Proton Therapy Physics

Series in Medical Physics and Biomedical Engineering

John G. Webster, E. Russell Ritenour, Slavik Tabakov, and Kwan-Hoong Ng

RECENT BOOKS IN THE SERIES

Radiation Protection in Medical Imaging and Radiation Oncology
Richard J. Vetter and Magdalena S. Stoeva (Eds)

Statistical Computing in Nuclear Imaging
Arkadiusz Sitek

The Physiological Measurement Handbook
John G. Webster (Ed)

Radiosensitizers and Radiochemotherapy in the Treatment of Cancer
Shirley Lehnert

Diagnostic Endoscopy
Haishan Zeng (Ed)

Medical Equipment Management
Keith Willson, Keith Ison, and Slavik Tabakov

Quantifying Morphology and Physiology of the Human Body Using MRI
L. Tugan Muftuler (Ed)

Monte Carlo Calculations in Nuclear Medicine, Second Edition: Applications in Diagnostic Imaging
Michael Ljungberg, Sven-Erik Strand, and Michael A. King (Eds)

Vibrational Spectroscopy for Tissue Analysis
Ihtesham ur Rehman, Zanyar Movasaghi, and Shazza Rehman

Webb's Physics of Medical Imaging, Second Edition
M. A. Flower (Ed)

Correction Techniques in Emission Tomography
Mohammad Dawood, Xiaoyi Jiang, and Klaus Schäfers (Eds)

Physiology, Biophysics, and Biomedical Engineering
Andrew Wood (Ed)

Proton Therapy Physics
Harald Paganetti (Ed)

Practical Biomedical Signal Analysis Using MATLAB®
K J Blinowska and J Żygierewicz (Ed)

Proton Therapy Physics

Second Edition

Edited by
Harald Paganetti

CRC Press
Taylor & Francis Group
Boca Raton London New York

CRC Press is an imprint of the
Taylor & Francis Group, an **informa** business

CRC Press
Taylor & Francis Group
6000 Broken Sound Parkway NW, Suite 300
Boca Raton, FL 33487-2742

First issued in paperback 2020

© 2019 by Taylor & Francis Group, LLC
CRC Press is an imprint of Taylor & Francis Group, an Informa business

No claim to original U.S. Government works

ISBN 13: 978-0-367-57078-1(pbk)
ISBN 13: 978-1-138-62650-8 (hbk)

Library of Congress Cataloging-in-Publication Data

Names: Paganetti, Harald, editor.
Title: Proton therapy physics / edited by Harald Paganetti.
Description: Second edition. | Boca Raton, Florida : CRC Press, [2019] |
Series: Series in medical physics and biomedical engineering |
Includes bibliographical references and index.
Identifiers: LCCN 2018024807 | ISBN 9781138626508 (hardback : alk. paper) |
ISBN 9781351855754 (ebook adobe reader) | ISBN 9781351855747 (ebook epub) |
ISBN 9781351855730 (ebook mobipocket)
Subjects: LCSH: Proton beams—Therapeutic use. | Medical physics. |
Cancer—Radiotherapy.
Classification: LCC RM862.P76 P765 2019 | DDC 610.1/53—dc23
LC record available at https://lccn.loc.gov/2018024807

Visit the Taylor & Francis Web site at
http://www.taylorandfrancis.com

and the CRC Press Web site at
http://www.crcpress.com

Contents

Editor

Dr. Harald Paganetti is the Director of Physics Research at the Department of Radiation Oncology at Massachusetts General Hospital and Professor of Radiation Oncology at Harvard Medical School, Boston, USA.

He received his PhD in experimental nuclear physics in 1992 from the Rheinische-Friedrich-Wilhelms University in Bonn, Germany, and has been working on experimental as well as theoretical projects in radiation therapy research since 1994. He has authored and co-authored more than 200 peer-reviewed publications and is renowned particularly for his work on proton therapy.

Dr. Paganetti has been awarded numerous research grants from the National Cancer Institute in the United States. He serves in several editorial boards and is a member of task-groups and committees for various associations, such as the American Society for Therapeutic Radiology and Oncology (ASTRO) and the American Association of Physicists in Medicine (AAPM).

List of Contributors

Christoph Bert
Department of Medical Physics
Universitätsklinikum Erlangen
Erlangen, Germany

Ben Clasie
Department of Radiation Oncology
Massachusetts General Hospital and
    Harvard Medical School
Boston, Massachusetts

David Craft
Department of Radiation Oncology
Massachusetts General Hospital and
    Harvard Medical School
Boston, Massachusetts

Juliane Daartz
Department of Radiation Oncology
Massachusetts General Hospital and
    Harvard Medical School
Boston, Massachusetts

Lei Dong
Department of Radiation Oncology
University of Pennsylvania
Philadelphia, Pennsylvania

Martijn Engelsman
Delft University of Technology
Delft, The Netherlands

Stella Flampouri
University of Florida Health Proton
    Therapy Institute
Jacksonville, Florida

Jacob Flanz
Department of Radiation Oncology
Massachusetts General Hospital and
    Harvard Medical School
Boston, Massachusetts

Bernard Gottschalk
Laboratory for Particle Physics and
    Cosmology
Harvard University
Cambridge, Massachusetts

Nisy Elizabeth Ipe
Shielding Design, Dosimetry &
    Radiation Protection
San Carlos, California

Antje C. Knopf
Faculty of Medical Sciences
University of Groningen
Groningen, The Netherlands

Hanne M. Kooy
Department of Radiation Oncology
Massachusetts General Hospital and
    Harvard Medical School
Boston, Massachusetts

Jon Kruse
Department of Radiation Oncology
Mayo Clinic
Rochester, Minnesota

Zuofeng Li
University of Florida Health Proton
    Therapy Institute
Jacksonville, Florida

Haibo Lin
New York Proton Center
New York City, New York

Anthony Lomax
Center for Proton Therapy (CPT)
Paul Scherrer Institute
Villigen, Switzerland

Hsiao-Ming Lu
Department of Radiation Oncology
Massachusetts General Hospital and
    Harvard Medical School
Boston, Massachusetts

Peter van Luijk
Department of Radiation Oncology
University Medical Center Groningen
University of Groningen
Groningen, The Netherlands

Harald Paganetti
Department of Radiation Oncology
Massachusetts General Hospital and
    Harvard Medical School
Boston, Massachusetts

Hugo Palmans
National Physical Laboratory
Medical Radiation Science
Teddington, United Kingdom
and
EBG MedAustron GmbH
Medical Physics
Wiener Neustadt, Austria

Jatinder R. Palta
Department of Radiation Oncology
Virginia Commonwealth University
Richmond, Virginia

Katia Parodi
Department of Experimental Physics -
    Medical Physics
Ludwig-Maximilians-Universität
    München (LMU Munich)
Garching b. München, Germany

Marco Schippers
Particle Accelerator Department
Paul Scherrer Institute
Villigen, Switzerland

Jiajian Shen
Department of Radiation Oncology
Mayo Clinic
Phoenix, Arizona

Roelf Slopsema
University of Florida Health Proton
    Therapy Institute
Jacksonville, Florida

Alexei V. Trofimov
Department of Radiation Oncology
Massachusetts General Hospital and
    Harvard Medical School
Boston, Massachusetts

Jan Unkelbach
Department of Medical Physics
Universitäts Spital Zürich
Zürich, Switzerland

Daniel K. Yeung
University of Florida Health Proton
    Therapy Institute
Jacksonville, Florida

Timothy C. Zhu
Department of Radiation Oncology
University of Pennsylvania
Philadelphia, Pennsylvania

Introduction

Harald Paganetti

Massachusetts General Hospital and Harvard Medical School

CONTENTS

PROTON THERAPY

According to the World Health Organization, cancer is the leading cause of death worldwide. A large portion of cancer patients, e.g., more than half of all cancer patients in the United States, receive radiation therapy during the course of treatment. Radiation therapy is used either as the sole treatment or, more typically, in combination with other therapies, including, for instance, surgery and chemotherapy.

Radiation interacts with tissue via atomic and nuclear interactions. The energy transferred to and deposited in the tissue in such interactions is quantified as "absorbed dose" and expressed in energy (Joules) absorbed per unit mass (kg), which has the units of Gray (Gy). Depending on the number and spatial correlation of such interactions, mainly with cellular DNA, they can result in mutations or complete functional disruption, i.e., cell death. Assessing radiation damage is a complex problem because the cell typically does have limited ability for repair. Furthermore, the translation from cellular to organ effects depends on many biological mechanisms and systemic reactions such as immune response are not fully understood.

The main focus in research and development of radiation therapy is on eradicating cancerous tissue while minimizing the irradiation of healthy tissue. The ideal scenario would be to treat the designated target without damaging any healthy structures. This is not possible for various reasons, e.g., uncertainties in defining the target volume as well as uncertainties in delivering the therapeutic dose as planned. Furthermore, applying external beam radiation therapy typically requires the beam to penetrate healthy tissue to reach the target.

There are many degrees of freedom when administering radiation, e.g., different radiation modalities, doses, or beam directions. Treatment planning in radiation therapy uses mathematical and physical formalisms to optimize the trade-off between delivering high and conformal dose to the target and limiting the doses to critical structures. The dose tolerance levels for critical structures, as well as the required doses for various tumor types, are typically defined based on decades of clinical experience.

When considering the trade-off between administering the prescribed target dose and the dose to healthy tissue, the term "therapeutic ratio" defines the maximally tolerated dose of a treatment to the minimally efficacious dose. Technological advances in beam delivery and treatment modality often focus on increasing the therapeutic ratio. Improvements can be achieved, for example, by applying advanced imaging techniques leading to improved patient setup or tumor localization.

The most prominent difference between photon and proton beams is the finite range of a proton beam. After a short buildup region, photon beams show an exponentially decreasing energy deposition with increasing depth in tissue. Except for superficial lesions, a higher dose to the tumor as compared to the organ at risk can only be achieved using multiple beam directions. Furthermore, a homogenous dose distribution can only be achieved by utilizing various different beam angles, not by delivering a single field. In contrast, the energy transferred to tissue by protons is inversely proportional to the proton velocity as protons lose their energy mainly in electromagnetic interactions with orbital electrons of atoms. The more the protons slow down, the higher the energy they transfer to tissue per track length, causing the maximum dose deposition at a certain depth in tissue. For a single proton the peak is very sharp. For a proton beam, it is broadened into a peak of typically a few millimeters width because of the statistical distribution of the proton tracks. The peak is called the "Bragg peak" (Figure 1). The depth and width of the Bragg peak is a function of the beam energy and the material (tissue) heterogeneity in the beam path. The peak depth can be influenced by changing the beam energy and can thus be positioned within the target for each beam direction. This feature allows pointing a beam towards a critical structure. Although protons from a single beam are able to deliver a homogeneous dose in the target, multiple beam angles are also used in proton therapy to even further optimize the dose distribution. Note that there is also a slight difference between photon and proton beams when considering the lateral penumbra. For large depths (more than 16 cm), the penumbra for proton beams is slightly wider than the one for photon beams by typically a few millimeters. Depending on the treatment depth, this can cause a slight dosimetric disadvantage of proton beams.

The rationale for using proton beams instead of photon beams is the feasibility of delivering higher doses to the tumor, while maintaining the total dose to critical structures.

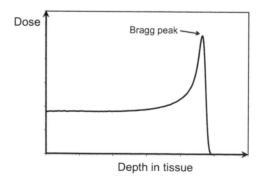

FIGURE 0.1   Energy deposition as a function of depth for a proton beam leading to the Bragg peak.

Or, maintaining the target dose while reducing the total dose to critical structures. Treatment plan comparisons show that protons offer potential gains for many sites. In some cases, the dose conformity that can be reached with photon techniques is comparable or even better to one that can be achieved with proton techniques. However, due to the fundamental difference in physics between photon beams and proton beams as outlined earlier, the total dose absorbed in the patient for any given treatment will always be higher with photons than with protons. The use of protons leads to a reduction of the total dose delivered to a patient when treating a given target by a factor of about two to seven compared to standard photon techniques and a factor of about two to three compared to intensity-modulated photon plans. Benefits can thus be expected particularly for pediatric patients, where the irradiation of large volumes is particularly critical in terms of long-term side effects.

MAJOR CHANGES FOR THIS SECOND EDITION COMPARED TO THE FIRST EDITION

This is the second edition of the successful book on *Proton Therapy Physics*. The book has been completely restructured and updated, including the addition of new chapters. Proton therapy is a very active are of research, and many aspects needed an update considering that the first edition was published almost six years ago. For instance, this new edition shifts focus more on beam scanning delivery (while still discussing passively scattered proton therapy). Another aspect where considerable research has been done in the last few years is on the relative biological effectiveness of proton beams, leading to a complete rewrite of this chapter. Generally, all chapters were updated based on new research and new publications since the last edition. Furthermore, it was deemed that certain aspects needed more in-depth discussion. This resulted in 23 chapters for this second edition as compared to the previous 20 chapters. The main structural changes are as follows:

The dosimetry discussion has been expanded by dividing the previous chapter on "Dosimetry" into two separate chapters titled "Dosimetry/Detectors and Relative Dosimetry" and "Dosimetry/Reference Dosimetry and Primary Standards." Also, the chapter on "Quality Assurance and Commissioning" was split into three, titled "Acceptance Testing and Commissioning," "Quality Assurance," and "Monitor Unit Calibration." A chapter on "Proton Image Guidance" was added. In addition to these expansions, two chapters have been combined, i.e., the chapter on "Late Effects from Scattered and Secondary Radiation" has been incorporated into "The Physics of Proton Biology."

The order of chapters was also altered to group chapters into "Background," "Beam Delivery," "Dosimetry," "Operation," "Treatment Planning/Delivery," "Imaging," and "Biological Effects."

BOOK OUTLINE

Under "Background," the book starts with an overview of the history of proton therapy. The pioneering work done in the early days of proton therapy is acknowledged, and the main developments up to the first hospital-based facilities are outlined. The chapter concludes with comments about the original and current clinical rationale for proton therapy.

The atomic and nuclear physics background necessary for understanding proton interactions with tissue is summarized in Chapter 2. The chapter covers the basic physics of protons slowing down in matter independent of their medical use. The ways in which protons can interact with materials/tissue is described from both macroscopic (e.g., dose) and microscopic (energy loss kinematics) points of view. Furthermore, Chapter 2 presents equations that can be used for estimating many characteristics of proton beams for daily use.

The next section is on "Beam Delivery." The first chapter, Chapter 3, describes the physics of proton accelerators, including currently used techniques (cyclotrons and synchrotrons) as well as a brief discussion of new developments. The chapter goes beyond simply summarizing the characteristics of such machines for proton therapy. It also describes some of the main principles of particle accelerator physics.

Chapter 4 outlines the characteristics of clinical proton beams and how clinical parameters are connected to the design features and the operational settings of the beam delivery system. Parameters such as dose rate, beam intensity, beam energy, beam range, distal falloff, and lateral penumbra are introduced.

The next two chapters describe in detail how to generate a conformal dose distribution in the patient. Passive scattered beam delivery systems are discussed in Chapter 5. Scattering techniques to create a broad beam as well as range modulation techniques to generate a clinically desired depth-dose distribution are outlined in detail. Next, Chapter 6 does focus on magnetic beam scanning systems. Scanning hardware as well as parameters that determine the scanning beam characteristic, e.g., its time structure, and performance are discussed.

The next group of chapters covers "Dosimetry," starting with a chapter on radiological aspects covering shielding considerations and measurement methods to ensure the safety of patients as well as operating personnel while operating a treatment beam. One aspect of increasing importance in the field of medical physics is the use of computer simulations to replace or assist experimental methods. After an introduction into the Monte Carlo particle tracking method, Chapter 8 demonstrates how Monte Carlo simulations can be used to address various clinical and research aspects in proton therapy. Chapters 9 and 10 cover detectors and relative dosimetry as well as reference dosimetry and primary standards, respectively. The underlying dosimetry formalism and technology are reviewed in detail in addition to the basic aspects of microdosimetry.

These chapters are then followed by three chapters on "Operation." Chapter 11 outlines all aspects of acceptance testing and commissioning, including treatment planning commissioning. Chapter 12 does focus on machine as well as patient-specific quality assurance, and Chapter 13 summarizes methods used to monitor unit calibration.

The next six chapters focus on "Treatment Planning/Delivery." Chapter 14 does focus on dose calculation concepts and algorithms. The formalism for pencil-beam algorithms is reviewed from a theoretical and practical implementation point of view. Further, the Monte Carlo dose calculation method and hybrid methods are outlined. Chapter 15 covers proton-specific aspects of treatment planning for passive scattering and scanning delivery for single field uniform dose, i.e., homogeneous dose distributions in the target from each beam direction. Proton-specific margin considerations and special treatment techniques are discussed. Chapter 16 describes treatment planning for multiple field uniform dose and intensity-modulated proton therapy using beam scanning. The challenges and the potential of intensity modulated treatments are described, including uncertainties and optimization strategies. A few case studies round up this chapter.

One of the advantages of proton therapy is the ability to precisely shape dose distributions, in particular, using the distal fall-off due to the finite beam range. Uncertainties in the proton beam range limit the utilization of the finite range of proton beams in the patient, because more precise dose distributions are less forgiving in terms of errors and uncertainties. Chapter 17 discusses precision and uncertainties for non-moving targets. Special emphasis is on the dosimetric consequences of heterogeneities. Chapter 18 deals with precision and uncertainties for moving targets, e.g., when treating lung cancer with proton beams. The clinical impacts of motion as well as methods of motion management to minimize motion effects are presented.

Computerized treatment planning relies on optimization algorithms to generate a clinically acceptable plan. Chapter 19 reviews the main aspects of treatment plan optimization, including the consideration of some of the uncertainties discussed in the previous chapters, e.g., robust and four-dimensional optimization strategies are described.

The section on "Imaging" starts with Chapter 20, which summarizes image-guided radiation therapy with special emphasis on proton therapy. Chapter 21 covers in vivo range verification. The methods include the detection of photons caused by nuclear excitations and of annihilation photons created after the generation of positron emitters by the primary proton beam. The new field of ionoacoustics is described as well.

The final two chapters discuss aspects of "Biological Effects" in proton therapy. While this book is mainly on proton therapy physics, biological implications are discussed briefly as they relate to physical aspects. The biological implications of using protons are outlined from a physics perspective in Chapter 22, particularly covering the issue of the relative biological effectiveness of proton beams. Furthermore, issues related to secondary doses and methods to estimate the risks for radiation-induced cancers are presented. Outcome modeling as it pertains to proton therapy is summarized in Chapter 23. This chapter illustrates the use of risk models for normal tissue complications in treatment optimization. Proton beams allow precise dose shaping and thus personalized treatment planning might become particularly important for proton therapy in the future.

THE SCOPE AND GOAL OF THIS BOOK

The share of patients treated with proton therapy as compared to photon therapy is currently low but is expected to increase significantly in the near future, as evidenced by the number of facilities currently planned or under construction. With the increasing use of protons as radiation therapy modality, there is a need for a better understanding of the characteristics of protons. Protons are not just heavy photons when it comes to treatment planning, quality assurance, delivery uncertainties, radiation monitoring, and biological considerations. To fully utilize the advantages of proton therapy and, just as importantly, to understand the uncertainties and limitations of precisely shaped dose distributions, proton therapy physics needs to be understood. Furthermore, the clinical impact and evidence for improved outcomes need to be studied. Proton therapy research has increased significantly

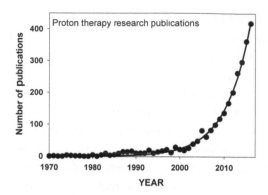

FIGURE 0.2   Number of publications listed in PubMed (a free database of citations on life sciences and biomedical topics) per year with the phrases "proton therapy," "proton radiation therapy," "proton beam therapy," or "proton beam radiation therapy" in title or abstract (closed circles). Also shown is an exponential fit of the form $Publications = a \cdot e^{(b*year)}$ (solid line.)

in the last few years. Figure 2 shows the number of proton therapy related publications in most relevant scientific journals over the years.

The goal of this book is to offer a coherent and instructive overview of proton therapy physics. It might serve as a practical guide for physicians, dosimetrists, radiation therapists, and physicists who already have some experience in radiation oncology. Furthermore, it can serve graduate students who are either in a medical physics program or considering a career in medical physics. Certainly, it is also of interest to physicians in their last year of medical school or residency with a desire to understand proton therapy physics. There are some overlaps between different chapters. This could not be avoided because each chapter should be largely independent.

Overall, the book covers most, but certainly not all, aspects of proton therapy physics.

I

Background

CHAPTER 1

Proton Therapy

History and Rationale

Harald Paganetti

Massachusetts General Hospital and Harvard Medical School

CONTENTS

1.1 THE ADVENT OF PROTONS IN CANCER THERAPY

The first medical application of ionizing radiation, using X-rays, was reported in 1895 [1,2]. In the following decades, radiation therapy became one of the main treatment options in oncology [3]. Many improvements have been made with respect to how radiation is administered considering biological effects, e.g., the introduction of fractionated radiation therapy in the 1920s and 1930s. In addition, technical advances have been aimed, for instance, at reducing dose to healthy tissue while maintaining prescribed doses to the target or increasing the dose to target structures with either no change or a reduction of dose to normal tissue. Computerized treatment planning, advanced imaging and patient setup, and the introduction of megavoltage X-rays are examples of new techniques that have impacted beam delivery precision during the history of radiation therapy. Another way of reducing dose to critical structures is to take advantage of dose deposition characteristics offered by different types of particles.

The advantages of proton radiation therapy, compared to "conventional" photon radiation therapy, were first outlined by Wilson in 1946 [4]. He presented the idea of utilizing the finite range and the Bragg peak of proton beams for treating targets deep within healthy tissue, and was thus the first to describe the potential of proton beams for medical use. Wilson's suggestion to use protons (in fact, he also extended his thoughts to heavy ions) was based on the well-known physics of protons as they slowed down during penetration of tissue.

1.2 HISTORY OF PROTON THERAPY FACILITIES

1.2.1 Early Days: Lawrence Berkeley Laboratory, Berkeley, USA

The idea of proton therapy was not immediately picked up at Wilson's home institution, Harvard University, but was adopted a couple of years later by the Lawrence Berkeley Laboratory (LBL) in California. Pioneering the medical use of protons, Tobias, Anger, and Lawrence [5] in 1952 published their work on biological studies on mice using protons, deuterons, and helium beams. Many more experiments with mice followed at LBL [6], and the first patient was treated in 1954 [7].

The early patients had metastatic breast cancer and received proton irradiation of their pituitary gland for hormone suppression. The bony landmarks made targeting of the beam feasible. The Bragg peak itself was not utilized. Instead, using a 340 MeV proton beam, patients were treated with a cross-firing technique, i.e., using only the plateau region of the depth-dose curve. This approximated a rotational treatment technique to concentrate the dose in the target. Protons as well as helium beams were applied. Between 1954 and 1957, 30 patients were treated with protons. Initially, large single doses were administered [7], and later, fractionated delivery treating three times a week was applied [8]. The first patient using the Bragg peak was treated in 1960 for a metastatic lesion in the deltoid muscle employing a helium beam [9]. The LBL program moved to heavier ions entirely in 1975, resulting in several developments that also benefited proton therapy.

1.2.2 Early Days: Gustav Werner Institute, Uppsala, Sweden

In 1955, shortly after the first proton treatments at LBL, radiation oncologists in Uppsala, Sweden, became interested in the medical use of protons. Initially, a series of animal (rabbits and goats) experiments were performed to study the biological effect of proton radiation [10–12]. The first patient was treated in 1957 using a 185 MeV cyclotron at the Gustav Werner Institute [12–14]. Subsequently, radiosurgery beams were used to treat intracranial lesions, and by 1968, 69 patients had been treated [15,16]. Because of limitations in beam time at the cyclotron, high doses per fraction were administered. Instead of the cross-firing technique, the use of the Bragg peak was adopted early on using large fields and range-modulated beams [14,17,18]. In fact, the Gustav Werner Institute was the first to use range modulation using a ridge filter, i.e., a spread-out Bragg peak (SOBP) with a homogeneous dose plateau at a certain depth in tissue [14], based on the original idea of Robert Wilson. Here, various monoenergetic proton beams resulting in Bragg peaks are combined to achieve a homogeneous dose distribution in the target. The proton therapy program ran from 1957 to 1976 and then reopened in 1988 [19].

1.2.3 Early Days: Harvard Cyclotron Laboratory, Cambridge, USA

Preclinical work on proton therapy at Harvard University (Harvard Cyclotron Laboratory, HCL) started in 1959 [20]. The cyclotron at the HCL had sufficient energy (160 MeV) to reach the majority of sites in the human body up to a depth of about 16 cm. The relative biological effectiveness (RBE) of proton beams was studied in the 1960s using experiments on chromosome aberrations in bean roots [21], mortality in mice [22], and skin reactions on primates [23]. Subsequently, the basis for today's practice of using a clinical RBE of 1.1 as a conservative estimate was established [24–27] (see Chapter 22).

The clinical program was based on a collaboration between the HCL and the neurosurgical department of the Massachusetts General Hospital (MGH). The first patients were treated in 1961 [28]. Intracranial targets needed only a small beam, which could be delivered using a single scattering technique to broaden the beam. As at LBL, pituitary irradiation was one of the main targets. Because of the maximum beam energy of 160 MeV, it was decided to focus on using the Bragg peak instead of applying a cross-fire technique. Until 1975, 732 patients had undergone pituitary irradiation at the HCL [29]. Based on the growing interest in biomedical research and proton therapy, the facility was expanded by constructing a biomedical annex in 1963, funded by the National Aeronautics and Space Administration (NASA). The research program was funded largely by the US Office of Naval Research, which also originally funded the cyclotron. This funding source was shut down in 1967, putting the proton therapy project in danger of being terminated. Extensive negotiations between the MGH and the HCL, as well as small grants by the National Cancer Institute (NCI) in 1971 and the National Science Foundation in 1972 helped, thus saving the program.

In 1973, the radiation oncology department commenced an extensive proton therapy program. The first patient was a 4-year-old boy with a posterior pelvic sarcoma. Subsequently, the potential of the HCL proton beam for treatment of skull base sarcomas, head-and-neck region carcinomas, and uveal melanomas was identified, and several studies on fractionated proton therapy were performed [30]. Furthermore, a series of radiobiological experiments were done [25]. Based on the development of a technique to treat choroidal melanomas at the MGH, the Massachusetts Eye and Ear Infirmary, and the HCL, melanoma treatments started in 1975 [31] after tests had been made using monkeys [32,33]. The first treatments for prostate patients were in the late 1970s [34]. A milestone for the operation at HCL as well as for proton therapy research in general was a large research grant by the NCI awarded in 1976 to MGH Radiation Oncology to allow extensive studies on various aspects of proton therapy. The HCL facility treated a total of 9,116 patients until 2002.

1.2.4 Second Generation: Proton Therapy in Russia

Proton therapy began early at three centers in Russia. Research on using proton beams in radiation oncology was started in Dubna (Joint Institute for Nuclear Research, JINR) and at the Institute of Theoretical and Experimental Physics (ITEP) in Moscow in 1967. The Dubna facility started treatments in April 1967 followed by ITEP in 1968 [35–39]. A joint project between the Petersburg Nuclear Physics Institute and the Central Research Institute of Roentgenology and Radiology in St Petersburg launched a proton therapy program in 1975 in Gatchina, a nuclear physics research facility near St Petersburg. The latter treated intracranial diseases using Bragg curve plateau irradiation with a 1 GeV beam [40].

The program at ITEP was the largest of these programs and was based on a 7.2 GeV proton synchrotron with a medical beam extraction of up to 200 MeV. Patients were treated with broad beams and a ridge filter to create depth-dose distributions. Starting in 1972, the majority of treatments irradiated the pituitary glands of breast and prostate cancer patients using the plateau of the Bragg curve [35,41]. By the end of 1981, 575 patients with various indications had also been treated with Bragg peak dose distributions [35].

1.2.5 Second Generation: Proton Therapy in Japan

The history of proton therapy treatments in Japan goes back to 1979 when the National Institute of Radiological Sciences (NIRS) at Chiba started treating at its 70 MeV facility [42]. Of the 29 patients treated between 1979 and 1984, only 11 received proton therapy alone, while 18 received a boost irradiation of protons following either photon beam or fast neutron therapy. The effort was followed by the use of a 250 MeV proton beam at the Particle Radiation Medical Science Center in Tsukuba in 1983 by degrading a 500 MeV beam from a booster synchrotron of the high-energy physics laboratory at the 'High Energy Accelerator Research Organization' (KEK) [43]. Japan has since emerged as one of the main users of proton and heavy ion therapy.

1.2.6 Second Generation: Proton Therapy Worldwide

The late 1980s and early 1990s saw a number of initiatives starting proton therapy programs on several continents, e.g., at the Paul Scherrer Institute (PSI) (Switzerland) in 1984, Clatterbridge (UK) in 1989, Orsay (France) in 1991, and iThemba Labs (South Africa) in 1993. In particular, the activities at PSI, starting with a 72 MeV beam for ocular melanoma treatments [44] and after 1996 using a 200 MeV beam, have led to many technical and treatment planning improvements in proton therapy.

1.2.7 Hospital-Based Proton Therapy

By the early 1990s, proton therapy had been based mainly in research institutions on a modest number of patients, due in part to very restricted beam time availability at some centers. Then, in 1990, the first hospital-based facility was built and started its operation at the Loma Linda University Medical Center in California [45]. The accelerator system, based on a synchrotron [46], was developed in collaboration with Fermilab. The gantries were designed by the HCL group [47]. By July 1993, 12,914 patients had been treated with protons worldwide, still roughly half of those at the HCL, and 25% in Russia [48]. Roughly 50% were radiosurgery patients treated with small fields. However, the facility at Loma Linda would soon treat the biggest share of proton therapy patients.

It took another few years before the first commercially available equipment was installed and in operation at the MGH, which transferred the program from the HCL to its main hospital campus in 2001. At the time when the facility was purchased, proton therapy was still considered mainly an experimental and research effort. In fact, the construction project was in part funded by the NCI. The commercial equipment sold to the MGH started the interest of different companies to offer proton therapy solutions and the interest of major hospitals to buy proton therapy facilities. Numerous other hospital-based facilities have been opened since then.

1.2.8 Facilities

Table 1.1 lists the facilities currently in operation. In total, more than 100,000 patients have been treated with proton radiation therapy worldwide.

TABLE 1.1   Proton Therapy Facilities in Operation as of August 2018

	Country	Who, Where	S/C/SC[a] Max. Energy (MeV)	Beam Directions	Start
1	Austria	MedAustron, Wiener Neustadt	S 250	2 fixed beams[b], 1 gantry[b] (under construction)	2017
2	Canada	TRIUMF, Vancouver	C 72	1 fixed beam	1995
3	Czech Republic	PTC Czech r.s.o., Prague	C 230	3 gantries[b], 1 fixed beam	2012
4	China	WPTC, Wanjie, Zi-Bo	C 230	2 gantries, 1 fixed beam	2004
5	China	SPHIC, Shanghai	S 250	3 fixed beams[b]	2014
6	England	Clatterbridge	C 62	1 fixed beam	1989
7	England	Proton Partner's Rutherford CC, Newport	C 230	1 gantry[b]	2018
8	France	CAL/IMPT, Nice	C165 SC 235	1 fixed beam, 1 gantry	1991, 2016
9	France	CPO, Orsay	C 230	1 gantry[b], 2 fixed beams	1991, 2014
10	France	CYCLHAD, Caen	C 230	1 gantry[b]	2018
11	Germany	HZB, Berlin	C 250	1 fixed beam	1998

(Continued)

TABLE 1.1 (*Continued*)   Proton Therapy Facilities in Operation as of August 2018

	Country	Who, Where	S/C/SC[a] Max. Energy (MeV)	Beam Directions	Start
12	Germany	RPTC, Munich	C 250	4 gantries[b], 1 fixed beam	2009
13	Germany	HIT, Heidelberg	S 250	2 fixed beams, 1 gantry[b]	2009, 2012
14	Germany	WPE, Essen	C 230	4 gantries[c], 1 fixed beam	2013
15	Germany	UPTD, Dresden	C 230	1 gantry[c]	2014
16	Germany	MIT, Marburg	S 250	3 horiz., 1 45 deg. fixed beams[b]	2015
17	Italy	INFN-LNS, Catania	C 60	1 fixed beam	2002
18	Italy	CNAO, Pavia	S 250	3 horiz., 1 vertical, fixed beams	2011
19	Italy	APSS, Trento	C 230	2 gantries[b], 1 fixed beam	2014
20	Japan	NCC, Kashiwa	C 235	2 gantries[c]	1998
21	Japan	HIBMC, Hyogo	S 230	1 gantry	2001
22	Japan	PMRC 2, Tsukuba	S 250	2 gantries[c]	2001
23	Japan	Shizuoka Cancer Center	S 235	3 gantries, 1 fixed beam	2003
24	Japan	STPTC, Koriyama-City	S 235	2 gantries[b], 1 fixed beam	2008
25	Japan	MPTRC, Ibusuki	S 250	3 gantries[c]	2011
26	Japan	Fukui Prefectural Hospital PTC, Fukui City	S 235	2 gantries[c], 1 fixed beam	2011
27	Japan	Nagoya PTC, Nagoya City, Aichi	S 250	2 gantries[c], 1 fixed beam	2013
28	Japan	Hokkaido Univ. Hospital PBTC, Hokkaido	S 220	1 gantry	2014
29	Japan	Aizawa Hospital PTC, Nagano	C 235	1 gantry	2014
30	Japan	Tsuyama Chuo Hospital, Okayama	S 235	1 gantry	2016
31	Japan	Hakuhokai Group Osaka PT Clinic, Osaka	S 235	1 gantry	2017
32	Japan	Kobe Proton Center, Kobe	S 235	1 gantry	2017
33	Netherlands	UMC PTC, Groningen	C 230	2 gantries[c]	2018
34	Poland	IFJ PAN, Krakow	C 230	1 fixed beam, 2 gantries	2011, 2016
35	Russia	ITEP, Moscow	S 250	1 fixed beam	1969
36	Russia	MIBS, Saint Petersburg	C 250	2 gantries[b]	2018
37	Russia	JINR 2, Dubna	C 200[d]	1 fixed beam	1999
38	South Africa	NRF – iThemba Labs	C 200	1 fixed beam	1993

(*Continued*)

TABLE 1.1 (*Continued*) Proton Therapy Facilities in Operation as of August 2018

	Country	Who, Where	S/C/SC[a] Max. Energy (MeV)	Beam Directions	Start
39	South Korea	KNCC, Ilsan	C 230	2 gantries, 1 fixed beam	2007
40	South Korea	Samsung PTC, Seoul	C 230	2 gantries	2015
41	Sweden	The Skandion Clinic, Uppsala	C 230	2 gantries[b]	2015
42	Switzerland	CPT, PSI, Villigen	C 250	3 gantries[b], 1 fixed beam	1984, 1996, 2013, 2018
43	Taiwan	Chang Gung Memorial Hospital	C 230	4 gantries[b], 1 fixed beam exp.	2015
44	USA, CA	J. Slater PTC, Loma Linda	S 250	3 gantries, 1 fixed beam	1990
45	USA, CA	UCSF-CNL, San Francisco	C 60	1 fixed beam	1994
46	USA, MA	MGH Francis H. Burr PTC, Boston	C 235	2 gantries[c], 1 fixed beam	2001
47	USA, TX	MD Anderson Cancer Center, Houston	S 250	3 gantries[c], 1 fixed beam	2006
48	USA, FL	UFHPTI, Jacksonville	C 230	3 gantries[c], 1 fixed beam	2006
49	USA, OK	ProCure PTC, Oklahoma City	C 230	1 gantry, 3 fixed beams	2009
50	USA, PA	Roberts PTC,UPenn, Philadelphia	C 230	4 gantries[c], 1 fixed beam	2010
51	USA, IL	Chicago Proton Center, Warrenville	C 230	1 gantry[b], 3 fixed beams	2010
52	USA, VA	HUPTI, Hampton	C 230	4 gantries, 1 fixed beam	2010
53	USA, NJ	ProCure Proton Therapy Center, Somerset	C 230	4 gantries[c]	2012
54	USA, WA	SCCA ProCure Proton Therapy Center, Seattle	C 230	4 gantries[c]	2013
55	USA, MO	S. Lee Kling PTC, Barnes Jewish Hospital, St Louis	SC 250	1 gantry	2013
56	USA, TN	Provision Center for Proton Therapy, Knoxville	C 230	3 gantries[b]	2014
57	USA, CA	California Proton Cancer Therapy Center, San Diego	C 250	3 gantries[b], 2 fixed beams[b]	2014
58	USA, LA	Willis Knighton Proton Therapy Cancer Center, Shreveport	C 230	1 gantry[b]	2014
59	USA, FL	Ackerman Cancer Center, Jacksonville	SC 250	1 gantry	2015
60	USA, MN	Mayo Clinic Proton Beam Therapy Center, Rochester	S 220	4 gantries[b]	2015

(*Continued*)

TABLE 1.1 (*Continued*)   Proton Therapy Facilities in Operation as of August 2018

	Country	Who, Where	S/C/SC[a] Max. Energy (MeV)	Beam Directions	Start
61	USA, NJ	Laurie Proton Center of Robert Wood Johnson Univ. Hospital, New Brunswick	SC 250	1 gantry	2015
62	USA, TX	Texas Center for Proton Therapy, Irving	C 230	2 gantries[b], 1 fixed beam	2015
63	USA, TN	St Jude Red Frog Events Proton Therapy Center, Memphis	S 220	2 gantries[b], 1 fixed beam	2015
64	USA, AZ	Mayo Clinic Proton Therapy Center, Phoenix	S 220	4 gantries[b]	2016
65	USA, MD	Maryland Proton Treatment Center, Baltimore	C 250	4 gantries[b], 1 fixed beam[b]	2016
66	USA, FL	Orlando Health PTC, Orlando	SC 250	1 gantry	2016
67	USA, OH	UH Sideman CC, Cleveland	SC 250	1 gantry	2016
68	USA, OH	Cincinnati Children's Proton Therapy Center, Cincinnati	C 250	3 gantries[b]	2016
69	USA, MI	Beaumont Health Proton Therapy Center, Detroit	C 230	1 gantry[b]	2017
70	USA, FL	Baptist Hospital's Cancer Institute PTC, Miami	C 230	3 gantries[b]	2017

Source: Based on information by the Particle Therapy Co-Operative Group (PTCOG), https://www.ptcog.ch/index.php/facilities-in-operation.
[a] S/C/SC = Synchrotron (S) or Cyclotron (C) or Synchrocyclotron (SC).
[b] With beam scanning.
[c] With passive scattering and beam scanning.
[d] Degraded beam.

1.3 HISTORY OF PROTON THERAPY DEVICES

1.3.1 Proton Accelerators

The concept of accelerating particles in a repetitive way with time-dependent varying potentials led to the invention of the cyclotron by Lawrence in 1929 [49]. Cyclotrons accelerate particles while they are circulating in a magnetic field and pass the same accelerating gap several times (see Chapter 3). Gaining energy, the particles are traveling in spirals and are eventually extracted. To overcome the energy limitation of a cyclotron, the principle of phase stability was invented in 1944 [50]. One was now able to accelerate particles of different energy on the same radius, leading to the synchrotron (see Chapter 3). The synchrotron concept was first suggested by Oliphant in 1943 [51]. Thus, both accelerator types were available when proton therapy was first envisaged.

1.3.2 Proton Facilities

Cyclotrons and synchrotrons were originally quite large, and the advent of superconducting technology allowed a significant reduction of not only accelerators but also gantry structures. While facilities were first built at large centers with the demand for multiple treatment rooms, nowadays more and more smaller cancer centers are interested in the use of proton therapy.

Single room facilities have been developed with a reduced footprint and reduced cost compared to the early hospital-based facilities. Recent developments even allow such systems to eventually occupy treatment rooms originally designed to host photon linear accelerator systems.

1.3.3 Mechanically Modulating Proton Beams to Extend the Dose in Depth

In his 1946 paper, Wilson introduced the idea of using a rotating wheel of variable thickness to cover an extended volume with an SOBP (although he did not define this term) [4,52]. This technique to produce SOBPs (see Chapter 5) was adopted by proton facilities, e.g., at the HCL [53–55]. Others have used a ridge filter design to shape an SOBP [14,56,57].

1.3.4 Scattering for Broad Beams to Extend the Dose Laterally

For treatment sites other than very small targets (e.g., in radiosurgery), proton beams produced by accelerators result in "pencil" beams that are too small to cover an extended target. Thus, scattering foils had to be used to increase their width. To produce a flat dose distribution in lateral direction, it was inefficient to use a single scattering foil because only a small area in the center of the beam would suffice beam flatness constraints. The double-scattering system, using two scatterers to achieve a parallel beam producing a flat dose distribution with high efficiency, was developed at the HCL in the late 1970s [58]. The idea was based on similar systems previously designed for heavy ion and electron beams [59]. The double-scattering concept was later improved using a contoured scatterer system (see Chapter 5) [60].

1.3.5 Magnetic Beam Scanning

The development of beam scanning was a major milestone in proton therapy. The clinical implications of beam scanning were analyzed in the late 1970s and early 1980s [61,62]. The advantage of scanning is not only the need for fewer beam-shaping absorbers in the treatment head (increasing the efficiency) but also the potential for delivering variable modulation and thus sparing structures proximal to the SOBP [61].

The concept of using magnets to deflect a proton beam (dynamic beam delivery) is as old as the double-scattering system. The idea to magnetically deflect proton beams for treatment was first published by Larsson in 1961 [14]. Continuous scanning using an aperture was done in the 1960s in Uppsala [14]. The aim was not to scan the tumor with individual pencil beams but to replace the scattering system using a sweeping magnetic field. A method using rotating dipoles instead of a scattering system to produce a uniform dose distribution was considered by Koehler et al. [58]. It can be considered as an intermediate between double-scattered broad beam delivery and beam scanning. Similarly, a technique called wobbling, using magnetic fields to broaden the beam without a double-scattering system, was developed at Berkeley for heavy ion therapy because here the material in the beam path when using a double-scattering system produces too much secondary radiation [63].

Full intensity modulation using beam scanning uses small proton beams of variable energy and intensity that are magnetically steered to precisely shape dose around critical structures (see Chapter 6). This concept of using beam scanning in three dimensions for clinical proton beam delivery was developed by Leemann et al. [64]. Many different types of beam scanning exist. Typical terms are spot, pixel, voxel, dynamic, and raster scanning. The terminology is not consistent. The main differences between scanning systems are whether the delivery is done in a step and shoot mode or continuously. Spot scanning, where the beam spots are delivered one by one with beam off time in between, thus covering the target volume instead of delivering a rectangular scanned field that has to be shaped by an aperture, was first introduced at the NIRS using a 70 MeV beam. Scanning was mainly done to improve the range of the beam by removing a scattering system. At first, two-dimensional scanning was applied in combination with a range modulating wheel [42]. Later, three-dimensional beam scanning was introduced by using a system with two scanning magnets and an automatic range degrader to change the spot energy [42,65–68].

Many studies on different scanning techniques (spot scanning, continuous scanning) were done in the early 1980s at LBL, and continuous scanning in three dimensions without collimator was first done in the early 1990s [69].

Beam scanning can be used in a similar way as passively scattered proton therapy, but it also allows creating fields delivering inhomogeneous dose distributions, where only a combination of several fields yields the desired dose distribution in the target (see Chapter 16). The method of intensity-modulated proton therapy (IMPT) is being used more routinely in proton centers, and a majority of new facilities are exclusively using IMPT.

1.3.6 Impact of Proton Technology in Other Areas of Radiation Therapy

Some of the developments in proton therapy have influenced the way radiation therapy is being conducted in conventional radiation therapy. External beam radiation therapy requires a geometric description of the internal patient anatomy. Until the advent of Computed Tomography (CT), this could only be obtained from X-ray images, which project the anatomy on a planar film. Since conventional radiation therapy uses photons, the imaging X-ray modality basically just replaces the treating photons. In the case of protons, which stop in the patient, this method does not suffice for treatment planning. At the start of proton treatment of cancer patients at the HCL, positioning of the target for each treatment field for each fraction was readily achieved by the simple use of bi-planar radiographs. The information was used to decide on potential beam approaches that covered the target in the lateral dimension for each beam path. Pituitary adenomas and arteriovenous malformation were the initial targets for proton therapy [16,70]. These lesions could be visualized on X-rays using contrast material to visualize the vasculature and thus could be treated without the use of CT imaging. It became clear that to utilize the superior dose distribution of proton beams, one needs to understand the impact of density variations for each beam path [30,71,72]. Thus, the treatment of other sites in the heterogeneous head-and-neck region, e.g., paranasal sinus or nasopharynx, required additional research on accurate imaging to visualize patient's geometry and densities in the beam path [71].

When CT imaging became available, proton radiation therapy was the early adaptor, i.e., using CT for treatment planning [73–75]. The proton therapy program at HCL, the heavy ion program at LBL, and the pi-meson program at the University of New Mexico were the first radiation therapy programs to install dedicated CT scanners. Some were modified to allow imaging in a seated position to mimic the treatment geometry.

Proton therapy paved the way for many other advances in radiation therapy. The proton therapy group at MGH developed the first computerized treatment planning program in the early 1980s, which was subsequently used clinically [76–79]. Other developments included the innovative concepts of beam's eye view and dose–volume histograms, the features that have become standard in radiation therapy. Sophisticated patient positioning was developed first in proton therapy because the finite range of proton beams required a more precise setup than in photon therapy [80].

1.4 THE CLINICAL RATIONALE FOR USING PROTONS IN CANCER THERAPY

1.4.1 Dose Distributions

Any new radiation treatment technology has to find acceptance amongst clinicians, e.g., by demonstrating improved dose distributions and suggesting a more favorable treatment outcome [30]. A more favorable dose distribution is a distribution that is more closely confined to the tumor volume. This allows reducing the dose to normal structures, decreasing the normal tissue complication probability, or increasing the dose to the tumor, increasing the tumor control probability, or both. When proton therapy became available, it was of interest mainly because it showed dose conformity far superior to any type of conventional photon radiation therapy at that time [72,81]. Nowadays, it is quite feasible for some tumor shapes to reach dose conformity to the target with photons that is comparable to the one achievable with protons, albeit at the expense of using a larger number of beams. The difference in dose conformity between protons and photons has certainly decreased since the early days of proton therapy (at least for regular-shaped targets), mainly due to the development of intensity-modulated photon therapy or rotational therapies. However, the integral dose (the total energy deposited in the patient) is always lower with proton beams.

There is a limit to further improving and shaping photon-generated dose distributions because the total energy deposited in the patient and thus to critical structures cannot be reduced but only distributed differently. Proton radiation therapy, on the other hand, can potentially achieve significant further physical improvements through the use of scanning-beam technology and IMPT. This will increase the advantage of proton therapy due to advanced dose-sculpting potential.

1.4.2 Early Clinical Implications

Protons are ideal for many targets, specifically if they are concave-shaped or close to critical structures. Target dose distributions can typically be shaped with proton beams by applying fewer beams than with photons. Proton therapy is of particular interest for tumors located close to serially organized tissues, where a small local overdose can cause significant complications. The advantages of proton therapy could not be utilized right from the start because of limitations in patient imaging and beam delivery (e.g., the absence of gantry systems). Proton treatments started with the cross-firing technique and the irradiation of pituitary targets. The proton therapy program at the HCL began with single fraction treatments of intracranial lesions [28]. In the early 1960s, the program of fractionated irradiation was commenced by the radiation oncologists at HCL and was used for a greatly expanded number of anatomic sites, e.g., skull base sarcomas, choroidal melanomas, and head and neck carcinomas. Choroidal melanomas quickly became the most commonly treated tumor at HCL [82]. Starting in 1973, all treatments for cancer patients were done by fractionated dose delivery [30]. By mid-1980s, roughly one-third of the treated patients received intracranial radiosurgery treatments, e.g., arteriovenous malformations [83,84].

Even with a limited number of indications, the distinct advantages of proton treatments compared to photon treatments were seen early on [85]. One was able to demonstrate clinical efficacy of proton radiation therapy in otherwise poorly manageable disease, e.g., for chordoma and chondrosarcoma of the skull base and the spine [86,87]. These present significant treatment challenges as they are often very close to critical structures, e.g., the brainstem, spinal cord, or optic nerves.

1.4.3 Current Clinical Implications

Today, proton therapy is a well-established treatment option for many tumor types and sites. Advantages when using protons in favor of photons have been shown in terms of tumor control probability and/or normal tissue complications probability. Various dosimetric studies clearly demonstrate superior normal tissue sparing with protons [88–99]. It is well recognized that protons are extremely valuable to treat tumors close to critical structures, e.g., for head-and-neck treatments [100]. However, there are circumstances and treatment sites where the advantage appears to be marginal at best [101].

In the pediatric patient population, the impact of the decreased total absorbed energy in the patient (by a factor of at least 2–3 [92]) with protons is most significant. The overall quality of life and reduction of secondary effects is of great importance, and the reduction in overall normal tissue dose is proven to be relevant [91]. Using protons for craniospinal cases can reduce the dose to the thyroid glands significantly. One prime example is the treatment of medulloblastoma, a malignant tumor that originates in the medulla and extends into the cerebellum. Treatment with photon radiation therapy invariably causes significant dose to the heart, lung, and abdominal tissues as well as organs at risk in the cranium, something that can largely be avoided using protons. These facts have boosted proton therapy, in particular, for pediatric patients. For example, at the MGH, about 90% of the pediatric patient population in radiation oncology is treated with proton therapy. About 60% of those are brain tumors.

Although the dose distributions achievable with protons are superior to those achievable with photons, it is debatable whether the dosimetric advantages of proton therapy are clinically significant for all treatment sites. Various randomized clinical trials comparing protons and photons are currently being conducted for sites such as the breast, prostate, lung, and many others. A review in early 2017 lists 122 active clinical trials involving proton therapy [102]. There is an ongoing discussion about the necessity for randomized clinical trials to show a significant advantage in outcome when using protons in favor of photons [103–106]. It is likely that for specific sites, proton therapy might be advantageous only for a subset of patient. This may lead to model-based trials to stratify patients into randomization [107]. In this approach, mathematical models are used to predict patient-specific outcomes, which then determine enrollment in a trial. Note that data on late morbidity are still scarce due to the follow-up of less than 20 years for most patients.

1.4.4 Economic Considerations

Related to the question of clinical trials mentioned earlier is the cost of health care, i.e., whether a potential gain in tumor control or reduced tissue complication is substantial enough to warrant the additional cost of proton therapy. This is one of the reasons why the treatment of prostate cancer with protons has been criticized [106,108], and it was argued that due to the limited availability of proton beams, proton therapy might be used predominantly for such cases where protons are expected to make the biggest difference, e.g., for the pediatric patient population [109]. The cost of a proton treatment is expected to decrease with the advent of more and more facilities. A detailed discussion on the economic aspects of proton therapy is beyond the scope of this book, and the reader may be referred to publications on this subject [110].

REFERENCES

1. Roentgen WC. Über eine neue Art von Strahlen. *Sitzungsberichte der Würzburger Physikalischmedicinischen Gesellschaft.* 1895;137:41.
2. Roentgen WC. On a new kind of rays. *Nature.* 1896;53(1369):274–6.
3. Hewitt HB. Rationalizing radiotherapy: Some historical aspects of the endeavour. *Br J Radiol.* 1973;46(550):917–26.
4. Wilson RR. Radiological use of fast protons. *Radiology.* 1946;47:487–91.
5. Tobias CA, Anger HO, Lawrence JH. Radiological use of high energy deuterons and alpha particles. *Am J Roentgenol Radium Ther Nucl Med.* 1952;67(1):1–27.
6. Ashikawa JK, Sondhaus CA, Tobias CA, Kayfetz LL, Stephens SO, Donovan M. Acute effects of high-energy protons and alpha particles in mice. *Radiat Res Suppl.* 1967;7:312–24.
7. Lawrence JH. Proton irradiation of the pituitary. *Cancer.* 1957;10(4):795–8.
8. Tobias CA, Lawrence JH, Born JL, McCombs R, Roberts JE, Anger HO, et al. Pituitary irradiation with high energy proton beams: A preliminary report. *Cancer Res.* 1958;18:121–34.
9. Lawrence JH, Tobias CA, Born JL, Wangcc, Linfoot JH. Heavy-particle irradiation in neoplastic and neurologic disease. *J Neurosurg.* 1962;19:717–22.
10. Falkmer S, Larsson B, Stenson S. Effects of single dose proton irradiation of normal skin and Vx2 carcinoma in rabbit ears: A comparative investigation with protons and roentgen rays. *Acta Radiol.* 1959;52:217–34.
11. Larsson B, Leksell L, Rexed B, Sourander P. Effect of high energy protons on the spinal cord. *Acta Radiol.* 1959;51(1):52–64.
12. Leksell L, Larsson B, Andersson B, Rexed B, Sourander P, Mair W. Lesions in the depth of the brain produced by a beam of high energy protons. *Acta Radiol.* 1960;54:251–64.
13. Larsson B. Blood vessel changes following local irradiation of the brain with high-energy protons. *Acta Soc Med Ups.* 1960;65:51–71.
14. Larsson B. Pre-therapeutic physical experiments with high energy protons. *Br J Radiol.* 1961;34:143–51.

15. Larsson B, Leksell L, Rexed B. The use of high-energy protons for cerebral surgery in man. *Acta Chir Scandinavia*. 1963;125:1–7.

16. Larsson B, Leksell L, Rexed B, Sourander P, Mair W, Andersson B. The high-energy proton beam as a neurosurgical tool. *Nature*. 1958;182(4644):1222–3.

17. Falkmer S, Fors B, Larsson B, Lindell A, Naeslund J, Stenson S. Pilot study on proton irradiation of human carcinoma. *Acta Radiol*. 1962;58:33–51.

18. Fors B, Larsson B, Lindell A, Naeslund J, Stenson S. Effect of high energy protons on human genital carcinoma. *Acta Radiol Ther Phys Biol*. 1964;2:384–98.

19. Montelius A, Blomquist E, Naeser P, Brahme A, Carlsson J, Carlsson A-C, et al. The narrow proton beam therapy unit at the Svedberg laboratory in Uppsala. *Acta Oncol*. 1991;30:739–45.

20. Kjellberg RN, Koehler AM, Preston WM, Sweet WH. Stereotaxic instrument for use with the Bragg peak of a proton beam. *Confin Neurol*. 1962;22:183–9.

21. Larsson B, Kihlman BA. Chromosome aberrations following irradiation with high-energy protons and their secondary radiation: A study of dose distribution and biological efficiency using root-tips of *Vicia faba* and *Allium cepa*. *Int J Radiat Biol*. 1960;2:8–19.

22. Dalrymple GV, Lindsay IR, Hall JD, Mitchell JC, Ghidoni JJ, Kundel HL, et al. The relative biological effectiveness of 138-MeV protons as compared to cobalt-60 gamma radiation. *Radiat Res*. 1966;28:489–506.

23. Dalrymple GV, Lindsay IR, Ghidoni JJ, Hall JD, Mitchell JC, Kundel HL, et al. Some effects of 138-Mev protons on primates. *Radiat Res*. 1966;28(2):471–88.

24. Hall EJ, Kellerer AM, Rossi HH, Yuk-Ming PL. The relative biological effectiveness of 160 MeV protons. II. Biological data and their interpretation in terms of microdosimetry. *Int J Radiat Oncol Biol Phys*. 1978;4:1009–13.

25. Robertson JB, Williams JR, Schmidt RA, Little JB, Flynn DF, Suit HD. Radiobiological studies of a high-energy modulated proton beam utilizing cultured mammalian cells. *Cancer*. 1975;35:1664–77.

26. Tepper J, Verhey L, Goitein M, Suit HD. In vivo determinations of RBE in a high energy modulated proton beam using normal tissue reactions and fractionated dose schedules. *Int J Radiat Oncol Biol Phys*. 1977;2:1115–22.

27. Todd P. Radiobiology with heavy charged particles directed at radiotherapy. *Eur J Cancer*. 1974;10(4):207–10.

28. Kjellberg RN, Sweet WH, Preston WM, Koehler AM. The Bragg peak of a proton beam in intracranial therapy of tumors. *Trans Am Neurol Assoc*. 1962;87:216–8.

29. Kjellberg RN, Kliman B. Bragg peak proton treatment for pituitary-related conditions. *Proc R Soc Med*. 1974;67(1):32–3.

30. Suit HD, Goitein M, Tepper J, Koehler AM, Schmidt RA, Schneider R. Exploratory study of proton radiation therapy using large field techniques and fractionated dose schedules. *Cancer*. 1975;35:1646–57.

31. Gragoudas ES, Goitein M, Koehler AM, Verhey L, Tepper J, Suit HD, et al. Proton irradiation of small choroidal malignant melanomas. *Am J Ophthalmol*. 1977;83(5):665–73.

32. Constable IJ, Goitein M, Koehler AM, Schmidt RA. Small-field irradiation of monkey eyes with protons and photons. *Radiat Res*. 1976;65:304–14.

33. Constable IJ, Roehler AM. Experimental ocular irradiation with accelerated protons. *Invest Ophthalmol*. 1974;13(4):280–7.

34. Shipley WU, Tepper JE, Prout GR, Jr., Verhey LJ, Mendiondo OA, Goitein M, et al. Proton radiation as boost therapy for localized prostatic carcinoma. *JAMA*. 1979;241(18):1912–5.

35. Chuvilo IV, Goldin LL, Khoroshkov VS, Blokhin SE, Breyev VM, Vorontsov IA, et al. ITEP synchrotron proton beam in radiotherapy. *Int J Radiat Oncol Biol Phys*. 1984;10(2):185–95.

36. Khoroshkov VS, Goldin LL. Medical proton accelerator facility. *Int J Radiat Oncol Biol Phys*. 1988;15(4):973–8.

37. Dzhelepov VP, Komarov VI, Savchenko OV. Development of a proton beam synchrocyclotron with energy from 100 to 200 MeV for medico-biological research. *Med Radiol (Mosk)*. 1969;14(4):54–8.

38. Khoroshkov VS, Barabash LZ, Barkhudarian AV, Gol'din LL, Lomanov MF, Pliashkevich LN, et al. A proton beam accelerator ITEF for radiation therapy. *Med Radiol (Mosk)*. 1969;14(4):58–62.

39. Dzhelepov VP, Savchenko OV, Komarov VI, Abasov VM, Goldin LL, Onossovsky KK, et al. Use of USSR proton accelerators for medical purposes. *IEEE Trans Nuclear Sci*. 1973;20:268–70.

40. Abrosimov NK, Gavrikov YA, Ivanov EM, Karlin DL, Khanzadeev AV, Yalynych NN, et al. 1000 MeV proton beam therapy facility at Petersburg Nuclear Physics Institute synchrocyclotron. *J Phys: Conf Ser*. 2006;41:424–32.

41. Savinskaia AP, Minakova EI. Proton hypophysectomy and the induction of mammary cancer. *Med Radiol (Mosk)*. 1979;24(2):53–7.

42. Kanai T, Kawachi K, Kumamoto Y, Ogawa H, Yamada T, Matsuzawa H, et al. Spot scanning system for proton radiotherapy. *Med Phys*. 1980;7:365–9.

43. Kurihara D, Suwa S, Tachikawa A, Takada Y, Takikawa K. A 300-MeV proton beam line with energy degrader for medical science. *Jap J Applied Phys*. 1983;22:1599–605.

44. Zografos L, Perret C, Egger E, Gailloud C, Greiner R. Proton beam irradiation of uveal melanomas at Paul Scherrer Institute (former SIN). *Strahlenther Onkol*. 1990;166(1):114.

45. Slater JM, Archambeau JO, Miller DW, Notarus MI, Preston W, Slater JD. The proton treatment center at Loma linda university medical center: Rationale for and description of its development. *Int J Radiat Oncol Biol Phys*. 1992;22:383–9.

46. Cole F, Livdahl PV, Mills F, Teng L. Design and application of a proton therapy accelerator. Proc 1987 IEEE Particle Accelerator Conference. 1987;Piscataway, NJ: IEEE Press. 1985–1987(Lindstrom E. R., Taylor L. S., eds.).

47. Koehler AM. Preliminary design study for a corkscrew gantry. Harvard Cyclotron Laboratory report. 1987.

48. Raju MR. Proton radiobiology, radiosurgery and radiotherapy. *Int J Radiat Biol*. 1995;67:237–59.

49. Lawrence EO, Edlefson NE. On the production of high speed protons. *Science*. 1930;72:376–7.

50. Veksler V. Concerning some new methods of acceleration of relativistic particles. Doklady Acad Sci USSR. 1944;43:444.

51. Oliphant MO. The acceleration of particles to very high energies. Classified memo submitted to DSIR, United Kingdom; now in University of Birmingham Archive. 1943.

52. Wilson RR. Range, Straggling, and multiple scattering of fast protons. *Physical Rev*. 1947;74:385–6.

53. Koehler AM. Dosimetry of proton beams using small silicon detectors. *Radiat Res*. 1967;7:s53–s63.

54. Koehler AM, Preston WM. Protons in radiation therapy. *Radiology*. 1972;104:191–5.

55. Koehler AM, Schneider RJ, Sisterson JM. Range modulators for protons and heavy ions. *Nucl Instrum Methods*. 1975;131:437–40.

56. Blokhin SI, Gol'din LL, Kleinbok Ia L, Lomanov MF, Onosovskii KK, Pavlonskii LM, et al. Dose field formation on proton beam accelerator ITEF. *Med Radiol (Mosk)*. 1970;15(5):64–8.

57. Karlsson BG. Methods for calculating and obtaining some favorable dosage distributions for deep therapy with high energy protons. *Strahlentherapie*. 1964;124:481–92.

58. Koehler AM, Schneider RJ, Sisterson JM. Flattening of proton dose distributions for large-field radiotherapy. *Med Phys*. 1977;4:297–301.

59. Sanberg G. Electron beam flattening with an annular scattering foil. *IEEE Trans Nuclear Sci*. 1973;20:1025.

60. Gottschalk B, Wagner M. Contoured scatterer for proton dose flattening. Harvard Cyclotron Laboratory report. 1989.

61. Goitein M, Chen GTY. Beam scanning for heavy charged particle radiotherapy. *Med Phys.* 1983;10:831–40.
62. Grunder HA, Leemann CW. Present and future sources of protons and heavy ions. *Int J Radiat Oncol Biol Phys.* 1977;3:71–80.
63. Chu WT, Curtis SB, LLacer J, Renner TR, Sorensen RW. Wobbler facility for biomedical experiments at the BEVALAC. *IEEE Trans Nuclear Sci.* 1985;NS-32:3321–3.
64. Leemann C, Alonso J, Grunder H, Hoyer E, Kalnins G, Rondeau D, et al. A 3-dimensional beam scanning system for particle radiation therapy. *IEEE Trans Nuclear Sci.* 1977;NS-24:1052–4.
65. Kanai T, Kawachi K, Matsuzawa H, Inada T. Three-dimensional beam scanning for proton therapy. *Nucl Instrum Methods.* 1983;214:491–6.
66. Kawachi K, Kanai T, Matsuzawa H, Inada T. Three dimensional spot beam scanning method for proton conformation radiation therapy. *Acta Radiol Suppl.* 1983;364:81–8.
67. Kawachi K, Kanai T, Matsuzawa H, Kutsutani-Nakamura Y, Inada T. Proton radiotherapy facility using a spot scanning method. *Nippon Igaku Hoshasen Gakkai Zasshi.* 1982;42(5):467–75.
68. Hiraoka T, Kawashima K, Hoshino K, Kawachi K, Kanai T, Matsuzawa H. Dose distributions for proton spot scanning beams: effect by range modulators. *Nippon Igaku Hoshasen Gakkai Zasshi.* 1983;43(10):1214–23.
69. Chu WT, Ludewigt BA, Renner TR. Instrumentation for treatment of cancer using proton and light-ion beams. *Rev Sci Instrum.* 1993;64:2055–122.
70. Kjellberg RN, Nguyen NC, Kliman B. The Bragg peak proton beam in stereotaxic neurosurgery. *Neurochirurgie.* 1972;18(3):235–65.
71. Goitein M. The measurement of tissue heterodensity to guide charged particle radiotherapy. *Int J Radiat Oncol Biol Phys.* 1977;3:27–33.
72. Suit HD, Goitein M, Tepper JE, Verhey L, Koehler AM, Schneider R, et al. Clinical experience and expectation with protons and heavy ions. *Int J Radiat Oncol Biol Phys.* 1977;3:115–25.
73. Goitein M. Compensation for inhomogeneities in charged particle radiotherapy using computed tomography. *Int J Radiat Oncol Biol Phys.* 1978;4(5–6):499–508.
74. Goitein M. Computed tomography in planning radiation therapy. *Int J Radiat Oncol Biol Phys.* 1979;5(3):445–7.
75. Munzenrider JE, Pilepich M, Rene-Ferrero JB, Tchakarova I, Carter BL. Use of body scanner in radiotherapy treatment planning. *Cancer.* 1977;40(1):170–9.
76. Goitein M, Abrams M, Gentry R, Urie M, Verhey L, Wagner M. Planning treatment with heavy charged particles. *Int J Radiat Oncol Biol Phys.* 1982;8:2065–70.
77. Goitein M, Abrams M. Multi-dimensional treatment planning: I. Delineation of anatomy. *Int J Radiat Oncol Biol Phys.* 1983;9(6):777–87.
78. Goitein M, Abrams M, Rowell D, Pollari H, Wiles J. Multi-dimensional treatment planning: II. Beam's eye-view, back projection, and projection through CT sections. *Int J Radiat Oncol Biol Phys.* 1983;9(6):789–97.
79. Goitein M, Miller T. Planning proton therapy of the eye. *Med. Phys.* 1983;10:275–83.
80. Verhey LJ, Goitein M, McNulty P, Munzenrider JE, Suit HD. Precise positioning of patients for radiation therapy. *Int J Radiat Oncol Biol Phys.* 1982;8:289–94.
81. Suit HD, Goitein M. Dose-limiting tissues in relation to types and location of tumours: implications for efforts to improve radiation dose distributions. *Eur J Cancer.* 1974;10(4):217–24.
82. Gragoudas ES, Seddon JM, Egan K, Glynn R, Munzenrider J, Austin-Seymour M, et al. Long-term results of proton beam irradiated uveal melanomas. *Ophthalmology.* 1987;94(4):349–53.
83. Kjellberg RN, Davis KR, Lyons S, Butler W, Adams RD. Bragg peak proton beam therapy for arteriovenous malformation of the brain. *Clin Neurosurg.* 1983;31:248–90.
84. Kjellberg RN, Hanamura T, Davis KR, Lyons SL, Adams RD. Bragg-peak proton-beam therapy for arteriovenous malformations of the brain. *N Engl J Med.* 1983;309(5):269–74.

85. Suit H, Goitein M, Munzenrider J, Verhey L, Blitzer P, Gragoudas E, et al. Evaluation of the clinical applicability of proton beams in definitive fractionated radiation therapy. *Int J Radiat Oncol Biol Phys.* 1982;8:2199–205.

86. Suit HD, Goitein M, Munzenrider J, Verhey L, Davis KR, Koehler A, et al. Definitive radiation therapy for chordoma and chondrosarcoma of base of skull and cervical spine. *J Neurosurg.* 1982;56(3):377–85.

87. Austin-Seymour M, Munzenrieder JE, Goitein M, Gentry R, Gragoudas E, Koehler AM, et al. Progress in low-LET heavy particle therapy: Intracranial and paracranial tumors and uveal melanomas. *Radiat Res.* 1985;104:S219–26.

88. Archambeau JO, Slater JD, Slater JM, Tangeman R. Role for proton beam irradiation in treatment of pediatric CNS malignancies. *Int J Radiat Oncol Biol Phys.* 1992;22:287–94.

89. Fuss M, Hug EB, Schaefer RA, Nevinny-Stickel M, Miller DW, Slater JM, et al. Proton radiation therapy (PRT) for pediatric optic pathway gliomas: Comparison with 3D planned conventional photons and a standard photon technique. *Int J Radiat Oncol Biol Phys.* 1999;45:1117–26.

90. Fuss M, Poljanc K, Miller DW, Archambeau JO, Slater JM, Slater JD, et al. Normal tissue complication probability (NTCP) calculations as a means to compare proton and photon plans and evaluation of clinical appropriateness of calculated values. *Int J Cancer (Radiat Oncol Invest).* 2000;90:351–8.

91. Lin R, Hug EB, Schaefer RA, Miller DW, Slater JM, Slater JD. Conformal proton radiation therapy of the posterior fossa: A study comparing protons with three-dimensional planned photons in limiting dose to auditory structures. *Int J Radiat Oncol Biol Phys.* 2000;48:1219–26.

92. Lomax AJ, Bortfeld T, Goitein G, Debus J, Dykstra C, Tercier P-A, et al. A treatment planning inter-comparison of proton and intensity modulated photon radiotherapy. *Radiother Oncol.* 1999;51:257–71.

93. St Clair WH, Adams JA, Bues M, Fullerton BC, La Shell S, Kooy HM, et al. Advantage of protons compared to conventional X-ray or IMRT in the treatment of a pediatric patient with medulloblastoma. *Int J Radiat Oncol Biol Phys.* 2004;58(3):727–34.

94. Suit HD, Goldberg S, Niemierko A, Trofimov A, Adams J, Paganetti H, et al. Proton beams to replace photon beams in radical dose treatments. *Acta Oncol.* 2003;42:800–8.

95. Yock T, Schneider R, Friedmann A, Adams J, Fullerton B, Tarbell N. Proton radiotherapy for orbital rhabdomyosarcoma: clinical outcome and a dosimetric comparison with photons. *Int J Radiat Oncol Biol Phys.* 2005;63(4):1161–8.

96. Isacsson U, Hagberg H, Johansson K-A, Montelius A, Jung B, Glimelius B. Potential advantages of protons over conventional radiation beams for paraspinal tumours. *Radiother Oncol.* 1997;45:63–70.

97. Isacsson U, Montelius A, Jung B, Glimelius B. Comparative treatment planning between proton and X-ray therapy in locally advanced rectal cancer. *Radiother Oncol.* 1996;41:263–72.

98. Miralbell R, Lomax A, Russo M. Potential role of proton therapy in the treatment of pediatric medulloblastoma/primitive neuro-ectodermal tumors: Spinal theca irradiation. *Int J Radiat Oncol Biol Phys.* 1997;38:805–11.

99. Weber DC, Trofimov AV, Delaney TF, Bortfeld T. A treatment plan comparison of intensity modulated photon and proton therapy for paraspinal sarcomas. *Int J Radiat Oncol Biol Phys.* 2004;58:1596–606.

100. Chan AW, Liebsch NJ. Proton radiation therapy for head and neck cancer. *J Surg Oncol.* 2008;97(8):697–700.

101. Lee M, Wynne C, Webb S, Nahum AE, Dearnaley D. A comparison of proton and megavoltage X-ray treatment planning for prostate cancer. *Radiother Oncol.* 1994;33:239–53.

102. Mishra MV, Aggarwal S, Bentzen SM, Knight N, Mehta MP, Regine WF. Establishing evidence-based indications for proton therapy: An overview of current clinical trials. *Int J Radiat Oncol Biol Phys.* 2017;97(2):228–35.

103. Glimelius B, Montelius A. Proton beam therapy – do we need the randomised trials and can we do them? *Radiother Oncol.* 2007;83(2):105–9.
104. Goitein M, Cox JD. Should randomized clinical trials be required for proton radiotherapy? *J Clin Oncol.* 2008;26(2):175–6.
105. Goitein M. Trials and tribulations in charged particle radiotherapy. *Radiother Oncol.* 2010;95(1):23–31.
106. Brada M, Pijls-Johannesma M, De Ruysscher D. Current clinical evidence for proton therapy. *Cancer J.* 2009;15(4):319–24.
107. Langendijk JA, Lambin P, De Ruysscher D, Widder J, Bos M, Verheij M. Selection of patients for radiotherapy with protons aiming at reduction of side effects: the model-based approach. *Radiother Oncol.* 2013;107(3):267–73.
108. Konski A, Speier W, Hanlon A, Beck JR, Pollack A. Is proton beam therapy cost effective in the treatment of adenocarcinoma of the prostate? *J Clin Oncol.* 2007;25(24):3603–8.
109. Jagsi R, DeLaney TF, Donelan K, Tarbell NJ. Real-time rationing of scarce resources: the Northeast Proton Therapy Center experience. *J Clin Oncol.* 2004;22:2246–50.
110. Verma V, Mishra MV, Mehta MP. A systematic review of the cost and cost-effectiveness studies of proton radiotherapy. *Cancer.* 2016;122(10):1483–501.

CHAPTER 2

Physics of Proton Interactions in Matter

Bernard Gottschalk

Harvard University

CONTENTS

2.1 INTRODUCTION

Radiotherapy protons (kinetic energy between 3 and 300 MeV) interact with matter in three ways. They lose energy and eventually stop by multiple electromagnetic (EM) collisions with atomic electrons (and to a much smaller degree, atomic nuclei). We call this process *stopping*. They scatter a few degrees by multiple EM collisions with atomic nuclei (and to a smaller degree, electrons). This is called *multiple Coulomb scattering* (MCS). Finally, they occasionally undergo single *hard scatters* either by the nucleus as a whole or by constituents of the nucleus (protons, neutrons, or clusters such as α particles). Hard scatters are also called *nuclear interactions*, though they may involve either the nuclear (strong) force or the EM force.

There exist well-tested and relatively simple theories for the first two processes: Bethe–Bloch theory for stopping and Molière theory for MCS. By contrast, the theory of hard scatters is complicated and largely phenomenological. We are saved by the fact that hard scatters are relatively infrequent, occurring at roughly 1%/cm in water. Typically, only 20% of the protons in a radiotherapy beam suffer a hard scatter before stopping. Many physics problems, such as beam line design, can be solved by considering only stopping and MCS. For others, such as predicting the dose distribution in a patient, hard scatters must eventually come in as a correction.

Those protons that escape hard scatters are called *primaries*, whereas particles emerging from a hard scatter are called *secondaries*. These include the incident proton, other protons, neutrons, clusters (such as α particles) ejected from the nucleus, and the residual nucleus itself.[*] The previous paragraph says, in other words, that the evolution of the primary beam in a homogeneous medium, or even in a heterogeneous medium, can be computed to a fair approximation ignoring hard scatters.

Absorbed dose (or just "dose" for short) is the energy per unit mass deposited by the beam in the stopping medium. It is measured in Gray (Gy): 1 Gy $\equiv$ 1 J/kg. Thus, dose is a mass density which varies from point to point in the patient, and the common expression "dose to the patient" makes no sense, taken literally. It usually means "maximum dose in the target volume." This may seem like nitpicking, but careful use of language is essential to analyzing a problem and reducing it to mathematics.

The archetypal problem of proton radiotherapy physics is finding the dose everywhere in some region of interest (ROI) exposed to a known proton beam or beams. The dose at any point of interest (POI) depends on the number of particles and their propensity to lose energy. Stated quantitatively, that leads to the concepts of *fluence* and *mass stopping power* and, eventually, to the fundamental formula for dose discussed in Section 2.2.

---

[*] Our definition is not universal. Some Monte Carlo programs regard elastically scattered protons as primaries. The rationale for our definition will be more obvious later.

After laying that groundwork, we cover the three processes by which fluence, mass stopping power, and, eventually, dose can be computed. Stopping is discussed in Section 2.3, MCS in Section 2.4, and hard scatters in Section 2.5. These processes underlie the *Bragg curve*, meaning the entire depth-dose distribution of a monoenergetic beam stopping in a uniform medium (usually water, the simplest proxy for tissue). *Bragg peak* (BP) refers to the peak in the Bragg curve near end-of-range. The BP is the defining characteristic of radiotherapy protons and other charged particle beams. Section 2.6 covers the Bragg curve and the closely related concept of *effective stopping power*.

That finishes the fundamentals. Although some interesting problems involve stopping or MCS separately, to compute dose we must recognize that stopping and MCS happen *simultaneously* in the stopping medium, thus requiring a *transport theory*. Luckily that exists in the form of *Fermi–Eyges* (FE) theory, and a comprehensive procedure for finding dose, a *dose algorithm*, can be built on it. For completeness, we sketch that development in Section 2.7. Proton dose algorithms will be covered more fully in later chapters.

Appendix A is a summary of symbol definitions, and Appendix B gives some useful kinematic relations.

2.2 THE FUNDAMENTAL FORMULA FOR DOSE

Suppose we wish to compute the dose $D(x, y, z) = D(\tilde{x})$ throughout a known heterogeneous terrain (e.g., a beam-spreading line followed by a water tank or patient) exposed to known proton beams. To start, consider a far simpler case: an infinitesimal volume of frontal area dA and thickness dz exposed to dN monoenergetic protons at normal incidence. Then,

$$D \equiv \frac{\text{energy}}{\text{mass}} = \frac{-dN(dT/dz)dz}{\rho\, dA\, dz} = \left(\frac{dN}{dA}\right) \times \left(-\frac{1}{\rho}\frac{dT}{dz}\right) = \Phi \times \frac{S}{\rho} \tag{2.1}$$

dose = fluence × mass stopping power

Equation 2.1 is the starting point of every dose calculation.[*] However, the terminology is often disguised. For instance, S/ρ is frequently called the "central axis term" and Φ the "off-axis term." That language obscures the essential point: in a particle beam, the dose at any point equals the areal density of particles times their rate, with z, of energy loss.

Equation 2.1 is inconvenient in MKS units. Also, it is easier to use total proton *charge* rather than the number of protons. In practical units, it follows that

$$D = \Phi_q(S/\rho) = (q/A)(S/\rho) \qquad \text{Gy} \tag{2.2}$$

with q/A in nC/cm^2 and S/ρ in MeV/(g/cm^2), as it is usually tabulated. Finally, if we are interested in dose *rate*, we take the derivative with respect to time obtaining

---

[*] The minus sign makes D positive even though energy *decreases* with z. Also, our definition of fluence assumes the protons are all more or less normal to dA as is often the case. Otherwise, a more general definition is required [1,2].

$$\dot{D} = \dot{\Phi}_q(S/\rho) = (i/A)(S/\rho) \qquad \text{Gy/s} \tag{2.3}$$

with i/A in nA/cm^2.

To illustrate, the dose rate at the surface of a water cylinder of radius 10 cm irradiated uniformly by 1 nA of 160 MeV protons is (anticipating Section 2.3.2 for S/ρ)

$$\frac{1 \text{ nA}}{\pi 10^2 \text{ cm}^2} \times 5.167 \frac{\text{MeV}}{\text{g/cm}^2} = 0.0164 \frac{\text{Gy}}{\text{s}} \approx 1 \frac{\text{Gy}}{\text{min}}$$

Since 1 Gy/min is a reasonable clinical dose rate, we have already learned something: proton radiotherapy currents are on the order of nA.

In more complicated cases, multiple proton bundles ("pencil beams" (PBs)) of different energies may pass near enough to the POI to contribute some dose. Fluence, stopping power, and dose must then be calculated separately for each PB, and the doses added.

If "dose" is not qualified, it means dose to the material at the POI. That is not always what we want to know. For instance, in a dosimeter calibration beam, the dosimeter might actually be in air. However, we are not interested in the dose to air at the POI, or even the dose to the materials making up the dosimeter, but rather the dose that would be delivered to a small imaginary volume of water (*dose to water*) at that same point. In that case, we will compute Φ and proton energy at the POI, assuming the actual beam line including air and a geometric model of the dosimeter, but we will use S/ρ for water at that energy.

2.3 STOPPING

Most radiotherapy protons traversing matter slow down and eventually stop by myriad soft EM collisions with atomic electrons. (These are the "primaries" defined earlier.) The present section concerns the *continuous slowing down approximation* (CSDA) theory of this process, range-energy tables, and some consequences for proton radiotherapy calculations.

2.3.1 Experimental Definition of Range

The *mean projected range* of primaries (which we will simply call *range*) can be measured as shown in Figure 2.1. A beam monitor is followed by a variable-thickness column of the material under test. The protons stop in a Faraday cup (FC) which measures total charge, acting as a proton counter. In the geometry shown (FC far from the stack), the FC catches all the primaries but few secondaries, because of their large angles and low energy. The measured charge (filled circles) at first falls slowly and almost linearly with stack thickness at $\approx 1\%/(\text{g/cm}^2)$ as primaries are lost to hard scatters. It then falls steeply to zero as the primaries range out. The range R (cm) or mass range ρR (g/cm^2) of the beam is defined as the halfway point of the steep part. It is the depth at which half the surviving primaries (n.b. not half the incident protons!) have stopped.

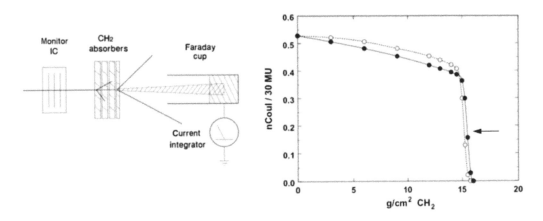

FIGURE 2.1 Measuring range with an FC. Filled circles correspond to the setup shown; open circles, to an FC placed much closer to the material under test. The arrow points to the mean projected range.

If the FC shown schematically in Figure 2.1 (see [3] for a complete description) is replaced by a vacuumless or "poor man's" FC (PMFC) [4] placed much nearer the stack, the curve changes to that shown in open circles because some secondaries now enter the FC.[*]

Alternatively, one can use a multilayer FC (MLFC) (see [5,6] for descriptions) obtaining the middle panel of Figure 2.2, where the range corresponds to the peak channel, or more precisely, the mean channel of the sharp peak. The MLFC allows the range to be verified to a fraction of a millimeter (water equivalent) in a matter of seconds [7].

In clinical facilities, range is frequently measured with a dosimeter and a water tank yielding a Bragg curve, which is dose vs. depth, *not* number of protons vs. depth. That raises a question: what point on the Bragg curve corresponds to the mean projected range just defined in terms of FC measurements? The answer

$$R = d_{80} \tag{2.4}$$

that is, the mean projected range equals the depth of the distal 80% point of the BP, was first found by A.M. Koehler [8]. It has been confirmed many times since then [9,10]. It follows from the fundamental Equation 2.1 as applied to a depth-dose measurement in water.[†,‡]

As Figure 2.2 shows, if we increase the energy spread of the proton beam, keeping the mean energy the same, the mean projected range measured all three ways, in particular using Equation 2.4, remains the same.

In the general physics literature, such as range-energy tables, "range" means precisely the quantity defined earlier, namely the depth at which half the surviving primaries have

---

[*] The sharp drop-off shifts slightly because the entrance window of this PMFC happened to be thicker.

[†] If we want the range of protons entering the proton nozzle, we must of course correct for the range loss in scatterers, air, tank wall, and dosimeter wall.

[‡] In particle beams, depth-dose measurements are best made with a plane-parallel ionization chamber (IC) rather than a cylindrical IC [11].

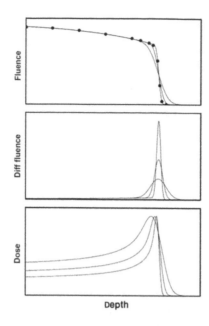

FIGURE 2.2   Illustrating $R = d_{80}$. Top panel: range measured with a simple FC. Middle panel: range measured with a multilayer FC. Bottom panel: range measured with a dosimeter in a water tank. Beams with the same mean energy but different energy spreads are shown. R is invariant in the FC curves, whereas d_{80} (= R) is invariant in the Bragg curve.

stopped. Unfortunately, in clinical practice, "range" is used rather loosely, often denoting d_{90} rather than d_{80}. Consistent use of the d_{xx} notation will help reduce confusion.

2.3.2 Theoretical CSDA Range

The theoretical CSDA rate of energy loss of fast charged particles in matter was found by Bethe and Bloch around 1933. Good modern accounts with many references can be found in introductions to the range-energy tables of Janni [12,13] and ICRU (International Commission on Radiation Units and Measurements) Report 49 [14]. If we restrict ourselves to 3–300 MeV protons (neither very low nor very high energy), most of the corrections described in those accounts are negligible, as is the energy loss to recoil nuclei. Taking advantage of that and applicable kinematic approximations, the mass stopping power in an elementary material of atomic number Z and atomic mass A is

$$\frac{S}{\rho} \equiv -\frac{1}{\rho}\frac{dT}{dz} = 0.3072\frac{Z}{A}\frac{1}{\beta^2}\left(\ln\frac{W_m}{I} - \beta^2 \right) \quad \frac{\text{MeV}}{\text{g/cm}^2} \tag{2.5}$$

where $\beta \equiv v/c$ of the proton, see Equation 2.48, and

$$W_m = \frac{2m_e c^2 \beta^2}{1-\beta^2} \tag{2.6}$$

is the largest possible proton energy loss in a single collision with a free electron (rest energy $m_e c^2 \approx 0.511$ MeV). I is the *mean excitation energy* of the target material, a critical parameter to which we will return.

Once we know S/ρ as a function of β (that is, of T), we can find the range of a proton by imagining that it enters the material at $T_{initial}$ and subtracting the energy lost in very thin slabs until the energy reaches some low value T_{final} (but not 0, because Equation 2.5 diverges there). The choice of T_{final} is not critical. Any very small value will do. Thus,

$$\rho R(T_{initial}) = \int_{T_{initial}}^{T_{final}} \left(\frac{1}{\rho} \frac{dT}{dx} \right)^{-1} dT = \int_{T_{finall}}^{T_{initial}} \frac{dT}{[S/\rho](T)} \tag{2.7}$$

is the theoretical CSDA range in g/cm². Because of MCS protons actually travel a wiggly path, so strictly speaking, the quantity we have computed is the total path length rather than mean projected range, ρR being smaller by a "detour factor." However, that correction is also negligible in the clinical regime (0.9988 for 100 MeV protons in water [14]). Here is a short range-energy table for water [13]:

kinetic energy (MeV)	1	3	10	30	100	300
Range (cm H_2O)	0.002	0.015	0.125	0.896	7.793	51.87

justifying our choice of 3–300 MeV as the clinical regime.

Having outlined the computation of ρR, we return to the mean excitation energy. I cannot be calculated accurately from first principles, so it is, in effect, an adjustable parameter of the theory. It depends on target material and is found by fitting measured ranges or stopping powers, when available, and by interpolation otherwise. It is roughly proportional to Z ($I \approx 10Z$ eV), but irregularities due to atomic shell structure make interpolation difficult [14]. Fortunately, S/ρ is logarithmic in I so that at 100 MeV for instance, a relative increase of 1% in I only causes an approximately 0.15% decrease in S/ρ.

If the stopping material is a mixture of elements, the atoms act separately and we can replace the mixture by a succession of thin sheets of each constituent element. That picture leads easily to the *Bragg additivity rule*

$$\frac{S}{\rho} = \sum_i w_i \left(\frac{S}{\rho} \right)_i \tag{2.8}$$

where w_i is the fraction by weight of the ith element.

Compounds are more complicated since their constituent atoms are bound in molecules and do not, strictly speaking, act separately. However, the Bragg rule seems to hold quite well [13] even then.

Water, often a proxy for tissue in radiotherapy and therefore of particular interest, is particularly complicated being a polar molecule. There is experimental evidence [15–17] that Janni [13] ($\approx$0.9% higher) is more accurate than ICRU Report 49 [14] for water.

To summarize, choosing I is a complicated business requiring considerable familiarity with the experimental literature. It is mainly for this reason that our practice is to obtain mass stopping power S/ρ and mass range ρR values by interpolating generally accepted range-energy tables rather than by computing them *ab initio* as outlined earlier. Tables can differ from each other by 1%–2% due solely to different choices of I. For the same reason, a given set of tables may be better for some materials than others.

One percent of range at 180 MeV corresponds to ≈2 mm water. Therefore, when the treatment depth itself depends on it, we must rely on *measured* ranges in water and *measured* water equivalents of other materials, rather than range-energy tables. Even so, the tables are invaluable in many other calculations such as beam line design.

A final note on compounds. If a compound is not listed in one of the standard tables, it might seem that we need to look up S/ρ for each constituent, combine them using the Bragg rule, and integrate the resulting stopping power table to obtain ρR. However, a shortcut is provided by Rasouli et al. [18] who combine ranges directly in the simple formula:

$$\frac{1}{\rho R} = \sum_i \frac{w_i}{(\rho R)_i} \tag{2.9}$$

which assumes the Bragg rule holds and is better than ±0.03% for an arbitrary collection of ten compounds over ranges from 0 to 45 cm [18]. Since Janni [13] gives the range-energy relation for all chemical elements, Equation 2.9 easily yields the proton range in any compound or mixture at any energy.

2.3.3 Interpolating Range-Energy Tables

Figure 2.3 shows the range-energy relation for some useful materials in the clinical regime. At a given energy, range is greater (stopping power is lower) for heavy materials. A plot of $\log R(\log T)$ is nearly linear. If it were exactly linear, the relation would be an exact power law $R = aT^b$, and if the lines were parallel, b would be the same for all materials. As it is, the power varies with material as well as T (the lines are not quite parallel or straight). For maximum accuracy, therefore, we parameterize the log-log graph for each material separately, using cubic spline interpolation of tabulated values at 1, 2, 5… 500, 1,000 MeV. Even this sparse data set yields ±0.1% accuracy in the clinical regime. This is the method used by LOOKUP [19].

For interpolation by hand, use *power-law* interpolation $R = aT^b$, particularly if the table steps are large. Linear interpolation always yields too large an answer because $R(T)$ is concave upwards.

When seeking answers in closed form as in [10], two-parameter approximations may be useful despite their reduced accuracy. We have already mentioned the power law $R = aT^b$. Another is the *Øverås approximation* [20], which can be written as $R = a(pv)^b$ and takes advantage of the close relation between pv and T, see Equation 2.49. The two have comparable accuracy (≈2%) but are seen (Figure 2.4) to be complementary. Øverås is better for light materials and $R = aT^b$ for heavy materials. Details depend on the fiducial energies (here 32 and 160 MeV), and it is possible to make one look considerably better than the

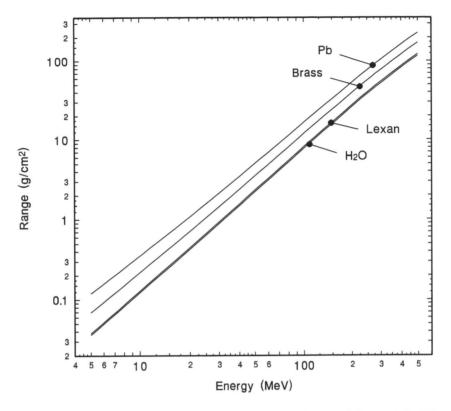

FIGURE 2.3   Range-energy relation of radiotherapy protons in four useful materials [14].

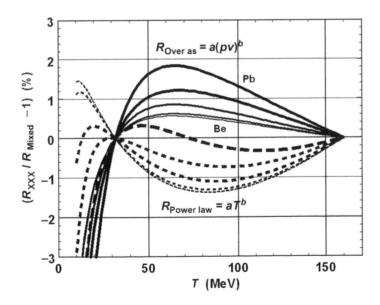

FIGURE 2.4   Comparison of the Øverås range-energy approximation (solid lines) to the more common $R = aT^b$ (dashed lines). Materials in order of increasing line weight are Be, water, Al, Cu, Pb. Parameters in this example are computed from exact fits at 32 and 160 MeV. The MIXED range-energy table corresponds to ICRU 49 [14] except for water which uses Janni82 [13].

other in particular cases. There are even cases [21,22] where the "weak" Øverås approximation $R = A(pv)^2$ (different A) is sufficiently good.

2.3.4 Range Straggling

Returning to Figure 2.1, note that the final falloff is steep but not infinitely so. Because EM stopping involves multiple discrete and random energy transfers, protons stop at slightly different depths. This *range straggling* yields a Gaussian distribution of stopping depths, characterized by σ_S. See [13] for theory and tables. Figure 2.5, a plot of Janni's σ_S/R vs. incident energy, is a summary. Roughly speaking, σ_S is a constant fraction of range. For light materials, the coefficient is ≈1.2%.

The fact that the coefficient is not very material-dependent is of practical importance. If materials other than water reduce the incident proton energy, the Bragg curve will not be very different from that observed in water alone. That greatly simplifies the design of passive range modulators or the computation of dose in a heterogeneous terrain.

Due to range straggling, at any depth in a water tank, protons will have a distribution of energies (*energy straggling*) even if the incident beam is monoenergetic. The same applies to protons leaving any finite slab of any material. Most pronounced for near-stopping depths or slab thicknesses, this is just another face of the same phenomenon. The energy distribution will be still broader if the incident beam also has an energy spread.

2.4 MULTIPLE COULOMB SCATTERING

MCS is the random deflection of protons by myriad EM interactions with atomic nuclei and (in low-Z materials) atomic electrons. Figure 2.6 illustrates a simple experiment. The transverse distribution on the *measuring plane* reflects the angular distribution emerging from the target, which is what MCS theory predicts.*

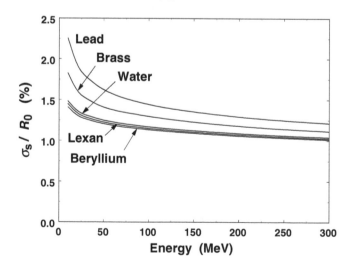

FIGURE 2.5 Range straggling in several materials. (From Janni [13].)

---

* Technically, finding the spatial distribution at the measuring plane is a transport problem, but it is a trivial one if energy loss and MCS in the air are ignored.

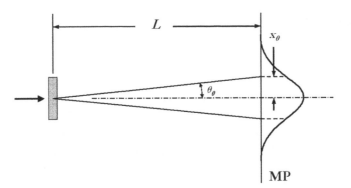

FIGURE 2.6 Multiple Coulomb scattering in a thin slab.

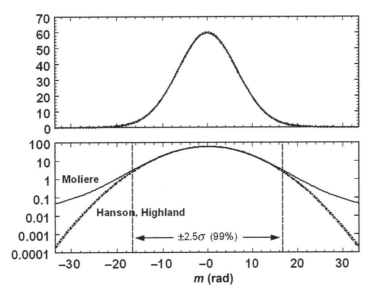

FIGURE 2.7   Angular distributions for 158.6   MeV protons traversing 1 cm of water. On a linear plot, the Molière/Fano distribution is indistinguishable from Gaussians using the Hanson or Highland θ_0. On a log plot, the correct distribution peels away at 2.5σ, and is more than $100\times$ higher at 5σ. 2.5σ encompasses 99% of the protons.

The angular distribution has a Gaussian core and a tail that falls much more slowly (Figure 2.7). The core contains some 99% of the protons so the Gaussian approximation suffices for the great majority of proton radiotherapy problems. Then the only thing that remains for MCS theory to predict is the Gaussian width parameter as a function of incident particle species and energy and the thickness and atomic properties of the target. We will focus on that rather than the complete theory, which also predicts the angular distribution in the non-Gaussian region (the "single scattering tail").

2.4.1 Suggested Reading

The good news is that you only need one theory, that of Molière. It has no adjustable parameters and is thought [23] to be accurate to 1%. For a review including experimental

evidence and many references but omitting derivations, see [24]. For a theoretical derivation in English and a good account of competing theories, see Bethe [23]. (However, the
thick target and compound-target aspects of Molière theory are missing.) The German
reader will find Molière's original papers [25,26] elegant and comprehensive, except that
experimental data were sparse at that early date.

The bad news is that much of the MCS literature is unhelpful and sometimes incorrect.
Ignore claims that Molière didn't cover energy loss by the incident particle (he did, and
the theory extends to near-stopping targets), that the more complicated "NSW" theory (by
Nigam, Sundaresan and Wu) is better (it isn't) or occasionally that Molière doesn't agree
with experiment (it does). At one time, complicated schemes were proposed to improve
Rossi's simple formula for the Gaussian approximation. These were all swept away by
Highland's elegant parameterization [27].

Finally, even though adding mixed slabs in quadrature is incorrect in principle, it works
in practice, and the alternative suggested by Lynch and Dahl [28] is unworkable. We'll
address that in connection with scattering power and transport theory.

2.4.2 Elements of Molière Theory

Molière theory is algebraically complicated.[*] Here we will only treat the simplest case,
a target consisting of a single element (atomic weight A, atomic number Z) sufficiently
thin that the proton (charge number z, momentum p, speed v) loses negligible energy.
We assume large Z so that scattering by atomic electrons is negligible. We wish to compute the distribution of proton space angle θ given protons of known energy entering a
target of mass thickness t (g/cm²), where t (for now) is much less than the proton mass
range ρR.

We first calculate a characteristic single scattering angle χ_c given by

$$\chi_c^2 = c_3 t (pv)^2 \tag{2.10}$$

where

$$c_3 \equiv 4\pi N_A \left(\frac{e^2}{\hbar c}\right)^2 (\hbar c)^2 \frac{z^2 Z^2}{A} \tag{2.11}$$

$N_A \approx 6.022 \times 10^{23}$ (g mol wt)$^{-1}$ is Avogadro's number, $(e^2/\hbar c) \approx 1/137$ is the fine structure
constant and $(\hbar c) \approx 197 \times 10^{-13}$ MeV cm is the usual conversion factor. The physical interpretation of χ_c is that, on average, a proton suffers exactly one single scatter greater than χ_c
in traversing the target.

Next, we compute a *screening angle* χ_a using

$$\chi_a^2 = \chi_0^2 (1.13 + 3.76\alpha^2) \tag{2.12}$$

---

[*] LOOKUP [19] computes MCS theory in all its variants for any target material and thickness.

where

$$\chi_0^2 = c_2 / (pc)^2 \tag{2.13}$$

and the *Born parameter* α is given by

$$\alpha^2 = c_1 / \beta^2 \tag{2.14}$$

where $\beta \equiv v/c$ of the proton, see Equation 2.48. The constants are

$$c_1 \equiv \left[\left(\frac{e^2}{\hbar c} \right) z\, Z \right]^2 \tag{2.15}$$

and

$$c_2 \equiv \left[\frac{1}{0.885} \left(\frac{e^2}{\hbar c} \right) \left(m_e c^2 \right) Z^{1/3} \right]^2 \tag{2.16}$$

The screening angle is that (very small) angle at which the single scattering cross section levels off (departs from Rutherford's $1/\theta^4$ law) because of the screening of the nuclear charge by atomic electrons. One of Molière's insights was that, though MCS depends on that angle, it is insensitive to the exact shape of the single scattering cross section near that angle. Next we compute a quantity

$$b = \ln \left(\frac{\chi_c^2}{1.167\, \chi_a^2} \right) \tag{2.17}$$

which is the natural logarithm of the effective number of collisions in the target. Next, the *reduced target thickness B* is defined as the root of the equation

$$B - \ln B = b \tag{2.18}$$

which can be solved by standard numerical methods. B is almost proportional to b in the mathematical region of interest.

Finally, Molière's characteristic multiple scattering angle

$$\theta_M = \frac{1}{\sqrt{2}} \left(\chi_c \sqrt{B} \right) \tag{2.19}$$

is analogous to θ_0 in the Gaussian approximation. Typically, it is about 6% larger.[*] Molière is now positioned to compute the distribution of θ. Defining a reduced angle

---

[*] The $1/\sqrt{2}$ is ours. It makes θ_M more or less equivalent to θ_0 in the Gaussian approximation.

$$\theta' \equiv \frac{\theta}{\chi_c \sqrt{B}} \tag{2.20}$$

he approximates the desired distribution function $f(\theta)$ by a power series in $1/B$

$$f(\theta) = \frac{1}{2\pi\theta_M^2} \frac{1}{2} \left[f^{(0)}(\theta') + \frac{f^{(1)}(\theta')}{B} + \frac{f^{(2)}(\theta')}{B^2} \right] \tag{2.21}$$

where

$$f^{(n)}(\theta') = \frac{1}{n!} \int_0^\infty y \, dy \, J_0(\theta' y) e^{y^2/4} \left(\frac{y^2}{4} \ln \frac{y^2}{4} \right)^n \tag{2.22}$$

$f^{(0)}$ is a Gaussian

$$f^{(0)}(\theta') = 2e^{-\theta'^2} \tag{2.23}$$

Molière [26] gives further formulas and tables for $f^{(1)}$ and $f^{(2)}$.

The foregoing equations, with Bethe's improved tables [23] for $f^{(1)}$ and $f^{(2)}$, permit one to evaluate the scattering probability density $f(\theta)$ if the target consists of a single chemical element with $Z \gg 1$ and the energy loss is small. Except for rearrangements of physical constants to conform to modern usage, and the normalization of $f(\theta)$, our equations are identical to Molière's. We repeat, however, that Molière [26] generalized them to *arbitrary energy loss* and to *compounds and mixtures* [24].

The generalization to low-Z elements, where scattering by atomic electrons (not just the screened nucleus) is appreciable, is handled two ways in the literature. Bethe's approach [23] is simply to substitute $Z(Z+1)$ for Molière's Z^2 wherever it appears. We call this Molière / Bethe. Fano's approach [24, 29, 30] is more complicated. He computes a correction to b. Molière /Fano theory fits experimental proton data from 1 MeV to 200 GeV for a wide variety of materials and thicknesses at the few percent level [24]. Since this is comparable to experimental error, we do not really know how good Molière theory is, but Bethe [23] claims 1%. Unlike stopping theory, there are no adjustable parameters!

Though the derivation of B is complicated, numerically B has a simple interpretation. It is proportional to the logarithm of normalized target thickness $t/(\rho R)$ with a coefficient depending on the material. Figure 2.8 shows that the angular distribution depends very weakly on B.

2.4.3 The Gaussian Approximation

The inset to Figure 2.8 shows that the best-fit Gaussian is somewhat narrower than that obtained by simply dropping $f^{(1)}$ and $f^{(2)}$. This was first pointed out by Hanson et al. [31]

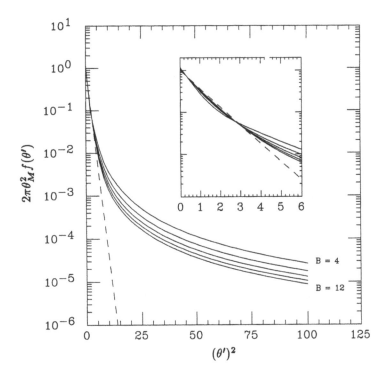

FIGURE 2.8 Molière angular distribution plotted so that a Gaussian becomes a straight line [24]. Dashed line: $f^{(0)}$ only. Inset: graph near the origin showing that the best-fit Gaussian is narrower than simply using $f^{(0)}$.

who, in the course of testing Molière theory with electrons[*], suggested using a Gaussian with a characteristic angle which we will call θ_{Hanson}:

$$\theta_0 = \theta_{\text{Hanson}} \equiv \frac{1}{\sqrt{2}}\left(\chi_c\sqrt{B-1.2}\right) \tag{2.24}$$

That's not a shortcut since one still has to evaluate the full Molière theory. In 1975, however, Highland parameterized Molière/Bethe/Hanson theory and obtained

$$\theta_0 = \theta_{\text{Highland}} \equiv \frac{14.1\ \text{MeV}}{pv}\sqrt{\frac{t}{\rho X_0}}\left[1+\frac{1}{9}\log_{10}\left(\frac{t}{\rho X_0}\right)\right]\quad \text{rad} \tag{2.25}$$

ρX_0 is the mass *radiation length* (g/cm²) of the target material, which can be found in tables or using a standard formula [33]. Equation 2.25 is a huge shortcut, avoiding Molière theory entirely. It also lays bare the structure of the full theory: the Gaussian width varies primarily as the square root of target thickness, but a correction factor proportional to the logarithm of target thickness is required. It is called the *single scattering correction* to θ_0 because it stems from the influence of single scattering at larger angles.

---

[*] Bichsel [32] later confirmed it with protons.

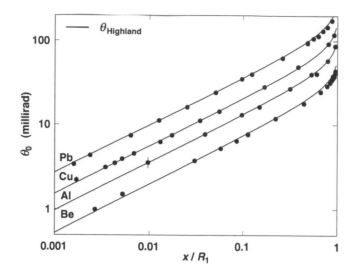

FIGURE 2.9   Accuracy of Highland's formula for four elements. The abscissa is target thickness divided by proton range. Points are experimental data at 158.6 MeV [24].

Equation 2.25 is obviously limited to thin targets, since the depth at which pv is to be evaluated is not specified. (Note that Equation 2.24 is *not* limited to thin targets.) A makeshift generalization to thick targets was proposed by us [24], namely

$$\theta_0 = 14.1\,\text{MeV}\, z \left[1 + \frac{1}{9} \log_{10} \left(\frac{t}{\rho X_0} \right) \right] \times \left(\int_0^t \left(\frac{1}{pv} \right)^2 \frac{dt'}{\rho X_0} \right)^{1/2} \tag{2.26}$$

We took Highland's logarithmic correction factor out of the integral, so that it is evaluated for the *entire* target thickness rather than each step in the integral. (Otherwise, the integral, approximated by a sum, decreases as the step size is decreased.) With that generalization, the Highland formula fits experimental data rather well (Figure 2.9). However, Equation 2.26 has been superseded by the concept of scattering power, to which we now turn.

2.4.4 Scattering Power

Scattering power was introduced by Rossi in 1952 [34], though he did not use that term. It was resurrected and named by Brahme [35] in connection with electron radiotherapy, and now appears regularly in discussions of proton transport. This section is rather technical. It may be skipped if you are only interested in computing multiple scattering in a single homogenous slab, for which we have already given several methods. Ref. [36] is a full discussion of scattering power. What follows is just a summary.

Brahme used an analogy with stopping power, but that is imperfect. Stopping power $S(z) \equiv -dT/dz$ depends solely on the proton speed and atomic properties of the target *at the current depth z* (Equation 2.5). It is a "local" function. We define *scattering power* by

$$T_{xx}(z) \equiv d < \theta_x^2 > / dz \tag{2.27}$$

that is, the rate of increase with z of the variance of the projected MCS angle.[*] To be consistent with what we already know we must recover θ_{Hanson} from Molière/Fano/Hanson theory if we integrate T_{xx} over the z of a full slab. For that to work for any target thickness, we find [36] that T_{xx} must be *nonlocal*. It must depend somehow on the *history* of the beam as well as its energy at the POI.

Before we go into that, we consider an obvious question: Who needs scattering power, since we already know the answer? The response to that is that Molière/Fano/Hanson theory tells us only one thing: the MCS angle out of a *given finite thickness* of a *given single material*. A general transport theory (whether deterministic or Monte Carlo) needs to do much more than that. It must treat both MCS and slowing down at the same time in mixed slabs, in a way that is independent of step size. For that we need a *differential* description of both processes. Stopping is easy: we already have Equation 2.5. But Molière theory treats only *finite* targets. Unlike stopping theory, it does not flow from a differential description. To answer our question: Scattering power can be viewed as a differential formulation of MCS theory, concocted after the fact.

Skipping the details, we find that six prescriptions for $T_{xx}(z)$ occur in the literature. Some are considerably better than others in agreeing with Molière/Fano/Hanson theory when integrated over target thickness. The best and most convenient is the "differential Molière" scattering power derived by the author, namely

$$T_{dM}(z) = f_{dM}(pv, p_1 v_1) \times \left(\frac{15.0\,\mathrm{MeV}}{pv(z)} \right)^2 \frac{1}{X_S(z)} \frac{\mathrm{rad}^2}{\mathrm{cm}} \tag{2.28}$$

where

$$f_{dM} \equiv 0.5244 + 0.1975\log_{10}\left(1 - (pv/p_1 v_1)^2\right) + 0.2320\log_{10}(pv/\mathrm{MeV})$$
$$- 0.0098\log_{10}(pv/\mathrm{MeV})\log_{10}\left(1 - (pv/p_1 v_1)^2\right) \tag{2.29}$$

and

$$\frac{1}{\rho X_S} \equiv \left(0.34896 \times 10^{-3}\right) \times \frac{Z^2}{A}\left\{2\ln\left(33219(AZ)^{-1/3}\right) - 1\right\} \frac{1}{\mathrm{g/cm}^2} \tag{2.30}$$

This reproduces Molière/Fano/Hanson theory to $\approx \pm 2\%$ for normalized target thicknesses from 0.001 to 0.97 (very thin to nearly stopping) over the full periodic table [36].

X_S is a *scattering length*, analogous to radiation length X_0 and of the same order of magnitude. ($X_S/X_0 = 1.42$ for Be decreasing monotonically to 1.04 for Pb.) In compounds and mixtures, X_S, like X_0, obeys a Bragg additivity rule:

$$\frac{1}{\rho X_S} = \sum_i \frac{w_i}{(\rho X_S)_i} \tag{2.31}$$

---

[*] The subscripts xx distinguish different formulas and differentiate scattering power from kinetic energy.

pv is the familiar kinematic quantity (Equation 2.49) at z, the POI, while p_1v_1 is the initial value of the same quantity. Equations 2.28–2.30 apply equally well to mixed homogeneous slabs. $pv(z)$ is computed using the appropriate range-energy relation in each slab, whereas X_S is a piecewise constant function of z.

Figures 2.10 and 2.11 demonstrate the accuracy of integrated T_{dM} for polystyrene and lead (Pb), respectively.*

An important characteristic of $T_{dM}(z)$ is that it is *nonlocal* via the correction factor f_{dM}. Scattering power depends not only on conditions (that is, pv and X_S) at z but also on

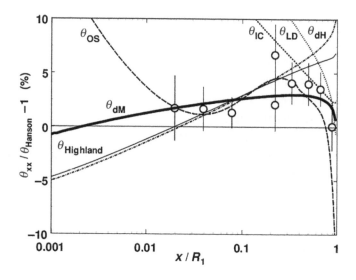

FIGURE 2.10 Compliance of experiment (points, [24]) and integrated T_{xx} (lines, [36]) with Molière / Fano/Hanson theory for polystyrene. The bold line represents T_{dM}.

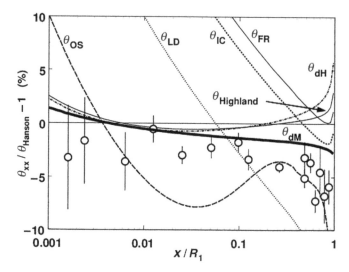

FIGURE 2.11 The same as Figure 2.10 for lead (Pb).

---

* Among inexpensive plastics, polystyrene is the best proxy for water. Thin-walled water targets are difficult to make.

how the protons started out ($p_1 v_1$). To take a simple example, for a 20-MeV proton in Be, $d < \theta_x^2 > /dz$ is smaller if the overlying thickness of Be is 0.1 cm (protons enter at 23.7 MeV) than if it is 5 cm (protons enter at 102 MeV). How can that be? How does the proton "know" what has gone before?

The answer is that, unlike stopping and single scattering, *MCS is not a primitive process.* It makes sense to speak of stopping and single scattering even in an atomic monolayer, and we would expect those to depend only on proton energy and atomic properties of the monolayer. By contrast, *multiple* scattering of a beam is a statistical statement about many protons, each undergoing many collisions, and we should not be too surprised if, in a given slab, that depends on what came before.

The factor f_{dM} measures the beam's progress towards "Gaussianity." Any $T_{xx}(z)$ that has some nonlocality built into it, some sense of the beam's history, is more accurate than any $T_{xx}(z)$ that does not [36]. pv is a particularly convenient proxy for "history," because a transport program needs to keep track of it (or kinetic energy T) anyway.

A final comment on step size. We mentioned that any computation, deterministic or Monte Carlo, should converge as a function of Δz. Over some reasonable range of Δz, the answer should remain the same. T_{dM} has this property, but some Monte Carlo MCS models do not [37], and using a nonlocal scattering power might remedy that.

2.5 NUCLEAR INTERACTIONS (HARD SCATTERS)

This section marks the greatest departure from the first edition of this book. There, we focused on the categorization of nuclear interactions and their relative probabilities. All of that is still correct, but here we focus instead on the effect of hard scatters on the dose distribution of a pencil beam, more directly relevant to pencil beam scanning (PBS).

In addition to multiple soft EM interactions, protons stopping in matter undergo single hard scatters. These throw dose out to large radii, creating what Pedroni et al. [38] called the *nuclear halo*. A more comprehensive terminology for the anatomy of the PB dose was proposed later by us [39]. The *core* is the compact central region due to primaries, the *halo* is the surrounding dose from charged secondaries, and the *aura* is the very large region due to neutral secondaries. The three regions overlap, but each has distinct characteristics as we will see.

There may also be *spray*, additional low dose outside the core which enters with the beam. Unlike the halo and aura, it depends on beam line design and can in principle be avoided. It may arise from hard scatters in upstream degraders [38], MCS in upstream beam profile monitors [40], or protons scraping the beam pipe [41]. We will ignore it here. Obviously, that is not possible in practice. Spray, if present, complicates parameterization of the PB.

The halo and aura comprise a significant fraction of the total proton energy (integrated dose), about 15% at 180 MeV. To see what that implies, suppose we build up a broad dose using PBS, starting at the center and working outwards. As we drop in PBs nearby, their halos overlap the central PB, increasing the dose slightly. That continues until the new PBs are so far out that their halos no longer contribute. Thus, the dose at the center increases by an amount depending on field size. The consequent need to parameterize the halo as well as the core was first pointed out by Pedroni et al. [38].

2.5.1 Contributing Reactions

Reactions leading to hard scatters can categorized in complementary ways. First of all, they may involve the EM or the nuclear force. Next, they may be *coherent* (interaction with the nucleus as a whole) or *incoherent* (interaction with a component of the nucleus). Finally, they may be *elastic* (kinetic energy conserved), *inelastic* (kinetic energy not conserved, recoil nuclide the same as the target, but excited) or *nonelastic* (recoil nuclide different, and possibly excited). We discuss a few such reactions, shown schematically in Figure 2.12.

Hard scattering on free hydrogen ^{1}H(p,p)p is elastic[*], resembling a billiard-ball collision between equal masses. The secondaries emerge 90° apart[†] and share the incident energy, the more forward proton being the more energetic.

Quasi-elastic p-p scattering ^{16}O(p,2p)^{15}N is nonelastic. Kinematics resemble scattering on free hydrogen, somewhat modified by the binding energy E_B required to extract the target proton and by the initial momentum of the target proton [43]. Quasi-elastic proton-neutron ^{16}O(p,pn)^{15}O scattering is similar, except, of course, that one secondary is neutral and travels much farther on average. (Also, the E_B of a neutron is greater.)

Finally, elastic proton-nucleus ^{16}O(p,p)^{16}O scattering to small angles is EM. In fact, it is identical to the Molière single scattering tail. Eventually, it goes over to nuclear scattering which falls far more slowly with angle. The two regimes are separated by a *Coulomb interference* dip. Figure 2.13 shows all this for C (O is very similar). Note that 96 MeV is a representative KE since hard scatters can occur anywhere along the primary's path.

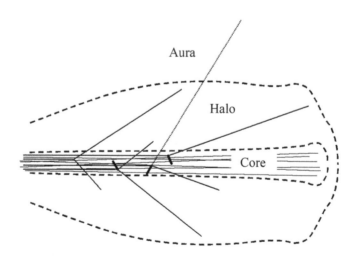

FIGURE 2.12 Core, halo, and aura with reactions ^{1}H(p,p)p, ^{16}O(p,2p)^{15}N, ^{16}O(p,pn)^{15}O and ^{16}O(p,p)^{16}O (from left). Ranges of recoil nuclei are exaggerated. The dashed lines are 10% and 0.01% isodoses drawn to scale.

---

[*] The reaction ^{1}H(p,pγ)p is permitted but is negligible at radiotherapy energies [42].
[†] Actually, slightly less due to a relativistic effect.

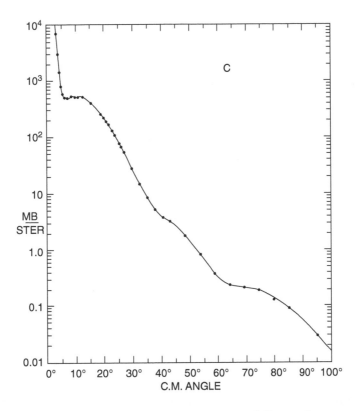

FIGURE 2.13   Gerstein et al. [60] Figure 2.3: Elastic scattering differential cross section of 96-MeV protons on carbon. Note (from left) the Molière single scattering tail, the Coulomb interference dip near 7° and scattering to large angles, with shallow diffraction minima, by the nuclear force.

2.5.2 Shape and Size of the Halo

Using conservation of energy and momentum, one can compute the energy, and therefore the range, of the more energetic secondary in scattering from free H. The quasi-elastic case is more complicated [39,44], but we have good values for E_B as well as the characteristic target proton momentum from the literature [43]. Figure 2.14 shows, for 180 MeV incident protons, stopping points of secondary protons from elastic scattering on free H as well as quasi-elastic scattering in O, assuming the reaction occurs at one of five depths along the primary track. The figure strongly suggests that a longitudinal scan at ≈6 cm will exhibit a dose bump around midrange.

From similar figures at other incident energies, we can conclude that the radius of the halo is roughly one-third the range of the incident beam.

2.5.3 Experiment

Pedroni et al. [38], and others since then, used a PBS facility itself to investigate the halo in water, laying down hollow frames of PBs and measuring the resulting depth-dose on the central axis. By contrast, Sawakuchi et al. [40] used a single PB on the axis of a water tank and took radial scans with a small IC at selected depths. We adopted that more direct

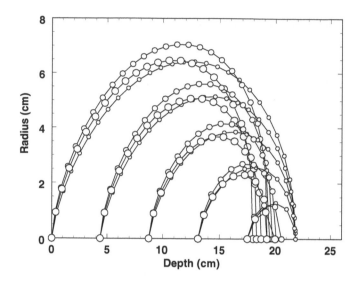

FIGURE 2.14   Stopping points in water of the outgoing proton, assuming a 180-MeV beam with reactions at five depths. Small, medium, and large circles represent elastic, QE with $E_B = 12.4$ MeV and QE with $E_B = 19.0$ MeV. Recoil nucleus parameters are $\theta = 300°$, $pc = 75$ MeV [43].

approach at 177 MeV, with two important changes. First, we used a beam monitor calibrated to measure the total number of incident protons, and second, we took depth scans at ten radii instead of radial scans at selected depths. The result [39,44] was the first absolute and comprehensive measurement of the halo.

Figure 2.15 shows two advantages of our method. Experimentally, it requires fewer adjustments of electrometer gain and beam intensity because the dose rate in any given

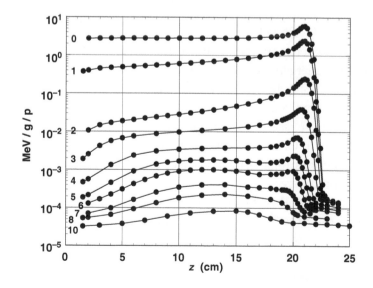

FIGURE 2.15   Measured depth-dose distributions at 10 radii [39]. Left-hand numbers are the distance (cm) of each scan from the beam center line. The lines serve only to guide the eye. 1 MeV/g = 0.1602 nGy.

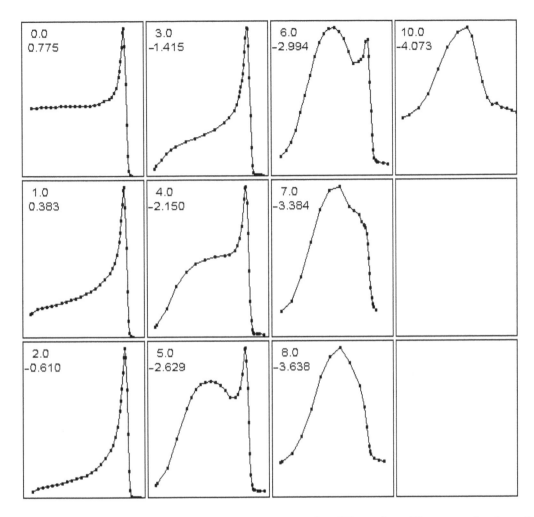

FIGURE 2.16 The same data as Figure 2.15 in autonormalized linear form. Upper number in each frame is r (cm), lower is $\log_{10}(\text{dose}/(\text{MeV/g/p}))$ for the highest point in the frame.

scan is more nearly constant. More important, it reveals features of the halo which are easily missed by selected radial scans. Linear Figure 2.16 shows these more clearly: the midrange bump at large radii predicted by elastic and quasi-elastic kinematics, the BP at radii far larger than the Molière single scattering tail, and lastly, evidence for the aura as dose well beyond the BP, and as a smooth background at large radii.

These data can be fit in a model-dependent fashion [44] which separates, at least approximately, the core, halo elastic, halo nonelastic, and aura. Figure 2.17 shows a transverse distribution at midrange. The transverse shape varies strongly with depth, making it difficult to fit with any single functional form.

2.5.4 Halo and Aura as Monte Carlo Tests

The experimental data just described provide an incisive test of Monte Carlo nuclear models because they are absolute and because the multiple EM, Molière tail, and nuclear coherent

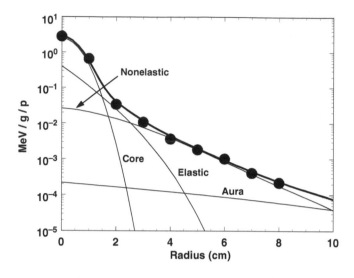

FIGURE 2.17 Bold line: model-dependent fit [39] to the transverse dose distribution at $z = 12\,\text{cm}$ (midrange) with experimental points (full circles). Light lines: contribution of each term in the model dependent fit.

and incoherent regions, each related to a different model, are experimentally separated as well as the underlying physics permits. At this writing, only Geant4 has been tested against these data. Overall, absolute agreement is remarkably good although some issues remain [45]. Whether these are experimental, defects in the Geant4 model, or both, is yet to be determined.

2.6 BRAGG CURVE AND EFFECTIVE STOPPING POWER

As previously noted, any dose calculation starts with

$$D = \Phi \times \frac{S}{\rho} \tag{2.32}$$

However, we must interpret S as the *effective* stopping power S_{eff} of a cohort of protons, not the theoretical CSDA value for a single proton, which is very large at low T.[*] The effective mass stopping power of a beam is much lower because the protons stop at slightly different depths due to range straggling, which smooths the dose out considerably.[†]

We will consider two limiting versions of S_{eff}, representing extreme cases. Both are required to understand and parameterize PBs. The first, S_{em}, is the effective stopping power of a proton beam with hard scatters turned off. As we will see, it is easy to compute, but it

---

[*] For instance, $S/\rho = 115\ \text{MeV}/(\text{g/cm}^2)$ at 3 MeV in water (Equation 2.5).

[†] To grasp "effective stopping power", consider this question: what is the stopping power of a 160 MeV beam in water at the Bragg peak? Answering that directly is difficult, since we need to average over a good model of the mix of energies and stopping points at the Bragg peak. However, observe that the peak/entrance dose ratio under typical circumstances is ≈4. Since tabulated $-(1/\rho)dT/dz$ in water at 160 MeV (the entrance energy) is ≈5.2 MeV/(g/cm²) the *effective* mass stopping power of that beam at the Bragg peak is ≈ 4 × 5.2 = 20.8 MeV/(g/cm²).

cannot be measured because hard scatters cannot, in fact, be turned off. The second, S_{mixed}, is the effective stopping power prevailing near the center of a broad beam. By contrast with S_{em} it can be measured (two ways, actually) but cannot be computed (except by Monte Carlo) because we have no simple mathematical model for hard scatters.

2.6.1 Electromagnetic Stopping Power S_{em}

S_{em} (suppressing ρ_{water}) is simply the convolution of Equation 2.5 with a Gaussian whose rms value σ_{sem} combines range straggling and initial beam energy spread in quadrature:

$$S_{em}(z) \equiv \int_{z-5\sigma}^{z+5\sigma} -\frac{dT}{dz}(z) \, G_{1D}(z-z';\sigma_{sem}) \, dz' \qquad (2.33)$$

G_{1D} is a 1D Gaussian of $\sigma = \sigma_{sem}$. Equation 2.33 assumes G_{1D} is negligible at $\pm 5\sigma_{sem}$.

We can perform the integral numerically using Simpson's rule [46], but we must deal with the singularity of dT/dz at end-of-range.[*] We therefore break the integral into two terms at some cutoff depth $z = z_c$, one term nonsingular and the other susceptible to approximation:

$$S_{em}(z) = \int_{z-5\sigma}^{zc} -\frac{dT}{dz}(z) \, G(z-z';\sigma_{sem}) \, dz' + \int_{zc}^{z+5\sigma} -\frac{dT}{dz}(z) \, G(z-z';\sigma_{sem}) \, dz' \quad (2.34)$$

The second integrand is dominated by the singularity at $R \approx z_c$, so we approximate it by setting $z' = z_c$ and find

$$S_{em}(z) = \int_{z-5\sigma}^{zc} -\frac{dT}{dz}(z) \, G(z-z';\sigma_{sem}) \, dz' + T(R-z_c) \, G(z-z_c;\sigma_{sem}) \qquad (2.35)$$

$T(R-z_c)$ is the residual kinetic energy at z_c, which we can obtain from range-energy tables. The overall result is insensitive to z_c.

S_{em} is plotted in Figure 2.18 (dot–dash line). The peak is greater than we observe in a real proton beam because the CSDA ignores hard scatters (nuclear interactions), assuming energy loss is entirely due to multiple soft EM interactions with atomic electrons. When hard scatters are turned on, fewer primaries reach the peak, a corresponding amount of energy is either moved upstream (charged secondaries) or entirely outside the ROI (neutral secondaries), and the solid line results. We discuss this next.

2.6.2 Transverse Equilibrium and S_{mixed}

Suppose we apply a large number of PBs close together and do a depth scan with a small IC (Figure 2.19, bottom). Because of both multiple EM and hard scatters, not all the protons initially aimed at the IC actually reach it. However, those that scatter away from the IC are exactly compensated by those from neighboring PBs that scatter into it.

---

[*] Alternatively, the integral can be expressed in closed form subject to some approximations [10].

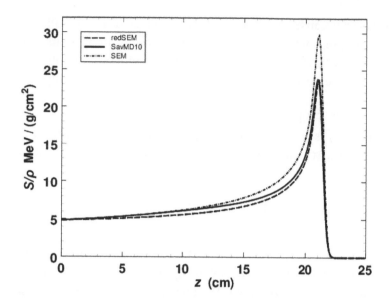

FIGURE 2.18   Bragg curves in water. Dot–dash: EM mass stopping power S_{em}/ρ. Solid: mixed mass stopping power. Dashed: S_{em}/ρ times a factor $(1-cz)$ with c adjusted to equalize the peak values.

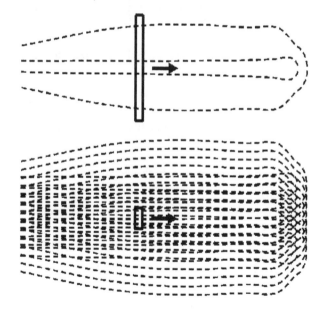

FIGURE 2.19   Top: depth scan of a single PB with a BPC. Bottom: depth scan of a broad beam with a small IC.

Therefore, near the center of a sufficiently broad beam in a uniform medium, the exact transverse coordinate of the IC does not matter. This condition is called *transverse equilibrium*. As we have seen, the characteristic radius R_{halo} of hard scatters is much greater than that of multiple EM scatters and governs the beam size required for transverse equilibrium.

If transverse equilibrium holds, the mix of particles at a given depth, both primaries and secondaries, is independent of the transverse coordinates. To calculate dose under those conditions, S_{mixed} is appropriate. As Figure 2.19 shows, it can be measured either by using a small IC in a sufficiently broad beam, or a sufficiently large "pancake" IC (a so-called "BP chamber," BPC) straddling a single PB. We now show that mathematically.

2.6.3 Measuring S_{mixed}

We wish to show that the integrated dose D to water per incident proton, measured by a BPC straddling a single PB, equals the local dose to water, divided by the fluence in air, on the axis of a sufficiently broad uniform beam.

The average energy dE_1 per proton deposited in a disk of thickness dz_1 and radius R_{halo} by many protons entering a large water tank near the axis (Figure 2.19 top) is

$$dE_1(z) = \left(\int_0^{2\pi} \int_0^{R_{\text{halo}}} D_1(r,z) r \, dr \, d\phi \right) \rho \, dz_1 \tag{2.36}$$

Now let the tank be exposed, instead, to a broad uniform parallel proton beam of fluence in air Φ. The energy $dE_n(z)$ deposited in a small disk of thickness dz_1 and radius R_{dosim} by n protons at the tank entrance directed at the disk (Figure 2.19, bottom) is

$$dE_n(z) = \left(\int_0^{2\pi} \int_0^{R_{\text{dosim}}} D_n(r,z) r \, dr \, d\phi \right) \rho \, dz_1 = \pi R_{\text{dosim}}^2 D_n(0,z) \rho \, dz_1 \tag{2.37}$$

In the second step, we have assumed transverse equilibrium (EM and nuclear) near the axis, so that D_n is independent of r there. Because the mix of stopping powers is the same in both discs, and they have the same thickness, it follows that

$$dE_1(z) = dE_n(z)/n \tag{2.38}$$

Transverse equilibrium also plays a role in Equation 2.38. Protons initially directed at the small dosimeter do not deposit all their energy in it. However, the shortfall is exactly compensated by energy from protons *not* directed at the dosimeter. We thus find

$$\int_0^{2\pi} \int_0^{R_{\text{halo}}} D_1(r,z) r \, dr \, d\phi = D_n(0,z)/\left(n/\left(\pi R_{\text{dosim}}^2 \right) \right) = D_n(0,Z)/\Phi \tag{2.39}$$

Dose per fluence is a mass stopping power (Equation 2.1), which we have called $S_{\text{mixed}}/\rho_{\text{water}}$.

In practice, either method has its problems. Commercial BPCs are not quite large enough at higher energies [39]. Scanning a BPC in a water tank can be unwieldy, so it is sometimes incorporated into a water column [47] which must of course be big enough. If the BPC is too small, Monte Carlo derived corrections can be applied.

The small IC/broad beam method with a single dosimeter suffers from the fact that *all* the PBs must be laid down to measure a single point on the depth-dose, which may be prohibitively slow. Multilayer ICs (MLICs) have been developed to address this [48], and there is at least one commercial version, the IBA "Zebra" [49].

2.6.4 Parameterizing the Nuclear Halo

The Paul Scherrer Institute (PSI) group first pointed out [38] the need to take the energy in the halo into account.* Their parameterization of the halo deserves special attention because it set the pattern for all subsequent PBS treatment planning systems and is, in our opinion, flawed. It calls for unnecessary measurements which have, unfortunately, become enshrined in practice.

We merely outline the argument here. See [39,44] for details and references. Their $T(w)$ is identical to our $S_{mixed}(z)/\rho_{water}$. Their parameterization [38] (Equation 2.7) in our notation and assuming cylindrical symmetry, reads

$$D(r,z)=\left[\left(1-f_{NI}(z)\right)\times G_{2D}^{P}\left(r;\sigma_{P}(z)\right)+f_{NI}(z)\times G_{2D}^{NI}\left(r;\sigma_{NI}(z)\right)\right]\times\frac{S_{mixed}}{\rho_{water}} \quad (2.40)$$

the familiar (dose) = (fluence)×(mass stopping power). The fluence, in square brackets, is divided into a primary or core term (P) and a secondary or "nuclear interaction" term (NI). f_{NI}, an adjustable positive function of z and incident energy, governs the attrition of primaries. It reduces the core and increases the halo by the same amount. $\sigma_{NI}(z)$, another adjustable function of z and energy, is the rms width of a two-dimensional (2D) Gaussian describing the fluence in the halo, whereas $\sigma_P(z)$ is the rms width of a 2D Gaussian describing the core, not adjustable but, instead, found by transport (FE) theory.

Associating $S_{mixed}(z)$ with the core is incorrect; it should be $S_{em}(z)$. Primaries suffer only multiple soft EM scatters, by definition, and not the hard scatters included in $S_{mixed}(z)$. Dose from hard scatters is in the halo, not the core! A red flag is the fact that $S_{mixed}(z)$ has the primary attrition function built-in, so the explicit $(1-f_{NI}(z))$ is duplicative, whereas S_{em} does not have a built-in primary attrition. What mass stopping power associates with the *halo* is a more complicated question, for which see [44].

Figure 2.20 compares our data [39] with the PSI fit [38] at 177 MeV, with σ_P adjusted to our beam size. Agreement in the core is acceptable, except for the excess at midrange ($r=0$) from using S_{mixed} instead of S_{em}, cf. Figure 2.18. A more obvious but less consequential problem is the Gaussian describing secondaries, which falls far too rapidly (Figure 2.20). PSI themselves characterized this as a "first preliminary estimate," and others (e.g., [50]) have used radial functions with longer tails. However, we caution that the transverse dose depends strongly on depth [39]. No single shape can be regarded as typical.

Our objection to Equation 2.40 is *not* that it causes errors in treatment planning, since any commercial treatment planning system has enough adjustable parameters to compensate for

---

* The contribution of the aura to the high-dose region is negligible.

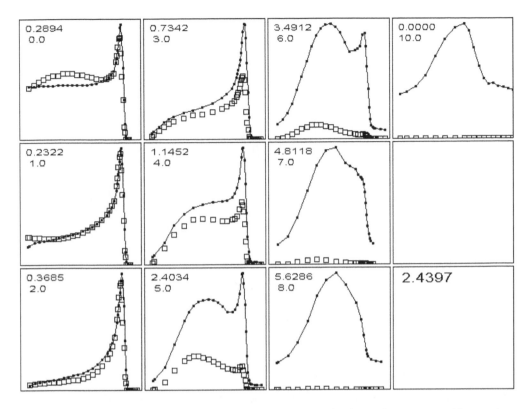

FIGURE 2.20   Our data with the PSI fit (absolute comparison) adjusted to our initial beam size.

the erroneous use of S_{mixed}. Rather, it is the misconception that it is necessary to measure S_{mixed}, leading in turn to commercial and unnecessary BPCs and MLICs. To compound the problem, the commercial BPCs are not quite large enough at higher energies. This has led to experimental [51] and Monte Carlo [52] papers to correct for lost signal in an unnecessary measurement.

In summary, the nuclear halo at a given energy is best mapped by measuring the depth dose of a single PB at selected radii with a small dosimeter, preferably with an absolute beam monitor. The radial integral of dose, $T(w)$ or IDD (Integral Depth Dose) or S_{mixed}/ρ_{water}, is unimportant. The most efficient parameterization of the core and halo for treatment planning purposes is, in our view, still to be determined.

2.6.5 Energy Dependence of Bragg Curves

Since straggling dominates the BP and is a constant fraction of range, it follows that BPs taken at lower energies are sharper (Figure 2.21). When we spread the dose in depth by adding BPs of different ranges (range modulation), we can either use degraders[*] or change the machine energy. In the former case, Section 2.3.4 shows that component BPs are just pulled back versions of the deepest one. In the latter case, the sharpening of the component BPs must be considered.

---

[*] A *degrader* is a slab of material whose primary purpose is to reduce energy.

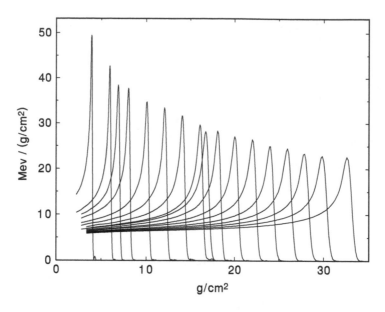

FIGURE 2.21 Measured Bragg curves from 69 to 231 MeV. (Data courtesy, D. Prieels, Ion Beam Applications s.a.)

2.6.6 The Disappearing Bragg Peak

Rather than our two limiting cases, large dosimeter/small beam and small dosimeter/large beam, what happens with a small dosimeter along the axis of a small beam? Again, we use $D = \Phi \times S / \rho$. As depth increases, energy decreases and stopping power increases. At the same time, fluence decreases because MCS spreads the beam out (the 2D Gaussian grows wider), and there is no compensating in-scatter from neighboring PBs. The decrease in fluence ultimately wins out, and the BP disappears. First predicted by Preston and Koehler [21] (Figure 2.22), this has been confirmed experimentally many times. It is a limiting factor in treating small deep fields.

2.6.7 Nuclear Buildup

If the depth dose in a broad beam entering a water tank from air or vacuum is measured with a small dosimeter, a nuclear buildup of a few percent in the first centimeter or so is observed. This was first reported by Carlsson and Carlsson [53], who noted that the buildup was smaller than expected.

It should be clear from the foregoing discussion that observed buildup depends on both beam size and detector size. With a small beam and a small detector, for instance, there is no buildup (Figure 2.16). Hard scatters occur, but their dose appears at larger radii. Full buildup will be observed only with a small detector in a beam sufficiently large for nuclear equilibrium. Probably the beam (size not given) in [53] was too small.

2.7 LOOKING AHEAD

As promised, we have discussed the three basic interactions of radiotherapy protons with matter: EM stopping, EM multiple scattering, and hard scatters. That still leaves us a long

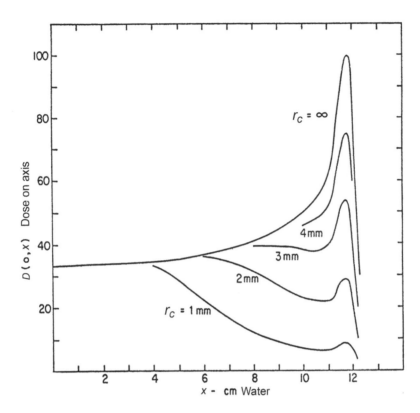

FIGURE 2.22 Figure 2.7 of the Preston and Koehler manuscript [21]: Relative dose on the axis of a uniform circular proton beam of initial range 12 cm of water and radius r_c at the collimator. The curve for $r_c = \infty$ is experimental, the others are calculated.

way from computing the dose in realistic situations. For that, we need a *transport theory*, namely a general way of computing fluence. Let us take a quick look ahead.

2.7.1 Fermi-Eyges Theory

Enrico Fermi, in unpublished lecture notes, computed the transverse displacement, the angle, and their correlation, for cosmic rays traversing the earth's atmosphere [54]. He used the Gaussian approximation to MCS. Leonard Eyges [55] added the effect of energy loss, which Fermi had ignored. Eyges assumed an ideal beam entering a single homogeneous slab, and built into his theory a scattering power (T_{FR} in [36]) which, it turns out, is the worst possible choice. Later, the theory found application in electron radiotherapy* and was generalized to nonideal incident beams, multiple homogeneous slabs, and arbitrary scattering powers. For a detailed account, see [22].

In FE theory, the terrain is discretized by "z-planes" perpendicular to the nominal beam direction. These may be actual slab boundaries or may simply be introduced for computational purposes. At each such plane, a cylindrically symmetric Gaussian PB is fully characterized by ten parameters: one for the total proton charge the PB carries, three for the

---

* Ironically, protons satisfy the approximations of FE theory far better than do electrons.

position *x,y,z* and two for the direction θ_x,θ_y of the PB's central axis, one for the nominal proton energy or an equivalent such as residual range in water, and three 'FE moments' related to the PB's transverse size, angular divergence, and emittance.*

To "transport" a PB through a step $z \rightarrow z'$ is to find these parameters at z', given their values at z. FE theory is simply a prescription for doing that. It follows that we can, in the Gaussian approximation, find the parameters of any PB at any depth in mixed-slab geometry by a relatively simple computation.

That is a big step forward. However, the resulting fluence distribution (and therefore the dose) is Gaussian, and only rarely does that correspond to a useful problem. (Think of the complicated dose distribution inside a patient.) Most problems involve *transverse* as well as longitudinal heterogeneities. Before tackling that, we take a short detour.

2.7.2 The Preston and Koehler Manuscript

William M. Preston (1910–1989) and Andreas M. ("Andy") Koehler (1930–2015) were among the pioneers of proton radiotherapy. Coming from a high energy rather than a radiotherapy background, they were unaware of FE theory. However, they used an analogous method in a manuscript [21] which to this day is the best theoretical and experimental study of the evolution of an ideal PB in a homogeneous slab. Unfortunately, it was rejected for publication. Proton radiotherapy was deemed uninteresting at the time (around 1968).

Because it solves the same problem as FE theory, their theory must in some way be equivalent. In [22], we rederive their most important results in FE language and extend them to heavy ions. Here we merely state them.

First, *the rms transverse spread $\sigma_x(R)$ at end of range is proportional to the range R.* The constant of proportionality is

$$\frac{\sigma_x(R)}{R} = \frac{E_s z}{2(pv)_{R/2}} \sqrt{\frac{R}{X_S}} \tag{2.41}$$

which despite appearances is nearly independent of R. $E_s = 15.0$ MeV, z is the particle charge number, pv is evaluated at the T value corresponding to $R/2$ (cf. Equation 2.49), and X_S is the scattering length of the material [36]. In Lexan, for instance, $\sigma_x(R) = 0.021R$. Values of $\sigma_x(R)/R$ and ρX_S for many other materials can be found in [22].

Second, at any lesser depth $z < R$,

$$\frac{\sigma_x(z)}{\sigma_x(R)} = \left[2(1-t)^2 \ln\left(\frac{1}{1-t}\right) + 3t^2 - 2t \right]^{1/2}, \quad t \equiv z/R \tag{2.42}$$

---

* The emittance, a measure of the correlation between position and angle, is the most subtle. If they are perfectly correlated, the emittance is zero, even if the beam has finite width and angular divergence. If they are completely uncorrelated, the emittance is maximal.

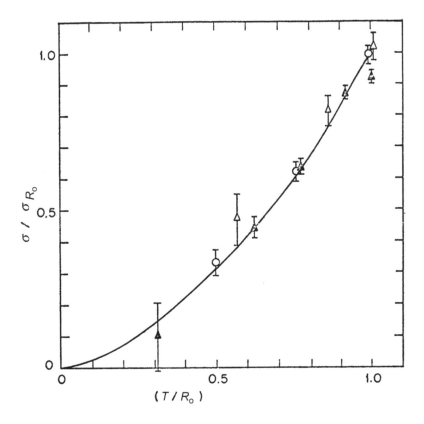

FIGURE 2.23   Facsimile of Preston and Koehler Figure 2.17 whose caption reads, "Dimensionless plot of the standard deviation due to scattering versus depth of penetration... Open triangles are experimental results for 112 MeV protons on aluminum; solid triangles for 158 MeV protons on aluminum; open circles for 127 MeV protons on water."

which, with Equation 2.41, *completely describes beam spreading in a homogeneous slab for any heavy charged particle at any energy in any material.* Equations 2.41 and 2.42 assume an ideal incident beam, so measurements of $\sigma_x(z)$ must be corrected for initial beam size, divergence and emittance, which are often significant in beams designed for PBS. Figure 2.23 is Preston and Koehler's experimental confirmation of Equation 2.42 for two materials and two energies.

Third, the *disappearing* BP, covered previously (Section 2.6.6 and Figure 2.22).

2.7.3 A Comprehensive PB Algorithm

The following description of a novel PB algorithm (PBA), using FE theory, is a brief outline. For a detailed description, see [56,57].

The archetypal forward problem of proton radiotherapy physics is that of computing the dose everywhere in a known heterogeneous terrain irradiated by known proton beams. For routine treatment planning, the computation needs to be fast and therefore deterministic. Like other deterministic algorithms, PBA relies on the fact that the high-dose region is well approximated even if we ignore hard scatters (nuclear interactions), that is, if we consider just the primary protons. Hard scatters are put in later as a correction.

Monte Carlo methods transport individual protons using models of basic interactions. Deterministic methods group protons into bundles or "PBs," transported using FE theory. As stated in Section 2.7.1, any PB can be transported correctly through any number of *longitudinal* heterogeneities. It is *transverse* heterogeneities that cause difficulties.

A brass collimator furnishes an extreme example. Depending on their coordinates, protons making up a PB may find themselves either in brass or in air. However, FE theory can only transport the *entire* PB, and only through one material or the other. If the PB central axis is just inside brass, FE will choose brass ("central axis" or CAX approximation), and *all* the protons represented by the PB will stop. Otherwise, FE will choose air, and *all* the protons will pass through, scattered and slowed only by a few centimeters of air. If the incident PB represents a great many protons, a serious error results either way.

We therefore break up the mother PB into smaller daughter PBs, separated transversely. That resolves the brass/air ambiguity for some of the daughters, and the others represent fewer protons, so the error is less serious. The breakup is repeated as often as necessary.

For rapid convergence, PBA combines two ways of breaking up PBs. In *redefinition* [58], the entire cohort of PBs is replaced at one or more selected z-planes. In *dynamic splitting* [59], individual PBs are split only when necessary.

Besides combining redefinition with recursive dynamic splitting, PBA has other novel features.

First and foremost, it is a calculation *from first principles*. That is, it starts directly at the level of established laws of physics and does not use empirical model and fitting parameters. All it requires is an accurate physical description of the incident beams, the beam line (if any), and the patient. Thus, in a passive beam spreading problem, it does not start with the beam's effective source position and size. Instead, these follow, as they should, from the physical parameters of the beam line.

Second, PBA makes no distinction between passive beam spreading and PBS. In passive spreading, the beam line is more complicated and there is only one incident beam. In PBS, the beam line is simpler or absent, and there are many incident beams.

Third, all objects (beam line, collimators, patient) are represented the same way, as stacks of homogeneous or heterogeneous slabs. There are no magic "beam defining" planes. Collimator scatter (protons interacting with the collimator without stopping in it) and collimator thickness effects emerge in a natural way.

Last, every material is associated with its correct stopping and scattering powers, rather than being treated as a variant of water. The concept of "radiological path length" is not used. That allows PBA to address problems involving, e.g., titanium implants.

As an example, Figure 2.24 is a PBA computation of collimator scatter (not handled at all by other PBAs). It takes 1.4 min on a single laptop computer (Lenovo T400 running Intel Fortran under Windows7). The same problem using TOPAS/Geant4 takes many hours on a multicore computer (see [56]).

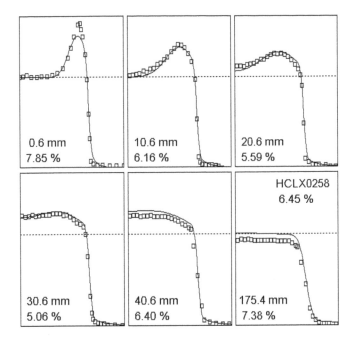

FIGURE 2.24 Transverse dose distributions in air at six distances (top number) from the downstream face of a brass collimator with a 9.88 mm radius hole. Percentages give the rms deviation of the PBA calculation (line) from experiment (squares). 1,665 K PBs were processed and 531 K left the collimator.

APPENDIX A: SYMBOLS

Symbol	Description	Units
A	area	cm^2
A_0, A_1, A_2	Fermi-Eyges moments	rad^2, cm rad, cm^2
c	speed of light	299,792,458 m/s = 0.984 ft/ns
D	dose	$Gy \equiv J/Kg$
		1 MeV/g = 0.1602 nGy
e	quantum of charge	1.602×10^{-10} nC
i	electric current	nA
mc^2	proton rest energy	938.272 MeV
N	number of protons	
N_A	Avogadro constant	6.022141×10^{23} mol^{-1}
pc	proton momentum	MeV
pv	proton momentum times speed	MeV
q	charge	nC
R	range	cm
ρR	mass range	g/cm^2
S	stopping power $= -dT/dz$	MeV/cm

(*Continued*)

Symbol	Description	Units
S/ρ	mass stopping power $= -dT/(\rho dz)$	MeV/(g/cm²)
t	time	s
	mass target thickness in MCS derivation	g/cm²
T	kinetic energy	MeV
T_{dM}	differential Moliere scattering power	rad²/cm
v	speed	cm/s
X_0	radiation length	cm
X_S	scattering length	cm
x, y, z	transverse coordinates	cm
α	fine structure constant	1/137.035
	Born parameter in Moliere theory	
β	v/c	
θ_0	Gaussian width parameter	rad
θ_x, θ_y	projected angles	rad
Φ	fluence $= dN/dA$	protons/cm²
Φ_q	charge fluence $= dq/dA$	nC/cm²
ρ	density	g/cm³

APPENDIX B: RELATIVISTIC SINGLE-PARTICLE KINEMATICS

At 160 MeV $v/c \approx 0.5$ so radiotherapy protons are relativistic. Although nonrelativistic equations are often adequate, the exact equations given here are preferable.

A proton beam is usually characterized by its kinetic energy T, but we occasionally need its speed βc or momentum pc or the quantity pv which governs MCS. The next three formulas can be found in any introductory book on special relativity:[*]

$$\beta \equiv \frac{v}{c} = \frac{pc}{E}$$

(2.43)

$$E \equiv T + mc^2$$

(2.44)

$$E^2 = \left(pc\right)^2 + \left(mc^2\right)^2$$

(2.45)

Define τ (reduced KE) and ξ (reduced pv) by

$$\tau \equiv T/\left(mc^2\right)$$

(2.46)

---

[*] E is *total* energy. Medical physicists often use E for *kinetic* energy but then $E = mc^2$, the most famous equation of the twentieth century, makes no sense. There is also precedent in the medical physics literature for T e.g. [14]. Equations in this section apply to any particle of rest energy mc^2.

$$\xi \equiv pv/(mc^2) \qquad (2.47)$$

From the first three equations we find

$$\beta^2 = \frac{2+\tau}{(1+\tau)^2}\tau \qquad (2.48)$$

$$pv = \frac{2+\tau}{1+\tau}T \qquad (2.49)$$

$$(pc)^2 = (2+\tau)mc^2T \qquad (2.50)$$

Over the radiotherapy range $3 \le T \le 300$ MeV the coefficient of T in Equation (2.49) varies from 2 to 1.76 so pv can be thought of as roughly $2T$. We occasionally require the inverse of Equation (2.49) which is

$$T = 0.5mc^2\left(\sqrt{(2-\xi)^2 + 4\xi} - (2-\xi)\right) \qquad (2.51)$$

REFERENCES

1. F. H. Attix, *Introduction to Radiological Physics and Radiation Dosimetry*, Wiley, Hoboken NJ 1986.
2. H. Palmans, Proton beam interactions: Basic, in *Principles and Practice of Proton Beam Therapy*, Ed. I.J. Das and H. Paganetti, AAPM Monograph #37, Medical Physics Publishing, Madison WI (2015).
3. L. J. Verhey, A. M. Koehler, J. C. McDonald, M. Goitein, I.-C. Ma, R. J. Schneider and M. Wagner, The determination of absorbed dose in a proton beam for purposes of charged-particle radiation therapy, *Radiat. Res.* 79 (1979) 34–54.
4. E. W. Cascio and B. Gottschalk, A simplified vacuumless Faraday cup for the experimental beamline at the Francis H. Burr proton therapy center, 2009 IEEE Radiation Effects Data Workshop, pp 161–155, digi ID 978-1-4244-5092-3/09/.
5. B. Gottschalk and R. Platais, Tests of a multilayer Faraday cup, Abstracts PTCOG22, San Francisco, CA, 1995, 13.
6. B. Gottschalk, R. Platais and H. Paganetti, Nuclear interactions of 160 MeV protons stopping in copper: a test of Monte Carlo nuclear models, *Med. Phys.* 26(12) (1999), 2597–2601.
7. B. Gottschalk, Calibration of the NPTC range verifier, technical report 16JAN01 available at http://users.physics.harvard.edu/~gottschalk.
8. A. M. Koehler, priv. comm. ca. 1982.
9. M. J. Berger, Penetration of proton beams through water I. Depth-dose distribution, spectra and LET distribution, NIST technical note NISTIR 5226 (1993) available from the National Technical Information Service, NTIS, U.S. Department of Commerce, Springfield, VA 22161.
10. T. Bortfeld, An analytical approximation of the Bragg curve for therapeutic proton beams, *Med. Phys.* 24(12) (1997) 2024–2033.
11. H. Bichsel, Calculated Bragg curves for ionization chambers of different shapes, *Med. Phys.* 22(11) (1995) 1721–1726.

12. J. F. Janni, Calculations of energy loss, range, pathlength, straggling, multiple scattering, and the probability of inelastic nuclear collisions for 0.1 to 1000 MeV protons, Air Force Weapons Laboratory Technical Report No. AFWL-TR-65-150 (1966).

13. J. F. Janni, Proton range-energy tables, 1KeV–10 GeV, Atomic Data and Nuclear Data Tables **27** parts 1 (compounds) and 2 (elements) (Academic Press, 1982).

14. M.J. Berger, M. Inokuti, H.H. Andersen, H. Bichsel, D. Powers, S.M. Seltzer, D. Thwaites, D.E. Watt, H. Paul and R.M. Sternheimer, Stopping powers and ranges for protons and alpha particles, ICRU Report 49 (1993).

15. M.F. Moyers, G.B. Coutrakon, A. Ghebremedhin, K. Shahnazi, P. Koss and E. Sanders, Calibration of a proton beam energy monitor, *Med. Phys.* 34(6) (2007) 1952–1966.

16. E. W. Cascio and S. Sarkar, A continuously variable water beam degrader for the radiation test beamline at the Francis H. Burr Proton Therapy Center, Proc. IEEE Radiation Effects Data Workshop (2007).

17. T. Siiskonen, H. Kettunen, K. Peräjärvi, A. Javanainen, M. Rossi, W.H. Trzaska, J. Turunen and A. Virtanen, Energy loss measurement of protons in liquid water, *Phys. Med. Biol.* 56 (2011) 2367–2374.

18. F. S. Rasouli, S. F. Masoudi and D. Jette, On the proton range and nuclear interactions in compounds and mixtures, *Med. Phys.* 42 (2015) 2364–2367.

19. LOOKUP, a proton desk calculator for IBM compatible PCs available for free download at http://users.physics.harvard.edu/~gottschalk.

20. H. Øveras, On small angle multiple scattering in confined bodies, CERN Yellow Report 60-18 (1960).

21. W. M. Preston and A. M. Koehler, The effects of scattering on small proton beams, unpublished manuscript (1968), Harvard Cyclotron Laboratory. A facsimile is available in \BGdocs. zip at http://users.physics.harvard.edu/˜gottschalk.

22. B. Gottschalk, Techniques of proton radiotherapy: Transport theory, arXiv:1204.4470v2 (2012).

23. H.A. Bethe, Molière's theory of multiple scattering, *Phys. Rev.* 89 (1953) 1256–1266. Four entries in the second column (the Gaussian) of Table II are slightly incorrect (A. Cormack, priv. comm.) but the error (corrected in our programs) is at worst 1%.

24. B. Gottschalk, A. M. Koehler, R. J. Schneider, J. M. Sisterson and M.S. Wagner, Multiple Coulomb scattering of 160 MeV protons, *Nucl. Instr. Meth.* B74 (1993) 467–490. We have discovered the following errors: Eq. (2) should read $\Xi(\chi) = \dfrac{1}{\pi} \dfrac{\chi_c^2}{\left(\chi^2 + \chi_a^2\right)^2}$ and in Table 1 the heading α should read α^2 and $\times 10^9$ under χ_c^2 should read $\times 10^6$.

25. G. Molière, Theorie der Streuung schneller geladener Teilchen I Einzelstreuung am abgeschirmten Coulomb-Feld, *Z. Naturforschg.* 2a (1947) 133–145. Available in \BGdocs.zip at http://users.physics.harvard.edu/~gottschalk.

26. G. Molière, Theorie der Streuung schneller geladenen Teilchen II Mehrfach- und Vielfachstreuung, *Z. Naturforschg.* 3a (1948) 78–97. Available in \BGdocs.zip at http://users. physics.harvard.edu/~gottschalk.

27. V. L. Highland, Some practical remarks on multiple scattering, *Nucl. Instr. Meth.* 129 (1975) 497–499 and Erratum, Nucl. Instr. Meth. **161** (1979) 171.

28. G. R. Lynch and O. I. Dahl, Approximations to multiple Coulomb scattering, *Nucl. Instr. Meth.* B58 (1991) 6.

29. U. Fano, Inelastic collisions and the Molière theory of multiple scattering, *Phys. Rev.* 93 (1954) 117–120.

30. W.T. Scott, The theory of small-angle multiple scattering of fast charged particles, *Rev. Mod. Phys.* 35 (1963) 231–313.

31. A. O. Hanson, L. H. Lanzl, E. M. Lyman and M. B. Scott, Measurement of multiple scattering of 15.7-MeV electrons, *Phys. Rev.* 84 (1951) 634–637.
32. H. Bichsel, Multiple scattering of protons, *Phys. Rev.* 112 (1958) 182–185.
33. S. Eidelman et al. Review of particle physics, *Phys. Lett.* B592 (2004). Tables, listings, reviews and errata are also available at the Particle Data Group website http://pdg.lbl.gov.
34. B. Rossi, *High-Energy Particles*, Prentice-Hall, New York, 1952.
35. A. Brahme, On the optimal choice of scattering foils for electron therapy, technical report TRITA-EPP-17, Royal Institute of Technology, Stockholm, Sweden (1972).
36. B. Gottschalk, On the scattering power of radiotherapy protons, arXiv:0908.1413 and *Med. Phys.* 37(1) (2010) 352–367.
37. W. Matysiak, D. Yeung, R. Slopsema and Z. Li, Evaluation of Geant4 for experimental data quality assessment in commissioning of treatment planning system for proton pencil beam scanning mode, *Intl. J. of Rad. Onc. Bio. Phys.* 87(2) (2013) S739 and accompanying poster, 011527 20130916 ASTRO poster.pdf.
38. E. Pedroni, S. Scheib, T. Böhringer, A. Coray, M. Grossmann, S. Lin and A. Lomax, Experimental characterization and physical modeling of the dose distribution of scanned proton pencil beams, *Phys. Med. Biol.* 50 (2005) 541–561.
39. B. Gottschalk, E. W. Cascio, J. Daartz and M. S. Wagner, 'On the nuclear halo of a proton pencil beam stopping in water,' arXiv:1412.0045v2 (2015) and *Phys. Med. Biol.* 60 (2015) 5627–5654.
40. G. O. Sawakuchi, X. R. Zhu, F. Poenisch, K. Suzuki, G. Ciangaru, U. Titt, A. Anand, R. Mohan, M. T. Gillin and N. Sahoo, Experimental characterization of the low-dose envelope of spot scanning proton beams, *Phys. Med. Biol.* 55 (2010) 3467–3478.
41. L. Lin, C. G. Ainsley, T. D. Solberg and J. E. McDonough, Experimental characterization of two-dimensional spot profiles for two proton pencil beam scanning nozzles, *Phys. Med. Biol.* 59 (2014) 493–504.
42. B. Gottschalk, W.J. Shlaer and K.H. Wang, Proton-proton bremsstrahlung at 158 MeV, *Nucl. Phys.* 75 (3) (1965) 549–560.
43. H. Tyrén, S. Kullander, O. Sundberg, R. Ramachandran, P. Isacsson and T. Berggren, Quasi-free proton-proton scattering in light nuclei at 460 MeV, *Nucl. Phys.* 79 (1966) 321–373.
44. B. Gottschalk, E. W. Cascio, J. Daartz and M. S. Wagner, Nuclear halo of a 177 MeV proton beam in water: theory, measurement and parameterization, arXiv:1409.1938v1 (2014).
45. D. C. Hall, A. Makarova, H. Paganetti and B. Gottschalk, Validation of nuclear models in Geant4 using the dose distribution of a 177 MeV proton pencil beam, *Phys. Med. Biol.* 61 (2016) N1–N10 and arXiv:1507.0815v3.
46. W. H. Press, B. P. Flannery, S. A. Teukolsky and W.T. Vetterling, *Numerical Recipes: the Art of Scientific Computing*, Cambridge University Press, Cambridge UK 1986.
47. P. T. W. Freiburg GMBH, Lörracher Strasse 7, 79115 Freiburg, Germany.
48. B. Gottschalk, Multi-layer ion chamber for eye depth-dose measurements, Proc. PTCOG40 (Paris, 2004) p.72.
49. Ion Beam Applications s.a., Chemin du Cyclotron 3, 1348 Louvain-la-Neuve, Belgium.
50. V. E. Bellinzona, M. Ciocca, A. Embriaco, A. Fontana, A. Mairani, M. Mori and K. Parodi, On the parameterization of lateral dose profiles in proton radiation therapy, *Physica Medica* 31 (2015) 484–492.
51. A. Anand, N. Sahoo, X. R. Zhu, G. O. Sawakuchi, F. Poenisch, R. A. Amos, G. Ciangaru, U. Titt, K. Suzuki, R. Mohan and M. T. Gillin, A procedure to determine the planar integral spot dose values of proton pencil beam spots, *Med. Phys.* 39 (2012) 891–900.
52. B. Clasie, N. Depauw, M. Fransen, C. Gomà, H. R. Panahandeh, J. Seco, J. B. Flanz and H. M. Kooy, Golden beam data for proton pencil-beam scanning, *Phys. Med. Biol.* 57 (2012) 1147–1158.

53. C.A. Carlsson and G.A. Carlsson, Proton dosimetry with 185 MeV protons: Dose buildup from secondary protons and recoil electrons, *Health Phys.* 33 (1977) 481–484.

54. B. Rossi and K. Greisen, Cosmic-ray theory, *Rev. Mod. Phys.* 12 (1941) 240–309.

55. L. Eyges, Multiple scattering with energy loss, *Phys. Rev.* 74 (1948) 1534–1535.

56. B. Gottschalk, Comprehensive proton dose algorithm using pencil beam redefinition and recursive dynamic splitting, arXiv:1610.00741v1 (2016).

57. B. Gottschalk, Deterministic proton dose calculation from first principles, *Phys. Med. Biol.* 63 (2018) 135016, doi: 10.1088/1361-6560/aacc96.

58. A. S. Shiu and K. R. Hogstrom, Pencil beam redefinition algorithm for electron dose distributions, *Med. Phys.* 18 (1991) 7–18.

59. N. Kanematsu, M. Komori, S. Yonai and A. Ishizaki, Dynamic splitting of Gaussian pencil beams in heterogeneity-correction algorithms for radiotherapy with heavy charged particles, *Phys. Med. Biol.* 54 (2009) 2015–2027.

60. G. Gerstein, J. Niederer and K. Strauch, Elastic scattering of 96-MeV protons, *Phys. Rev.* 108 (1957) 427–432.

II

Beam Delivery

CHAPTER 3

Proton Accelerators

Marco Schippers

Paul Scherrer Institute

CONTENTS

3.1 INTRODUCTION

Proton therapy has been developed and was performed initially in nuclear physics laboratories that were equipped with a particle accelerator, such as a (synchro)cyclotron or a synchrotron, like in Berkeley (CA, USA) [1,2], Cambridge (MA, USA) [3], Paul Scherrer Institute (PSI; Switzerland) [4–6], and Uppsala (Sweden) [7,26]. The first hospital-based facility with gantries was built in Loma Linda (CA, USA) [8] in the 1990s. Around that time, also commercial companies started to develop accelerators and offered complete treatment facilities, including gantries. Nowadays, the cyclotron and the synchrotron are the two typical types of accelerators that are offered by companies and are proven to be reliable machines in clinical facilities. Many good textbooks and proceedings of accelerator schools exist on accelerator design, see for example [9,10], but in this chapter, the emphasis will be on relevant issues of proton therapy to understand the reason for the typical design choices and to become aware of the important technical and accelerator physics issues that should be discussed in a selection and acquisition procedure.

In this chapter, first the relation between certain accelerator specifications and the quality and type of the dose delivery method will be discussed, followed by a detailed description of the currently used accelerators: cyclotron and synchrotron. Then, an overview is given of the devices that rotate the beam delivery system around the patient, to aim the proton beam at any direction to the tumor. It will be made clear why these devices ("gantries") are quite big compared to the ones used in conventional photon therapy. Finally, after a short overview of new developments in accelerator physics that may be applied in proton therapy in the coming one or two decades, some words of caution will be given in the conclusions.

3.2 CONSEQUENCES OF THE DOSE APPLICATION TECHNIQUE

Several important specifications of the accelerator and beam delivery system depend on the chosen technique to apply the proton dose to the tumor. As these techniques [11] will be described in detail in Chapters 4–6, only aspects relevant for the accelerator and beam delivery will be discussed here.

3.2.1 Dose Spreading in Depth

To distribute the dose in depth, the energy of the protons has to be adjusted accordingly before they enter the patient. In some accelerator types, one can accelerate protons up to the needed energy and transport the protons to the patient. Such accelerators are, for example, synchrotrons, linear accelerators, and Fixed Field Alternating Gradient (*FFAG*) accelerators. It is important that the change to another beam energy is sufficiently fast to limit the treatment time and to allow fast switching between treatment rooms and very precise to set the range in the patient with sufficient accuracy.

In accelerators such as cyclotrons and synchrocyclotrons this is not possible, since the machines developed for proton therapy can work at one specific proton energy only, which is the maximum to be used in the facility. To obtain a lower energy, the accelerated protons are slowed down in an adjustable amount of material, a so-called *degrader*. However, the beam coming out of the degrader is subject to some spread due to the statistical process of energy loss and multiple scattering in the degrader material. This would increase the beam size in the following systems too much. So, to limit this spread, the degrader is followed by a collimator system and an energy selection system (ESS). This is a magnetic bending system, which will spread out the beam with a strong correlation between energy and lateral position. Protons with a too much deviating energy will be intercepted by a dedicated slit system. Due to the neutrons created by protons lost in the collimators and this slit (see Chapter 8), the ESS needs sufficient shielding. Only the protons within the correct energy range (1–2% spread) are then guided into the following beam transport system. In case the degrader is somewhere in the beam transport system or in case the accelerator can accelerate to the needed energies, all following beam line magnets must be adjusted according to the chosen energy. Also the read-out systems of all downstream monitors and beam diagnostics must use the correct energy-dependent factors in their conversion from a measured signal to beam intensity. Proton energy changes need some time, which is specified by the energy-varying systems. If it is only for setting the maximum range used in a certain gantry angle, it is quite acceptable if an energy change takes a few seconds. However, range modulation to cover the target thickness must be done much faster (<0.5 s per energy step) to limit the treatment time.

If the energy adjustment is performed in the *nozzle*, so just before the patient, a stack with a variable number of plates (*range shifter*), a plate with specially shaped ripples (*ridge filter*), or a rotating wheel with an azimuthally changing thickness (*range modulation wheel*) is used. In all of these systems, the energy change is very fast (<0.1 s), but beam spreading can hardly be compensated.

Therefore, the choice of the location of the energy variation has important consequences on issues like transmission, neutron background radiation at the patient, energy spectrum, and beam spreading or emittance [11]. In this respect, relevant parameters of the accelerator choice and the beam transport system design are the needed speed of the energy change; the accuracy of the obtained energy (range); the effect on beam parameters such as intensity, energy spread, and beam broadening; and last but not least, the background dose to the patient from neutrons created by proton interactions with materials, for example, in collimators.

3.2.2 Lateral Dose Spreading

Beam spreading in lateral direction is necessary because the typical 1 cm size of a proton beam is much smaller than typical tumor dimensions. To irradiate target volumes of sizes between a few centimeters (e.g., eye treatments) up to 30–40 cm (e.g., sarcoma), one needs dedicated beam spreading systems adapted to the needed size.

Until now, the most commonly used method is *passive scattering* (Chapter 5) at which the beam is broadened by multiple scattering of the protons in a (set of) foil(s) or thin plate(s) and collimated just before the patient. When crossing the scatter foil, the beam will also lose a few MeV. When specifying the accelerator energy, one should be aware of this energy loss. Also one should keep in mind that, if no dedicated beam focusing measures are taken, up to 90% of the intensity can be lost at the collimator(s) in the nozzle, which will generate neutrons. With respect to the scattering system, accurate beam alignment is very important to prevent a lateral shift of the dose distribution and an asymmetry (tilt) in the lateral dose profile at the tumor. If the necessary beam size is larger than what can reasonably be accomplished with (double) scattering, *beam wobbling* can be added to the system. Here, a fast steering magnet sweeps the scattered beam along a certain trajectory over the target, so that an additional area is covered by the beam. The trajectory is fixed, for example a circular path or a set of parallel lines (*raster scanning*). This is repeated several times a second. For the accelerator, this implies that an eventual periodic time structure in the beam (e.g., pulses) should not interfere with this periodic motion; otherwise, locally severe under- and overdosage will occur in the dose pattern. In general, it is important to evaluate such interference risks for any periodic change in the beam characteristics, for example those due to a range modulator wheel.

The best coverage of the target volume in combination with the lowest dose in the surrounding normal tissue is obtained with the *pencil beam scanning* technique (see Chapter 6). Here fast steering magnets (*scanning magnets*) are used to aim a narrow beam at the volume elements (*voxels*) in the target volume sequentially. At each voxel, a specific dose is deposited. This is usually being done on a discrete grid (*spot scanning*), using a "*step and shoot*" method [6], in which the beam is only switched "on" if the pencil beam is aimed at the correct voxel. In some systems, the beam is not switched off when shifting to the next voxel, and of course, this is taken care of in the treatment planning. Currently, one is also developing *continuous scanning*, in which the pencil beam moves along a certain trajectory within the target volume [77]. The beam is kept "on" continuously. The dose is set by the combination of pencil-beam intensity and scanning speed. For spot scanning, the specifications on the accelerator are rather relaxed. The beam intensity should be sufficient and the on–off switching must be fast and reproducible. Beam positioning is controlled by the scanning magnets and must be fast and typically correct within a millimeter.

Continuous scanning techniques can be grouped into two categories: *time driven* and *event driven*. In the time-driven category, the scanning speed is fixed and the beam intensity is set as a function of the position of the pencil beam. In event-driven systems, the beam intensity is more or less fixed. The speed of the pencil beam motion is adjusted according to the necessary voxel dose and eventually corrected for the actual beam intensity.

For time-driven systems, the intensity of the beam must be adjustable within, say, a fraction of a millisecond (depends on scanning speed) and set to the desired value with an accuracy of a few percent. Unexpected fluctuations or interruptions in the beam intensity or a pulsed beam are not desired. In event-driven systems, the stability of the beam intensity is a bit less critical, if they can be compensated by the speed of the scanning magnets. Of course, a combination of the time- and event-driven methods is possible as well. More details are discussed in Chapter 6.

An important problem to be dealt with in radiation therapy is the motion of the tumor and/or nearby critical healthy tissue during the dose administration (see Chapter 18). Especially for time-dependent dose administration techniques like pencil beam scanning, having typical time constants in the same order of magnitude as those describing the motion, this is a problem. Therefore, various strategies are being considered to deal with this problem: *beam gating* [12–14], in which one suppresses the beam when the target is not at its correct place, a very fast multiple *rescanning* of the target volume [15,16], or *tumor tracking* or *adaptive scanning* [17] in which an online correction of the beam position, the intensity, or the energy is applied to "follow" the motion. All three methods require an extremely accurate and fast control of the (preferably continuous) beam intensity of the beam from the accelerator. Therefore, pulsed machines are less suitable for scanning applications, unless the repetition rate is sufficiently high (kHz) and the dose per pulse can be controlled with sufficient accuracy (less than few percent).

3.2.3 Timing Considerations

The characteristics of a proton beam from any accelerator will be subject to time-dependent variations. Slow variations are usually due to, e.g., temperature changes of magnets or sputtering of components in the ion source. These can easily be compensated or corrected for in control loops. For faster variations, one should consider the possibility how these changes interfere with the dose delivery.

Intensity variations are the most occurring "spontaneous" variations in a proton beam. One distinguishes regular oscillations or random fluctuations (beam noise). Also the beam is pulsed. As will be described later, a synchrotron operates in a so-called spill mode. During a spill, the beam is extracted from the synchrotron during a period lasting between a second and a fraction of a minute. The time between two spills is in the order of a few seconds. During the spill, the beam intensity pulses at a rate of several tens of MHz. In all accelerators, the MHz-frequency pulses originate from the acceleration voltages in the accelerator, which oscillate with a so-called radio frequency (*RF*). From a cyclotron, the extracted beam is continuous, but it is pulsed at a frequency in the range of 50–100 MHz. This is also in the beam from a synchrocyclotron, but here, these pulses occur in groups with a repetition rate of a few hundred Hz.

In most treatments and techniques, repetition rates of higher MHz frequency do not play a role, since few microseconds time structures in the beam are much faster than the time constants in the scanning process, which are in the range of milliseconds to seconds. However, in dosimetry devices like ionization chambers, one deals with processes with microsecond timescales in the charge collection processes. In such studies, it may be necessary to consider this "burst" nature of the proton beam, since it may, for example, cause space-charge effects.

The passive scattering technique is a robust technique with respect to timing issues in the proton beam. The only issue on timing is the range modulation. The rotation of the wheel causes energy changes at a frequency of a few hundred Hz. Processes occurring at this (or related) frequency may cause distortion in the dose application and measurement data. One should, for example, take care of intensity oscillations (pulses) from the accelerator or data read-out sequences at these frequencies. Especially, when using a synchrocyclotron, one should take care of such timing interferences.

The spot scan technique is relatively insensitive to time-dependent intensity variations and periodic effects. However, the timing of the dose monitors and the beam on–off switching are critical. Continuous scanning is quite sensitive, however. In this technique, the intensity is controlled to be at a certain value during a fraction of a millisecond. When the applied dose is controlled by means of the scanning speed, low speed and low beam intensity are needed to prevent dose errors due to a very slow matching of the scan speed. Currently, the spill character of a beam from a synchrotron and the kHz pulses from a synchrocyclotron make their application for continuous scanning impossible. However, spot scanning is certainly possible during the spill of a synchrotron beam. In current synchrocyclotrons, spot scanning is applied at the pulse rate of the synchrocyclotron. The pulse length cannot be varied, so the intensity per spot is controlled by the intensity at each pulse. However, since the intensity regulation per pulse is usually not accurate enough, it is difficult to apply the prescribed spot dose with a single pulse, especially if the needed dose varies per spot. Therefore, the dose per spot is applied in several low-dose pulses. This, however, may be disadvantageous for the total treatment time.

For a typical beam line magnet, it takes time to change its strength. The magnet power supply must give an extra voltage over the magnet coils to change the field strength of a magnet. The faster one changes, the higher this voltage must be. This voltage is limited, however, by the insulation between the different turns in the coil of the magnet. In addition to this, the change in magnetic field is partly compensated by eddy currents, which are induced in the iron yoke and pole of the magnet. These currents usually decay within a few seconds, but this limits the speed of the energy variations. By making the yoke and pole out of laminated iron (iron sheets of only a few mm thickness and insulated from each other) orientated parallel to the magnetic field lines, the effect of the eddy currents can be reduced dramatically. For example, in the beam lines at PSI, a typical energy change of 2% for range modulation is reached in less than 0.1 s. This is the fastest system currently in operation.

As briefly mentioned earlier, the timing issues in the beam are important to deal with in relation to organ/tumor motion during the dose delivery (see Chapter 18).

3.3 CYCLOTRONS

Modern cyclotrons dedicated for proton therapy accelerate protons to a fixed energy of 230 or 250 MeV [18–21]. Compared to the classical cyclotrons in accelerator laboratories, the new cyclotrons (see Figure 3.1 for two examples) have been made much simpler. Many operating parameters of the cyclotron have been fixed to a more or less optimal value for the given beam energy and the typical intensities needed. This simplified the system as well as its operation. These machines are rather compact with a magnet height of approximately 1.5 m and a typical diameter between 3.5 m (100 tons) and 5 m (200 tons), when equipped with superconducting coils or with room temperature coils, respectively. Usually, some extra space is needed above and/or below the cyclotron for the support devices of the ion source, eventual liquid helium supply, and space for access and equipment to open the machine.

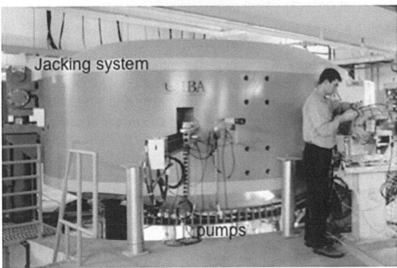

FIGURE 3.1 The 250 MeV superconducting cyclotron of Varian, of which the first one has been installed by ACCEL at PSI, Switzerland [20,21], and the 230 MeV proton cyclotron of IBA (Louvain la Neuve, Belgium), of which the first one has been installed in Boston, MA (USA) [18].

The most important advantages of a cyclotron are the continuous character (continuous wave, CW) of the beam and that its intensity can be adjusted quickly to any typically desired value. Although the cyclotron has a fixed energy, the beam energy at the patient can be adjusted fastly and accurately by means of a fast degrader and an appropriate beam line design, as discussed in the previous section. In addition, the simplicity of the design concept and the relatively low number of components are often considered as advantages for the reliability and availability of the accelerator.

The major components of a typical compact cyclotron are (Figure 3.2)

- an RF system, which provides oscillating strong electric fields by which the protons are accelerated,

- a strong magnet of few Tesla, confining the particle trajectories into a spirally shaped orbit, so that they can be accelerated repeatedly by the RF voltage in the order of 30–100 kV,

- a proton source in the center of the cyclotron, in which hydrogen gas is ionized and from which the protons are extracted,

- an extraction system that guides the particles that have reached their maximum energy out of the cyclotron's magnetic field into a beam transport system.

Although tilted cyclotron orientations are in use, in this chapter, it is assumed that the particle orbits in the accelerator are in the horizontal plane.

The RF system consists of two to four electrodes (due to their shape in the first cyclotrons built, often called "*Dee*"), which are connected to an RF generator, driving the oscillating voltage with a fixed frequency somewhere in the range of 50–100 MHz (so in the RF domain). Each Dee consists of a pair of copper plates on top of each other with a few centimeters in between. The top and bottom plate are connected to each other near the center of the cyclotron and at the outer radius of the cyclotron. The Dees are placed between the magnet poles. The magnet iron and environment of the Dees are at ground potential. When a proton crosses the gap between the Dee and the grounded region, it experiences acceleration towards the grounded region when the Dee voltage is positive. When it approaches the Dee at the negative voltage phase, the proton is accelerated into the gap between the two plates. During its trajectory within the electrode (so between the top and bottom plate) or in the ground-potential region, the proton is in a region free of electric

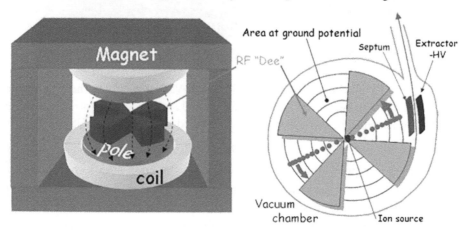

FIGURE 3.2 Schematic view of the major components of a cyclotron: The magnet, the RF system (Dees), ion source, and extraction elements. The protons being accelerated are schematically indicated on their "spikes."

fields from the Dees, and at those moments, the voltages on the electrodes change sign, without effecting the proton in the field-free region. The magnetic field forces the particle trajectory along a circular orbit, so that it crosses a gap between Dee and ground several times during one circumference. In the example shown in Figure 3.2, there are four Dees, so that a proton is accelerated eight times during one turn. When the electrode voltage is, for example, 60 kV at the moment of gap crossing, the proton gains $\Delta E = 0.48$ MeV per turn. Due to the energy gain, the radius of the proton orbit increases so that it spirals outward. The maximum energy E_{max} (typically, 230 or 250 MeV) is reached at the outer radius of the cyclotron's magnetic field, after approximately $E_{max}/\Delta E$ turns (530 in the example).

3.3.1 RF System of a Cyclotron

The RF system is the most challenging subsystem in a cyclotron since many contradicting requirements need to be dealt with. Important operational parameters are the RF voltage and frequency. A minimum value of the RF voltage is needed to make the first turn. This starts at the ion source in the center and, in the first turn, apertures are passed, and the protons have to go around the ion source to pass connections between the top and bottom halves of the Dees, see Figure 3.12 in the section dealing with the ion source. A high Dee voltage is advantageous. This yields a large ΔE, which enhances turn separation, so that the beam passes all areas more quickly. This makes the beam less sensitive to small local errors in the magnetic field, and in addition to this, it is a prerequisite to obtain high extraction efficiency. The combination of a high voltage and a high frequency needs a high power. This and the increasing risk of discharges at high field strengths limit the maximum Dee voltage.

Further, the period (1/frequency) of the RF voltage at each Dee must be synchronous to the azimuthal location of the protons at all radii. The time T a proton (with electric charge q and mass m) needs to make one turn with radius r, depends on its velocity v and the strength of magnetic field B. For a circular orbit, the Lorentz force Bqv acts as the centripetal force:

$$\frac{mv^2}{r} = Bqv \qquad (3.1)$$

The proton velocity can be written as $v = 2\pi r/T$, where T is the time it takes a proton to make one turn. With Eq. (3.1), this time yields

$$T = \frac{2\pi m}{qB} \qquad (3.2)$$

Note that this time T does not depend on the radius or the velocity of the particle. This means, that all particles that are properly accelerated, are at the same azimuthal angle in the cyclotron. Other particles will be lost, since they do not cross the acceleration gaps at the right moment, as determined by the RF frequency, which is equal to an integer × 1/T. As indicated in Figure 3.2, they are all within a cloud resembling a rotating spike of a wheel. Due to the energy-independent matching of the orbit frequency 1/T with the RF frequency, this "normal" type of cyclotron is also called *isochronous* cyclotron.

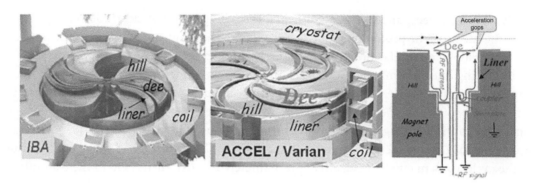

FIGURE 3.3 The IBA/SHI cyclotron (11 and www.striba.com) with two Dees, the Varian cyclotron with four Dees, and a sketch of the RF currents in the cavity, creating a voltage across the acceleration gap.

Although pulsed at the RF frequency of the accelerating voltage, this frequency is so high that the beam intensity extracted from the cyclotron can be considered continuous (*CW*) for almost all processes in the dose application in particle therapy. The ratio between the RF frequency and the orbit frequency of the protons must be an integer number, the *harmonic number h*. In case of one electrode covering 180° of the pole, one typically uses $h = 1$ and in case of two or four electrodes of 45° (as used in Figure 3.2), $h = 2$ can be used. In Figure 3.3, two typical RF electrode configurations (with 2 and 4 Dees, respectively) are shown.

The magnet pole consists of *hills* and *valleys,* and the Dees can be mounted in (some of) the valleys, so that the gap between the upper and lower hill can be minimized. A small gap needs less electric current in the magnet coils and also has a large effect on the exact shape of the magnetic field. The Dee is mounted on a copper pillar (*stem*), and the valley wall is covered with a grounded copper sheet (*liner*). At the bottom of the valley, a *short plate* connects the stem with the liner. The combination of Dee, stem, and liner acts as a resonant *cavity.* This means that an RF current can flow back and forth along the stem, with a frequency determined by the resonance frequency of this cavity. The Dee will get a negative potential when the electrons flow to the edge of the Dee, and when the electrons flow to the grounded liner, the Dee will get a positive potential. A quality factor Q of the cavity is defined in terms of the ratio of the energy stored in the cavity to the energy being dissipated in one RF cycle and can have a typical value in the order of 3,000–7,000.

The RF current is induced by a *coupler,* which is an antenna that emits the RF power from the external generator into the cavity. By slightly modifying the volume of the cavity, for example by shifting the stem-liner short plate, the resonance frequency of the cavity slightly changes. Since the oscillation frequency is enforced by the RF generator, this detuning of the cavity yields a change in Q factor of the cavity and thus of the power absorbed in the cavity. This then changes the Dee voltage amplitude. This is typically used for fine regulation of the Dee voltages.

The RF currents float along the inner surface of the cavity and can be on the order of kA. Also high voltages are obtained across small gaps. These conditions imply rigorous

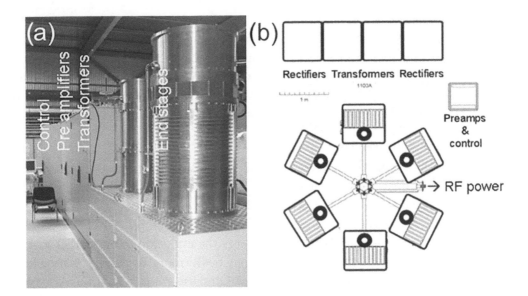

FIGURE 3.4 (a) Picture of a 150 kW, 72 MHz, RF amplifier (Courtesy: PSI) and (b) top view of a design developed by Cryolectra, consisting of 6 racks filled with 24 solid-state amplifiers of 1.45 kW [22].

cooling, good vacuum, and clean surfaces. A lot of power (60–120 kW) dissipates as small components in the Dees and in the transfer lines, making the RF system usually one of the most vulnerable subsystems of a cyclotron.

The RF generator operates in the FM-frequency range and uses typical techniques for radio transmitters. Between 100 and 200 kW, RF power is needed in total. Most cyclotrons still work with a radio-tube-driven RF-amplifier system. In those cases, the amplifier consists of several stages in series, of which the final stage is a high-power vacuum tube (typically, a tetrode or triode), see Figure 3.4. A recently developed RF amplifier [22] consists of many parallel–coupled, solid-state amplifiers. This might potentially offer higher reliability due to the possibility of redundant amplifier units.

3.3.2 Cyclotron Magnet

Specifications of the magnetic field are determined by the beam dynamics. First of all, the field must be *isochronous*: at each radius r, it must have the appropriate strength to match the time T a proton needs to make one turn to the RF frequency, as given in Eq. (3.2). Second, the shape of the field lines must provide a focusing force in the vertical direction, to confine the vertical space in which the protons are moving. However, a small gap between the upper and lower pole (hills) helps to limit the current in the main magnet coil, and a small gap is also advantageous for field shaping and the reduction of RF fields outside the Dee.

The magnetic field must be correct within approximately 10^{-5}. Although small local deviations can be accepted at some locations, a repetitive encounter of the beam with such a distortion often leads to systematic trajectory distortions, yielding instabilities and beam losses. Therefore, careful selection and shaping of the iron, positioning of the coils, and field mapping are essential steps in the production phase of a cyclotron [23,24].

Once the cyclotron has been commissioned and the field optimized, one usually does not need to care much anymore about the magnetic field. Sometimes, small adjustments of the current through the magnet coil are necessary to compensate temperature changes of the iron or changes in component positions after a service. But since the magnetic field specifications are quite critical, the magnet design may have strong implications for the operational quality and eventual future upgrades.

3.3.3 Superconducting Magnet Coils

The magnetic field strength in the commercially available cyclotrons is between 2 and 3.5 T. Both conventional copper magnet coils [18,19] and, since a few years, superconducting coils are used for therapy cyclotrons. Advantages of superconducting coils are low power consumption (20 kW, mostly for the liquid helium cooling, versus 300–350 kW) and, especially, stronger magnetic field. This allows the cyclotron to be smaller and less heavy, but more importantly, since the iron is magnetically more saturated at strong fields. This makes the magnetic field less sensitive to small imperfections in the iron. Also, when switching on, cycling (a ramping procedure to erase the "magnetic history," in which the magnet is first set at a stronger field than needed) is not needed. In the superconducting (SC) cyclotron at PSI (Varian) [20], the coil is mounted in a closed ring and cooled by means of liquid helium. As is also visible in Figure 3.3, the coil ring is suspended in a vacuum cryostat providing the thermal insulation. In addition, the coil is surrounded by a heat shield (at 40–70 K) and many layers of super insulation. The outside of the cryostat is at room temperature. Therefore, opening of the cyclotron does not require warming up of the coil. Due to the insulation between coil and magnet iron, temperature effects in the iron, which occur in cyclotrons with normal conducting coils, are not present. Disadvantages of the SC coil are the risk of a quench due to operational errors (which are normally prevented by the control system) or when a high-intensity beam is lost or stopped close to the coil.

3.3.4 The Average Magnetic Field

According to Eq. (3.2), the magnetic field in the cyclotron should be homogeneous. However, when the energy of the protons becomes larger than 20–30 MeV, their velocity becomes a considerable fraction of the speed of light c ($v/c = 0.2$ at 20 MeV). Due to relativistic effects, this yields an increase of the proton mass m with respect to its rest mass m_0:

$$m(r) = \gamma(r)m_0 = \frac{1}{\sqrt{1-v(r)^2/c^2}} m_0 \qquad (3.3)$$

At 20 MeV, this is a 2% effect, but at 250 MeV, with $v/c = 0.61$, this yields a mass increase of $\gamma = 1.27$. In a homogeneous field, this would imply (see Eq. 3.2) that the orbit frequency would drop with energy (or radius). With increasing energy (i.e. radius), the protons would lose pace with the acceleration voltage and would no longer be accelerated.

Until the late 1950s, the method to accelerate protons in a cyclotron to energies above about 30 MeV was by adapting the RF to the radius of the proton orbit. The RF is then modulated *synchronous* to the mass increase of the protons. The system that performs this frequency modulation has a typical repetition rate in the range of 0.2–1 kHz. A similar frequency modulation can be used to compensate for a decrease of the magnetic field strength with radius (see Figure 3.7, top), which may occur in cyclotrons with strong fields. This technique in the so-called *synchrocyclotrons* has experienced a revival in recent designs of compact high-energy SC cyclotrons [25,83]. As a consequence of RF modulation, the beam intensity extracted from these synchrocyclotrons is pulsed at the repetition rate of the modulation, so at typically 0.2–1 kHz.

Initially, synchrocyclotrons of 160–200 MeV have been used for proton therapy at Harvard, Berkeley, and Uppsala. However, the more modern cyclotrons employ an increase of the magnetic field with radius to cope with relativistic effects and keep the cyclotron *isochronous*:

$$B(r) = \gamma(r) B_0 \tag{3.4}$$

Here B_0 is the field strength that would be needed if relativity were ignored. As an example, the field strengths of the 250 MeV cyclotron at PSI (Varian [23]) and the 230 MeV cyclotron of Ion Beam Applications (IBA), Louvain-La-Neuve, Belgium [27] are plotted as a function of radius in Figure 3.5. With this field shape, the protons do not lose pace with respect to the RF signal and the beam remains CW.

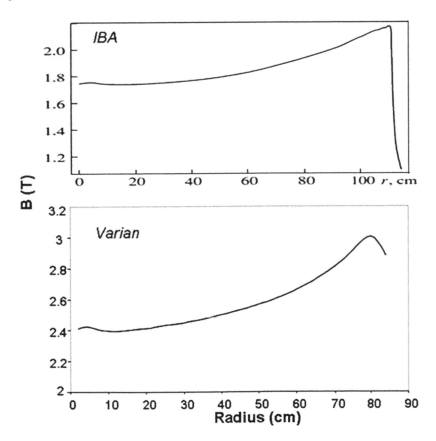

FIGURE 3.5 The average magnetic field strength as a function of radius for the IBA cyclotron [27] and de Varian SC cyclotron [23].

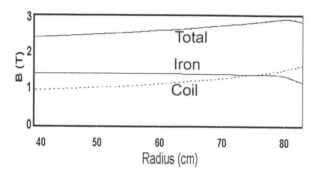

FIGURE 3.6   Contributions of the (saturated) iron and the coil to the average magnetic field.

Two methods are commonly used to increase the field with radius. In the IBA and Sumitomo (SHI) cyclotrons [11], the gap between the magnet poles is decreasing with radius, as can be seen in Figure 3.3. The method employed in the SC cyclotron of Varian takes advantage of the strong magnetic field close to the SC coils. At fields above 2.5 T, the iron is saturated and gives an almost constant (i.e., independent of the current through the coil) contribution of 40–60% to the total magnetic field. The rest of the field comes directly from the coil [28–30], and this part increases with radius, as shown in Figure 3.6.

3.3.5 Focusing Properties

A magnetic field that increases with radius, however, would cause a lack of vertical beam stability. Contrary to a field that decreases its strength with radius, in which particles are pushed back to the median plane by the Lorentz force after a vertical excursion, particles are pushed towards the poles in a field that increases with radius (see Figure 3.7).

The radial variation of a magnetic field can be expressed by means of the field index n:

$$n(r) = -\frac{r}{B(r)} \cdot \frac{dB(r)}{dr} \tag{3.5}$$

If B decreases with radius (i.e., dB/dr is negative), n has a positive value and the field has vertically focusing properties. The relativistic correction of Eq. (3.4) yields an increase of B with radius, so a positive dB/dr. In that case, the field index is negative, which is thus causing defocusing in the vertical plane. The field index Eq. (3.5) varies from 0 in the center to approximately −0.5 at extraction in typical isochronous cyclotrons for proton therapy. To compensate this defocusing, additional vertical focusing power must be added. This is achieved by an azimuthally varying field (AVF) by adding pie-shaped pieces (*hills*) to the pole, as shown in Figure 3.8.

Along a turn, a proton thus experiences stronger magnetic fields when crossing the pole hills and weaker fields when crossing the valleys. The average value of the field along a turn is, of course, equal to the isochronous one. In the strong field, the orbit is slightly more curved than in the weaker field. Therefore, the protons do not cross the boundary between the two regions perpendicularly. When a proton is not traveling in the median plane, it will

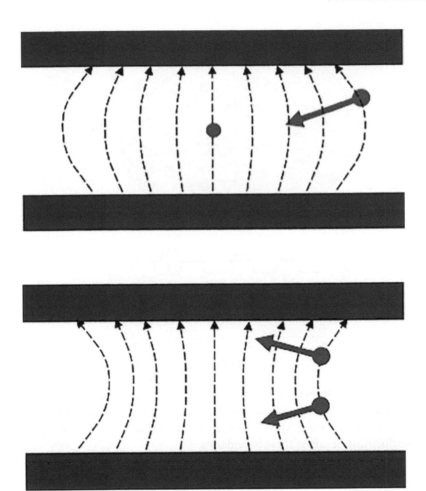

FIGURE 3.7   Vertically focusing and defocusing in a cyclotron with a magnetic field that is decreasing with radius (top) and in a cyclotron with a field that is increasing with radius (bottom).

therefore experience an azimuthal component of the magnetic field, which is not anymore parallel to the beam. This component creates a vertical component of the Lorentz force. The magnitude of this force is proportional to the difference between the hill and valley field strengths, the distance of the proton to the median plane and $\tan(\varphi)/r$, φ being the angle between the trajectory and the normal to the hill–valley boundary. Depending on the traveling direction, the magnitude of φ and whether the proton track is above or below the median plane, this force is directed either towards the median plane or towards the pole. This repetitive focusing and (depending on φ) defocusing results in a net focusing force. The focusing strength of these field steps depends on the variation of the field, as expressed by the flutter $F(r)$:

$$F(r) = \frac{\left\langle B(r)^2 \right\rangle - \left\langle B(r) \right\rangle^2}{\left\langle B(r) \right\rangle^2} \tag{3.6}$$

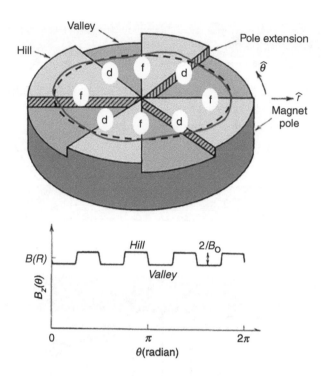

FIGURE 3.8 Pie-shaped hills and valleys on a pole give an azimuthally (θ) varying magnetic field strength. In the hill, the orbit has a smaller radius of curvature due to the stronger field. Since the boundary between hill and valley is not crossed perpendicularly due to this, alternating vertical focusing and defocusing occurs at these boundaries.

where <....> denotes the average over one turn. Typical values of flutter can be in the order of 0.05. Starting at a certain radius, the flutter will decrease with radius. In addition the vertical Lorentz force decreases with increasing r. Therefore, the focusing effect of the flutter must be enhanced with increasing radius in the cyclotron. This is done by increasing the angle between the proton trajectory and the field step with radius; the hill must be twisted to a spiral with spiral angle $\psi(r)$, as shown in Figure 3.9 and as was also visible in Figure 3.3.

The restoring forces ensure vertical stability. The proton tracks perform a vertical *betatron oscillation* around the equilibrium orbit in the median plane with frequency or tune Q_z (instead of Q, the symbol ν is also frequently used). The strength of the focusing is given by Q_z^2, where a high betatron frequency yields strong focusing. The total focusing power in the vertical plane can then be written as the sum of above contributions. So, for such an *AVF* cyclotron, consisting of N sectors (hills), one gets

$$Q_z^2(r) \approx n(r) + \frac{N^2}{N^2-1} F(r)(1+2\tan^2 \psi(r)) \tag{3.7}$$

Typical values of Q_z are around 0.2, which suffices for the (weak) vertical focusing.

At the extraction of the cyclotron, the vertical focusing can become strong due to the grazing crossing of the fringe field at the outer edge of the cyclotron pole. Therefore, it is

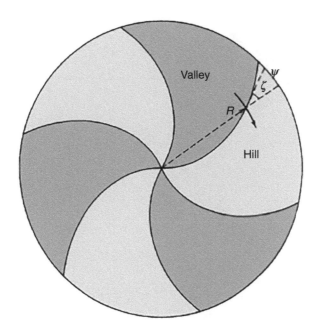

FIGURE 3.9   By twisting the hill into a spiral shape, the angle ψ is increasing with radius, so that the vertical focusing is maintained at large radius.

important to guide the protons as fast as possible through this region to prevent a too dramatic increase of the vertical beam size.

In a cyclotron, focusing in the horizontal plane is rather straightforward and is automatically achieved since the isochronicity requirement of Eq. (3.4) yields a fixed relation between the horizontal (=radial) betatron frequency Q_r and the relativistic mass increase $\gamma(r)$ (defined in Eq. (3.3)):

$$Q_r(r)^2 = 1 - n(r) = \gamma(r)^2 \qquad (3.8)$$

In addition, there is a contribution of a few percent from the spiral-shaped hills and valleys. The quantity Q_r is always greater than unity, so radial focusing is strong. Note that the second equality in (3.8) is valid until the field starts to flatten off just before extraction, since there the isochronicity criterion need not be maintained anymore in these last few turns. The horizontal betatron frequency plays an important role in the central region and during extraction. An oscillation with the associated $Q_r \approx 1$ yields a radial shift of the orbit, as illustrated in Figure 3.10.

In the central region, the initial centering is quite sensitive to small field distortions. These can cause an accumulation of orbit shifts and lead to large *precessions* (changes in the position of the orbit center) of beam orbits. In a working machine beam, centering corrections may be necessary after a service involves exchange of components. Slight shifts of Dee positions or of the ion source may result in such centering errors. However, on the other hand, corrections can be made easily with small trim coils attached to the pole or by changing the insertion depth of iron *trim rods* in pockets in the central pole region.

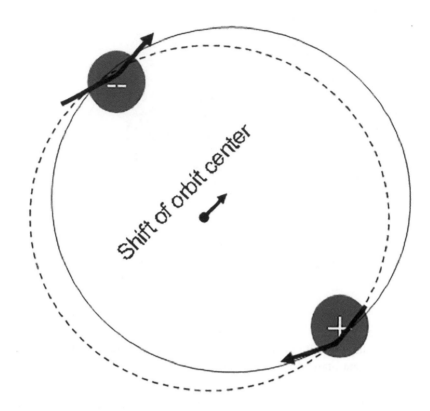

FIGURE 3.10   At the $Q_r = 1$ betatron resonance, the center of the orbit is shifted by a small magnetic bump.

Just before the extraction radius, the $Q_r = 1$ resonance is crossed again, and this is used to extract the beam, as will be discussed later. Although, as in this example, in some designs, a betatron resonance is used in advantage, in general, one must be very careful. Excitation of a resonance will cause an increase of the oscillation amplitude and may have beam loss as a consequence. Also coupling resonances, in which there is an exchange of amplitude between radial and vertical oscillations, are possible. In general, resonances occur at radii in the cyclotron, where

$$mQ_r(r) + nQ_z(r) = p, \tag{3.9}$$

in which m, n, and p are integer numbers. In the design phase of the machine, one tries to adjust the spiral and flutter to avoid such resonances. If a resonance cannot be avoided, one tries to tune the field (and RF voltage) such that it is crossed in a minimum number of turns, so that small errors in the magnetic field do not have the chance to excite such resonances.

After commissioning of the cyclotron, the tunes are fixed, and one only needs to be aware of resonances in case of small changes in orbit positions due to machine changes (e.g., upgrades).

3.3.6 Proton Source

Currently, all commercially available cyclotrons are equipped with a proton source in the center of the cyclotron. External sources are of interest if other particles need to be accelerated. The internal sources operate by exploiting the Penning effect [31,32]: the ionization of gas by energetic electrons in an electrical discharge.

The ion source consists of two cathodes at a negative voltage of a few kV, located at each end of a vertical hollow cylinder at ground potential (*chimney*), into which the H_2 gas is flushed (about a cm³/minute), see Figure 3.11. Free electrons are created by heating a filament or, in case of a "cold-cathode source," by spontaneous emission from the cathode in the strong electric field between cathode and ground or by emissions after impact of a proton at the cathode. In the electric field between cathode and ground, the electrons are accelerated and will ionize the gas. They are confined along the magnetic field lines and bounce up and down between the cathodes, thus increasing the ionization of the gas.

The ions (H^+, H_2^+, H^-, etc.) and electrons form a plasma that fills the chimney volume. Since a plasma is an electric conductor, external electric fields hardly penetrate the plasma. However, protons and other ions that diffuse to a little hole in the chimney wall will experience electric field from the nearest Dee (the *puller*). When the Dee is at a negative potential, protons that escape from the plasma will accelerate towards the Dee. If they arrive at the right phase, they will cross the distance to the puller and will be further accelerated, as shown in Figure 3.12. Due to the narrow acceptance windows in time and in the further acceleration path, only a few percent of the protons leaving the source is actually being accelerated.

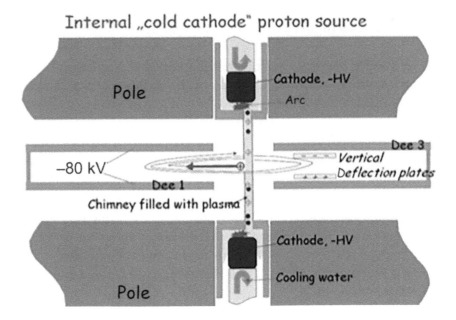

FIGURE 3.11   Schematic view of the ion source of the "cold cathode" type (=no filament used) and the first few turns.

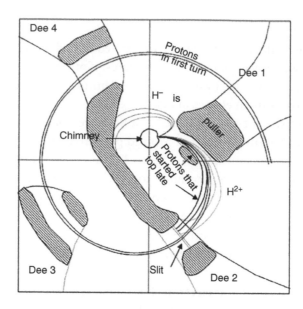

FIGURE 3.12   Orbits in the central region of different ion species coming out of the source. To prevent acceleration of unwanted particles, a slit is selecting the protons that travel at the correct RF phase. The radii of the orbits are determined by the amplitude of the RF voltage, which should be high enough to get around the obstacles in the central region.

3.3.7 Practical Issues

Important specifications of the ion source are: the total proton current extracted from the source within moderate emittance, the stability of the intensity, and the time between services (goes on cost of operational time of the cyclotron). At an ion source service, one typically exchanges filaments (if used), cathodes (worn out by sputtering), and the chimney (its extraction hole increases by sputtering). Modest operational conditions and a careful material choice (heat properties, electron emission, sputtering resistance) are of importance to obtain a source lifetime of 1–2 weeks or even up to a month in a typical proton therapy cyclotron.

3.3.8 Beam Intensity Control

The beam intensity is regulated by adjusting the arc current between the cathodes and ground, the gas flow, and the current through the filament (if present). Usually, the proton current reacts rather slow (milliseconds to seconds) on changing these parameters. In these kinds of sources, it has been observed that an intensity increase also leads to an increase of the emittance (beam divergence) of the beam coming out of the source [33]. By selecting a fraction of the beam around the maximum of the emittance, a very stable beam intensity of several percent (at kHz band width) can be obtained routinely [34]. This is essential for fast line scanning.

In the IBA cyclotron, the beam intensity is regulated at the source by a feedback loop from the measured beam intensity. At PSI, the ion source is operated at a fixed setting. In this cyclotron, and recently also in the Sumitomo cyclotron, a pair of adjustable radial

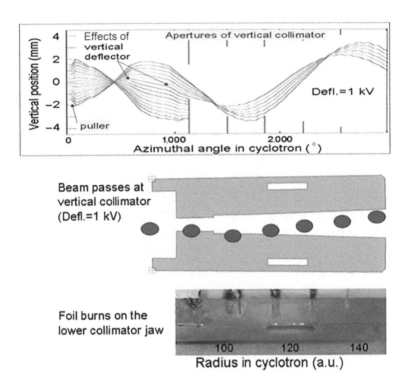

FIGURE 3.13 In the PSI cyclotron, a fast regulation of the beam current is performed by means of vertical deflection of the beam, shown in the top figure. The aperture (middle figure) is crossed six times. The bottom figure shows beam passages at the lower aperture limiter by means of foil burns. Due to the vertical betatron oscillations, the beam is mostly cut at the first three crossings.

apertures (*phase slits*) close to the cyclotron center sets the maximum intensity of the accelerated beam. Between 0 and this maximum and within 50 μs, the intensity is regulated with an accuracy of 5% by a vertically deflecting electric field and a vertically limiting collimator, cutting the beam in the first few turns, as shown in Figure 3.13. This fast intensity regulation at the vertical collimator by the vertical deflector (see Figure 3.14) is used for intensity control during line scanning.

3.3.9 Beam Extraction

The *extraction efficiency* is an important specification of a cyclotron. A low efficiency means severe beam losses, which will cause radioactive components in the cyclotron [35] and enhanced wear and dirt deposition on insulators. This reduces the availability of the cyclotron and increases the radiation dose to the maintenance staff. One should therefore strive for extraction efficiencies of at least 60–70%. Mind that an extraction efficiency of, e.g., 25% implies nine times (!) more lost protons per extracted proton, compared to a cyclotron with 75% extraction efficiency.

The major problems to overcome with the extraction of the beam from a compact cyclotron are to separate the protons to be extracted from those who still have to make one or more turns and to prevent destruction of the extracted beam in the grazing passage of

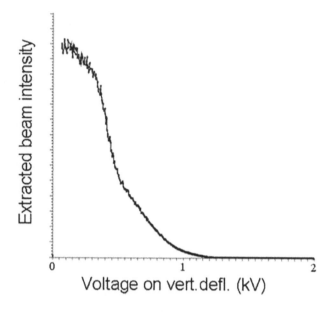

FIGURE 3.14 The extracted beam intensity as a function of voltage on the vertically deflecting plates, mounted in Dee 3 of the PSI cyclotron.

the rapidly decreasing fringe field. The design of the extraction is therefore aimed at an increase of the orbit separation at the appropriate location and to cross the fringe field as quickly as possible, followed by a reduction of the vertical focusing to compensate for a vertical overfocusing and a horizontal focusing to prevent a too large beam width.

The extraction is performed by several *extraction elements*, of which the first one is a *septum* (see Figure 3.2). This thin vertical blade is aligned parallel to the beam orbit and separates the extracted beam from the ones that still have to make one or more turns. The septum should be able to withstand power dissipation from beam particles that hit the septum. A bar-shaped cathode is positioned parallel to the septum at a larger radial position. The electric field between this cathode (at several tens of kV) and the septum deflects the beam more outward into a channel in between the coils and through the yoke. In this *extraction channel*, additional steering and (de)focusing are performed by permanent magnets and/or field-shaping iron blocks.

The major problem to deal with at extraction is the small orbit separation. During acceleration, this is decreasing due to two effects. First, the radius of each orbit increases linearly with the momentum p and thus approximately with $\sqrt{E}$. A constant ΔE per turn thus leads to smaller distances between sequential orbits. In addition, ΔE will decrease due to loss of isochronism if the magnetic field becomes too weak at extraction radius (see Figure 3.5 and inset at Figure 3.15).

The orbit separation can be increased by exciting the $Q_r = 1$ resonance to shift the center of the orbits. This already creates a larger separation between two turns, as shown in Figure 3.10. But a stronger effect can be achieved when a precession of the orbits is used, which becomes effective a few turns later. Due to the decreasing magnetic field beyond the maximum, Q_r starts to decrease. For $Q_r < 1$, the phase of the betatron resonance starts to

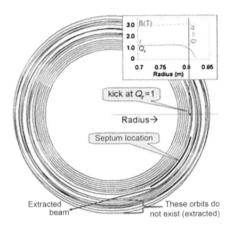
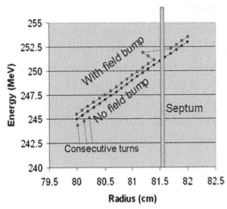

FIGURE 3.15 To obtain a large turn separation, one uses the precession of the orbits, after exciting the $Q_r = 1$ resonance at the radius where the field index is $n = 0$. The precession is caused by a phase advance of the betatron oscillation due to $Q_r < 1$ beyond the resonance. To illustrate the precession, orbits are drawn, which do not exist when the beam is extracted. Along the radial direction, the precession shifts consecutive turns to smaller radii, followed by a jump to large radius. This is the optimal location for a septum to split off the extracted beam.

shift to a larger azimuth and an orbit precession starts, as illustrated in Figure 3.15. When looking along one radial line, one then observes that the orbit with next higher energy is at smaller radius than the previous one. However, when continuing the precession with a few turns, this will lead to an orbit shift giving an increase of distance between two sequential orbits, as demonstrated in Figure 3.15. This is an optimal location for a septum that splits the extracted beam off.

The resonance is excited by kicking the beam with a magnetic bump, created with little coils or with iron trim rods. The orbit separation is further increased by a high-energy gain per turn and by using a small horizontal beam emittance at the beginning of the acceleration. A high energy gain per turn is accomplished by high Dee voltages and many Dees and by limiting the phase slip due to loss of isochronicity after crossing the field maximum. A small horizontal emittance can be achieved by cutting the beam sufficiently with slits and collimators in the central region (low energy: low power loss and low/no activation). Application of this *resonant extraction* method in the PSI cyclotron (Varian) has resulted in a routinely obtained extraction efficiency of 80%.

In the IBA and SHI cyclotrons, another approach, *self-extraction*, is pursued [27]. In these cyclotrons, the acceleration continues to a maximum possible radius, and the field is kept isochronous by employing an elliptical pole gap (visible in Figure 3.3). This gap is very narrow at the pole edge, so that the field drops to zero for a very short radial distance (see Figure 3.5). With this technique, one strives to guide the beam out of the main field and through the fringe field very quickly. Just before extraction radius, a groove is made in one hill to get the extraction at a well-specified azimuthal angle. The beam is deflected further by means of a septum and extraction cathode. To limit the beam size in these cyclotrons, several (de)focusing elements are following the septum.

3.3.10 Matching the Beam Energy of a Cyclotron Beam

Since current cyclotrons for therapy have been optimized for one energy, energy changes can only be made after extraction from the cyclotron. In systems using the passive scattering technique, one can usually set a maximum energy (e.g., per field) by a degrader at the beginning of the beam transport system. Energy (or range) modulation is then done in the nozzle, just before the beam enters the patient. In general, when degrading just before the patient, one should take care of the lateral beam size increase due to multiple scattering in the degrader. This would be quite bad in the case of pencil beam scanning. So, when applying the pencil beam scanning technique, an upstream degrader is preferred.

The upstream energy setting necessitates all following beam line magnets to be set such that they are matched to the beam energy. As discussed in Section 3.2.1. An upstream degrader is usually followed by an ESS to limit the spread of energy.

3.3.11 Developments in Cyclotrons for Therapy

Of course, many small details and technologies in cyclotrons have been improved in the last years. However, initially proton therapy has been performed by cyclotrons in physics laboratories, next to the physics program. The modification from an accelerator for physics experiments into a cyclotron dedicated for proton therapy has been stimulated by the commercial companies developing cyclotrons for isotope production. These developments have initiated the search for simplicity and removal of all "unnecessary knobs." A next step was the introduction of superconducting coils, which reduced the size and weight, but kept at least the same performance level. The use of superconducting coils has been essential for a further reduction of size and weight of the cyclotron. However, these small synchrocyclotrons have a pulsed beam, so that their application possibilities are less than those of the "normal" isochronous cyclotrons. But it is interesting to see how these developments stimulate industries as well as laboratories to continue their search for an ideal small cyclotron.

3.4 SYNCHROTRONS

The first hospital-based proton therapy facility, at Loma Linda Medical Center (CA, USA), has been equipped with a synchrotron built by Fermilab (Figure 3.16) and started its operation in 1992 [8]. Equipment based on the design used at Loma Linda has become commercially available at the Optivus company [36]. The proton synchrotron designed by Hitachi [12], see Figure 3.16, is in use at centers in Japan and in the United States and the one of Mitsubishi in several centers, Japan [37]. All machines provide protons between 70 and 250 MeV.

For proton therapy, synchrotrons and cyclotrons are quite competitive, but for heavy ion therapy, synchrotrons are currently the only machines in operation. Recognized advantages of a synchrotron are that the protons are accelerated until the desired energy, that almost no radioactivity is created due to beam losses [39] and that low-energy protons have the same intensity as high-energy protons, since there is no transmission loss in a degrader [40].

The space required for a proton synchrotron is larger than for a cyclotron: the synchrotron itself has a diameter of 6–8 m and the injection system has a length of 6–10 m. The

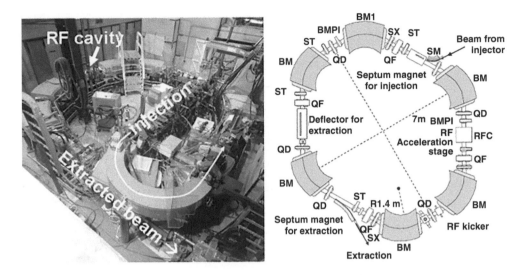

FIGURE 3.16 The synchrotron used at Loma Linda [38] with the 2 MeV injector mounted on top of the synchrotron ring and the synchrotron of Hitachi (injector not shown) [12].

injection system consists of an ion source, one or two linear accelerators in series (*RFQ* (RF quadrupole) and *DTL* (drift tube linac)) and a beam transport system. At the Loma Linda facility, a relatively small footprint has been achieved by mounting the injector on top of the synchrotron. A synchrotron (+ injection) consists of many small components that can be built in series, which gives relatively easy access to the machine parts.

3.4.1 Operation

The acceleration process in a synchrotron is in cycles (*spills*), each consisting of filling of the ring with a bunch of 2×10^{10} protons of 2 MeV (Loma Linda) or 10^{11} protons of 7 MeV (Hitachi, Mitsubishi), acceleration until the desired energy between 70 and 250 MeV, slow extraction of the protons into the beam line, finally followed by ramping down to the initial situation, eventually with deceleration and dumping unused protons at low energy, see Figure 3.17. In general, this sequence takes rather long to be useful for energy modulation, so the energy to be extracted is chosen for each gantry angle and equal to the maximum energy needed at that angle. If the spill repetition rate is considered too slow, energy modulation is done in the nozzle by means of a modulator wheel or ridge filters.

A synchrotron itself consists mainly of a lattice with bending magnets and focusing elements. Quadrupole magnets are used to focus the beam and sextupole magnets are used to increase the acceptance of beam energy spread. In some synchrotrons, the bending magnets are shaped such that the focusing properties are added to the bending field. By applying strong focusing schemes (field index $n \gg 1$), small beam diameters can be achieved. The periodicity of the lattice imposes a periodic shape of the beam envelope. However, in the design, care must be taken that the period of the betatron oscillation of individual protons does not coincide with a characteristic dimension of the machine, such as the circumference, since that could give a resonance condition. The huge number ($\sim 10^9$) of revolutions a proton makes in the ring yields an extreme sensitivity to small errors in

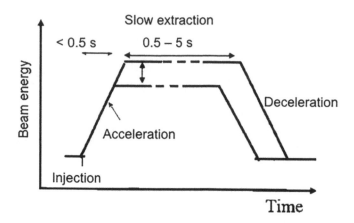

FIGURE 3.17 A typical spill from a synchrotron: The machine is filled with protons of 2 or 7 MeV, and the protons are accelerated until the desired energy and are slowly extracted. The unused remaining protons are decelerated and dumped.

magnet alignment, magnetic fields, or power supply ripples, which easily induce a periodic distortion. In the design phase, the focusing lattice (quadrupole strengths) is designed such that the working points (the *tunes*) are far away from a betatron resonance condition, Eq. (3.9); otherwise, this will unavoidably lead to beam loss.

3.4.2 Ion Source and Injector

For the proton ion sources, there is a wide choice of commercially available types, typically based on ionization by microwaves in a special configuration of coils or permanent magnet to confine the electrons [32,41]. The source is usually set at a positive potential, to preaccelerate the protons towards an RFQ, acting as first linear accelerator. An RFQ consists of four rods parallel to the beam direction. Each rod is shaped as a wave-like structure along its length, as shown in Figure 3.18. The rods are mounted in a resonant cavity such that the pairs of opposing rods have 180° phase shift of the RF voltage with respect to the orthogonal pair. The component of the electric field along the beam direction provides the acceleration and the radial components provide focus. Acceleration up to 2–3 MeV is achieved, and subsequently, the protons can be accelerated to 7 MeV in a DTL (see Figure 3.18).

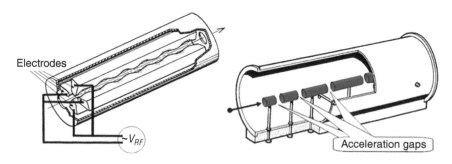

FIGURE 3.18 Typical RFQ and DTL linac configurations, for a preaccelerator to inject the beam into the synchrotron.

Also, the operation of a DTL is based on electromagnetic oscillations in tuned structures. The structures support a traveling wave of alternating voltages on cylindrical electrodes, between which the protons are accelerated. The electrode lengths increase along the tube, in accordance with the velocity of the accelerated particles.

Injection in the ring must be done at the correct phase with respect to the RF of the ring. One can inject all particles at once (*single turn injection*) or gradually add particles to the circulating beam at the moment the bunch passes the injection point. Since the injection system should not touch the already circulating beam, the new protons are added next to the protons that are already in the ring. The emittance of the circulating beam therefore increases until the acceptance of the ring is filled.

To reduce treatment time, it is important to fill the ring with as many protons as possible. This gives possibilities to lengthen the extraction phase and to reduce dead time between spills [48]. The maximum intensity in the machine is limited by space charge or Coulomb repulsion forces. The higher the injection energy, the lower the effect of these defocusing space-charge forces and thus the higher the amount of protons that can be stored in the ring. For single turn injection, intensity may also be limited by the maximum beam current from the injector to fill the ring in one turn. The maximum intensity can be increased by allowing more circulating bunches behind each other in the ring and by modifying the time structure of the RF voltage across the acceleration gap, to spread the bunch in the longitudinal direction [12].

3.4.3 Acceleration and RF

The acceleration phase usually lasts approximately 0.5 s and thus takes place over approximately 10^6 turns. The energy of the circulating particle bunch is increased in an RF cavity located in the ring. The increasing proton momentum p and thus also the relativistic γ factor need a synchronous increase of the magnet strengths in the ring, since the protons must remain in an orbit with constant average radius. Following Eq. (3.1), this yields

$$\frac{p}{Bq} = r = \text{const.} \tag{3.10}$$

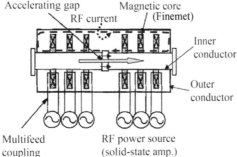

FIGURE 3.19 The RF acceleration device used in a synchrotron [42] consists of an induction cavity filled with magnetic cores. The cores drive an RF current (dashed line) that causes an RF voltage across the acceleration gap in the center of the cavity.

The increase of the magnetic field drives the frequency of the RF voltage: the frequency must remain synchronous to the increasing revolution frequency (~1→8 MHz) and increases (nonlinearly) in time:

$$f(t) = \frac{1}{2\pi r} \frac{p(t)}{m_0 \gamma(t)} \tag{3.11}$$

Therefore, the RF cavities are nonresonant (quality factor $Q < 1$) wide band structures and of moderate power [12]. A widely used type of cavity is the *induction cell*, for example the one developed by Hitachi [42], as shown in Figure 3.19.

An induction cavity consists of ferrite rings (*magnetic cores*) that surround the beam pipe. Around each core, a coil has been wound to induce a magnetic field in the ring. The electric current through the coil is driven with the RF frequency and induces an RF magnetic field in the ferrite ring. On its turn, this varying field induces an electric current in the beam pipe, which acts as inner conductor. An outer conductor surrounding the cores is closing the loop for this driven current. However, in the center of the device, the inner conductor is interrupted. Across this gap, an RF voltage of a few hundred Volts is built up by the RF current, so that the protons that cross the gap at the correct RF phase will be accelerated. This moderate RF voltage is sufficient, since the protons pass this gap many times during the acceleration phase. The applied frequency and voltage need to be controlled as a function of the magnetic field in the ring magnets as given in Eq. (3.11). The system is much simpler than the high-power, narrow bandwidth systems used for cyclotrons. The RF power can be generated with reliable solid-state amplifiers.

3.4.4 Extraction

Instead of fast extraction in a single turn, a slow extraction scheme is necessary for accurate dose application at which the beam is spread over the tumor with scanning techniques or with additional range modulation. The time during which the protons are extracted varies between 0.5 and 5 s, depending on the amount needed at the extracted energy. The extracted beam intensity can be regulated and is constant on average, but at several kHz, a ripple or noise of up to 50% is present. This ripple depends on the extraction method and is caused by the extreme sensitivity of the beam orbit to small misalignments and to ripples in power supplies.

Several methods to extract the beam are currently applied (see Figure 3.20). *Resonant extraction* based schemes, in which the machine tune is slowly shifted towards a resonance (e.g., $Q_r = 1/2$ or $1/3$). Effectively, this narrows the stable region around the phase space of the beam, since resonance bands are approached by the set tune. Particles with large oscillation amplitude will then slide into the unstable region of the phase space, so that their oscillation amplitude will grow until they pass the extraction septum at the external side. The shift of the beam tune is performed by changing the fields of specific quadrupoles and/or sextupoles in the ring [38]. However, while the phase space of the circulating beam gets more and more empty, the quadrupoles need to shift the tune

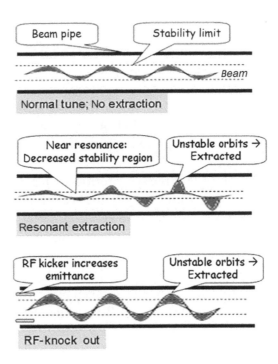

FIGURE 3.20 The betatron oscillation amplitude of the unperturbed beam tune (top) remains below the stability limit. In a resonant extraction method (middle), the tune is continuously shifted towards a resonance, which brings the stability limits tighter. The particles exceeding their amplitude outside the stability region will be extracted. In the resonant extraction scheme (bottom), the beam is kicked randomly and the amplitude is increasing. Particles leaving the original stability region will be extracted.

closer and closer to the resonance to catch the protons with the smallest amplitudes as well. A disadvantageous consequence of this method is that the position and/or size and/ or energy of the extracted beam are varying at the entrance of the beam line during the extraction period.

Alternatively, one can give quasi-random transversal kicks to the beam with a dedicated RF kicker. The *RF knock-out method* increases the circulating beam emittance, and particles that are kicked out of the stability region are guided to the extraction septum. With this method, the extracted beam position, size, and momentum remain constant during the spill [12]. This method also easily allows a fast on/off switching of the extracted beam intensity, which is conveniently used for *gating* on the respiration motion of the patient [43,44]. Its fast response can also be used to stabilize the intensity of the extracted beam, as has been demonstrated at the synchrotron in Heidelberg [81]. In both methods, the extracted intensity is controlled by a feedback of a beam intensity monitor on the acting extraction elements.

The horizontal emittance and momentum spread of a beam from a synchrotron is typically up to a factor 10 smaller than the emittance of a cyclotron beam. However, the emittance may show large asymmetries, which must be taken into account for preserving an angular independence of the beam characteristics when using rotating gantry systems.

3.4.5 Recent Developments in Therapy Synchrotrons

Scale reduction is of interest for synchrotron systems. Balakin et al. [45] have built a small ring of 5 m in diameter for proton acceleration up to 330 MeV. The protons are injected by a 1-MeV linear accelerator. Similar studies of a "tabletop" proton/carbon ion synchrotron are reported by Endo et al. [46]. The relatively low cost of such accelerators is expected to be the major advantage of these small machines. In Hokkaido (Japan), Hitachi has introduced a smaller version (5.1 m diameter) of its proton, a 7.8 m diameter synchrotron. It started its operation in 2013 [82].

Efforts towards a more rapid cycle of the synchrotron are reported in the last few years. In a proposal for a rapid cycling synchrotron [47], the beam is kicked out in a single turn extraction after single turn beam injection and acceleration to the desired energy. This sequence is repeated at ~30 Hz. Single turn extraction implies, however, that the amount of protons per pulse can be adjusted carefully. A slow extraction combined with fast cycling has been demonstrated in the Hitachi synchrotron [48] (see Figure 3.21a).

In recent years, an interesting development has taken place in the Carbon synchrotron at National Institute of Radiological Sciences (NIRS), Chiba, Japan. A method to reduce the energy during the extraction process [78] has been developed, so that energy modulation can be done within a spill. Of course, use of this method implies sufficiently high initial beam intensities in the ring as well as a good control of the extracted beam intensity. The energy steps (Figure 3.21b) are as fast as cyclotron-based systems, such that a considerable amount of time is gained. This technique is now routinely used at NIRS.

Summarizing, one can expect that the developments on beam intensity and energy change during extraction will bring the technical performance of the synchrotron at a

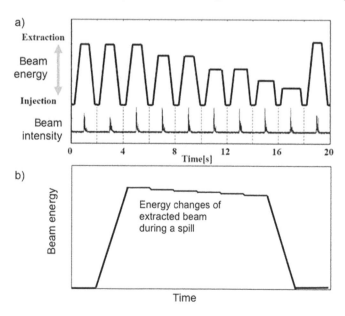

FIGURE 3.21 (a) Spill structure of a rapid cycling synchrotron [48]. After single-turn beam injection and acceleration to the desired energy, the beam is used after a slow extraction. (b) Energy variation during a spill [78]. The beam energy is decreased in small steps during the slow extraction process.

comparative level as the cyclotron. The choice between these two will then mostly depend on nontechnical aspects, such as space, service, and price.

3.5 GANTRIES

To direct the particle beam from different directions to the tumor in a patient lying on a treatment table, a gantry is used. This is a mechanical system that rotates the magnets of the last part of the beam transport system around the patient. Gantries for particle therapy are of considerable size. To bend and focus the proton beam, large (and thus heavy) magnets are needed. This is due to the limited bending power of the magnets. Bending magnets bend the proton tracks with a curvature radius of approximately 1.5 m. Typically, a distance of at least 2 m is needed to distribute the beam over a sufficiently large field size. Also between the last magnet and the patient, space is needed for additional equipment, such as scanning magnets, dose monitors, range shifter, and collimation systems. Therefore, a typical diameter of these 100–200 ton gantries is 10–12 m, both for scattered beams and for scanned beams.

The Harvard group [84] has designed the first gantries for protons. Three of these gantries have been installed at the facility in Loma Linda [8]. The beam transport system layout on this type of gantry is usually referred to as a corkscrew, since it is a combination of two perpendicular bending planes, as shown in Figure 3.22. It consists of two achromatic

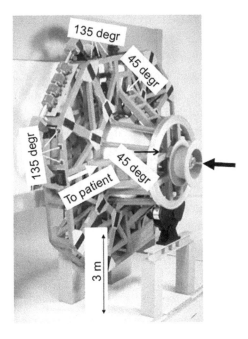

FIGURE 3.22   A model of the gantry at the Loma Linda proton therapy facility. The arrow indicates the proton beam entering the gantry from the right. At the gantry position shown, the beam is bent upwards by 45° after entering the gantry. Then, it is bent another 45° into the vertical direction (magnets are indicated). From the vertical direction, the beam is bent to the left and down by a 135° magnet, and a second 135° magnet aims the beam in a horizontal direction towards the patient position.

bending systems. If the gantry rotation angle is such that the beam is bent upwards first, the beam points vertically upwards when leaving the first achromatic bending section. In the second achromat, the beam is bent leftwards from the vertical direction by a 135° magnet and into a horizontal direction towards the patient position by the second 135° magnet. This layout has the advantage of being relatively short, as seen in the original beam direction.

Achromaticity is an essential characteristic of the beam optics of a gantry. This implies that the beam position and beam size at the isocenter are, within a certain range, not sensitive to the beam energy. This provides some robustness against small energy errors. The double achromatic bending system of the corkscrew (first: 2 × 45°, second: 2 × 135°) is an excellent example of such a system. Gantries developed later bend within one plan and use only two magnets; for example 45° + 135°. Designs have been made, both for scattering and for scanning. In some gantries, the 135° magnet has been split up into several shorter bending magnets with quadrupoles in between for focusing and to achieve achromaticity.

With the increasing demand for pencil beam scanning, dedicated nozzle designs have been made to allow (also or only) for scanning. In that case, the scanning magnets are usually located after the last bending magnet and act as the virtual source of the pencil beams. After this point, an Source to Axis Distance (SAD) of 1.5–2 m is needed to obtain a sufficiently large field size. However, this is in a complicated conflict with the need for small pencil beam diameters at isocenter. Dedicated beam focusing and extension of the beam vacuum until shortly before the patient are the usual techniques to deal with this.

At PSI, a compact proton gantry ("Gantry1") [85] has been optimized for pencil beam scanning. It was the first gantry in the world applying the scanning technique. In this gantry, a scanning magnet has been mounted before the 90° magnet that bends the beam towards the isocenter. The beam optics of this last bending magnet has been designed such that excitation of the scanning magnet causes a parallel shift of the pencil beam at the isocenter. In this way, the virtual source of the pencil beams is located at infinity and an orthogonal pencil beam displacement is obtained. The other two orthogonal displacements are performed by inserting range-shifter plates in the nozzle to shift the Bragg peak and by shifting the table in the direction orthogonal to the magnetic scanning direction. The thus-obtained orthogonal grid and parallel beam configuration have the advantage that it is relatively easy to obtain large field sizes by shifting the patient table ("field patching"). However, the total treatment time is rather long due to the slow motion of the patient table. Fast magnetic parallel scanning in two orthogonal directions covering a field of 12 × 20 cm² has been developed in "Gantry2," see Figure 3.23. Since both scanning magnets are located before the last 90° magnet, this magnet needs a relatively large gap. To benefit from the fast (<0.2 s energy steps) upstream degrader, magnets of Gantry 2 are laminated.

The optics of the beam transport system to the gantry is subject to several constraints to match the beam to the gantry at the gantry entrance. At the rotation coupling point, it is of utmost importance that the beam is well aligned. The beam must be symmetric in size, and in some gantries, also symmetric in divergence and it must be dispersion free at the coupling point. These constraints are preventing a gantry-angle dependence of the beam shape, position, or intensity at the isocenter. Ideally, at the isocenter, the lateral shape of

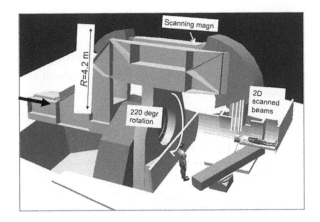

FIGURE 3.23   Design of "Gantry2" at PSI. This gantry is designed for parallel beam scanning in two dimensions and fast energy scanning. It rotates 220° around the patient, to have good access to the patient at any gantry angle.

the pencil beam is a circular Gaussian, but in any case, it should not change with gantry angle. This is especially important for gantries that use the pencil beam scanning technique. At the PSI gantries, the beam optics has been designed such that the entrance of the gantry is imaged to the isocenter, with a magnification factor close to unity. A collimator has been mounted at the entrance of the gantry (=coupling point), which thus defines the size of the pencil beam size at the isocenter, making the pencil beam size at isocenter to be rather independent of the beam divergence and incoming angle at the gantry entrance. So, because of this collimator imaging, a misalignment of the fixed beam or a focusing error at the coupling point only results in a lower beam intensity at the isocenter.

In case a synchrotron is used to accelerate the beam, one has to deal with the rather asymmetric emittance of the extracted beam, as mentioned in the section on synchrotrons. This difference of up to a factor 10 between the horizontal and vertical emittance must be taken into account for preserving an angular independence of the beam characteristics when using rotating gantry systems. This can be done by a system of quadrupoles that rotates with the gantry, or by letting the beam go through a scatter foil to make a symmetric emittance.

Currently, there is a strong interest to reduce the gantry size or to include energy degrading to reduce the facility footprint. In fact, to obtain a dramatic decrease of gantry radius, one has to configure the gantry for pencil beam scanning with "upstream" scanning, with scanning magnets before the last bending section before the isocenter. When achieving almost parallel pencil beam scanning at isocenter in such a design, one can work with a relatively short distance between the nozzle exit. With "downstream" scanning, an additional distance of 1.5–2 m is needed at the isocenter.

When replacing the bending magnets by superconducting magnets with a much stronger magnetic field, one can achieve a minimum proton bending radius of approximately 0.3–0.5 m. Due to the remaining necessary space, the radial size reduction is limited, but the proton gantry weight can be reduced to 20–40 tons [86]. At the moment, such magnets are being developed in several groups as well as at several companies, and interesting designs are in progress. In 2016, the first gantry with SC magnets has started its clinical operation with carbon ions at NIRS [87].

A reduction of the footprint of a gantry is achieved when the gantry is rotated over a bit more than 180° instead of 360°. This geometry allows easy patient access as well as an easy combination with an in-room imaging device (e.g., Computed Tomography (CT) and/or Positron Emission Tomography (PET)). A horizontal rotation of the patient table over 180° is necessary for a full treatment. This combination has been applied for the first time in PSI's Gantry-2 [16], and this idea has been taken over by several companies.

3.6 NOVEL ACCELERATOR TECHNOLOGIES

Most of the acceleration principles being investigated by different groups are aiming at scale reduction, with an affordable single treatment room facility as an ultimate goal. This would decrease the financial difference with conventional photon therapy. Although initial investment costs will of course be lower for a single-room proton therapy facility, it is not expected that the costs per treatment room will become lower than those of a multiroom facility. However, for certain regions, such facilities may be the only possibility to have proton therapy at all. For extended regions with more homogeneously spread populations like in Europe, the large centers with multiple treatment rooms are expected to operate more economical.

3.6.1 FFAG Accelerators

Since several years, it is being investigated whether an *FFAG* accelerator would be a suitable accelerator for proton therapy [49–55]. In this concept, the AVF cyclotron described above is split up in separate sector magnets, with strong focusing edges or with alternating signs of magnetic fields (Figure 3.24). The RF is varying to match the revolution frequency of the protons. The magnet design and large gradients of opposing signs are complicating the design process. Also, the cavity and RF generator are complicated due to the opposing requirements of very strong electric fields at a varying frequency.

For proton therapy, a possible advantage of an FFAG may be its capacity to have much faster repetition rates than synchrocyclotrons. Also, the optics and magnetic field setting may change from pulse to pulse, allowing fast energy changes between each pulse [51]. On the other hand, the FFAGs do not have a small radius (7–8 m) or a low weight (150–200 tons). In general, FFAGs are very complex to design, build, and operate and need very complex RF amplifiers and magnet power supplies of high power. An FFAG also needs a ~10

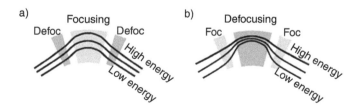

FIGURE 3.24 Schematic configuration and orbit structure of (a) a scaling FFAG and (b) a nonscaling FFAG.

MeV injection system, which can be a cyclotron [50]. For proton therapy, a promising spin-off from FFAG technology is the use of very strong gradients, alternating along a beam line, which has been proposed for a new type of gantry design [56].

3.6.2 Linear Accelerators

Linear accelerators (*linacs*) for electron acceleration are the most widely used accelerators in radiation therapy. However, compared to electrons, the higher mass of protons makes it much more difficult to use linear accelerators. This is because electrons quickly reach a constant velocity (almost the speed of light) when accelerated, allowing repetitive accelerator structures of equal dimensions. But due to the higher mass of protons, the energy increases with an increase of speed. This needs varying dimensions along the accelerator structure. Therefore, especially, the low-energy part of a proton linac needs to consist of different stages.

The time structure of the beam is of importance for the direct use of the beam for therapy. In linacs, a very high frequency (GHz) is needed to enable very strong electric acceleration fields. However, to limit the power dissipation, the RF power must be pulsed, and currently, one is working on systems that have sufficiently high pulse rates (100–200 Hz) for spot scanning. The RF power is generated by amplifiers, each driving its section of the linac. This offers the natural possibility for rapid energy variation of the accelerated beam, by adjusting the power of some amplifiers. A higher repetition rate and sufficient accuracy of the dose per pulse could make a linac suitable for rapid spot scanning.

3.6.3 Linac-Based Systems for Proton Therapy

To overcome the linac problems at low energies, the *Cyclinac* concept has been developed [57]. A cyclotron of 60 or 30 MeV is used as an injector for a linear accelerator with side-coupled cavities (*CCL*) in which very strong electric fields are created at 3 GHz, so that the length of the linac is minimized. In a recent cavity design, a gradient of up to 27 MV/m was reached [58]. The linac will consist of several RF-klystron driven tanks with groups of accurately machined cavities. A possible layout of a clinical facility is shown in Figure 3.25. Recently, a new cavity for the energy range of 15–66 MeV has been presented [59].

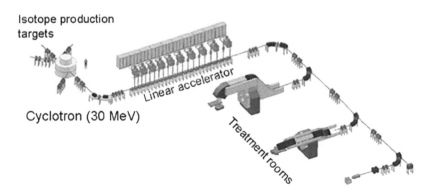

FIGURE 3.25 Layout of the cyclotron-driven linac concept (59). A 30 MeV cyclotron is used for isotope production or injects protons into a linear accelerator of the LIBO (Linac Booster) type.

Higher gradients of the accelerating electric field and thus shorter structures are a natural line of development. But there are limits in the peak power that can be injected into a linac tank and in the maximum electric field at the cavity wall. Current data on breakdown phenomena shows a limit just above 30 MV/m at 3 GHz in a CCL [60]. Higher frequencies (e.g., 5.7 GHz) will allow stronger acceleration fields, but at the cost of reduced transversal acceptance. In the future, superconducting cavities may be of help to increase the field strengths.

3.6.4 Dielectric Wall Accelerator

Recently, one has demonstrated that one can increase the acceleration fields in induction cavities by replacing the usual ferrite core by a ring of specially developed [62] *high-gradient insulator (HGI)* material, surrounded by a dielectric, which is sandwiched between two conducting sheets. By connecting a high voltage and ground potential sequentially to the stack of different sheets in this *Dielectric Wall Accelerator (DWA)*, a traveling wave of a strong electric field is generated to accelerate the protons [61,63].

The alternating fast (de)connection of the conducting sheets to the high voltage is performed by a special switching circuit (*Blumleins*). Energy variation per pulse can be achieved by selecting the appropriate amount of switches. To obtain a very strong accelerating field, one shortens the time that the field is present to only a few ns. Experiments on small HGI samples have demonstrated [63] that an accelerating field of more than 100 MV/m can be possible for 3 ns pulses. As shown in Figure 3.26, this could lead to a linear accelerator of only 2 m for 200 MeV protons. Although the system is pulsed with a repetition frequency of only several tens of Hz, this accelerator type has been of interest for commercial companies [64].

However, in the last few years, it has become clear that this challenging project has suffered from reaching the limits of current technology: high-gradient vacuum insulators, breakdown strength dielectrics for pulse-forming circuits, a high number of high-voltage switches that have to operate within a ns, the safe and accurate control of very high amount of protons per pulse as part of the developments of the proton source [66] and cooling requirements [65]. Therefore, it is expected that the developments on DWAs for proton therapy will not be continued.

3.6.5 Laser-Driven Accelerators

When the light beam from a very high-power laser hits a target, protons will be emitted from the target. For proton therapy, this can be of interest since the laser and light

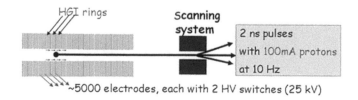

FIGURE 3.26 The DWA consists of a stack of special insulators between which a traveling electric field of 100 MV/m is traveling. Protons are accelerated and aimed at a voxel in the target [62–64].

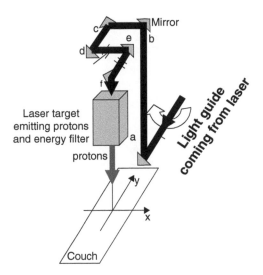

FIGURE 3.27   The concept of a treatment facility using protons that are generated by a laser [67].

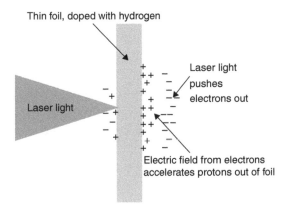

FIGURE 3.28   An intensive laser accelerates electrons from a proton-enriched area (polymer) at the far side of a titanium foil. This creates a strong electric field that accelerates protons out of the surface [75].

transmission components do not need heavy concrete shielding. Further, heavy and big magnets are not needed anymore. Also, scanning the light beam could be used for pencil beam scanning in principle. These ideas are illustrated in Figure 3.27 [67].

Currently, a lot of research is done in the field of particle acceleration by means of laser pulses [68]. At the moment, most experience has been obtained with the target normal sheet acceleration (*TNSA*) method [69]. As shown in Figure 3.28, a high-intensity laser irradiates the front side of a solid target. At the front surface, a plasma is created due to the energy absorption in the foil. The electrons in this plasma are heated and emerge from the rear surface. This induces strong electric fields, which pull ions and protons out of the evaporating target.

In this method, typical proton energies of about 20 MeV are observed [71] with a laser power intensity of 6×10^{19} W/cm^2 during a pulse length of 320 fs. The observed agreements [71]

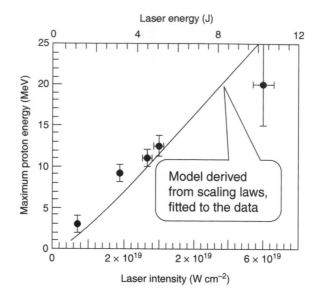

FIGURE 3.29   The measured and modeled proton intensities as a function of laser power [71].

between experimental data and scaling laws (Figure 2.29) have yielded an accurate description of the acceleration of proton beams for a large range of laser and target parameters [76].

Extrapolation of this model to the delivery of 200 MeV protons indicates that the needed laser power is as high as 10^{22} W/cm². Apart from the quest to obtain higher proton energies, the obtained energy spectrum (Figure 3.30) is of concern, since it shows a broad continuum [72] that is not usable for proton therapy. Several filter and energy compression techniques are proposed [67], but these will yield an intensive neutron production.

Even though the field is developing continuously [70], it is expected that it will still take many years to develop a laser-driven medical facility for proton/ion therapy [73]. First of all, one currently relies on a huge extrapolation to high power from the values on which the current models are based. Also, there are many technical difficulties in the transport of

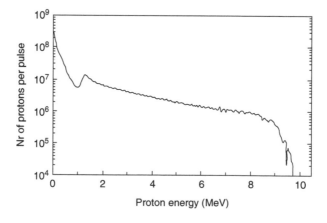

FIGURE 3.30   Measured proton energy spectrum, obtained with a peak laser power of 6×10^{19} W/cm², focused at a thin aluminum foil [72].

such a high power light beam and to develop appropriate target technologies. To date, these problems make it uncertain that a reproducible pulse of protons having the desired energy and the required intensity [74] can be made. Second, the obtained proton beam intensities are still far too low, e.g., 10^9 protons per pulse [71] or 10^8 protons of <10 MeV per pulse [75]. Also, the reached repetition rate of the pulses (1–100 Hz has been reached in some systems) is far too low.

A method alternative to the *TNSA* method is the radiation pressure acceleration (*RPA*) or the laser piston regime [79]. Here, the light pressure of a laser pulse accelerates the target foil as a plasma slab. Simulations predict higher proton energies and less energy spread. However, the *RPA* method faces even more technological challenges.

In a recent review [80], it was concluded that although significant progress is reported in the theoretical developments, today's best lasers are far from reaching the required performance levels.

3.7 CONCLUDING REMARKS

The isochronous cyclotron and the synchrotron will remain the working horses in proton therapy for the coming decade, especially scanning technology and its developments to continuous scanning are of more interest. These machines work reliably but are still being improved, and a huge amount of experience exists and is shared. These systems have undergone certification processes from various authorities. Although one was dealing with existing and well-developed technology, this process has taken about two decades.

Development of new acceleration techniques, such as those discussed in this chapter, needs to be encouraged and is essential to make proton therapy more accessible for patients. However, a few words of caution should be addressed here. Most currently proposed new accelerators emphasize one specific advantage (e.g., fast energy change) with respect to the cyclotron and synchrotron. However, that is not sufficient for operating a patient treatment facility. *All parameters* defining beam characteristics at the patient are of importance and the safety requirements to prevent a wrong dose (location) in the patient should not be underestimated.

Therefore, before a claim is made that a new technology will soon outdate the currently used systems, it first needs to be proven that the new technology can *at least provide the same quality of treatment* as achieved today. Next, one should realize that the claimed advantage very often comes with compromises in other parameters.

The consequences of "cheap" protons providing nonoptimal treatments should not be underestimated: suboptimal results will have a catastrophic impact on proton therapy in general. This will endanger the operation of centers in which good, high-quality treatments are being given. Due to the urgent need to search for cheaper new techniques to provide proton therapy, dramatic expectations have been created for certain developments. Apart from the quality, the reliability of new technology should be evaluated seriously. Does the new technique allow patient treatments at least 14 h/ day, 5–6 days a week, without major interruptions, and with professional support? It is also very important that an honest and fair estimate is given of the time it will take to have the new technology available, safely and routinely usable in a clinical environment. It is simply misleading to announce the release of the first new system to customers 2–3 years after a successful first proof of principle. Although 15–20 years is more realistic, such announcements have seriously delayed critical decisions in several planned proton therapy facilities.

ACKNOWLEDGMENTS

I would like to thank my colleagues from the Division of Large Scale Research Facilities and the Center of Proton Therapy at the PSI for sharing their valuable and long years of experience with me. Knowing how an accelerator works in principle is one thing, but the organization and detailed knowledge needed to develop, use, and maintain a reliable and safe treatment facility is another thing. Also, I would like to acknowledge the numerous discussions and interesting details I have learned from my colleagues at proton therapy centers, various companies and the international community of accelerator physicists during conferences, site visits, and personal discussions.

REFERENCES

1. M. C. Pirruccello and C. A. Tobias, eds. *Biological and Medical Research with Accelerated Heavy Ions at the Bevalac, 1977–1980*. Lawrence Berkeley Laboratory Report LBL-11220, p. 423, 1980.
2. W. T. Chu, B. A. Ludewigt and T. R. Renner. Instrumentation for treatment of cancer using proton and light-ion beams. *Rev. Sci. Instr.* 64 (1993), 2055–2122.
3. R. Wilson. *A Brief History of the Harvard University Cyclotrons*. Cambridge, MA: Harvard University Press, 2004.
4. C. Von Essen et al. The Piotron: II. Methods and initial results of dynamic pion therapy in phase II studies. *Int. J. Rad. Onc. Biol. Phys.* 11 (1985), 217–226.
5. E. Egger, A. Schalenbourg, L. Zografos, others. Maximizing local tumor control and survival after proton beam radiotherapy of uveal melanoma. *Int. J. Radiat. Oncol. Biol. Phys.* 51 (2001), 138–147.
6. E. Pedroni et al., The 200 MeV proton therapy project at the Paul Scherrer Institute: Conceptual design and practical realisation. *Med. Phys.* 22(1) (1995), 37-53.
7. B. Larsson. Application of a 185 MeV proton beam to experimental cancer therapy and neurosurgery: A biophysical study. PhD Thesis, University of Uppsala, 1962.
8. J. M. Slater, J. O. Archambeau, D. W. Miller et al. The proton treatment center at Loma Linda university medical center: Rationale for and description of its development. *Int J Radiat Oncol Biol Phys.* 22 (1992), 383–389S.
9. S. Humphries. *Principles of Charged Particle Acceleration*. John Wiley and Sons, 1986. (ISBN 0-471-87878-2, QC787.P3H86). http://www.fieldp.com/cpa.html.
10. Proceedings of the Cern Accelerator Schools CAS, Zeegse (2005). http://cas.web.cern.ch/cas/Holland/Zeegse-lectures.htm.
11. J.M. Schippers. Beam delivery systems for particle radiation therapy: Current status and recent developments. *Rev. Acc. Sci. Technol.*, II (2009), 179–200.
12. K. Hiramoto et al. The synchrotron and its related technology for ion beam therapy. *Nucl. Instr. Meth.* B 261 (2007), 786–790.
13. S. Yamada et al. HIMAC and medical accelerator projects in Japan, Asian Particle Accelerator Conference 1998, KEK Proceedings 98–10.
14. Y. Tsunashima et al. Efficiency of respiratory-gated delivery of synchrotron-based pulsed proton irradiation. *Phys. Med. Biol.* 53 (2008), 1947–1959.
15. M. H. Phillips et al. Effects of respiratory motion on dose uniformity with a charged particle scanning method. *Phys. Med. Biol.* 37 (1992), 223–233.

16. E. Pedroni et al. The PSI gantry 2: A second generation proton scanning gantry. *Z. Med. Phys.* 14 (2004), 25–34.

17. N. Saito et al. Speed and accuracy of a beam tracking system for treatment of moving targets with scanned ion beams. *Phys. Med. Biol.* 54 (2009), 4849–4862.

18. IBA. Website: http://www.iba.be/healthcare/radiotherapy/particle-therapy/particle-accelerators.php.

19. Sumitomo Heavy Industries ltd. Website: http://www.shi.co.jp/quantum/eng/product/proton/proton.html

20. M. Schillo et al. Compact superconducting 250 MeV proton cyclotron for the PSI proton therapy project. *Proceedings of the 16th International Conference on Cyclotrons and Applications* (2001), 37–39.

21. J.M. Schippers et al. First year of operation of PSI's new SC cyclotron and beam lines for proton therapy. *Proceedings of the 18th International Conference on Cyclotrons and Applications*, Catania, Italy,1–5 October 2007, ed. D. Rifuggiato, L.A. Piazza, INFN-LNS Catania, Italy (2008), 15–17.

22. M. Getta et al. Modular High Power Solid State RF Amplifiers for Particle Accelerators. *Proceedings of PAC09*, Vancouver (2009), TU5PFP081

23. A. Geisler et al. Status report of the ACCEL 250 MeV medical cyclotron. *Proceedings of the 17th International Conference on Cyclotrons and Applications*, Tokyo (2004) 18A3

24. Y. Jongen et al. Progress report on the IBA-SHI small cyclotron for cancer therapy. *Nucl. Instr. Meth. B* 79 (1993), 885–889.

25. Website of Mevion: www.mevion.com

26. S. Graffman et al. Proton radiotherapy with the Uppsala cyclotron. Experience and plans. *Strahlentherapie* 161 (1985) 764–70.

27. G. A. Karamysheva and S. A. Kostromin. Simulation of beam extraction from C235 cyclotron for proton therapy. *Phys. of Part. and Nucl. Lett.* 6 (2009), 84–90.

28. C. Baumgarten et al. Isochronism of the ACCEL 250 MeV medical proton cyclotron. *Nucl. Instrum. Meth. Phys. Res. A* 570 (2007), 10–14.

29. J.M. Schippers et al. Beam-dynamics studies in a 250 MeV superconducting cyclotron with a particle tracking program. *Nukleonika* 48 (2003), S145–S147.

30. J.M. Schippers et al. Results of 3D dynamic studies in distorted fields of a 250 MeV superconducting cyclotron. *Proceedings of the 17th International Conference on Cyclotrons and Applications*, Tokyo, October 2004, p 435

31. B. Wolf (ed.) *Handbook of Ion Sources*. CRC press, 1995. ISBN 0-8493-2502-1

32. I.G. Brown (ed.) *The Physics and Technology of Ion Sources*. Wiley-VHC, Freiburg, 2004. ISBN 3-527-40410-4.

33. E. Forringer et al. A cold cathode ion source. *Proceedings of the 16th International Conference on Cyclotrons and Applications* (2001), 277–279, and PhD Thesis (NSCL, unpublished)

34. J.M. Schippers et al. Beam intensity stability of a 250 MeV SC Cyclotron equipped with an internal cold-cathode ion source. *Proceedings of the 18th International Conference on Cyclotrons and Applications*, Catania, Italy, 2007, ed. D. Rifuggiato, L.A.C. Piazza, INFN-LNS Catania, Italy, 2008, 300–302.

35. J.M. Schippers et al. Activation of a 250 MeV SC-cyclotron for proton therapy. *Proceedings of the Cyclotrons 2010, 19th International Conference on cyclotrons and their applications*, 6–10 September 2010, Lanzhou, China

36. Optivus Proton Therapy, Inc.: http://www.optivus.com/index.html

37. Website of Mitsubishi: http://global.mitsubishielectric.com/bu/particlebeam/index.html

38. G. Coutrakon. Synchrotrons for proton therapy, PTCOG 47 and educational session PTCOG 49

39. M.F. Moyers and D.A. Lesyna, Exposure from residual radiation after synchrotron shutdown. *Radiat Meas.* 44 (2009), 176–181.

40. M. J. van Goethem et al. Geant4 simulations of proton beam transport through a carbon or beryllium degrader and following beam line. *Phys. Med. Biol.* 54 (2009), 5831–5846.
41. S. Hara et al. Development of a permanent magnet microwave ion source for medical accelerators. *Proc. of the 2006 EPAC, Edinburgh* (2006), 1723.
42. K. Saito et al. FINEMET-core loaded untuned RF cavity. *Nucl. Instr. Meth.* A 402 (1998), 1–13.
43. S. Yamada et al. HIMAC and medical accelerator projects in Japan. Asian Particle Accelerator Conference 1998, KEK Proceedings 98–10.
44. Y. Tsunashima et al. Efficiency of respiratory-gated delivery of synchrotron-based pulsed proton irradiation. *Phys. Med. Biol.* 53 (2008), 1947–1959.
45. V.E. Balakin et al. TRAPP-Facility for proton therapy of cancer. *Proc. of the 1988 EPAC, Rome,* 2 (1988), 1505.
46. K. Endo et al. Compact proton and carbon ion synchrotrons for radiation therapy. *Proc. of the 2002 EPAC, Paris* (2002), 2733–2735.
47. S. Peggs et al. The rapid cycling medical synchrotron, RCMS. *Proc. of the 2002 EPAC, Paris* (2002), 2754–2756.
48. K. Hiramoto. Synchrotrons, educational session PTCOG 49.
49. K. R. Symon et al., Fixed field alternating gradient particle accelerators. *Phys. Rev.* 103 (1956), 1837–1859.
50. E. Keil et al., Field Alternating Gradient Accelerators (FFAG) for Fast Hadron Cancer Therapy, PAC 2005 – Proceedings Knoxville, Tennessee, pp. 1667–1669.
51. S. Antoine et al. Principle design of a proton therapy, rapid-cycling, variable energy spiral FFAG. *Nucl. Instr. Meth.* A 602 (2009), 293–305.
52. I. Blumenfeld, et al. Energy doubling of 42 GeV electrons in a metre-scale plasma wakefield accelerator. *Nature* 445 (2007), 741–744.
53. C.R. Prior (ed.), chapter 4, Theme section: FFAG accelerators, beam dynamics. *Newslett.* 43 (2007), 19.
54. M.K. Craddock, K.R. Symon. Cyclotrons and fixed-field alternating-gradient accelerators. *Rev. Acc. Sci Technol.* I (2008), 65–97.
55. D. Trbojevic. FFAGs as accelerators and beam delivery devices for ion cancer therapy. *Rev. Acc. Sci Technol.* II (2009), 229–251.
56. D. Trbojevic et al. Superconducting non-scaling FFAG gantry for carbon/proton cancer therapy. *Proceedings of the PAC07*, Albuquerque, New Mexico, THPMS092 3199–3201.
57. K. Crandall and M. Weiss, Preliminary design of a compact linac for TERA, TERA 94/34, ACC 20 (1994)
58. C. De Martinis et al. Beam tests on a proton linac booster for hadron therapy. *Proceedings of the 2002 EPAC, Paris* (2002), 2727–2729.
59. U. Amaldi et al. CLUSTER: A high-frequency H-mode coupled cavity linac for low and medium energies. *Nucl. Instrum. Meth. Phys. Res.* A 579 (2007), 924–936.
60. W. Wuensch. High gradient breakdown in normal-conducting RF cavities. *Proceedings of the EPAC02*, 2002, pp. 134–138.
61. A. Pavlovski et al. A coreless induction accelerator. *Sov. J. At. En.* 28 (1970), 549.
62. G.J. Caporaso et al. Compact accelerator concept for proton therapy. *Nucl Instr Meth B* 261 (2007), 777.
63. G.J. Caporaso et al. Status of the dielectric wall accelerator. *Proceedings of the PAC 2009* TH3GAI02.
64. Website of TomoTherapy http://www.tomotherapy.com/news/view/20070614_tomo_proton/
65. G.J. Caporaso et al. The dielectric wall accelerator. *Rev. Acc. Sci. Technol.* II (2009), 253–263.
66. Y.J. Chen et al. Compact proton injector and first accelerator system test for compact proton dielectric wall cancer therapy accelerator. *Proceedings of the PAC* 2009, TU6PFP094
67. C.M. Ma et al., Development of a laser-driven proton accelerator for cancer therapy. Laser Phys. 16 (2006), 639–646.

68. T. Tajima et al. Laser acceleration of ions for radiation therapy, *Rev. Acc. Sci Technol.* II, (2009), 201–228.
69. S.C. Wilks et al. Energetic proton generation in ultra-intense laser–solid interactions. *Phys. Plasmas* 8 (2001), 542–549.
70. K. Ledingham, Desktop accelerators going up? *Nature Phys.* 2 (2006), 11–12.
71. J. Fuchs et al. Laser-driven proton scaling laws and new paths towards energy increase. *Nature Phys.* 2 (2006), 48–54.
72. V. Malka et al. Practicability of protontherapy using compact laser systems. *Med. Phys.* 31 (2004), 1587–1592.
73. C.M.C. Ma and R.L. Maughan. Point/Counterpoint, within the next decade conventional cyclotrons for proton radiotherapy will become obsolete and replaced by far less expensive machines using compact laser systems for the acceleration of the protons. *Med. Phys.* 33 (2006), 571–573.
74. U. Linz and J. Alonso. What will it take for laser driven proton accelerators to be applied to tumor therapy? *Phys. Rev. ST Accel. Beams* 10 (2007), 094801.
75. H. Schwoerer et al. Laser-plasma acceleration of quasi-monoenergetic protons from micro-structured targets. *Nature* (London) 439, 445 (2006).
76. K. Zeil et al. The scaling of proton energies in ultrashort pulse laser plasma acceleration. *New. Phys.* 12 (2010), 045015.
77. T. Haberer et al. Magnetic scanning system for heavy ion therapy. *Nucl. Instr. Meth. A* 330 (1993), 296–305.
78. Y. Iwata et al. Multiple-energy operation with Quasi-DC extension of flattops at HIMAC, *Proc. IPAC10*, MOPEA008 (2010), 79–81.
79. A.P.L. Robinson et al. Radiation pressure acceleration of thin foils with circularly polarized laser pulses. *New J. Phys.* 10 (2008) 013021, 1–13.
80. U. Linz and J. Alonso. Laser-driven ion accelerators for tumor therapy revisited. *Rev. ST Accel. Beams* 19 (2016), 124802.
81. C. Schoemers et al. Intensity feedback system at Heidelberg ion-beam therapy centre. *Nucl. Instr. Meth. A* 795(2015), 92–99.
82. Hitachi Website: http://www.hitachi.com/New/cnews/month/2014/03/140317.html
83. W. Kleeven et al. The IBA superconducting synchrocyclotron project S2C2. Proc. of "Cyclotrons 2013", Vancouver, BC, Canada, MO4PB02, pp 115–119.
84. A.M. Koehler. Proceedings of the 5th PTCOG meeting: International workshop on biomedical accelerators. Berkeley, CA: Lawrence Berkeley Lab., 1987, 147–158.
85. E. Pedroni, et al. The 200-MeV proton therapy project at the Paul Scherrer Institute: Conceptual design and practical realization. *Med. Phys.* 22 (1995), 37–53.
86. A. Gerbershagen, et al. Novel beam optics concepts in particle therapy gantries utilizing the advantages of superconducting magnets. *Z. Med. Phys.* 26 (2016), 224–237.
87. Y. Iwata, et al. Design of superconducting rotating-gantry for heavy-ion therapy. *Proc. IPAC2012*, THPPR047 (2012).

CHAPTER 4

Characteristics of Clinical Proton Beams[*]

Hsiao-Ming Lu and Jacob Flanz

Massachusetts General Hospital and Harvard Medical School

CONTENTS

4.1 INTRODUCTION

The clinical characteristics of a proton beam are largely determined by the intended therapeutic use. The intrinsic physical properties of the protons that originate from the accelerators generating the beam are modified by the devices used to control the beam. Different beam delivery system designs can affect the beam properties substantially. Before one can design beam-shaping systems, such as those discussed in the next couple of chapters, it is useful to explore the basic beam properties from a clinical point of view and how the clinical requirement are related to the design features and the operational settings of a delivery system. While

---

[*] Colour figures available online at www.crcpress.com/9781138626508

a majority of proton therapy treatments in the past were delivered by scattered beam systems, the use of scanning beams has spread rapidly in the last few years, with many new centers now using only scanning; hence, our discussions will consider both such systems.

4.2 PROTON DOSE DELIVERY

When considering how a proton beam can be used for clinical purposes, one must factor in both the biological and physical effects. Despite the distinguished physical properties, protons are biologically used in therapy in more or less the same manner as photons are used in conventional radiotherapy. Protons and photons have different physical energy-loss behaviors and biological behaviors. Indeed, a proton's relative biological effectiveness (RBE) does depend on the linear energy transfer and can become substantially large at the falling edge the Bragg peak. However, as discussed in Chapter 22, the overall effect when combining multiple Bragg peaks can be approximated by a constant RBE value of 1.1 in most situations. As a result, one can essentially prescribe proton treatment with the same amount of Cobalt equivalent dose, i.e., Gy(RBE), for the same level of expected local control as in a conventional therapy treatment. This is often viewed as an advantage of protons over heavier particles with much more complicated radiological effects, given the rich and invaluable clinical experiences about local control and normal tissue toxicity obtained from conventional radiotherapy over many years and from many patients. For this reason, we will consider only the physical dose in the following discussions.

In the current practice of radiotherapy, a treatment is specified as a prescribed dose to a volume of tissue in the patient, i.e., the target volume. A number of volumes with different definitions are often used to guide the treatment planning and delivery processes. These include gross target volume, clinical target volume, and planning target volume. For a more detailed explanation of these volumes and their implications to treatment process, please see Chapters 15 and 16.

The objective of the treatment is that every tissue voxel in a given target volume should receive a prescribed amount of dose to achieve the appropriate level of cell response based on the biological models. It could be that the entire volume requires a homogeneous dose. However, for many cancer treatments, different regions of tissue may require different amount of radiation doses in consideration of cancer cell distribution probability and other medical issues. Traditionally, this is achieved by the patient receiving multiple courses of treatments sequentially, each with a different target volume prescribed to a different dose, although for each fraction, the goal is to deliver a homogeneous dose distribution to the target volume of the treatment course. In recent years, however, the technique of so-called "simultaneous boosting" is gaining recognition, where multiple target volumes are treated with different dose levels in the same fraction, to reduce treatment fractions with possibly improved outcomes. This is made possible entirely by the power of intensity modulation for either photon [intensity modulated radiation therapy treatment (IMRT)/volumetric modulated arc therapy (VMAT)] or particle beam scanning (PBS*) beams, although protons will inevitably be more suited given its superb dose-shaping capabilities.

---

* The abbreviation PBS is not used consistently in the field. It may beam particle beam scanning, proton beam scanning, or pencil beam scanning.

It is the challenge of the beam delivery system to deliver the beam to the target volume with an appropriate dose distribution. Various types of beam delivery constraints will affect the beam characteristics, based upon the method of beam delivery chosen.

4.3 BEAM CHARACTERIZATION

A number of parameters are needed to characterize the physical properties of the proton beam for clinical use. Different beam delivery systems produce different beam characteristics, and therefore, the specifications and how these beams are used clinically may be defined in relation to the manner in which the beam is produced and characterized by measurements. Treatment planning requires a model of the beam, and this model may also have a parameterization based on the beam delivery modality. It is critical from the clinical point of view that the treatment planning parameters and measured beam properties be unified to avoid any potential confusion or misrepresentation. That is, the same set of parameters is used to describe the beam model in the treatment planning system, specify the desired beam production in the beam control system, and specify the quality assurance measurements. For this reason, we will discuss only those specifications currently adopted clinically. Readers interested in the historical evolution of the beam specifications for various purposes should see the references provided herein.

During proton treatment, the accelerator and the beam transport system transport to the patient a monoenergetic beam with a small lateral cross section. This beam must be modified for patient treatment.

The depth-dose distribution of such a pencil-like beam is the Bragg peak introduced in Chapter 2 with a sharp peak in depth. To obtain the desired depth-dose distribution throughout the target volume for the rationale discussed earlier, one must build a superposition of many Bragg peaks with proper intensities and locations. As an example, Figure 4.1 shows a specific case of a near-uniform depth-dose distribution (solid line) in a

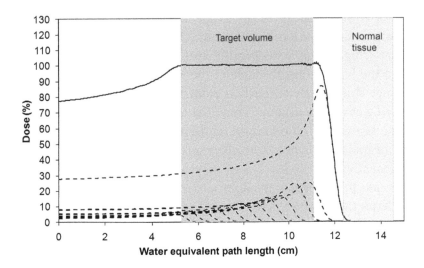

FIGURE 4.1   A schematic view of SOBP construction, showing the SOBP depth-dose distribution (solid line) and the component Bragg peaks (dashed lines).

homogeneous media achieved in a single field. It contains a number of Bragg peaks (dotted lines) in the same beam direction, but with different incident proton energies and thus different Bragg peak locations spread out in depth, therefore the term "spread-out Bragg peaks (SOBP)." Two main categories of methods are used to produce a superposition of Bragg peaks for clinical use. They either use materials in the beam path to modify the beam energy or the energy coming from the accelerator without additional material, as described in detail in Chapters 5 and 6.

This type of dose distribution, derived in this way, in Figure 4.1 exhibits several characteristic features. First, it delivers a near-uniform dose distribution in depth along the target volume (darker gray area). Second, it preserves the sharp distal fall of the Bragg peak, and therefore, the ability of the proton beam to spare normal tissue behind the target volume (light gray area). On the proximal side of the target volume, the dose changes gradually, i.e., a soft knee. Third, the total entrance dose has increased from about 30% due to the deepest Bragg peak along to nearly 80% due to the additional shallower peaks. These features are the main clinically relevant properties of the proton beam in the longitudinal direction.

While all beam delivery systems spread out the Bragg peak as described earlier, the lateral distribution of the beam must also be modified to match the shape of the desired target volume. The method to do this depends on the beam-shaping modality used. There are two main types of beam delivery systems, named for the method used to spread out the lateral beam distribution: scattering and scanning. In the following sections, we briefly describe the principles of these methods to identify the characteristics of the beam and potential clinical use of the beams they produce.

4.3.1 Scattering Systems

In systems that use beam scattering to spread the lateral extent of the beam, the small beam coming to the nozzle is scattered to a large area, and the scatterers are specially designed such that the beam has uniform penetration and uniform intensity across the scattered area specified for clinical use. At the same time, the energy of the beam is modulated, as noted above, to spread out the location of the Bragg peaks over the target volume in depth. Typically, a scattering system is designed in such a way as to very quickly, or instantaneously, spread the beam range throughout the target. In this case, in current systems, mechanical range shifting devices such as ridge filters and range modulator wheels are used. A scattering system is usually configured to produce a homogeneous dose distribution with the same penetration across the beam, as shown in Figure 4.1. For patient treatment, the beam is collimated by an aperture to match the target volume and a range compensator (usually, a Lucite block with varying thicknesses) to "pull back" the most distant Bragg peaks to conform to the distal surface of the target volume (see Chapter 5).

Figure 4.2 shows the parameters used to describe the SOBP dose distribution. The distribution is normalized to 100% at the dose plateau. The parameters are defined in terms of positions in depth at the given dose levels, i.e., d20, d80, and d90 at the distal end, with p90 and p98 on the proximal side.

The distal margin of the SOBP is given by the distance between d20 and d80, corresponding to the 20% and 80% dose levels. The quantity has been termed "distal dose

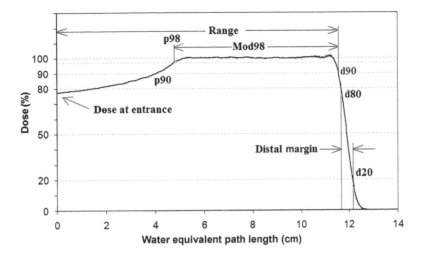

FIGURE 4.2   Specification of SOBP depth-dose distribution produced by scattering.

falloff" (DDF) [1]. The dose at the surface entrance is also a useful parameter for characterizing the dosimetric property.

Therefore, in the depth-dose direction, the most clinically relevant parameters are the beam range, modulation width of the SOBP, and the distal falloff. The beam range is defined as the depth of penetration at distal 90%, that is, d90. Historically, the modulation width was defined as the width of the dose plateau at the 90% level, i.e., the distance in depth from p90 to d90. In our institution, however, we recently changed the definition of modulation width to m98, as indicated in Figure 4.2, for the following advantages [2].

1. For SOBP distributions with large modulation width, the magnitude of the upstream "plateau" rises making the proximal "knee" much less noticeable. As a result, the p90 value becomes overly sensitive in that small changes in the beam delivery can make large changes in the location of p90. A small difference in dose normalization, or measurement, for the same SOBP, could result in a large difference in p90. The p98 point, on the other hand, is at the steepest part of the "knee" and is therefore more stably defined.

2. For cases where the target volume extends close to the patient body surface, the dose plateau of the SOBP must extend close to the surface as well to provide full dose coverage. In that case, the p90 point would go outside of the body surface and become totally undefined, while the p98 is still valid.

3. m98 clearly reflects the extent of the high-dose region required to cover the target volume.

Figure 4.3 shows the lateral dose distribution at the middle of the SOBP dose plateau produced by a scattering system with a beam range of 13 cm and a modulation width of 5 cm (m98). Therefore, in the lateral direction, the most clinically relevant parameters include

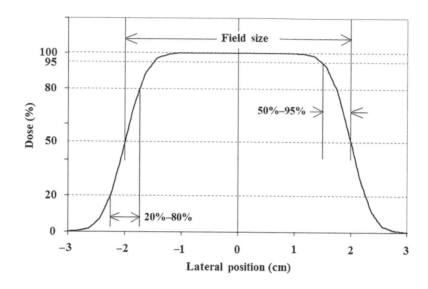

FIGURE 4.3   Specification of lateral dose distribution produced by a scattering.

the field size; defined at the 50% level, as for a photon field and the lateral penumbra or the lateral falloff at either end; both 20%–80% and 50%–95% values are used for that specification. While the former is used traditionally to reflect the general quality of the penumbra, the latter is particularly needed for determining the margins of aperture for a given treatment beam (see Chapter 17).

The clinical parameters of the beam in three dimensions (3D) have been identified for a scattered beam. What remains is to characterize the dose. The absolute dose delivered by an SOBP field is controlled by a monitor unit ion chamber located in the beam, downstream from the scatterers and range modulators. The SOBP beam has a well-defined output factor, expressed as the dose measured at the dose plateau of the SOBP in centigray per monitor unit reading (cGy/MU), as it is for the photon beam (see Chapter 13). The difference here is that the output factor depends on both the beam range and modulator width. For a given SOBP treatment field with a specific combination of range and modulation width, the output factor must be first determined either by measurement or by modeling [3,4]. The required monitor unit for the treatment delivery can then be computed for the prescribed dose. The calibration of the monitor unit chamber follows the procedure described in International Atomic Energy Agency report 398 [5]. Note that the dose calibration is usually based on the physical dose, while the prescription is given in Colt Gray Equivalent. The proper conversion of 1.1 must be included in the monitor unit computation (Chapter 22).

4.3.2 Pencil Beam Scanning

In systems that use beam scanning to spread the lateral extent of the beam, the small beam coming through the nozzle is directly sent into the patient without interacting with any scattering or energy modulation devices (Chapter 6). Single or multiple magnetic elements are used to steer the thin beam across the target profile to reach the full lateral extent of the target volume. The depth-dose distribution is delivered by placing the Bragg peak in the patient one depth at a time (known as a layer) and then varying the beam energy (this is dependent on the time needed to change the system's energy). In this way, there is full freedom to put any amount of dose, anywhere within the target. A consequence of this, without material inter-cepting the beam, is that the beam energy can be effectively modified from the accelerator and beamline. In a current operation, the transition from one beam energy to another takes some time (seconds to tenths of seconds). Therefore, in the depth-dose direction, the most clinically relevant beam characterization is the Bragg peak (as shown in upper graph of Figure 4.4), and in the lateral direction, the most clinically relevant beam characteristics include the beam width (usually specified in terms of the sigma (σ) parameter of a Gaussian curve representing the beam and lateral falloff, as shown in the lower graph of Figure 4.4. There is no a priori dose distribution such as uniform homogenous dose as in the case of a scattering system, so more degrees of freedom and flexibility are possible. The "pencil" beam spot can be positioned in a spot-by-spot manner or can be moved continuously across the target.

Three main categories of dose delivery have been used in beam scanning, uniform scan-ning (US), single-field uniform dose (SFUD), and multifield uniform dose (MFUD).

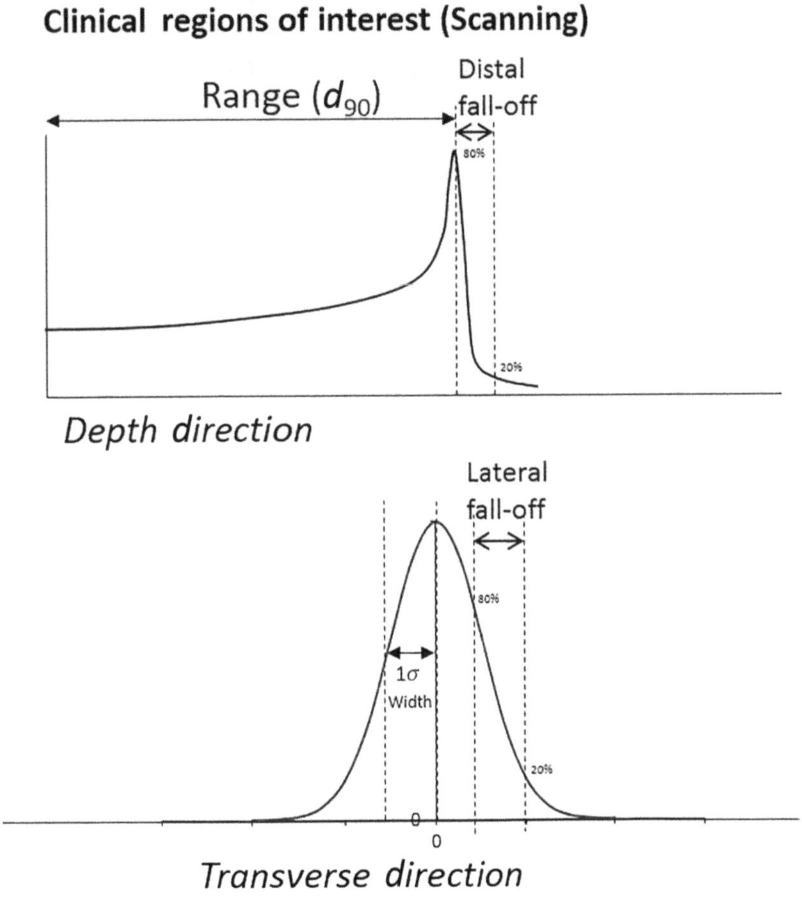

FIGURE 4.4 Specification of the clinical parameters of a scanned beam. (The upper graph rep-resents the Bragg peak/depth-dose distribution, and the lower graph represents the lateral dose distribution of a single beam spot.)

US employs a fixed scanning pattern with constant beam intensity for each layer. The relative intensities of the layers are fixed to produce a flat dose plateau longitudinally for a homogeneous medium. It is also generally a larger, lightly scattered beam and uses an aperture for beam collimation and a range compensator for distal conformity, just as for beams produced by scattering. However, the size of the laterally spread-out beam is still smaller than the aperture and can result in lower secondary radiation when compared to a nonoptimized scattered beam field. The dose distributions produced by US are largely the same as those by scattering, except that the maximum field size is no longer limited by the scattering system. Because of this, the beam is treated in the same manner as scattering in treatment planning as well as in delivery. That is, the beam is specified by the range and modulation width for the overall dose distribution as those given in Figure 4.2, rather than the energy and intensity of each individual Bragg peak [6].

For SFUD, both the scanning pattern and intensity distribution of the scanned beam are customized for a treatment field, but the resultant dose distribution at the end of each field is still uniform over the target volume. This is not required in MFUD, or incorrectly termed intensity-modulated proton therapy (IMPT), where the homogeneous dose distribution over the target volume is constructed only by the combination of two or more treatment fields (Chapter 16).

For both SFUD and MFUD, the treatment planning system considers each pencil beam Bragg peak explicitly, rather than their combinations of any type as for US or scattering, thus being consistent with the beam parameter characterizations defined in this section. The concept of modulation width becomes irrelevant. The specification of a treatment beam is basically a list of Bragg peaks, each with the energy of the proton, the lateral location of the peak projected unto the isocenter plane, and the number of protons, often given as units of gigaprotons (10^9). In this case, the quality of the beam is determined largely by the quality of the individual pencil beam and their relative weights. The dose distribution of the pencil beam is essentially a Bragg peak longitudinally and a Gaussian transversely. The width of the peak reflects the energy spread of the protons, while the sigma of the Gaussian gives the spot size of the beam. Note that the spot size is defined as the width of the Gaussian in air at the isocenter (Chapter 6).

4.4 CLINICALLY MOTIVATED BEAM PROPERTIES

The beam characterizations arising from the beam delivery systems are indications of the types of beams that can be delivered. Exactly what the parameter values need to be, to appropriately treat clinical targets, depends on the clinical indications and constraints. The following sections discuss aspects of these beam properties, as used in particle radiotherapy for clinical treatments.

4.4.1 Beam Range

For a given beam direction, the deepest Bragg peak must reach the deepest point of the target volume with sufficient margins, considering the various effects due to the distal penumbra and uncertainties in treatment planning and patient setup.

At the Massachusetts General Hospital (MGH)facility, the beam is supplied by cyclotron at 230 MeV, with the beam range in water at about 34 cm. In scattering mode, the maximum treatment depth achievable is 29.3 cm [2]. Although this is deep enough for most patients, occasionally some deep-seated tumors, prostate or pancreas, for example, for an exceptionally large patient could require a deeper range to reach the target. . The minimum range achievable (with acceptable dose rate) in our scatting system is 4.6 cm. The need for a lower value could occur, for example, for pediatric patients, or superficial target volumes, like post mastectomy chest wall. In these cases, one could shorten the range by increasing the minimum thickness of the compensator, although it could increase the distance between the patient and the aperture, thereby degrading the lateral penumbra of the beam. In pencil beam scanning mode at our facility, the maximum beam range is 32 cm, while the minimum is 7 cm. Again, range shifter must be used for superficial target volumes.

As mentioned earlier, range uncertainty in patients has been a challenging issue for proton therapy (Chapter 17). Some techniques have shown promising results (Chapter 21), but they have not been used widely in the clinic. As a result, most of the treatments do not utilize the distal falloff for narrow margin sparing. The exception is the patching technique where the distal end of the patch field meets the lateral penumbra of through beam (Chapter 15). In that situation, a large-range uncertainty can substantially distort the designed treatment dose distribution. Even in a homogenous medium, a millimeter change of the beam range could create hot/cold spots of 5%–10%. For this reason, the patching technique is used for shallower target volumes, mostly in head and neck area, requiring only short beam paths and thus relatively small-range uncertainty in the patient. Moreover, alternate patch lines are used and rotated daily to even out the potential hot/cold spots. In any case, it is critical that the delivery system can maintain an accurate beam range. Note that such patching technique is needed only for treatment by passive scattering and is no longer necessary with MFUD (or IMPT) of pencil beam scanning.

To date, all proton accelerators for proton therapy use have been built to produce proton energy of 235 MeV or higher. But, is this really necessary? We have collected beam information for the total of 4,033 patients treated by passive scattering in the two gantry rooms at the Francis H. Burr Proton Therapy Center (FHBPTC) over the 10-year period from 2005 to 2015. Figure 4.5 shows a collection of 23,603 beams used with the range in patient and modulation width. A total of 21,811 fields were used for nonprostate treatment, and only 5 of them for 4 patients (1 pancreas, 1 lumbar chordoma, 2 spine metastasis) required beam range greater than 25 cm (group A). Among the 446 prostate patients, 386 had at least one of the lateral treatment fields with beam ranges greater than 25 cm (group B). One can safely say that the beam ranges greater than 25 cm were primarily for prostate treatments.

Obviously, the need for long beam range for prostate treatment is dictated by the lateral beam technique. If anterior or anterior oblique beams were used instead, the beam range would be well below 25 cm, as was verified for the ten patients with the longest lateral beam ranges (group C in Figure 4.5). In that case, a maximum beam range in a patient of 25 cm should be sufficient to treat all treatment sites with very rare exceptions. For passive scattering, the maximum accelerator energy needs to be substantially higher due to energy loss in scatterers, range modulators, compensators, etc. For PBS, however, such losses are

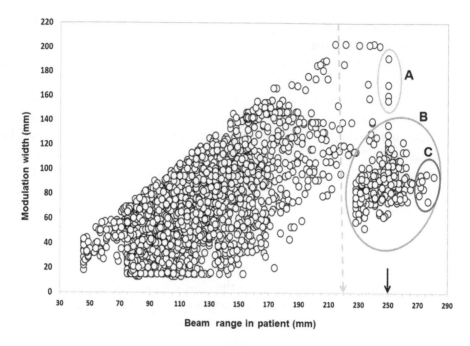

FIGURE 4.5   Ranges and modulations widths for the beams (small circles) used for the 4,033 patients treated by passive scattering in the two gantry rooms at FHBPTV over the 10-year period from 2005 to 2015. Groups A–C include largest ranges for nonprostate treatments, all prostate treatments, and ten prostate treatment with largest range, respectively.

minimal, and a maximum beam energy of 200 MeV would be sufficient. This reduction of maximum beam energy from 235 to 200 MeV could mean low field strength in the accelerator, in beamline magnets, smaller gantry size, etc., and with associated lower cost for the overall proton therapy system.

4.4.2 Field Size

The majority of cases treated by protons have relatively small target volumes, with a few exceptions including medulloblastoma and large sarcomas. As proton treatment facilities become more accessible, proton treatment expands to more indications including late-stage diseases, often with nodal involvement and thus much larger target volumes or multiple targets. Although one can, in principle, design the system to provide the largest field size possible, practical considerations must be included, like nozzle size, gantry size, dose rate, etc. In scattering systems, large field size requires large and heavy apertures and compensators that are difficult to handle (Chapter 5). For pencil beam scanning, larger field size requires stronger or longer scanning magnets.

The Ion Beam Applications (IBA) scattering system used at MGH has a field size up to 25 cm in diameter, although the effectively usable field with up to 2% dose heterogeneity is around 22 cm. The system has three sizes of snouts, 12, 18 and 25 cm in diameter. The smaller snouts allow for lighter apertures and compensators, and most importantly, closer patient contact leaving to smaller air gaps and thus better lateral beam penumbra preservation. At our facility, the 12 cm snout is used for close to 70% of treatment fields, with 18 and 25 snouts at about 15% each. Note that we currently limit prostate treatment to less than ten per day. The usage of the small field size could be substantially larger at centers with significantly more prostate patients.

At our facility, the field size for pencil beam scanning is a rectangle of 30×40 cm at the isocenter. The substantially larger field size is particularly useful for the treatment of medulloblastoma, one of the treatments that substantially benefits from proton therapy. The target volume in this case is the entire central nervous system (CNS), including the whole brain, and the spinal cavity extending inferiorly nearly to the coccyx and could be more than 80 cm, particularly with young adult patients. With scattering, the treatment is often broken into four parts using five fields, two laterals for brain and three abutting spinal fields, each covering a portion of the spinal cavity. Feathering must be used to even out the hot/cold spots at the junction. With repeated setup verifications between different fields due to the isocenter move, the entire treatment could take up to 30–40 min. Any increase of field size allowing for a smaller number of fields will be substantially appreciated by both the patient and staff. Note that with pencil beam scanning, feathering is no longer necessary. Each of the two abutting fields is designed to deliver a tapered dose distribution across the "match line." It completely eliminates hot/cold spots caused by geometric uncertainties from matching.

4.4.3 Dose Rate

An external beam radiotherapy system must be able to produce a high-enough dose rate so that the treatment can be delivered in a reasonably short amount of time. This is not only in consideration of the efficiency in facility utilization. It is directly related to treatment quality. In most of the treatment procedures today, the patient is first set up by image guidance (two-dimensional radiography, cone-beam computed tomography (CBCT), portal imaging, etc.; see Chapter 20), and the treatment is then delivered using one or a few treatment fields, under the assumption that the patient does not change his/her body configuration from the time of the last imaging to the completion of dose delivery. While this assumption is more valid for certain types of treatment with appropriate patient immobilization devices than for others, it is clear that the longer the treatment takes, the more likely the patient will change the body configuration and affect the quality of treatment. Typically, proton treatment setup used to take more time than photon treatment given the required accuracy for the treatment, as mentioned earlier; however, with the advent of image guidance in photon treatments, similar setup times can occur.

It was found in a recent study that, during prostate treatment, the target volume can stay within 5 mm margin only for 5 min. If an endorectal balloon is used, the margin is reduced to 3 mm but still only for 5 min [7]. In this case, the lack of sufficiently high dose rate would require increasing the planning margins of the target volume and delivering more doses to normal tissues nearby. Patient fatigue is another consideration for reducing the treatment time, particularly for elderly or pediatric patients. However, this does not mean that one should use the highest dose rate possible. Instead, it should be low enough to give the operator a reasonable amount of time to stop the treatment in response to sudden patient movement or any anticipated equipment problems.

The majority of therapy patients are currently treated with a regular fractionation schedule with a daily dose of 1.8–2.0 Gy. The most commonly used dose rate for conformal photon treatment is 2–4 Gy/min. For IMRT, even higher dose rate, e.g., 6 Gy/min, is often used, since each segment of the Multi-Leaf Collimator (MLC) pattern could irradiate only a small portion of the target volume. Note that even with such high dose rate, the treatment may still take a long time, due to the large number of segments in each field and the large number of field directions used. This is the main reason for the recent trend towards arc delivery together with intensity modulation, e.g., RapidArc, VMAT, and Tomotherapy.

Proton treatment, on the other hand, does not require a large number of treatment fields for each treatment fraction, due to its superiority in minimizing dose to normal tissues around the target volume. A large number of fields may be needed to satisfy the total dose constraints on the critical organs over the whole treatment course, but only a subset of these fields are sufficient to deliver the daily prescription with acceptable dose tolerance for the surrounding normal tissues. Most of treatment uses two fields a day, alternating between different field combinations.

For regular fractionated treatment, i.e., 1.8–2.0 Gy per fraction, this means 0.9–1.0 Gy per field. Dose rates consistent with delivering a field in up to a few minutes are consistent with the targeting issues identified above and still much less than the overall setup time. A dose rate of 1.0–4.0 Gy/min will be reasonable, but the intensity required for the accelerator to achieve this dose rate is highly dependent on the target field size. Naturally, for hypofractionated treatment, a much higher dose rate will be desirable. The extreme case is stereotactic radiosurgery treatment, where each field could deliver up to 8 Gy. Another type of treatment where a higher dose rate may be appreciated is respiratory gating, where the beam is turned on only for a portion of the respiration cycle, usually 30% centered on the end of the respiration phase [8]. Ideally, the dose rate should be three times more than normal if the usual amount of time is used for the treatment. However, the dose rate should not be too high, given that the treatment is to be delivered over a sufficient number of respiratory cycles to average out the uncertainties due to breathing irregularities.

The dose rate ultimately depends on the beam current transported to the nozzle entrance, where it is to be scattered in scattering mode or to be guided into the patient directly in scanning mode. This largely depends on the capability of the accelerator and the energy selection system. For some cyclotrons, the beam current can be continuous and generally has a high operating current, e.g., 300 nA at the cyclotron exit. However, the fixed energy of the beam must be reduced by the energy selection system to the appropriate value before sending it to the patient, and this could reduce the beam current significantly. As a result, the smaller the beam range required, the lower the dose rate. For synchrotron-based accelerators, dose rate does not depend on the beam energy as much, given the absence of energy selection system. However, the beam is not delivered continuously but in "spills," and the overall peak beam current reaching the nozzle must be adjusted accordingly to allow adequate average beam intensity.

In the scattering mode, the dose rate also depends on the scattering design, particularly the intended field size. Obviously, the larger the field size, the lower the dose rate for a given beam current intensity at the nozzle entrance. In fact, in some situations, the scattering system was designed to give a small field size for a particular type of treatment, for example, for prostate treatment only, where the maximum field size is 12 cm in diameter.

For our system in scattering mode, the dose rate for a 4.6 cm beam range is 1.5 Gy/min at the maximum beam current, i.e., 300 nA at the cyclotron exit. At a beam range of 16 cm, the dose rate can be as high as 10 Gy/min.

For pencil beam scanning, the dose can be delivered spot by spot (in step-n-shoot mode) or line by line (in continuous scanning) and then layer by layer. The overall dose rate is usually specified by the time to deliver a uniform dose to a 10 cm cube or 1 Liter of tissue equivalent material. Note that this depends not only on the beam current intensity for each scan layer but also the time between the layers for beam energy change and the corresponding adjustments of the beam transport system. For the same incident beam current, the effective dose rate will also depend significantly on the target size to be treated.

4.4.4 Lateral Penumbra

A sharp lateral penumbra is essential for sparing critical organs adjacent to the target volume. This happens to be one of the most attractive features of the proton beam. The lateral penumbra achievable in the patient depends on the design of the beam delivery system and also the nature of interaction between protons and tissues in the patient. The beam nozzle is generally designed to keep the penumbra as sharp as possible, although in some situations, a less sharp penumbra may be beneficial, for example, for beam patching (Chapter 15). Another example is for treatments where large patient setup uncertainties must be tolerated.

For scattering, the lateral beam penumbra is affected by the source size and source position, the position of the aperture, the range compensator, and the air gap between the compensator and patient's body surface, and naturally, the depth of tissue that the beam must penetrate before reaching the target volume (Chapter 5). The scatterer and the modulator determine the source size, and they are positioned far upstream, as far as possible from the aperture, resulting in a much longer Source to Axis Distance (SAD) (>200 cm) than that for photon beams (100 cm). The aperture should be as close to the patient as possible to reduce the effect of source size. This also reduces the air gap that degrades the penumbra substantially (Chapter 5).

In the patient, a proton beam interacts with tissues very differently from megavoltage photon beams (Chapter 2). For the later, the main mechanism is attenuation due to Compton scattering, where the scattered photons essentially escape from the beam. As a result, the lateral penumbra is mainly determined by the beam source size and source position, and increases only moderately as the beam goes through the patient. For protons, on the other hand, the main interacting mechanism is multiple Coulomb scattering. A proton changes its direction very little after each interaction but stays in the beam. The change accumulates rapidly and increases the beam penumbra much faster than in the case of a photon beam.

Figure 4.6 plots the lateral penumbra for a 6-MV photon beam and that for a scattered proton beam (range 14 cm/modulation width 10 cm), at both 4 and 10 cm depths. Clearly, the proton penumbras are much sharper than those of the photon counterparts. However, it increases drastically as depth increases from 4 to 10, while the photon penumbra increase

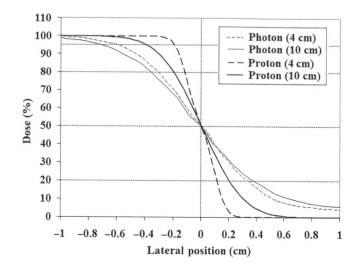

FIGURE 4.6 Lateral beam profiles in the penumbra region for scattered beam with a range of 14 cm and a modulation width of 10 cm (m98) at both 4 and 10 cm depths in water. The profiles of a 6 MV photon beam at the two depths are shown for comparison.

is moderate. Figure 4.7 shows the increase of beam penumbra of protons over deeper depths, again together with those for higher-energy photon beams.

Overall, at shallower depths, the proton penumbra (both 20%–80% and 50%–95%) is smaller than with photon beams, but it increases rapidly with depth and becomes larger than the 15-MV photon beam for depths greater than 17 cm for 20%–80%, and 22 cm for 50%–95%. It is interesting to note that this is the typical treatment depth in current prostate treatment, where only bilateral fields are used. It is exactly the reason why proton plans do not demonstrate any substantial dosimetric benefit over IMRT in terms of dose to anterior part of the rectal wall situated right next to the prostate target volume [9].

Safai et al. [10] compared the properties of the lateral beam penumbra between the scattered beam and a pencil beam. Figure 4.8 shows the measured and modeled penumbra [20%–80%] as a function of depth in water for two broad pristine Bragg peaks with ranges 22.1 cm (T_0 = 183 MeV) and 7.85 cm (T_0 = 102 MeV), where T_0 denotes the initial beam energy. Two sets of data were obtained for each beam range, with and without a 4 cm thick plate of PolyMethyl MethAcrylate (Acrylic/PMMA) to simulate the contribution of the range compensator. The air gap was 10 cm without the PMMA and 6 cm with it. For all configurations, the penumbra increases with depth, as was the case shown in Figure 4.7. The lower energy beam starts with a slightly larger value at the surface, but increases much more rapidly than the higher energy beam. The PMMA plate broadens the penumbra for both beam energies, more at depth than near the surface due to reduced beam energy. At the end of the beam range, the penumbras are comparable with or without the PMMA plate. Note that although the data given here is from pristine Bragg peaks without range modulation, it was shown that the lateral penumbra of SOBP fields is almost independent of the beam modulation width and therefore follows the same trend described here [11,12].

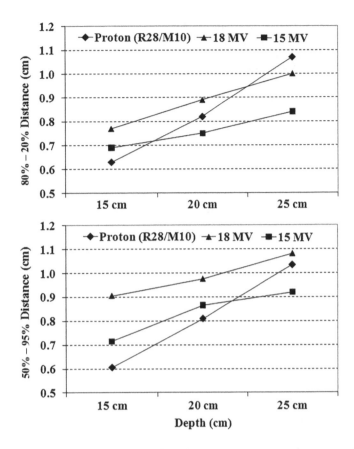

FIGURE 4.7 Measured lateral beam penumbra (20%–80% upper and 50%–95% lower) as functions of depth in water for 15 and 18 MV photon beam, and for scattering beam with a range of 28 cm and a modulation width of 10 cm.

In pencil beam scanning, the main source of penumbra broadening before the patient is the scattering by the air column extending from the end of the nozzle to the patient's body surface. Safai et al. [10] modeled this effect for a single Gaussian beam ($\sigma = 3$ mm) and compared the results with a broad Gaussian beam having an aperture and compensator. It was found that if the vacuum window is at the same upstream location as in a scattering system, the penumbra for the uncollimated pencil beam is significantly larger than that for the collimated beam at the surface. Of course, no scanning system would have an air column that long. In depth of water, however, the uncollimated penumbra increased much slower than the collimated, and became less than the later at larger depths greater than 18 cm. This is partly due to the different effective source positions (i.e., divergence) of the beam. Overall, the collimated penumbra is superior at shallow depths but is inferior at greater depths. The authors concluded that for most of the clinical sites (e.g., head and neck) the penumbra of a pencil beam is inferior to that of a collimated divergent beam, unless the vacuum window is moved downstream substantially, or the beam spot size is reduced to 5 mm or less [11].

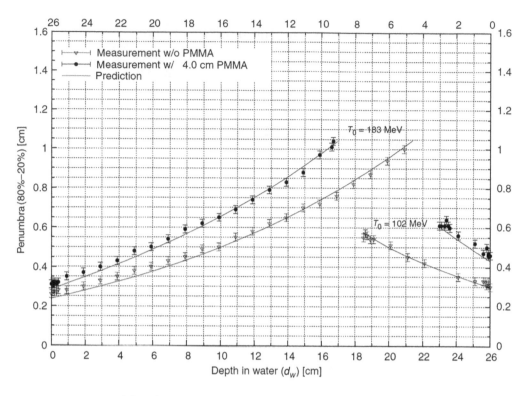

FIGURE 4.8 Measured (symbols) and modeled (lines) lateral beam penumbra (20%–80%) as functions of depth in water for two scattered beams with beam range of 21.1 cm (left) and 7.85 cm (right), with and without a 4.0 cm plate of PMMA.

The investigation reported in Safai's work focused on the penumbra of a single pencil beam. Clinically, what is more relevant in terms of organ sparing is the lateral penumbra of the composite dose distribution as the superposition by all individual pencil beams used for a treatment field. As discussed in Chapter 6, such a superposition with a uniform dose plateau can be built using evenly spaced pencil beams with a constant intensity, but the resultant composite lateral penumbra will be broader than that for the individual pencil beam. However, composite doses with equally acceptable dose plateau can also be obtained using a nonuniform spacing together with modulated intensities optimized to produce a much sharper penumbra, nearly the same as for a single pencil beam. For either case, though, the smaller the spot size for the individual pencil beam, the narrower the lateral penumbra of the total dose distribution.

Is it always better to have a smaller spot size for PBS? Trofimov and Bortfeld [13] studied the issue in their efforts to develop a set of clinically relevant specifications for pencil beam scanning. They pointed out that, for deep tumors, a finer pencil beam would not necessarily lead to any significant improvement in dose conformity, simply because the main contribution to the spot size at the tumor arises from multiple Coulomb scattering along the beam path in patient. For shallower target volumes, they performed treatment planning using a range of spot sizes for a head-and-neck case where this perhaps matters the most. It was found that reducing the spot size from 8 to 5 mm lead to a marked improvement in dose conformality for the target volume, whereas the difference was not as dramatic from 5 to 3 mm. They concluded that for most clinical cases, pencil beams of widths $\sigma = 5$ mm will be sufficient for delivery of the target-conformal planned dose with a high precision. Reducing the beam spot size below 5 mm does not lead to a substantial improvement in the target coverage or sparing of healthy tissue.

4.4.5 Distal Penumbra

The distal penumbra of the SOBP dose distribution is determined primarily by the deepest few Bragg peaks. The penumbra (50%–95%) is the smallest with only the deepest peak, but increases up to a millimeter as more peaks are added to build up the dose plateau, as shown in Figure 4.1. In relatively homogeneous medium without high gradient in tissue density, the distal beam penumbra is always much sharper than the lateral one. It increases moderately with energy due to range straggling in the patient and also by scattering and range modulation components in the nozzle if scattering is used. The former is unavoidable, but the latter can be eliminated using pencil beam scanning or minimized by reducing the water equivalent thickness of those components. For the scattering system at MGH, the distal penumbra increases monotonically from 3.5 to 5.0 mm (20%–80%) over the beam range of 4.8–25 cm. But for a beam range of 25–28 cm, the penumbra is only 4.7 mm, actually smaller due to the use of a very thin second scatter.

When high-gradient tissue inhomogeneity is present, range mixing can occur. The distal penumbra can be degraded substantially and could be much larger than the lateral penumbra. The extreme case is when the beam passes along the interface between a high density and lower density mediums, where the DDF can be seriously distorted. The situation has been investigated by Schaffner et al. [14]. Needless to say that the compensator causes additional range straggling, which will also increase the distal penumbra. Thick compensator and, in particular, those with sharp variations in thickness should be avoided if possible.

It should be pointed out that although the distal beam penumbra is much sharper than the lateral, it is not always used clinically for tight margin sparing, due to uncertainties in predicting the beam range in the patient (Chapter 17). The recent development using posttreatment Positron Emission Tomography (PET) imaging based on treatment-induced isotope activities has shown promising results for in vivo range verification, as described in Chapter 21. While it works well mainly for coregistered bony structures in head and neck, for soft tissue target volumes in other parts of the body, the accuracy drops substantially due to various reasons. In any case, the method is still at the research stage and is not widely available.

In current practice, the range uncertainty issue is managed by adding an additional amount to the beam range in treatment planning, usually 3.5%, to head off the potential "undershooting" (see Chapters 15 and 17). While this guarantees the coverage of the distal aspect of the target volume, it also risks overdosing the normal tissue behind the target volume. A good example is the treatment of prostate cancer mentioned above. Anterior or anterior oblique fields have never been used, despite the fact that such fields can utilize the sharp distal penumbra (~4 mm for 50%–95%) to separate the prostate and the rectum behind. Instead, only lateral fields are used, relying solely on the much broader lateral beam penumbra (>10 mm for 50%–95%) to spare the rectum. The reason is because when a water-filled endorectal balloon is used to immobilize the prostate, as widely practiced, the anterior rectal wall is only about 5 mm thick and is situated right next to the posterior side of the prostate in many areas. If an anterior field is used for treatment, the typical beam range is about 15 cm and its 3.5% will be 5 mm, just the thickness of the anterior rectal wall. Therefore, the usual method of adding extra range will risk delivering the full dose to the anterior rectal wall, clearly unacceptable given rectal bleeding as the leading treatment toxicity for prostate treatment.

4.4.6 Dose Uniformity

A proton beam has the unique ability that a single beam can be used to deliver a homogeneous dose across the entire target volume, as opposed to a photon beam where multiple fields must be used to achieve a similar coverage through superposition. In scattering, this is achieved by carefully designing the scattering components including scatterers and range modulators (or ridge filters in some systems) so that the dose in the SOBP is uniform in the dose plateau both in depth and laterally across the field size required. Alternatively, a uniform dose distribution can also be achieved with a scanning system. Naturally, this specification is for a homogeneous phantom, while in patients, it could be larger due to the effect of tissue inhomogeneity.

The details of the scattering systems will be discussed in Chapter 5. We only comment here that, given the sensitive dependence of the scattering property on the beam energy, it is a formidable task to design a scattering system that accommodates all clinical relevant beam ranges with sufficient quality and at the same time provides convenience and efficiency that are critical in day-to-day clinical practice. With a method developed only recently, it was possible to optimize a critical component of the system so that, for all the beam ranges, the dose uniformity in depth is within 2% [2].

Note that the double scattering system is very sensitive to beam steering, and a slight change in the beam spot position on the scatters can seriously affect the field flatness and symmetry. This requires an effective monitoring system and frequent quality assurance verifications by independent means.

For pencil beam scanning, the total dose distribution is constructed by individually delivered Bragg peaks, and its uniformity will directly depend on the accuracy of the delivery, i.e., the beam energy, the spot location, etc. Inaccuracies in the beam energy will create uneven distances between adjacent layers in depth and thus cause ripples in depth that could exceed the 2% requirement, particularly at the shallower part where the Bragg peak is distinctively narrow. Across the field, when a spot is not accurately positioned as planned, the superposed Gaussian distribution will contain spikes and valleys, which could exceed the uniformity requirement. The accuracy requirement in delivery for keeping an acceptable level of dose uniformity and achieved with the current technological capability will be discussed in Chapter 6.

4.4.7 Characteristics of Proton Therapy Treatment

The distinguished physical characteristics of the proton beam discussed above also result in some special characteristics of the proton therapy treatment itself in comparison to treatment by photon and/or electron beams. Proton beams have been used to treat nearly all major types of cancer at all treatment sites, e.g., CNS, head and neck, lung, esophagus, liver/pancreas, prostate, rectum, and sarcomas, and the list is ever increasing. Recently, proton beams were used for locally advanced breast cancer. While these conditions are also treated by photon/electron beams as the only choice for the majority of patients today, treatment using protons can be quite different, given its unique physical properties of the beam.

Undoubtedly, the sharp DDF gives proton the most important advantage in sparing normal tissues near the target volume. The clinical implication of this is best illustrated by the spine fields used in the treatment of medulloblastoma, a condition occurring most often in pediatric and/or young adult patients. The protons from a posterior beam stop right behind the spinal target volume, leaving no dose to the rest of the body, while a photon treatment would give up to 50% of the prescription dose the anterior portion of the body [15]. Another unique feature made possible by the sharp distal falloff is the patching technique widely used in scattering and US (Chapter 15).

We must note that while the sharp DDF can offer great potentials for normal tissue sparing, it also makes the dose distribution extremely sensitive to uncertainties in treatment planning and patient setup (Chapter 17). The range of the proton beam in patients depends largely on the water equivalent path length (WEPL) along the beam. If the WEPL value changes by 1 cm, the location of the distal falloff will also change by 1 cm, causing either an "undershoot" missing the distal portion of the target volume by one full centimeter or an "overshoot" delivering full dose to a centimeter of normal tissue behind the target volume. In contrast, the same magnitude of change in WEPL in a photon treatment will only change the dose at most by 3%–4% for 10 MV beam, for example. A patient setup error in an IMRT treatment may cause a shift of the so-called "dose cloud," missing some peripheral regions of the target volume. In a proton treatment, the same error may also distort the dose cloud substantially, due to the mismatch between the beam energy distribution and the tissue heterogeneities along the beam path.

The presence of a dose plateau in Figure 4.1 means that a single proton field can deliver an uniform dose to the target volume, which is not possible with photon or electron beams. This allows for the use of only one or two fields for each treatment fraction to deliver the prescription dose, although the whole treatment course may use a large number of treatment fields to spread the dose to normal tissues. Many patients are treated in this manner, i.e., with only one or two fields per fraction, but with different fields or field combinations on different days. This is very different compared to a photon treatment where each fraction uses the same number of fields throughout the treatment course.

When the total prescription dose for a fraction is delivered by only one or two fields, the uncertainties involved in each field become much more important. This and the sensitive nature of the proton dose distribution mentioned above determine that proton treatment requires highly accurate patient setup. Note that since the treatment beam stops in the patient completely, a photon-like portal imaging is not possible, and one must rely entirely on X-ray imaging. Usually, the patient is first setup with an orthogonal pair of images to produce the correct anatomical configuration, and then for each treatment field, the patient is imaged again along the beam direction, often with the aperture when available, to ensure target volume coverage and normal structure avoidance. As a result, the time for patient setup is generally longer than with a photon treatment. Recently, volumetric imaging for patient setup has made its way into proton treatment rooms, either with CT on rail or CBCT. It is expected that this will become the standard of practice with improved setup accuracy, higher operation efficiency, and the potential for adaptive therapy.

4.4.8 Beam Directions

Ideally, the treatment beam should take the shortest path to reach the tumor to minimize the volume of normal tissue on the beam path. However, the entrance dose of the proton beam, while much lower than that of photon beams is nonzero, and it is advantageous to spread it out over the normal tissue around the target volume by using multiple beam directions including those with longer beam paths. Use of the scanning beam delivery modality offers the possibility of proximal dose sparing and even fewer beam angles may be required. Beam direction may also be restricted by the geometric position of the target volume in relation to the nearby critical organs. Prostate treatment is such an example where the rectum is situated right next to the prostate on the posterior side. When considering the relative sharpness of the distal falloff compared to the lateral penumbra, naturally, the best approach is from the anterior so that the sharp distal falloff of the proton beam can be used to cover the target volume while sparing the rectum behind. This would require, however, a precise control of the beam range in the patient with millimeter accuracy, which is not currently possible. As a result, one can only use lateral fields with a substantially long beam path going through more than half of the patient's body [9].

Can treatment be delivered with only fixed beamline rather than a gantry system? This seems to be a question relevant only in particle therapy. A gantry system is large and expensive. Moreover, the beam transport system on the gantry adds another layer of complexity to the whole system. The answer to this question really depends on the specific treatment site and treatment techniques involved. Treatments of ocular melanoma have always used fixed beamlines. Current prostate treatment uses only lateral beams as mentioned earlier, and therefore can be treated by only fixed horizontal beamlines as indeed practiced at some centers. For other treatment approaches using anterior and anterior oblique beams currently under development, the gantry will be required. For some treatment sites like nasopharynx, skull base chordoma, however, it is clearly difficult without a gantry. Figure 4.9 shows the number of fields at each gantry angle for all the patients treated during a 12-month period for one gantry treatment room at the FHBPTC. Of all the fields, 23% used only lateral beams, 41% for just the four normal angles (0, 90, 180, 270), and 59% used other gantry angles. Interestingly, the treatment planners only varied the gantry angle at 5° increments.

In principle, the beam orientation relative to the patient provided by all the coplanar beams on the gantry can be achieved equally by a fixed horizontal beamline with the patient sitting on a treatment chair rotatable around a vertical axis. The fact is that most of the treatment sites require only coplanar beams. Figure 4.10 shows the distribution over gantry and table angles for the beam used over a 12-month period for the major noncranial

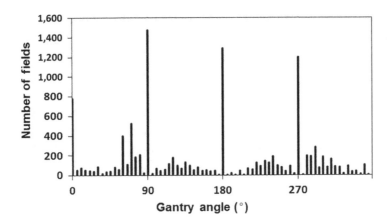

FIGURE 4.9 Distribution of the number of treatment fields over gantry angles for all patients treated in a gantry room over a 12-month period.

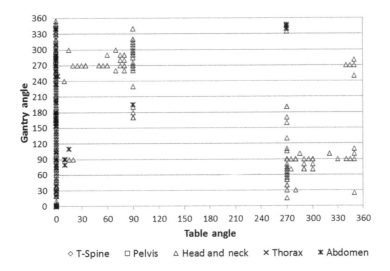

FIGURE 4.10 Gantry and table angles for the beams used over a 12-month period for the major noncranial sites on one of the gantry treatment rooms at FHBPTC.

sites on one of the gantry treatment rooms at the FHBPTC. Nearly all the noncoplanar beams were used only for head and neck treatments.

The combination of a chair with a fixed horizontal beam was used in the early days of proton therapy for a broad range of treatment sites, given that it was the only option at the time. The practice largely stopped as soon as gantries became available. However, reconsideration of this technique appeared recently [16]. While the efforts were mostly aimed for a potential cost reduction of proton therapy equipment, they were also motivated by possibilities offered by new technologies in two main areas. One is patient position control with significant improvements in accuracy and setup speed supported by modern robotics, volumetric imaging, surface imaging and monitoring, etc. The other is the availability of pencil beam scanning. The power of intensity modulation offered by IMPT allows one to obtain the desired dose distribution with less number of beams and more flexible beam angles compared with scattered beams. It was found in a recent study that IMPT plans could achieve satisfactory dose distributions for head and neck treatments without the noncoplanar beams that can only be provided by a gantry. This is, in fact, not surprising considering the impact of intensity modulation on beam geometry in the photon world. While noncoplanar beams are always needed by 3D conformal treatments for certain group of treatment sites, IMRT/VMAT can provide equal or improved dose distributions for these sites by coplanar beam directions only. A good example is the Tomotherapy system that simply does not provide any noncoplanar beam directions but can treat all sites including cranial target volumes. This new effort in reexploring fixed beamline treatment technique for proton therapy could come to fruition in the near future.

4.5 SUMMARY

A survey of the properties of proton beam and how they are used in clinical treatment has been presented. Although proton therapy has existed for decades, the more widespread use started only in the last ten years, and current clinical practice is now being augmented by investigating the use of the beam for a wider range of clinical sites and of the effects of modifying the beam properties. It is interesting to point out that one of the characteristics of proton beams is that their properties can, in fact, be modified. Thus, it becomes even more important to understand what works best clinically and to evaluate the design of any proposed beam spreading system with an eye towards how it will be used clinically. One issue may be the construction of multipurpose systems as opposed to single-use systems. In much the same way as a gantry is more expensive than a fixed beamline, the use of a beam for multiple treatment sites may also be more expensive. However, with the increase in the use of scanning beams, this may not be an issue any longer. The properties of the beam must be matched to the characteristics of the clinical target and the beam delivery modality, and understanding how the beam characteristics can be used and modified is an essential part of optimal treatment.

REFERENCES

1. ICRU Report 78. Prescribing, recording, and reporting proton-beam therapy. International Commission on Radiation Units and Measurements 2007.
2. Engelsman M, Lu HM, Herrup D, Bussiere M, Kooy HM. Commissioning a passive-scattering proton therapy nozzle for accurate SOBP delivery. *Med Phys.* 2009 Jun; 36(6):2172–80.
3. Kooy HM, Schaefer M, Rosenthal S, Bortfeld T. Monitor unit calculations for range-modulated spread-out Bragg peak fields. *Phys Med Biol.* 2003; 4, 2797–2808.
4. Kooy HM, Rosenthal SJ, Engelsman M, Mazal A, Slopsema R, Paganetti H, Flanz, JB. The prediction of output factors for spread-out proton Bragg peak fields in clinical practice. *Phys Med Biol.* 2005; 50, 5847–56.
5. IAEA Report 398. Absorbed dose determination in external beam radiotherapy: A international code of practice for dosimetry based on standards of absorbed dose to water. International Atomic Energy Agency (2000).
6. Farr J, Mascia AE, His WC, Allgower CE, Jesseph F, Schreuder AN, Wolanski M. Clinical characterization of a proton beam continuous uniform scanning system with dose layer tacking. *Med Phys.* 2008; 35, 4945–54.

7. Both S. Private communications.
8. Lu HM, Brett R, Sharp G, Safai S, Jiang S, Flanz J, Kooy H. A respiratory-gated treatment system for proton therapy. *Med Phys.* 2007; 34, 3273–78.
9. Trofimov A, Nguyen PL, Coen JJ, Doppke KP, Schneider RJ, Adams JA, et al. Radiotherapy treatment of early-stage prostate cancer with IMRT and protons: A treatment planning comparison. *Int J Radiat Oncol Biol Phys.* 2007 Oct 1; 69(2), 444–53.
10. Safai S, Bortfeld T, Engelsman M. Comparison between the lateral penumbra of a collimated double-scattered beam and uncollimated scanning beam in proton radiotherapy. *Phys Med Biol.* 2008; 53, 1729–50.
11. Oozeer R, Mazal A, Rosenwald J C, Belshi R, Nauraye C, Ferrand R and Biensan S . A model for the lateral penumbra in water of a 200 MeV proton beam devoted to clinical applications. *Med Phys.* 1997; 24, 1599–604.
12. Urie MM, Sisterson JM, Koehler AM, Goitein M, Zoesman J. Proton beam penumbra: Effects of separation between patient and beam modifying devices. *Med Phys.* 1986; 13, 734–41.
13. Trofimov A, Bortfeld T. Optimization of beam parameters and treatment planning for intensity modulated proton therapy. *Technol Cancer Res Treatm.* 2003; 2, 437–444.
14. Schaffner, B., Pedroni, E., Lomax, AJ. Dose calculation models for proton treatment planning using a dynamic beam delivery system: An attempt to include density heterogeneity effects in the analytical dose calculation. *Phys Med Biol.* 1999; 44: 27–42.
15. Clair WH, Adams JA, Bues M, Fullerton BC, Shell SL, Kooy HM, Loeffler JS, Tarbell NJ. Advantages of protons compared to conventional X-ray or IMRT in the treatment of pediatric patient with medulloblastoma. *Int J Radiat Oncol Biol Phys.* 2004; 58, 727–34.
16. Yan S, Lu HM, Flanz J, Adam J, Trofimov A, Bortfeld T. Reassessment of the necessity of the proton gantry: Analysis of beam orientations from 4332 treatments at the Massachusetts General Hospital Proton Center over the past 10 years. Particle Therapy Special Edition. *Int J Radiat Oncol Biol Phys.* 2016; 95: 224–33.
17. Yan S, Depauw N, Flanz J, Adams J, Gorissen B, Shih H, Bortfeld T, Lu H. Does the greater flexibility of pencil beam scanning reduce the need for a proton gantry? *Med Phys.* 2016; 43: 3509–3510. doi:10.1118/1.4956345 (abstract).
18. Bentefour, E. H. and Lu H. Re-thinking the useful clinical beam energy in proton therapy: An opportunity for cost reduction. *Med Phys.* 2016; 43: 3511. doi:10.1118/1.4956353 (abstract).

CHAPTER 5

Beam Delivery Using Passive Scattering*

Roelf Slopsema

University of Florida Health Proton Therapy Institute

CONTENTS

5.1 INTRODUCTION

Passive scattering is a delivery technique in which scattering and range-shifting materials spread the proton beam. After the protons are accelerated, either by a cyclotron or synchrotron, they are transported into the treatment room through the beamline

---

* Colour figures available online at www.crcpress.com/9781138626508

(see Chapter 3). The proton beam that reaches the treatment room is monoenergetic and has a lateral spread of only a few millimeters. Without modification, this beam would give a dose distribution that is clinically not very useful. Along the beam axis, the dose is initially fairly constant, but peaks sharply towards the end of travel of the protons (*Bragg peak*). In the lateral direction, the profile would be a Gaussian with a spread in the order of centimeters. Clinical use of the proton beam requires both spreading of the beam to a useful uniform area in the lateral direction as well creating a uniform dose distribution in the depth direction. The main function of the treatment head, or *nozzle*, is shaping the proton beam into a clinically useful three-dimensional (3D) dose distribution. In general, two methods of lateral beam spreading are applied: *passive scattering*, in which high-Z materials scatter the proton beam to the desired dimension, and *magnetic beam scanning*, in which magnetic fields sweep the proton beam over a desired area. Scattering systems are described in this chapter; scanning systems are the topic of Chapter 6. In the depth direction, a uniform dose region is created by adding Bragg peaks that are shifted in depth and given an appropriate weight to obtain a flat dose region called *spread-out Bragg peak* (SOBP). This method of adding pristine peaks is called *range modulation*. The number of peaks that is added proximally can be varied, varying the extent of the uniform region in depth. Combining scattering and range modulation gives a uniform dose distribution shaped like a cylinder. Field-specific *apertures* and *range compensators* conform the dose to the target. The aperture blocks the beam outside the target and conforms the beam laterally. The range compensator is a variable range shifter that conforms the beam to the distal end of the target. In this chapter, we will first describe the basic techniques of scattering and range modulation. Next, the design and application of apertures and range compensators is discussed. In the final section, several complete scattering systems are discussed, combining scattering technique, range modulation method, and conforming devices. Finally, we discuss the current and future role of passive scattering in proton therapy, given the recent ascendance of pencil beam scanning (PBS) systems.

5.2 SCATTERING TECHNIQUES

5.2.1 Flat Scatterer

The simplest scattering system is a single, flat scatterer that spreads a small proton beam into a Gaussian-like profile (Figure 5.1a). A collimator (aperture) blocks the beam outside the central high-dose region. To keep the dose variation over the profile within clinically acceptable limits, most of the beam will need to be blocked. For a Gaussian beam profile with spread σ, the efficiency η defined as the proportion of protons inside a useful radius R is given by [1]

$$\eta = 1 - e^{-\frac{1}{2}(R/\sigma)^2} \tag{5.1}$$

It follows that the fraction of protons outside the useful radius is equal to the relative dose at the useful radius (when normalizing to the central axis). Allowing for a dose variation

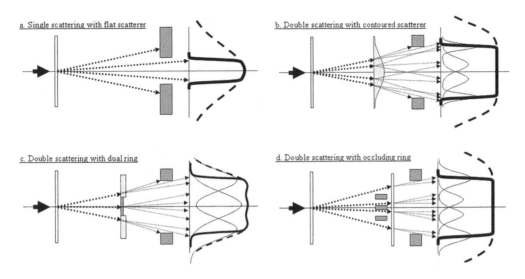

FIGURE 5.1   A schematic representation of the single-scattering technique using a flat scatterer (a) and double-scattering techniques using a contoured scatterer (b), dual ring (c), or occluding ring (d). Dashed lines indicate lateral profile without aperture; solid lines with aperture.

of ±2.5% over the profile and setting the useful radius at the 95% dose level results in an efficiency of only 100% − 95% = 5%. Because of this low efficiency, requiring relatively large beam currents and generating high production of secondary neutrons, spreading using a flat scatterer (single scattering) is limited to small fields with a diameter typically not exceeding 7 cm. Besides its simplicity, the advantage of a single flat scatterer over more complex scattering techniques is the potential for a very sharp lateral penumbra. Most of the scattering occurs in a single location along the beam axis, limiting the angular diffusion of the beam. Especially if the scatterer is placed far upstream of the final collimator, a very sharp lateral penumbra can be achieved. The field size limitation and sharp dose falloff make a single scattering ideal for eye treatments [2–9] and intracranial radio surgery [10].

Typically, the scatters are made of high-Z materials, like lead or tantalum, providing that largest amount of scattering for the lowest energy (range) loss. A scattering system that allows variation of the thickness of the scattering material can be used to maintain scattering power for varying proton energy. An example of such a system has a binary set of scatterers (with each scattering foil double the thickness of the previous) that can be independently moved in or out of the beam path.

5.2.2 Contoured Scatterer

A better efficiency can be achieved by scattering more of the central protons to the outside and creating a flat profile (Figure 5.1b). The shape of a contoured scatterer, thick in the center and thin on the outside, has been optimized to do this [11,12]. Typically a flat scatterer (first scatterer) spreads the beam onto the contoured scatterer (second scatterer) that flattens out the profile at some distance. This type of system is called a double-scattering system. Mathematically, the lateral dose distribution $\Phi(r)$ created by a double-scattering system can be described as

$$\Phi(r) = \frac{1}{\left(2\pi\, z_{FS}\theta_{FS}\right)^2} \int_0^{2\pi} \int_0^{R} \exp\left[-\frac{\vec{r}'^2}{\left(z_{FS}\theta_{FS}\right)^2}\right] \cdot \frac{1}{\left(z_{CS}\theta_{CS}(\vec{r}')\right)^2} \cdot \exp\left[-\frac{(\vec{r}-\vec{r}')^2}{\left(z_{CS}\theta_{CS}(\vec{r}')\right)^2}\right] \vec{r}'\, d\vec{r}'\, d\phi \quad (5.2)$$

Here z_{FS} and z_{CS} are the distances from the first and second scatterer to the plane of interest (typically the isocenter plane); θ_{FS} is the characteristic scattering angle of the first flat scatterer which is constant, θ_{CS} is the characteristic scattering angle of the contoured scatterer which depends on radial position r'. R is the radius of the contoured scatterer, assuming all protons outside R are blocked. The radial coordinates r' and R of the contoured scatterer are projected from the first scatterer onto the plane of interest. The first exponential in the equation gives the fluence of the beam hitting the contoured scatterer at position r'. Without contoured scatterer, the dose distribution would be equal to this term and be a single-scattering system (Section 5.2.1). The second exponential describes the operation of the contoured scatterer. It describes the proportion of protons hitting the second scatterer at position r' and ending up in position r. This depends on the distance between r' and r and the angular spread added by a contoured scatterer, which is a function of the characteristic scattering angle (i.e., thickness) at location r'. By integrating over all positions r', the dose in point r is found. (Note that we have made some simplifications here like Gaussian scattering, rotational symmetry, thin scatterers, and small parallel entrance beam. More realistic properties will complicate the formulation, but can typically still be described analytically.) Given a desired flat dose distribution $\Phi(r)$, it is not possible to analytically solve Equation 5.2 and find the required shape of the contoured scatterer $\theta_{CS}(r')$. Instead, the shape of the contoured scatterer is determined using numerical methods. The scattering shape is described by a parameterized function like a cubic spline through a limited number of points whose thickness is optimized [12] or a modified cosine with four independent variables [11]. The variables of the contoured scatterer are optimized in combination with the scattering power of the first scatterer to obtain a dose distribution of desired size and acceptable uniformity. Efficiency can be made an additional objective in the optimization. Efficiencies of up to 45% can be obtained, significantly larger than in single scattering.

Protons hitting the center of the contoured scatterer lose more energy than those going through the thinner parts at the periphery. To avoid a concave distortion of the distal isodose plane, with the range increasing away from the beam axis, energy compensation is applied to the contoured scatterer. A high-Z scattering material (lead, brass) is combined with a low-Z compensation material (plastic). The thickness of the two materials is designed to provide constant energy loss, while maintaining the appropriate scattering power variation. The thickness of the high-Z material decreases with distance from the axis, while the thickness of the compensating low Z material increases. Figure 5.2 shows a schematic of an energy-compensated scatterer. Note that energy compensation will increase the total water-equivalent thickness of the scatterer because the compensation material increases the scattering power on the outside of the scatterer, which needs to be

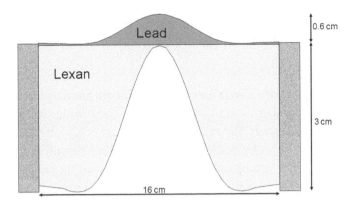

FIGURE 5.2 Schematic cross section of an energy-compensated contoured scatterer in the IBA universal nozzle.

compensated for by adding additional scattering material in the center. The energy of the protons entering the nozzle needs to be increased to achieve the same range in patient as with an uncompensated scatterer.

The dose distribution is sensitive to misalignments of the beam with respect to the second scatterer. A displacement of the beam increases the fluence on one side of the second scatterer and reduces it on the other side, causing a tilt in the dose profile at the isocenter. To keep the symmetry within clinical tolerance, the alignment of the beam typically needs to be better than ~1 mm. The large distance between the final steering magnet of the beamline and the second scatterer makes this difficult to achieve without a feedback mechanism. By monitoring the profile symmetry downstream of the second scatterer (e.g., using strip ionization chambers), the final steering magnets can be controlled maintaining a flat profile [13]. In addition to misalignment, the profile is sensitive to variations in beam size at the second scatterer level. If the beam hitting the second scatterer is too large, the dose profile at isocenter will have "horns"; if it is too small, the dose profile will be "domed." If the second scatterer is placed far enough downstream, meaning that its physical size is large enough, the width of the beam onto it will be dominated by the scattering of the first scatterer and variations in beam spot size at nozzle entrance (the beam's phase space) will not play a role.

5.2.3 Dual-Ring Scatterer

An alternative to the contoured scatterer is the dual-ring scatterer [14]. It consists of a central disk made of a high-Z material (lead, tungsten) and a surrounding ring of a lower Z material (aluminum, Lucite). The physical thickness of the outer ring is chosen such that the energy loss is equal (or close) to the energy loss in central disk. A first, flat scatterer spreads the beam onto the dual-ring scatterer. The central disk produces a Gaussian-like profile and the surrounding ring produces an annulus-shaped profile that combine to produce a uniform profile at isocenter. (Figure 5.1c) In the design, the projected scattering radius of the first scatterer and of the two dual-ring materials, together with the diameter of the central disk, are optimized to generate a flat dose profile of desired size [15]. Because of the binary nature of the dual ring, the dose distribution is not perfectly flat. A small cold spot is allowed at the level of the interface between the two materials. Like the contoured scatterer, the dual-ring system is sensitive to both beam alignment and beam size changes.

5.2.4 Occluding Rings

Most double-scattering systems use a contoured or dual-ring scatterer, but occluding rings combined with a flat second scatterer can also flatten the profile [16]. Instead of scattering the central protons outwards, they are blocked (Figure 5.1d). The "hole" created in the fluence distribution is filled in by scattering through a flat, second scatterer. Larger field sizes can be obtained by not just blocking the center but by adding one or more occluding rings. A flat dose distribution is achieved by simultaneously optimizing the ring diameters and scattering power of the flat scatterers. Because the protons are not redistributed but blocked, the efficiency of an occluding ring system drops significantly as the number of rings and maximum field size are increased. It is significantly lower than for the contoured scatterers. The energy loss is smaller though, because a relative thin second scatterer foil is needed to spread out the beam. The geometry of the occluding rings makes them just as sensitive to beam misalignment as the contoured scatterers.

5.3 RANGE MODULATION TECHNIQUES

When looking at the depth-dose curve of a monoenergetic proton beam, it is obvious that the Bragg peak is too sharp to cover a target of any reasonable size. By combining proton beams of decreasing energy, range modulation transforms the pristine Bragg peak into a uniform depth-dose region called the SOBP. Addition of Bragg peaks shifted in depth and weighted appropriately yields a uniform dose. Depending on the size of the target to be covered the extent of the uniform region can be adjusted by changing the number of added peaks (Figure 5.3). Several range modulation techniques are applied in proton therapy:

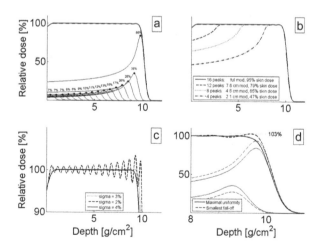

FIGURE 5.3 Creation of the SOBP with a range of 10 g/cm². Subplot A shows the weights of the pristine peaks when creating a full-modulation SOBP. In subplot B, SOBP of various modulation widths are shown. Subplot C shows the effect of a change in pristine peak energy spread on the SOBP. In subplot D, the alternative methods of optimization at the distal end of the SOBP are shown.

energy stacking, range modulator (RM) wheels, and ridge filters. After exploring some of the general principles of range modulation and SOBP construction, we discuss each of these techniques in detail.

5.3.1 Range Modulation Principles

Using a power-law approximation of proton stopping power, it is possible to analytically describe the Bragg peaks and calculate the optimal weights for an SOBP [17]. In reality, the shape of the Bragg peak is complex and depends on the energy spread and scattering properties of the delivery system. Measured Bragg curves are used, and the weights are determined numerically using simple optimization algorithm [18–20]. Mathematically, the problem can be described as follows:

$$\text{SOBP}(R,d) = \sum_{i=1}^{N} w_i \cdot \text{PP}(R_i,d) \tag{5.3}$$

where $\text{PP}(R_i,d)$ is the pristine peak depth-dose curve with range R_i, w_i is the relative contribution of peak i to the SOBP given by the ratio between the maximum dose in the peak and the SOBP plateau dose, N is the number of peaks summed, and $\text{SOBP}(R,d)$ is the resulting spread-out depth-dose curve with range R. The weights w_i are optimized minimizing the difference between $\text{SOBP}(R,d)$ and an ideal, presumably uniform, dose distribution. Figure 5.3a shows the optimization of an SOBP with a range of 10 g/cm^2 and full dose up to skin. The range shift between the pristine peaks is fixed at 6 mm, and 16 pristine peaks are required to obtain full skin dose. The peak weights drop exponentially from 85% of the plateau dose for the distal layer to 36% for the second layer, down to 7% for the most proximal layers. The increasing contribution of dose from the more distal layers reduces the dose required of more proximal layers to reach the plateau. The extent of the uniform dose region can be varied by changing the number of pristine peaks delivered. Figure 5.3b shows the SOBP when the distal 12, 8, or 4 layers are delivered, resulting in a 90%–90% modulation width of 7.6, 4.8, and 2.1 g/cm^2.

All peaks add dose to the skin. The distal peaks have a small skin dose due to the rising shape of the Bragg curve but have a large weight. Proximal peaks have a larger skin-to-peak dose ratio, but their contribution to the SOBP is less. These competing effects result in an almost linear increase of skin dose with modulation width. The skin dose also depends on the SOBP range, with the skin dose decreasing as the SOBP range increases.

Although the range of the individual pristine peaks can be made a variable in the SOBP optimization process, good results are obtained by keeping the range shift (pullback) between the peaks constant and setting it equal to the 80% width of the individual peaks [12]. A larger pullback between the peaks results in a ripple in the dose distribution, and a smaller pullback does not improve uniformity. In Figure 5.3c, the SOBP is shown when the width of the peak is changed from 6 to 5 or 7 mm, but the pullback is kept at 6 mm. The sharper peaks generate a strong ripple in the SOBP, while the wider peaks do not change the uniformity of the plateau and just deteriorate the sharpness of the corners. It is interesting to consider that we have not reoptimized the weights here. The sharper peaks generate a ripple but not a tilt in the plateau, suggesting the weights are still optimal. This effect has also been observed experimentally when changing the energy spread of the proton beam entering the nozzle [21].

A choice needs to be made at the distal end of the SOBP when optimizing the weights. One can either make the dose as flat as possible or make the distal falloff as sharp as possible (Figure 5.3d). If the extent of the desired uniform dose region is limited to a point halfway between the distal two peaks, the dose distribution can be made completely flat, but the distal dose falloff exhibits a fairly large shoulder. The falloff can be made sharper by increasing the weight of the distal peak while decreasing the weight of the second peak. The resulting dose distribution acquires a hot spot at the distal end of the uniform region and a cold spot more proximal. For the SOBP in Figure 5.3d, the 95%–20% distal falloff is decreased from 6.8 to 5.5 mm when limiting the hot spot to 103%. An additional consideration here is the biological effect. Several studies have shown that the radiological bioeffectiveness increases in the shoulder and distal falloff of the SOBP (see Chapter 22). Optimizing for the sharpest falloff might result in an even larger hot spot in biological effect. Whichever approach is taken, it should be mimicked in the treatment planning algorithm [22].

5.3.2 Energy Stacking

Changing the energy of the protons entering the nozzle is conceptually the easiest method of range modulation as it requires no dedicated nozzle elements. Between layers, the energy is changed at the accelerator level, either by changing the extracted synchrotron energy or the energy-selection system setting at the exit of the cyclotron (Chapter 3). By accurately controlling the number of protons delivered for a given energy, for example by terminating the beam once a preset number of monitor units in the reference ionization chamber is reached, the appropriate SOBP weighting is realized. A major advantage of this form of range modulation is that the protons do not need to interact with a range shifter inside the nozzle. Range-shifting material scatters the protons increasing the lateral penumbra, straggles the protons increasing the energy spread and distal falloff, and generates neutrons [23]. However, energy stacking with range-shifting upstream is not applied in current clinical scattering systems because of the relatively long energy switching time (Chapter 3), increasing the overall irradiation time, and the interplay between the energy switching and organ motion creating inhomogeneities in dose to the target (Chapter 18).

The energy switching time can be improved by applying range shifting inside the nozzle. The accelerator energy is set to the appropriate energy for the distal pristine peak only and kept constant during the irradiation. After delivering the dose for the distal layer, an absorber with a water-equivalent thickness equal to the desired pullback is inserted in the beam path and the next layer is delivered. The absorber thickness is increased sequentially while stepping through the energy layers. Low-Z materials, like plastic or water, are the preferred range-shifter materials because they provide the least amount of scattering per unit of range shift. Several types of variable range shifters have been implemented, including a variable water column that uses a movable piston to accurately adjust the amount of water in a cylinder, a binary set of plastic plates that can be moved independently into the beam path, and a double-wedge variable absorber [24]. Energy stacking inside the nozzle does not solve the problem of interplay effects with organ motion, which is one of the reasons why energy stacking in the nozzle is also not much used in clinical scattering systems.

5.3.3 Range Modulator Wheels

In his original paper on proton therapy, Wilson proposed an RM wheel as a method to spread the dose: "This can easily be accomplished by interposing a rotating wheel of variable thickness, corresponding to the tumor thickness, between the source and the patient" [25]. Koehler et al. [26] were the first to report the implementation of an RM wheel. The RM wheel has been and continues to be the method of range modulation in most clinical proton-scattering systems.

An RM wheel has steps of varying thickness, each step corresponding to a pristine peak in the SOBP. When the wheel rotates in the beam, the steps are sequentially irradiated. The thickness of a step determines the range shift of that pristine peak; the angular width of the step determines the number of protons hitting the step and thus the weight of the pristine peak. By progressively increasing the step thickness while making the angular width smaller, a flat SOBP can be constructed. Like the range shifters used in energy stacking, modulator wheels are preferentially made of low-Z materials to limit scattering. Plastics (Plexiglas, Lexan) are often used, but for wheels that need to provide large-range shifts and are mounted in nozzles where space is limited, carbon [27] and aluminum [28] have been applied.

Figure 5.4 shows an RM that was used at the now decommissioned Harvard Cyclotron Laboratory. Symmetrically cut, fan-shaped sheets of Plexiglas are stacked together

FIGURE 5.4 Range modulator wheel used at the Harvard Cyclotron Laboratory. (Picture courtesy of B. Gottschalk.)

repeating the modulation pattern four times per wheel revolution. Because the wheel is mounted close to the patient where the beam is already spread out, the dimension of the wheel is relatively large (~85 cm diameter). The size of the RM wheel can be made smaller by moving it further upstream. A smaller RM wheel allows higher rotational speeds, and for easier, even automatic, exchange of wheels. Figure 5.5 shows an example of such an "upstream propeller" from the Ion Beam Applications (IBA) Universal Nozzle [27]. It is located 2.8 m from isocenter, where the size of the beam is in the order of 8 mm full width at half maximum. Here, the RM pattern can be made small enough so that three separate 3 cm RM tracks are combined on one 34 cm diameter wheel. On this wheel, the range modulation pattern is not repeated. The high rotational speed (10 Hz) requires that the wheel is accurately balanced and counterweights are mounted on the outside of the wheel to compensate the uneven weight distribution in the tracks. (Note that the angular width of the steps in the outer track is constant. This track is used as a variable range shifter with the wheel in stationary position when energy stacking in uniform scanning (US) mode.)

The drawback of an "upstream propeller" is a larger dependence on energy. The steps of an RM wheel are optimized to give a flat dose distribution for a specific energy. When the incident energy, that is the range of SOBP, is changed, the weights are no longer optimal and the SOBP no longer flat. The main reason for the change in SOBP is the change in scattering power of the RM wheel steps with energy. (Change in energy spread and source-to-skin distance also play a role but are less important.) Combined with the large drift distance to isocenter, small changes in RM scattering power can cause large changes in fluence at isocenter. When the energy of the proton beam increases, the scattering in the RM steps will decrease. For the thinnest step, the scattering in the RM wheel will not contribute much to the total spread at isocenter, which is dominated by scattering in other nozzle elements (like the second scatterer). With increasing step thickness, RM scattering

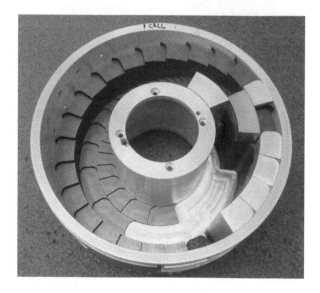

FIGURE 5.5   Range modulator wheel combining three range modulation tracks as installed in the IBA universal nozzle. (Note that the outer track is used in range shifting for US.)

will contribute progressively more to the spread at isocenter. A change in scattering power will affect the spread of the proximal layers more than the distal layers. Consequently, the increase in fluence on the beam axis associated with the decreased spread will be larger for the thicker than for the thinner steps. The increased relative weight of the proximal layers tilts the SOBP, which acquires a negative slope. The spread of the beam at isocenter is not only proportional to the scattering angle of the RM steps, but to the drift space between RM and iso. For large downstream propellers, the energy dependence is less because of the shorter drift space between the RM wheel and patient.

The way to avoid this problem is using a scatter-compensated RM wheel. By combining a low-Z (plastic, carbon, aluminum) and high-Z material (lead, tungsten), each step gives the appropriate range shift but with constant scattering power. Starting with only high-Z material for the thinnest step, the thickness of high-Z is progressively reduced while the thickness of low-Z material is increased for thicker steps. Like with the energy-compensated contoured scatterer, the scatter-compensated RM wheel has larger water-equivalent thickness than an uncompensated wheel.

Even with scatter compensation, an upstream RM wheel is usable within a limited range span. The ability of cyclotrons to accurately and quickly vary the beam current (by manipulating the ion source current) allows for a further extension of the range over which an RM wheel can be used. By changing the beam current as function of RM wheel position, the number of protons hitting a step can be adjusted and the intrinsic weight of the step, defined by its angular width, can be adjusted [29]. Figure 5.6 shows how an SOBP with a negative slope (~1.3%/cm) is made flat by a beam current modulation profile that decreases the beam current by a factor of two when moving from the thin to the thick RM steps. Beam current modulation requires a complex feedback control system that needs to be closely monitored during irradiation. Drifts in ion source output, feedback ionization chamber response, and RM wheel timings can cause large dose deviations [30].

Beam current modulation is a powerful tool that in theory can be used for other purposes than creating a flat SOBP at various energies. In one application, beam current modulation is used to vary the range of SOBP without changing the energy (range) at nozzle entrance, avoiding the time-consuming change in accelerator or energy-selection system energy [31]. Instead of always turning on the beam on the first step of the RM wheel, the starting position of the beam is delayed until thicker RM steps, decreasing the range of the distal layer in the SOBP. When not irradiating the first RM step(s), the weights of the pristine peaks are no longer optimal and the reduced-range SOBP is no longer flat. Beam current modulation is used to readjust the weights, creating a flat SOBP at lower range. It has been shown experimentally that with this technique the range of a 15 cm SOBP can be reduced from 9 to 4 cm. The major drawbacks of this approach are a significant increase in the SOBP distal falloff and a reduction in dose rate caused by the shortened gating window. Another application of beam current modulation adjusts the relative weight of the distal peaks of the SOBP to increase the distal falloff [32]. This technique can be used in the patch–match planning technique (Chapter 15) optimizing the distal falloff of the patch field to match the lateral falloff of the match field to reduce the dose variation in the overlap region.

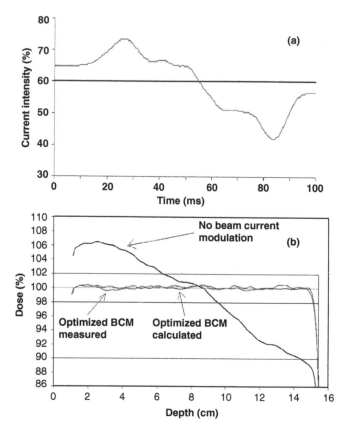

FIGURE 5.6   By varying the beam current (top) synchronized with the RM wheel position, a tilted SOBP can be made flat (bottom). (Reproduced from Lu 2006 with permission.)

We have seen that the range span an RM wheel can cover varies from a single range for an upstream wheel that is not scatter compensated and covers a large modulation, up to a significant range span for a scatter-compensated wheel with beam current modulation. How many different modulation widths can an RM wheel deliver? The thickest step determines the maximum shift in range between the most distal and proximal Bragg peaks and defines the maximum modulation width of a wheel. If the whole track is irradiated, the maximum modulation width is always delivered and a library of wheels is required to cover different target sizes (modulation widths). Alternatively, the beam can be turned on and off (*gated*) synchronized with the RM wheel position. Every wheel revolution the beam is turned on at the first step and off at the step corresponding to the most proximal layer needed to cover the target. By varying the number of irradiated steps, the modulation width can be varied, up to the maximum modulation width for which a wheel is designed when all steps are irradiated. Modulation width control using beam gating is applied in both cyclotron and synchrotron systems. In cyclotron systems, beam gating is performed by cutting the ion source current or dephasing the accelerating radio frequency [27], while in synchrotron systems, by turning off the extraction RF power [33]. Instead of switching on and off the beam, a block can be used to cover the part of the modulator wheel that

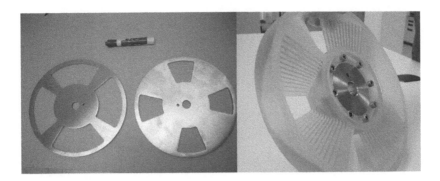

FIGURE 5.7 Two modulator wheel blocks (left) determining the modulation width for an RM wheel (right) in the IBA eyeline.

should not be irradiated. Figure 5.7 shows an example of two modulator blocks from an ocular scattering system [9]. The same blocks can be used on different RM wheels and each wheel–block combination corresponds to a specific range and modulation width. The limited overall range and modulation span required for eye treatments result in a manageable wheel and block library.

For every revolution of the RM wheel, the SOBP is delivered. If the rotational speed is high enough, the delivery can be considered quasi-instantaneous and the issues related to interplay effects with organ motion disappear. Typical rotational speeds of 6–10 Hz are larger than organ motion frequencies. Repetition of the range modulation pattern as seen earlier (Figures 5.4 and 5.7) increases the frequency of SOBP delivery even more. Synchrotron systems have an additional speed requirement, because the beam spill structure is generally not synchronized with the wheel rotation. The start and end of each spill occur at random wheel positions (within the gating window). Sufficiently complete SOBPs need to be delivered per beam spill to avoid unacceptable perturbations by partial SOBP delivered at the start and end of each spill [24]. The dose variation ε for a wheel with n repetitions spinning at f Hz in a beam with a spill time of τ s, is equal to $1/(nf\tau)$ [24]. Limiting the dose variation to 1% for a rotational speed of 10 Hz and a spill time of 2.5 s requires four repetitions.

Making the range modulation pattern smaller, by moving the wheel upstream and repeating the pattern, will start to affect the delivered dose distribution once the size of the steps becomes smaller than the beam spot [34]. When the beam is gated, the beam spot not only covers the gating step but spills onto neighboring steps. The steps thicker than the gating step receive some protons and thinner steps receive less than their full weight, resulting in a softening of the shoulder at the proximal end of the uniform region. Figure 5.8 shows the effect of "partial shining" for an RM wheel designed for a range of 15 g/cm² and full modulation width (30 steps). The track radius is 12.5 cm, and the beam spot is Gaussian with a sigma of 0.7 cm. The beam is gated on step 20, which has an arc length of 1.1 cm when there is no repetition. Repeating the pattern 1, 4, or 6 times reduces the extent of the 100% dose region by 0.2, 1.8, or 2.8 cm compared to an infinitesimal small beam. Note that proximal to the 90% dose level, the dose hardly varies and that the integral dose decreases.

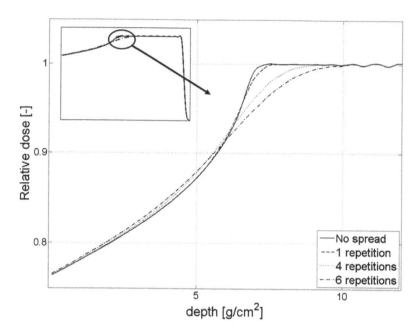

FIGURE 5.8 The effect of beam spilling onto multiple steps of the RM wheel when it is gated off. Proximal region of the SOBP is shown for an infinitesimal small beam; a beam with a 7-mm sigma; and 1, 4, or 6 repetitions of the modulation pattern on a 25 cm diameter wheel.

This can be explained by the fact that although the protons not delivered on the thinner steps are all delivered on the thicker steps, these steps contribute less dose because they are pulled back and deposit more dose in the RM wheel.

5.3.4 Ridge Filters

Ridge filters have been applied in proton therapy for at least as long as modulator wheels [35,36] and, although less wide spread than RM wheels, they are used in several clinical scattering systems [37–39]. The principle of the ridge filter is the same as the modulator wheel: the thickness of the ridge filter steps determines the pullback of the peaks and the width of the steps sets the weight of the peaks. Figure 5.9 shows a ridge filter designed for a modulation width of 6 cm. Protons hitting the tip of the ridge will form the most proximal peak in the SOBP; protons passing outside the ridge will form the distal peak. The thickness and width of steps in-between is optimized to provide a flat SOBP. The energy lost depends on the location where the protons hit the filter. To avoid the dependence of the SOBP shape on lateral position, the ridge is made small enough that the incident proton angular diffusion and scattering in the ridge filter smooth out any positional dependence. In the optimization of the ridge shape, the scattering can be taken into account [40,41]. The width of the ridges is typically 5 mm in systems that have the ridge filter downstream of the second scatterer. By arranging many bar-shaped ridges in parallel, a large beam area can be covered. Spiral ridge filters rely on the same principles as bar ridge filters, but the ridges are arranged in a circular pattern [42].

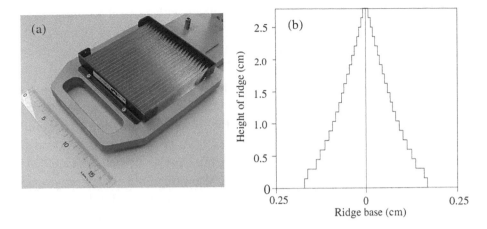

FIGURE 5.9   Bar ridge filter designed to modulate the beam to a 6 cm SOBP. On the right hand side the cross section of a ridge is shown. (Reproduced from Akagi 2003 with permission.)

Manufacturing constraints need to be taken into account when designing ridge filters. The total height needs to be limited to avoid ridges that are too sharp to be accurately machined. This limits the maximum modulation width that can be achieved. Traditionally, ridge filters are made of high-Z materials, like brass, that can be machined accurately. The large stopping power of brass limits the height and gradient of the ridges, but its scattering power has a negative effect on the lateral penumbra. With the improvement of machining technology, low-Z materials like aluminum [40] and even plastics have been used.

As we have seen, energy stacking delivers the pristine peaks sequentially in time. Spinning RM wheels deliver them sequentially as well, but at the frequency of the wheel rotation, delivering the SOBP quasi-instantaneously. The ridge filter delivers all pristine peaks at the same time making SOBP delivery truly instantaneous. This makes ridge filters suitable for range modulation in systems in which the beam itself has a time structure, like synchrotron-based scattering systems with a pulsed beam of low duty cycle or US systems in which the beam is scanned over a ridge filter [43,44].

The main drawback of ridge filters is the fact that they can only be used for a single modulation width. A possible solution is tilting the bar ridge filter [44 and references within]. Tilting the ridge filter by $\theta°$ in the plane of the beam axis and the long axis of the ridges increases the thickness of all ridge steps by $1/\cos(\theta)$. Unlike the gated modulator wheel that increases modulation by adding more pristine peaks, the tilted ridge filter keeps the same number of peaks but increases the pullback between the peaks. The number of steps in the ridge needs to be large enough to avoid ripple in the SOBP at maximum tilt. Feasibility has been shown experimentally for a ridge filter, increasing modulation width from 10 to 14.5 cm when tilted at 45° [44].

The miniature ridge filter is designed to be used in combination with energy stacking [45]. It spreads the monoenergetic Bragg peak into a wider peak; energy stacking combines multiple range-shifted wide peaks into an SOBP. In this hybrid delivery technique, the number of energy layers is reduced compared to pure energy stacking. The delivery efficiency is increased, while interplay effects with respect to organ motion are reduced. The modulation width can be varied, although with a worse resolution than can be obtained with pure energy stacking or gated RM wheels. On the downside, the broadening of the peaks deteriorates the distal falloff of the SOBP. In another method of combining ridge filter and energy stacking, the miniature ridge filter is used to combine the distal pristine peaks into a small-modulation SOBP width, and energy stacking is used to vary the modulation by adding monoenergetic pristine peaks proximally (N. Schreuder and G. Mathot, private communication, December 2010). Application of these forms of range modulation has only been reported in US systems, but could in principle be applied in scattering systems as well. A miniature ridge filter can also extend the energy range over which a ridge filter can be applied [46]. If the beam energy is decreased below the energy for which a ridge filter is designed, the width of the peaks will become sharper (specifically, for synchrotron systems that have a small energy spread at low energies), the pullback between peaks is no longer optimal, and a ripple appears in the SOBP. To counter the sharpening of the peaks, a miniature ripple filter increases the energy spread of the protons and widens the peaks. A library of few ridge filters and mini ridge filters can cover the complete clinical range span.

5.4 CONFORMING TECHNIQUES

The second scatterer spreads the beam to a uniform lateral distribution, and the RM wheel spreads the dose to a uniform depth-dose distribution. Clearly, something is needed to conform the dose to the target. Like in conventional external radiotherapy, a block, called aperture in proton therapy, shapes the dose laterally. Alternatively, a multileaf collimator (MLC) can be used. Unlike conventional radiotherapy, the sharp distal falloff of the proton beam allows to conform the beam to the distal end of the target. A range compensator is designed to perform this function.

5.4.1 Aperture

The shape of the aperture is defined by the shape of the target projected along the beam axis, with added margins accounting for penumbra width and setup uncertainty. Because of the large geometric source size of a proton-scattering system, it is important to bring the aperture close to the patient. Large air gaps inflate the lateral penumbra undesirably. Aperture and range compensator are mounted together on a snout. The snout travels along the beam axis and brings the aperture close to patient skin. Based on the snout position, the virtual source-to-axis distance (SAD) of the scattering system and the projection of the target onto the isocenter plane, a simple backprojection gives the physical shape of the aperture.

The primary physical consideration in selecting aperture material is stopping power. High-Z materials are the obvious choice as they stop the protons in the shortest physical distance. From a practical point of view, we want a material that can be manufactured easily and cheaply. The two materials that are commonly used are brass and cerrobend. Brass apertures are cut with a milling machine, and cerroband apertures are poured into a mold. In a proposed method to reduce cost, the inner shaped part of the aperture is fabricated from brass and the outer ring from cheaper stainless steel [47].

It should be pointed out that such an aperture will *completely* stop the beam. Due to the finite range of protons, the chance is zero that a primary proton will traverse the complete aperture. Unlike X-ray therapy, leakage of primary protons through the aperture does not exist. This does not mean that all protons hitting the aperture will be absorbed. Protons hitting the upstream face close to the aperture opening can escape through the inner surface of the opening; protons hitting the inner surface can scatter out of the aperture through the inner surface or escape through the downstream face. These slit-scattered protons perturb the dose distribution specifically at shallow depth and near the field boundary [48–50]. Cutting the aperture edge divergent, parallel to the divergent proton beam, is a method to reduce the slit-scattering dose perturbations [51].

The aperture and snout are the largest contributors of neutrons to the patient [52] (see also Chapter 7). Not only do they completely absorb high-energy protons in material with large cross section for neutron production, but also they do it close to the patient. Several approaches to reduce neutrons have been investigated. Changing aperture material would not reduce the neutron dose dramatically. Stopping 235 MeV protons in nickel instead of brass reduces neutron dose by about 15% [53]. A bigger reduction can be obtained by using a precollimator, limiting the number of protons hitting the aperture [53–55]. A related radiation safety issue is the activation of the apertures [56,57]. Activation levels are low enough to avoid special procedures in the daily handling of the apertures by staff. For disposal of the apertures, activation levels need to be below the background. Irradiation of brass yields several isotopes with relevant activation and half lives (^{58}Co–$T_{1/2}$ = 71 d, ^{57}Co–$T_{1/2}$ = 271 d). In typical clinical practice, apertures are stored for about 2 weeks when activation has reached background levels.

5.4.2 Multileaf Collimator

In conventional radiotherapy, an MLC has become the standard field-shaping device. Perhaps surprisingly, MLCs are currently not much applied in proton therapy. The vast majority of passive-scattering proton treatments today use custom-made apertures. The lack of MLCs in proton therapy has several reasons. Unlike conventional radiotherapy where the MLC is used in IMRT to create highly conformal dose distributions, an MLC in a passive-scattering beam is inferior to PBS when delivering intensity-modulated proton therapy (IMPT) plans. The MLC blocks the majority of protons, making it very inefficient compared to PBS. As a result, the irradiation time, neutron dose, and activation are negatively impacted. Even if an MLC replaces the aperture block in 3D conformal proton therapy, the field-specific range compensators still needs to be manufactured and installed in-between treatment fields, reducing the efficiency gains obtained when eliminating the custom aperture. Finally, the versatility of an MLC compared with a custom-made aperture installed in snout comes at the expense of increased physical size. An MLC is bulkier, limiting the ability to bring it close to the patient to reduce air gap and penumbra. Still, the MLC has received some attention as beam-shaping device in passive-scattering systems, and several centers have successfully implemented an MLC in their clinical system [10,58].

The main issues when commissioning an MLC for proton therapy are not different from conventional radiotherapy: penumbra and conformity, leakage, neutron production, and activation. In collaboration with Varian, University of Philadelphia developed a proton MLC to be used with their IBA universal nozzle [58]. It consists of two banks of 50 tungsten leafs. The leaf width is 4.4 mm, and leaf height is 6 cm. The parallel and abutting sides have steps of 0.45 and 0.30 mm, respectively, to limit leakage. For a 30 cm snout position, the maximum projected opening is 25×18 cm^2. Measurements show that the scalloping effect of the leaves on the dose distribution is clinically acceptable. Multiple scattering of the protons in the target smear out the leaf effects with depth. Although tungsten has a higher neutron yield than brass, maximum neutron dose levels for a 230 MeV proton beam are about 20% lower for the MLC compared with a 6.5 cm brass aperture. This is presumably because of the self-shielding effect of the additional tungsten on top of the 3.5 cm required to stop a 230 MeV proton beam. (About 6 cm brass is required.) This self-shielding also limits the measured activation on the downstream face to levels far below regulatory limits. In addition to the MLC, a "range compensator loader" has been installed so that it can hold up to two compensators and hence eliminates the need to go into the room between fields. Massachusetts General Hospital had implemented a mini-MLC in its radiosurgery beamline [10]. Although this MLC was designed for stereotactic treatments on a linear accelerator, it behaves well in the proton beam. For a typical field, agreement between MLC-shaped and custom aperture-shaped field is within 1.5 mm or 2%. The leakage dose is below the measurement threshold (0.3%). Neutron dose for the MLC is 1.5–1.8 times higher than for the brass aperture. This increased neutron dose, when using an MLC instead of a brass aperture, is contrary to the results obtained with the University of Pennsylvania MLC but agrees with a study by the University of Indiana [59].

Alternative applications of the MLC in proton therapy have been considered. Tayama [54] has described the MLC as a precollimator to a custom-made aperture. By reducing the proton flux onto the final aperture, the neutron dose to the patient is significantly reduced. Full-blown IMPT using MLCs is not practicable as discussed above. Still the flexibility of the MLC can be used to improve dose conformity compared to a single-aperture field. In systems that apply energy-stacking to create the SOBP, the MLC opening can be optimized per energy layer [60]. Because each point of the target needs to get the dose contributions of all "upstream" layers to achieve full dose, the collimator opening for a layer can never be larger than the opening of the previous layers. The collimator can only be progressively closed while stepping from the distal to the proximal layers. Improved proximal conformity will only be achieved in convex targets.

5.4.3 Range Compensator

The range compensator conforms the dose to the distal end of the target. Figure 5.10 shows a schematic representation of the application of range compensation. The water-equivalent depth of the distal end of the target varies with lateral position. It is a function of the shape of both external body contour and target, as well as of the composition of the tissue in between. The range compensator is designed to remove the depth variation by adding more absorbing material in areas where the depth is small and less where it is large. For example, the energy loss in the high-density structure in the figure is compensated for by removing more material from the compensator. It is obvious that each treatment field has a unique

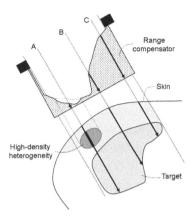

FIGURE 5.10 Schematic representation of the application of a range compensator that compensates for the shape of the body entrance, the distal target shape, and inhomogeneities.

compensator. Designing the range compensator is an important part of the treatment planning process. Most algorithms use a ray tracing algorithm to determine the water-equivalent depth of selected points on the distal surface of the target [61–63]. The deepest point will determine the required proton range. For each of the other points, the difference in depth with the deepest point determines the required pullback. The matrix of pullbacks is divided by the compensator's relative stopping power to obtain physical thickness. To accurately model the effect of the real compensator, the milling pattern described below can be applied to the compensator model before the final dose calculation.

Correct compensation is only achieved if the range compensator is exactly aligned to the geometry and heterogeneities for which it compensates. Misalignment causes the dose to fall short, resulting in underdosage of the target, and/or to overshoot, resulting in dose to normal tissue distal to the target. Smearing is a geometrical operation applied to the range compensator to account for uncertainties (Chapter 15).

Ideally, the range compensator provides pullback with as little scattering as possible. Scattering decreases the compensator's conforming ability and generates undesirable cold and hot spots inside the target. Low-Z, high-density materials give the least scattering per energy loss. Lucite and wax are the two materials most commonly used. Proponents of wax favor it because it is easier (faster) to mill, has lower cost, and can be recycled. Lucite users value its transparency, which allows for visual screening for air pockets and validation of the isoheight lines by placing it on a paper printout during quality assurance. A standard milling machine mills the desired profile into a blank compensator. Resolution needs to be weighed against speed when selecting drill bit size and spacing. The smallest drill provides the best lateral resolution in conforming the dose, but it takes the longest time to drill. Because of scattering, there is a lower limit below which reduction of the drill bit size does not significantly improve conformity anymore. A typical drilling pattern will use a 5-mm diameter drill bit and a spacing of 5 mm. For such a pattern, machining times (for Lucite) range from 10 min for a small brain lesion, 45 min for a prostate, and up to 4 h for a 2 l sarcoma field.

The tapering (angle) of the drill bit affects the magnitude of the dose perturbation created by large-range compensator gradients [64]. A gradient in the range compensator will scatter more protons from the thick part to the thin part than vice versa, creating a cold spot beyond the thicker part and a hot spot beyond the thinner part. If the gradient is exactly parallel to the beam axis, the magnitude of the dose perturbation is maximal. When irradiating a 4 cm deep, 0.5 cm diameter hole in acrylic with 160 MeV protons, ±20% dose perturbations are observed 1 cm downstream. By applying a tapered drill bit, the gradient becomes less steep, reducing the dose perturbation to ±15%/ ±5% for a 1.5°/3° tapering, respectively. A 3° tapering is common because it reduces the scattering to an acceptable level without compromising the lateral resolution too much in most clinical situations.

An interesting approach to limit the compensator-scattering perturbations is the bimaterial range compensator [65]. By combining the low-Z compensating material with a high-Z scattering material, a compensator can not only be designed with the desired range compensation pattern but also with constant scattering power. (This is similar to the approach taken with the scatter-compensated RM wheel described above.) The drawback of such an approach, besides added complexity in design, fabrication, and quality assurance, is that it increases the overall compensator thickness. As a result, a bimaterial compensator requires larger proton range and has worse lateral penumbra because of increased scattering.

In recent years, 3D printing technology has become popular in many different applications. Ju and colleagues [66] have shown that a range compensator printed on a commercial 3D printer has the same mechanical and dosimetric properties as a range compensator fabricated on a milling machine. 3D printing can potentially be a faster and cheaper method of range compensator fabrication, eliminating the need for the large and expensive milling machines.

5.5 SCATTERING SYSTEMS

In the previous sections, we have discussed the individual elements of passive-scattering systems. In this section, we will focus on integrated scattering systems. A larger number of different scattering systems exist. This makes a general description of scattering systems complicated. The approach taken here is to discuss a few representative scattering systems in more or less detail. We just describe the configuration of these systems and refer to Gottschalk [12] for a detailed discussion of the design methods and tools. First we will discuss several "general-purpose" scattering systems. These large-field double-scattering systems can be seen as the proton equivalent to the standard X-ray linear accelerators. They have been designed to treat a large variety of target sizes and depths, and are typically gantry mounted. Next we will discuss a single-scattering system that has been designed to treat a specific target, the eye. This system is mounted at the end of a fixed beamline. It should be emphasized that the choice of scattering systems discussed is purely based on familiarity to the author and availability in the literature, not on presumed superiority over other systems.

5.5.1 Large-Field Double-Scattering Systems

Figure 5.11 gives an overview of four commercially available, turn-key proton therapy systems that use double scattering on a gantry [27,32,33,67]. Two of the systems have a synchrotron and the other two a cyclotron. The Mevion accelerator stands out as it is a small superconducting cyclotron that is mounted on a 190° gantry, just upstream of the treatment nozzle. The four nozzles are similar in design, all applying range-modulator wheels to create the SOBP and a contoured second scatterer to flatten the lateral profile. The Optivus system has the RM wheel downstream of the second scatterer, while the Hitachi, IBA, and Mevion systems have an upstream RM.

Figure 5.12 shows the layout of the IBA universal nozzle in detail. It is called "universal," because it permits irradiations not only in double-scattering mode, but also in single-scattering, uniform-scanning, and PBS mode. The Hitachi nozzle, which has only the double-scattering delivery mode, is shown in Figure 5.13. While both the IBA and Hitachi systems vary the energy at nozzle entrance, the Mevion system (shown in Figure 5.14) has

Manufacturer	Installations	Accelerator	Range modulation	Lateral spreading
Optivus	Loma Linda University Medical Center (1991)	Synchrotron (250 MeV)	RM wheel downstream	Contoured scatterer
Ion Beam Applications	Massachusetts General Hospital, Boston (2001) Wanjie Hospital, Wanjie, China (2004) University of Florida, Jacksonville (2006) National Cancer Center, Ilsan, Korea (2006) University of Pennsylvania, Philadelphia (2009) Institut Curie, Paris, France (2010) HUPTI, Hampton (2010) WPE, Essen, Germany (2013) UPTD, Dresden, Germany (2014)	Isochronous cyclotron (235 MeV)	RM wheel upstream	Contoured scatterer
Hitachi	M.D. Anderson Cancer Center (2006)	Synchrotron (250 MeV)	RM wheel upstream	Contoured scatterer
Mevion Medical Systems	Barnes Jewish Hospital, St. Louis (2013) Robert Wood Johnson, New Brunswick (2015) Ackerman Cancer Center, Jacksonville (2015) Orlando Health PTC, Orlando (2016) Sideman Cancer Center, Cleveland (2016)	Superconducting synchrocyclotron (250 MeV)	RM wheel upstream	Contoured scatterer

FIGURE 5.11 Commercially available turn-key proton therapy systems with double scattering.

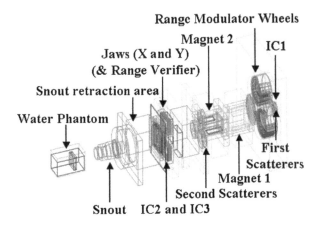

FIGURE 5.12 Schematic layout of the IBA universal nozzle. (Reproduced from Paganetti 2004 with permission.)

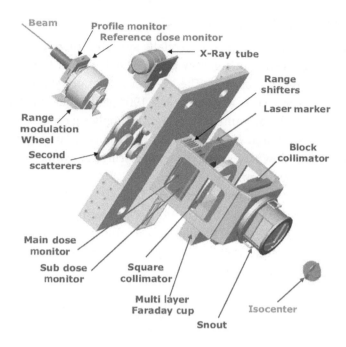

FIGURE 5.13   Schematic layout of the Hitachi large-field scattering nozzle. (Reproduced from Smith 2009 with permission.)

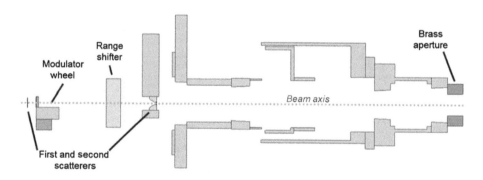

FIGURE 5.14   Schematic layout of the Mevion double-scattering nozzle. (Reproduced from Hill et al., 2013, with permission.)

a fixed energy at nozzle entrance, set by the gantry-mounted cyclotron at 250 MeV, and requires additional range shifting in the nozzle supplied by the range shifter between the RM wheel and second scatterer.

The IBA nozzle uses two contoured scatterers to spread the beam to a uniform field diameter of 24 cm for ranges from 4.6 to 23.9 g/cm² in water. The amount of scattering material (first and second scatterer) needed to scatter the beam to this field size reduces the maximum cyclotron range (~34 g/cm²) significantly. To treat deeper-seated targets, a third, thinner second scatterer is added that allows treatments up to 28.4 g/cm² depth, but with a limited field diameter of up to 14 cm. All three contoured scatterers are energy compensated, combining Lexan and lead. They are mounted together on a large wheel that is located 178 cm from the isocenter. Before start of irradiation, the appropriate scatterer is rotated into the beam path. The variable collimators are set to block the beam outside the aperture opening. Unlike the IBA nozzle that always spreads the beam to the maximum field diameter, both the Hitachi and Mevion nozzles have contoured scatterers optimized for different field sizes. By scattering the beam less for smaller targets, the efficiency of the system is improved, increasing maximum dose rate and range, and reducing production of secondary neutrons in the nozzle. In addition, the lateral penumbra is improved for small fields. The Mevion system uses 3 second scatterers to generate two field sizes of 14 and 25 cm diameter over the complete clinical range span [68]. The Hitachi system provides three field sizes: 25×25, 18×18, and 10×10 cm^2. For each field size, 3 second scatterers cover the whole range span, resulting in a total of 9 second scatterers. The maximum range for each of the field sizes is 25.0, 28.5, and 32.4 g/cm^2, respectively. It is interesting to see that the maximum range for the 10×10 cm^2 field size is 4.0 g/cm^2 larger than the maximum range for the equivalent field size ($\sqrt{2} \cdot 10 = 14$ cm diameter) in the IBA nozzle. This is equal to the difference in maximum accelerator range. (The synchrotron energy of 250 MeV corresponds to a range in water of 38 g/cm^2, the cyclotron energy of 235 MeV to a range of 34 g/cm^2.) The thickness of the scattering material required is similar due to the similar position of the scattering elements inside both nozzles.

The Hitachi system extracts eight energies from the synchrotron. An RM wheel has been designed for each of the 24 energy and field-size combinations. The steps of the RM wheel are scatter compensated, resulting in a constant scattering power over the steps. As range-shifting material, plastic or aluminum is used; for scattering compensation, tungsten is used. The wheel not only acts as a modulator but also as the first scatterer in the double-scattering system. Additional tungsten is added to provide, combined with the second scatterer, the desired uniform field size at isocenter. The fine range adjustment in the Hitachi nozzle is done with a variable range shifter located downstream of the second scatterer. In the IBA nozzle, the range adjustment is done by changing the energy of the protons entering the nozzle (using the energy-selection system at the exit of the cyclotron). As a result, the RM wheels are not used at a single energy, but for a range of energies. As the beam energy increases, the scattering power of both RM wheel and second scatterer decreases, resulting in a nonflat lateral profile at the isocenter. This can be compensated for by adding additional scattering material to the first scatterer. In the IBA nozzle, adjustment of scattering is done by the *fixed scatterer*, a binary set of lead foils that can be inserted independently into the beam path and is located upstream of the RM wheel. RM wheel and first scatterer combine to form a first scatterer with variable scattering power. The RM steps are scatter compensated, limiting the effect of change in scattering power on the pristine peak weights and increasing the range of energies over which a flat depth-dose curve is generated. Still, an RM wheel track can only be used for a range span of 0.4 g/cm^2 for the lowest ranges and up to 2.0 g/cm^2 for the highest ranges. By applying beam current modulation, adjusting the beam current as function of modulator wheel position, the number of modulator wheels can be limited. A total of five modulator wheel tracks cover the complete energy range. Unlike the Hitachi system where the RM wheels are loaded manually, the IBA nozzle has an automated system. Three RM tracks are combined on a single wheel (Figure 5.5) and three wheels are mounted on a large wheel, whose position determines which track is in the beam path. (The four remaining tracks are used in single scattering and US.) The Mevion RM-loading system is also automatic, with its 14 RM wheels mounted on a tray that can be moved to position the desired wheel in the beam path. Like the IBA system, the Mevion system applies beam current modulation varying the cyclotron current as function of RM wheel position.

Because the cyclotron generates a continuous beam, the speed requirements are not very stringent for the IBA and Mevion RM wheels. The IBA wheels spin at 600 rpm and the modulation pattern is not repeated. The Hitachi RM wheels spin at 400 rpm and the modulation pattern is repeated six times per revolution, washing out the effects of the beam pulse structure on the SOBP shape. In all systems, the beam is turned on and off synchronized with the wheel rotation, allowing for variation of the modulation width. The repetition of the pattern on the Hitachi wheel causes more beam spilling over the steps and a softer shoulder on the proximal side of the SOBP compared to the single pattern of the IBA track.

All three nozzles have ionization chambers at the entrance and exit of the nozzle to monitor the beam properties and terminate the beam once the prescribed dose has been reached. A snout that can move along the beam axis holds the aperture and range compensator and collimates the beam outside the aperture. Both the Hitachi and IBA systems have a library of three snouts. The Mevion system has two sizes of "applicators" to hold the aperture and range compensator that have been designed minimizing weight and can be easily switched by the therapists.

The dosimetric properties of the delivery system depend mostly on the design of the scattering system. Figure 5.15 shows the virtual SAD of the IBA nozzle, determined by backprojecting the 50% field width in air, as measured in several planes along the beam axis. For all ranges, the source position falls between the RM (270 cm) and second scatterer

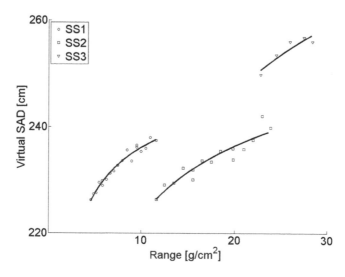

FIGURE 5.15 Virtual SAD as function of range in patient for the double-scattering options of the IBA universal nozzle.

(178 cm) as expected. For options that use the same second scatterer, the SAD increases continuously with range. Given all that has been said before, it is not that difficult to explain the observed behavior. The scattering power of the second scatterer decreases as the range increases. To compensate, the scattering power of the first scatterer is increased either by adding additional scattering material to the RM wheel (between options) or by increasing the fixed scatterer thickness (within an option). (An option is a combination of an RM wheel track and second scatterer.) The increased scattering power of the first scatterer pulls the source towards it, increasing the SAD. The small-field second scatterer is thinner than the two large-field second scatterers, resulting in a source position closer to the first scatterer and a larger SAD. Figure 5.16 shows the source size as a function of range. The source size is determined by measuring the lateral penumbra in air at several distances from a square aperture. The measured 80%–20% penumbra is backprojected to a nominal source position of 230 cm. The source size of the large field options (options 1–7) is significantly smaller than the small field option (option 8). The additional scattering to spread the beam to a larger diameter increases the angular confusion of the beam. Within the large field options, the source size decreases continuously with range.

The fact that the dosimetric properties are mostly determined by the scattering inside the nozzle elements, and only to a small extent by the phase space of the beam entering the nozzle, raises the question whether different installations of the same scattering nozzle have the same dosimetric properties. Comparing the SAD, effective source size, and pristine peak shape of the IBA universal nozzle installed in various gantries at the University of Florida and the University of Pennsylvania, Slopsema et al. [69] show that variations between these nozzles are minimal. A single set of beam data (golden beam data) could potentially be used to commissioning treatment planning for any installation of this nozzle.

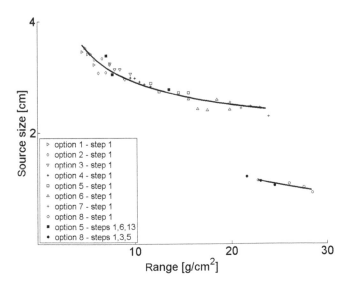

FIGURE 5.16 Effective source size as function of range in patient for the double-scattering options in the IBA universal nozzle.

5.5.2 Single-Scattering System for Eye Treatments

Ocular tumors have been successfully treated with protons for decades. Most eye lines spread the beam with single scattering [2,4,5,7,8,9] or an occluding beam stopper [3,6], although contoured second scatterers [33,70] have been applied as well. Figure 5.17 shows the IBA eye line, whose design is based on the eye line at the Centre de Protonthérapie Orsay in France. The beam is brought into the room at a fixed energy of 105 MeV. After passing through a beam monitor, the protons hit the RM wheel spinning at 1,200 rpm. The RM wheels are not scatter compensated and can be used within a very small range span (0.2–0.4 g/cm²). Eleven wheels are required to cover ranges from 0.3 to 3.4 g/cm². The wheels are designed for full modulation, and blocks are used to vary the modulation width (Figure 5.7). The blocks are made of brass and are 1.2 cm thick. Both RM wheel and block are loaded manually. Next, the beam passes through a variable range-shifter and scatterer system. For a given range and field size, the appropriate Lucite range-shifter plates and lead scattering foils are selected. A brass collimator blocks the majority of protons. The neutron shield downstream of the collimator is intended to absorb most of the neutrons generated in this collimation. The virtual source position of this system is located between the RM wheel and variable range-shifter system, where most of the scattering takes place. The resulting SAD is about 150 cm. Because the distance between the RM wheel and variable range shifter is small, the angular diffusion of the beam and thus the effective source size are small. Given the large SAD and small source size, the penumbra of this system is very sharp. The 80%–20% penumbra in air at 7 cm from the final aperture is 1.2 mm.

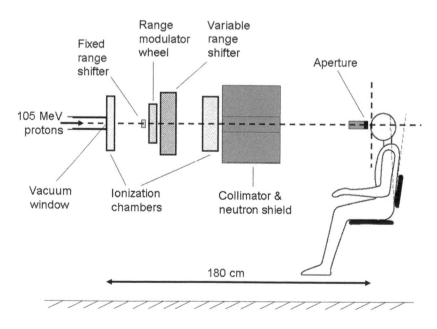

FIGURE 5.17 Schematic representation of the IBA eye line that applies single scattering to spread the beam.

5.6 OUTLOOK

Proton-scattering systems have some major drawbacks compared to PBS systems (Chapter 6). When protons interact with the scattering and range modulation material in the nozzle, they lose energy, decreasing the maximum penetration depth of the beam, and gain angular diffusion, increasing the lateral penumbra. Nuclear interactions will cause activation and create unwanted secondary particles like neutrons. Apertures and range compensators need to be made for every treatment field requiring an expensive, labor-intensive, and time-consuming fabrication and quality assurance process. In a state-of-the-art proton scanning system, no field-specific hardware is required and no interaction of the proton beam with nozzle material occurs. Depending on size and shape of target, a PBS system also allows for better dose conformity reducing integral nontarget dose. When optimizing the PBS spot map over all fields simultaneously (IMPT), highly conformal dose distribution can be obtained. And the absence of field-specific hardware allows for rapid replanning, potentially on a Computed Tomography (CT) scan taken just before the day's treatment.

Despite these long-known advantages of PBS, passive scattering was the main delivery technique during the first two-and-a-half decades of "modern" proton therapy [71]. Up until 2009, the majority of centers only applied passive scattering (or US, a delivery technique that uses magnets to spread the beam but still has an aperture and range compensator) (Figure 5.18). Between 2009 and 2015, most centers had PBS in addition to a passive-scattering system. Starting in 2014, the majority of new centers opening their doors

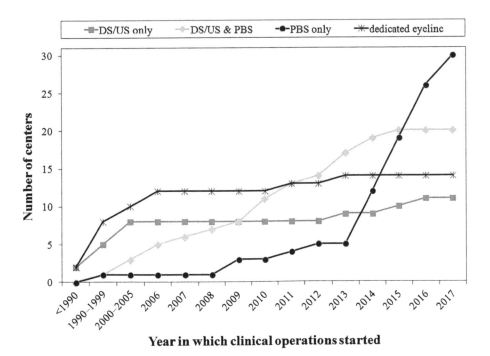

Year in which clinical operations started

FIGURE 5.18 Number of clinical proton centers over the years, sorted by delivery technique (DS: double scattering, US: uniform scanning, PBS: pencil beam scanning).

only had PBS, leading to a rapid increase in the number of PBS centers. In 2017, half the centers in operation still used a form of passive scattering or US, while the other half was purely PBS. These numbers ignore the proton systems dedicated to eye treatments that all apply a form of passive scattering and have delivered a substantial proportion of the proton treatments for decades.

For many years, the spread of PBS was limited by technological challenges. The accurate control of both position and dose of the pencil beam as well as the rapid switching between energy layers required control processes and hardware that took more than a decade to develop into reliable commercial systems. In contrast, the hardware elements of the passive-scattering systems had in essence already been developed in the early days of proton therapy in the 1960s and 1970s, and could readily be implemented into the first commercial proton therapy systems. Looking at the trends in Figure 5.18, it is clear that the technological challenges of PBS have mostly been overcome. Most, if not all, of the centers starting clinical operations in the coming years are likely to opt for PBS, taking advantage of its benefits over the passive-scattering systems. Still many of the centers currently in operation use passive scattering and are likely to keep doing so in the coming years.

REFERENCES

1. Preston WM, Koehler AM. The effects of scattering on small proton beams. [Online]. 1968 [cited 2010 December 10]; Available from: http://huhepl.harvard.edu/˜gottschalk/BGDocs. zip
2. Montelius A, Blomquist E, Naeser P, Brahme A, Carlsson J, Carlsson AC et al. The narrow proton beam therapy unit at the the Svedberg laboratory in Uppsala. *Acta Oncol.* 1991;30(6):739–45.
3. Bonnett DE, Kacperek A, Sheen MA, Goodall R, Saxton TE. The 62 MeV proton beam for the treatment of ocular melanoma at Clatterbridge. *Br J Radiol.* 1993 Oct;66(790):907–14.
4. Moyers MF, Galindo RA, Yonemoto LT, Loredo L, Friedrichsen EJ, Kirby MA et al. Treatment of macular degeneration with proton beams. *Med Phys.* 1999 May;26(5):777–82.
5. Newhauser WD, Burns J, Smith AR. Dosimetry for ocular proton beam therapy at the Harvard Cyclotron Laboratory based on the ICRU Report 59. *Med Phys.* 2002 Sep;29(9):1953–61.
6. Cirrone GAP, Cuttone G, Lojacono PA, Lo Nigro S, Mongelli V, Patti IV, et al. A 62-MeV proton beam for the treatment of ocular melanoma at Laboratori Nazionali del Sud-INFN. *IEEE Trans Nucl Sci.* 2004;51(3):860–5.
7. Hérault J, Iborra N, Serrano B, Chauvel P. Monte Carlo simulation of a protontherapy platform devoted to ocular melanoma. *Med Phys.* 2005 Apr;32(4):910–9.
8. Michalec B, Swakoń J, Sowa U, Ptaszkiewicz M, Cywicka-Jakiel T, Olko P. Proton radiotherapy facility for ocular tumors at the IFJ PAN in Kraków Poland. *Appl Radiat Isot.* 2010 Apr–May;68(4–5):738–42.
9. Slopsema RL, Mamalui M, Zhao T, Yeung D, Malyapa R, Li Z. Dosimetric properties of a proton beamline dedicated to the treatment of ocular disease. *Med Phys.* 2014 Jan;41(1):011707.
10. Daartz J, Bangert M, Bussière MR, Engelsman M, Kooy HM. Characterization of a mini-multileaf collimator in a proton beamline. *Med Phys.* 2009 May;36(5):1886–94.
11. Grusell E, Montelius A, Brahme A, Rikner G, Russell K. A general solution to charged particle beam flattening using an optimized dual-scattering-foil technique, with application to proton therapy beams. *Phys Med Biol.* 2005 Mar 7;50(5):755–67.
12. Gottschalk B. Passive beam spreading in proton radiation therapy. [Online]. 2004 October 1 [cited 2010 December 3]; Available from: http://huhepl.harvard.edu/~gottschalk

13. Nishio T, Kataoka S, Tachibana M, Matsumura K, Uzawa N, Saito H et al. Development of a simple control system for uniform proton dose distribution in a dual-ring double scattering method. *Phys Med Biol.* 2006 Mar 7;51(5):1249–60.
14. Takada Y. Dual-ring double scattering method for proton beam spreading. *Jpn J Appl Phys.* 1994;33:353–9.
15. Takada Y. Optimum solution of dual-ring double-scattering system for an incident beam with given phase space for proton beam spreading. *Nucl Instrum Methods Phys Res A.* 2002;485(3):255–76.
16. Koehler AM, Schneider RJ, Sisterson JM. Flattening of proton dose distributions for large-field radiotherapy. *Med Phys.* 1977 Jul–Aug;4(4):297–301.
17. Bortfeld T, Schlegel W. An analytical approximation of depth-dose distributions for therapeutic proton beams. *Phys Med Biol.* 1996 Aug;41(8):1331–9.
18. Koehler AM, Preston WM. Protons in radiation therapy. Comparative dose distributions for protons, photons, and electrons. *Radiology.* 1972 Jul;104(1):191–5.
19. Petti PL, Lyman JT, Castro JR. Design of beam-modulating devices for charged-particle therapy. *Med Phys.* 1991 May–Jun;18(3):513–8.
20. Gardey KU, Oelfke U, Lam GK. Range modulation in proton therapy—an optimization technique for clinical and experimental applications. *Phys Med Biol.* 1999 Jun;44(6):N81–8.
21. Hsi WC, Moyers MF, Nichiporov D, Anferov V, Wolanski M, Allgower CE et al. Energy spectrum control for modulated proton beams. *Med Phys.* 2009 Jun;36(6):2297–308.
22. Engelsman M, Lu HM, Herrup D, Bussiere M, Kooy HM. Commissioning a passive-scattering proton therapy nozzle for accurate SOBP delivery. *Med Phys.* 2009 Jun;36(6):2172–80.
23. Harvey MC, Polf JC, Smith AR, Mohan R. Feasibility studies of a passive scatter proton therapy nozzle without a range modulator wheel. *Med Phys.* 2008 Jun;35(6):2243–52.
24. Chu WT, Ludewigt BA, Renner TR. Instrumentation for treatment of cancer using proton and light-ion beams. *Rev. Scu. Instrum.* 1993;64:2055–121.
25. Wilson RR. Radiological use of fast protons. *Radiology* 1946 Nov;47(5):487–91.
26. Koehler AM, Schneider RJ, Sisterson JM. Range modulators for protons and heavy ions. *Nucl Instrum Methods* 1975:131:437–40.
27. Paganetti H, Jiang H, Lee SY, Kooy HM. Accurate Monte Carlo simulations for nozzle design, commissioning and quality assurance for a proton radiation therapy facility. *Med Phys.* 2004 Jul;31(7):2107–18.
28. Polf JC, Harvey MC, Titt U, Newhauser WD, Smith AR. Initial beam size study for passive scatter proton therapy. I. Monte Carlo verification. *Med Phys.* 2007 Nov;34(11):4213–8.
29. Lu HM, Kooy H. Optimization of current modulation function for proton spread-out Bragg peak fields. *Med Phys.* 2006 May;33(5):1281–7.
30. Lu HM, Brett R, Engelsman M, Slopsema R, Kooy H, Flanz J. Sensitivities in the production of spread-out Bragg peak dose distributions by passive scattering with beam current modulation. *Med Phys.* 2007 Oct;34(10):3844–53.
31. Sánchez-Parcerisa D, Pourbaix JC, Ainsley CG, Dolney D, Carabe A. Fast range switching of passively scattered proton beams using a modulation wheel and dynamic beam current modulation. *Phys Med Biol.* 2014 Apr 7;59(7):N19–26.
32. Hill PM, Klein EE, Bloch C. Optimizing field patching in passively scattered proton therapy with the use of beam current modulation. *Phys Med Biol.* 2013 Aug 21;58(16):5527–39.
33. Smith A, Gillin M, Bues M et al. The M. D. Anderson proton therapy system. *Med Phys.* 2009 Sep;36(9):4068–83.
34. Li Y, Zhang X, Lii M, et al. Incorporating partial shining effects in proton pencil-beam dose calculation. *Phys Med Biol.* 2008 Feb 7;53(3):605–16.
35. Larsson B. Pre-therapeutic physical experiments with high energy protons. *Br J Radiol.* 1961 Mar;34:143–51.

36. Lomanov M. The Bragg curve transformation into a prescribed depth dose distribution. *Med. Radiol.* 1975;11:64–69.
37. Inada T, Hayakawa Y, Tada J, Takada Y, Maruhashi A. Characteristics of proton beams after field shaping at PMRC. *Med Biol Eng Comput.* 1993 Jul;31 Suppl:S44–8.
38. Kostjuchenko V, Nichiporov D, Luckjashin V. A compact ridge filter for spread out Bragg peak production in pulsed proton clinical beams. *Med Phys.* 2001 Jul;28(7):1427–30.
39. Ando K, Furusawa Y, Suzuki M, Nojima K, Majima H, Koike S et al. Relative biological effectiveness of the 235 MeV proton beams at the National cancer center hospital east. *J Radiat Res (Tokyo).* 2001 Mar;42(1):79–89.
40. Akagi T, Higashi A, Tsugami H, Sakamoto H, Masuda Y, Hishikawa Y. Ridge filter design for proton therapy at Hyogo Ion Beam Medical Center. *Phys Me d Biol.* 2003 Nov 21;48(22):N301–12.
41. Fujimoto R, Takayanagi T, Fujitaka S. Design of a ridge filter structure based on the analysis of dose distributions. *Phys Med Biol.* 2009 Jul 7;54(13):N273–82.
42. Khoroshkov VS, Breev VM, Zolotov VA, Luk'iashin VE, Shimchuk GG. Spiral comb filter. *Med Radiol (Mosk).* 1987 Aug;32(8):76–80.
43. Akagi T, Higashi A, Tsugami H, Sakamoto H, Masuda Y, Hishikawa Y. Ridge filter design for proton therapy at Hyogo Ion Beam Medical Center. *Phys Med Biol.* 2003 Nov 21;48(22):N301–12.
44. Nakagawa T, Yoda K. A method for achieving variable widths of the spread-out Bragg peak using a ridge filter. *Med Phys.* 2000 Apr;27(4):712–5.
45. Fujitaka S, Takayanagi T, Fujimoto R, Fujii Y, Nishiuchi H, Ebina F et al. Reduction of the number of stacking layers in proton uniform scanning. *Phys Med Biol.* 2009 May 21;54(10):3101–11.
46. Takada Y, Kobayashi Y, Yasuoka K, Terunuma T. A miniature ripple filter for filtering a ripple found in the distal part of a proton SOBP. *Nucl. Instrum. Methods Phys. Res. A.* 2004 Ajn 29;524:366–73.
47. Chen H, Matysiak W, Flampouri S, Slopsema R, Li Z. Dosimetric evaluation of hybrid brass/stainless-steel apertures for proton therapy. *Phys Med Biol.* 2014 Sep 7;59(17):5043–60.
48. van Luijk P, van t' Veld AA, Zelle HD, Schippers JM. Collimator scatter and 2D dosimetry in small proton beams. *Phys Med Biol.* 2001 Mar;46(3):653–70.
49. Titt U, Zheng Y, Vassiliev ON, Newhauser WD. Monte Carlo investigation of collimator scatter of proton-therapy beams produced using the passive scattering method. *Phys Med Biol.* 2008 Jan 21;53(2):487–504.
50. Vidal M, De Marzi L, Szymanowski H, Guinement L, Nauraye C, Hierso E, Freud N, Ferrand R, François P, Sarrut D. An empirical model for calculation of the collimator contamination dose in therapeutic proton beams. *Phys Med Biol.* 2016 Feb 21;61(4):1532–45.
51. Zhao T, Cai B, Sun B, Grantham K, Mutic S, Klein E. Use of diverging apertures to minimize the edge scatter in passive scattering proton therapy. *J Appl Clin Med Phys.* 2015 Sep 8;16(5):5675.
52. Pérez-Andújar A, Newhauser WD, Deluca PM. Neutron production from beam-modifying devices in a modern double scattering proton therapy beam delivery system. *Phys Med Biol.* 2009 Feb 21;54(4):993–1008.
53. Brenner DJ, Elliston CD, Hall EJ, Paganetti H. Reduction of the secondary neutron dose in passively scattered proton radiotherapy, using an optimized pre-collimator/collimator. *Phys Med Biol.* 2009 Oct 21;54(20):6065–78.
54. Tayama R, Fujita Y, Tadokoro M, Fujimaki H, Sakae T, Terunuma T. Measurement of neutron dose distribution for a passive scattering nozzle at the Proton Medical Research Center (PMRC). *Nucl Instrum Methods Phys Res A.* 2006;564:532–6.
55. Taddei PJ, Fontenot JD, Zheng Y, Mirkovic D, Lee AK, Titt U et al. Reducing stray radiation dose to patients receiving passively scattered proton radiotherapy for prostate cancer. *Phys Med Biol.* 2008 Apr 21;53(8):2131–47.

56. Faßbender M, Shubin YN, Lunev VP, Qaim SM. Experimental studies and nuclear model calculations on the formation of radioactive products in interactions of medium energy protons with copper, zinc and brass: Estimation of collimator activation in proton therapy facilities. *Appl Radiat Isot.* 1997;9:1221–30.

57. Sisterson JM. Selected radiation safety issues at proton therapy facilities. 12th Biennial Topical Meeting of the Radiation Protection and Shielding Division of the American Nuclear Society (Santa Fe, NM) [Online]. 2002 [cited 2010 Dec 5]; Available from: URL: http://gray. mgh.harvard.edu/content/dmdocuments/Janet2002.pdf

58. Diffenderfer ES, Ainsley CG, Kirk ML, McDonough JE, Maughan RL. Comparison of secondary neutron dose in proton therapy resulting from the use of a tungsten alloy MLC or a brass collimator system. *Med Phys.* 2011 Nov;38(11):6248–56.

59. Moskvin V, Cheng CW, Das IJ. Pitfalls of tungsten multileaf collimator in proton beam therapy. *Med Phys.* 2011 Dec;38(12):6395–406.

60. Kanai T, Kawachi K, Matsuzawa H, Inada T. Broad beam three-dimensional irradiation for proton radiotherapy. *Med Phys.* 1983 May–Jun;10(3):344–6.

61. Goitein M. Compensation for inhomogeneities in charged particle radiotherapy using computed tomography. *Int J Radiat Oncol Biol Phys.* 1978 May–Jun;4(5–6):499–508.

62. Urie M, Goitein M, Wagner M. Compensating for heterogeneities in proton radiation therapy. *Phys Med Biol.* 1984 May;29(5):553–66.

63. Petti PL. New compensator design options for charged-particle radiotherapy. *Phys Med Biol.* 1997 Jul;42(7):1289–300.

64. Wagner MS. Automated range compensation for proton therapy. *Med Phys.* 1982 Sep–Oct;9(5):749–52.

65. Takada Y, Himukai T, Takizawa K, Terashita Y, Kamimura S, Matsuda H et al. The basic study of a bi-material range compensator for improving dose uniformity for proton therapy. *Phys Med Biol.* 2008 Oct 7;53(19):5555–69.

66. Ju SG, Kim MK, Hong CS, Kim JS, Han Y, Choi DH, Shin D, Lee SB. New technique for developing a proton range compensator with use of a 3-dimensional printer. *Int J Radiat Oncol Biol Phys.* 2014 Feb 1;88(2):453–8.

67. Moyers MF. Proton Therapy. In: Van Dyk J, ed. *The Modern Technology of Radiation Oncology.* 1st ed. Madison, WI: Medical Physics Publishing; 1999. p. 823–69.

68. Chen KL. Neutron exposure from electron linear accelerators and a proton accelerator: Measurements and simulations. PhD Thesis, University of Missouri, Columbia, OH, 2011.

69. Slopsema RL, Lin L, Flampouri S, Yeung D, Li Z, McDonough JE, Palta J. Development of a golden beam data set for the commissioning of a proton double-scattering system in a pencil-beam dose calculation algorithm. *Med Phys.* 2014 Sep;41(9):091710.

70. Titt U, Suzuki K, Li Y, Sahoo N, Gillin MT, Zhu XR. Technical note: Dosimetric characteristics of the ocular beam line and commissioning data for an ocular proton therapy planning system at the Proton Therapy Center Houston. *Med Phys.* 2017 Dec;44(12):6661–71.

71. Jermann M. Hadron Therapy Patient Statistics. [Online]. March 2010 [cited 2010 Dec 10]; Available from: URL: http://ptcog.web.psi.ch/Archive/Patientenzahlen-updateMar2010.pdf

CHAPTER **6**

Particle Beam Scanning*

Jacob Flanz

*Massachusetts General Hospital and
Harvard Medical School*

CONTENTS

---

* Color figures available online at www.crcpress.com/9781138626508

6.1 INTRODUCTION

The dimensions of a clinical target are typically different than the dimensions of an unmodified particle beam. The beam extracted and transported from a typical accelerator will have dimensions on the order of millimeters and will have a narrow energy spectrum, which results in a narrow spread of ranges (also millimeters) in a target. Therefore, that beam has to be spread out in three dimensions (3D) to match the target volume. In this chapter, the method of spreading called beam scanning will be described. Beam scanning is quite a general technique, and while it has acquired many acronyms such as particle beam scanning (PBS), intensity-modulated proton therapy (IMPT), and spot scanning (SS), coming from specific implementations of the technology and limitations of accelerator beam properties; these acronyms only serve to minimize the power and generality of this beam spreading approach.

Beam scanning can be defined as the act of moving a charged particle beam of particular properties and perhaps changing one or more of the properties of that beam for the purpose of spreading the dose deposited by a beam throughout the target volume. Some examples (nonexhaustive) of these properties include position, size, range, and intensity. They are all adjusted in such a way as to deposit the appropriate dose at the correct location and time and minimize the dose outside of the target. Physical equipment in the system is used to control these properties. For example, the beam position and size on target can be controlled using magnetic fields or other mechanical motion techniques. When a beam penetrates the target, it delivers dose to the intercepting volume along the beam trajectory until it stops. The goal of this beam delivery is to deliver dose according to a prescription. This prescription provides a map of the dose that is necessary to deliver at each region in the target. The beam parameters can, and should, be able to change on a location-by-location basis, for example two locations can have different ranges or beam sizes.

In the transverse dimension, there are a variety of ways of moving the beam across the target. Some of these methods include

- Scanning by mechanical motions

  - Physically moving the target with respect to a fixed beam position

- Mechanically moving a bending magnet to change the position of the transported beam

- Using an adjustable collimator to effectively adjust the location/edges of the beam

- Scanning by magnetic field variation to bend the beam trajectory

  - Scanning an unmodified beam (sometimes called a pencil beam)

  - Scanning a slightly scattered beam, so that the beam scanned on the target is a larger dimension. This is called "wobbling," or a version of this can be called uniform scanning

- Aforesaid combinations

  - Two-dimensional "Ribbon" scanning of a beam wide in the dimension perpendicular to the direction of motion, with the beam extent adjusted by a variable collimator

  - Scanning the beam magnetically in one dimension and moving the target mechanically in the other dimension

- Other combinations are possible

In fact, the first implementation of a scanning beam was demonstrated in Japan using a novel system, including a range modulator wheel to modulate the beam range while scanning the beam transversely with magnetic dipoles [1]. As implied earlier, the beam size used in scanning can be varied. An unmodified beam (nonscattered or tightly focused) beam is sometimes called a "Pencil Beam", although sometimes this term is used for a beam that has a dimension on the order of a few millimeters. It is possible to obtain a raw, unmodified (unscattered and uncollimated) beam which is on the order of several millimeters or even a centimeter, in which case, one can consider using the term "crayon beam," owing to the larger size. In any case, it is more important to define the terms and understand the regime being considered than to rely on unclear acronyms. The longest (period of time) implementations of the scanning modality have been ongoing at research institutions such as the Paul Scherrer Institute (PSI) [2] and until it closed for clinical application, at the Gesellschaft fur Schwerionenforschung (GSI) [3]. Implementation outside the research environment began with commercial and academic hospital collaborations at MD Anderson and Massachusetts General Hospital (MGH) in 2008. Currently, many particle therapy facilities are using beam scanning as the primary beam delivery modality.

6.1.1 General Description of Scanning

Beam scanning is the process of spreading the beam over the target volume by moving the beam throughout the target. Pictorially, Figure 6.1 describes the scanning process. It is important to understand that true beam scanning could involve the variation of many parameters of the beam while it is being scanned. A beam at position A can be characterized by a variety of parameters including the vector transverse coordinate $\mathbf{X}_A$, $(x_A y_A)$ its energy

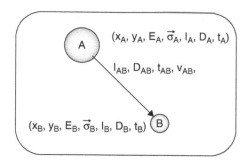

FIGURE 6.1 Parameters useful in describing the scanning process. A beam A with the indicated parameters is modified to the beam B, with modified parameters.

E_A, which determines its depth (the third dimension), the current I_A, the beam size (a vector since it may be different in x and y) $\boldsymbol{\sigma}_A$ the time it is in a given location, e.g. τ_A and others. The beam deposits a dose D_A in the voxel around A. After that dose is deposited, having stayed at A for a time $\mathbf{t_A}$, the beam is moved to location B. The time it takes to move from location A to B is $\mathbf{t_{AB}}$. The beam current during that movement is $\mathbf{I_{AB}}$, which could be a function of position (or time). The velocity that the beam moves from position A to B is $\mathbf{v_{AB}} = (\mathbf{x_B} - \mathbf{x_A})/\mathbf{t_{AB}}$, and the average current change rate between A and B is $\mathbf{dI/dt} = (\mathbf{I_B} - \mathbf{I_A})/\mathbf{t_{AB}}$ (which could be time-dependent functions). The charge delivered (related to the dose absorbed) is determined by the integral of the beam current and the time spent in a given location. In this way, we have defined all the terms that are necessary in the delivery of beam scanning.

6.1.2 Limits of Scanning Implementations

The earlier description appears intrinsically discrete or digital. There are usually, two interpretations of this description related to one extreme or the other. This has unfortunately caused confusion in the specification of the system and in the terminology. In the limit when $\mathbf{t_A} = 0$, the beam does not stop at a particular location, and its motion is characterized by $\mathbf{v_{AB}}$. Also, in such a case, the concept of $\mathbf{I_A}$ is undefined, but rather the quantity $\mathbf{I_{AB}}$ is defined. This extreme has been called "continuous or raster or line scanning." It is the equivalent beam motion as that used in the old cathode ray tube televisions. In the case when this extreme limit is **not** valid, it has been called "SS" or even sometimes a form of "raster scanning" (different than the previously identified raster scanning). However, the distinction between these extremes in the case when $\mathbf{t_A}$ is sufficiently small and/or $\mathbf{x_B} - \mathbf{x_A} \ll \boldsymbol{\sigma}_A$ is not relevant. Whether or not the $\mathbf{t_A} = 0$ limit is reached, is an implementation decision or based upon physically realizable quantities. Thus, the level of discreteness vs. continuousness will all lead to the measurements and control to be used. These may all be determined by the capabilities of the hardware and software.

In the case when the beam fully stops at a given point, it may be necessary to measure $\mathbf{D_A}$, if the integrated current is not predictable; however, even when the beam does not "stop" at a given point, it is necessary to measure some form of the quantity $\mathbf{D_A}$ at places on a 3D grid in the target volume to compare with the desired dose distribution. A simplified way of looking at this is to identify two extremes of implementation, which we will call "Dose-Driven"

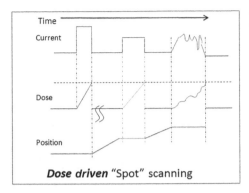
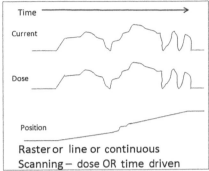

FIGURE 6.2   Graphs of beam current, position, and dose delivered in the dose-driven and time-driven techniques during a one-line scan.

(DD) and "Time-Driven" (TD) scanning, as depicted in Figure 6.2. Note that so-called "SS" can be implemented in a dose-driven or time-driven mode, but the figure to the left (in Figure 6.2) depicts the dose-driven mode of SS. This figure examines three scenarios, represented by three pulses of current. The first one is a larger current than the second one, but both are well regulated with a constant value when on. The Dose increases linearly until the desired dose is reached. The second pulse takes longer to deliver the desired dose. The third pulse shows a not well regulated beam current. Even in this case, although the rate of dose accumulation is nonlinear, dose can be measured and the current is stopped when the dose has been reached. After the dose has been reached, the beam is moved (bottom graph). This DD mode allows for both well regulated and unregulated beam current from the system. Even the case in which the beam is not stopped during dose accumulation, (moving continuously—sometimes called continuous, line or raster scanning) the process can be approximated by a series of small moves and can be implemented both in the time- and dose-driven modes. Also note that the implementation chosen normally depends upon the ability of the accelerator to deliver beams of well-controlled intensity and whether the beam is continuous or pulsed. In the TD mode, instead of the dose being measured, the time the beam is on, is measured. One assumes that the current is well regulated and the total charge delivered is the integral of the regulated dose rate over time. The current can be varied, or the position speed can be varied (see page 183). Alternatively the dose can be measured on the fly and the same continuous scanning beam would then be DD.

The control of these quantities, whether open loop, or closed loop is a subject that must be discussed. In an ideal world, all quantities would be perfect and no control is needed. In another idealistic world, all parameters would be controlled in a closed-loop mode and verified with very fast response times. In the physical world, wherein the measurements made require finite periods of times, open-loop delivery and subsequent correction or feed forward process is needed. In the case when $t_A \neq 0$, such as in the left side of Figure 6.2, this just means that the dose is delivered in quantized doses, and there might be a correction for the next quantum if the previous quantum was not as expected. This allows the beam delivery to be possible even with a fluctuating beam current (assuming its maximum value is limited). The beam is stopped (or moved) when the dose is reached independent of its time dependence. However, if $t_A = 0$, then the position of the beam at any given small time interval is not known (due to the Heisenberg uncertainty principle), owing to the time it takes to make the beam measurement, and therefore, an analysis and dose distribution correction after the fact may be necessary. Of course, this assumes that any fluctuations are of a magnitude that does not lead to an unsafe dose delivery. One could, for example, stop the beam before the desired dose is reached and resume with a lower current if the beam current fluctuation at the higher level, together with the finite measurement time, could lead to an overdose. This possibility would be reduced with a lower beam current. In any case, one will have to identify the timescale for measurement and control for each parameter and/or device and define the "control" strategy to obtain the appropriate dose distribution.

6.1.3 Safety

The highest importance must be given to a safe and accurate delivery of dose to the patient. The scanning system must ensure that the correct dose fidelity is achieved, which includes giving the right dose to the right place, and not giving the wrong dose to anyplace. This could be a voxel-by-voxel consideration or a more global view such as an online gamma index. The method to use to ensure this represents severe constraints on the implementation of the scanning system. This will impact the control strategy mentioned earlier, given the time constants built into the system. This includes, based on the time for measurements, limitations of beam current (i.e. dose rate). Also, a safety plan should contain the definitions of the required types of redundancy and sensors to be used, and therefore, helps define the interfaces with which the scanning system will interact both for measurements and beam control.

6.2 PARAMETERS THAT AFFECT THE BEAM AND DOSE DELIVERY

The goal of radiotherapy implementation is to deliver the prescribed dose to the target with the prescribed dose distribution. In general, the dose fidelity (conformality to the prescription) is given by properly controlling the beam position, size, range, current, and gradient at any given time or integrated over any given time interval. Each of the devices affecting or measuring the beam properties can be controlled and/or measured in a finite time period or continuously depending upon the measuring device (e.g. power supply current vs. beam position), which should become part of the specification. The scanning system will control equipment parameters, read back instrumentation parameters, and make decisions about the settings of the equipment parameters based upon the instrumentation parameters.

It is useful to separate the discussion of beam parameters into static and motion regimes. The former is the unperturbed property of the beam when it is not in motion and the latter are effects arising from the fact of motion.

6.2.1 Static Beam Parameters

6.2.1.1 Depth-Dose Distribution

The depth-dose distribution will be determined by the superposition of the Bragg peaks used in the delivery of dose volume. Unlike a spread-out Bragg peak (SOBP) as used in the scattering technique (see Chapter 5), the delivery of a scanned beam can be general. An example of a

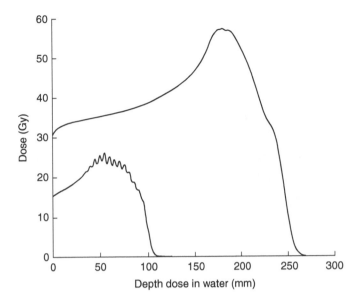

FIGURE 6.3 Two nonuniform, but realistic longitudinal dose distributions. The one at shallower range (lower curve) suffers from the narrower Bragg peak and is less smooth when the Bragg peak spacing is too far apart.

nonuniform depth (single field)-dose distribution required for a prostate carcinoma is shown in the upper curve of Figure 6.3. However, without any degraders (material) in the beam path, the width and distal falloff of a Bragg peak for a low-energy (shallow) beam is small, and creating a smooth dose distribution by superposition of these depth doses is difficult, as shown in the lower curve for a base of skull ependymoma of Figure 6.3 [4]. A range shifter or ridge filter is generally used for this situation by increasing the beam range spread. (This will also increase the beam transverse spreading.) Thus even the scanning beam may need to be modified. It should be added that while scanning is considered beam delivery without modifying devices, use of a range compensator can provide advantages in some situations and the use of a collimator can provide certain advantages sometimes, as well. The compensator would allow one beam range to be used to irradiation an irregular shape distally and thus reduce the number of energy levels required for the dose delivery.

6.2.1.2 Transverse-Dose Distribution and Modulation
The transverse-dose distribution is given by a superposition of the transverse raw beam profiles. Much depends on the shape of that raw beam. A beam is a collection of particles. The distribution of particles in a beam is typically statistical and, owing to the physics of the source of ions and the accelerator, the two-dimensional (position and angle) phase space, in each plane, results in a distribution as shown in the upper left of Figure 6.4. An integrated projection along either the position (vertical) labeled "y" or angular (horizontal) labeled "φ_y" axes result in a Gaussian distribution as shown in the darker curve in the projections. The more irregular curve is the distribution with limited statistics. The Gaussian has a special shape with magic properties. A superposition of Gaussians results in a flat

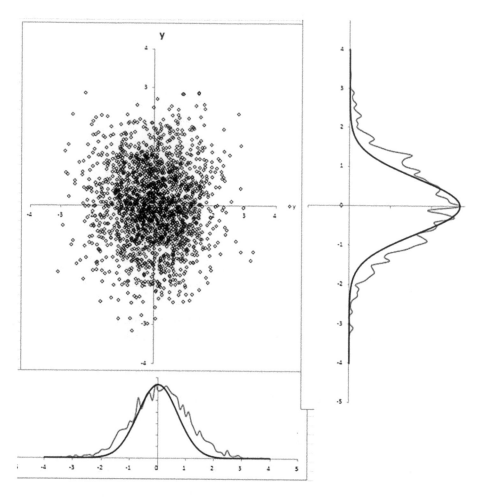

FIGURE 6.4 Statistical coordinates of protons in a beam defined by position and angle. The side graphs are histogram projections of the two dimensions, showing the Gaussian distribution.

distribution over some range of distance between the superimposed Gaussians. There is also considerable tolerance to the relative positions of these beams before their spacing affects the overall distribution. On the other hand, an asymmetric beam, or the one that has a transverse falloff with a different shape than a Gaussian will lead to much tighter beam-positioning tolerances.

The sharpness of the Gaussian beam falloff will determine the sharpest dose falloff possible for a clinical beam. Figure 6.5 shows some of the features of a Gaussian beam which is characterized by the equation

$$y = e^{-\frac{1}{2}\left(\frac{x}{\sigma}\right)^2}$$

(6.1)

The beam sigma (σ) is the single parameter characterizing the Gaussian shape. The full-width-half-maximum is given by 2.35 * σ. In particular, if the Gaussian is unmodified (no collimator to produce sharper edges), and one wants to separate the target (left dashed box in Figure 6.5)

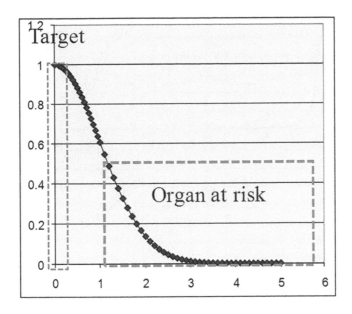

FIGURE 6.5   A graph of the intensity versus distance of a Gaussian beam. An example of regions of target and organ at risk are shown to relate the spacing to the Gaussian parameters.

from a critical structure in such a way that the critical structure does not receive more than, say, 50% of the target dose (right, lower dashed box in Figure 6.5), the distance between the target and this critical structure has to be at least 0.85σ. This sets the scale of the beam size needed for different treatment sites. For example, if the target and organ at risk are separated by 5 mm, then the beam sigma should be smaller than approximately 6 mm. Alternatively one might modify the beam, or apply an aperture to sharpen an edge. Apertures are generally thought of as an inconvenience and expense when confused with the type used in scattering systems. However, one can conceive of a beam optics solution that allows a sharper edge to be created upstream of the target, and the beam on target is imaged from this aperture so that the beam that is scanned has a sharper edge at the target, first proposed by Flanz at a meeting of the Particle Therapy CoOperative Group (PTCOG) in 2002 and subsequently implemented by Pedroni [5]. Of course, if the beam traverses too much material on the way to the target, the multiple scattering in this material will broaden the beam again.

The clinical utility of such a pencil beam is yet to be determined. Some initial treatment planning studies have been conducted to evaluate the dose-volume histograms and dose to critical structures as a function of the size of the beam [6]. Examples of these comparisons are shown in Figure 6.6. Note the relative values of the dose-volume histograms for various beam sizes and for comparison purposes, also for a highly conformal photon plan. Since some of these types of plans are a bit subjective, in that different planners may achieve different results, one has to be careful, however, there is an interesting indication of the relative importance of the beam sigma in this particular case.

If the goal, however, is to achieve the smallest transverse beam penumbra, one can only do almost as good as achieved with a collimator if the unmodified beam sigma is very small. Very small beam sigma in a scanning beam system is expensive to achieve, commands very

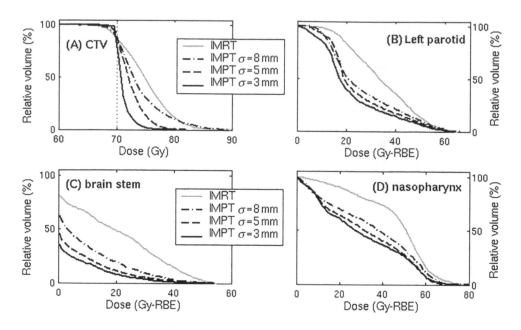

FIGURE 6.6 Dose-volume histograms for one treatment site planned with different size Gaussian beams and intensity-modulated photons. (a) CTV, (b) left parotid, (c) brainstem, (d) nasopharynx. The clinical impact of these differences is not yet known, but the relative quantitative importance of these parameters can be seen. (Courtesy of Alex Trofimov.)

tight tolerances on beam delivery, and can add significantly to the beam delivery time for several reasons. The use of a collimator, on the other hand, invokes several perceived disadvantages such as air gap minimization requirements, bulky equipment, therapist involvement installing and removing the equipment, and the time involved for manual work. The use of a multileaf collimator (MLC) as normally envisioned to improve the workflow and reduce the time for manual handling has been regarded as undesirable owing to the large size of a typical MLC and air gap requirements. However, the use of a scanning beam does not require a full-size MLC. First, the scanned beam motion can stop on the collimator, and therefore, the collimator only needs to have a transverse extent of a few beam sigmas, instead of the full length of the maximum delivered field size. Also, it is possible to conceive of a collimator bar that is synchronized with the beam motion and dynamically collimates the beam where needed. Such a system has been designed by the Iowa group [7], and an equivalent system is included in the new scanning system used by MeVIon. Figure 6.7a shows a drawing of the device that uses what are called "beam trimmers," not collimators. It has the capability of inclusion of a range shifter. Figure 6.7b shows the substantial reduction in penumbra (note the upper part of the beam spot compared to the lower part) that can be obtained in a Monte Carlo simulation.

If one is scanning the unmodified Gaussian beam, the superposition of beams affects the falloff at the edge of the field, as shown in Figure 6.8a. The addition of Gaussians results in a larger than optimal edge falloff distance. However, in much the same way that one sharpens the edge of an SOBP by emphasizing the Bragg peak at the distal edge, it is possible to achieve similar results by modifying the distribution of the number of protons across the

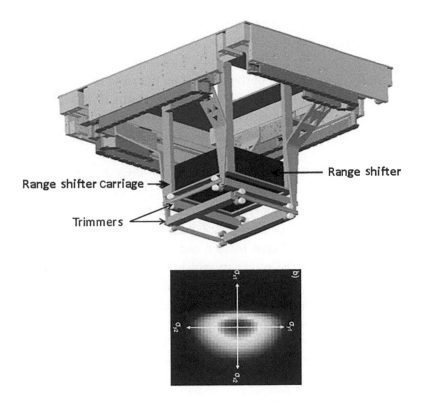

FIGURE 6.7   Dynamic collimator/trimmer for PBS. Upper figure: The trimmer bars move along with the scanned beam to "trim" an edge of the beam as programmed. Lower figure: Shapes that depend on how the trimmers are used can be used to sharpen arbitrary portions of the beam.

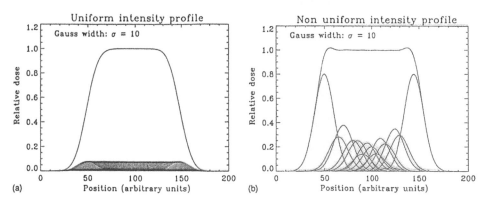

FIGURE 6.8   (a) A superposition of equally space and equal amplitude Gaussian beams. (b) A non-uniformly spaced and modulated dose distribution resulting in an acceptable uniform overall dose with a sharper penumbra.

field, as shown in Figure 6.8b [2,8]. Thus, to achieve a flat distribution with optimal edges, it is necessary to modulate the dose delivered across the target even for a single field delivery.

Finally, it is important to realize that the transverse dose delivered at any given depth will depend on the overall depth dose. Consider, for example, a dose distribution that is desired to be uniform after the delivery of a single field. This is called single-field uniform dose (SFUD) (see Chapter 15). Figure 6.9a (left) shows an actual example of a treated field under such conditions [9]. Figure 6.9b (right, lower) shows the transverse-dose distribution along the transverse section A at this depth. Looking, now at a shallower depth B, one has to account for the proximal tail of the depth-dose distribution that was delivered to A. The resulting transverse-dose distribution is shown in Figure 6.9b (right upper), and one can see a typical "Island-of-dose" pattern for these fields. Thus, the dose across the shallower depth is also a highly modulated distribution, to obtain an optimized uniform field throughout the volume.

These last two examples highlight the fact that the dose distribution for most fields delivered will naturally be characterized by a dose modulation pattern. In common usage today, the term IMPT is used as contrasting SFUD to indicate a single field delivery that is not uniform, vs. a single field delivery that is uniform. In the former, subsequent fields are delivered to finish the desired dose pattern and, for example, results in a dose-uniform overall volume. However, in the mode in which Multiple Fields are superimposed to achieve a Uniform Dose (MFUD) (sometimes, mistakenly called IMPT), especially in the dose-driven case, the intensity is NOT necessarily modulated. Furthermore, in the case of the SFUD delivery, the dose pattern is modulated, as it can be in what has been called IMPT. Therefore, the beam delivery methods are NOT different, and are just Particle Beam Scanning. Understanding these distinctions and care to not use an incorrect term, such as IMPT, is vital in properly specifying a scanning beam system.

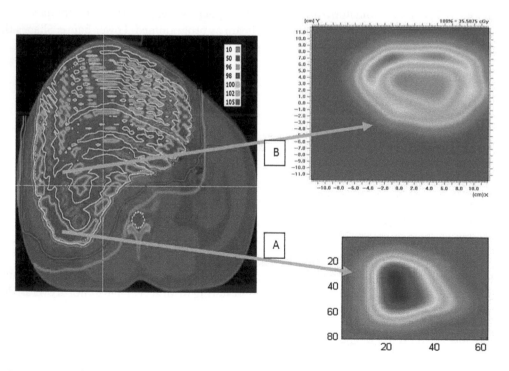

FIGURE 6.9 Left: an actual treatment plan for an SFUD delivery. Right: cross sections of the dose distribution required to achieve the overall uniform dose at the two depths indicated by the magenta arrows are labeled as A (deeper depth) and B (more proximal depth).

6.2.2 Motion Effects

It is useful to note that the dose delivered to the target can be affected by motion. This includes the motion of the beam and the motion of the target [10]. The latter will be covered in Chapter 18, and the former will be mentioned here. In the case of dose-driven SS, the dose delivered is what one would expect from a static beam. Indeed, in the extreme example, the beam is only turned on when it is at the correct location. If, however, the beam is not turned off before moving to the next spot, the dose delivered during the motion must be accounted for. Therefore, the beam will start to be moved before the full dose is delivered (thus delivering the rest as it is moving), and the final dose will depend on the stability and predictability of the current delivered.

In the extreme example, wherein the beam is continuously moving, there are a variety of effects to consider. One may desire to change the dose from one location to another, and this change can be as extreme as turning off. If one does not stop the motion of the beam (which is possible, but also takes time), one has to account for the distance that the beam travels while the intensity is changing, as illustrated in Figure 6.10 (upper). In the figure,

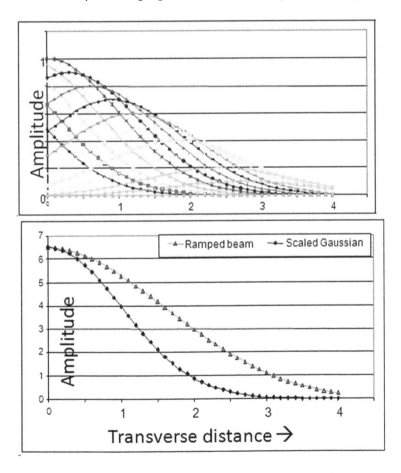

FIGURE 6.10 Upper: reducing amplitude of a Gaussian beam at specific locations (multiexposure snapshots) along its path as the beam intensity is being reduced. Lower: difference between the effective penumbra (upper curve) and the unmodified pure Gaussian penumbra for this case.

there are several Gaussians, each one displaced a given distance (as time evolves and the beam moves) and each one with a smaller amplitude as the beam is being turned off. Take, for example, a moderate scanning dipole with an effective 30 Hz frequency that sweeps over 30 cm. This results in an effective sweep speed of 18 m/s. If one wants to effect the desired intensity change before the beam moves 1 mm, this requires that change to be done within 55 μs. Figure 6.10 (lower) shows an example of the effective increase in beam size owing to the motion. At 20 m/s, with a 3 mm sigma, beam off in 500 μs produces a penumbra growth, from the undisturbed Gaussian of the lower curve to the ramped beam above, of 70% (from 3 to 5 mm sigma). Similarly, the effect is only 10% for a beam with a 1 cm sigma, since the absolute distance traveled does not change and makes a relatively smaller effect.

6.3 TIME SEQUENCE OF BEAM SCANNING TASKS

The events that take place during a scanning sequence will determine the time required to deliver the treatment plan, and depending upon motion effects, will determine the actually delivered dose distribution. One can consider that the treatment plan is a 3D map (it will probably grow into a four-dimensional map as adaptive therapy evolves). How this map is to be delivered can have NOTHING to do with the treatment plan, other than delivering the correct dose to the correct voxel within tolerance, but is related to the scanning implementation. For example, in some situations it is advantageous to vary the beam energy first and the position second or to mix the two. Figure 6.11 illustrates these

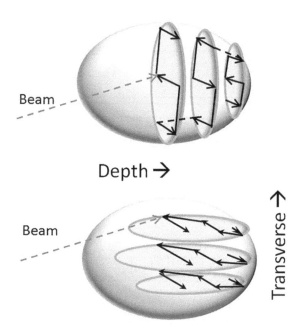

FIGURE 6.11 Two styles of scanning with different time sequences. The upper sketch shows the beam painting the transverse cross section first (left, right, up, and down), then changing the range, while the lower sketch shows changing the range and one coordinate (left and right) before changing the third coordinate (up or down).

different sequences. In the upper figure, the ellipsoidal volume is scanned first in the transverse direction, as indicated by the transverse planes shown, and then the beam energy is modified to move the beam in the third dimension. Scanning may be used to conform to a distal layer without a range compensator by adjusting the field size as a function of the distal depth. Alternatively, the depth direction can be combined with one transverse direction first, followed by a motion to the other transverse direction. Scanning can also be employed to deliver dose to each of these layers, but less than the total required dose, so that the layers may be repainted ([11] or see Chapter 18) repeatedly to compensate for organ motion, or just to deliver a time-dependent field position with a variation of range. Focusing on the transverse positions, the pattern delivered can be optimized and will be based upon the implementation specifics. All of the variations can be independent of the treatment plan. Therefore, there will be the equivalent of an intensity-modulated X-ray therapy sequencer (a series of MLC settings), or more accurately, for scanning, a trajectory manipulator that converts the treatment plan into something closer to what should be delivered (e.g. contour scanning [12]). Of course, it should be possible for the treatment planning software to read this manipulated map and be sure that the treatment plan is unaffected (especially if interpolation is required). Furthermore, it may be that voxel by voxel the tolerances on the beam parameters are different (e.g. small dose may have a higher tolerance), depending on the sensitivities of the treatment plan. This can also be accounted for.

While some accelerators have certain beam delivery limitations with respect to the time-optimized order of the steps described earlier, new accelerators being developed may not and may be able to vary the beam energy, spot size, and position and current quickly (Chapter 3). For example, multiple synchrotron accelerators have the capability to change the extracted beam energy within one extraction period. Therefore, it is possible to vary the energy from the accelerator in milliseconds instead of what used to take seconds.

6.3.1 Scanning Techniques

Various types of scanning have been identified. For example, there is the dose-driven SS technique employed at PSI [2], a variation of this in which the beam is not turned off between spots was used at GSI [13] and is used by Varian, and continuous scanning, which in recent years has been developed at both research institutions such as PSI and the National Institute for Radiological Sciences (NIRS) as well as commercially for clinical use by Sumitomo machines. There is the general feeling that the main advantage of a continuous raster technique would be that of speed, not necessarily speed in delivering all the dose in a field, but speed in covering a dose layer with a fraction of the dose and the possibility to repeat that several times. This could be an advantage to mitigate organ motion, but the details of the timing are yet to be definitively worked out.

One type of continuous scanning method can be characterized by a constant beam deflection speed on the target. The beam intensity is modulated to adjust the dose per voxel. The constant beam deflection rate translates into benign requirements for the deflection magnet power supply (essentially a constant voltage).

In time-driven SS, the beam intensity is attempted to be held constant. The dose per voxel is controlled by the time the beam spends on each spot. Between spots, the beam is deflected as fast as possible by applying the maximum power supply voltage to the deflection magnet. This method requires a high-compliance power supply and more sophisticated control circuitry. This is not used in practice yet.

Assuming the motion is similar (same average speed) in both situations, it is clear that the shortest time to deliver the dose is by a method that can deliver the beam with a maximum dose rate. In fact, the velocity of the scan can be used to control the dose deposited during continuous scanning, instead of modulating the beam intensity and possibly allowing the maximum dose rate to continue, or a spot-by-spot approach can always allow the maximum dose rate. There are different limitations to each of these approaches.

Figure 6.12 contains two graphs in which the upper curve is dose rate and the lower curve is scan speed. The one on the left represents a constant velocity scan and the one on the right shows a constant dose rate scan, whereby the dose in a voxel is modulated by the scan speed. Both deliver the same modulated dose to the target. It is instructive to compare these two cases.

In the case of constant speed, one can examine several scenarios. Let us assume, in arbitrary units that we have a dose rate less than 1. If the velocity (in arbitrary units) is 1, then the time to deliver this scan is, say, t1. If we increase the speed by a factor of 2, but do not increase the dose rate, for each scan, since the time it took to cover the area was reduced by ½, the dose delivered is also ½. Therefore, to achieve the desired dose one would have to rescan twice and the overall time is the same or a little longer. If we double the dose rate, then we must increase the scan speed by a factor of 2, or the dose delivered during time t1 would be twice the required dose. In this way, we can reduce the time by ½. Again, increasing the scan speed without increasing the dose rate does not gain any time. However, increase in the dose rate requires instrumentation that can safely detect the beam properties with commensurate speed.

In the case of constant dose rate, one can examine similar scenarios. Using a scan speed of 1 is not good, since, as seen in the graphs of Figure 6.12, it was desired to reduce the

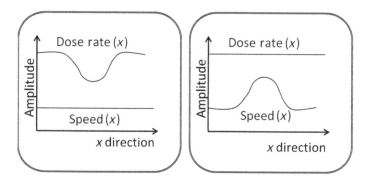

FIGURE 6.12 Graphs of the amplitudes of either dose rate or speed as a function of position during a transverse scan. This shows an example of two extremes of scanning dose delivery. On the left is a constant speed scan with an irregular desired dose distribution, and on the right is a constant dose scan, with the same desired irregular dose distribution.

dose in the center of the scan; therefore, the speed has to be increased in the center of the scan. If, for example, the ratio of maximum to minimum dose during the scan is a factor of 2, we can consider a solution with the speed starting at 1 and rising to 2 in the middle of the scan. This will take approximately 0.75 times the original time owing to the higher dose rate (constant) in the middle of the scan compared to the constant speed scenario. However, increasing the dose rate will further reduce the time since the speed must be increased to compensate.

In conclusion, while it may seem trivial, the dose rate limitation is the main issue. Given a dose rate, faster speed only buys you repainting, not time. However, if you have higher dose rate, you need scan speed to use it.

This brings up one of the safety issues. In the process of radiotherapy, one cannot prevent an overdose because one does not stop the beam (normally) before the dose has been reached, but can react in sufficient time to stop the increasing dose, once the desired dose has been reached, from becoming clinically relevant. Therefore, depending upon the system parameters, including dose readout time and beam control time, one is limited in dose rate by these instrumentation time constants.

Recently, systems with exceptionally fast scanning speeds have been developed. These will be mentioned later.

6.3.2 Contributions to Time

It is helpful to consider, within safety limits, how to best deliver an efficient scanned beam. One can identify the contributions to the scanning time per layer.

- Beam control

  - Time to change beam intensity or to turn it on and off. An example of this is shown in the current modulated scope trace shown in Figure 6.13a.

  - Time to irradiate a location (dose rate)

  - As discussed earlier with respect to the Dose Rate—the faster the better, but for safety reasons, the maximum safe dose rate will be inversely proportional the time that it takes measure and stop the beam. If the time taken to detect the beam is long, then the dose rate must be lower to ensure that the maximum unwanted dose delivered is at an acceptable level.

  - Time to move from spot to spot

  - Note that in time-driven scanning mode, the maximum dose rate depends on maximum scan speed, and the instrumentation time constants and the maximum scanning speed depend upon the time it takes for a beam current change and the desired effective penumbra.

- Scanning Magnets

  - Time to change the magnetic field is a balance between speed and accuracy. Some practical limitations such as the voltage available also come into play given the

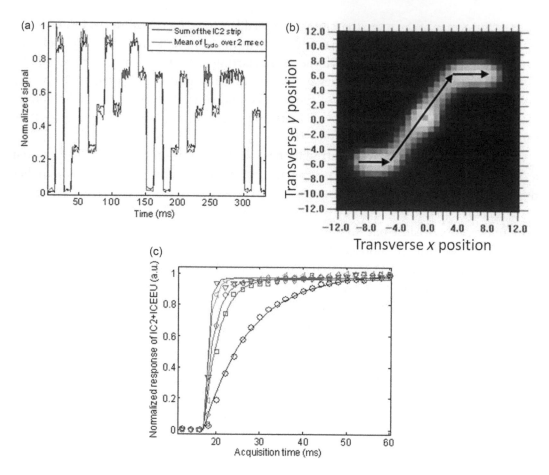

FIGURE 6.13 (a) A graph of beam intensity (signal) vs. time showing an example of beam current modulation. (b) A scintillator screen image of a beam position change requiring a finite time (and distance). (c) The time response of an ionization chamber for different electronic configurations.

finite inductance of a scanning magnet. An example of a moving beam is shown in Figure 6.13b. Also important is the time to detect that the magnets are settled, this can sometimes be longer than the time to move the beam.

• Instrumentation

    • As in all the earlier contributions, instrumentation plays a crucial role. For example, the time to read dose is determined by a number of factors, including the ion drift time in the ionization chambers (ICs), to the speed of the electronics readout. The graph in Figure 6.13c shows multiple rise times as the resistors of an ionization chamber electronic unit and was modified to achieve faster response times.

It is useful to separate the timing information into that required to deposit the dose, the time needed by the scanning magnets to move the beam between adjacent locations, and the time taken by other equipment such as the dosimetry system. Once this is done, there are a variety of parameters that can be explored. Among these are the relative time of

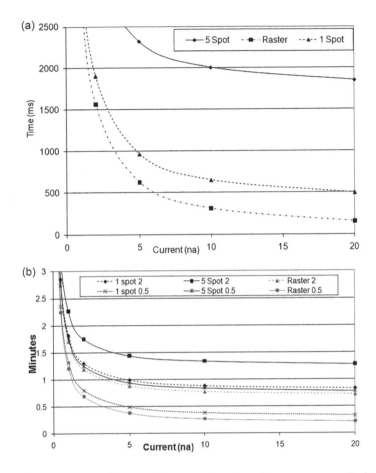

FIGURE 6.14   (a) The time to irradiate one layer as a function of beam current. (Safety issues are not included, or there would be a hard cut off with an upper limit of beam current.) (b) Curve includes the time for energy change for a full volumetric delivery with different currents and different beam sizes. For example "5, Spot 0.5" is short for 5 repaintings of spot scanning with a beam size sigma of 0.5 cm. Raster beam is scanned as many times as needed to deliver the appropriate dose, as fast as allowed.

the extreme of SS compared to continuous scanning, as a function of dose rate (or beam current) as shown in Figure 6.14. For the specific conditions explored here, and for ONE energy layer, the highest curve in Figure 6.14a includes five repaintings of SS, the middle is SS for one repainting, and the lower is raster scanning which, due to the high current used, requires multiple repaintings anyway. Owing to the extra steps in an SS method in which the beam is turned on and off (not all dose-driven methods require this!), there is indeed some increase in irradiation time; however, for some realistic parameters for a given layer, the difference may only be a fraction of a second, and when considering the overall time of the irradiation (including the time to change the energy which is normally a few seconds, but can be as low as fractions of a second), the difference can be less than half a minute assuming repainting with both methods, as shown in Figure 6.14b. This plot is typical for the case in which the time to change the energy is relatively significant (e.g. seconds).

6.4 SCANNING HARDWARE

The unmodified accelerator beam must be spread out throughout the target volume, and the appropriate dose must be delivered. The hardware required to do this can be divided into two main categories.

1. Equipment to adjust the beam properties

2. Instrumentation to measure the beam properties.

The equipment used for these manipulations at the MGH system developed jointly by Ion Beam Applications (IBA) and MGH is summarized in Figure 6.15. The equipment in these figures, from left to right, include a quadrupole doublet to control the beam size, a pair of scanning dipoles to deflect the beam to the desired position, and ionization chambers to measure the dose, position, and beam profiles. Other commercial scanning systems are depicted in Figure 6.16.

6.4.1 Adjust the Beam Properties

The equipment that is used to deliver the dose introduces physical constraints and therefore limits the types of patterns that can be applied and the timing that is possible. The scanning patterns that will be allowed are determined by the ability to adjust the relevant beam properties.

6.4.1.1 Energy

In the case of scanning, this is simple. The beam range is normally given by the accelerated beam energy or the beam energy resulting from a degrader system, and perhaps, slightly further modified by a minimum of material in the beam path. Any material in the beam path will scatter the beam (see Chapter 2) and increase the beam emittance.

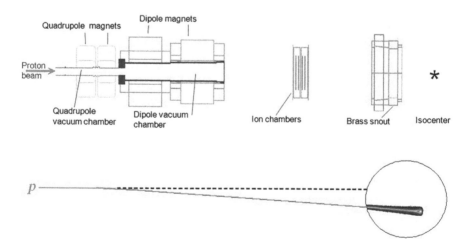

FIGURE 6.15 Upper: The equipment used to deflect and measure the beam for the MGH/IBA system. Lower: The beam deflections resulting from the magnetic fields.

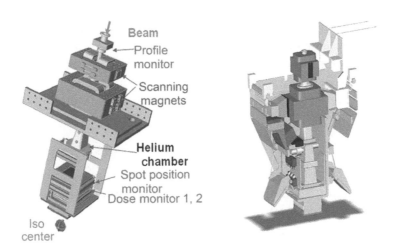

FIGURE 6.16   Left: The Scanning Nozzle of Hitachi as implemented at MD Anderson. (Courtesy of Hitachi.) Right: The Scanning Nozzle of Accel/Varian as implemented at the Renniker facility. (Courtesy of Accel.)

6.4.1.2 Size

The beam size is determined by the intrinsic emittance of the beam, as modified by the beam-focusing elements in the beamline. Sometimes, a set of final quadrupoles can be used to fine-tune the beam size in the nozzle.

6.4.1.3 Position on Target

It takes two parameters to define a trajectory. One must control the position and angle of the beam. Assuming one already knows the position and angle of the beam entering the scanning dipoles (to the left of Figure 6.15), the excitation of the magnetic field in the scanning dipoles will determine the final beam path angle and thus the location of the beam in the target, as shown in the lower part of Figure 6.15.

The overall system of power supply and dipole will determine the speed of the scan. The relation $V = L\, di/dt$ indicates that the faster the scan (di/dt) the higher the voltage (V) requirements, which also depend on the magnet design such as the inductance (L).

6.4.1.4 Scan Patterns

We can call "field area" as the area in which we expect beam to be delivered according to a particular pattern of scan lines with a certain distance between scan lines. If one chooses a pattern with constant frequencies along X and Y axes (f_x and f_y, respectively), a Lissijoux pattern results with a spacing between the painted lines. To reach the required line spacing (δ), which depends on the beam size and desired overlap, the frequency ratios could be adjusted. If you change the frequency ratio $\alpha = \dfrac{f_y}{f_x}$ while keeping nearly the same sizes for the scanning area, one can theoretically adapt the pattern to reach any required line spacing with a repeating pattern. However, the limitations on frequencies and speeds along both axes due to the hardware inherent to any power supply will limit the lowest reachable spacing between scan lines. One way to solve this is to work around those hardware limitations and modify the relative phasing of the frequencies. This will reduce the distance between the scanning lines since the scan will not repeat for a to-be-defined number of cycles.

In the ideal case, the beam trajectory is arbitrary and one is not dependent on a set of fixed frequencies; however, there will always be a dependence on the hardware constraints of the system.

6.4.1.5 Dose Rate

The highest dose rate available is almost completely determined by the extracted current from the accelerator, assuming there are no losses in the beam path, or in the case of a degraded energy beam, the beam intensity will be reduced by the scattering resulting from a degrader system (see Chapter 3) and the collimation system which contains a fixed amount of beam losses. The highest dose rate allowable will be determined by the parameters of the instrumentation and beam control. Most accelerators have the capacity to deliver a higher dose rate than is currently used. However, the limitation is mainly the typical 100 μs response time of an ionization chamber. Such a response time can be reduced through optimization of the voltage gap and voltage amplitude. Consideration must be given to recombination effects as well. The speed with which one can detect and turn off the beam will determine how much dose is delivered during that time and will therefore limit the dose rate.

6.4.1.6 Dose Limitations

While there can be a maximum dose rate, one can, in principle, deliver as much dose as desired to a given voxel. On the other hand, there are several considerations that lead towards a limit on the minimum dose that can be delivered.

As noted in other sections in this chapter, there is generally a finite time required to detect the beam delivered, initiate a request to turn the beam off, and time to turn the beam off. During all of this time, some beam is delivered to the target. This amount of beam should either be insignificant when compared to the prescription, or should be predictable so that one can calculate the dose delivered while it is turning off and thus begin the turnoff process a bit early counting on that extra dose to meet the prescription requirements. The amount of dose delivered during this time depends both on the time it takes and on the dose rate.

If one desires a small dose in a given region, it could be that the desired dose is equivalent or smaller than the dose that would be delivered during the time the beam is turning off, so one would effectively have to turn off the beam before it is turned on. That is, unless one reduces the dose rate. However, at a certain point, the signal from the ionization chamber for a reduced dose rate beam is low enough to be at the level of noise in the system electronics or below. Therefore, if one cannot decrease the time it takes to detect and turn off the beam, then one has to lower the dose rate for low-dose spots and one reaches a lower limit.

Assume a beam current of 1 nA which is consistent with a healthy dose rate for scanning. Let us also assume that the entire detection and turnoff time takes 100 μs (a pretty fast time). Then, during the turnoff time period, 0.1 picoCoulombs (pC) or 0.625×10^6 protons have been transported to the target. Some electronics are designed to count the charge in 10 pC units. Let us not forget the amplification of the ionization chamber, which increases the signal from the beam current through the chamber on the order of 200, so the electronics will detect 20 pC or 2 counts. At MGH, the arbitrary parameter called monitor units (MUs) is set to be 300 counts. Therefore, the minimum MU that can be delivered with 1 nA beam current is 0.0067 MU. One can reduce the beam current by a factor of two and reach 1 count in the time interval. The noise in the system should be a fraction of a count and is sometimes in the neighborhood of 1×10^6 protons. Thus, we have the basis for a minimum deliverable quantity of dose.

One should note that this must be considered for treatment planning. Regions in the target should not be designed to have less dose than this. One can take all such regions and modify the plan to deliver either 0 or the minimum dose, and the appropriate rounding will have to be devised so as not to compromise the treatment plan.

6.4.1.7 Faster Scanning Systems

Recently, within the last couple of years, what some may consider the next generation of scanning systems is starting to be used clinically. Actually, these systems were designed several (up to 10) years ago, but that started about the time that some commercial systems were already being introduced for clinical use. One can imagine what might be possible if one started now. The primary attribute of these systems is speed. While it may be that the time it takes to irradiate an average volume in a hospital setting these days is from 30 s to a couple of minutes, the newer systems seem to be able to irradiate an entire volume in a somewhat reduced time (e.g. several seconds). If a full target volume were to be able to be irradiated in about 1 s, this would, for most cases, make the issue of organ motion irrelevant. It can also be the case, with appropriate beam delivery timing spread over the course of a minute, the effects of target motion can be averaged out, however more healthy tissue is irradiated. One can see that the evolution to a few seconds of time to irradiate a volume is going in a useful direction, but it has not reached that point yet. One might also note that, for example, a 3 to 5 s irradiation would be very similar to a respiration cycle and therefore not a good timing for averaging the effects of organ motion [14] without breath hold or equivalent techniques.

How have these systems accomplished this feat when a score of commercial systems currently installed are slower? These systems at PSI and NIRS use a cyclotron and synchrotron, respectively, and therefore that is not the issue, although they have implemented some unique aspects to both to achieve the goals. It should also be understood that each of the system parameters identified in this section have been addressed from the perspective of time and safety, so systems that provide safety have been improved.

Energy: The time to change the beam energy has been reduced from seconds to fractions of a second, as low as, or lower than 100 ms. In the case of the NIRS system, it is done through multienergy extraction [15] and fast beamline magnetic field changes. In the case of PSI, it has been done through the use of a fast linear motion mechanical degrader and beamline magnetic field changes [16] and compensation of eddy current effects. Note that for 25 energy layers, a 0.1 s energy change time contributes 2.5 s, so the energy change time is still a dominant factor.

TABLE 6.1   Parameters of Recent Faster Scanning Systems

Parameter	NIRS System	PSI System
Beam intensity	3×10^7 to 1×10^9 particles/s	>0.5 nA (3×10^9 protons/s)
Scan speed	$V_x > 100$ mm/ms	20 m/ms (shorter direction)
(At maximum energy)	$V_y > 50$ mm/ms	5 mm/ms (larger direction)
(Lower energy can be faster)		
Energy change time	<300 ms	<100 ms
Scanning amplitude		12 cm×20 cm
Minimum time/spot	25 μs	3–5 ms
Average time/spot	200 μs	
Time to delivery one 2D layer	~40 ms (10 cm×10 cm)	~80 ms (12 cm×20 cm)

Dose Rate: Until recently, the parameters used in clinical ion chambers have been relatively standard, but now these values are better optimized for speed, linearity, and accuracy. The NIRS instrumentation can operate at 200 kHz (5 μs), and the instrumentation used as PSI is focused on low current precision and has a 500 μs charge collection time.

Scanning Magnet Speed: While the speed with which the current in an electromagnet can be changed is related to the amount of voltage applied $V = L * dI/dt$, optimized design of the scanning dipole can reduce the Inductance (L) and thus increase the current change speed for the same voltage applied. In addition, the effects of magnetic field variation such as eddy currents are controlled by care and detailed design of the laminations and electrocoil designs of these systems. Taking care of such parameters has resulted in these systems being capable of increasing scanning magnet, magnetic field change speed from 4–40 to 100–200 Hz.

Table 6.1 illustrates the performance of these new scanning systems.

6.5 SCANNING INSTRUMENTATION AND CALIBRATION

We can divide the situation for which we must measure the beam properties into online and offline measurements. The latter includes calibration and quality assurance (QA; see Chapter 12). It is important to decide how to verify that the beam is entering the target with the correct parameters. Introducing a monitor that intercepts the beam will affect the beam profile in a potentially adverse way. Weighing the optimized beam parameters with the impact of measurements on safety is a complicated process. For example, one can consider that the measurement of the current and voltage and magnetic field of a scanning magnet is a triply redundant way to know the effect of that magnet on the beam trajectory. Others may believe that only a direct measurement of the beam parameters before the target can definitively tell that the beam is entering the target correctly. Still others may desire an ionization chamber to determine the beam position entering the scanning dipoles in addition to the downstream (before the target) monitors to ascertain the beam angle. One can argue that downstream ionization chambers can measure the position of the beam before the patient is sufficient; however, this can depend on the distance of these chambers to the patient. If they are close, then the beam angle will not play a significant role, but if they are far, the angle could be a cause for consideration. The overall system design is important in determining the optimum configuration of instrumentation.

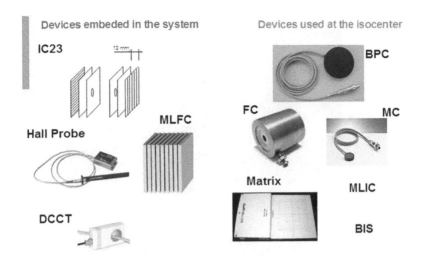

FIGURE 6.17   Some instrumentation used including: MLFC (Multilayer Faraday Cup), DCCT (DC Current Transformer), FC (Faraday Cup), MC (Monitor Chamber), BIS (Beam Imaging Scintillator), BPC (Bragg Peak Chamber), and IC23 (Ionization Chambers in the MGH System).

6.5.1 Offline Measurements

6.5.1.1 Calibration

A significant amount of measurements are necessary to both characterize the system performance and calibrate the system settings so that the desired performances could be achieved. Some of the tools that can be used are depicted in Figure 6.17. Depending upon the system, some of these devices might be embedded in the system and some might require external implementation, for example at isocenter.

6.5.1.2 Calibration of the Beam Position at Isocenter

It is necessary to ensure that the scanning dipoles can position the beam to the desired location on the target. Given the safety nature of this quantity, multiple levels of redundancy can be used to monitor the status of the scanning dipole, including redundant current measurements, voltage measurements, magnetic field measurements as well as the status of the power supply. This, together with input trajectory information fully characterizes the deflection of the beam. Still, providing a functional redundancy with ionization chamber information can also be used to further augment the redundancy in beam position; however, this is usually installed upstream of the isocenter. Thus, a device mounted at isocenter is useful for calibration. Figure 6.18 shows the MatriXX [17] used at isocenter to measure the deflection as a function of set point voltage and to calibrate the magnet settings as well as the upstream ionization chamber readout. These calibrations result in position accuracy of ±0.7 mm, under the conditions of this measurement.

6.5.1.3 Calibration of the Beam Size at Isocenter

The beam size performance has been previously discussed in the section on beam performance. Measurements can be made using a scintillator screen and camera system (BIS, Lynx, etc.) or the MatriXX. Figure 6.18 shows some MatriXX results.

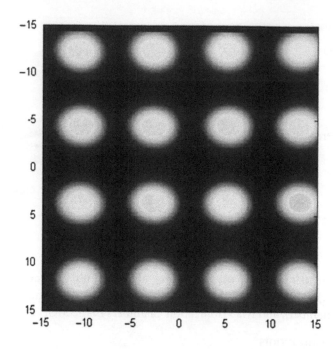

FIGURE 6.18 Beam spots at Isocenter using a MatriXX (12) ionization chamber system. The lower axis is the horizontal position and the abscissa is the vertical beam position.

6.5.1.4 Calibration of the Dose Delivery at Isocenter

There is a series of steps that are necessary to perform all the calibrations required to enable accurate dose delivery. These include

- Calibration from charge delivered to counts to MUs

- Characterization of the beam turnoff time

- Ionization chamber effects

With respect to the beam turnoff time, a plot of the difference between the expected dose and actual dose as a function of dose rate will help determine the time. The slope of this curve is related to the beam turnoff time. Measurements of this sort may result in a parameter that requires the beam to be shut off before reaching the desired dose at a spot, if the beam intensity during the turnoff time can be predicted within reason. If fluctuations are large, the beam turnoff must begin earlier with the risk of delivering a lower than desired dose.

The signal from the IC itself can depend upon the beam current due to recombination effects. The maximum beam current will be limited by the ionization chamber response as well as the beam turnoff time, as discussed earlier.

6.5.2 Online Verification of Beam Position

The tolerances applied during beam position are dictated by the clinical tolerance. However, a beam position cannot be measured with sufficient accuracy by an ionization chamber until sufficient charge is collected. Therefore, the beam position (and size) measurement may have to be delayed. However, other measurements may be quicker.

A good scanning system will record the online measurement data and compare it with the expected values within the specified tolerances. An example of this is shown in Figure 6.19. These are plots of the scanning magnet current values as a function of spot positions. Included on these plots are the measured spots of the expected values, and the high and low thresholds. Similar plots can be made of the scanning magnetic field and power supply parameters. It is a fact that much calibration must be done to optimize the speed when it is important and to focus on the accuracy when it is important. Adjustment of tolerance is a very important and time-consuming part of the calibration process.

6.5.3 Beam Steering Corrections

While significant amount of time can be spent on calibrating the position and in trying to understand the effects that contribute to mispositioning, not all the effects may be understood, and it is necessary to consider how to best deliver an irradiation if not all parameters are a priori perfect. Given the desire to reduce the time for an irradiation and continue an irradiation once started, assuming all the parameters are within tolerances, it is possible to apply more robust techniques in correcting the beam position if deviations occur.

Four methods have been developed.

1. Dead reckoning via the algorithmic tables or formulae developed.

2. Beam tuning on the first layer (if an energy layer approach is used).

3. Adaptive beam tuning that applies the position offset corrected in the first energy layer to all subsequent layers.

4. Adaptive scanning magnet correction, which corrects the beam position within a scan.

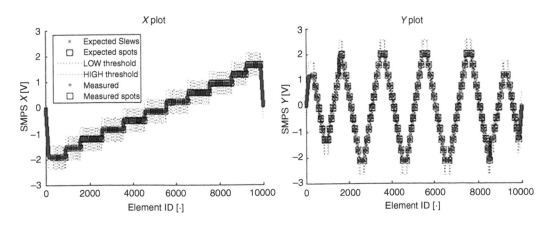

FIGURE 6.19 An example of an online data analysis system monitoring each of the axes of the scanning dipole power supply (left and right) as a function of an element of the treatment plan. Also contained is the actual measured *x* and *y* power supply current behavior.

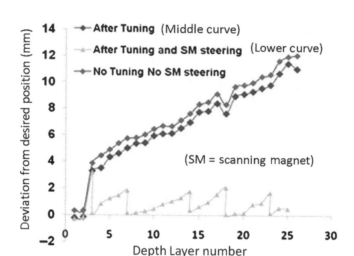

FIGURE 6.20 Data (offset vs. time) obtained from a simulated beam position correction process. The upper curve shows the raw (purposely misaligned position both during the first and third depth layers). The middle curve is corrected from the first position (but purposely modified subsequently. The lower curve is corrected each time the beam exceeds the maximum offset.

Together, these approaches significantly enhance the precision of the absolute beam position and provide a degree of robustness so that the irradiation is less sensitive to potential beam-positioning fluctuations.

Figure 6.20 shows an example of the beam position correction processes. The vertical axis is the beam position in mm and the horizontal axis is the layer number for a 3D irradiation. The upper curve is a measured position of the beam without corrections, for a contrived case. (Note in particular the discontinuity between layers 2 and 3.) The middle (just below the upper) curve shows the result from adaptive beam tuning that was implemented during the first layer. Thus, all subsequent layers are corrected by the same amount; however, there was a discontinuity between the second and third layer. Finally, the lower curve shows the result of the adaptive scanning magnet correction. In this example the threshold is set at 1.6 mm. Note that in layers 3, 7, etc. the measured beam position exceeds the threshold and the scanning magnet is used for correction. This correction is applied to the entire map. The mispositioning is measured and reacted to before dose of any clinical significance is delivered with the beam out of tolerance.

6.6 SCANNING GANTRIES

The delivery of any beam will be characterized by the relative orientation of the beam and the patient target. While, in general, either the patient or the beam orientation can be manipulated, for the most part, nonplanar beams are delivered by rotating Gantries that manipulate the beam from the horizontal orientation, to an orientation perpendicular to a patient in a prone or supine, horizontal position (see also Chapter 3). While the subject of Gantry design and implementation is large, it may be worthwhile herein to note that the beam delivery modality can have a significant impact on their design. For the most part, a Gantry that contains a scattering beam delivery system will have that scattering system immediately after the last bending dipole of the Gantry. The scattered beam becomes very large and uniform, and further, transport by magnets will require very large magnet apertures and may negatively affect particle distribution. However, in the case of particle beam scanning, with an unmodified beam, the potential geometries expand. The first Gantry to take advantage of this possibility is shown in Figure 6.21 [18]. This gantry is also compact in the transverse dimension by setting the patient position away from the gantry rotation axis. Inserting the scanning dipoles before the last gantry dipole allows one to minimize the space between the last gantry dipole and the target, at a cost of a large aperture in the gantry dipole to accept the diverging scanned beam. It is possible to obtain a variety of optical solutions that will control the angle of the scanned beam at the target. For example, the beam angle can be anything from parallel to the actual angle the beam was bent by the scanning dipole and magnified beyond that. Additional geometries include positioning the scanning dipoles before or after the last gantry dipoles, and the inclusion of focusing elements, which will affect both the beam size and effective scanning beam angle (or effective source-to-axis distance). Some of these geometries are shown in the sketches of Figure 6.21b, as implemented by Hitachi, for example, and the one in Figure 6.21c in which the scanning dipoles are split before and after the gantry dipole with the appropriate beam optics in between. In cases where the dipoles are together, they can be replaced by a single combined dipole [19], if there is sufficient space to achieve the desired field size. Many more combinations are possible.

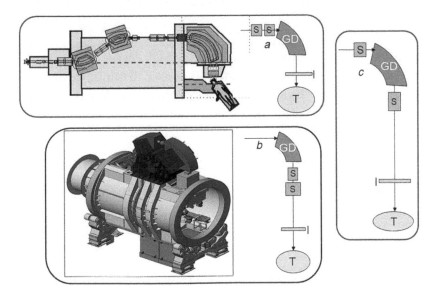

FIGURE 6.21   Depictions of three types of gantry geometries capable of providing scanning beams. Geometry a (courtesy of E. Pedroni) is the first scanning proton gantry with both scanning dipoles upstream of the last gantry bending dipole. Geometry b (courtesy of Hitachi) is the Hitachi scanning gantry with scanning dipoles downstream of the last gantry dipole. A drawing of the physical gantries are in the left of each box, and the line drawing to the right shows the scanning dipole(s) (lighter colored blocks with S inscribed), gantry dipole (darker wedge with GD), the instrumentation (I), and an ellipse representing the target (T). The geometry in box c shows the scanning dipoles split upstream and downstream of the gantry dipole.

6.7 BEAM PROPERTY QA

One can have a friendly discussion about all the aspects of the beam that have to be correct to achieve the desired dose within a 3D field that contains nonuniform field distributions. However, one way to cut that discussion short is to find a way to measure some of the important beam properties explicitly and to do so in a very short time.

One such method is a test pattern that can measure many important properties of the beam in a very short time. It takes a few seconds to deliver that pattern, and the dose delivered is displayed on a MatriXX or Scintillator screen monitor. This pattern has the capability, with analysis, to give the beam position, size, and dose.

An example of this with an automated analysis tool is shown in Figure 6.22. The gray-scale in the left of Figure 6.22a shows the pattern, assuming perfect beam resolution. Figure 6.22b, in living color, shows the results when this pattern is delivered with the MGH scanning beam. In this case, an unfocused beam with a sigma of about 1cm was used. Also indicated are some of the regions in the pattern which can be used to analyze beam position, dose, and beam size. It is envisioned to repeat this test pattern daily, after the scanning system hardware is configured and to use this data as QA testing.

Finally it is necessary to determine the correspondence of the beam position and the imaging fiducials. One way to do this is to generate an exposure with the X-ray tube and produce an image of the X-ray cross hairs on the same screen as that which is measuring the QA pattern. This completes the correlation between the beam and the patient setup data.

6.8 SAFETY

PBS is a modality to deliver a proton beam achieving a high degree of irradiation conformality to the desired target. As a form of radiation, it is desired to minimize the radiation delivered outside the target. Thus, the level of criticality increases if dose is delivered to healthy tissue. Underdosing the target can also have undesirable results. However, if the dose is accurately monitored, full dose can be delivered as required. Dose outside the target may

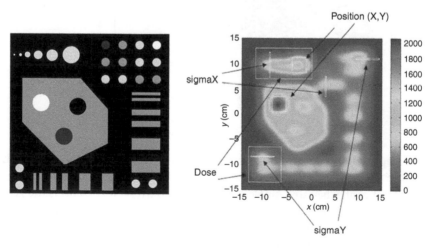

FIGURE 6.22   Left: A theoretically designed scan test pattern. Right: The result of a measurement with a larger beam size and some of the parameters that can be obtained through analysis using this pattern.

be acceptable up to certain limits, as defined by the medical prescription. The equipment must be capable of ensuring that this unwanted dose is not delivered, or of detecting this unwanted dose and stopping the irradiation in sufficient time to prevent too high a quantity of undesirable dose. Thus, the level of severity in some cases will in part be determined by the specifications of the instrumentation used to detect beam properties. One must ensure

1. The appropriate dose is delivered

2. The dose distribution is correctly delivered

3. The dose is delivered to the correct location.

One way to ensure that a system operates in a safe manner is to evaluate the failure modes and develop a strategy that leads to a safe and consistent mitigation of each and every conceived failure. For the PBS, a comprehensive Failure Mode and Effects Analysis (FMEA) and hazard analysis should be done. Elements of the process include

- Possible hazards/effect on beam
- Risk mitigation method
    - Hardware detection
    - Software mitigation
    - Functional redundancy
- Clinical protocol
    - What has to be done for QA?
    - What should a therapist/physicist do?

6.8.1 Safety Strategy

To develop an appropriate strategy, a philosophy of risk mitigation is needed. The following questions are posed:

- What needs to be set to determine a property? (*e.g. magnets and scattering for beam size*)
- Are there other things (that we do not control) that can determine a property? (*e.g. accelerator-extracted beam trajectory*)
- What can be monitored to determine that the setting of the DEVICE is correct? Are there redundant ways to monitor it? (*e.g. limit switches, current transformers*)
- Is there a direct way to measure that beam property? (*e.g. ionization chambers, range verifiers*)
- What can be done with Software and what can be done with hardware?
- What additional QA can/should be done?

6.8.2 Beyond Safety

An old adage says that "the safest system is one that does not work." We cannot always expect everything to be perfect, and therefore, we look for some ways to be insensitive to errors or to correct errors in a fast and accurate way that allows treatment to continue. For example, the system should accept a Pause, Stop, and Resume command. Also note, it should be possible to make vector corrections to adjust the beam position. This should allow the beam to be properly located, and the data that is uploaded after a plan is delivered would then include the appropriate beam data, even though the physical magnet settings may not be nominal due to the vector correction made to compensate for machine fluctuations.

If one has paused or stopped the beam before the irradiation was completed and it is desired to complete that irradiation, this process must account for all the dose that has been previously delivered. Unlike scattering, where the entire volume is irradiated almost simultaneously, in the case of scanning, there is a time dependence to the beam delivery. In the case of recovering a treatment after it has been stopped or paused, the measurement of dose already delivered is crucial. Figure 6.23 shows the case of a particular layer irradiation. The numbered sections are voxels, and by the time of the pause, the gray-filled boxes have been irradiated, while box number 11 is only partially irradiated. The rightmost figure represents the desired irradiation of the remaining voxels upon resuming the irradiation.

Treatment planning simulations have been carried out to understand the effect on different types of fields. For example, in a plan, shown in Figure 6.24 with a 34 Gy highly modulated field, a scenario could be created in which there are localized changes of ~6 Gy ~ 16%. This does not show up in the dose volume histogram (DVH), but does contribute to the desired accuracy of the already delivered dose. Note that in this case, if thousands of spots are used, 0.1% can result in missing ten spots. If those ten spots, are located in regions of steep gradient in a highly nonuniform distribution, the results can be noticeable. Much depends on the details of the treatment plan and therefore the results are a bit unpredictable or at least nonintuitive. At MGH, it has been determined that 0.7 MU accuracy or limitation on the noise that contributes to the dose measurement is required. This gives an acceptable dose distribution for the cases examined.

It is also important to realize that we do not live in an ideal world. In our world, we have electronics noise, glitches and microphonics, to name a few. It would be disadvantageous to continually pause the perfectly accurate beam when spurious issues arise. It is not

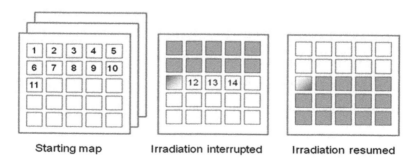

FIGURE 6.23   Illustration of an interrupted scanning map and the recovery and continuation.

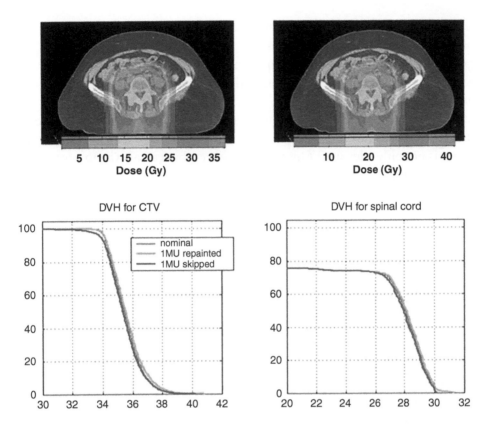

FIGURE 6.24 Treatment planning example of an interrupted scanning beam delivery to determine the accuracy required of knowing the actually delivered MUs, to accurately resume the treatment. The left treatment plan shows a local underdose.

unsafe to account for reality and allow the beam delivery to continue through minor signal noise excursions. However, it would potentially be unsafe to stop the beam and prolong the treatment time. Therefore, error detection techniques that are not simply binary (inside or outside tolerances) are considered. One can use a method of allowing a certain number of out-of-tolerance measurements below a certain threshold before taking action. A more statistical approach applies the exponentially weighted mean average [20] to account for systematic and random statistical variations separately. In all cases, it is necessary to be convinced that the best things are being done to ensure safe and efficient patient treatments.

6.9 SUMMARY

Particle beam scanning represents a technical improvement in obtaining the most conformal dose distributions possible with particles. Additionally, it allows for dose delivery without patient-specific hardware. This allows for the most efficient beam delivery modality. One can envision multiple field irradiations being delivered without entering the treatment room between fields.

The depth-dose distribution together with the conformality of this approach leads to a minimization of beam fields necessary for an excellent dose distribution, although there may be clinical considerations in these numbers. The number of fields can be so reduced that consideration of gantryless solutions can potentially be considered [21].

The overall system is in fact simpler in the number of elements required, than a scattering system; however, the implementation of the system with respect to timing and tolerances is more challenging.

Given the fact that for a scanned beam it is not normally necessary to prepare elements that will control the beam delivery in advance, it is ideally suited for adaptive radiotherapy. Adaptation is not only relevant for intrafraction effects but can also be important for interfraction effects. Indeed, the largest obstacle to widespread use of scanning in all body sites is due to organ motion. In general, the scanning technique contains knobs to compensate for all types of organ motion effects if the treatment planning can be adapted to account for these effects and the accelerator system is nimble enough to compensate.

As is the case in all particle beam spreading methods, the tolerances associated with the delivery of the particle beam are tighter than those associated with photon beam delivery; however, the gains in dose conformality, efficiency, and compensatory power are all significant with the Beam Scanning modality.

Accelerator, beamline, and scanning systems have previously been built for scattering beam delivery; however, newer systems have now been optimized for scanning and provide more flexibility for adaptive and fast beam delivery.

On the whole, the cost of delivering a beam using scanning can be lower than other alternatives, and the treatment is more efficient, thus positively contributing to the cost-effectiveness of particle therapy.

The Beam Scanning modality is now integrated into hospital-based clinical treatment facilities, and is not just in the laboratory environment anymore. This represents a new phase in the evolution of particle beam therapy and research.

ACKNOWLEDGMENTS

The author would like to thank Dr. Paganetti for the opportunity to update this chapter in the second edition of the book. The author would also like to acknowledge IBA, Hitachi, Sumitomo, Varian, MeVIon, ProTom, and others who are continuing to make advances in scanning beam systems for medical therapy and the National Cancer Institute for their support of some of the work reported herein. Finally, thanks are owed to the author's wife, Nancy, and although his children, Adam and Scott, are no longer at home watching this chapter being edited, their encouragement is much appreciated.

REFERENCES

1. Kanai T, Kawachi K, Kumamoto Y, Ogawa H, Yamada T, Matsuzawa H, and Inada T, Spot scanning system for proton radiotherapy, *Med Phys* 7, 365–369 (1980).
2. Pedroni E, Bacher R, Blattmann H, Bohringer T, Coray A, Lomax A, Lin S, Munkel G, Scheib S, and Schneider U, The 200-MeV proton therapy project at the Paul Scherrer Institute: conceptual design and practical realization, *Med Phys* 22, 37–53 (1995).
3. Haberer T, Becher W, Schardt D, and Kraft G, Magnetic scanning system for heavy ion therapy, *Nucl Instrum Meth Phys Res A* 330, 296–305 (1993).

4. J. Hubeau, Private communication.
5. Pedroni E, Bearpark R, Bohringer T, Coray A, Duppich J, Forss S, George D, Grossmann M, Goitein G, Hibes C, Jermann M, Lin S, Lomax A, Negrazus M, Schippers H, kotle G, The PSI Gantry 2: a second generation proton scanning gantry, *Z Med Phys* 14(1), 25–34 (2004).
6. Trofimov A and Bortfeld T, Optimization of beam parameters and treatment planning for intensity modulated proton therapy, *Technol Cancer Res Treat* 2(5), 437–444 (2003).
7. Moignier A, Gelover E, Smith B, Wang D, Flynn R, Kirk M, Lin L, Solverg T, Lin A, and Hyer D, Toward improved target conformity for two spot scanning proton therapy delivery systems using dynamic collimation, *Med Phys* 43(3), 1421–1427 (2016).
8. Staples JW and Ludewigt BA, Proceedings Particle Accelerator Conference. 1759–1761, (1993).
9. Kooy HM, Clasie BM, Lu HM, Madden TM, Bentefour H, Depauw N, Adams JA, Trofimov AV, Demaret D, Delaney TF, and Flanz JB. A case study in proton pencil-beam scanning delivery, *Int J Radiat Oncol Biol Phys* 76(2), 624–630 (2010).
10. Saito N, Bert C, Chaudhri N, Gemmel A, Schardt D, Durante M, and Rietzel E, Speed and accuracy of a beam tracking system for treatment of moving targets with scanned ion beams, *Phys Med Biol* 54(16), 4849–4862, (2009).
11. Seco J, Robertson D, Trofimov A, and Paganetti H, Breathing interplay effects during proton beam scanning: simulation and statistical analysis. *Phys Med Biol* 54(14), N283–N294, (2009).
12. Meier G, Leiser D, Besson R, Mayor A, Safai S, Weber D, and Lomax A, Contour scanning for penumbra improvement in pencil beam scanned proton therapy, *Phys Med Biol* 62(6), 2398, (2017).
13. Eickhoff H, Haberer T, Kraft G, Krause U, Richter M, Steiner R, and Debus J. The GSI Cancer Therapy Project, *Strahlenther Onkol* 175 (Suppl2), 21–24, (1999).
14. Dowdell S, Grassberger C, Sharp GC, and Paganetti H, Interplay effects in proton scanning for lung: a 4D Monte Carlo study assessing the impact of tumor and beam delivery parameters, *Phys Med Biol.* 58(12), 4137–4156 (2013).
15. Furukawa T et al., Commissioning of NIRS fast scanning system for heavy-ion therapy, Proceedings of IPAC2011, San Sebastian, Spain, THPS072; 3595–3697; Furukawa T, Inaniwa T, Sato S, Shirai T, Takei Y, Takeshita E, Mizushima K, Iwata Y, Himukai T, Mori S, Fukuda S, Minohara S, Takada E, Murakami T, and Noda K. Performance of the NIRS fast scanning system for heavy-ion radiotherapy, Med Phys. 2010 Nov;37(11): 5672–5682.
16. Pedroni E, Meer D, Bula C, Safai S, and Zenklusen S, Pencil beam characteristics of the next-generation proton scanning gantry of PSI: design issues and initial commissioning results, *Eur Phys J Plus* 126, 1–27, (2011).
17. Herzen J, Todorovic M, Cremers F, Platz V, Alvers D, Bartels A, and Schmidt R, Dosimetric evaluation of a 2D pixel ionization chamber for implementation in clinical routine, *Phys Med Biol* 52(4): 1197–1208, (2007).
18. Pedroni E and Enge H, Beam optics design of compact gantry for proton therapy, *Med Biol Eng Comput* 33(3), 271–277, (1995).
19. Anferov V, Combined x-y scanning magnet for conformal proton radiation therapy, *Med Phys* 32(3), 15–18, (2005).
20. Clasie BM, Kooy HM, and Flanz JB, PBS machine interlocks using EWMA, *Phys Med Biol* 61(1), 400–412, (2016).
21. Yan S, Lu HM, Flanz J, Adam J, Trofimov A, and Bortfeld T, Reassessment of the necessity of the proton gantry: analysis of beam orientations from 4332 treatments at the Massachusetts General Hospital Proton Center over the past 10 years, *Intl J Rad Oncol*, 95(1), 224–233, (2016).

III

Dosimetry

CHAPTER 7

Secondary Radiation Production and Shielding at Proton Therapy Facilities

Nisy Elizabeth Ipe

Consultant, Shielding Design, Dosimetry & Radiation Protection

CONTENTS

7.1 INTRODUCTION

Radiological considerations for proton accelerators have played an important role, since the construction and operation of particle accelerators first occurred in the 1930s. Proton therapy accelerators operate in the intermediate energy range. Proton energies are classified as follows (where E_p is the energy of the incident proton):

Very Low Energy: $E_p < 10$ MeV;

Low Energy: 10 MeV $\leq E_p < 50$ MeV;

Intermediate Energy: 50 MeV $\leq E_p \leq 1,000$ MeV (1 GeV);

High Energy: $E_p > 1$ GeV.

The prompt radiation field produced by protons in the therapeutic energy range of interest, i.e., 67–330 MeV is comprised of a mixture of charged and neutral particles as well as photons, with neutrons being the dominant component. The neutrons have energies as high as the incident proton energy and dictate the shielding for proton accelerators. The purpose of structural shielding is to reduce the prompt secondary radiation to dose levels that are within regulatory or design limits for individual exposure.

This chapter is an updated version of the chapter titled "Basics of Shielding" in the first edition of *Proton Therapy Physics* [1]. The primary focus of this chapter is secondary radiation production and shielding. The reader is referred to various other literature [2,3] for an extensive coverage of radiological considerations at charged particle therapy facilities.

7.2 SECONDARY RADIATION PRODUCTION

The interaction of protons with any material in their path results in the production of secondary radiation. Losses of the primary proton beam may occur along the beamline or in beamline components. Beam losses occur in the synchrotron and cyclotron during injection, during energy degradation in the cyclotron, during beam transport to the treatment room, in passive scattering systems, range degraders, and modulators, and in beam-shaping devices such as collimators, apertures, and range compensators placed near the patient [2]. Secondary radiation is also produced in the patient and in the dosimetric phantom. Thus, the proton therapy facility will require structural shielding to reduce individual radiation exposure below regulatory dose limits.

Secondary radiation consists of both prompt and residual radiation. Prompt radiation is produced *only* while the machine is on. Residual radiation is produced by materials that have become radioactive during beam operation (referred to as "activated material"). Residual radiation remains, even after the machine is turned off, for a time period that is determined by the half-life of the activated material. Short-lived radioactive materials decay away rapidly, while, long-lived radioactive materials continue to emit radiation for a much longer period of time. The large concrete (structural shielding) thicknesses for the facility are determined by the prompt radiation, while the residual radiation can be shielded with considerably less-localized shielding.

7.2.1 Physics of Secondary Radiation Production

Nuclear interactions are the major mechanisms through which secondary radiation is produced. They can be elastic, nonelastic, or inelastic. In elastic interactions, kinetic energy is conserved in the interaction between the proton and the nucleus, while in nonelastic scattering, kinetic energy is not conserved. Inelastic interactions are a subset of nonelastic reactions, where the final nucleus is the same as the target nucleus. Evaporation and the intranuclear cascade are the two important nuclear processes in secondary radiation production for protons in the therapeutic range [1–5].

The interaction of low-energy protons ($E_p < 10$ MeV) with a nucleus can be described by the compound nucleus model [5]. The incident particle is absorbed by the target nucleus, resulting in the formation of a compound nucleus. The compound nucleus is in an excited state with a number of allowed decay channels. The preferred decay channel is the entrance channel. As the energy of the incident particle increases, the number of levels available to the incident channel increases considerably. Instead of discrete levels in the quasi-stationary states of the compound nucleus, there is a complete overlapping of levels inside the nucleus.

The interaction of protons in the energy range of 50–1000 MeV with matter results in the production of an intranuclear cascade (spray of particles), in which neutrons have energies as high as the incident proton. Thus, the intranuclear cascade is an important consideration for therapeutic protons. There are five distinct and independent stages to be considered, as shown in Figure 7.1 [4,6].

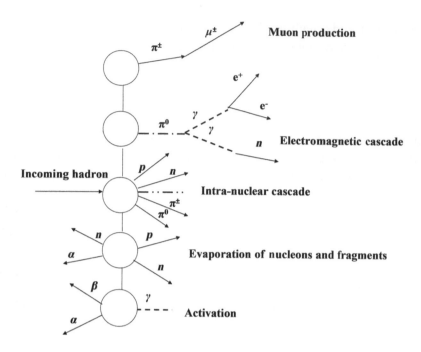

FIGURE 7.1 A schematic representation of various stages of an intranuclear cascade. (Adapted from Ref. [6] with permission of the International Commission on Radiation Units and Measurements, http://ICRU.org.)

7.2.2 Intranuclear Cascade

A qualitative description of the intranuclear cascade is as follows. An intranuclear cascade is produced when an incoming hadron (proton, neutron, etc.) with energy less than ~2 GeV, interacts with individual nucleons in a nucleus, producing a spray of particles, such as protons, neutrons, and π mesons (pions). Pions can be charged ($\pi^\pm$) or neutral. At particle energies below ~100 MeV, all interactions occur just between nucleons and the process is called nuclear cascade [7]. As the incident particle energy increases, the threshold energies for particle production in nucleon–nucleon collisions are exceeded.

The scattered and recoiling nucleons resulting from the interaction proceed through the nucleus. Each nucleon may in turn interact with other nucleons in the nucleus, leading to the development of a cascade. Only those cascade particles that have sufficiently high energy escape from the nucleus, and they are emitted in the direction of the incident particle. The rest of the energy is equally distributed among nucleons in the nucleus which is left in a highly excited state [7]. During the creation of a compound nucleus, the energy which is initially concentrated on a few nucleons spreads through the composite nucleus. The latter evolves towards a state of statistical equilibrium. Nucleons or fragments, having considerable energy, may be ejected during this precompound stage. These nucleons and fragments are referred to as "pre-equilibrium" or "precompound" particles. They may be emitted after each interaction between the incident particle or another cascade particle, and a nucleon inside the nucleus. The pre-equilibrium particles have higher energies than the particles emitted during the equilibrium decay.

The residual nucleus evaporates particles such as alpha particles and other nucleons, which contribute only to local energy absorption. When particle emission is no longer energetically possible, the remaining excitation energy is emitted in the form of gamma rays.

A large fraction of the energy in the cascade is transferred to a single nucleon. This nucleon with energy greater than 150 MeV propagates the cascade. The cascade neutrons that arise from individual nuclear interactions are forward-peaked and have longer attenuation lengths than evaporation neutrons. However, it is important to note that while the high-energy neutrons transport the cascade, the low-energy neutrons deposit a major fraction of the absorbed dose, even outside thick shields. The attenuation length is defined later in this section.

7.2.2.1 Cross Section

For a specific particle interaction, the cross section, σ, represents the "effective size" of the target atom or nucleus [8]. If a particle beam of fluence Φ (particles cm^{-2}) is incident on a thin slab of shielding material of thickness dx and atomic density N (atoms cm^{-3}), the number of incident particles that interact and are removed from the original fluence, $-d\Phi$, is given by

$$-d\Phi = \sigma N \Phi dx \tag{7.1}$$

where σ is the cross section in cm^{-2}.

$N = \rho N_A / A$, where ρ is the material density (g cm^{-3})

N_A = Avogadro's number = 6.02×10^{23} atoms g mole^{-1}

A is the atomic weight (i.e., the mass number).

Cross sections are usually expressed in units of barns (1.0 barn is 10^{-24} cm^2). A millibarn (mb) = 10^{-27} cm^2. The fluence after traversing a distance x is obtained by integration of Equation 7.1:

$$\Phi(x) = \Phi_0 e^{-N\sigma x} = \Phi_0 e^{-\mu x} \tag{7.2}$$

where Φ_0 is the initial fluence and μ is the linear attenuation coefficient.
$\mu = N\sigma$ (cm^{-1})
   The attenuation length (defined later) is given by

$$\lambda = 1/N\sigma \tag{7.3}$$

Figure 7.2 shows the neutron inelastic cross sections of various materials [6].
   The inelastic cross sections are represented by the following equation:

$$\sigma_{in} = 43.1 A_N^{0.70} mb \tag{7.4}$$

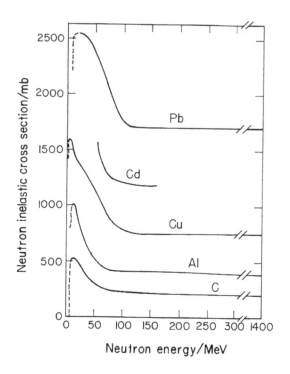

FIGURE 7.2 Neutron inelastic cross sections as a function of neutron energy for various materials. (Reprinted from Ref. [6] with permission of the International Commission on Radiation Units and Measurement, http://ICRU.org.)

This equation is valid for $A_N \geq 3$ and $E_p \geq 150$ MeV regardless of whether the particle is a proton or neutron. The cross sections increase with increasing mass number and decrease with increasing energy to constant values above ~150 MeV. Thus, cascade neutrons with energy >150 MeV will eventually control the radiation field for protons with energy above 150 MeV. The macroscopic inelastic cross section Σ_{in} is given by

$$\Sigma_{in} = \frac{\sigma_{in} N_A}{A} \, \text{cm}^{-1} \tag{7.5}$$

where N_A is the Avogadro constant and A is the molar mass of the medium.

The neutron absorption length, λ_{in}, is given by

$$\lambda_{in} = \frac{1}{\Sigma_{in}} \, \text{cm} \tag{7.6}$$

Radiation transmission can be approximated by an exponential function over a limited range of thickness [5]. The attenuation length is defined as the distance traveled in the medium through which the intensity of the radiation is reduced to 37% of its original value. The neutron attenuation length, λ, is given by

$$\lambda = \frac{1}{\Sigma} \, \text{cm} \tag{7.7}$$

where Σ is the total neutron macroscopic cross section.

It is important to note that while the high-energy neutrons transport the cascade, the low-energy neutrons deposit a major fraction of the absorbed dose, even outside thick shields, except in the forward direction.

Nucleons with energies between 20 and 150 MeV transfer energy to several nucleons. Therefore, on an average, each nucleon receives energy of about 10 MeV. Charged particles at these low energies are quickly stopped by ionization. Thus, at low energies, there is a preponderance of neutrons.

7.2.3 Production of Muons and Electromagnetic Cascade

Charged pions decay into muons and neutrinos. Muons have a mass of 105.7 MeV/c^2. They are very penetrating particles compared to other charged particles. They deposit energy by ionization. Photonuclear reactions are also possible. Protons and pions with energy less than 450 MeV have a high rate of energy loss. Thus, neutrons are the principal propagators of the cascade with increasing depth in the shielding.

Neutral pions decay into two energetic gamma rays which initiate electromagnetic cascades. The photons that are produced interact through pair production or Compton collisions, resulting in the production of electrons. These electrons radiate high-energy photons (bremsstrahlung), which in turn interact to produce more electrons. At each step in the cascade, the number of particles increases and the average energy decreases. This process continues until the electrons fall into the energy range where collision losses dominate over radiative losses, and the energy of the primary electron is completely dissipated in excitation and ionization of the atoms, resulting in heat production. This entire process resulting in a cascade of photons, electrons, and positrons is called an electromagnetic cascade. A very small fraction of the bremsstrahlung energy in the cascade goes into the production of hadrons, such as neutrons, protons, and pions. The energy that is transferred is mostly deposited by ionization within a few radiation lengths. However, the attenuation length of these cascades is much shorter than the neutron absorption length, which is the reciprocal of the inelastic cross section, and therefore, does not include elastic scattering of neutrons. Thus, the electromagnetic cascade does not contribute significantly to energy transport. It is important to note that the intranuclear cascade dominates for protons in the therapeutic energy range of interest.

7.2.4 Evaporation Process and Activation

The energy of those particles that do not escape is assumed to be distributed among the remaining nucleons in the nucleus, leaving it in an excited state. It then de-excites by emitting particles, mainly neutrons and protons, which are referred to as "evaporation nucleons," alpha particles, and some fragments. The evaporation nucleons are so called, because they can be considered as boiling off a nucleus which is heated by the absorption of energy from the incident particle. The energy distribution of emitted neutrons can be described by the following equation:

$$n(E)dE = aEe^{-E/\tau} \tag{7.8}$$

where a is a constant, E is the energy of the neutron, and τ is the nuclear temperature which has the dimensions of energy with a value that lies between 0.5 and 5 MeV. The evaporated particles are emitted isotropically in the laboratory system, and the energy of the evaporation neutrons extends to 8 MeV in shielding materials such as concrete [4,5]. The emission of low-energy charged particles is suppressed by the Coulomb barrier. Therefore, charged particles produced by evaporation are unimportant in shielding considerations.

The evaporation neutrons travel long distances, continuously depositing energy. Evaporation neutrons produced by interactions near the source contribute to dose inside the shield and to leakage dose through doors and openings. However, because they are strongly attenuated in the shield, they do not contribute to dose outside the shield. The dose outside the shield is dominated by evaporation neutrons produced near the outer surface of the shield. The remaining excitation energy may be emitted in the form of gammas. The de-excited nucleus may be radioactive, thus leading to residual radiation.

7.2.5 Spallation

Spallation is the emission of a large number of nucleons and fragments caused by the interaction of a high-energy proton or neutron with kinetic energies from ~100 MeV to several GeV, with a target nucleus. It includes the intranuclear cascade and de-excitation and plays an important role in activation.

7.3 PROMPT RADIATION AND SHIELDING

The prompt radiation field produced by protons of energies up to 330 MeV encountered in proton therapy is quite complex. It consists of a mixture of charged and neutral particles as well as photons. Neutrons are the dominant component of the prompt radiation field and dictate the structural shielding thicknesses. However, f or mazes and penetrations, low-energy neutrons and capture gamma rays also contribute to dose.

7.3.1 Neutron Energy Classification and Interactions

Neutrons are classified according to their energy as follows:

Thermal: $\bar{E}_n = 0.025\,\text{eV}$ at 20°C. Typically, $E_n \leq 0.5\,\text{eV}$

Intermediate: $0.5\,\text{eV} < E_n \leq 10\,\text{keV}$

Fast: $10\,\text{keV} < E_n \leq 20\,\text{MeV}$

Relativistic: $E_n > 20\,\text{MeV}$

High-Energy Neutrons: $E_n > 100\,\text{MeV}$

where E_n is the energy of the neutron and $\bar{E}_n$ is the average energy of the neutron.

Because neutrons are uncharged, they can travel appreciable distances in matter without undergoing interactions. A neutron can undergo an elastic or an inelastic reaction upon collision with an atom [9]. An elastic reaction is one in which the total kinetic energy of the incoming particle is conserved. In an inelastic reaction, the nucleus absorbs some energy and is left in an excited state. Inelastic scattering can occur only at energies above the lowest excited state (or inelastic scattering threshold) of the material. The neutron can also be captured or absorbed by a nucleus in reactions such as $(n, 2n)$, (n, p), (n, α), or (n, γ). These reactions result in the change of the atomic mass number and/or the atomic number of the target nucleus.

7.3.1.1 Thermal Neutrons

Thermal neutrons (n_{th}) are in approximate thermal equilibrium with their surroundings. They gain and lose only small amounts of energy through elastic scattering, but they diffuse about until captured by atomic nuclei. Thermal neutrons undergo radiative capture, i.e., neutron absorption leads to the emission of a gamma ray, such as in the $^1H(n_{th}, \gamma)^2H$ reaction. The capture cross section for this reaction is 0.33×10^{-24} cm^2, and the emitted gamma ray energy is 2.22 MeV. This reaction occurs in hydrogenous shielding materials such as polyethylene and concrete. For shielding maze doors, borated polyethylene is used instead of polyethylene, because the cross section for capture in boron is much higher $(3,480 \times 10^{-24}$ cm$^2)$ and the subsequent capture gamma ray from the ^{10}B $(n_{th}, \alpha)^7Li$ has a much lower energy of 0.48 MeV.

7.3.1.2 Intermediate Neutrons

The capture cross sections for low-energy neutrons (<1 keV) decrease as the reciprocal of the velocity or the reciprocal of the square root of the neutron energy. As neutron energy increases, the capture cross section decreases. Intermediate energy neutrons lose energy by scattering and are absorbed.

7.3.1.3 Fast Neutrons

Fast neutrons include evaporation neutrons and neutrons from specific nuclear reactions such as (p, n) reactions. They interact with matter mainly through a series of elastic and inelastic scattering, and are finally absorbed after giving up their energy [6]. Approximately, 7 MeV is given up to gamma rays, on an average, during the slowing down and capture process. Therefore, the shielding calculations must take capture gamma rays into account for both structural shielding (if it is thick enough) as well as scattering through mazes and penetrations. Inelastic scattering is the dominant process in all materials at neutron energies above 10 MeV. Elastic scattering dominates at lower energies. Below 1 MeV, elastic scattering is the principle process by which neutrons interact in hydrogenous materials. A large amount of elastic scattering takes place in high-Z materials, but results in negligible energy loss. However, elastic scattering increases the path length of the neutrons in the shielding material, thus providing more opportunities for inelastic and $(n, 2n)$ reactions to occur. The mean free path is the average distance traveled by the particle in the material between two interactions. When high-Z material such as steel or lead is used for shielding, it must always be followed by a hydrogenous material such as concrete or polyethylene. The reason being that the energy of the neutrons may be reduced by inelastic scattering in the high-Z material to a lower energy, at which the neutrons may be transparent to the nonhydrogenous material. For example, lead is virtually transparent to neutrons with energy below 0.57 MeV [10].

The sum of the inelastic and $(n, 2n)$ cross sections in the energy range <20 MeV is called the nonelastic cross section [10]. The inelastic scattering dominates at lower energies, while the $(n, 2n)$ reactions dominate at higher energies. In an inelastic collision, there is a minimum energy loss that equals the energy of the lowest excited state; however, the energy loss in any inelastic collision cannot be determined exactly. The binding energy of a nucleus is the energy that would be needed to take it apart into its individual protons and neutrons. Typically, there is a large energy loss in a single collision, resulting in the excitation of energy states above the ground state, followed by the emission of gamma rays. The minimum energy loss is equal to the binding energy of the neutron in the $(n, 2n)$ reaction, which produces a large number of lower-energy neutrons, because the energies of the two neutrons that are produced are similar. In high-Z materials, a large amount of elastic scattering takes place, but results in negligible energy loss. However, the mean free path or path length of the neutrons in the shielding material increases, thus providing more opportunities for inelastic and $(n, 2n)$ reactions to occur.

7.3.1.4 Relativistic Neutrons

Relativistic or "cascade" neutrons arise from cascade processes in proton accelerators. They are important in propagating the radiation field. The high-energy component of the cascade with neutron energies above 100 MeV propagates the neutrons through the shielding; and continuously regenerates lower-energy neutrons and charged particles at all depths in the shield via inelastic reactions with the shielding material [11]. Neutrons with energies above ~100 MeV produce copious amounts of neutrons through spallation.

The reactions occur in three stages for neutrons with energies between 50 and 100 MeV. In the first stage, an intranuclear cascade develops, where the incident high-energy neutron interacts with an individual nucleon in the nucleus [12]. In the second stage, the residual nucleus is left in an excited state and evaporates particles such as alpha particles and other nucleons. In the third stage, after particle emission is no longer energetically possible, the remaining excitation energy is emitted in the form of gamma rays. The de-excited nucleus may be radioactive. For neutrons with energy below 50 MeV, only the second and third stages are assumed to be operative.

7.3.2 Neutron Yield and Angular Distribution

As the proton energy increases, the threshold for nuclear reactions is exceeded and more nuclear interactions can occur. At energies above 50 MeV, the intranuclear cascade process becomes important. The neutron yield of a target is defined as the number of neutrons emitted per incident primary particle. Calculations using Monte Carlo codes and measurements of neutron yields, energy spectra, and angular distributions for protons of various energies incident on different types of materials have been reported in the literature [5,13–19]. Between proton energies of 50 and 500 MeV, the neutron yields increase as approximately E_{p^2} for all target materials, where E_p is the energy of the incident proton [5]. Neutron yields increase with increasing proton energy because the thresholds for neutron production in the target material are exceeded.

Thick targets are targets in which the protons are completely stopped, i.e., the thickness is greater than or equal to the particle range. They are also referred to as "stopping targets." By contrast, thin targets are targets with thicknesses that are significantly less than the particle range. Thus, for example, the protons lose an insignificant amount of energy in the target, and the kinetic energy available for neutron production in the target is the full incident proton energy [5]. The yield from a thin target will be proportional to the target thickness. A stopping target can be used to determine dose from the beam incident on the patient, but, the use of a stopping target is not necessarily conservative in other cases where thin targets are involved. The dose equivalent (Sv m^2 p^{-1}) in the forward direction as a function of depth in concrete for protons with energies from 100 to 400 MeV is typically higher for the thin copper target than for the thick copper target [4]. The effective attenuation length for a thin copper target is higher than for a thicker copper target. Ipe [3] has shown that the forward-directed neutrons are more penetrating for a thin graphite target than a thick graphite target.

According to data provided by Tesch [13], as the mass number (not atomic number, Z) increases, the neutron yield increases. For shielding purposes, the angular neutron yield and spectra (usually expressed as double differential neutron yield) are more meaningful than the total neutron yield, because targets encountered in proton therapy facilities are not always thick targets. Therefore, the target material impacts the shielding thickness.

7.3.3 Neutron Spectra

The characteristics of the unshielded and shielded neutron spectra are discussed in this section.

Figure 7.3 shows the unshielded neutron spectra for neutrons at various emission angles, produced by 250 MeV protons incident on a thick iron target (without any concrete shielding) [14].

The double differential neutron spectra shown as lethargy plots were calculated with the Monte Carlo radiation transport code, FLUKA (FLUktuierende KAskade) [18,19]. Lethargy plots are plots in which the differential neutron spectra are multiplied by the neutron energy E, to account for the logarithmic energy scale. Neutron lethargy or logarithmic energy decrement, u, is a dimensionless logarithm of the ratio of the energy of source neutrons (E_0) to the energy of neutrons (E) after a collision:

$$u = \ln \frac{E_0}{E} \tag{7.9}$$

$$E = E_0 \exp(-u) \tag{7.10}$$

A plot of E vs. u will show an exponential decay of energy per unit collision, indicating that the greatest changes of energy (ΔE) result from the early collisions. The energy distributions in these figures are typically characterized by two peaks: a high-energy peak (produced by the scattered primary beam particle) and an evaporation peak at ~2 MeV. The high-energy peaks shift to higher energies with increasing proton energies, which

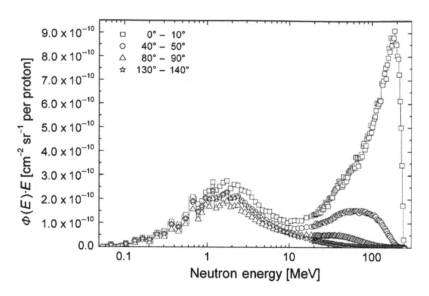

FIGURE 7.3 Unshielded neutron spectra for neutrons at various emission angles, produced by 250-MeV protons incident on a thick iron target (without any concrete shielding). (Reprinted from Ref. [14] with permission from Elsevier.)

are particularly evident in the forward direction (0°–10°). The high-energy peak for the unshielded target is not the usual 100 MeV peak that is observed outside thick concrete shielding, which will be discussed in the next section.

The high-energy component of the cascade with neutron energies (E_n) above 100 MeV propagates the neutrons through the shielding and continuously regenerates lower-energy neutrons and charged particles at all depths in the shield via inelastic reactions with the shielding material [11]. However, the greater yield of low-energy neutrons is more than compensated for by greater attenuation in the shield due to a higher cross section at low energy. Shielding studies indicate that the radiation field reaches an equilibrium condition beyond a few mean-free paths within the shield. Neutrons with energies greater than 150 MeV regenerate the cascade even though they are present in relatively small numbers. They are accompanied by numerous low-energy neutrons produced in the interactions. The typical neutron spectrum observed outside a thick concrete shield consists of peaks at a few MeV and at ~100 MeV. The high-energy neutrons are anisotropic and forward peaked, but the evaporation neutrons are isotropic. The highest-energy neutrons detected outside the shielding are those that arrive without interaction, or that have undergone only elastic scattering or direct inelastic scattering with little loss of energy, and a small change in direction. Low-energy neutrons and charged particles detected outside the shielding are those that have been generated at the outer surface of the shield. Thus, the yield of high-energy neutrons ($E_n > 100$ MeV) in the primary collision of the protons with the target material determines the magnitude of the prompt radiation field outside the shield in the therapeutic proton energy range of interest. The charged particles produced by the protons will be absorbed in shielding that is sufficiently thick to protect against neutrons. Thus, neutrons dominate the radiation field outside the shielding. Degraded neutrons might undergo capture reactions in the shielding, giving rise to neutron-capture gamma rays.

Figure 7.4 shows the normalized neutron spectra in the transverse direction at the surface of the concrete and at various depths in the concrete, for 250 MeV protons, incident on a thick iron target [14].

The low-energy neutron component is attenuated up to about a depth of 100 cm with a short attenuation length, thus giving rise to a less intense but more penetrating spectrum with a longer attenuation length. Beyond a depth of 100 cm, the spectrum reaches equilibrium.

Thus, the neutron energy distribution consists of two components, high-energy neutrons produced by the cascade and evaporation neutrons with energy peaked at ~2 MeV. As previously mentioned, high-energy neutrons are anisotropic and forward peaked, but the evaporation neutrons are isotropic. Because the neutron energy spectra extend to the energy of the incident proton, it is important to use wide-energy range instruments for neutron monitoring [20,21].

7.3.4 Shielding Calculations

Shielding calculations are performed to ensure that the facility is designed and constructed so that exposures of personnel and public are well within regulatory limits. The primary purpose of radiation monitoring is to demonstrate compliance with design or regulatory limits, thus permitting safe operation. Naturally, this implies that calculations and

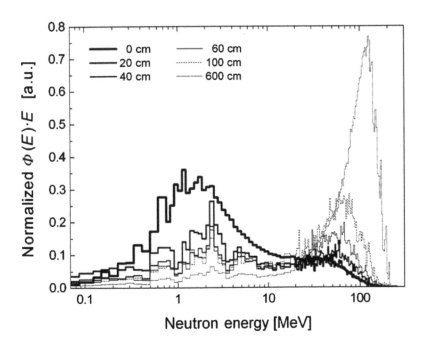

FIGURE 7.4 Normalized neutron spectra in transverse direction at surface of a concrete shield, and at various depths, produced by 250 MeV protons incident on a thick iron target. (Reprinted from Ref. [14] with permission from Elsevier.)

measurements must be expressed in terms of quantities in which the regulatory dose limits are defined.

7.3.4.1 Dose Quantities

Dose limits that are expressed in terms of protection quantities are defined by the International Commission on Radiological Protection (ICRP) [22,23]. An example of such a dose limit is the effective dose (E) that is measured in the human body and is not directly measurable. Therefore, for external individual exposure, the operational quantity-ambient dose equivalent $H^*(d)$, defined by the International Commission on Radiological Units (ICRU), can be used [24]. Frequently, radiation fields are characterized in terms of absorbed fluence [25].

The protection quantities and operational quantities can be related to the particle fluence and, in turn, by conversion coefficients to each other.

The weighting factor, w_R for the protection quantity, equivalent dose, recommended by ICRP Publication 103 [23], is shown in Table 7.1. In general, for radiation protection, the operational dose quantities provide a good and conservative estimate of effective dose under most exposure conditions, when the w_R values from Table 7.1 are used for neutrons. However, there are exceptions as discussed in the next section.

7.3.4.2 Conversion Coefficients

Conversion coefficients are used to relate the protection and operational quantities to physical quantities characterizing the radiation field [25]. Thus, for example, the effective dose, E, can be obtained by multiplying the fluence with the fluence-to-effective dose conversion coefficient. The ambient dose equivalent, $H^*(d)$, can be obtained by multiplying the fluence with the fluence-to-ambient dose equivalent conversion coefficient.

Conversion coefficients have been calculated by various authors using the FLUKA [18,19] for many types of radiation (photons, electrons, positrons, protons, neutrons, muons, charged pions, kaons) and incident energies up to 10 TeV [26]. Because the conversion coefficient for $H^*(10)$ for neutrons becomes smaller than that for E(AP) (where AP refers to Anterior–Posterior) above 50 MeV, use of E(AP) may be considered more conservative for

TABLE 7.1 Radiation Weighting Factors Recommended by ICRP Publication 103 [23]

Radiation Type	Energy Range	W_R
Photons, electrons, and muons	All energies	1
Neutrons	<1 MeV	$W_R = 2.5 + 18.2\exp[-\frac{(\ln(E))^2}{6}]$
Neutrons	1–50 MeV	$W_R = 5 + 17\exp[-\frac{(\ln(2E))^2}{6}]$
Neutrons	>50 MeV	$W_R = 2.5 + 3.5\exp[-\frac{(\ln(0.04E))^2}{6}]$
Protons, other than recoil protons	>2 MeV	2
Alpha particles, fission fragments, and heavy nuclei	All energies	20

Source: Reference [3]. With permission from Medical Physics Publishing, Inc.

high-energy neutrons. The conversion coefficient for E(AP) becomes smaller than that for Posterior–Anterior (PA) irradiation geometry, E(PA), at neutron energies above 50 MeV. However, the integrated dose from thermal neutrons to high-energy neutrons is highest for AP geometry, and therefore, the choice of E(AP) is more conservative than the choice of E(PA).

For mazes where low-energy neutrons and photons are encountered, the ambient dose equivalent is more conservative than E(AP). However, the effective dose is the legal quantity in regulatory requirements. The term "dose" is used in a generic sense for the dose equivalent or effective dose in this chapter.

7.3.4.3 Methodology

The methodology for shielding calculations includes analytical methods, Monte Carlo calculations, or use of computational models.

7.3.4.3.1 Analytical Methods

Most analytical models consist of line-of-sight or "point kernel" models. They are limited in their use because they are based on simplistic assumptions and geometry. Many of the models are restricted to transverse shielding. Further, they do not account for changes in energy, angle of production, target material and dimensions, and concrete material composition and density. They tend to grossly overestimate or underestimate the true doses, and are therefore *not* recommended for proton therapy shielding calculations.

7.3.4.3.1.1 Point Source

A radiation source can be considered a point source if the distance at which the dose is determined is at least five times the dimension of the radiation source. For a point source, the dose decreases as the square of the distance. Because protons have a defined range in the target material, the dimension of the source can be considered to be the range of the proton in the target material. For example, the range of a 230 MeV proton in tissue is 32 cm, while the distance from the target to the point of interest outside the shielding walls is typically greater than 1.6 m (5 × 32 cm).

7.3.4.3.1.2 Attenuation Length

Radiation transmission can be approximated by an exponential function over a limited range of thickness [5]. Attenuation length was defined in Section 7.2.2.1. The attenuation length, λ, is usually expressed in cm (or m) and in g cm^{-2} (or kg m^{-2}) when multiplied by density (ρ). The value of λ changes with increasing depth in the shield for thicknesses (ρd) that are less than ~100 g cm^{-2}, because the "softer" radiations are more easily attenuated, and the neutron spectrum "hardens." The attenuation length ($\rho\lambda$) for monoenergetic neutrons increases with increasing neutron energy at energies greater than ~20 MeV [4]. In the past, it has typically been assumed that the attenuation length reaches a high-energy limiting value of about 120 g cm^{-2}, even though the data [4] shows a slightly increasing trend above 200 MeV. According to Nakamura [27], the measured neutron dose attenuation length (thermal to maximum energy) for concrete, in the forward direction, gradually increases above 100 MeV to a maximum value of about 130 g cm^{-2}, which may be considered the high-energy limit. It is important to note that, in addition to particle type and energy, λ also depends on the production angle (θ), material composition, and density.

The effective attenuation length in the forward direction as a function of proton energy for thin and thick copper targets is reported in the literature [5]. The values of λ_{eff} in the forward direction for neutrons from a thin copper target were higher than those from a thick target, because of the softer spectrum emitted from the thick target compared to the thin target.

It is important to note that, in addition to particle type and energy, λ also depends on the production angle (θ), material composition, and density.

7.3.4.3.1.3 Removal Cross Section For fast neutrons (E_n < 20 MeV), the most common approach to calculate shielding thicknesses is that of the removal cross-section theory. Neutron shielding is complicated by significant nuclear structure effects that result in resonances associated with compound nucleus that can be excited [8]. There are also several nuclear reaction channels that lead to a large number of nuclear excited states. These excited states have a wide variety of nuclear structure quantum numbers and very narrow widths in energy. Clark [28] has provided some general rules for fast neutron shielding.

The removal cross section can be applied to dose calculations as follows: The dose equivalent, $H(d)$, as a function of shield thickness, is approximately given by the line-of-sight equation:

$$H(d)=\frac{H_0 e^{-\Sigma_r d}}{r^2} \tag{7.11}$$

where H_0 is the unshielded dose at a distance of 1 m from the source, r is the distance from the source to the point of interest outside the shield, and d is the thickness of the shield in cm, and Σ_r is the macroscopic removal cross section in cm^{-1}.

$$\Sigma_r = \frac{N_0 \sigma_r \rho}{A} \tag{7.12}$$

where σ_r is the microscopic removal cross section in cm, N_0 is Avogadro's number, ρ is the density (g cm^{-3}), and A is the mass number. The attenuation data for fast neutrons can be adequately described by removal cross sections, even though neutrons of energy lower than a few MeV dominate at greater depths in the shielding.

7.3.4.3.1.4 Moyer Model Burton Moyer developed a semiempirical method for the shield design of the 6 GeV proton Bevatron [4] in 1961. This model is only applicable to production angles close to 90° and the transverse shielding for a high-energy proton accelerator is determined using the following simple form of the Moyer model [29]:

$$H = \frac{H_0}{r^2}\left[\frac{E_P}{E_0}\right]^\alpha \exp\left[-\frac{d}{\lambda}\right] \tag{7.13}$$

where H = maximum dose equivalent rate at a given radial distance (r) from the target, d = shield thickness, E_P = proton energy, E_0 = 1 GeV, H_0 = 2.6×10^{-14} Sv m^2, and α is about 0.8.

The Moyer model is effective in the GeV region, because the neutron dose attenuation length (λ) is nearly constant regardless of energy. However, the model is restricted to the determination of neutron dose equivalent produced at an angle between 60° and 120°. At proton energies in the therapeutic range of interest, the neutron attenuation length increases considerably with energy. Therefore, the Moyer model is ineffective and inappropriate for use in proton therapy shielding.

Kato and Nakamura [30] have developed a modified version of the Moyer model, which includes changes in attenuation length with shield thickness and also includes a correction for oblique penetration through the shield.

7.3.4.3.2 Monte Carlo Calculations   The Monte Carlo radiation transport codes such as FLUKA [19,20,22,23] and MCNP (Monte Carlo N-Particle)/ MCNPX (Monte Carlo N-Particle eXtended) [31] have been used extensively in shielding calculations for particle accelerators of all energies. An extensive coverage of other codes used in shielding can be found in PTCOG (Particle Therapy Co-operative Group) Report 1 [2]. These codes can be used to perform a full simulation, modeling the accelerator and room geometry in its entirety. They can also be used to derive computational models, as discussed in the next section. An important difference between FLUKA and MCNP/MCNPX is their treatment of low-energy neutrons. Transport of low-energy neutrons is performed in FLUKA by use of energy multigroups, whereas the MCNP/MCNPX simulations are based on evaluated continuous-energy cross-section tables in the code. Therefore, neutron transport in MCNP/MCNPX is immune from multigroup approximations and self-shielding effects. Thus, MCNP/MCNPX may be more suitable for maze and penetration scatter calculations [32].

There are also variations between codes, cross-section libraries, and models. Oh et al. [33] determined double differential neutron yields from Be, C, Al, and Fe targets for incident proton energies of 113 and 256 MeV protons. They compared these yields with Meier's experimental data [34]. Neutron yields from protons with energies of 100, 150, 200, and 230 MeV incident on thick Al, Fe, Cu, and Pb targets were also studied. The results calculated using FLUKA and PHITS (Particle and Heavy Ion Transport **Code** System) [35] agreed well with Meier's experimental results. However, the MCNPX results were not in agreement with experimental results. When the LA150 cross-section library was used in MCNPX, this discrepancy disappeared. They also found that neutron yields calculated using MCNPX and PHITS increase continuously for neutron energies lower than 1 MeV. However, FLUKA results showed a decrease of yields at lower neutron energies. A comparison of the results between the three codes indicated reasonable agreement at 90°, but discrepancies were found at forward and large angles with factors as large as 5.

Sheu et al. [36] performed benchmark calculations for double differential neutron yields based on experimental data from Meier et al. [34] for a 256 MeV proton beam incident on a thick iron target. Calculations were performed with FLUKA v2011.2 and MCNPX v2.7.0. The results indicate that at large production angles (120° and 150°), MCNPX overestimates the neutron yields above 20 MeV compared to FLUKA and experimental data. Therefore, one expects MCNPX to overestimate the dose at large angles compared to FLUKA, at least for a thick iron target.

The code GEANT4 (GEometry ANd Tracking) [37,38] is a toolkit that allows free definition of complex geometries, scoring options, and physical parameters. However, great care must be exercised in its use, because the improper choice of physical parameters can lead to inconsistent results [39].

Monte Carlo codes have been used in the shielding design of several particle therapy facilities [39–46]. Results of simulations performed for an IBA (Ion Beam Applications, Louvain-La-Neuve, Belgium) cyclotron show that at forward angles, doses calculated using GEANT4 are two times smaller than MCNPX values, while at backward angles, doses from the GEANT4 simulations are larger than MCNPX doses, up to a factor of 3 for tissue (at a depth of 3 m in the concrete). For the transmission through the walls, discrepancies of up to a factor of 3 were noted between MCNPX and GEANT4 [39].

For comparison of Monte Carlo calculations to experimental data, the actual experimental configuration should be modeled, including the beamline components, the instrument response, and the site-specific concrete composition. Appropriate calibration correction factors need to be applied to the instrument response. Further, the experiment should have been performed using the appropriate instrumentation such as wide-energy neutron rem-meters. Any deviations from the earlier conditions will result in large discrepancies between measurements and simulations. Unfortunately, there is a paucity of published data for charged proton therapy facilities that meets all these conditions. Further, several authors have used theoretical concrete composition and densities in their Monte Carlo calculations, which make their results invalid. Satoh et al. [47] measured neutron doses behind the concrete shields and at the maze of the Fukui Prefectural Hospital Proton Therapy Center, Japan, with three types of radiation monitors (DARWIN, Wendi-2, and a rem-meter) and solid-state nuclear track detectors. The measured data were compared with estimations using analytical models and the Monte Carlo code, PHITS. The analytical model resulted in considerably higher values than the measured data, indicating that the facility was designed with a sufficient margin of safety. The results calculated by PHITS were lower than those determined with the analytical model but were about three times larger than the measured data.

De Smet et al. [44] have reported a comparison of simulated results ($H^*(10)$) using FLUKA and GEANT4) with measured results using Wendi 2 for an IBA cyclotron facility. At all positions located outside the vaults, the simulated $H^*(10)$ values systematically overestimated the WENDI-2 measurements. For most positions located inside the access mazes of the rooms, the simulated results match the measured WENDI-2 responses within a factor 2.5. Outside the shielded rooms, all simulation results overestimated the WENDI-2 measurements by approximately a factor of 2–7, except for one position where the factor was 12 (with MCNPX). The MCNPX values were higher than the GEANT4 values by up to 209%. It is important to note that appropriate calibration correction factors were not applied to their measurements. It is also not clear whether site-specific concrete composition and density were used in the calculations. Since then, De Smet [46] has published new data that includes calibration correction factors. Neutron spectrometry measurements were performed with an extended-range Bonner Sphere Spectrometer. Neutron $H^*(10)$ measurements were made using an extended-range rem-meter WENDI-2, a conventional rem-meter LB 6411, and a tissue-equivalent proportional counter. All measurements were performed inside and around the Fixed-Beam Treatment Room at the proton therapy facility of Essen, in Germany. The measurements were also compared to simulation results obtained with MCNPX 2.7.0 using two different physics models for the hadron interactions above 150 MeV: the Bertini & Dresner models and the CEM03 model. The Bertini & Dresner simulations resulted in dose rates that were 1.6–1.8 higher than the CEM03 simulations. The latter underestimated total neutron production rate in water by a factor of 1.3, while the Bertini & Dresner models reproduce the neutron production in the water relatively well. The Bertini model appeared to overestimate the production of neutrons with energies above 100 MeV at the most forward angle. Outside the shielding, the simulated $H^*(10)$ overestimated the WENDI-2 measurements by factors of 2–3 when compared to the Bertini & Dresner models, and 1.1–1.7 when compared to the CEM03 model. Thus, both simulations were conservative. The authors conclude that the conservative behavior is probably caused by a combination of uncertainties in cross sections and uncertainties in concrete composition and density. Clearly, concrete composition and density play a significant role in the attenuation of neutrons.

In the early stages of design, the facility undergoes several iterations of changes in layout, and therefore, a full Monte Carlo simulation is not practical or cost effective. According to Sunil [48], the cumulative CPU (central processing unit) time taken for the simulation of doses outside the shielding for four sources in the cyclotron room, was 900 days. Full simulations should be performed only after the layout has been finalized. Full Monte Carlo simulations should be used for special issues such as maze design, penetration shielding, skyshine, and groundshine.

7.3.4.3.3 Computational Models   The use of computational models to evaluate structural shielding is a commonly accepted industry practice [1–4,14,36,49–51] because of the deep penetration problems associated with thick shields encountered in proton therapy and the requirement for extremely cumbersome runs to get good statistics. Further, the entire process will have to be repeated if there are any changes to the shielding configuration. The use of computational models should be a conservative approach, which is what is desired for shielding calculations. However, this assumption should be validated prior to the use of the computational models. The entire room geometry is not modeled, but usually spherical shells of shielding material are placed around the target. Monte Carlo codes are used to score total dose from all particles, at given angular intervals, and in each shell of shielding material. Plots of dose vs. shielding thickness can be fitted to obtain source terms and attenuation lengths as a function of angle in increments of 10°, and at the energies of interest, with the appropriate target. Site-specific concrete composition, proton parameters, and target material/dimensions are used.

Computational models (derived using Monte Carlo codes) that are independent of geometry typically consist of a source term and an exponential term that describes the attenuation of the radiation. Both the source term and the attenuation length are dependent on particle type, incident proton energy, and angle. Shielding can be estimated over a wide range of thicknesses by the following equation for a point source, which combines the inverse square law and an exponential attenuation through the shield and is independent of geometry [14].

$$H\left(E_p, \theta, d / \lambda(\theta)\right) = \frac{H_0(E_p, \theta)}{r^2} \exp\left[-\frac{d}{\lambda(\theta)g(\theta)}\right] \tag{7.14}$$

where H is the dose equivalent outside the shielding, H_0 is the source term at a production angle θ with respect to the incident beam and is assumed to be geometry independent, E_p is the energy of the incident particle, r is the distance between the target and the point at which the dose equivalent is scored, d is the thickness of the shield, $d/g(\theta)$ is the slant thickness of the shield at an angle θ, $\lambda(\theta)$ is the attenuation length for dose equivalent at an angle θ and is defined as the penetration distance in which the intensity of the radiation is attenuated by a factor of e, $g(\theta) = \cos\theta$ for forward shielding, $g(\theta) = \sin\theta$ for lateral shielding, and $g(\theta) = 1$ for spherical geometry. In some cases, the data may require a multiple exponential fit.

Computational models are especially useful during the schematic phase of the facility design, when the design undergoes several changes, to determine the barrier shielding [51]. Most of the published computational models provide source terms and attenuation lengths for only neutrons. For shielding purposes, the total dose from all particles should be considered instead of only neutron dose, as first proposed by Ipe [50]; further, the effective dose should be calculated. The source terms and attenuation lengths will depend on the combination of composition and density of the shielding material. Ray traces can be performed at various angles, and the source terms and attenuation lengths can be used for dose calculations. These models are also useful in identifying thin shielding and facilitate improved shield design. The strength of the computational model is its simplicity and independence of geometry. However, this strength can result in deviations from the true dose when there is oblique incidence or increased scatter.

Lai et al. [49] compared the results of a full Monte Carlo simulation to computational models for 100–300 MeV protons incident on a thick copper target. According to them, the computational model provided rather accurate predictions on the transmitted doses for a wide range of angles, except for those at angles from 250° to 310°, where the three-leg maze resides. However, the geometry of the treatment room that was used in the simulation was a simple rectangular room, with normal incidence of the proton beam on the target relative to the walls. For oblique incidence, one would expect an overestimate of the dose with the computational models. Further, the cyclotron room geometries are complicated, because they have a lot of intervening structures (both shielding and otherwise) that can cause scatter. A comparison of computational models and full simulations for such a room may not yield such favorable results.

Several authors have published computational models [14,36,49–51]. Such models are only of academic interest and should *not* be used for calculations, since they are typically based on stopping targets (whose material and dimensions can be quite different from actual targets), and some theoretical concrete composition and density (which, can be significantly different from the site-specific concrete composition and density). Computational models for proton energies, target material, and dimensions—as well as concrete composition and density that are facility-specific—should be derived on a case-by-case basis.

Care must be exercised in the use of computational models. There are differences in codes and differences in nuclear models and cross-section libraries that can result in large uncertainties. Depending upon the code and nuclear model that is used, the doses can be overestimated or underestimated.

7.3.4.4 Angular Dose Profile and Dependence on Target Materials and Dimensions

Figure 7.5 shows the angular dose profile from thick unshielded ICRU tissue targets for various proton energies calculated using the Monte Carlo code, FLUKA [3]. The total ambient dose equivalent from all particles is normalized to a distance of 1 m and expressed in pSv per proton. For a given production angle, as the proton energy increases, the dose increases, because the thresholds for nuclear interactions are exceeded. At a given energy, the dose decreases significantly with increasing angle initially and then levels off at larger angles. The decrease in dose is due to the softer spectra at larger angles. Therefore, more shielding will be required in the forward direction.

Table 7.2 shows the dimensions of the ICRU tissue targets that were used in the calculations earlier for various proton beam energies.

Figures 7.6 and 7.7 show the total ambient dose equivalent attenuation curves in concrete in the 0°–10° direction and in the 80°–90° direction, respectively. Simulations were

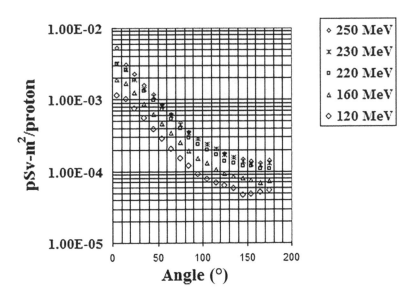

FIGURE 7.5 Angular dose profile from unshielded thick ICRU tissue targets for various proton energies calculated using the Monte Carlo code, FLUKA. (Reprinted from Ref. [3] with permission from Medical Physics Publishing, Inc.)

TABLE 7.2   ICRU Tissue Dimensions as a Function of Energy

Energy (MeV)	250	230	220	160	120
Target length (cm)	38	34	31	18	11
Target radius (cm)	20	17	16.5	10	6.5

Source: Reference [3]. With permission from Medical Physics Publishing, Inc.

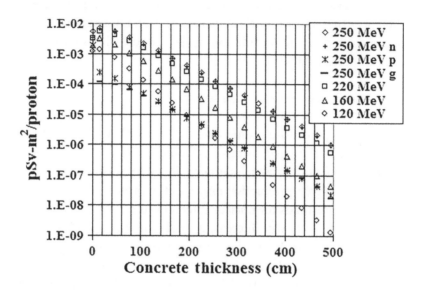

FIGURE 7.6   Dose as a function of shielding thickness in the 0°–10° direction for protons incident on ICRU tissue target. Total dose is shown for all energies. At 250 MeV, neutron (*n*), proton (*p*), and gamma (*g*) dose contributions are also shown. (Reprinted from Ref. [3] with permission from Medical Physics Publishing, Inc.)

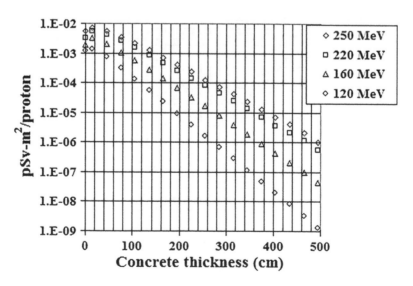

FIGURE 7.7   Dose as a function of shielding thickness in the 80°–90° direction for protons incident on ICRU tissue target. (Reprinted from Ref. [3] with permission form Medical Physics Publishing, Inc.)

performed for a pencil beam of protons incident on thick ICRU tissue targets for various proton energies [3].

A dose buildup in the forward direction can be seen at small depths in Figure 7.6. For thicknesses greater than 75 cm, the source term and attenuation length in the forward direction (0°–10°) produced by 250 MeV protons can be fitted as follows: $H_0 = 1.66 \times 10^{-14}$ Sv m^2 per proton and $\lambda_{tot} = 122.4$ g cm^{-2}, respectively. This fit is only valid for the specific concrete used in calculations. For thin targets, a single exponential fit may not be sufficient over the entire range of concrete thickness.

The neutron (n), proton (p), and gamma (g) components of dose at 250 MeV in the forward direction are also shown. In the forward direction, more than 95% of the dose is from neutrons. The proton dose is higher than the photon dose for concrete thicknesses up to 165 cm. At concrete thicknesses greater than 165 cm, the photon dose is greater than the proton dose. The ratio of neutron dose to total dose decreases at large angles. Therefore, it is prudent to use total dose from all particles for shielding calculations.

From Figure 7.7, it appears that at large angles, at least a two-component fit may be required for thick targets. Plots of dose equivalent vs. shielding indicate that there is a dose buildup in the forward direction at small depths. The low-energy component of the radiation is attenuated quickly in small thicknesses, at large angles, due to the change in neutron spectrum with depth in the shield. The spectrum hardens with depth and reaches equilibrium after a depth greater than about 100 cm. In this case, there are at least two attenuation lengths. The second attenuation length is referred to as the effective attenuation length and is valid for thicknesses greater than 100 cm of concrete.

7.4 SHIELDING DESIGN CONSIDERATIONS

In radiation protection, the primary purpose of shielding is to protect workers and the public from radiation by attenuating secondary radiation to levels that are within regulatory or design limits. Proper shielding design requires the consideration of various factors such as beam parameters and losses (including location, target material, and dimensions), treatment parameters, beam delivery system, facility layout, adjacent occupancies, shielding material composition, and density and regulatory dose limits [1–4,50,51].

7.4.1 Beam Parameters and Losses

The shielding thicknesses for various parts of the proton therapy facility may range from about 60 cm to 7 m of concrete. Cyclotron rooms have the thickest shielding. Beam losses or beam interception by any material, including the patient, result in the production of secondary radiation. Therefore, to design effective shielding, the beam parameters (energy, intensity, beam dimensions, etc.), beam losses, and sources of radiation for proton therapy facilities must be well understood. This requires knowledge of how the accelerators operate and deliver beam to the treatment rooms. In addition to treatment, there are other checks and procedures that can also impact the shielding. Further, status of the beam when there are no treatments, or in-between treatments, should also be considered. The equipment vendor should provide specific details of beam losses, duration, frequency, targets (material and dimensions), and locations. Higher beam losses will occur during startup and commissioning as the beam is tuned and delivered to the final destination, and this should be planned for.

7.4.1.1 Synchrotron- and Cyclotron-Based Systems

Synchrotrons are designed to accelerate protons to the exact energy required for therapy, thus eliminating the need for energy degraders. This results in less local shielding and less activation of beamline components when compared to cyclotrons. Sources of radiation for the synchrotron injector include X-rays from the ion source, X-rays produced by back-streaming electrons striking the linac (linear accelerator) structure, and neutrons produced by the interaction of ions with the linac structure towards the end of the linac. The target material is typically copper or iron. The production of X-rays from backstreaming electrons will depend upon the vacuum conditions and the design of the accelerator [52]. The use of a Faraday cup to intercept the beam downstream of the linac must also be considered. Typically for synchrotrons, beam losses can occur during the injection process, radio frequency (RF) capture and acceleration, and during extraction. Some of these losses may be distributed in the synchrotron, while others may occur locally. Losses will be machine-specific and therefore the equipment vendor should provide this information. Particles that are not used in a spill may be deflected on to a beam dump or stopper and will need to be considered in the shielding design and activation analysis. If the particles are decelerated before being dumped, they are not of concern in the shielding design or activation analysis. X-rays may be produced at the injection and extraction septa due to the voltage applied across electrostatic deflectors and may need to be considered in special cases.

About 20%–50% of the accelerated beam particles can be lost continuously in the cyclotron. Significant self-shielding is provided by the steel in the magnet yoke, except in regions where there are holes through the yoke. These holes need to be considered in the shielding design. Losses at very low proton energies are not of concern for prompt radiation shielding, but can contribute to activation of the cyclotron. The shielding is determined by beam losses that occur at higher energies and those due to protons that are close to their extraction energy (230–250 MeV depending upon the cyclotron type) striking the dees and the extraction septum that is made of copper. These beam losses also result in activation of the cyclotron.

The Energy Selection System (ESS) consists of an energy degrader, collimators, spectrometer, energy slits, and a beam stop. The ESS allows the proton energy to be lowered after extraction. The intensity from the cyclotron is increased as the degraded energy is decreased to maintain the same dose rate at the patient. Thus, copious amounts of neutrons are produced in the degrader, especially at lower energies, resulting in thicker local shielding requirements in this area. The degrader scatters the protons and increases the energy spread. A collimator is used to reduce the beam emittance. A magnetic spectrometer and energy slits are used to reduce the energy spread. Beam stops are used to tune the beam. Neutrons are also produced in the collimator, slits, and beam stop. Losses in the ESS are large, and they also result in activation.

Losses occur in the beam transport line for synchrotron- and cyclotron-based systems. Although these losses are usually very low (a few %) and distributed along the beamline, they need to be considered for shielding design. The target material is typically copper or iron. During operation, the beam is steered onto Faraday cups and beam stoppers. Beam incident on these components should also be considered in the shielding design.

7.4.1.2 Treatment Rooms
The radiation produced from the beam impinging on the patient (or phantom) is a dominant source for treatment rooms. Thus, a thick tissue or water target should be assumed in computer simulations for shielding calculations. In addition, losses in the nozzle, beam-shaping, and range-shifting devices must also be considered in the shielding design. The contributions from adjacent areas, such as the beam transport and other treatment rooms, should also be taken into account. Typically, the large facility treatment rooms do not have shielded doors, and therefore, the effectiveness of the maze design is critical. In such cases, a full computer simulation for the maze is recommended. Some of the smaller single-room facilities have shielded doors. Treatment rooms either have fixed beam rooms or gantries.

7.4.1.3 Fixed Beam and Gantry Rooms
Either a single horizontal fixed beam or dual (horizontal and vertical or oblique) beams are used in fixed beam rooms. Shielding walls in the forward direction are much thicker than the lateral walls and the walls in the backward direction. The Use Factor (U) is defined as the fraction of time that the primary proton beam is directed towards the barrier. For rooms with dual beams, the Use Factor for the wall in the forward (0°) direction for each beam should be considered. This may be either one-half for both beams, or two-thirds for one beam and one third for the other. For a single beam, the Use Factor is one for the wall in the forward direction.

The beam is rotated about the patient in gantry rooms. In some cases, the gantries may rotate completely (360°), while in other cases only partial rotation is possible (~180°). For full rotation, on average, it can be assumed that the Use Factor for each of the four barriers (two walls, floor, and ceiling) is 0.25. In some designs, the gantry counterweight (made of large thicknesses of steel) acts as a stopper in the forward direction. However, it usually covers a small angle and is asymmetric. The ceiling, lateral walls, and floor are exposed to the forward-directed radiation. The walls in the forward direction can be thinner than for fixed beams, because of the lower Use Factor. The primary proton beam should never be allowed to directly strike the lateral shielding walls, roof, or floor. Administrative controls should be in place to ensure that either the patient or a dosimetric phantom always intercepts the primary beam.

7.4.1.4 Beam-Shaping Techniques
The various techniques used to shape and deliver the beam to the patient can be divided into two categories: passive scattering and pencil beam scanning.

In passive scattering, a spread-out Bragg peak is produced by a range modulation wheel or a ridge filter located in the nozzle [53]. Lateral spread of the beam is achieved by scatterers located downstream. Typically, for small fields, a single scatterer is used, while for large fields, a double scatterer is used. A collimator (specific to the treatment field) located between the nozzle exit and the patient is used to shape the field laterally. A range compensator is used to correct for the shape of the patient surface, inhomogeneities in the tissues traversed by the beam, and the shape of the distal target volume. A much higher beam current is required at the nozzle entrance when compared to the other delivery techniques, because there are losses due to the incidence of the primary beam on various delivery and shaping devices. The typical maximum efficiency of a passive scattering system with a patient field is about 45%. Thus, more shielding is required for passive scattering when compared to pencil beam scanning.

7.4.2 Workload

The term "workload" is used in a generic sense to include for each treatment room, each proton energy, the beam-shaping method, the number of fractions per week and the time per fraction, the dose per fraction, and the proton current required to deliver a specific dose rate. Once the workload for the treatment room has been established, one must determine the corresponding energies and currents from the cyclotron or the synchrotron. The workload for each facility will be facility-specific and equipment-specific. Therefore, the workload will vary from facility to facility and from one equipment vendor to the other. An example of a workload can be found in PTCOG Report 1 [2].

7.4.3 Regulatory Dose Limits

The use of protons for therapy purposes is associated with the generation of secondary radiation. Therefore, protection of the occupationally exposed workers and members of the public must be considered. Most of the national radiation protection regulations are based on international guidelines or standards. In the United States, medical facilities are subject to state regulations. These regulations are based on standards of protection issued by the US Nuclear Regulatory Commission [54].

The dose limits for the countries vary. Further, there are few countries where there are instantaneous dose rate limits. Some countries do not use Occupancy Factors. Therefore, a cookie-cutter design originating in one country could potentially underestimate or overestimate the shielding in some areas for a proton therapy facility in another country, assuming similar patient workload, usage, and beam parameters. Therefore, country-specific dose limits and dose rates should be considered in the shielding design.

7.4.3.1 Occupancy Factor

According to the National Council on Radiation Protection and Measurements (NCRP) [55], the occupancy factor (T) for an area is the average fraction of time that the maximally exposed individual is present in the area while the beam is on [35]. If the use of the machine is spread out uniformly during the week, the occupancy factor is the fraction of the working hours in the week during which the individual occupies the area. For instance, corridors, stairways, bathrooms, or outside areas have lower occupancy factors than offices, nurse's stations, wards, staff, or control rooms. The occupancy factor for controlled areas is typically assumed to be 1, and is based on the premise that a radiation worker works 100% of the time in one controlled area or another. In the United States, the regulatory agencies allow the use of Occupancy Factors.

7.4.4 Shielding Materials

Earth, concrete, and steel are typically used for particle accelerator shielding [4]. Other materials such as polyethylene and lead are used to a limited extent. For particle therapy facilities, neutrons are the dominant secondary radiation, and when using steel, a layer of hydrogenous material must be used in conjunction with steel.

7.4.4.1 Earth

Earth is often used as shielding material at underground accelerator facilities. However, it must be compacted to minimize cracks and voids, and attain a consistent density. Because earth is primarily composed of silicon dioxide (SiO_2), it is suitable for shielding of both gamma radiation and neutrons [4,55]. Its water content improves the shielding of neutrons. However, the water content can vary from 0% to 30%. The density of earth typically ranges from 1.7 to 2.2 g cm^{-3} and depends on the soil type, water content, and degree of compaction. The activation of the ground water must also be considered for underground facilities. Floor slabs of treatment rooms and accelerator rooms must be sufficiently thick to minimize groundshine and activation.

7.4.4.2 Concrete and Heavy Concretes

Concrete is a mixture of cement, coarse and fine aggregates, water, and sometimes supplementary cementing materials and/or chemical admixtures [4]. The density of concrete depends on the amount and density of the aggregate, the amount of air that is entrapped or purposely entrained, and the water and cement contents. Typically, ordinary concrete has a density that varies between 2.2 and 2.4 g cm^{-3}. The hydrogen content of concrete is important for its effectiveness in neutron shielding. Almost all of the hydrogen in concrete is in the form of water, which is present as bound water (i.e., water of hydration in the cement and aggregate) as well as free water in the pores of the cement. Both forms of water may be lost at elevated temperatures and over a period of time, thus reducing the neutron attenuation of concrete. The water content of the concrete is important in the shielding of neutrons with energies between 1 and 15 MeV [56]. However, the concrete absorbs moisture from the surrounding environment until it reaches some equilibrium. Therefore, all shielding calculations should be performed using the oven-dry density, and not the wet density. In the United States, ordinary concrete is usually considered to have a density of 2.35 g cm^{-3} (147 lb ft^{-3}). Concrete has many advantages compared to other shielding materials NCRP [4]. For example, it can be poured in almost any configuration and provides shielding for both photons and neutrons. It is relatively inexpensive. The poured-in-place concrete is usually reinforced with steel rebar, which makes it more effective for neutrons.

Concrete used for floor slabs in buildings are typically lightweight, with a density that varies between 1.6 and 1.7 g cm^{-3}. The disadvantage of concrete is that it takes months to pour. Concrete is also available in the form of blocks. If concrete blocks are used, they should be interlocking or staggered both horizontally and vertically to minimize gaps.

Heavy concretes contain high-Z aggregates or small pieces of scrap steel or iron, which increase its density and effective Z. Densities as high as 4.8 g cm^{-3} can be achieved.

The pouring of such high-Z enhanced concrete is a special skill and should not be undertaken by an ordinary concrete contractor because of settling, handling, and structural issues. The high-Z aggregate enhanced concrete is also sold in the form of prefabricated interlocking or noninterlocking modular blocks. It is preferable to use the interlocking blocks to avoid the streaming of radiation. Concrete enhanced with iron ore is particularly effective for the shielding of relativistic neutrons but is susceptible to higher activation. The shielding effectiveness of the concrete depends both on the concrete composition and density.

De Smet et al. [57] found that the effectiveness of the neutron shielding depends on a complex interplay of the different elements and the mass density that can result in differences between a factor of 2.7 in the forward direction and 3.8 at large angles.

Brandl et al. [58] found that shielding concretes with varying compositions and densities ranging from 1.6 to 5.38 g cm^{-3} resulted in variations in neutron ambient dose equivalent ranging from −50 to +30%, at a proton energy of 250 MeV. Therefore, all shielding calculations should be performed using the density and composition of the site-specific concrete and <u>not</u> some theoretical composition.

Unlike in photon therapy shielding, where the concrete thicknesses can be scaled inversely with the density; a given thickness of concrete *cannot be replaced* with concrete of twice the density and half the thickness. Dose attenuation by 250 MeV protons on a tissue target for heavy concrete and composite shielding (concrete and steel) have been reported by Ipe [59]. One important consideration in the choice of shielding materials is their susceptibility to radioactivation by neutrons, which can last for decades.

7.4.4.3 Steel and Iron

Steel is an alloy of iron and is used for shielding photons and high-energy neutrons. The high density of steel (~7.4 g cm^{-3}) together with its physical properties leads to tenth-value thickness for high-energy neutrons of about 41 cm [60]. The attenuation of steel can be considered equivalent to the attenuation of iron for photon shielding but not for neutron shielding. Steel is often used in conjunction with concrete (composite shielding) when space is at a premium. Steel or iron are usually available in the form of blocks. Iron has an important deficiency in shielding neutrons because it contains no hydrogen. Natural iron is composed of 91.7% ^{56}Fe, 2.2% ^{57}Fe, and 0.3% ^{58}Fe. The lowest inelastic energy level of ^{56}Fe is 0.847 MeV (5). Neutrons with energy above 0.847 MeV will lose energy by inelastic scattering in ^{56}Fe, but below this energy, neutrons can only lose energy by elastic scattering, which is a very inefficient process. Therefore, there is a buildup of neutrons below this energy. Furthermore, the neutron quality factor is at a maximum near 0.7 MeV. In addition, natural iron has two regions where the total cross section is very low because of resonances in ^{56}Fe. There is one resonance at 27.7 keV (minimum cross section = 0.5 barn) and another at 73.9 keV (minimum cross section = 0.6 barn). The net result is an increased attenuation length. Thus, large fluxes of low-energy neutrons are found outside steel or iron shielding. If steel is used for the shielding of high-energy neutrons, it must be followed by a hydrogenous material for shielding the low-energy neutrons that are generated. Dose attenuation by 200 MeV protons on a thick iron target for iron shielding has been reported by Sheu et al. [61]. Due to the large variety of nuclear processes, including neutron capture reactions of thermalized neutrons, steel can be highly activated [2].

7.4.4.4 Polyethylene

Polyethylene $(CH_2)_n$ is used for neutron shielding. Attenuation curves in polyethylene of neutrons from 72 MeV protons incident on a thick iron target have been published by Teichmann [62]. The thermal neutron capture in polyethylene yields a 2.2 MeV gamma ray which is quite penetrating. Therefore, boron-loaded polyethylene can be used. Thermal neutron capture in boron yields a 0.475 MeV gamma ray. Borated polyethylene can be used for shielding of doors and ducts and other penetrations.

7.4.4.5 Lead

Lead is used primarily for the shielding of photons. It has a very high density (11.35 g cm^{-3}) and is available in bricks, sheets, and plates. Lead is malleable [5] and therefore cannot support its own weight when stacked to large heights. Therefore, it will require a secondary support system. Lead is transparent to fast neutrons, and therefore, it should not be used for door sills or thresholds for proton therapy facilities where secondary neutrons dominate the radiation field. However, it does decrease the energy of higher energy neutrons by inelastic scattering down to about 5 MeV, making the hydrogenous material following it more effective. Below 5 MeV, the inelastic cross section for neutrons drops sharply. Lead is toxic and should be protected by paint or encased in other materials.

7.4.4.6 Transmission

The transmission of a given thickness of shielding material is defined as the ratio of the dose at a given angle with shielding to the dose at the same angle without shielding. Transmission curves are useful for comparing different shielding materials. The total particle dose equivalent transmission (based on FLUKA calculations) of composite shields (iron and concrete) and iron and concrete as a function of shielding thickness for 250 MeV incident on a 30 cm ICRU tissue sphere have been reported by Ipe [59].

7.5 SPECIAL TOPICS

Skyshine, groundshine, direct-shielded doors, mazes, and penetrations are briefly discussed.

7.5.1 Skyshine and Groundshine

The characteristics of the prompt radiation field in the immediate vicinity of the radiation source are strongly dependent upon the proton energy. At large distances from the radiation source, the prompt radiation field consists of two components: direct and scattered [4]. Secondary radiation may escape the roof and then be scattered down by the atmosphere or be scattered by the roof to the ground level; if the roof shielding is thin compared to the lateral walls, secondary radiation may escape the roof and then be scattered down by the atmosphere; or be scattered by the roof to the ground level. This process is loosely referred to as "skyshine." Similarly, "groundshine" refers to the radiation that escapes the floor slab, reaches the earth, and scatters upwards or radiation scattered upwards from the floor slab. Roof and floor slabs shielding should be thick enough to prevent both skyshine and groundshine. Neutron scattering is more important than photon scattering because neutrons are not easily absorbed above thermal energies. The neutron component is much higher than the photon component at proton therapy facilities. For skyshine, it is also important to consider the presence of adjacent elevated floors and multistoried buildings. In general, skyshine and groundshine are not of concern at facilities that have well-shielded roofs and floor slabs.

7.5.2 Mazes and Direct-Shielded Doors

Typically, most cyclotron, synchrotron, and treatment rooms have mazes. The radiation at the maze entrance consists of neutrons that scatter through the maze and get thermalized. The thermal neutrons are captured, resulting in the production of capture gamma rays. The forward-directed radiation from the target should never be aimed toward the mouth of the maze. The maze attenuation increases with reduced cross-sectional area and increased number of legs. The legs should be perpendicular to each other, thus forcing the radiation to scatter at least twice before reaching the maze entrance. The maze walls should be thick enough to minimize the direct radiation transmitted through the walls to the maze entrance. The sum of the individual maze wall thicknesses between the source of radiation and the maze entrance should be at least equal to the direct shield wall thickness that would be required if there was no maze [4].

Typically, direct-shielded doors are used when space is a premium. They must provide the same attenuation as the adjacent shield wall. Prefabricated high-Z aggregate concrete blocks or a combination of steel, lead, and borated polyethylene can be used to minimize the thickness, and hence, the weight of the direct-shielded door. Shielding will be required for air conditioning ducts or any other conduits that penetrate the wall shielding above the door.

7.5.3 Penetrations

Penetrations in the shielding walls are required for the routing of various utilities such as air conditioning, RF, cooling water, electrical conduits, etc. Various options can be used to reduce the amount of radiation scattered through penetration, such as extension of the duct length, use of single and double bends shadowing the penetration with a shield, etc. However, one must always ensure that there is no direct external radiation from the source pointing in the direction of penetration or a portion of penetration, resulting in an unshielded path.

7.6 CONCLUSIONS

Secondary radiation production as well the physical processes that govern radiation transport through accelerator shielding have been described. Shielding design considerations have been addressed.

REFERENCES

1. Ipe NE. Basic aspects of shielding. In: Paganetti H, editor. *Proton Therapy Physics*. New York: CRC Press; 2012. pp. 525–554.
2. PTCOG. Shielding design and radiation safety of charged particle therapy facilities. In: Ipe NE, editor. *PTCOG Report 1*. Switzerland: Particle Therapy Cooperative Group; 2010. http://www.ptcog.ch/archive/Software_and_Docs/Shielding_radiation_protection.pdf Accessed on September 19, 2017.
3. Ipe NE and Sunil C. Secondary radiation production and shielding. In: Paganetti H and Das IJ, editors. *Medical Physics Monograph No. 37, American Association of Physicists in Medicine, 2015 Summer School Proceedings, Colorado College*. Colorado Springs, CO; Medical Physics Publishing; 2015.
4. NCRP. *Radiation Protection for Particle Accelerator Facilities, NCRP Report*. Bethesda, MD: National Council on Radiation Protection and Measurements; 2003. p. 144.
5. IAEA. *Radiological Safety Aspects of the Operation of Proton Accelerators, IAEA Technical Reports Series No. 283*. Vienna: International Atomic Energy Agency; 1988.
6. ICRU. *Basic Aspects of High Energy Particle Interactions and Radiation Dosimetry, ICRU Report*. Bethesda, MD: International Commission on Radiation Measurements and Units; 1978; p. 28.
7. Krasa A. Spallation reaction physics. Lecture notes for students of the Faculty of Nuclear Sciences and Physical Engineering, Czech Technical University in Prague. May 2010, http://ojs.ujf.cas.cz/~krasa/ZNTT/SpallationReactions-text.pdf Accessed 29 September 2014.
8. Cossairt JD. Radiation physics for personnel and environmental protection. Fermilab Report TM-1834 Revision 15, April 2016.
9. Turner JE. *Atoms, Radiation, and Radiation Protection*. New York: Pergamon Press; 1986.
10. NCRP. *Neutron Contamination from Medical Electron Accelerators, NCRP Report*. Bethesda, MD: National Council on Radiation Protection and Measurements Report; 1984; p. 79.
11. Moritz LE. Radiation protection at low energy proton accelerators. *Radiat Prot Dosim* 2001; 96(4): 297–309.
12. NCRP. *Protection against neutron radiation, NCRP Report*. Bethesda, MD: National Council on Radiation Protection and Measurements; 1971; p. 38.
13. Tesch K. A simple estimation of the lateral shielding for proton accelerators in the energy range 50 to 1000 MeV. *Radiat Prot Dosim* 1985; 11(3): 165–172.
14. Agosteo S, Magistris M, Mereghetti A, Silari M, and Zajacova Z. Shielding data for 100 to 250 MeV proton accelerators: double differential neutron distributions and attenuation in concrete. *Nucl Instrum Methods Phys Res B* 2007; 265: 581–589.
15. Kato T, Kurosawa K, and Nakamura T. Systematic analysis of neutron yields from thick targets bombarded by heavy ions and protons with moving source mode. *NIM Phys Res A* 2002; 480: 571–590.
16. Nakashima H, Takada H, Meigo S, Maekawa F, Fukahori T, Chiba S, Sakamoteo Y, Sasamota N, Tanaka Su, Hayashi K, Odano N, Yoshizawa N, Sato O, Suzuoki Y, Iwai S, Uehara T, Takahashi H, Uwanmino Y, Namito Y, Ban S, Hirayama H, Shin K and Nakamura T. Accelerator shielding benchmark experiment analyses. In: *Proceedings of SATIF-2 Shielding Aspects of Accelerators, Targets and Irradiation Facilities*; 1995 Oct 12–13; Geneva, Switzerland. Paris: Nuclear Energy Agency, Organization for Economic Co-operation and Development; 1996. pp. 115–145.
17. Tayama R, Handa, H, Hayashi, H, Nakano, H, Sasmoto, N, Nakashima, H, and Masukawa, F. Benchmark calculations of neutron yields and dose equivalent from thick iron target for 52-256 MeV protons. *Nucl Eng and Design* 2002; 213: 119–131.
18. Ferrari A, Sala P R, Fasso A, and Ranft J. FLUKA: a multi-particle transport code. CERN Yellow Report CERN 2005-10; INFN/TC 05/11, SLAC-R-773. CERN, Geneva, Switzerland; 2005.

19. Battiston G, Cerutti F, Fasso A, Ferrari A, Muraro S, Ranft J, Roesler S and Sala PR *. The FLUKA code: description and benchmarking. In: Albrow M, Raja R, editors. *Proceedings of the Hadronic Shower Simulation Workshop*; 2006 Sep 6–8; Fermilab, Battavia, IL. AIP Conference Proceedings 2007; 896 pp. 31–499.

20. Birattari C, Ferrari A, Nuccetelli C, Pelliccioni M and Silari M. An extended range neutron rem counter. *NIM* 1990; A297: 250–257.

21. Olsher RH, Hsu H, Beverding A, Kleck JH, Casson WH, Vasilik DG and Devine RT. WENDI: an improved neutron rem meter. *Health Phys* 2000; 70: 171–181.

22. ICRP. Recommendations of the International Commission on Radiological Protection. ICRP Publication 1991; 60. Annals of ICRP 21(1–3), Oxford: Pergamon Press.

23. ICRP. Recommendations of the International Council on Radiological Protection. ICRP Publication 2007; 103. Annals of the ICRP, Oxford: Elsevier Science.

24. ICRU. *Quantities and Units in Radiation Protection Dosimetry, ICRU Report*. Bethesda, MD: International Commission on Radiation Units and Measurements; 1993, p. 51.

25. ICRU. *Conversion Coefficients for Use in Radiological Protection against External Radiation, ICRU Report*, Bethesda, MD: International Commission on Radiation Units and Measurements; 1998, p. 57.

26. Pelliccioni M. Overview of fluence-to-effective dose and fluence-to-ambient dose equivalent conversion coefficients for high energy radiation calculated using the FLUKA Code. *Radiat Prot Dosim* 2000; 88(4): 277–297.

27. Nakamura T. Summarized experimental results of neutron shielding and attenuation length. In: *SATIF7 Proceedings Shielding Aspects of Accelerators, Targets and Irradiation Facilities*; 2004 May 17–18, Portugal, Nuclear Energy Agency. Paris: Nuclear Energy Agency, Organization for Economic Co-operation and Development; 2004. pp. 129–146.

28. Clark FH. *Shielding Data Appendix E, in Protection Against Neutron Radiation, NCRP Report*. Bethesda, MD: National Council on Radiation Protection and Measurements; 1971, p. 38.

29. Thomas RH. *Practical Aspects of Shielding High-Energy Particle Accelerators*. Washington DC: U.S. Department of Energy. Report UCRL-JC-115068; 1993.

30. Kato T and Nakamura T. Analytical method for calculating neutron bulk shielding in a medium-energy facility. *NIM Phys Res B* 2001; 174: 482–490.

31. Goorley JT. MCNP6.1.1-Beta Release Notes, LA-UR-14-24680 2014.

32. Sheu RJ. Neutron deep-penetration calculations & shielding design of proton therapy accelerators. *Paper presented at the 7th International Workshop on Radiation Safety*, Synchrotron Radiation Sources (RADSYNCH2013).

33. Oh J, Hee-Seock L, Park S, Kim M, Hong G S, Koi S and Cho W. Comparison of the FLUKA, MCNPX, and PHITS codes in yield calculation of secondary particles produced by intermediate energy proton beam. *Prog Nucl Sci Technol* 2011; 1: 85–88.

34. Meier MM, Goulding, CA, Morgan G and Ullmann JL. Neutron yields from stopping-length and near-stopping-length targets for 256-MeV protons. *Nucl Sci Eng* 1990; 104: 339–363.

35. Iwase H, Nita K, and Nakamura T. Development of general-purpose particle and heavy ion transport Monte Carlo code. *J Nucl Sci Technol* 2002; 39: 1142–1151.

36. Sheu RJ, Bo-Lun L, Uei-Tyng L and Jiang, SH. Source terms and attenuation lengths for estimating shielding requirements or dose analyses of proton therapy accelerators. *Health Phys* 2013; 105: 128–139.

37. Allison J, Amako K, Apostolakis J et al. Recent developments in Geant4. *NIM Phys Res A* 2016: 835: 186–225.

38. GEANT4—a simulation toolkit. *NIM Phys Res A* 2003: 506: 250–303.

39. Vanaudenhovea T, Stichelbaut, F, Dubus A, Pauly N and De Smet V. Monte Carlo calculations with MCNPX and GEANT4 for general shielding study—application to a proton therapy center. *Prog Nucl Sci Technol* 2014; 4: 422–426.

40. Kim J. Proton therapy facility project in national cancer center, Republic of Korea. *J Repub Korean Phys Soc* 2003; 43: 50–54.
41. Porta A, Agosteo S, and Campi F. Monte Carlo simulations for the design of the treatment rooms and synchrotron access mazes in the CNAO hadron therapy facility. *Radiat Prot Dosim* 2005; 113(3): 266–274.
42. Hofmann W and Dittrich W. Use of isodose rate pictures for the shielding design of a proton therapy centre. In: *SATIF7 Proceedings of SATIF7 Shielding Aspects of Accelerators, Targets and Irradiation Facilities*; 2004 May 17–18; Portugal, Paris: Nuclear Energy Agency, Organization for Economic Co-operation and Development; 2005. pp. 181–187.
43. Dittrich W and Hansmann T. Radiation measurements at the RPTC in Munich for verification of shielding measurements around the cyclotron area. In: *SATIF8 Proceedings of Shielding Aspects of Accelerators, Targets and Irradiation Facilities*; 2006 May 22–24; Gyongbuk, Republic of Korea; Paris: Nuclear Energy Agency, Organization for Economic Co-operation and Development; 2007. pp. 345–349.
44. De Smet V, Stichelbaut F, Vanaudenhove T, Mathot G, De Lentdecker G, Dubus A, Pauly N and Gerardy, I.H*(10) inside a proton therapy facility: comparison between Monte Carlo simulations and WENDI-2 measurements. *Radiat Prot Dosimetry* 2014; 161: 417–421.
45. Ferraro, D and Brizuela M. Shielding verification for a 230 MeV proton therapy cyclotron bunker. *Paper presented at Conference: Asociación Argentina de Tecnología Nuclear - XLII Reunión Anual*, Buenos Aires – Argentina. 2015.
46. De Smet V, De Saint-Hubert M, Dinar N, Manessi GM, Aza E, Cassell C, Vargas CS, Hoey OV, Mathot G and Stichelbaut F. Secondary neutrons inside a proton therapy facility: MCNPX simulations compared to measurements performed with a Bonner sphere spectrometer and neutron H*(10) monitors. *Radiat Meas* 2017; 99: 25–40.
47. Satoh D, Yoshikazu, M, Tameshige Y, Nakashima H, Shibata T, Endo A, Tsuda S, Sasaki M, Maekawa M, Shimizu Y, Yamazaki M, Katayose T and Niita K. Shielding study at the Fukui Prefectural Hospital Proton Therapy Center. *J Nucl Sci Technol* 2012; 49: 1097–1109.
48. Sunil C. Analysis of the radiation shielding of the bunker of a 230 MeV proton cyclotron therapy facility: comparison of analytical and Monte Carlo techniques. *Appl Radiat Isot* 2016; 110: 205–211.
49. Lai BL. Sheu RJ, and Lin UT. Shielding analysis of proton therapy accelerators: a demonstration using Monte Carlo-generated source terms and attenuation lengths. *Health Phy* 2015: 108(2 Suppl 2): S84–S93.
50. Ipe NE and Fasso A. Preliminary computational models for shielding design of particle therapy facilities. In: *Proceedings of SATIF8 Shielding Aspects of Accelerators, Targets and Irradiation Facilities*; 2006 May 22–24; Gyongbuk, Republic of Korea. Paris: Nuclear Energy Agency, Organization for Economic Co-operation and Development; 2007. pp. 351–359.
51. Ipe NE. Particle accelerators in particle therapy: the new wave. In: *Proceedings of the 2008 Mid-Year Meeting of the Health Physics Society on Radiation Generating Devices*; Oakland, CA; 2008.
52. NCRP. Radiation protection design guidelines for 0.1–100 MeV particle accelerator facilities. National Council on Radiation Protection and Measurements Report; 1977: p. 51.
53. Smith AR. Vision 20/20: proton therapy. *Med Phys* 2009; 36(2): 556–568.
54. USNRC United States Nuclear Regulatory Commission. Standards for Protection Against Radiation 10CFR20, Code of Federal Regulations. https://www.nrc.gov/reading-rm/doc-collections/cfr/part020/ Accessed 22 September 2017.
55. NCRP. *Structural Shielding Design and Evaluation for Megavoltage X- and Gamma-Ray Radiotherapy Facilities, NCRP Report*, Bethesda, MD: National Council on Radiation Protection and Measurements; 2005; p. 151.
56. Chilton AB, Schultis JK, and Faw RE. *Principles of Radiation Shielding*. Upper Saddle River, NJ: Prentice Hall; 1984.

57. De Smet V. Neutron measurements in a proton therapy facility and comparison with Monte Carlo shielding simulations. *PhD Thesis*, Université Catholique de Louvain, Belgium.
58. Brandl A, Hranitzky C, and Rollet S. Shielding variation effects for 250 MeV protons on tissue targets. *Radiat Prot Dosim* 2005; 115(1–4): 195–199.
59. Ipe NE. Transmission of shielding materials for particle therapy facilities. *Nucl Technol* 2009; 168(2): 559–563.
60. Sullivan AH. *Guide to Radiation and Radioactivity Levels near High-Energy Particle Accelerators*. Kent: Nuclear Technology Publishing; 1992.
61. Sheu RJ, Chen YF, Lin UT, and Jiang SH. Deep penetration calculations in concrete and iron for shielding of proton therapy accelerators. *NIM Phys Res B* 2012; 280: 10–17.
62. Teichmann S. Shielding parameters of concrete and polyethylene for the PSI proton accelerator facilities. In: *Proceedings of SATIF8 Shielding Aspects of Accelerators, Targets and Irradiation Facilities*; 2006 May 22–24, Gyongbuk, Republic of Korea; Paris: Nuclear Energy Agency, Organization for Economic Co-operation and Development: 2007; pp. 45–54.

CHAPTER 8

Monte Carlo Simulations*

Harald Paganetti

Massachusetts General Hospital and Harvard Medical School

CONTENTS

---

* Colour figures available online at www.crcpress.com/9781138626508.

8.1 THE MONTE CARLO METHOD

The strategy applied in Monte Carlo methods is to repeat a certain process many times until the result converges to an expectation value with a certain variance. To solve a problem, a process is divided into subprocesses with randomly varying outcome based on stochastic sampling. A simple example to demonstrate the use of Monte Carlo is the determination of the value of π. It is based on drawing a circle of radius r enclosed in a square and then throwing darts at the square n times at random without aiming. The number of darts hitting the inside of the circle divided by the number hitting the inside of the square equals the area of the circle divided by the area of the square, i.e., $\pi r^2/(2r)^2$. To obtain the numerical value of π with high accuracy, one needs to throw a considerable amount of darts. The example illustrates three key features of the Monte Carlo method, i.e., a large number of events need to be simulated, the accuracy of the result depends on the number of events, and the desired accuracy affects the calculation time.

Solving problems using the Monte Carlo method is typically slower compared to the use of analytical algorithms, because subprocesses have to be simulated many times to reduce statistical variance, whereas analytical methods require solving a set of equations only once. On the other hand, the latter method depends on the complexity of the problem, while Monte Carlo methods are typically not affected by macroscopic complexities. Thus, Monte Carlo simulations often simplify the computational task because microscopic physics processes are typically well understood while phenomena in complex systems are more difficult to parameterize. The relationship between calculation time and complexity is illustrated in Figure 8.1 [1].

The use of the Monte Carlo method requires a sequence of random numbers, i.e., a set of numbers where each number is independent from the other numbers in the sequence. Random numbers generated by computer algorithms typically provide a certain distribution of random numbers within a certain interval. The sequence then has to be adopted for the distribution of random numbers according to the underlying process being simulated.

A probability density function (PDF) expresses the relative likelihood that a variable will have a certain value, as determined by a random number. In mathematical terms, $PDF(x)$ represents the probability of finding the random variable x' within dx of a given value x. Considering a $PDF(x)$ defined for a certain interval $[a,b]$, the goal is to sample randomly between a and b according to the $PDF(x)$ (a and b being finite and $b > a$). In case of a continuous uniform distribution, one obtains $PDF(x) = 1/(b-a)$ for $a \leq x \leq b$ and 0 for $x < a$ and $x > b$. Assuming that random numbers R_i generated by a random

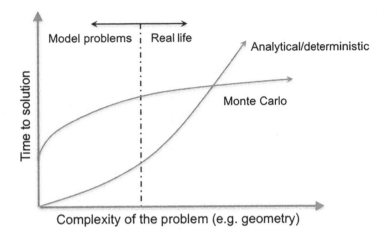

FIGURE 8.1 Time to solve a problem as a function of geometrical complexity using Monte Carlo versus a deterministic analytic approach. (Adopted from Ref. [1].)

number generator are uniformly distributed in [0,1], one can obtain random events via PDF(x) $dx = d$CDF. Thus,

$$\text{CDF}(x) = \int_a^x \text{PDF}(x')dx' \tag{8.1}$$

with CDF(x) = ($x - a$)/($b - a$) being the cumulative distribution function (CDF; i.e., the integral of the PDF), which leads to $x = a + (b - a)R$. In the above example, the mean value would be $\mu = a + (b - a)/2$ with a variance of $\sigma^2 = x^2 - \mu^2 = (b - a)^2/12$.

Another approach applied frequently is the sampling from a normal (Gaussian) distribution, where the PDF with the mean μ and the standard deviation σ is

$$\text{PDF}(x, \alpha, \sigma) = \frac{1}{\sqrt{2\pi\sigma}} e^{-\frac{(x-\alpha)^2}{2\sigma}} \tag{8.2}$$

The inversion method using a CDF is applicable whenever a distribution is invertible. Thus, if the physics can be parameterized as a PDF, the random action probability can be generated with an inverse function.

A key feature of Monte Carlo simulations is that the result of the calculation is subject to statistical uncertainties. For a large number of histories N, the central limit theorem shows that the uncertainty is proportional to $1/\sqrt{N}$. The standard error of the mean SE (i.e., the standard deviation σ of the mean estimate by the sample) can be given as

$$\text{SE} = \frac{1}{\sqrt{N}} \sqrt{\frac{\sum_{i=1}^{N}(x_i - \bar{x})^2}{N-1}} \tag{8.3}$$

8.2 MONTE CARLO PARTICLE TRANSPORT ALGORITHMS

8.2.1 Proton Transport Simulations Using Monte Carlo

8.2.1.1 Particle Tracking

The Monte Carlo method is the most accurate method of simulating particle interactions within a medium. A particle is simulated undergoing consecutive steps. At each step, the statistical probability of certain physics events based on the underlying material is determined. For instance, the process of simulating the energy deposited in a detector can be based on simulating the path of particles traveling through the detector and undergoing microscopic energy deposition events that may be randomly selected based on the probability of certain interactions determined by the underlying physics and the probability per unit path length of having an interaction. Figure 8.2 illustrates a particle being tracked through a medium. The tracking of one primary particle, including its secondary particles until their complete absorption or escape from the region of interest, is termed as history. A simulation with a finite amount of histories is termed a run. Sometimes, simulations are divided into several runs if, for instance, simulations are performed in parallel on several computers.

The simulation of a particle history starts by sampling a number of events from a starting source distribution. This can be a mathematical function or a parameter list resembling an initial particle source, such as the flux from an accelerator. Next, the passage of particles through a geometry is simulated, one particle at a time, one small step at a time, sampling one or more probability distributions at each step to choose how the particle might interact (be absorbed, be annihilated, change direction, or change energy) under random distributions consistent with the laws of physics. This particle tracking geometry is a well-defined geometrical model, such as a treatment head or a patient geometry. Materials are characterized by their physical properties, e.g., elemental composition, electron density, mass density, or mean excitation energy.

At each step, the algorithm samples a random distance to the next physics interaction from a PDF. The determined step length might be shortened if the particle has to cross a boundary to a new geometrical object (with a different underlying physics characteristics).

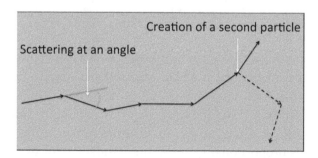

FIGURE 8.2 Schematic illustration of a particle tracking using Monte Carlo. The particle is being scattered at certain angles for the first five steps, before a physics interaction causes a second particle to be created (dashed line). Consequently, afterwards, two particles need to be tracked independently. Each step (straight line) might resemble several interactions in the condensed history method (see Section 8.2.1.2).

In general, the step length is the distance of a particle between two interaction sites and might be determined via a random number and an interaction coefficient. The step size has to be determined in three-dimensional space by sampling direction cosines. Typically, the maximum allowed step size is a user parameter so that a certain geometrical resolution of the physics interactions is ensured.

The outcome of Monte Carlo simulations depends on the chosen step size. It should be small, so that the difference of the cross sections at the beginning and the end of the step is small. On the other hand, a large step size decreases the computing time. Monte Carlo codes often use various methods to ensure proper step sizes, particularly near boundaries (changes in material) [2–4]. Depending on the flexibility of the code, the user may be allowed to define the maximum permitted step size.

The type of interaction is determined based on probabilities after each step. The interaction could lead to scattering where the particle changes direction for the next step to take, or it could lead to terminating the track because of an absorption event. There could also be secondary particles that need to be tracked separately. The latter would typically be stored on a memory stack and considered once the primary particle steps are finished. Not all secondary particles are considered. A threshold is defined as a low-energy limit to avoid having to track particles that, because of their low energy, would not significantly impact the result of the simulation. Sometimes, depending on the application, secondary particles may not be explicitly tracked at all (i.e., their energy may be deposited locally) depending on a user setting.

The reason for the increased accuracy of Monte Carlo simulations compared to analytical algorithms lies in the way the underlying physics is modeled. Monte Carlo simulations are able to take into account the physics of interactions on a particle-by-particle basis. This is done using theoretical models, parameterizations, and/or experimental cross-section data for electromagnetic and nuclear interactions. Monte Carlo accuracy thus depends on the detailed knowledge of physics for a particular particle, energy region, or material.

The simulation is terminated if the particle's energy is below a user-defined threshold or if the particle leaves the region of interest. The process can then be repeated with additional particles until a statistically valid result is achieved. Monte Carlo simulations are based on first principles, and considering that the underlying physics is well known, it should be considered the gold standard. However, often there are uncertainties in the underlying physics because of the lack of experimental data or models.

8.2.1.2 Proton Physics

If a particle is tracked through a medium, there are probabilities for certain physics interactions to occur at a given point during a step, as shown in Figure 8.3 for proton physics interactions. For uncharged particles, it is feasible to simulate all physics interactions. However, for charged particles, like protons, this is computationally ineffective because they interact very frequently. For example, the simulation of each elastic Coulomb interaction would cause a large number of small angle scattering events. This observation leads to the development of condensed history algorithms [5]. Rather than considering every discrete

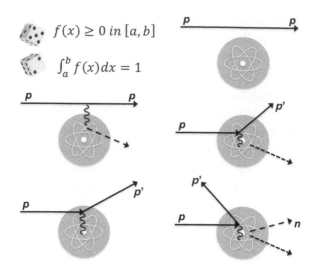

FIGURE 8.3 Illustration of the Monte Carlo method. Rolling the dice determines which of the many different physics interactions might happen for a proton interaction with a nucleus (including on the upper right the potential that no interaction takes place). Each interaction has a finite probability $f(x)$, and the total probability for all events equals 1.

interaction, individual events can be combined or "condensed." For instance, many single-scattering events causing a slight deflection of a charged particle can be grouped into one scattering event with a bigger scattering angle. Multiple scattering theories provide PDFs that represent the net result of several single-scattering events (see Chapter 2). This method can only be applied for events happening frequently so that Monte Carlo codes need to be capable of distinguishing between events that can be treated in the condensed history approach and those that need explicit consideration.

There are two different condensed history techniques, class I and class II. In the class I method, all interactions of a certain type are grouped in one condensed history step using a predetermined set of path lengths. The simplest choice is a constant path length. The condensed history class II method groups only "minor" collisions where energy losses or deflections are small, but considers the individual sampling of major events, where a large energy loss or deviation occurs. Those interactions that have to be treated individually are often called "catastrophic" or "hard" interactions. Typically, a maximum step size is defined, up to which continuous energy loss and a certain multiple scattering angle is assumed, unless a catastrophic event occurs. For certain Monte Carlo applications, the threshold for considering interaction explicitly can be critical. One often used class I algorithm, which is the continuous slowing down approximation for charged particles where energy loss is described using the unrestricted total stopping power. In general, class II algorithms are more accurate than class I, and most Monte Carlo codes use a combination of continuous processes based on condensed history and discrete processes based on an explicit model of each interaction. In proton therapy, discrete processes are typically nuclear interactions, secondary particle production (including δ-electrons), and large angle Coulomb scattering. Monte Carlo results are affected by user parameters such as the step size or the energy cutoffs, in particular in condensed history algorithms. Such information is thus important and should be stated when showing Monte Carlo simulations results. Furthermore, the condensed history approach has to be abandoned for some applications (e.g., for microdosimetry or small geometrical regions such as biological systems).

Typically, the energy loss of protons is calculated by the Bethe-Bloch equation down to 2 MeV. Below 2 MeV, a parameterization based on stopping power formalism, for example based on the International Commission on Radiation Units and Measurements (ICRU) [6], might be used. Multiple scattering is realized in condensed history class II implementations [7]. The Moliere theory (see Chapter 2) predicts the scattering angle distribution but does not give information about the spatial displacement of the particle. The Lewis method [8] allows calculation of moments of lateral displacement, angular deflection, and correlations of these quantities. Multiple-scattering algorithms used in Monte Carlo codes may differ but are typically variations of Lewis' theory.

Although nuclear interactions are not responsible for the shape of the Bragg peak (the majority of dose is deposited via electromagnetic ionization and excitation), they do have a significant impact on the depth-dose distribution because they cause a reduction in the proton fluence as a function of depth (~1% of primary protons undergo a nuclear interaction per cm range of the beam) [9]. An interaction between the projectile and the nucleus can be modeled as an intranuclear cascade with the probability of secondary particle emission. Once the energy of the particles in a cascade has reached a lower limit, a pre-equilibrium model can be applied. To accurately account for secondary particles from nuclear interactions, the nuclear interaction probability and the secondary particle emission characteristics must be known.

Nuclear interactions are typically parameterized using cross sections, i.e., interaction probabilities. Cross sections as a function of proton energy may not be available for all reaction channels. In these cases, models, parameterizations, or a combination of models, parameterizations, and experimental data must be used. The specific choice may depend not only on the particle or energy region but also on the required accuracy (versus efficiency) of a particular application. The cross section for a specific atomic or nuclear interaction caused by an incident particle is defined as the probability for the occurrence of the event for one target nucleus divided by the particle fluence. Cross sections are divided into elastic cross sections (scattering of the incident particle of the nucleus with conservation of kinetic energy) and nonelastic cross sections (nuclear excitation with potential creation of secondary particles). Inelastic cross sections are a subset of nonelastic cross sections, without changing the nuclear composition. Cross sections can be single differential or double differential. The latter would describe the probability for energy loss with the primary particle deflected under a specific angle. Cross sections for proton–nucleus interactions for applications in proton beam therapy are summarized by the ICRU [10].

For pencil beam scanning, small uncertainties in the "nuclear halo" (secondary particles emitted in nuclear interactions surrounding the primary beam) or multiple Coulomb scattering can cause large uncertainties when adding multiple pencils [11]. The dose distribution is small for each pencil but can be significant for a set of pencils delivering dose to the target volume, or in the sharp dose gradient at the distal falloff, as has been studied

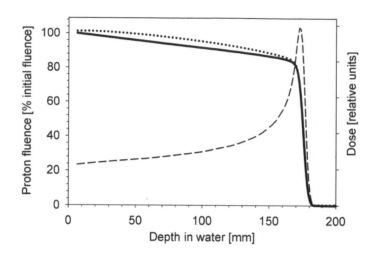

FIGURE 8.4 Monte Carlo simulated proton fluence as a function of depth for a 160 MeV proton beam. The dose is shown as dashed line, while the solid and dotted lines illustrate the primary proton and the total (primary and secondary) proton fluence.

using Monte Carlo [12]. Figure 8.4 shows the primary and secondary proton fluence as a function of depth for a 160 MeV beam. If one considers simulations solely for proton dose calculation, primary and secondary protons are mainly important, because they account for roughly 98% of the dose, depending on the beam energy [9]. This includes the energy lost via secondary electrons created by ionizations. The δ-electron energy E_δ can be calculated assuming maximum energy transfer, which a point-charge particle can impart to a stationary unbound electron (masses of the proton and electron are denoted as m_p and m_e, respectively):

$$E_\delta = 2m_e c^2 \frac{\beta^2}{\left(1-\beta^2\right)}\cos^2\theta\left[1+\frac{2m_e}{m_p\sqrt{\left(1-\beta^2\right)}}+\left(m_e\big/m_p\right)^2\right]^{-1} \tag{8.4}$$

The maximum energy of the δ rays can be approximated as a function of proton energy

$$E_\delta^{\mathrm{max}} \cong 4\,m_e\big/m_p\,E_p \cong E_p\big/500 \tag{8.5}$$

This corresponds to a maximum range of delta electrons of about 2.5 mm for a 250 MeV proton. The highest energy electrons are preferentially ejected in a forward direction. The energy of most electrons in a proton beam is much less than 300 keV, which corresponds to a range of 1 mm in water. Thus, explicit tracking of electrons is not necessarily required for dose calculations on a typical computed tomography (CT) grid but is certainly required for microscopic simulations, e.g., to study radiobiological properties, and for microdosimetry. For applications other than dose calculation, additional particles might need to be considered, e.g., neutrons (see Chapter 7).

8.2.2 Efficient Monte Carlo Techniques

There are various methods to improve the computational efficiency of Monte Carlo simulations. One can consider limiting simulations to certain particles and/or energies. Particles not being tracked should deposit their energy locally to ensure energy conservation. For instance, one might define a threshold for the production of δ-electrons. Above an energy threshold, δ-electrons are produced explicitly, but below the threshold, continuous energy loss of the primary particle might be assumed. Note that production cuts for secondary particles can influence energy loss and thus the simulation results [4].

Variance reduction techniques aim at improving computational efficiency by giving more emphasis to certain physical quantities of interest. For example, the splitting of primary particles in regions of high interest is often used. By splitting at specific locations, one predominantly considers particles that have a high likelihood of contributing in regions of interest. If primary particles are split, their simulated effects are subsequently treated with a weighting factor <1 (which will also be inherited by subsequent secondary particles). A related technique is Russian roulette, which, instead of splitting, combines several particles into one particle. Another method is the importance sampling, i.e., a method to oversample certain regions of interest while reducing the weight of events to maintain the statistical balance. Some of these techniques have been implemented specifically for proton therapy applications [13–15].

The efficiency in Monte Carlo simulations, ε, can be defined with an estimate s^2 of the true variance σ^2 of the quantity of interest and the computing time needed to reach this variance, T:

$$\varepsilon = \frac{1}{\left(Ts^2\right)} \tag{8.6}$$

With N being the number of histories, both Ns^2 and T/N are approximately constant so that the efficiency becomes independent of N. Variance reduction techniques try to improve the efficiency for a given N.

Track repeating algorithms have also been introduced, mainly for improving the efficiency of dose calculation. Here, pregenerated tracks are being reused during Monte Carlo transport [16,17]. In addition to using computational methods, Monte Carlo simulations can also be speeded up using advanced hardware solutions, such as graphics processing units [18–22].

8.2.3 User Input/Output in Monte Carlo Simulations

The simulation of a particle history using Monte Carlo typically starts by selecting a particle from a starting source distribution. This sampling is based on a mathematical function describing the distribution of particle characteristics. For instance, each particle in the source can be characterized by its type T, momentum Ω, and position Σ, i.e., $\phi(T, \Omega_x, \Omega_y, \Omega_z, \Sigma_x, \Sigma_y, \Sigma_z)$. This relationship can be defined analytically or, alternatively, the sampling can be performed from a list of particles that characterizes a radiation field. Monte Carlo simulations often use a phase space, which is a data file containing characteristics of a large number of particles with defined type, energy, direction, and other user-defined identifiers. Phase spaces are typically defined at a specific plane perpendicular to the central axis of a beam and contain the information for typically many millions of particles. They are used to increase computational efficiency if parts of the simulation are generic, allowing a phase space to be reused for subsequent simulations.

In addition to the initial particle spectrum, a Monte Carlo simulation setup requires the tracking geometry, the materials involved, the physics processes governing particle interactions, the response of sensitive detector components, the generation of event data, the storage of events and tracks, and the analysis of simulation data at different levels of detail (including the potential visualization of the detector and particle trajectories). Depending upon the code, some of these tasks may be taken care of by default settings. Most Monte Carlo codes would expect a more or less complex definition of geometries in an input file, while others might expect a definition of geometry using a programming language. If one uses an existing executable that is already tailored to a specific application, input parameters might be limited to defining, for example, the proton beam energy and the settings of a specific device in a bigger geometry assembly.

The results of Monte Carlo simulations are typically analyzed from 1D, 2D, or 3D histograms. These are filled during the simulation if certain conditions for a histogram bin are fulfilled, e.g., a particle has deposited a specific amount of energy in a specific area of the geometry. One can also store entire particle histories for retrospective analysis. Caution is warranted when dealing with dose scoring in a Monte Carlo system. A Monte Carlo system typically provides information about the status of a tracked particle, either at the beginning or at the end of an individual step. Typically, Monte Carlo systems require a particle to stop at a geometrical boundary to adjust for the change in physics. One has to make sure that energy is deposited in the volume according to the path of the particle track to avoid dosimetric artifacts at boundary crossings.

8.2.4 Monte Carlo Codes

There are various Monte Carlo codes used in proton therapy, e.g., FLUKA [23], Geant4 [24,25], MCNPX [26,27], VMCpro [28], and Shield-Hit [29]. Typically, a Monte Carlo program is a software executable for which the user has to write an input file depending on the specific problem. There are also other approaches, like Geant4, where the code provides only an assembly of object-oriented toolkit libraries with a functionality to simulate different processes organized in different functions within a C++ class structure. The ability to program is prerequisite for designing simulations using these types of codes. Monte Carlo codes also differ in the level of control over tracking parameters. The ability to control every parameter (e.g., physics settings, step sizes, and material constants) adds flexibility but may also make the code difficult to use and prone to inaccurate results due to user error.

Most of these codes are multipurpose Monte Carlo codes that were not necessarily designed for radiation therapy applications. There are frameworks that tailor these codes to radiation therapy applications such as GAMOS [30], GATE [31,32], PTsim [33], TOPAS [34] (all based on GEANT4), and FICTION [35] (based on FLUKA). These systems extend the functionality of the basic codes and provide user-friendly interfaces. For instance, TOPAS was particularly designed to model physics and biology for proton therapy to allow nonexperts and nonphysicists to perform complex Monte Carlo simulations without the need to engage in programming [36].

8.3 MONTE CARLO CODE VALIDATION

8.3.1 Uncertainties Due to Physics Models

The accuracy of Monte Carlo simulations depends on the accuracy of the underlying physics. A Monte Carlo code might allow different physics settings from which the user can choose. There might even be different settings for different energy domains. Models might not be tailored for proton therapy simulations because they were originally designed for other applications and therefore require adjustment. Thus, different Monte Carlo implementations can result in discrepancies. For instance, the implementation of the multiple-scattering theory can differ slightly from Molière theory [37,38], e.g., for multiple scattering, the code Geant4 uses a condensed history algorithm that utilizes functions to calculate the angular and spatial distributions of the scattered particle implementations [39].

The significance of uncertainties depends on the application. For instance, for shielding calculations, accurate double differential cross sections (in energy and emission angle) for proton–neutron interactions are important (see Chapter 7). According to the ICRU [10], angle-integrated emission spectra for neutron and proton interactions are known only to within 20%–30% uncertainty. Total nonelastic and elastic cross sections have uncertainties of <10%. This also impacts the difference in results amongst different Monte Carlo systems [40].

Uncertainties in stopping power parameters also influence calculation uncertainties, specifically the proton beam range (see Chapter 17) [4,41,42]. Values for the mean excitation energy might have uncertainties in the order of 5%–15%, and differences in published mean excitation energies for given elements can exceed 10%. Such an uncertainty for beam-shaping materials can lead to up to 1 mm uncertainty in the predicted beam range in water [4,41,43]. Typically, mean excitation energies are adjusted to agree with measurement for elements where data exist, and otherwise interpolated based on theory.

8.3.2 Validation Measurements for Proton Dose Calculations

The validation of Monte Carlo codes with respect to settings of different physics parameters (e.g., cross-section data, model parameterizations, cutoff values for particle production, consideration of secondary particles) is typically done by code developers. Direct experimental validation of cross sections is often not feasible in proton therapy institutions where the main users are located. Here, benchmarking is mostly based on Monte Carlo simulations of less fundamental quantities, e.g., dose in a phantom [44–47]. Benchmarking studies should not be too complicated in terms of the underlying geometry so that discrepancies can be attributed to differences in physics and not to shortcoming in simulating the geometry. Comparison with experimental data should always be done, particularly if the code is used for applications not anticipated by developers.

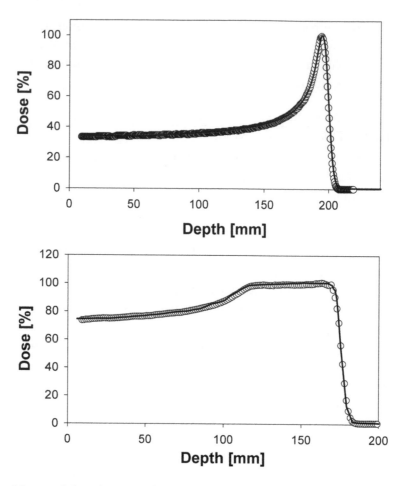

FIGURE 8.5 Measured data (open circles) and Monte Carlo generated data (solid lines) for two depth-dose curves from the Francis H Burr Proton Therapy Center at Massachusetts General Hospital in Boston, USA. The upper graph shows a Bragg peak with a nominal beam energy at the treatment head entrance of ~190 MeV. The SOBP (lower graph) is based on a clinical field with the prescription of 17.2 cm range and 6.8 cm modulation width. The Monte Carlo simulation included the entire treatment head geometry. (Adopted from Ref. [49].)

Three of the most commonly used codes in proton therapy (Geant4, FLUKA, and MCNPX) were compared with each other and with measurements by Kimstrand et al. [2]. The study aimed at the simulation of the scattering contribution at aperture edges, where protons scatter at inner surfaces of apertures at very small angles, i.e., a situation where small uncertainties in scattering power would be measurable. A significant impact of user-defined parameters was found. Benchmarking studies comparing experimental depth-dose distributions and beam profiles have been conducted for complex treatment heads [48–52]. As examples, Figures 8.5 and 8.6 show a comparison of Monte Carlo simulations using the Geant4 code with experimental data measured with an ionization chamber. Studying heterogeneous half-beam blocks and measuring/simulating the dose downstream can be a valuable test of scattering models.

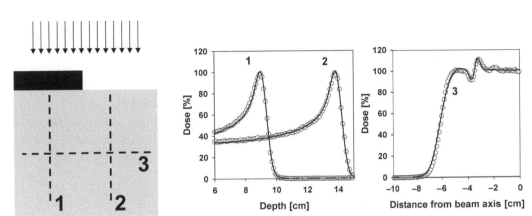

FIGURE 8.6 Measured (open circles) and simulated (solid lines) dose distribution. The left figure shows the experimental setup with the beam impinging on a half block of bone equivalent material (black) into a water phantom (gray). The dashed lines indicate the directions of the scans shown in the middle and right figure. The beam profile [3] was measured in an SOBP plateau. (Adopted from Ref. [51].)

8.3.3 Validation Measurements for Proton Nuclear Interactions

The correct modeling of nuclear interactions is essential for simulating the primary and secondary particle spectrum and for dose calculation, especially if Monte Carlo is used for absolute or relative dosimetry [53]. In general, nuclear interaction components are difficult to study separately in an experiment solely measuring dose, but large-angle scattering and thus the contribution of nuclear interaction products to dose distributions can be measured and compared with Monte Carlo [46,47]. An experimental tool particularly useful for testing proton nuclear interaction data is the multilayer Faraday cup [54–57]. It is sensitive to electromagnetic and nuclear inelastic reactions and measures the longitudinal charge distribution of primary and secondary particles. Because it relies on charge rather than dose, it is capable of separating the nuclear interaction component from the electromagnetic component. This is because nuclear stopping of protons takes place predominantly in the plateau region of a Bragg peak, while electromagnetic stopping takes place around the Bragg peak. The Faraday cup can only validate total cross sections.

For treatment head simulations and beam characterization, total and differential cross sections for materials of beam-shaping devices are required to compensate for fluence loss due to nuclear interactions. For primary standards and reference dosimetry, these cross sections with high accuracy are needed for a limited set of detector materials. Another example is the simulation of nuclear activation of tissues relying on isotope production cross sections for human tissues [58–60] (see Chapter 21). There are also considerable uncertainties when simulating neutron production. A precise modeling of neutron yields is needed when simulating both scattered neutron doses to assess potential risks for patients and neutron production for protection and shielding (see Chapter 7). These simulations require double differential production cross sections for tissues, beam-shaping devices, and shielding materials. There are various nuclear interactions channels for neutron production, as neutron and secondary charged particle emissions can be the result of complex interactions. There are often insufficient experimental data of inelastic nuclear cross sections in the energy region of interest in proton therapy so that parameterized models for Monte Carlo are used despite uncertainties in the physics of intranuclear cascade mechanisms. Agreements of Monte Carlo results and measured data on neutron doses have been reported to be only between 10% and 340% [48,61–64].

It would be desirable to have a set of benchmarking experiments whose results could be used to standardize the validation of Monte Carlo codes [45]. It is difficult for code developers to anticipate all potential applications and comprehensive validation measurements are thus difficult to design.

8.4 THE USE OF MONTE CARLO TO STUDY PROTON-SCATTERING EFFECTS

The understanding of proton scattering is particularly important for dose calculation in patient geometries. Analytical methods for dose calculation in many commercial planning systems use approximations (see Chapter 14). If a proton beam passes through complex heterogeneous geometries, scattering causes range degradation [65,66]. Another effect that is best studied with Monte Carlo is scattering at sharp edges, e.g., at apertures [67–69]. Further, markers implanted in the patient for setup or motion tracking do have an impact on dose calculations. They are typically not modeled accurately in pencil beam algorithms due to their high-Z nature. Monte Carlo has been used to study the impact of such markers on the dose distribution [70].

8.5 THE USE OF MONTE CARLO FOR BEAMLINE DESIGN

Physicists in proton facilities are rarely involved in designing beamlines, as this is normally done by scientists working for machine vendors. The basic beamline elements between the accelerator and the treatment room that might be simulated using Monte Carlo are bending magnets and energy degraders. Energy degraders are needed in cyclotron-based facilities because cyclotrons extract a single energy. Degraders are built as single or multiple wedges that can move in and out of the beam. Monte Carlo beam transport through Carbon and Beryllium degraders has been performed with the goal of improving beam characteristics [4].

An important part in beamline simulations deals with beam steering through magnetic fields. Although this can be done with many Monte Carlo codes, tracking through magnetic fields is usually quite slow, because it might be based on using Runge–Kutta algorithms and parameterized field maps. The curved particle in a specific field is broken up into linear chord segments. Beam optics calculations are therefore often done numerically, although there are specialized Monte Carlo codes that simulate magnetic beam steering [71].

8.6 THE USE OF MONTE CARLO FOR TREATMENT HEAD SIMULATIONS
8.6.1 Characterizing the Beam Entering the Treatment Head
For typical Monte Carlo applications in proton therapy, a simulation would start at the treatment head entrance. A parameterization of the phase space at treatment head entrance might in principle be defined from first principles, i.e., based on the knowledge

of the magnetic beam steering system or by fitting measured data [51]. Measuring some of these parameters directly can be difficult, and a user might have to rely on the manufacturer's information. Typically, the beam is characterized using the beam energy, the energy spread, the beamlet size, and the angular distribution [49,72].

The angular spread is not easily measurable and is typically on the order of 2–5 mm mrad. It can also be parameterized using the emittance of the beam, defined as the product of the size and angular divergence of the beam in a plane perpendicular to the beam direction. The size of the proton beamlet is typically well known because a segmented transmission ionization chamber is typically located at the treatment head entrance for beam monitoring purposes. For passive-scattering delivery, the impact of simulating the correct beamlet size depends on the width of each step used on the modulator wheel, i.e., the number of absorber steps covered by the beam at a given time [49,73]. For beam scanning, the exact beamlet size and its shape are, of course, vital, as it directly influences the spot size and shape at treatment head exit as prescribed by the planning system. The initial energy spread and beamlet size at nozzle entrance might influence the flatness of a spread-out Bragg peak (SOBP) because these parameters influence the peak-to-plateau ratio of the individual Bragg peaks that form an SOBP.

Clinically, one of the most important parameters is certainly the beam energy, as fields are prescribed using the beam range in water. The energy and energy spread can be obtained either directly, using an elastic scattering technique [74], or indirectly, by measuring the range and shape of Bragg peaks in water [73,75].

The parameters listed above are typically correlated. This correlation is insignificant when modeling a passive-scattering system, mainly because the amount of scattering material in the beam's path uncouples the parameters at the treatment head exit [49]. For beam scanning simulations, such a correlation, for example, between the particle's position within the extended beamlet and its angular momentum, needs to be taken into account, as it might affect the size of the pencil beam exiting the treatment head. Due to the energy spread of the beam, a deflection in a magnetic field will lead to a correlation of particle energy and position.

8.6.2 Modeling of Beam Monitoring Devices

Beam monitoring devices are typically transmission ionization chambers. To simulate a realistic dose distribution, simulating these chambers in detail might not be necessary, because they cause only little scattering and energy loss of the beam. Simulating plain or segmented ionization chambers is done for the purpose of designing ionization chambers or studying beam steering, as well as calculating the absolute dose in machine monitor units. The explicit tracking of secondary electrons should not be neglected for ion chamber simulations [53,76,77].

For treatment delivery, the prescribed dose is converted into machine monitor units (see Chapter 13). A monitor unit typically corresponds to a fixed amount of charge collected in a transmission ionization chamber in the treatment head. This reading is related to a dose at a reference point in water [78,79]. Absolute dose can thus be simulated if a detailed model of the treatment head, including the ionization chamber geometry and readout, is available [53]. In a segmented ionization chamber, the volume used for absolute dosimetry can be quite small. This causes low statistics when simulating the chamber response (from energy deposition events) and thus requires a large number of histories to be simulated. For instance, as an approximate approach, one might simulate the energy deposited in the ionization chamber without predicting the actual output charge, i.e., simulate a relative number that has to be normalized against a reference charge [80,81].

8.6.3 Modeling of Beam-Shaping Devices in Passive Scattering

Monte Carlo simulations are valuable in treatment head design studies, as they can reduce the required number of experiments considerably. Small design changes can be tested computationally before building or modifying hardware components. Furthermore, for research studies requiring accurate characterization of the radiation field, treatment heads are modeled to characterize the beam at treatment head exit. Such simulations are also useful when commissioning planning systems. There are various reports on Monte Carlo treatment head simulations (e.g., [49,52,73,80,82–85]). Most Monte Carlo models attempt to model machine-specific components in the treatment head using manufacture blueprints [49–51]. Creating a Monte Carlo model of a treatment head could also be based on a computer aided design interface to the Monte Carlo code [86]. If Monte Carlo codes only allow the definition of geometrical objects out of a library of standard objects, a complex geometry can be represented as a combination of regular geometrical objects.

Treatment heads in passive-scattering proton therapy can be rather complex, and the position or state of certain devices can change depending upon the specified field. For example, scattering foils might be inserted or different modulator wheels are evoked for a certain combination of range and modulation width. The Monte Carlo model of a treatment head has to accommodate possible variations of geometrical settings. One might generate a generic treatment head in the Monte Carlo code that is initialized when the code is set up (or compiled) and then modify the generic geometry using parameters provided via an input file. To simulate a specific field, one has to define specific treatment head parameters or the range and modulation width, which can subsequently be converted into treatment head parameters. The parameters are provided either by a treatment planning system (if it prescribes the treatment head settings for a patient field), by the treatment control system (if the treatment head settings are defined by an interface to the planning system or manually by an operator), or the user can set them.

The desired accuracy when modeling the treatment head in a Monte Carlo system depends on the purpose of the simulation. For calculating phase space distributions for dose calculation in the patient, it is often sufficient to have only beam-shaping devices included in the simulation. Thus, passive-scattering simulations require the double- or single-scattering system, the modulator wheel, an aperture, and a compensator. For other applications, e.g., ionization chambers for detector studies for absolute dosimetry or housing of devices to study scattering or shielding effects, one might need a more realistic description of other treatment head components. The following paragraphs briefly describe the modeling of the most important devices in a treatment head.

A contoured scatterer as part of a double-scattering system typically consists of two components, one made out of a high-Z material and the other made out of a low-Z material (see Chapter 2). The thickness of the high-Z material decreases radially with distance to the field center, while the thickness of the low-Z material increases as a function of radial distance. The bimaterial design ensures that the scattering power is independent of the beam energy. Such a complex geometry might be modeled by combining regular objects, e.g., by combining cones [49]. A modulator wheel can be simulated by using segments out of a circular structure. Each of these segments is characterized by thickness, material, minimum and maximum radius, as well as the angle covered. Because beam spots often overlap with several wheel steps, it is not sufficient to use one simulation per absorber step and combine simulations. Figure 8.7 depicts Monte Carlo models of a contoured scatterer and a modulator wheel.

The geometries of patient field specific apertures and compensators are typically provided by the planning system. Apertures are often studied using Monte Carlo because of the secondary radiation they produce, the effects of edge scattering, as well as their impact on the beam penumbra [87,88]. An aperture opening might be characterized by a set of points following the inner shape of the drilled opening. If the Monte Carlo code can translate this information into a 3D object representing the aperture, the information can

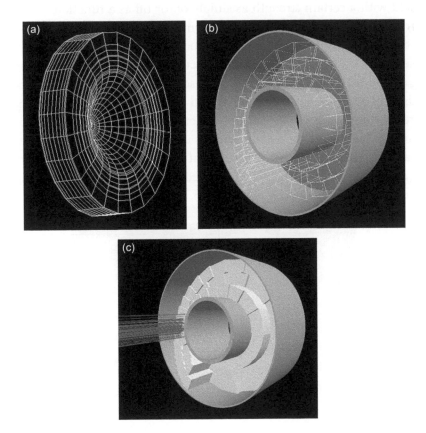

FIGURE 8.7   Monte Carlo model of a contoured scatterer used in a double-scattering system (a; mesh-type image) and a modulator wheel based [(b) showing the transparent steps of the wheel and (c) showing actual simulated proton tracks] at the Francis H Burr Proton Therapy Center at the Massachusetts General Hospital, Boston [49]. Simulation done with TOPAS [34].

be used directly. Otherwise, different parameterizations need to be applied [49]. For a compensator, the geometry can be described by a set of points in space defining the position and depth of drill bits for use in a milling machine. This information can be applied in a Monte Carlo code to model the devices using standard regular shapes (such as tubes) [49].

8.6.4 Modeling of Scanned Beam Delivery

Treatment heads for beam scanning are, geometrically, less complicated to model because there are fewer objects in the beam path. Essential to model might only be the scanning magnets and perhaps an aperture [87]. Treatment head simulations for scanned beams have shown good agreement with measurements [82]. To fully model a scanned beam treatment head, the Monte Carlo code has to be able to simulate magnetic beam steering. Depending on the beam steering, a scanned proton beam can be convergent or divergent. Magnetic fields are typically not modeled physically but rather geometrically by defining an area within the geometry in which particle tracking is affected by a magnetic force of certain strength [49]. Particle tracking is then performed according to field equations. As an approximation, instead of implementing the field lines, one might approximate a magnetic field with a certain strength as simply on or off as a function of positioning in the geometry (i.e., assume perfect dipoles). Attention needs to be given to the maximum step size while tracking protons through the magnetic field because large steps lead to considerable uncertainties when simulating the curved path of particles through the field. Depending on the scanning system, accurate description of the field lines as compared to binary fields, and accurate beam emittance is important when simulating dose distributions. Figure 8.8 shows an example of protons steered through a magnetic field. To simulate a full scanning pattern, time-dependent settings in the Monte Carlo are required [89].

If the scanning pattern for a prescribed treatment for pencil beam scanning is simulated, the magnetic field settings are prescribed by either the planning system or by a treatment control system, based on information by the planning system [90]. These field

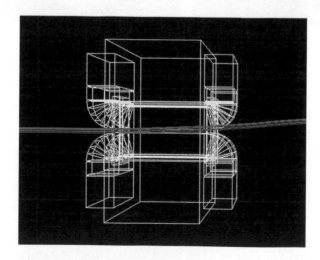

FIGURE 8.8 Proton tracks steered though a magnet with a specific field strength as modeled within Geant4. Simulation done with TOPAS [34].

characterizations are typically beam spot positions at a plane upstream of the patient and are parameterized as x and y coordinates and spot energy. If the spots are being modeled using Monte Carlo tracking through the nozzle, this information needs to be translated into magnetic strengths in tesla. The relationship is typically known from results of commissioning measurements.

8.6.5 Time-Dependent Geometries

For passive scattering (modulator wheel rotation) as well as beam scanning (changing magnetic field), the treatment head geometry involves devices with a time-dependent setting. Considering time-dependent settings by adding numerous individual runs can be cumbersome, as separate executables or input files might be needed for individual runs, e.g., for rotational steps of a modulator wheel. Alternatively, one can change the geometry dynamically by applying a four-dimensional Monte Carlo technique [89,91]. While still discrete, the technique allows geometry changes after each particle history. To simulate a beam scanning pattern, four-dimensional Monte Carlo techniques can be used to constantly update the magnetic field strength, allowing the assessment of beam scanning delivery parameters [82,89–92].

Another time-dependent feature in passively scattered beam delivery might be the modulation of the beam current. Because a unique wheel design for each field is impractical, one uses a finite set of modulator wheels, each resembling a unique step design to create full modulation to the patient's skin. To deliver a certain modulation width as well as meet specifications for SOBP flatness, the beam current might be continuously modulated as a function of rotation angle at the accelerator (beam source) level to fine-tune the shape of the SOBP depth-dose distribution (see Chapter 5) [93].

8.6.6 Treatment Head Simulation Accuracy

Figure 8.9 shows the simulated geometry of a proton therapy treatment head. The accuracy has to be validated experimentally, at least by comparing the flatness of the SOBP and the shape of the lateral beam profile (see Figures 8.3 and 8.4) [45]. For beam scanning, one might validate the simulation by analyzing the beamlet size and Bragg peak width [75].

For a passive-scattering treatment head, typically consisting of many beam-shaping devices, the accuracy of the simulation depends on the available information regarding the geometry of treatment head elements. In addition, it might depend on the knowledge of exact material compositions and material properties [73]. For example, the SOBP range and modulation width might be very sensitive (up to 1–2 mm range variations) to changes in the density of materials in the modulator wheel or scatterers [73]. Materials commonly used for these devices are polyethylene (lexan), lead, and carbon. Specifically, in the case of carbon, the nominal density as specified by the manufacturer can vary from the actual density because carbon is available in various specifications.

8.6.7 Phase Space Distributions

The results of particle tracking through a treatment head are typically stored in a phase space distribution where the kinetic parameters are recorded for each particle at a geometrical

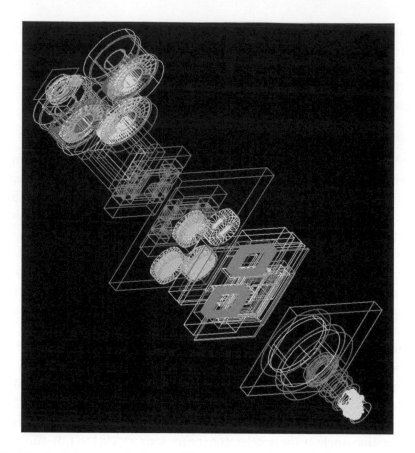

FIGURE 8.9 Monte Carlo model of one of the treatment heads at the Francis H Burr Proton Therapy Center at Massachusetts General Hospital. Simulation done with TOPAS [34].

location (see Section 8.2.3). Phase spaces thus serve as a full beam characterization at a given position [94]. It can also be defined as a particular surface of any given shape recording all particles that enter a particular device. For dose calculations, the phase space is typically defined at a plane perpendicular to the beam axis between the treatment head and the patient or phantom. Phase space files may only contain the most relevant parameters, such as the energy, directional cosine, and particle type. However, for specific studies, one is often interested in the history of particles, e.g., if the particle is a secondary or primary particle or if the particle was scattered at a specific device. In this case, the phase space may contain additional parameters or binary flags to allow partial reconstruction of a particle's history. Reusing the phase space can save calculation time if multiple scenarios with similar field characteristics are to be studied.

In photon therapy, phase space distributions play a big role for dose calculations. For dose calculations in passive-scattered proton therapy, phase spaces are less useful because each field typically has a unique setting of the treatment head and beam energy, thus making it unlikely that the phase space can be reused. This is true in particular if the patient-specific aperture and compensator are included in the treatment head simulation.

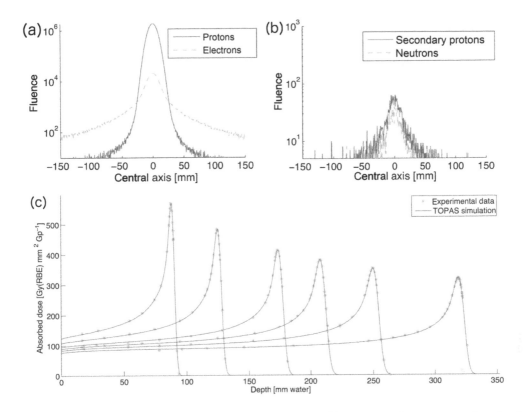

FIGURE 8.10 Simulated fluence profiles of a 230 MeV pencil beam scored at treatment head exit. (a) Primary protons and electrons, (b) secondary protons and neutrons. The simulations are plotted in units of Gy(RBE) mm² Gp⁻¹ and the measurements are normalized to the maximum dose (dashed) and neutrons from nuclear interactions (dotted). (c) Monte Carlo simulated Bragg curves (blue lines) compared with measurements (red stars). (Adapted from Ref. [75].)

This is different for beam scanning, where a beam parameterization and a generic phase space can typically be generated and reused because of the limited variations in beam settings (energy, energy spread, spot position). Note that even for scanning, the phase space cannot be limited to contain just primary protons. Nuclear interactions do play a role as they cause a loss of primary protons along the beam path through the treatment head. Figure 8.10 shows energy distributions of protons, electrons, and neutrons at the exit of a treatment head for a scanned beam delivery.

8.6.8 Beam Models

Beam models (parameterized phase space distributions) are a standard feature in Monte Carlo simulations for photon or electron beams [95]. As discussed earlier, passive-scattering treatment head settings are highly field dependent, and thus beam models are difficult to define. It is possible to deconvolve an SOBP into its pristine peak contributions for optimizing beam current modulation [93]. This method, in combination with assumptions about the angular spread of the field, could also serve to construct a beam model depending on the complexity of the double-scattering system. Note that similar dose distributions can be delivered with different underlying proton energy distributions (but different fluences). The creation of a given SOBP might not be unique in terms of the underlying pristine Bragg peaks.

The situation is different for beam scanning. Here, beam models are feasible because a field can be characterized by a fluence map of pencil beams (x, y, beam energy, weight, divergence, and angular spread). The parameters can be obtained by Monte Carlo simulations of the entire treatment head, including the magnetic fields [49] or by experiments, e.g., fluence distributions of pencil beams in air and depth-dose distribution measured in water [75,96]. For some scanned beam deliveries, one might want to use an aperture to reduce the beam penumbra. In this case, it is essential to consider aperture scattering in the beam model.

The fact that realistic beam models can be constructed for beam scanning has important implications when using Monte Carlo for clinical dose calculation. For passive-scattering simulations, the majority of calculation time is spent tracking particles through the treatment head due to the low efficiency of proton therapy treatment heads for passive scattering (typically between 2% and 40%, depending on the field). The use of beam models allows the use of fast Monte Carlo for beam scanning [51].

8.7 THE USE OF MONTE CARLO FOR QUALITY ASSURANCE

Even though quality assurance is based on routine measurements, Monte Carlo simulations can assist clinical quality assurance procedures (see Chapter 12). By simulating dose distributions and varying beam input parameters, tolerance levels for beam parameters can be defined [49]. Monte Carlo simulations are particularly valuable for studying scenarios that cannot be created easily in reality, e.g., slight uncertainties or misalignments in the treatment head geometry might occur over time.

8.8 OTHER MONTE CARLO APPLICATIONS

8.8.1 Organ Motion Studies

Monte Carlo simulations in radiation therapy are mostly used for dose calculation purposes (see Chapter 14). Given its accuracy, especially for low-density materials and in the presence of heterogeneities, Monte Carlo dose calculation is attractive, particularly, for studying motion effects in lung [97–102].

For studying dosimetric effects in time-dependent geometries, the results of individual three-dimensional calculations are usually combined. This method can become cumbersome, in particular, when assessing double-dynamic systems, e.g., when investigating the influence of time-dependent beam delivery (i.e., magnetically moving beam spots in proton beam scanning) on the dose deposition in a moving target. In four-dimensional Monte Carlo simulations, the time parameter is translated into the number of histories per updated geometry [89]. The use of this technique when modeling time-dependent structures in the treatment head has been described above. It has been applied to model respiratory patient motion [91,103] and to study interplay effects between respiratory motion and beam scanning speed [90]. If dose is to be calculated based on a time-dependent patient geometry, it cannot be accumulated based on a fixed coordinate system. Deformable image registration has to be applied to map dose distributions to a reference frame [103].

8.8.2 Modeling of Detector Systems

Monte Carlo studies play also a major role in designing detectors for range verification (see Chapter 21), particularly in the field of prompt gamma imaging, where, depending on the design of the method, no off-the-shelve detector solution exists [104–107]. Another example is the use of Monte Carlo simulations to optimize image reconstruction for proton computed tomography [108,109]. Monte Carlo methods are increasingly being used to assist in routine quality assurance method with the need for modeling ionization chamber responses [110,111] or optimizing microdosimetric detector systems [112]. Some detectors (e.g., gel-based dosimeters) experience a quenching effect, i.e., a change in response as a function of linear energy transfer (LET), which can be studied using Monte Carlo (e.g., [113–115]). Simulations also play a role in understanding uncertainties in calorimetry [116,117].

8.8.3 Simulating Proton-Induced Photon Emission for Range Verification

Protons undergo nuclear interactions in the patient, which can lead to the formation of positron emitters or excited nuclei emitting prompt gamma rays (see Chapter 21). Monte Carlo simulations play a vital role for this purpose [118]. For instance, they are used to generate a theoretical positron emission tomography (PET) image based on the prescribed radiation field, which can then be compared to the measured PET distribution for treatment verification [59,60]. The problem when simulating these images is the low statistics due to the relatively low cross section for generating a positron emitter. Fluence-to-yield conversion methods are therefore being used where positron emitter distributions are calculated by internally combining the proton fluence at a given voxel with experimental and evaluated cross sections for yielding ^{11}C, ^{15}O, and other positron emitters [58].

Both PET and prompt gamma-based range verification suffer from uncertainties in cross sections for Monte Carlo simulations with substantial uncertainties in cross-section data [119–124]. The considerable uncertainties are due to lack of nuclear physics experiments in the relevant energy range and for the relevant materials. Experiments at physics laboratories typically focus on thin targets, whereas, in a patient, particle energy distributions have to be considered. Most cross-section data are gathered at a few energies only [125–128], emphasizing the need of more accurate measurement of the cross-section values of the reaction channels, contributing to the production of PET isotopes by proton beams before PET in vivo range verification method can achieve mm accuracy.

8.8.4 Simulating Secondary Neutron Doses

Monte Carlo simulations have been used for shielding design studies (see Chapter 7). Secondary neutrons reaching the patient are of concern because of potential long-term toxicities (see Chapter 22). The neutron-generated dose cannot be calculated using an analytical dose calculation method implemented in a treatment planning system because the dose calculation is not commissioned for low doses. Neutron doses are very low (typically <0.1% of the target dose) and are thus negligible for treatment planning. Furthermore, secondary neutron doses are difficult to measure because neutrons are indirectly ionizing and interact sparsely. Consequently, Monte Carlo simulations have been extensively used for assessing neutron dose or scattered dose in patients (see, for instance, [129–135]).

For uncharged particles, like neutrons, interacting less frequently, Monte Carlo simulations might be time consuming because more histories are required to achieve a reasonable statistical accuracy. One approach to overcoming this problem is the use of tabulated energy-dependent fluence-to-equivalent dose conversion coefficients [136–138] in combination with calculating particle fluences at the surface of a region of interest (organ) [62,139,140].

8.8.5 Computational Phantoms

Organs not directly considered in the treatment planning process are typically not imaged. Consequently, whole-body computational phantoms can play an important role when combined with Monte Carlo dose calculations to simulate scattered or secondary doses (e.g., neutron doses) or imaging doses to organs [141]. Initially, the radiation protection community defined stylized phantoms that are based on simple geometrical shapes, e.g., an elliptical cylinder representing the arm, torso, and hips, a truncated elliptical cone representing the legs and feet, and an elliptical cylinder representing the head and neck. A more realistic representation of the human body can be achieved using voxel phantoms, where each voxel is identified in terms of tissue type (soft tissue, hard bone, etc.) and organ identification [142]. In a Monte Carlo environment, each phantom voxel is usually tagged with a specific material composition and density. Voxel phantoms are mostly based on CT images and manually segmented organ contours. For each organ and model, age- and gender-dependent densities, as well as age-dependent material compositions, can be adopted based on ICRU [143], and organ-specific material composition as a function of age can be based on individuals at the International Commission on Radiological Protection (ICRP) reference ages [144,145]. Dosimetric differences between the use of stylized phantoms and the use of voxel phantoms can be up to 150% [146]. To match a particular patient as closely as possible using a voxel phantom in a Monte Carlo simulation, one might have to interpolate between two different phantoms of a specific age doing uniform scaling [147]. Hybrid phantoms have been developed that can even consider differences in the distribution of subcutaneous fat when trying to create an individual from a reference phantom [148,149]. These phantoms are typically based on combinations of polygon mesh and nonuniform rational B-spline surfaces. They provide the flexibility to model thin tissue layers and allow for free-form phantom deformations for selected body regions and internal organs. A series of reference (i.e., 50th height/weight percentile) pediatric hybrid models has been developed [150]. Figure 8.11 shows a hybrid phantom and how it can be shaped to resemble a specific patient.

8.8.6 Simulating LET Distributions for Radiobiological Considerations

Proton doses are related to photon doses using the relative biological effectiveness (see Chapter 22). Biophysical modeling is far from being able to simulate all radiation effects (both damage and repair) in subcellular structures. The physics, however, can be simulated reasonably well. To interpret biological experiments, one needs to know the characteristics of the radiation beam, as the absorbed dose can be described as the integral of the particle fluence times the total mass stopping power over the particle energy distributions [151,152]. One parameter determining the biological effectiveness of proton beams is the LET. One

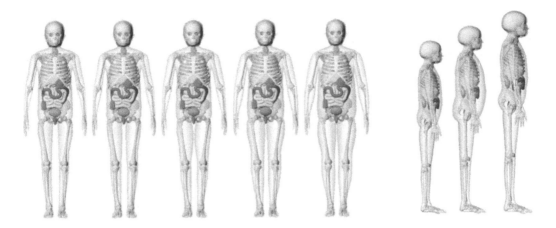

FIGURE 8.11   Frontal views of patient-dependent pediatric female phantoms at a specific targeted standing height and at their 10th, 25th, 15th, 75th, and 19th percentile body masses. Also shown are lateral views for different targeted standing heights. From Ref. [149] with permission (© 2009 IEEE).

can simulate dose-averaged LET distributions in a patient geometry to identify potential hot spots of biological effectiveness and use this information for biologically driven optimization (see Chapter 19) [153–155].

For instance, when calculating the dose-average LET during a Monte Carlo run, one needs to record each energy loss, dE, of a particle and the length of the particle step that leads to the energy deposition event, dx. In a CT geometry, all values can be scored voxel by voxel (v):

$$\text{LET}_d(v) = \frac{\sum\limits_{\text{events}} dE \cdot \left(\frac{dE}{dx}\right) \cdot \frac{1}{\rho}}{\sum\limits_{\text{events}} dE} \tag{8.7}$$

Note that when simulating LET_d using Monte Carlo simulations, the cutoff defined to stop further tracking of the proton, as well as the inclusion of secondary particles generated during a step, can have a significant influence [156,157]. Strictly speaking, the LET concept is incompatible with the Monte Carlo technique as it is based on a two-dimensional quantity.

8.8.7 Biology: Track Structure Simulations

Microdosimetric quantities considering small volumes are biologically more relevant than LET (see Chapter 22). In addition to model energy deposition events in microdosimetric volumes, the particle track structure describes the pattern of energy deposition events of proton tracks, including the secondary electrons. Simulations may assume a proportionality between ionization frequency and lesion type in the DNA so that predictions can be made on the likelihood of DNA damage and damage clustering [158], which in turn can be used to make assumptions about repair probability for certain DNA damages. Analyzing the distance of two energy deposition events or clustering of events on a nanometer scale can give insight into lesion complexity [159–161].

To understand the biological effect of radiation in more detail, Monte Carlo simulations can even be used to study the interactions of particles with biological structures like the DNA directly by modeling subcellular geometries (e.g., [162–165]). Radiation-induced clustered damage other than double-strand breaks can be simulated with Monte Carlo codes as well, but the scarcity of experimental validation data makes it difficult to validate the results [166].

Track structure can be simulated with the same codes that are used for macroscopic dose simulations. However, many specific codes have been developed particularly to deal with low-energy particle tracks and with the δ-electrons produced (e.g., [167,168]). Accurate single and double differential cross sections for inelastic and elastic interactions, including ionization, leading to δ-electron emission are needed but are mostly available only for water [169]. Specifically designed for Monte Carlo simulations on subcellular level for radiobiology is the Geant4DNA code, which extends the physics of Geant4 to encompass very low-energy interactions for organic materials [170–172]. Similar to TOPAS as a wrapper around Geant4 (see above), there now also is a wrapper for Geant4DNA (i.e., TOPAS_nBio [164]) to simplify Monte Carlo simulations for radiobiology.

Currently, Monte Carlo simulations are unable to simulate radiation action and radiation effects on living cells considering all processes involved. In order to model cellular radiation effects, it is necessary to predict the relationship between the lesion distribution and the kinetics of damage processing. The Monte Carlo code PHITS has been modified to allow the simulation of radiolysis [173] to be combined with the MKM model (see Chapter 22). More importantly, recent developments also suggest the use of Monte Carlo techniques to model more detailed radiochemistry mechanistically [174].

REFERENCES

1. Bielajew AF. History of Monte Carlo. In: Joao Seco and Frank Verhaegen, editors. *Monte Carlo Techniques in Radiation Therapy*. 2013, CRC Press, pp. 3–16.
2. Kimstrand P, Tilly N, Ahnesjo A, Traneus E. Experimental test of Monte Carlo proton transport at grazing incidence in GEANT4, FLUKA and MCNPX. *Phys Med Biol* 2008; 53(4): 1115–129.
3. Poon E, Seuntjens J, Verhaegen F. Consistency test of the electron transport algorithm in the GEANT4 Monte Carlo code. *Phys Med Biol* 2005; 50: 681–694.
4. van Goethem MJ, van der Meer R, Reist HW, Schippers JM. Geant4 simulations of proton beam transport through a carbon or beryllium degrader and following a beam line. *Phys Med Biol* 2009; 54(19): 5831–46.
5. Berger MJ. Monte Carlo Calculation of the penetration and diffusion of fast charged particles. Methods in Computational, Physics, edited by B Alder, S Fernbach, and M Rotenberg (Academic, New York). *Inspire* 1963; 1: 135–215.
6. ICRU. Stopping powers and ranges for protons and alpha particles. International Commission on Radiation Units and Measurements, Bethesda, MD, 1993; Report No. 49.
7. Kawrakow I, Bielajew AF. On the condensed history technique for electron transport. *Nucl Instrum Methods Phys Res B*. 1998; 142(3): 253–80.
8. Lewis HW. Multiple scattering in an infinite medium. *Phys Rev* 1950; 78: 526–9.
9. Paganetti H. Nuclear interactions in proton therapy: dose and relative biological effect distributions originating from primary and secondary particles. *Phys Med Biol* 2002; 47: 747–64.

10. ICRU. Nuclear data for neutron and proton radiotherapy and for radiation protection. International Commission on Radiation Units and Measurements, Bethesda, MD, 2000; Report No. 63.

11. Soukup M, Alber M. Influence of dose engine accuracy on the optimum dose distribution in intensity-modulated proton therapy treatment plans. *Phys Med Biol* 2007; 52: 725–40.

12. Sawakuchi GO, Titt U, Mirkovic D, Ciangaru G, Zhu XR, Sahoo N et al. Monte Carlo investigation of the low-dose envelope from scanned proton pencil beams. *Phys Med Biol* 2010; 55(3): 711–21.

13. Ramos-Mendez J, Perl J, Faddegon B, Schumann J, Paganetti H. Geometrical splitting technique to improve the computational efficiency in Monte Carlo calculations for proton therapy. *Med Phys.* 2013; 40(4): 041718.

14. Ramos-Mendez J, Schuemann J, Incerti S, Paganetti H, Schulte R, Faddegon B. Flagged uniform particle splitting for variance reduction in proton and carbon ion track-structure simulations. *Phys Med Biol* 2017; 62(15): 5908–25.

15. Mendez JR, Perl J, Schumann J, Shin J, Paganetti H, Faddegon B. Improved efficiency in Monte Carlo simulation for passive-scattering proton therapy. *Phys Med Biol* 2015; 60(13): 5019–35.

16. Yepes P, Randeniya S, Taddei PJ, Newhauser WD. A track-repeating algorithm for fast Monte Carlo dose calculations of proton radiotherapy. *Nucl Technol.* 2009; 168(3): 736–40.

17. Fix MK, Frei D, Volken W, Born EJ, Aebersold DM, Manser P. Macro Monte Carlo for dose calculation of proton beams. *Phys Med Biol* 2013; 58(7): 2027–44.

18. Giantsoudi D, Schuemann J, Jia X, Dowdell S, Jiang S, Paganetti H. Validation of a GPU-based Monte Carlo code (gPMC) for proton radiation therapy: clinical cases study. *Phys Med Biol* 2015; 60(6): 2257–69.

19. Jia X, Schumann J, Paganetti H, Jiang SB. GPU-based fast Monte Carlo dose calculation for proton therapy. *Phys Med Biol* 2012; 57(23): 7783–97.

20. Ma J, Beltran C, Seum Wan Chan Tseung H, Herman MG. A GPU-accelerated and Monte Carlo-based intensity modulated proton therapy optimization system. *Med Phys* 2014; 41(12): 121707.

21. Qin N, Botas P, Giantsoudi D, Schuemann J, Tian Z, Jiang SB et al. Recent developments and comprehensive evaluations of a GPU-based Monte Carlo package for proton therapy. *Phys Med Biol* 2016; 61(20): 7347–62.

22. Wan Chan Tseung H, Ma J, Beltran C. A fast GPU-based Monte Carlo simulation of proton transport with detailed modeling of nonelastic interactions. *Med Phys* 2015; 42(6): 2967–78.

23. Battistoni G, Bauer J, Boehlen TT, Cerutti F, Chin MP, Dos Santos Augusto R et al. The FLUKA Code: an accurate simulation tool for particle therapy. *Front Oncol* 2016; 6: 116.

24. Agostinelli S, Allison J, Amako K, Apostolakis J, Araujo H, Arce P et al. GEANT4—a simulation toolkit. *Nucl Instrum Methods Phys Res, A* 2003; 506: 250–303.

25. Allison J, Amako K, Apostolakis J, Araujo H, Arce Dubois P, Asai M et al. Geant4 developments and applications. *IEEE Trans Nucl Sci* 2006; 53: 270–8.

26. Pelowitz DBE. MCNPX User's Manual, Version 2.5.0. Los Alamos National Laboratory, 2005; LA-CP-05-0369.

27. Waters L. MCNPX User's Manual. Los Alamos National Laboratory, 2002.

28. Fippel M, Soukup M. A Monte Carlo dose calculation algorithm for proton therapy. *Med Phys* 2004; 31(8): 2263–73.

29. Dementyev AV, Sobolevsky NM. SHIELD-universal Monte Carlo hadron transport code: scope and applications. *Radiat Meas* 1999; 30: 553–7.

30. Arce P, Rato P, Canadas M, Lagares JI. GAMOS: a GEANT4-based easy and flexible framework for nuclear medicine applications. *2008 IEEE Nuclear Science Symposium and Medical Imaging Conference (2008 NSS/MIC)*. 2008: 3162–8.

31. Jan S, Benoit D, Becheva E, Carlier T, Cassol F, Descourt P et al. GATE V6: a major enhancement of the GATE simulation platform enabling modelling of CT and radiotherapy. *Phys Med Biol* 2011; 56(4): 881–901.

32. Jan S, Santin G, Strul D, Staelens S, Assie K, Autret D et al. GATE: a simulation toolkit for PET and SPECT. *Phys Med Biol* 2004; 49(19): 4543–61.
33. Akagi T, Aso T, Faddegon B, Kimura A, Matsufuji N, Nishio T, et al. The PTSim and TOPAS projects, bringing Geant4 to the particle therapy clinic. *Prog Nucl Sci Technol* 2011; 2: 912–7.
34. Perl J, Shin J, Schumann J, Faddegon B, Paganetti H. TOPAS: an innovative proton Monte Carlo platform for research and clinical applications. *Med Phy* 2012; 39(11): 6818–37.
35. Bohlen TT, Bauer J, Dosanjh M, Ferrari A, Haberer T, Parodi K, et al. A Monte Carlo-based treatment-planning tool for ion beam therapy. *J Radiat Res* 2013; 54 (Suppl 1): i77–81.
36. Polster L, Schuemann J, Rinaldi I, Burigo L, McNamara AL, Stewart RD, et al. Extension of TOPAS for the simulation of proton radiation effects considering molecular and cellular endpoints. *Phys Med Biol* 2015; 60(13): 5053–70.
37. Andreo P, Medin J, Bielajew AF. Constraints of the multiple-scattering theory of Moliere in Monte Carlo simulations of the transport of charged particles. *Med Phys* 1993; 20: 1315–25.
38. Gottschalk B, Koehler AM, Schneider RJ, Sisterson JM, Wagner MS. Multiple Coulomb scattering of 160 MeV protons. *Nucl Instrum Methods Phys Res, B* 1993; 74: 467–90.
39. Urban L. Multiple scattering model in Geant4. *CERN Report.* 2002;CERN-OPEN-2002-070.
40. Robert C, Dedes G, Battistoni G, Bohlen TT, Buvat I, Cerutti F, et al. Distributions of secondary particles in proton and carbon-ion therapy: a comparison between GATE/Geant4 and FLUKA Monte Carlo codes. *Phys Med Biol* 2013; 58(9): 2879–99.
41. Andreo P. On the clinical spatial resolution achievable with protons and heavier charged particle radiotherapy beams. *Phys Med Biol* 2009; 54(11): N205–15.
42. Gottschalk B. On the scattering power of radiotherapy protons. *Med Phys* 2010; 37(1): 352–67.
43. Moyers MF, Coutrakon GB, Ghebremedhin A, Shahnazi K, Koss P, Sanders E. Calibration of a proton beam energy monitor. *Med Phys* 2007; 34(6): 1952–66.
44. Faddegon BA, Shin J, Castenada CM, Ramos-Mendez J, Daftari IK. Experimental depth dose curves of a 67.5 MeV proton beam for benchmarking and validation of Monte Carlo simulation. *Med Phys* 2015; 42(7): 4199–210.
45. Testa M, Schumann J, Lu HM, Shin J, Faddegon B, Perl J, et al. Experimental validation of the TOPAS Monte Carlo system for passive scattering proton therapy. *Med Phys* 2013; 40(12): 121719.
46. Hall DC, Makarova A, Paganetti H, Gottschalk B. Validation of nuclear models in Geant4 using the dose distribution of a 177 MeV proton pencil beam. *Phys Med Biol* 2016; 61(1): N1–N10.
47. Lin L, Kang M, Solberg TD, Ainsley CG, McDonough JE. Experimentally validated pencil beam scanning source model in TOPAS. *Phys Med Biol* 2014; 59(22): 6859–73.
48. Clasie B, Wroe A, Kooy H, Depauw N, Flanz J, Paganetti H, et al. Assessment of out-of-field absorbed dose and equivalent dose in proton fields. *Med Phys* 2010; 37(1): 311–21.
49. Paganetti H, Jiang H, Lee S-Y, Kooy H. Accurate Monte Carlo for nozzle design, commissioning, and quality assurance in proton therapy. *Med Phys* 2004; 31: 2107–18.
50. Titt U, Sahoo N, Ding X, Zheng Y, Newhauser WD, Zhu XR, et al. Assessment of the accuracy of an MCNPX-based Monte Carlo simulation model for predicting three-dimensional absorbed dose distributions. *Phys Med Biol* 2008; 53(16): 4455–70.
51. Paganetti H, Jiang H, Parodi K, Slopsema R, Engelsman M. Clinical implementation of full Monte Carlo dose calculation in proton beam therapy. *Phys Med Biol* 2008; 53(17): 4825–53.
52. Stankovskiy A, Kerhoas-Cavata S, Ferrand R, Nauraye C, Demarzi L. Monte Carlo modelling of the treatment line of the Proton Therapy Center in Orsay. *Phys Med Biol* 2009; 54(8): 2377–94.
53. Paganetti H. Monte Carlo calculations for absolute dosimetry to determine output factors for proton therapy treatments. *Phys Med Biol* 2006; 51: 2801–12.
54. Gottschalk B, Platais R, Paganetti H. Nuclear interactions of 160 MeV protons stopping in copper: a test of Monte Carlo nuclear models. *Med Phys* 1999; 26: 2597–601.
55. Paganetti H, Gottschalk B. Test of Geant3 and Geant4 nuclear models for 160 MeV protons stopping in CH_2. *Med Phys* 2003; 30: 1926–31.

56. Zacharatou-Jarlskog C, Paganetti H. Physics settings for using the Geant4 toolkit in proton therapy. *IEEE Trans Nucl Sci* 2008; 55: 1018–25.

57. Rinaldi I, Ferrari A, Mairani A, Paganetti H, Parodi K, Sala P. An integral test of FLUKA nuclear models with 160 MeV proton beams in multi-layer Faraday cups. *Phys Med Biol* 2011; 56(13): 4001–11.

58. Parodi K, Ferrari A, Sommerer F, Paganetti H. Clinical CT-based calculations of dose and positron emitter distributions in proton therapy using the FLUKA Monte Carlo code. *Phys Med Biol* 2007; 52(12): 3369–87.

59. Parodi K, Paganetti H, Cascio E, Flanz JB, Bonab AA, Alpert NM, et al. PET/CT imaging for treatment verification after proton therapy: a study with plastic phantoms and metallic implants. *Med Phys* 2007; 34(2): 419–35.

60. Parodi K, Paganetti H, Shih HA, Michaud S, Loeffler JS, DeLaney TF, et al. Patient study of in vivo verification of beam delivery and range, using positron emission tomography and computed tomography imaging after proton therapy. *Int J Radiat Oncol Biol Phys* 2007; 68(3): 920–34.

61. Schneider U, Agosteo S, Pedroni E, Besserer J. Secondary neutron dose during proton therapy using spot scanning. *Int J Radiat Oncol, Biol, Phys* 2002; 53: 244–51.

62. Polf JC, Newhauser WD. Calculations of neutron dose equivalent exposures from range-modulated proton therapy beams. *Phys Med Biol* 2005; 50(16): 3859–73.

63. Moyers MF, Benton ER, Ghebremedhin A, Coutrakon G. Leakage and scatter radiation from a double scattering based proton beamline. *Med Phys* 2008; 35(1): 128–44.

64. Zheng Y, Fontenot J, Taddei P, Mirkovic D, Newhauser W. Monte Carlo simulations of neutron spectral fluence, radiation weighting factor and ambient dose equivalent for a passively scattered proton therapy unit. *Phys Med Biol* 2008; 53(1): 187–201.

65. Sawakuchi GO, Titt U, Mirkovic D, Mohan R. Density heterogeneities and the influence of multiple Coulomb and nuclear scatterings on the Bragg peak distal edge of proton therapy beams. *Phys Med Biol* 2008; 53(17): 4605–19.

66. Urie M, Goitein M, Holley WR, Chen GTY. Degradation of the Bragg peak due to inhomogeneities. *Phys Med Biol* 1986; 31: 1–15.

67. Kimstrand P, Traneus E, Ahnesjo A, Tilly N. Parametrization and application of scatter kernels for modelling scanned proton beam collimator scatter dose. *Phys Med Biol* 2008; 53(13): 3405–29.

68. Titt U, Zheng Y, Vassiliev ON, Newhauser WD. Monte Carlo investigation of collimator scatter of proton-therapy beams produced using the passive scattering method. *Phys Med Biol* 2008; 53(2): 487–504.

69. van Luijk P, van t' Veld AA, Zelle HD, Schippers JM. Collimator scatter and 2D dosimetry in small proton beams. *Phys Med Biol* 2001; 46(3): 653–70.

70. Newhauser W, Fontenot J, Koch N, Dong L, Lee A, Zheng Y, et al. Monte Carlo simulations of the dosimetric impact of radiopaque fiducial markers for proton radiotherapy of the prostate. *Phys Med Biol* 2007; 52(11): 2937–52.

71. Brown KL, Carey DC, Iselin C, Rothacker F. Transport: a computer program for designing charged particle beam transport systems. CERN 73-16 (1973) & CERN 80-04. 1980.

72. Hsi WC, Moyers MF, Nichiporov D, Anferov V, Wolanski M, Allgower CE, et al. Energy spectrum control for modulated proton beams. *Med Phys* 2009; 36(6): 2297–308.

73. Bednarz B, Lu HM, Engelsman M, Paganetti H. Uncertainties and correction methods when modeling passive scattering proton therapy treatment heads with Monte Carlo. *Phys Med Biol* 2011; 56(9): 2837–54.

74. Brooks FD, Jones DTL, Bowley CC, Symons JE, Buffler A, Allie MS. Energy spectra in the NAC proton therapy beam. *Radiat Prot Dosim* 1997; 70: 477–80.

75. Grassberger C, Lomax A, Paganetti H. Characterizing a proton beam scanning system for Monte Carlo dose calculation in patients. *Phys Med Biol* 2015; 60(2): 633–45.

76. Verhaegen F, Palmans H. Secondary electron fluence perturbation by high-Z interfaces in clinical proton beams: a Monte Carlo study. *Phy Med Biol* 1999; 44: 167–83.
77. Verhaegen F, Palmans H. A systematic Monte Carlo study of secondary electron fluence perturbation in clinical proton beams (70–250 MeV) for cylindrical and spherical ion chambers. *Med Phys* 2001; 28(10): 2088–95.
78. Kooy H, Schaefer M, Rosenthal S, Bortfeld T. Monitor unit calculations for range-modulated spread-out Bragg peak fields. *Phys Med Biol* 2003; 48: 2797–808.
79. Kooy HM, Rosenthal SJ, Engelsman M, Mazal A, Slopsema RL, Paganetti H, et al. The prediction of output factors for spread-out proton Bragg peak fields in clinical practice. *Phys Med Biol* 2005; 50: 5847–56.
80. Herault J, Iborra N, Serrano B, Chauvel P. Spread-out Bragg peak and monitor units calculation with the Monte Carlo code MCNPX. *Med Phys* 2007; 34(2): 680–8.
81. Koch N, Newhauser WD, Titt U, Gombos D, Coombes K, Starkschall G. Monte Carlo calculations and measurements of absorbed dose per monitor unit for the treatment of uveal melanoma with proton therapy. *Phys Med Biol* 2008; 53(6): 1581–94.
82. Peterson SW, Polf J, Bues M, Ciangaru G, Archambault L, Beddar S, et al. Experimental validation of a Monte Carlo proton therapy nozzle model incorporating magnetically steered protons. *Phys Med Biol* 2009; 54(10): 3217–29.
83. Fontenot JD, Newhauser WD, Titt U. Design tools for proton therapy nozzles based on the double-scattering foil technique. *Radiat Prot Dosim* 2005; 116(1–4 Pt 2): 211–5.
84. Newhauser W, Koch N, Hummel S, Ziegler M, Titt U. Monte Carlo simulations of a nozzle for the treatment of ocular tumours with high-energy proton beams. *Phys Med Biol* 2005; 50: 5229–49.
85. Prusator M, Ahmad S, Chen Y. TOPAS Simulation of the Mevion S250 compact proton therapy unit. *J Appl Clin Med Phys* 2017; 18(3): 88–95.
86. Constantin M, Constantin DE, Keall PJ, Narula A, Svatos M, Perl J. Linking computer-aided design (CAD) to Geant4-based Monte Carlo simulations for precise implementation of complex treatment head geometries. *Phys Med Biol* 2010; 55(8): N211–20.
87. Yasui K, Toshito T, Omachi C, Kibe Y, Hayashi K, Shibata H, et al. A patient-specific aperture system with an energy absorber for spot scanning proton beams: verification for clinical application. *Med Phys* 2015; 42(12): 6999–7010.
88. Charlwood FC, Aitkenhead AH, Mackay RI. A Monte Carlo study on the collimation of pencil beam scanning proton therapy beams. *Med Phys* 2016; 43(3): 1462–72.
89. Shin J, Perl J, Schumann J, Paganetti H, Faddegon BA. A modular method to handle multiple time-dependent quantities in Monte Carlo simulations. *Phys Med Biol* 2012; 57(11): 3295–308.
90. Paganetti H, Jiang H, Trofimov A. 4D Monte Carlo simulation of proton beam scanning: Modeling of variations in time and space to study the interplay between scanning pattern and time-dependent patient geometry. *Phys Med Biol* 2005; 50: 983–90.
91. Paganetti H. Four-dimensional Monte Carlo simulation of time dependent geometries. *Phys Med Biol* 2004; 49: N75–N81.
92. Peterson S, Polf J, Ciangaru G, Frank SJ, Bues M, Smith A. Variations in proton scanned beam dose delivery due to uncertainties in magnetic beam steering. *Med Phys* 2009; 36(8): 3693–702.
93. Lu HM, Kooy H. Optimization of current modulation function for proton spread-out Bragg peak fields. *Med Phys* 2006; 33: 1281–7.
94. Tessonnier T, Marcelos T, Mairani A, Brons S, Parodi K. Phase space generation for proton and carbon ion beams for external users' applications at the Heidelberg ion therapy center. *Front Oncol* 2015; 5: 297.
95. Ma CM, Faddegon BA, Rogers DWO, Mackie TR. Accurate characterization of Monte Carlo calculated electron beams for radiotherapy. *Med Phys* 1997; 24: 401–16.
96. Kimstrand P, Traneus E, Ahnesjo A, Grusell E, Glimelius B, Tilly N. A beam source model for scanned proton beams. *Phys Med Biol* 2007; 52(11): 3151–68.

97. Dowdell S, Grassberger C, Paganetti H. Four-dimensional Monte Carlo simulations demonstrating how the extent of intensity-modulation impacts motion effects in proton therapy lung treatments. *Med Phys* 2013; 40(12): 121713.

98. Dowdell S, Grassberger C, Sharp G, Paganetti H. Fractionated lung IMPT treatments: sensitivity to setup uncertainties and motion effects based on single-field homogeneity. *Technol Cancer Res Treat* 2016; 15(5): 689–96.

99. Dowdell S, Grassberger C, Sharp GC, Paganetti H. Interplay effects in proton scanning for lung: a 4D Monte Carlo study assessing the impact of tumor and beam delivery parameters. *Phys Med Biol* 2013; 58(12): 4137–56.

100. Grassberger C, Daartz J, Dowdell S, Ruggieri T, Sharp G, Paganetti H. Quantification of proton dose calculation accuracy in the lung. *Int J Radiat Oncol Biol Phys* 2014; 89(2): 424–30.

101. Grassberger C, Dowdell S, Lomax A, Sharp G, Shackleford J, Choi N, et al. Motion interplay as a function of patient parameters and spot size in spot scanning proton therapy for lung cancer. *Int J Radiat Oncol Biol Phys* 2013; 86(2): 380–6.

102. Grassberger C, Dowdell S, Sharp G, Paganetti H. Motion mitigation for lung cancer patients treated with active scanning proton therapy. *Med Phys* 2015; 42(5): 2462–9.

103. Paganetti H, Jiang H, Adams JA, Chen GT, Rietzel E. Monte Carlo simulations with time-dependent geometries to investigate organ motion with high temporal resolution. *Int J Radiat Oncol, Biol, Phys* 2004; 60: 942–50.

104. Ortega PG, Torres-Espallardo I, Cerutti F, Ferrari A, Gillam JE, Lacasta C, et al. Noise evaluation of Compton camera imaging for proton therapy. *Phys Med Biol* 2015; 60(5): 1845–63.

105. Pinto M, Dauvergne D, Freud N, Krimmer J, Letang JM, Ray C, et al. Design optimisation of a TOF-based collimated camera prototype for online hadrontherapy monitoring. *Phys Med Biol* 2014; 59(24): 7653–74.

106. Biegun AK, Seravalli E, Lopes PC, Rinaldi I, Pinto M, Oxley DC, et al. Time-of-flight neutron rejection to improve prompt gamma imaging for proton range verification: a simulation study. *Phys Med Biol* 2012; 57(20): 6429–44.

107. Janssen FM, Landry G, Cambraia Lopes P, Dedes G, Smeets J, Schaart DR, et al. Factors influencing the accuracy of beam range estimation in proton therapy using prompt gamma emission. *Phys Med Biol* 2014; 59(15): 4427–41.

108. Li T, Liang Z, Singanallur JV, Satogata TJ, Williams DC, Schulte RW. Reconstruction for proton computed tomography by tracing proton trajectories: a Monte Carlo study. *Med Phys* 2006; 33(3): 699–706.

109. Schulte RW, Penfold SN, Tafas JT, Schubert KE. A maximum likelihood proton path formalism for application in proton computed tomography. *Med Phys* 2008; 35(11): 4849–56.

110. Goma C, Andreo P, Sempau J. Monte Carlo calculation of beam quality correction factors in proton beams using detailed simulation of ionization chambers. *Phys Med Biol* 2016; 61(6): 2389–406.

111. Sorriaux J, Testa M, Paganetti H, Bertrand D, Lee JA, Palmans H, et al. Consistency in quality correction factors for ionization chamber dosimetry in scanned proton beam therapy. *Med Phys* 2017; 44(9): 4919–27.

112. Solevi P, Magrin G, Moro D, Mayer R. Monte Carlo study of microdosimetric diamond detectors. *Phys Med Biol* 2015; 60(18): 7069–83.

113. Wang LL, Perles LA, Archambault L, Sahoo N, Mirkovic D, Beddar S. Determination of the quenching correction factors for plastic scintillation detectors in therapeutic high-energy proton beams. *Phys Med Biol* 2012; 57(23): 7767–81.

114. Perles LA, Mirkovic D, Anand A, Titt U, Mohan R. LET dependence of the response of EBT2 films in proton dosimetry modeled as a bimolecular chemical reaction. *Phys Med Biol* 2013; 58(23): 8477–91.

115. Hoye EM, Skyt PS, Balling P, Muren LP, Taasti VT, Swakon J, et al. Chemically tuned linear energy transfer dependent quenching in a deformable, radiochromic 3D dosimeter. *Phys Med Biol* 2017; 62(4): N73–N89.

116. Lourenco A, Thomas R, Homer M, Bouchard H, Rossomme S, Renaud J, et al. Fluence correction factor for graphite calorimetry in a clinical high-energy carbon-ion beam. *Phys Med Biol* 2017; 62(7): N134–N46.

117. Palmans H, Al-Sulaiti L, Andreo P, Shipley D, Luhr A, Bassler N, et al. Fluence correction factors for graphite calorimetry in a low-energy clinical proton beam: I. Analytical and Monte Carlo simulations. *Phys Med Biol* 2013; 58(10): 3481–99.

118. Kraan AC. Range verification methods in particle therapy: underlying physics and Monte Carlo modeling. *Front Oncol* 2015; 5: 150.

119. Verburg JM, Shih HA, Seco J. Simulation of prompt gamma-ray emission during proton radiotherapy. *Phys Med Biol* 2012; 57(17): 5459–72.

120. Dedes G, Pinto M, Dauvergne D, Freud N, Krimmer J, Letang JM, et al. Assessment and improvements of Geant4 hadronic models in the context of prompt-gamma hadrontherapy monitoring. *Phys Med Biol* 2014; 59(7): 1747–72.

121. Jeyasugiththan J, Peterson SW. Evaluation of proton inelastic reaction models in Geant4 for prompt gamma production during proton radiotherapy. *Phys Med Biol* 2015; 60(19): 7617–35.

122. Pinto M, Dauvergne D, Freud N, Krimmer J, Letang JM, Testa E. Assessment of Geant4 prompt-gamma emission yields in the context of proton therapy monitoring. *Front Oncol* 2016; 6: 10.

123. Testa M, Min CH, Verburg JM, Schumann J, Lu HM, Paganetti H. Range verification of passively scattered proton beams based on prompt gamma time patterns. *Phys Med Biol* 2014; 59(15): 4181–95.

124. Schumann A, Petzoldt J, Dendooven P, Enghardt W, Golnik C, Hueso-Gonzalez F, et al. Simulation and experimental verification of prompt gamma-ray emissions during proton irradiation. *Phys Med Biol* 2015; 60(10): 4197–207.

125. Espana S, Zhu X, Daartz J, El Fakhri G, Bortfeld T, Paganetti H. The reliability of proton-nuclear interaction cross-section data to predict proton-induced PET images in proton therapy. *Phys Med Biol* 2011; 56(9): 2687–98.

126. Bauer J, Unholtz D, Kurz C, Parodi K. An experimental approach to improve the Monte Carlo modelling of offline PET/CT-imaging of positron emitters induced by scanned proton beams. *Phys Med Biol* 2013; 58(15): 5193–213.

127. Rohling H, Sihver L, Priegnitz M, Enghardt W, Fiedler F. Comparison of PHITS, GEANT4, and HIBRAC simulations of depth-dependent yields of beta(+)-emitting nuclei during therapeutic particle irradiation to measured data. *Phys Med Biol* 2013; 58(18): 6355–68.

128. Seravalli E, Robert C, Bauer J, Stichelbaut F, Kurz C, Smeets J, et al. Monte Carlo calculations of positron emitter yields in proton radiotherapy. *Phys Med Biol* 2012; 57(6): 1659–73.

129. Fontenot J, Taddei P, Zheng Y, Mirkovic D, Jordan T, Newhauser W. Equivalent dose and effective dose from stray radiation during passively scattered proton radiotherapy for prostate cancer. *Phys Med Biol* 2008; 53(6): 1677–88.

130. Zacharatou Jarlskog C, Lee C, Bolch W, Xu XG, Paganetti H. Assessment of organ specific neutron doses in proton therapy using whole-body age-dependent voxel phantoms. *Phys Med Biol* 2008; 53: 693–714.

131. Jiang H, Wang B, Xu XG, Suit HD, Paganetti H. Simulation of organ specific patient effective dose due to secondary neutrons in proton radiation treatment. *Phys Med Biol* 2005; 50: 4337–53.

132. Athar BS, Bednarz B, Seco J, Hancox C, Paganetti H. Comparison of out-of-field photon doses in 6-MV IMRT and neutron doses in proton therapy for adult and pediatric patients. *Phys Med Biol* 2010; 55: 2879–92.

133. Athar BS, Paganetti H. Neutron equivalent doses and associated lifetime cancer incidence risks for head & neck and spinal proton therapy. *Phys Med Biol* 2009; 54(16): 4907–26.

134. Geng C, Moteabbed M, Seco J, Gao Y, Xu XG, Ramos-Mendez J, et al. Dose assessment for the fetus considering scattered and secondary radiation from photon and proton therapy when treating a brain tumor of the mother. *Phys Med Biol* 2016; 61(2): 683–95.

135. Dowdell SJ, Clasie B, Depauw N, Metcalfe P, Rosenfeld AB, Kooy HM, et al. Monte Carlo study of the potential reduction in out-of-field dose using a patient-specific aperture in pencil beam scanning proton therapy. *Phys Med Biol* 2012; 57(10): 2829–42.
136. Bozkurt A, Chao TC, Xu XG. Fluence-to-dose conversion coefficients from monoenergetic neutrons below 20 MeV based on the VIP-man anatomical model. *Phys Med Biol* 2000; 45: 3059–79.
137. Chen J. Fluence-to-absorbed dose conversion coefficients for use in radiological protection of embryo and foetus against external exposure to protons from 100 MeV to 100 GeV. *Radiat Prot Dosim* 2006; 118(4): 378–83.
138. Alghamdi AA, Ma A, Tzortzis M, Spyrou NM. Neutron-fluence-to-dose conversion coefficients in an anthropomorphic phantom. *Radiat Prot Dosim* 2005; 115(1–4): 606–11.
139. ICRU. Conversion coefficients for use in radiological protection against external radiation. *International Commission on Radiation Units and Measurements*, Bethesda, MD, 1998; 57.
140. Zheng Y, Newhauser W, Fontenot J, Taddei P, Mohan R. Monte Carlo study of neutron dose equivalent during passive scattering proton therapy. *Phys Med Biol* 2007; 52(15): 4481–96.
141. Paganetti H. The use of computational patient models to assess the risk of developing radiation-induced cancers from radiation therapy of the primary cancer. *Proc IEEE.* 2009; 97: 1977–87.
142. Zaidi H, Xu XG. Computational anthropomorphic models of the human anatomy: the path to realistic Monte Carlo modeling in radiological sciences. *Annu Rev Biomed Eng* 2007; 9: 471–500.
143. ICRU. Photon, electron, proton and neutron interaction data for body tissues. *International Commision on Radiation Units and Measurements*, Bethesda, MD, 1992; Report No. 46.
144. ICRP. Basic anatomical and physiological data for use in radiological protection: reference values. *International Commission on Radiological Protection (Pergamon Press).* 2003; 89.
145. ICRU. Tissue substitutes in radiation dosimetry and measurement. *International Commission on Radiation Units and Measurements*, Bethesda, MD, 1989; Report No. 44.
146. Lee C, Lee C, Williams JL, Bolch WE. Whole-body voxel phantoms of paediatric patients: UF Series B. *Phys Med Biol* 2006; 51(18): 4649–61.
147. Moteabbed M, Geyer A, Drenkhahn R, Bolch WE, Paganetti H. Comparison of whole-body phantom designs to estimate organ equivalent neutron doses for secondary cancer risk assessment in proton therapy. *Phys Med Biol* 2012; 57(2): 499–515.
148. Lee C, Lodwick D, Hurtado J, Pafundi D, Williams JL, Bolch WE. The UF family of reference hybrid phantoms for computational radiation dosimetry. *Phys Med Biol* 2010; 55(2): 339–63.
149. Johnson PB, Whalen SR, Wayson M, Juneja B, Lee C, Bolch WE. Hybrid patient-dependent phantoms covering statistical distributions of body morphometry in the U.S. adult and pediatric population. *Proc IEEE* 2009; 97(12): 2060–75.
150. Lee C, Lodwick D, Williams JL, Bolch W. Hybrid computational phantoms of the 15-year male and female adolescent: applications to CT organ dosimetry for patients of variable morphometry. *Med Phys* 2008; 35: 2366–82.
151. Paganetti H, Goitein M. Radiobiological significance of beam line dependent proton energy distributions in a spread-out Bragg peak. *Med Phys* 2000; 27: 1119–26.
152. Paganetti H, Schmitz T. The influence of the beam modulation method on dose and RBE in proton radiation therapy. *Phys Med Biol* 1996; 41: 1649–63.
153. Grassberger C, Trofimov A, Lomax A, Paganetti H. Variations in linear energy transfer within clinical proton therapy fields and the potential for biological treatment planning. *Int J Radiat Oncol Biol Phys* 2011; 80(5): 1559–66.
154. Giantsoudi D, Grassberger C, Craft D, Niemierko A, Trofimov A, Paganetti H. Linear energy transfer-guided optimization in intensity modulated proton therapy: feasibility study and clinical potential. *Int J Radiat Oncol Biol Phys* 2013; 87(1): 216–22.
155. Unkelbach J, Botas P, Giantsoudi D, Gorissen B, Paganetti H. Reoptimization of intensity-modulated proton therapy plans based on linear energy transfer. *Int J Radiat Oncol Biol Phys* 2016; 96(5): 1097–106.

156. Granville DA, Sawakuchi GO. Comparison of linear energy transfer scoring techniques in Monte Carlo simulations of proton beams. *Phys Med Biol* 2015; 60(14): N283–91.
157. Cortes-Giraldo MA, Carabe A. A critical study of different Monte Carlo scoring methods of dose average linear-energy-transfer maps calculated in voxelized geometries irradiated with clinical proton beams. *Phys Med Biol* 2015; 60(7): 2645–69.
158. Dos Santos M, Villagrasa C, Clairand I, Incerti S. Influence of the DNA density on the number of clustered damages created by protons of different energies. *Nucl Instrum Methods Phys Res, B* 2013; 298: 47–54.
159. Gonzalez-Munoz G, Tilly N, Fernandez-Varea JM, Ahnesjo A. Monte Carlo simulation and analysis of proton energy-deposition patterns in the Bragg peak. *Phys Med Biol* 2008; 53(11): 2857–75.
160. Backstrom G, Galassi ME, Tilly N, Ahnesjo A, Fernandez-Varea JM. Track structure of protons and other light ions in liquid water: applications of the LIonTrack code at the nanometer scale. *Med Phys* 2013; 40(6): 064101.
161. Bueno M, Schulte R, Meylan S, Villagrasa C. Influence of the geometrical detail in the description of DNA and the scoring method of ionization clustering on nanodosimetric parameters of track structure: a Monte Carlo study using Geant4-DNA. *Phys Med Biol* 2015; 60(21): 8583–99.
162. Ottolenghi A, Merzagora M, Tallone L, Durante M, Paretzke HG, Wilson WE. The quality of DNA double-strand breaks: a Monte Carlo simulation of the end-structure of strand breaks produced by protons and alpha particles. *Radiat Environ Biophys* 1995; 34: 239–44.
163. Friedland W, Jacob P, Bernhardt P, Paretzke HG, Dingfelder M. Simulation of DNA damage after proton irradiation. *Radiat Res* 2003; 159: 401–10.
164. McNamara A, Geng C, Turner R, Mendez JR, Perl J, Held K, et al. Validation of the radiobiology toolkit TOPAS-nBio in simple DNA geometries. *Phys Med* 2017; 61(16): 5993–6010.
165. Lazarakis P, Bug MU, Gargioni E, Guatelli S, Rabus H, Rosenfeld AB. Comparison of nanodosimetric parameters of track structure calculated by the Monte Carlo codes Geant4-DNA and PTra. *Phys Med Biol* 2012; 57(5): 1231–50.
166. Watanabe R, Rahmanian S, Nikjoo H. Spectrum of radiation-induced clustered non-DSB damage: a Monte Carlo track structure modeling and calculations. *Radiat Res* 2015; 183(5): 525–40.
167. Nikjoo H, Uehara S, Emfietzoglou D, Cucinotta FA. Track-structure codes in radiation research. *Radiat Meas* 2006; 41(9–10): 1052–74.
168. Semenenko VA, Stewart RD. A fast Monte Carlo algorithm to simulate the spectrum of DNA damages formed by ionizing radiation. *Radiat Res* 2004; 161(4): 451–7.
169. Incerti S, Psaltaki M, Gillet P, Barberet P, Bardiès M, Bernal MA, et al. Simulating radial dose of ion tracks in liquid water simulated with Geant4-DNA: A comparative study. *Nucl Instrum Methods Phys Res, B.* 2014; 333: 92–8.
170. Incerti S, Baldacchino G, Bernal M, Capra R, Champion C, Francis Z, et al. The Geant4-DNA project. *Int J Model Simul Sci Comput* 2010; 1: 157–78.
171. Bernal MA, Bordage MC, Brown JMC, Davidkova M, Delage E, El Bitar Z, et al. Track structure modeling in liquid water: a review of the Geant4-DNA very low energy extension of the Geant4 Monte Carlo simulation toolkit. *Phys Med* 2015; 31(8): 861–74.
172. Incerti S, Douglass M, Penfold S, Guatelli S, Bezak E. Review of Geant4-DNA applications for micro and nanoscale simulations. *Phys Med* 2016; 32(10): 1187–200.
173. Tomita H, Kai M, Kusama T, Ito A. Monte Carlo simulation of physicochemical processes of liquid water radiolysis. The effects of dissolved oxygen and OH scavenger. *Radiat Environ Biophys* 1997; 36(2): 105–16.
174. Tian Z, Jiang SB, Jia X. Accelerated Monte Carlo simulation on the chemical stage in water radiolysis using GPU. *Phys Med Biol* 2017; 62(8): 3081–96.

CHAPTER 9

Detectors, Relative Dosimetry, and Microdosimetry

Hugo Palmans

Medical Radiation Science EBG MedAustron GmbH

CONTENTS

9.1 INTRODUCTION—DETECTOR REQUIREMENTS FOR RELATIVE DOSIMETRY AND MICRODOSIMETRY

Relative dosimetry concerns the measurement of relative distributions of absorbed dose to water or medium. This can be with respect to the absorbed dose to water determined in reference conditions, treated in Chapter 10, or to an intermediate dose point. Reference dosimetry aims to obtain relative doses in discrete points in different positions or different radiation conditions (e.g., field output factors) or in a continuous way (1D, 2D, 3D, or 4D dose distributions). For proton beams, the detectors used are to a large extent similar as those used for photon and electron beams, but there also exist dedicated systems. Apart from universal requirements such as sufficient spatial resolution, minimal volume averaging, and, for 4D dosimetry, adequate time resolution, there are some requirements for proton beam dosimetry that are specific and different from those for photon and electron beam dosimetry. These are related to the different interaction and scatter properties of protons as well as to the dependence of the response of many detectors on the charged particle spectrum at the measurement point. Another aspect of importance to protons is that it is well known that the cellular response to proton beam irradiation is not proportional to the absorbed dose, since there is an additional dependence on the spectral characteristics of the local distribution of energy deposition (see Chapter 22). While there is at present no complete understanding of the relation between the local stochastic fluctuations of energy deposition and the biological effects, it is generally assumed that these effects can be well correlated with the microdosimetric spectra, i.e., the distribution of energy deposition on a microscopic scale coinciding with the dimensions of the cell, cell nucleus, and DNA. This chapter reviews the specific characteristics and requirements for relative dosimetry and microdosimetry in proton beams. Other sources reviewing information on these subjects can be found in the literature and complement the information in this chapter [1–8].

9.1.1 Requirements for Relative Dosimetry

The requirements of detectors for the measurement of relative values or distributions of absorbed dose to water depend on the application, but in general, an ideal detector is water equivalent and has a response that does not depend on energy or dose rate and which is linear as a function of absorbed dose to water at a point. Issues like dose and dose rate linearity are universal criteria for detector systems and are not discussed here in the context of proton dosimetry.

Water equivalence refers to the condition that the interaction properties such as the stopping power, scattering power, and nuclear interaction cross sections are the same as those of water. It also implies that the charged particle fluence in the detector is the same as the charged particle fluence present at the point of measurement in water or, in other words, that the detector is perturbation free. Another implication is that volume-averaging effects are absent which, if steep local gradients are present, requires the detector to be small. Nevertheless, the latter criterion may not always be suitable. Indeed, the interest to obtain absorbed dose to water at a point stems from the fact that absorbed dose is a macroscopic quantity defined at a point. But if one has to measure absorbed dose or dose distributions in a heavily modulated intensity-modulated proton-therapy (IMPT) field, for example, a uniformity index or gamma index can determine if a dose distribution within the target is clinically acceptable, and it may be more convenient to measure an average dose in a larger volume rather than a point dose that would be extremely sensitive to local dose fluctuations.

Detector criteria will also be different depending on the type of profiles being measured. For lateral profiles, spatial resolution and volume averaging are important issues to be considered in steep penumbrae and small-field profiles used in stereotactic radiosurgery and in scanned beams. The variation of charged particle energy spectra over the field area is in most cases a minor problem, unless the modulation is very complex. The influence of energy spectra is much more pronounced on the measurement of depth-dose distributions, since many detectors exhibit a response that is energy dependent. For many detectors, this results in a quenching effect of the response in the Bragg peak.

9.1.1.1 Spatial Resolution and Volume Averaging

Criteria of spatial resolution for proton beam dosimetry are similar as for other types of radiotherapy; the dimensions of a detector have to be small compared to the detail one wants to resolve in the dose distributions, and the distance between measurement points has to be adapted to the local gradient. Given that protons scatter very little over the dimensions of small detectors, these criteria can be assessed on pure geometrical considerations.

For lateral dose profiles, the most critical features in the dose distributions are penumbrae, cold and hot spots in the target region, and behind nonuniformities. Lateral volume averaging of a finite detector will yield wider penumbrae, reduced cold and hot spots, and reduced dose fluctuations behind nonuniformities compared to the real values. Deconvolution approaches are possible if an accurate kernel for the detector is available, but the most common approach is to resort to detectors that have a small dimension in the scanning direction. For static lateral dose profiles, microdiamonds and diodes are suitable for this purpose and are attractive because of their online readout. Two-dimensional (2D) scintillating screens also have a high spatial resolution (limited by the resolution of the charge-coupled device (CCD) camera) and can also be used for scanned beams. Films (both radiographic and radiochromic) are very high-resolution 2D devices but have the disadvantage that the readout is offline.

Concerning depth-dose profiles, resolution in the depth direction is obviously an important criterion, but in addition, most nonwater detectors will exhibit a shift of the depth-dose curve which can be characterized by an effective point of measurement. The following paragraph discusses this aspect for ionization chambers, which have a density much lower than water, and solid detectors, which have a density that is close to or often larger than water.

For cylindrical ionization chambers, the mean energy of the protons entering the cavity is determined by the distance of water traversed by the proton before reaching the cavity. For conditions where the local depth-dose gradient is approximately constant, an analytical tracing of protons along straight lines through the ionization chamber geometry, thus neglecting scatter over the dimensions of the ionization chamber, has been proven to be adequate and in good agreement with Monte Carlo simulations, taking into account full modeling of the scatter [9]. The results are also confirmed by comparison with experimental data. The shift of the effective point of measurement from the center of the cavity towards the radiation source is according to the analytical model $0.85 \times R_{cav}$, where R_{cav} is the inner radius of the cavity. The presence of a thin wall material with a higher density than water and a metallic central electrode shifts this slightly back to the center of the chamber, such that the overall effective point of measurement of Farmer-type chambers is about $0.75 \times R_{cav}$.

For plane-parallel solid-state detectors, the shift of the effective point of measurement from the center of the detector can be quantified in terms of the thickness and the water equivalence ratio (WER) of the detector. WER is defined as the ratio of the continuous slowing-down approximation (CSDA) ranges of the detector material and water. If the local depth-dose gradient is approximately constant, the shift of the effective point from the center of the detector volume away from the beam is given by

$$P_{eff} - P_{centre} = \left(\frac{\rho_{det} r_{0,w}}{\rho_w r_{0,det}} - 1 \right) \frac{t_{det}}{2}, \tag{9.1}$$

where ρ_i is the mass density of material i, $r_{0,i}$ is the CSDA range (in g cm^{-2}) of material i and t_{det} is the thickness of the detector.

9.1.1.2 Stopping Power

Under the assumption that when using a small detector to determine dose to water, the charged particle fluence is not altered by the presence of the detector, the dose to water is derived from the dose to the detector medium by the Bragg–Gray cavity theory using Bragg–Gray (unrestricted) water-to-medium mass stopping power ratios or, if the detector is smaller than the range of delta electrons, by Spencer–Attix (restricted) water-to-medium mass stopping power ratios.

For air-filled ionization chambers, which, due to the low mass density of the detector medium have to be considered as very small, Spencer–Attix water-to-air mass stopping power ratios, $s_{w,air}$, are required. For monoenergetic protons of energy T, $s_{w,air}$ can be calculated as

$$s_{w,air} = \frac{\left(S^{\Delta}(T)/\rho\right)_w}{\left(S^{\Delta}(T)/\rho\right)_{air}}, \tag{9.2}$$

where $\left(S^{\Delta}(T)/\rho\right)_{med}$ is the restricted mass stopping power in medium "med" for protons of energy T with energy cutoff Δ. In case of a spectrum, the numerator and denominator in Equation 9.2 become a convolution of charged particle fluence in water with restricted stopping powers:

$$s_{w,air} = \frac{\sum_i\left(\int_{\Delta}^{E_{max,i}} \Phi_{E,i,w}(z_{ref})\left(S^{\Delta}(E,i)/\rho\right)_w dE + TE_{i,w}\right)}{\sum_i\left(\int_{\Delta}^{E_{max,i}} \Phi_{E,i,w}(z_{ref})\left(S^{\Delta}(E,i)/\rho\right)_{air} dE + TE_{i,air}\right)}, \tag{9.3}$$

where $\Phi_{E,i,w}(z_{ref})$ is the fluence, differential in energy, of charged particle type i in water at the reference depth, $\left(S^{\Delta}(E,i)/\rho\right)_w$ and $\left(S^{\Delta}(E,i)/\rho\right)_{air}$ are the restricted mass stopping powers of charged particle type i with energy E in water and air, respectively, and $TE_{i,w}$ and $TE_{i,air}$ are the track-end terms in water and air, respectively, accounting for energy deposition by charged particles of which the energy has fallen below the cutoff energy Δ.

The restricted mass stopping power can be calculated from the unrestricted mass stopping power $\left(S(T)/\rho\right)_{med}$, as, for example, tabulated in the International Commission on Radiation Units and Measurements (ICRU) Report 90 [10] by

$$\left(S^{\Delta}(T)/\rho\right)_{med} = \left(S(T)/\rho\right)_{med} - \frac{K}{2\beta^2}\cdot(Z/A)_{med}\cdot z^2\cdot\ln\left(W_{max}/\Delta\right), \tag{9.4}$$

where the constant $K = 0.307075$ MeV cm^2 g^{-1} and W_{max} is the maximum energy that a secondary electron can obtain from a proton–electron collision. For energies from 1 to 300 MeV, restricted mass stopping power ratios water-to-air with cutoff energy $\Delta = 10$ keV derived from ICRU Report 90 [10] using Equations 9.2 and 9.4 can be approximated by

$$s_{w,air}(T) = \frac{a\cdot T}{(T-b)^n}, \tag{9.5}$$

where the constants $a = 1.1337$, $b = 0.02$, and $n = 1.0009$.

Given that the maximum range of delta electrons in water is of the order of 2 mm for the highest proton energies (and that the vast majority of delta electrons has much smaller ranges), it is reasonable to assume that for liquid or solid detectors, the mass density and

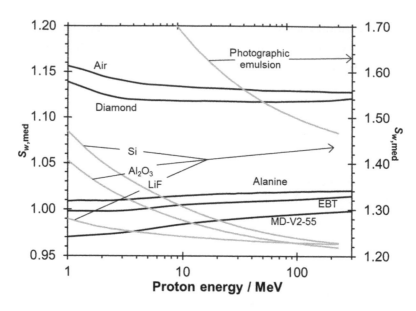

FIGURE 9.1 Mass stopping power ratios of water-to-medium for a variety of detector materials calculated using ICRU Report 90 data for water, air, and diamond and ICRU Report 49 data for all other materials. The black curves for air, diamond, alanine, GafChromic EBT film, and GafChromic MD-V2-55 film are represented on the left-hand-side vertical axis, whereas the gray curves for silicon, aluminum oxide, lithium fluoride, and photographic emulsion are represented on the right-hand-side vertical axis. The figure is updated from Palmans [4] with new data from ICRU Report 90.

size are usually such that Bragg–Gray stopping power ratios can be used. Bragg–Gray mass stopping power ratios water-to-medium for a number of detector materials and Spencer–Attix water-to-air mass stopping power ratios are shown in Figure 9.1, using data from ICRU Reports 49 [11] and 90. For materials consisting of elements with charge number, Z, close to water, such as alanine, radiochromic films, and diamond detectors, these ratios are almost constant and will thus have little influence on the measurement of depth-dose profiles. For detectors containing higher-Z materials such as silicon diodes and Al_2O_3-based luminescent crystals, the variation can be substantial and thus affect depth-dose distributions. For photographic emulsions containing silver, the variation becomes very large and induces a large uncertainty.

9.1.1.3 Energy Dependence

Apart from the energy dependence of the stopping power ratio between water and the detector material discussed in the previous section, most, if not all, detectors exhibit a response to a given absorbed dose to the detector medium that is dependent on the energy of the protons or other charged particles present in the radiation field. This dependence is often referred to as a linear-energy transfer (LET) dependence, but this is not strictly correct since charged particle spectra with the same average LET could lead to a different response.

While for ionization chambers, a beam quality correction factor (defined in Chapter 10) is usually employed to correct for their energy dependence, for solid detectors, the relative effectiveness is normally used. It is defined as the ratio of absorbed dose to the detector medium in a ^{60}Co beam, and the absorbed dose to the detector medium in the proton beam that results in the same detector signal. For a one-hit detector system or in the absence of significant track overlap effects, the relative effectiveness of a detector in a mixed particle field is calculated as a dose-weighted average of the energy-dependent relative effectiveness for monoenergetic particles over the particle spectrum [12]:

$$\bar{\eta}_{det} = \frac{\sum_{i=1}^{n_{proj}} \int_0^{E_{max,i}} (S/\rho)_{det} \; \eta_{det,i}(E)\phi_{E,i} dE}{\sum_{i=1}^{n_{proj}} \int_0^{E_{max,i}} (S/\rho)_{det} \; \phi_{E,i} dE}, \tag{9.6}$$

where i refers to the charged particle type and $\phi_{E,i}$ is the fluence differential in energy for charged particle type i.

The equivalent property for ionization chambers is the mean energy required to produce an ion pair in air by protons, $W_{air,p}$, compared to that by the electron spectrum in a ^{60}Co beam. This quantity relates the energy deposition to the amount of ions created and depends also on proton energy [13,14]. A model for energies above 1 MeV, based on the assumption that for high-energy protons the value should evolve asymptotically to the value for high-energy electrons, is given by

$$W_{air,p}(T) = W_{air,e} \cdot \frac{T}{T-k}, \tag{9.7}$$

where $W_{air,e} = 33.97$ J C^{-1} is the mean energy required to produce an ion pair in air by relativistic electrons and the constant $k = 0.08513$ is derived from the data from Dennis [13]. A fit to the more recent data from Grosswendt and Baek [14] results in $k = 0.05264$. These two values for k represent quite a large difference in the variation of $W_{air,p}$ with energy (9.3% and 5.5% from 1 to 300 MeV, respectively) mainly due to the structure of the functional dependence at lower energies, but this variation is mainly significant for protons energies under 10 MeV (hence the last couple of mm of the proton range).

The equivalent property to the relative effectiveness for chemical dosimeters is the chemical yield, G, compared to that in a ^{60}Co beam. Considering, for example, the ferrous sulphate (Fricke) dosimeter, the amount of ferrous ions oxidized to ferric ions is determined by the amount of initial reactive species formed after the radiolysis of water. Given that the yields of these initial species is energy dependent (see Chapter 10 regarding the discussion of the chemical heat defect in water calorimetry), the resulting chemical yields will be energy dependent.

It should be noted that even calorimeters, which are discussed in detail as the primary measurement instruments of absorbed dose in Chapter 10, are surprisingly prone to energy dependence. Given that they are instruments to measure absorbed dose from first principles by determining the increase of internal energy by radiation, they could be considered as ideal gold standard instruments for comparing dose distributions obtained with any other detector (in practice, calorimetric measurements are too time consuming to perform such benchmarks on a substantial scale). However, the assumptions to make this principle work and the correction factors to be applied to calorimeters make that their response is energy dependent. Indeed, a basic assumption of calorimetric measurement of absorbed dose is that all the energy deposited by radiation appears as heat and thus that there is no change of the physicochemical state of the medium. For water calorimeters, this assumption is potentially violated by the energy balance of the chain of chemical reactions that follows the radiolysis of water. Both experimental work [15] and simulations of the chemical energy balance of all reactions involved [16] show that pure water exhibits a particle energy-dependent chemical heat defect with corrections varying by up to 4%. For clinical protons, which have still relatively low LET, the effects are found to be smaller than 1%. For graphite calorimeters, a potential physical heat defect is caused by the creation and annihilation of interstitial lattice defects. This is generally assumed to be small but may be larger in proton beams than in photon beams, due to the higher probability of a sufficient energy transfer to a recoil nucleus. A detailed discussion of the heat defect in water and graphite calorimeters is provided in Chapter 10.

The relative effectiveness of alanine, radiochromic film, Al_2O_3:C based optically stimulated luminescence (OSL) dosimeter and thermoluminescent dosimeter (TLD), the relative chemical yields of Fricke and the relative $W_{air,p}$ values for ionization chambers are illustrated in Figure 9.2 to provide a sense of how much energy dependence these properties induce, which is important to consider for the determination of depth-dose distributions. The alanine data were taken from Herrmann [17], the radiochromic film data from Fiorini et al. [18], the OSL data from Yukihara et al. [19], the Fricke data from LaVerne and Schuler [20], and the TLD data from Geiss et al. [21]. Also, the quenching effect on the depth-dose curve on a low-energy proton beam is illustrated showing that alanine and EBT3 radiochromic film exhibit a similar under response of about 10%, OSL and Fricke exhibit about a 20% under response, and TLD more than 30%. This quenching effect can also have a substantial impact on the range determination when using those detectors, as is obvious in the figure.

9.1.1.4 Fluence Perturbation Factors

For small detectors, it is generally assumed that the proton fluence is not perturbed by the detector medium so that differences in scattering properties between the detector medium and water have little impact. The secondary particle production after target fragmentation can also be different and thus affect the fluence of those fragments within the detector. In distributed detectors, e.g., gel dosimeters, liquid or solid scintillators, an additional issue is that the phantom medium itself is usually not water equivalent, and fluence correction factors are required to account for the different secondary particle fluence in the phantom as in water due to differences in nuclear attenuation and production cross sections. The following paragraphs first discuss the fluence perturbations with ionization chamber and solid detectors and then a discussion of the second type of fluence corrections for phantom materials.

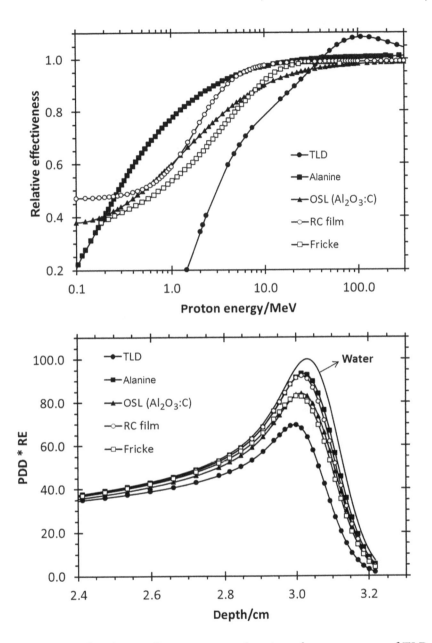

FIGURE 9.2 Upper graph: relative effectiveness as a function of proton energy of TLD, alanine, Al_2O_3:C-based OSL, radiochromic film and Fricke. Lower graph: the percentage depth dose (PDD) in water for a 60 MeV nonmodulated proton beam (indicated with an arrow) multiplied what the relative effectiveness for each detector, as calculated from the local proton spectra using Equation 9.6.

9.1.1.4.1 Ionization Chambers The wall perturbation factor for ionization chambers has been demonstrated to be slightly energy dependent due to the perturbation of the secondary electron fluence [22,23]. Given the low density of air, the range of the highest energy secondary electrons can be substantially larger than the cavity dimensions.

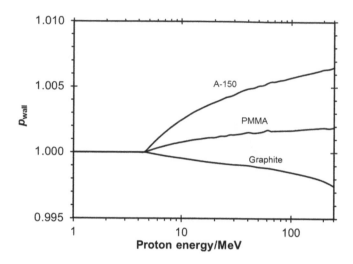

FIGURE 9.3 p_{wall} for three ionization chamber wall materials calculated using Equation 9.8.

This contribution to the wall perturbation correction factor can be approximated based on cavity theory as [24]

$$p_{\text{wall}} = \left(s_{w,\text{wall}}^{\text{BG}} s_{\text{wall,air}}^{\text{SA}} \right) / s_{w,\text{air}}^{\text{SA}}, \tag{9.8}$$

where the superindices BG and SA have been introduced to indicate Bragg–Gray stopping power ratios (using unrestricted stopping powers) and Spencer–Attix stopping power ratios, respectively. Values of the wall perturbation correction factor as a function of energy for monoenergetic protons calculated using Equation 9.8 are shown in Figure 9.3 for graphite, A-150, and polymethyl methacrylate as wall materials. This shows that the correction factor can differ by up to 1% for two chamber types with a different wall material. While these factors are generally ignored, they could be significant when carefully comparing reference dose measurements and depth-dose curves measured with ionization chambers having these two wall materials.

9.1.1.4.2 Solid-State Detectors   There is little research done on fluence perturbations for solid-state detectors. The range of the majority of secondary electrons is assumed to be sufficiently small compared to the detector dimensions so that Bragg–Gray proton stopping power ratios can be used. For solid-state detectors with very small dimensions, such as a microdiamond, there is nevertheless the potential that, similarly as for ionization chambers, small secondary electron fluence perturbation effects are present.

For the measurement of depth-dose curves using a stack of solid detectors in a plastic phantom of lower density than the detector material, one must consider that dose tails due to primary ions (which are normally stopped around the distal edge of the Bragg peak) can occur beyond the Bragg peak due to in-scatter from the surrounding phantom material. For protons, this tail signal would be obvious but for heavier light ions could complicate the interpretation and quantification of the fragmentation tail. Tunneling of ions within air gaps between detectors and phantom could also result in a tail signal that can be avoided by orienting the stack under a small angle with respect to the beam axis. A similar effect takes place when film is used for depth-dose or range measurements orienting the film between two phantom plates parallel with the beam axis. It has been demonstrated that an angle of as little as 1° can avoid this tunneling effect [25,26], but in general, larger angles are recommended (the limiting angle usually determined by the field size and the range that needs to be covered).

9.1.1.4.3 Fluence Correction Factors for Phantom Materials   Something which is often over-looked for proton beam dosimetry is that for measurements in phantom materials that are substitutes of water or tissues, fluence correction factors are required. These are the result of a difference in charged particle fluence at equivalent depths in two phantom materials due to differences in nuclear interaction cross sections between both materials. Two types of related correction factors can be distinguished:

- those to account for difference between the ratio of doses to medium at equivalent depths for both materials and the mass stopping power ratio for both materials, generally called fluence correction factors;

- those to account for the different fluence present in the sensitive volume of detector to determine dose to water when positioned at equivalent depths in both materials, generally called fluence scaling factor [27].

Both are not necessarily the same since short-ranged secondary particles such as recoiling nuclei and the heaviest target fragments may contribute to the dose in a medium but do not reach the detector sensitive volume through encapsulation materials [28].

Equivalence depths in this context are defined as depths in both phantom materials where the energy degradation of the primary protons is equal. For water, tissues and low-Z phantom materials for which stopping power ratios are approximately constant in the clinical energy range, this can be to a good approximation represented as depths that scale with the continuous slowing down (CSDA) ranges.

As an example, we consider here the difference between a plastic phantom material and water. The water-equivalent depth, z_{w-eq}, of a depth of interest in the phantom material, z_{pl}, is given by

$$z_{w-eq} = z_{pl} \frac{r_{0,w}}{r_{0,pl}}, \tag{9.9}$$

where $r_{0,w}$ and $r_{0,pl}$ are the CSDA ranges in water and the plastic material, respectively. The ratio of dose to water and dose to plastic at equivalent depths can be derived from the charged particle spectra differential in energy, $\Phi_{E,w,i}$ and $\Phi_{E,pl,i}$ at those equivalent depths in both phantoms as

$$\frac{D_w\left(z_{w-eq}\right)}{D_{pl}\left(z_{pl}\right)} = \frac{\sum_i\left[\int_0^{E_{max,i}} \Phi_{E,w,i}\left(S/\rho\right)_{w,i} dE\right]}{\sum_i\left[\int_0^{E_{max,i}} \Phi_{E,pl,i}\left(S/\rho\right)_{pl,i} dE\right]},$$ (9.10)

where i refers to the charged particle type, S/ρ is the mass stopping power, and $E_{max,i}$ is the maximum energy of particle type i in the fluence distribution.

If the charged particle spectra are identical at equivalent depths in both phantoms, i.e., $\Phi_{E,w,i} = \Phi_{E,pl,i}$, then the dose conversion is adequately described by the mass stopping power ratio of water to plastic, $s_{w,pl}$, for the charged particle spectrum in either phantom, e.g.,

$$\frac{D_w\left(z_{w-eq}\right)}{D_{pl}\left(z_{pl}\right)} = \frac{\sum_i\left[\int_0^{E_{max,i}} \Phi_{E,w,i}\left(S/\rho\right)_{w,i} dE\right]}{\sum_i\left[\int_0^{E_{max,i}} \Phi_{E,w,i}\left(S/\rho\right)_{pl,i} dE\right]} = s_{w,pl}.$$ (9.11)

If the charged particle spectra at the measurement depth in graphite and at the water-equivalent depth in water are not the same, then a fluence correction factor, k_{fl}, has to be introduced:

$$\frac{D_w\left(z_{w-eq}\right)}{D_{pl}\left(z_{pl}\right)} = s_{w,pl}k_{fl},$$ (9.12)

which can be derived from comparing Equations (9.10) and (9.11):

$$k_{fl} = \frac{\sum_i\left[\int_0^{E_{max,i}} \Phi_{E,w,i}\left(S/\rho\right)_{w,i} dE\right]}{\sum_i\left[\int_0^{E_{max,i}} \Phi_{E,pl,i}\left(S/\rho\right)_{w,i} dE\right]},$$ (9.13)

demonstrating that this correction factor accounts purely for the difference in charged particle fluence since the stopping power is the same in numerator and denominator. The differences in fluence are due to the differences in the absorption of primary protons in nonelastic nuclear interactions and differences in the production of secondary charged particles in water and plastic material.

A similar factor, somehow related but with a different function, is needed when a detector is used in a plastic phantom with the aim of determining absorbed dose to water in a water phantom. This factor, called the fluence scaling factor, is determined as the ratio of detector readings in water and in the plastic phantom at the same equivalent depth (as defined by Equation 9.9) and can be expressed as

$$h_{pl} = \frac{M_{det,w}}{M_{det,pl}} = \frac{\sum_i \left[\int_0^{E_{max,i}} \Phi_{E,w,i}(S/\rho)_{w,i}\, dE \right]}{\sum_i \left[\int_0^{E_{max,i}} \Phi_{E,pl,i}(S/\rho)_{pl,i}\, dE \right]} \cdot \frac{s_{pl,det}\, p_{pl}}{s_{w,det}\, p_w}, \qquad (9.14)$$

where $s_{pl,det}$ and $s_{w,det}$ are the mass stopping power ratios of plastic to detector medium and water to detector medium, respectively, and p_{pl} and p_w are the detector perturbation correction factors in plastic and water to account for the perturbation of the charged particle fluence by the presence of the detector, i.e., the difference in fluence within the detector volume as compared with the fluence present at the measurement point in homogeneous phantom material.

If we can assume that the stopping power ratios are Bragg–Gray stopping power ratios,

e.g., $s_{pl,det} = \dfrac{\sum_i \left[\int_0^{E_{max,i}} \Phi_{E,pl,i}(S/\rho)_{pl,i}\, dE \right]}{\sum_i \left[\int_0^{E_{max,i}} \Phi_{E,pl,i}(S/\rho)_{det,i}\, dE \right]}$, then we can write the fluence scaling factor as

$$h_{pl} = \frac{\sum_i \left[\int_0^{E_{max,i}} \Phi_{E,w,i}(S/\rho)_{det,i}\, dE \right]}{\sum_i \left[\int_0^{E_{max,i}} \Phi_{E,pl,i}(S/\rho)_{det,i}\, dE \right]} \cdot \frac{p_{pl}}{p_w} \approx k_{fl} \cdot \frac{p_{pl}}{p_w}, \qquad (9.15)$$

because the fluence correction factor depends very weakly on which low-Z stopping power is used to calculate it [29], showing a simple relation between the fluence scaling factor and the fluence correction factor via the ratio of detector perturbation factors in both materials. Figure 9.4 shows fluence correction factors and fluence scaling factors for a number of common plastic materials used as water substitutes calculated by FLUKA [28]. It is striking that the fluence scaling factors vary almost linear with depth while this is not the case for the fluence correction factors. The h_{pl} values are also unity at the phantom surface, which is consistent with the idea that in the absence of significant backscatter effects ionization chamber measurements at the phantom surface should be independent of the phantom material. This is not the case for the k_{fl} values that can be explained by the dose contribution from alpha particles generated in nonelastic nuclear interactions and that do not scale with the water-to-medium mass stopping power ratio. The more complex behavior of the fluence correction factors as a function of depth is related to the complexity of the secondary particle production cross sections (mainly secondary protons) and the finite ranges of those particles.

9.1.2 Different Requirements for Scanned and Scattered Beams

A suitable detector choice has to consider the delivery method and time structure of the beam. For passively scattered proton beams, shaped by static devices only, or in some cases, a dynamic modulator wheel with a high spinning frequency, the dose distribution is static

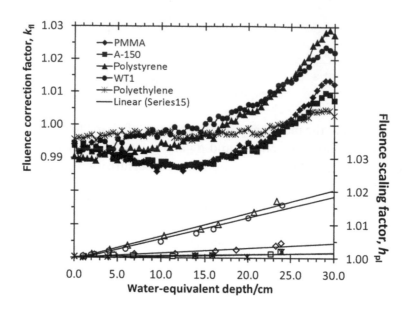

FIGURE 9.4 Monte Carlo calculated fluence correction factors (full symbols; represented on the left-hand side vertical axis) and fluence scaling factors (hollow symbols; represented on the right-hand side vertical axis) for a number of plastic materials using FLUKA. The straight lines are linear fits to the h_{pl} data to show that they vary almost linear as a function of depth. The data are extracted from Lourenço et al. [28].

or quasi-static (on the timescale of the acquisition of the dose distribution; see Chapter 5). For such systems, a plotting tank with a point detector can be used to measure the relative distribution of interest, although multidimensional detector systems will obviously be beneficial by reducing the measurement time. For scanned proton beams, on the other hand, it is essential to use multidimensional detector systems since the dose at a single point in the radiation field is composed of contributions from a large number of spots (see Chapter 6) [3,5]. Using a point detector, the measurement of each dose point would require the entire beam delivery sequence to be completed, resulting in an impractically large measurement time. This can be compared with complex intensity-modulated radiotherapy fields, where for the same reason, multidimensional detectors are a necessity.

Dose calculation algorithms for scanned beams are based on a full characterization of the individual pencil beams (see Chapter 14). This means that lateral profiles, depth-dose profiles, and spot positions and alignment need to be quantified. Another important aspect is the nuclear halo (see Chapter 2), the dose deposition by secondary particles (mainly protons) emitted after nuclear interactions, which has a lateral distribution that is distinct from the primary proton contribution and for dose calculations should be quantified separately. For lateral profiles, the considerations mentioned in Section 9.1.1.1 should be taken into account. For depth-dose profiles, it must be realized that individual beam-lets can be very small, and their lateral profile exhibits a sharp maximum, making alignment of a scanning detector with the central axis very challenging. The general approach adopted is to measure the depth dependence of the laterally integrated dose for which large-area plane-parallel ionization chambers can be used [30] or arrays of small volume ionization chambers [31]. The accurate determination of spot positions may also require different criteria on the lateral positioning accuracy as needed for profile measurements [32]. For the nuclear halo measurements, the most adequate method currently employed is the use of small-volume ionization chambers [31], although measurements using concentric ionization chambers [33] or large-area plane-parallel ionization chambers with varying diameters are a potential route as well as indirect methods such as the frame experiment [34].

9.1.3 Requirements for Small-Field Dosimetry

In photon beams, small fields are mainly associated with a loss of lateral charged particle equilibrium. A similar criterion can be applied to protons, but it is not so much associated with secondary electrons. The loss of lateral charged particle equilibrium is rather due to multiple Coulomb scattering (MCS) in the treatment field when the field size becomes smaller than the scattering distance of the protons [35]. Similarly as in photon beams, the loss of lateral charged particle equilibrium can also be associated with the overlap of penumbrae, which provides a practical way to assess if small field conditions exist. Given the scattering properties of protons, it is clear that it will depend on the incident proton beam energy, the depth of measurement as well as on the presence of air gaps and collimators between scatterers or range shifters and the phantom below what field size small field conditions exist. Single spots used in scanned beams are in principle small fields, and their individual characterization involves all problems of small-field dosimetry. The issue of small fields is, however, rather associated with stereotactic radiosurgery, since they require a full characterization of the beam profiles, which can only be achieved with detectors that have a high spatial resolution. For central axis profiles, in addition, very accurate alignment of the detector with the beam axis needs to be realized. Since in scanned beams many spots are used to create single-energy layers, the detailed dosimetric characteristics are often less critical, and it is sufficient to know the laterally integrated profile of the spot that can be measured with a large-area plane-parallel ionization chamber. Ocular beams, which are small, have usually low energy and very sharp penumbrae and will mostly not exhibit small field conditions but will nevertheless require detectors with high spatial resolution to enable a full characterization of the beam profiles.

9.1.4 Requirements for Microdosimetry

While radiotherapy prescription is based on the macroscopic quantity absorbed dose to water, multiplied with a quality factor, the effect of radiation on individual cells is governed by the local energy deposition within the cell or parts of the cell (see Chapter 22). Microdosimetry concerns the study of the spatial and temporal distribution of energy deposition within volumes having typical dimensions of a few micrometers to tens of micrometers corresponding with the sizes of cell nuclei or entire cells. In spatial microdosimetry, the clustering of interactions is quantified, while in regional microdosimetry, the fluctuations of energy deposition in the target is determined without considering the spatial distribution of energy deposition within this target. Microdosimetric quantities account for the stochastic nature of energy losses and track structure and therefore provide the physical information that is more directly linked with the biological effects of radiation. The microdosimetric distribution is often used to represent the radiation quality of the investigated beam that affects the radiation response of a biological system [36].

For spatial microdosimetry, it is of course important that the spatial resolution of the detector system is sufficient to resolve individual ionizations or ionization clusters. This is usually achieved in low-density gas environments, such as in a cloud chamber, but can also be achieved in solids in combination with confocal or super resolution microscopy to visualize the reaction products from ionization clusters.

For regional microdosimetry, the density and dimensions of the detector need to be adapted such that the mean energy imparted is the same as in the microscopic tissue site of interest (TSOI), for example, the cell nucleus, i.e., $\bar{\varepsilon}_{det} = \bar{\varepsilon}_{TSOI}$. The same equation can be written as $(S/\rho)_{det}\, \rho_{det}\, \bar{l}_{det} = (S/\rho)_{TSOI}\, \rho_{TSOI}\, \bar{l}_{TSOI}$, where S/ρ are the mass stopping powers, ρ the mass densities, and $\bar{l}$ the mean chord lengths of the sites. If the detector is tissue equivalent, i.e., $(S/\rho)_{det} = (S/\rho)_{TSOI}$ (in other words, it has the same interaction properties as the TSOI), then the scaling can be expressed by mass densities only: $\rho_{det}\, \bar{l}_{det} = \rho_{TSOI}\, \bar{l}_{TSOI}$. This is, for example, achieved in tissue-equivalent proportional counters (TEPCs) by adjusting the gas pressure of the tissue-equivalent gas mixture.

9.2 DETECTORS FOR RELATIVE DOSIMETRY

A range of detectors can be used for lateral or depth-dose profile measurements in proton beams, and as for photon beams, one can distinguish small, point-like, detectors that are used for scanning one-dimensional profiles and continuous or matrix detectors that form rigid structures and which can measure two- or three-dimensional profiles in one exposure with a high or relevant resolution.

As already mentioned, even water calorimeters exhibit an LET-dependent dose response due to the chemical heat defect, apart from the fact that they are impractical instruments for measuring depth-dose data. Ionization chambers remain the instrument of choice for depth-dose measurements, since they exhibit a rather modest LET-dependent dose response which is usually ignored. Other detectors may be preferred for a variety of reasons such as higher spatial resolution, higher time resolution, higher signal, or practicality of handling and their use in solid phantoms.

9.2.1 Ionization Chambers

The well-known ionization chamber is a gas-filled detector in which all the charges created by ionization are collected by the application of an electric field. The most commonly used type in proton dosimetry is the air-filled ionization chamber, with the air in the cavity in direct contact with the environment via a small vent tube or hole. They are usually thimble-shaped or plane-parallel individual detectors used as scanning probes or arranged in arrays for simultaneous measurement of entire profiles. For lateral measurements, 2D arrays exist. An interesting array configuration for fast checking of depth-dose profiles and ranges is a multilayer ionization chamber consisting of a stack of plane-parallel ionization chambers interspaced by slabs of a water-equivalent material. Plane-parallel ionization chambers are the instrument of choice for depth-dose measurements, although small

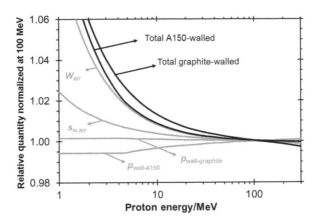

FIGURE 9.5 Relative variation as a function of proton energy of the mean energy required to produce and ion pair in dry air, W_{air}, the mass collision stopping power ratio water-to-air, $s_{w,air}$, and the ionization chamber wall perturbation correction factor due to secondary electron effects, p_{wall}, for two wall materials according to the data and expressions discussed in the text as well as the product of the three. The figure is updated from Palmans [4] with new data from ICRU Report 90.

thimble ionization chambers can be used within limits as well, since they exhibit a rather modest energy-dependent dose response that is usually ignored but nevertheless worth to be considered.

The energy dependences of the water-to-air stopping power ratio, the mean energy required to produce an ion pair, and the perturbation correction factor have been discussed in Section 9.1.1. The combined effect of those three contributions is illustrated in Figure 9.5 for graphite- and A-150-walled ionization chambers using the $W_{air,p}$ variation from the Dennis [13] data. All contributions were normalized to unity at 100 MeV.

Convolving this with proton spectra and dose contributions as a function of depth leads to a reduction of peak-to-entrance ratio of about 1.5%–2% for air-filled ionization chambers in high-energy proton beams and 3%–4% for low-energy beams typically used for the treatment of eye melanoma, where the lower numbers are obtained using the Grosswendt and Baek [14] data and the higher numbers using the Dennis [13] data. If no variation of the $W_{air,p}$ value for protons with energy is assumed, the reduction of the peak to entrance ratio is about 0.7% for high-energy beams and 1.2% for low-energy beams.

9.2.2 Alanine

The L-isomer of alpha-alanine, a natural amino acid, is used in crystalline powder form as a dosimeter. Free radicals formed by ionization remain in a stable position in the crystal, and the number of radicals present in a sample of alanine can be quantified by electron spin resonance. Alanine is an attractive detector for reference or relative dosimetry, because the signal that is produced by ionizing radiation is stable over a long period of time (the radicals are locked in the crystal), the readout of the signal is nondestructive, the response is linear over a large range of dose levels and dose rates, the signal exhibits low temperature dependence, and the composition and radiological properties of alanine are close to those of tissue. The crystalline powder can be used in a small container, embedded in a thin film layer or compressed in pellets using a binding agent, such as paraffin wax. The relatively low sensitivity of the technique puts a lower limit on the size of the sample for a given dose level; for radiotherapeutic dose levels, pellets containing about 90% alanine of 5 mm diameter and 2.5 mm thickness can be used, which is perfectly suitable for dosimetric audit or end-to-end tests. At low dose levels, the assumption can be made that alanine behaves as a one-hit detector (each molecule can only form a single free radical) and extensive experimental investigation [17] has shown that the energy dependence of alanine for protons can be well described by amorphous track structure models [37].

9.2.3 Semiconductor Devices: Diodes and Diamonds

In diodes, the energy levels defining the band gap vary between two regions in the device, either by a difference in doping (p–n junction) or by the presence of a metallic contact (Schottky diode). As a result, the negative and positive charge carriers will drift in opposite directions within an intermediate region, creating a depletion zone without charge carriers over which an internal electric field exists. Radiation-induced charges (electrons and holes) will diffuse in one direction by this internal electric field, creating a measureable current. Silicon diode detectors are most commonly realized by implanting electron donor and acceptor atoms in different regions of the silicon chip, thus creating a p–n junction. In natural diamond detectors, the depletion layer is created by applying an external bias voltage, while for chemical vapor deposition (CVD) diamond detectors, a Schottky diode is usually created. The energy dependence of the mean energy required to produce an ion pair in silicon diode detectors has been found to be extremely small, and the same can be assumed for diamond detectors [38]. Apart from the water-to-silicon or water-to-diamond mass stopping power ratio, the main mechanism resulting in energy dependence is initial recombination. This explains why the energy dependence is very small in very thin CVD diamonds, as observed by Marinelli et al. [39] for the 1 μm thick PTW-30019 microdiamond, since a very high electric field exists across the thin depletion layer. For such detectors, only in very high-LET beams, a small under response is observed [40,41].

9.2.4 Luminescence Devices: Scintillators, TLD, OSL
9.2.4.1 Scintillating Detectors
Scintillating detectors are based on the detection of prompt light emission due to molecular or crystalline transitions after ionization. Organic scintillators can be embedded in a polymer matrix but can also be available in liquid form or gaseous phase, making the application of these detectors as online dosimeters very versatile. A small sample of a polyvinyl toluene containing organic scintillators, mounted on an optical fiber has been used as a point detector for protons [42]. Gaseous scintillators in a gas electron multiplier system [43], and scintillating screens [44] have been developed as a 2D detector for beam profile monitoring. In both cases, CCD cameras are used for 2D mapping of the scintillating signal. Liquid scintillators have been used for 4D dosimetry of

scanned proton beam IMPT [45]. Inorganic scintillators, which are usually not water or tissue equivalent, can also be embedded in a polymer to improve the equivalence. Both organic and inorganic scintillators exhibit, in general, a substantial energy-dependent response, but Safai et al. [46] showed that this differs from material to material, and some materials even overrespond in low-energy protons. This prompted the design of a mixture that enabled the measurement of a Bragg peak of a particular energy without response quenching.

9.2.4.2 Thermoluminescent Dosimeter

TLDs consist of a crystalline solid-state material containing impurities or interstitial atoms exhibiting energy levels in the bandgap. These energy levels form trapping centers for electrons or holes, which can be populated by radiation-induced charges and subsequently emptied by heating. The emitted luminescence signal can be measured by a photomultiplier tube. Their behavior is complicated by energy levels of shallow electron and hole traps in thermal equilibrium with the conduction and valence band, the distribution of energy levels of luminescence centers and the different mobilities of electrons and holes. This makes the modeling of their response as a function of ionization cluster distributions or LET difficult, and the complexity can depend substantially on the type of crystal used, the distribution of luminescence centers, and the level of control on the purity and energy levels of trapping centers [21]. The most common TLD materials are based on LiF crystals doped with various high-Z elements.

9.2.4.3 OSL Dosimeter

The physical principles of OSL dosimeter (OSLD) are similar as for TLD. The difference is that readout of the trapped charges is performed by laser stimulation. Al_2O_3:C based OSLDs have been shown to exhibit a more unique dependence on LET, regardless of the ion type, than TLDs [19].

9.2.5 Chemical Dosimeters: Fricke, Radiochromic Film, Polymer Gels

9.2.5.1 Radiochromic Dosimeters

Radiochromic dosimeters are based on a change of color as a result of radiation-induced chemical reactions. Various chemicals have been used as radiochromic dosimeters, such as aminotriphenyl-methane dyes and leuco dyes. Commercial radiochromic films, such as MD-55, EBT, EBT2, EBT3, and EBT-XD (Ashland, NJ), are 2D detectors that are widely used in radiotherapy in recent years. These all contain pentacosa-10,12-diyonic acid (PCDA) or its lithium salt LiPCDA in the sensitive layer. The crystalline PCDA or LiPCDA polymerize after irradiation, forming an intense blue darkening [47] without a requirement for any postirradiation development process. One or more thin layers of the radiochromic material are embedded in polyester films, and the overall composition of the films is nearly tissue equivalent. This dosimeter exhibits energy dependence due to the LET dependence of the chemical yields (number of molecules formed per 100 eV of energy imparted) of initial reactive species formed and, in addition, due to recombination or termination reactions that depend on the ionization density. For very low-energy protons, response quenching can be up to 50% [18], while in the Bragg peak of clinical low-energy proton beams, the under response is typically between 10% and 20%.

Another radiochromic dosimeter that has received considerable attention recently is the 3D PRESAGE™ (Heuris Inc., Skillman, NJ) dosimeter, consisting of a free radical initiator and leuco dye locked in a polyurethane matrix. The response quenching in proton beams has not been as thoroughly quantified as for radiochromic film but is considerably larger, and the detector may be more suitable to verify range, for example, in anatomic phantoms.

9.2.5.2 Gel Dosimeters

Another class of chemical dosimeters are formed of those of which the radiosensitive agents can be distributed within an aqueous gelatin matrix providing a large flexibility in shaping the dosimeter and resulting in a nearly water- or tissue-equivalent medium. Two main concepts of gel dosimeters can be distinguished: Fricke gels are based on the principles of the Fricke dosimeter, and polymer gel dosimeters are based on the polymerization of organic monomers. Their LET dependence is determined by the energy-dependent production yields of primary species (see Chapter 10).

The well-known ferrous sulphate or Fricke dosimeter consists of an aerated acidic ferrous sulfate solution in which radiation-induced reactive species induce chemical reactions that result in ferrous ions to be converted to ferric ions. The main reactive species involved in this conversion are hydroxyl, hydroperoxyl, and hydrogen peroxide. The light absorption peak at wavelengths of 224 and 303 nm is commonly used for the quantification in 3D of the concentration of ferric ions using optical tomography. Given the magnetic susceptibility change, the distribution of ferric ions can also be quantified by magnetic resonance imaging.

Another example is the use of polymer gels in which radicals, such as hydroxyl, react with monomers that are locked up in the gel matrix in a suitable aqueous environment [48]. The general reaction can be denoted $R\bullet + M \rightarrow R + M\bullet$, where the radical is denoted $R\bullet$ and the monomer M. The radicalized monomers, $M\bullet$, initiate propagation reactions, resulting in a polymerization reaction, $M_m\bullet + M_n \rightarrow M_{m+n}\bullet$, which propagates until a termination reaction of the format $M_m\bullet + M_n\bullet \rightarrow M_{m+n}$ or $R\bullet + M_n\bullet \rightarrow RM_n$ takes place. The polymers resulting from this process form opaque clusters that can be quantified in 3D by optical tomography or by magnetic resonance imaging, given that the polymers have a different magnetic susceptibility than the original solution. The energy dependence can be understood as that of a one-hit, two-component detector [49], i.e., the under response is not only due to the reduced amount of radicals produced per unit of energy imparted at higher ionization density but also by the increased probability of termination reactions. This explains in general a more complex behavior as a function of energy, absorbed dose, and dose rate as for simple one-hit detectors.

9.3 APPLICATIONS

In this section, a number of example applications are taken from the literature which is not meant to be an exhaustive review; the examples are chosen to illustrate certain peculiarities one can encounter in relative dosimetry of proton beams.

9.3.1 Lateral Profile Measurements

For lateral profile measurements and verification of field uniformity and penumbral widths, radiochromic film and scintillating screens are adequate as they exhibit a very high resolution. Any nonlinearity of their response should of course be accounted for. However, for lateral profiles of small fields or single spots in scanning systems, these detector types do not necessarily exhibit the large dynamic or sensitivity range to distinguish all the important features in the profile. Indeed, Gottschalk et al. [31] pointed out four distinct dose contributions to the lateral profile: a *core* deposited by primary protons that has been widened due to MCS in the phantom, a *halo* of dose contributions from secondary particles generated in nonelastic nuclear interactions, an *aura* due to indirectly ionizing particles (neutrons and gammas generated in nonelastic nuclear interactions) and a *spray* of particles originating from the beam-shaping components. For in-beam dosimetry of single spots in scanning systems, only the core and the halo are important contributions, with the core contribution named the high-dose region and the halo contribution the low-dose region. For collimated small beams, the spray may contribute substantially to the low-dose region. The high-dose and low-dose regions are in many planning systems modeled as two separate contributions, both as Gaussian lateral distributions with different amplitudes and variances. It is thus important to quantify both contributions accurately, but the range of doses covered by the core and the halo spans four orders of magnitude. While ionization chambers and diodes have a sufficient range of sensitivity, in array configurations, they usually lack the lateral resolution to model the narrow profiles. Gottschalk et al. [31] proposed the use of microionization chambers scanned parallel to the beam axis at different off-axis positions as an efficient way of mapping the entire 3D dose profile of a single spot. Grevillot et al. [32] not only used a linear configuration of 24 individual microionization chambers interspaced by 1 cm (by repeatedly shifting this array, a higher resolution of measurement points could be obtained) but also showed that a scintillating screen can be used if the dose distribution is delivered in three different dose levels, each time evaluating a different section of the profile. In this work, measurements with a very thin microdiamond used as a scanning probe showed very good agreement, yielding slightly lower penumbral widths as microionization chambers in agreement with the results from a scintillating screen. A comparison of a measurement with a linear array of microionization chamber and a scintillating screen is shown in Figure 9.6.

9.3.2 Depth-Dose Profiles in Small Fields

For depth-dose profiles, plane-parallel ionization chambers are usually preferred while multilayer ionization chambers can be used for regular verification [50]. Small thimble ionization chambers can be used if the Bragg peak is not too sharp, i.e., for high-energy beams [9]. While for scattered beams, small-area ionization chambers would be used, for single-spot characterization (and in principle for any small field), a depth distribution of laterally integrated profiles is often preferred, especially for single-spot characterization in scanned beams. In principle, the information in a central axis depth-dose curve plus lateral profiles contains the same information as the depth distribution of laterally integrated profiles plus the lateral profiles.

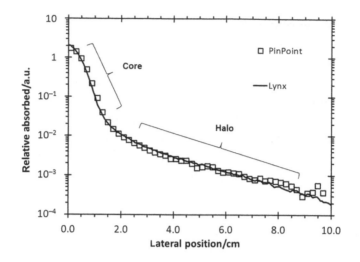

FIGURE 9.6 Lateral profile with an array of thimble microionization chambers (PTW-31015 PinPoint ionization chambers, PTW Freiburg, Germany) and a scintillating screen (Lynx, IBA dosimetry, Schwarzenbruck, Germany) with the indication of the core and the halo region. The data are extracted from Grevillot et al. [32] and renormalized to the laterally integrated profile.

For stereotactic radiosurgery applications, central axis depth-dose curves are often relevant to demonstrate the reduced dose in the distal region beyond the target that is beneficial for normal tissue sparing. Vatnitsky et al. [51] showed that for the smallest fields (below 5 mm diameter) even the smallest point detectors, such as diodes, are too large and a continuous dosimeter, such as radiochromic film, from which the average readout over a very small area can be derived could be preferred. Figure 9.7 illustrates the large reduction in the Bragg peak that results from lateral scatter deficiency in small radiosurgery beams; for the smallest fields, the Bragg peak even completely disappears.

For single-spot profiles depth distributions of laterally integrated profiles are commonly measured using large-area plane-parallel ionization chambers mounted in a conventional scanning water tank (dedicated solutions also exist). One problem is that treatment planning systems (TPSs) usually rely on a value of the integrated profile over the entire lateral space while typical large-area ionization chambers have limited diameters because of the requirement to sustain a constant spacing between the electrodes over the entire area. Given that the halo of a single spot has a maximum extension around the middle of the range of monoenergetic protons that amounts to about one-third of the range of incident protons [31], a significant fraction of the dose is deposited outside a typical large-area chamber diameter of 8 cm [30]. Figure 9.8 illustrates for two single-energy proton beams the fraction of dose that is deposited outside a diameter of 8 cm and which thus corresponds with the percentage correction that needs to be applied to the measurement to obtain the full integral of the lateral profile using a large-area plane-parallel ionization chamber with a collecting electrode diameter of 8 cm. An additional consideration is that the nonuniformity of the response over the area of such chambers may amount to 10% or more, which will affect the depth distribution of the laterally integrated profile because of

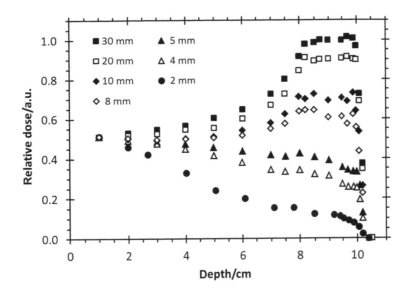

FIGURE 9.7 Central axis depth-dose profiles as a function of field diameter in a stereotactic radio-surgery system, measured using a diode (except for the two smallest fields for which radiochromic film was used) in a 126 MeV modulated proton beam with a range of 10 cm in water and a spread-out Bragg peak (SOBP) of 2 cm. The data are extracted from Vatnitsky et al. [51] and renormalized to the shallowest data point to optimally visualize the reduction of the Bragg peak with decreasing field diameter (the output factors in the SOBP do, consequently, not coincide with those reported in the paper).

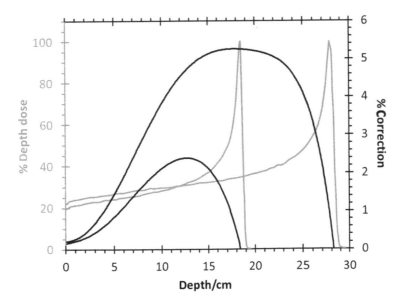

FIGURE 9.8 The fraction of dose that is deposited outside a diameter of 8 cm around the central axis of the single spot for two single-energy proton beams (represented on the right-hand-side vertical axis. For reference, the depth distributions of the laterally integrated profiles are shown as well (represented on the left-hand-side vertical axis). The data are extracted from Clasie et al. [30].

the increasing area of the chamber that receives a significant dose with increasing depth due to beam widening [52].

9.3.3 Field Output Factors

Field output factors (see Chapter 13) are commonly employed in the modeling of scattered beams and stereotactic radiosurgery beams but less commonly in scanned systems. A similar rule as for central axis depth-dose curves is that the detector always has to be small enough in comparison to the field size, but since the measurement always concerns a ratio between two doses at the same depth (the absorbed dose to water determined for a clinical field divided by the reference absorbed dose to water determined for a reference field as discussed in Chapter 10) with the same modulation, the concerns about the energy dependence of the detector are not important, making that a larger range of detectors can be employed. Small plane-parallel ionization chambers, small thimble ionization chambers, silicon diodes, diamonds, TLDs, and radiochromic films have, for example, been shown to be all suitable and consistent for field output factor measurements in stereotactic radiosurgery fields, each with their own lower limit of field size [51]. One must realize that output factors are not just a function of field size but, for a given residual range, also a function of the incident energy [53], as illustrated in Figure 9.9.

9.3.4 Dosimetric End-to-End Test

Given the complexity of scanned beam deliveries, a dosimetric end-to-end test is an efficient way of verifying the integrity of the entire dose delivery chain. During an end-to-end

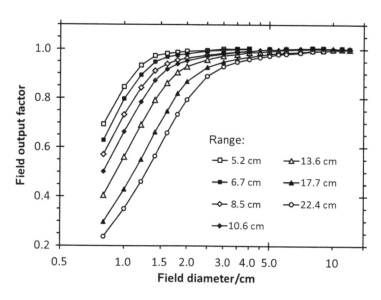

FIGURE 9.9 Field output factors as a function of field diameter obtained with a microionization chamber (PTW-31006 PinPoint ionization chamber, PTW Freiburg, Germany) for a gantry proton beamline in the center of a 2 cm SOBP for different incident energies (characterized by the different ranges in the legend). The data are extracted from Daartz et al. [53] and combine data obtained for two primary collimation systems.

test, a sufficiently representative phantom (e.g., an anthropomorphic phantom) is moved through the entire clinical workflow just as a real patient, and the dose distribution is verified directly within this phantom. One of the complications is that corrections for the energy dependence of many suitable detectors, as discussed in Section 9.1.1, are needed, which can be challenging given that for the application of Equation 9.6 the full charged particle fluence spectrum differential in energy needs to be known. This information cannot be extracted from pencil beam algorithms, for example, and rather requires the availability of a Monte Carlo tool for dose calculations.

Carlino et al. [54] developed a dosimetric end-to-end test using alanine, ionization chambers, and radiochromic film as dosimeters, of which the former two served to determine absorbed dose to water in absolute terms. The relative effectiveness of alanine, $\bar{\eta}_{Aln}$, was calculated inline within the Monte Carlo simulation according to Equation 9.6. Given that the alanine was calibrated in terms of absorbed dose to water in a ^{60}Co calibration beam and the results are compared in terms of absorbed dose to water in the proton beam calculated with the TPS, an additional correction is required since $\bar{\eta}_{Aln}$ is specified in terms of absorbed dose to alanine in both beam qualities. The corrected absorbed dose to water determined by the alanine pellet in the proton beam, D_w^p, is obtained from the ^{60}Co-equivalent absorbed dose to water, $D_w^{^{60}Co}$, readout by the alanine calibration service, as

$$D_w^p = \frac{D_w^{^{60}Co}}{\bar{\eta}_{Aln}} \left[\left(\frac{S}{\rho} \right)_{Aln}^w \right]^p \left[\left(\frac{\mu_{en}}{\rho} \right)_w^{Aln} \right]^{^{60}Co}, \tag{9.16}$$

where $\left[\left(\frac{\mu_{en}}{\rho} \right)_w^{Aln} \right]^{^{60}Co}$ is the alanine-to-water ratio of mass–energy absorption coefficients in the ^{60}Co calibration beam. Tests were performed in a homogeneous polystyrene phantom, a customized head and neck phantom and a pelvis phantom. Figure 9.10 illustrates the result for one of the test cases. The ^{60}Co-equivalent absorbed dose to water, $D_w^{^{60}Co}$, was systematically lower than the absorbed dose to water calculated by the TPS (which was in agreement with ionization measurements) and the difference was larger for pellets close to the distal edge of the target volume (lower pellet numbers). The values of D_w^p were closer to the TPS prediction and increased the consistency of the result substantially showing that the correction of Equation 9.16 is necessary. The remaining difference of about 2%, which is within the uncertainties of the comparison, was nevertheless systematically observed in all the performed tests and must be associated with the interaction data in Equation 9.16 and/or the beam quality correction factors for the ionization chamber used to calibrate the beam monitor.

Dosimetric end-to-end tests can also be developed as auditing capabilities that could contribute to dosimetric harmonization among particle therapy centers. A pilot study, preceding the study performed by Carlino et al. [54], already investigated the feasibility of a dosimetric end-to-end audit for scanned proton beams using the same methodology [55].

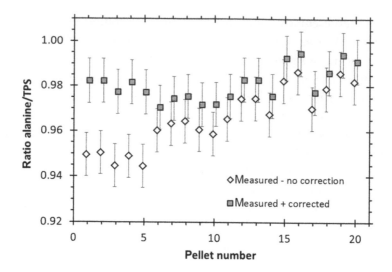

FIGURE 9.10   Ratios of dose values obtained by Carlino et al. [54] from alanine pellets in the poly-styrene phantom for the beam delivery sequence described by Ableitinger et al. [55] and those pre-dicted by the TPS. Diamond symbols represent those ratios for the ^{60}Co-equivalent absorbed dose to water, as obtained from the alanine calibration service (indicated as "measured – no correction"), and squares represent those for the corrected dose values obtained with Equation 9.16 (indicated as "measured + correction"). The uncertainty bars represent a 1% relative standard uncertainty of the ^{60}Co-equivalent readout as specified by the alanine calibration service. Data are extracted and reinterpreted from Carlino et al. [54].

TLDs were used for absolute dosimetry in another dosimetric end-to-end test in an anthropomorphic spine phantom by Cho et al. [56]. Due to the more complex and more pronounced energy dependence of TLDs for low-energy protons, however, these detectors can only be used for accurate dosimetry in the entrance region.

9.4 MICRODOSIMETRY

9.4.1 Microdosimetric Quantities

Contrary to the quantity absorbed dose, microdosimetric quantities are stochastic, and therefore, microdosimetric information is normally present as probability density functions of the following microdosimetric quantities:

- The energy imparted in a single event, ε_S, to the matter in the volume, also called the energy deposit, is the difference of the radiative energy entering the volume and the radiative energy leaving the volume plus any decrease of the rest mass within the volume.

- The specific energy, z, is defined as the energy imparted divided by the mass of matter in the volume, i.e., $z = \varepsilon_S/m$. Note that absorbed dose is theoretically defined as the expectation value of the sum of all specific energy contributions in an infinitesimally small domain.

- The lineal energy, y, of a single event is defined as the energy imparted in the microscopic volume divided by the mean effective chord length, $\bar{l}$, of particles crossing the volume, i.e., $y = \varepsilon_S / \bar{l}$. For isotropic radiation, the mean effective chord length can be approximated by the mean chord length of the volume, which, for concave volumes is given by $\bar{l} = 4V/S$ where V is the volume and S is the outer surface of the detecting volume. For spherical detector, this is the case for any incidence of radiation. For plane-parallel detectors oriented perpendicularly to the direction of the incident radiation, the mean effective chord length can be approximated by the detector thickness.

The probability density function of the lineal energy, $f(y)$, is referred to as the microdosimetric spectrum. Its ordinary representation is a semilogarithmic plot with the lineal energy on the horizontal axis and the product $y^2 f(y)$ on the vertical axis. That way, the area segment under the curve defined by a horizontal interval equals the contribution to the absorbed dose from the lineal energy values in that interval. Figure 9.11 gives an example of the latter representation of a microdosimetric spectrum of a proton beam compared to that of a ^{60}Co beam together with a biological weighting function for the acute effects in the intestinal crypt cells of mice as endpoint as a function of lineal energy, indicating why the biological effect of protons is slightly enhanced as compared to ^{60}Co [57].

The first moment of $f(y)$ is called the frequency mean lineal energy $\bar{y}_F = \int\limits_0^\infty y f(y) dy$. The

second moment of $f(y)$ divided by $\bar{y}_F$ is the dose mean lineal energy, $\bar{y}_D = \dfrac{1}{\bar{y}_F} \int\limits_0^\infty y^2 f(y) dy$.

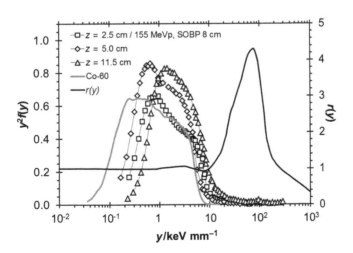

FIGURE 9.11 Microdosimetric spectrum of a 155 MeV modulated proton beam at three different depths (the larger two in the SOBP) hollow symbols compared with the microdosimetric spectrum of a ^{60}Co beam (gray line) and a biological weighting function for intestinal crypt regeneration in mice (black line). Data are extracted from Coutrakon et al. [57].

Similarly, the frequency mean specific energy and the dose mean specific energy are calculated from the probability density function of the specific energy, $f(z)$ as $\bar{z}_F = \int_0^\infty z f(z) dz$ and $\bar{z}_D = \dfrac{1}{\bar{z}_F} \int_0^\infty z^2 f(z) dz$. $\bar{z}_D$ is an important quantity in the relation between microdosimetry and radiobiology, since according the theory of dual radiation action, which assumes that the elementary lesions within the cell nucleus are the result of pairwise interaction of molecular products (called sublesions), the relation between the yield of lesions, $\bar{\epsilon}$, and absorbed dose, D, is $\bar{\epsilon} \sim \bar{z}_D D + D^2$ [58]. The theory of dual radiation action forms the basis of the linear-quadratic model that is the most commonly used radiobiological model in fractionated radiotherapy (see Chapter 22) [59].

9.4.2 Detectors for Microdosimetry

A distinction has to be made between regional microdosimetry, the most common focus of experimental microdosimetry, aiming at measuring $f(y)$ and $f(z)$ for a particular site of interest, and structural microdosimetry, which aims at deriving microdosimetric quantities and actions from detailed 3D distributions of energy transfer points. For regional dosimetry, gas-filled proportional counters, silicon diodes, and diamonds are commonly used, while some new devices having been described that use calorimetric methods.

9.4.2.1 Gaseous Detectors for Regional Microdosimetry: Tissue-Equivalent Proportional Counters

Proportional counters are cylindrical or spherical ionization chambers operated with an electric field between anode and cathode that is chosen such that radiation-induced ions drift for most of the distance due to the electric force and induce an avalanche in the vicinity of the collecting electrode. The charge collected at the electrode is then proportional to the amount of ionization in the gas and large enough to be accurately quantified for each individual event (i.e., for a single ionizing particle crossing the ionization chamber). Under the condition that the mean energy expended per ion pair in the gas, W_{gas}, is a constant, the amount of charge collected is also proportional to the energy deposited within the drift volume. It is obviously difficult to determine what the collecting volume exactly is and the boundary between the drift region and the avalanche region is not well defined, such that the lineal energy scale of microdosimetric spectra obtained from proportional counters needs to be calibrated by the use of characteristic edges of known particles or a known source of radiation that is totally absorbed such as an alpha particle from a radioactive source. The former method is based on the highest lineal energy values occurring near the end of the range, resulting in a reasonably sharp edge in the lineal energy distribution (only blurred by energy and range straggling). The edge lineal energy for protons in a $2\,\mu m$ in diameter water-equivalent sphere was determined to be $136\ keV\ \mu m^{-1}$ [60].

The oldest and most common proportional counters are large volume chambers filled with a tissue-equivalent gas operated at low pressure so as to measure ionization in a macroscopic volume containing a mass of tissue-equivalent material similar as the microscopic site of interest [7]. These are denoted as TEPC. Corrections are required for the delta-ray effect, the V-effect, and the scatter effect, all three of which concern two charged particles from the same event crossing the large cavity, whereas in the real tissue environment, only one of those would enter the microscopic volume of interest [61,62]. Another correction is required for backscatter also called the re-entry effect; this occurs when a particle re-enters the gas volume after having crossed the cavity and entered the wall, while in real tissue, the backscattered particle would pass to the side of the small volume of interest. At high count rates, corrections for pile up also have to be considered. Electromagnetic noise present in the measurement environment usually sets a lower detectable lineal energy limit or threshold. Since the low lineal energy part contributes little to dose-averaged microdosimetric quantities, this is normally not problematic, and the same can be said about their contribution to radiobiological effectiveness. More recently, a mini-TEPC operating under similar principles, but having a much smaller volume than conventional TEPCs, has been described [63]. The corrections mentioned earlier are much smaller for this device, and it can be operated at higher gas pressure.

The gas pressure in a TEPC is adapted such that the mean energy imparted is the same as in the microscopic TSOI, for example, the cell nucleus, i.e., $\bar{\varepsilon}_{gas} = \bar{\varepsilon}_{TSOI}$. The same equation can be written as $(S/\rho)_{gas}\,\rho_{gas}\,\bar{l}_{gas} = (S/\rho)_{TSOI}\,\rho_{TSOI}\,\bar{l}_{TSOI}$, where S/ρ are the mass stopping powers, ρ the mass densities, and $\bar{l}$ the mean chord lengths of the sites. If the gas has the same atomic composition of the TSOI, i.e., $(S/\rho)_{gas} = (S/\rho)_{TSOI}$, then the scaling can be expressed by the mass densities only: $\rho_{gas}\,\bar{l}_{gas} = \rho_{TSOI}\,\bar{l}_{TSOI}$. The last equation is generally used in experimental microdosimetry for properly adjusting the gas pressure when filling the TEPC with TE gas mixtures.

9.4.2.2 Solid-State Detectors for Regional Microdosimetry

Silicon detectors are attractive as microdosimeters because of the possibility to create sensitive volumes that are similar in size with the biological sites of interest, avoiding many of the corrections typical for TEPCs, as well as their simplicity, low cost, and transportability. However, due to the different radiological properties as compared to tissue, the measurements performed using silicon microdosimeters have to be corrected, as explained in Section 9.1.4. Other disadvantages are that the collecting volume is often not well defined either, and silicon devices are prone to radiation damage. Silicon devices based on silicon-on-insulator technology have been used for in-field and out-of-field microdosimetric characterization of proton beams [64,65]. A newer developed device consists of a microdosimetric diode with underneath a larger total-absorbing layer that is used to determine the total energy and type of the incident particle. Such a $\Delta E - E$ silicon telescope provides a coincidence measurement of the lineal energy and the total particle energy under the condition that it is totally absorbed in the thick layer. Since different particle types occupy different regions in a $\Delta E - E$ map, it is also possible to determine with a high probability the type of particle, enabling an improved silicon-to-tissue conversion procedure on a particle-by-particle basis [66], which is especially of interest in mixed particle fields. Such devices have been used for a detailed microdosimetric characterization of proton beams [67], and pixelated versions of such device have been developed as well [68].

Diamond-based microdosimeters operating as a Schottky diode have been developed [69] and are particularly promising because of their better tissue equivalence as compared to silicon. Both silicon- and diamond-based microdosimeters have the disadvantage of higher electronic noise levels as compared to TEPC, limiting their application to the lower lineal energy range.

Another novel solid-state detector that has been proposed is based on the Inductive Superconductive Transition Edge Sensor (ISTED), which itself is based on a Superconducting Quantum Interference Device, or SQUID. An ISTED contains the signal-generating superconducting layer within the SQUID loop, and by depositing a tissue-equivalent layer on top of the superconducting layer, it can act as a tissue-equivalent microdosimeter with an identical size and shape as the biological site of interest [70]. The energy deposition by an ionizing event in the absorber heats up the superconducting layer, which in turn causes a change in the inductive coupling between the superconducting layer and the SQUID loop, which can be measured as a change in voltage across the branches of the loop. The main drawback is that it requires operation in a cryostat, increasing the external dimensions and limiting its portability. The contamination signal from direct ionizations within the superconducting layers also needs to be corrected for Ref. [71].

9.4.2.3 Track Structure Detectors

Some track structure measurement devices have been used in proton beams that are mentioned, of which a more extended review can be found in Ref. [62]. In cloud chamber microdosimetry, a low-pressure supercooled gas is used in which the individual ionizations of the proton and its secondary particles create a 3D pattern of droplets that can be resolved by stereoscopic photography providing extremely high-resolution detail on the location of individual ionizations within the gas. In an optical ionization chamber, electrons in the particle track are made to oscillate rapidly by the application of an external, short-duration, high-voltage electric field. The excited electrons produce additional ionization and electronic excitation of the gas molecules in their immediate vicinity, leading to fluorescent light emission from the gas allowing the location of the electrons to be determined with a resolution of 10 nm. In three-dimensional optical random access memories, the energy deposited along the proton track changes a bistable photochromic material from the stable nonfluorescent form to a quasistable fluorescent form. The location of the fluorescent sites can be read out by confocal laser microscopy.

9.5 CONCLUSION

This chapter reviews the detector requirements for relative dosimetry in clinical proton beams, including those for profile measurements, depth-dose distributions, and field output factors. Specific discussion is given on the particularities of small fields used in scanned beams and stereotactic radiosurgery. Apart from the choice of detector size related to achieving sufficient spatial resolution, the main issues that are discussed are related to the water equivalence and the energy-dependent response of many detector systems, an important property to consider in proton beam dosimetry, and the use of plastic phantoms for relative dosimetry. An overview of the specific behavior of a number of suitable detectors is given and their application illustrated in a few examples. Finally, a brief introduction to microdosimetry is given and the requirements on microdosimeters and the various principles to establish lineal energy measurements are discussed.

REFERENCES

1. Chu, W. T., Ludewigt, B. A., and Renner, T. R. 1993. Instrumentation for treatment of cancer using proton and light-ion beams. *Rev. Sci. Instrum.* 64:2055–122.
2. Palmans, H., Kacperek, A., and Jäkel, O. 2009. Hadron dosimetry. In *Clinical Dosimetry Measurements in Radiotherapy*, eds. D. W. O. Rogers and J. E. Cygler, 669–722. Madison, WI: Medical Physics Publishing.
3. Karger, C. P., Jäkel, O., Palmans, H., and Kanai, T. 2010. Dosimetry for ion beam radiotherapy. *Phys. Med. Biol.* 55:R193–234.
4. Palmans, H. 2011. Dosimetry. In *Proton Therapy Physics*, ed. H. Paganetti, 191–219. London: CRC Press.
5. Vatnitsky, S. M., and Palmans, H. 2015. Detector systems. In *Principles and Practice of Proton Beam Therapy (AAPM 2015 Summer School)*, eds. I. J. Das and H. Paganetti, 275–316. Madison, WI: Medical Physics Publishing.
6. Palmans, H. 2017. Light-ion beam dosimetry. In *Clinical 3D Dosimetry in Modern Radiation Therapy*, ed. B. Mijnheer, 301–27. Boca Raton, FL: CRC Press.
7. Rossi, H. H. and Zaider, M. 1996. *Microdosimetry and Its Applications*. London: Springer.
8. Palmans, H., Rabus, H., Belchior, A. L., Bug, M. U., Galer, S., Giesen, U., Gonon, G., Gruel, G., Hilgers, G., Moro, D., Nettelbeck, H., Pinto, M., Pola, A., Pszona, S., Schettino, G., Sharpe, P. H. G., Teles, P., C. Villagrasa, C., and Wilkens, J. J. 2015. Future development of biologically relevant dosimetry. *Br. J. Radiol.* 87: 20140392.
9. Palmans, H. 2006. Perturbation factors for cylindrical ionization chambers in proton beams. Part I: corrections for gradients. *Phys. Med. Biol.* 51:3483–501.
10. ICRU. 2016. *Key Data for Ionizing-Radiation Dosimetry: Measurement Standards and Applications.* ICRU Report 90. Oxford: Oxford University Press.
11. ICRU. 1993. *Stopping Powers and Ranges for Protons and Alpha Particles.* ICRU Report 49. Bethesda, MD: International Commission on Radiation Units and Measurements.
12. Bassler, N., Hansen, J. W., Palmans, H., Holzscheiter, M. H., Kovacevic, S., and AD-4/ACE Collaboration. 2008. The antiproton depth dose curve measured with alanine detectors. *Nucl. Instrum. Methods Phys. Res. B* 266:929–36.
13. Dennis, J. A. 1973. Computed ionization and kerma values in neutron irradiated gases. *Phys. Med. Biol.* 18:379–95.
14. Grosswendt, B. and Baek, W. Y. 1998. W values and radial dose distributions for protons in TE-gas and air at energies up to 500 MeV. *Phys. Med. Biol.* 43:325–37.
15. Brede, H. J., K. D. Greif, O. Hecker, P. Heeg, J. Heese, D. T. L. Jones, H. Kluge, and D. Schardt. 2006. Absorbed dose to water determination with ionization chamber dosimetry and calorimetry in restricted neutron, photon, proton and heavy-ion radiation fields. *Phys. Med. Biol.* 51:3667–82.
16. Sassowsky, M. and Pedroni, E. 2005. On the feasibility of water calorimetry with scanned proton radiation. *Phys. Med. Biol.* 50:5381–400.
17. Herrmann, R. 2012. Prediction of the response behaviour of one-hit detectors in particle beams. *PhD Thesis*, Aarhus University, Denmark.

18. Fiorini, F., Kirby, D., Thompson, J., Green, S., Parker, D. J., Jones, B., and Hill, M. A. 2014. Under-response correction for EBT3 films in the presence of proton spread out Bragg peaks. *Phys. Med.* 30:454–61.

19. Yukihara, E. G., Doull, B. A., Ahmed, M., Brons, S., Tessonnier, T., Jäkel, O., and Greilich, S. 2015. Time-resolved optically stimulated luminescence of Al_2O_3:C for ion beam therapy dosimetry. *Phys. Med. Biol.* 60:6613–38.

20. LaVerne, J. A. and Schuler, R. H. 1987. Radiation chemical studies with heavy ions: oxidation of ferrous ion in the Fricke dosimeter. *J. Phys. Chem.* 91:5770–6.

21. Geiss, O. B., Krämer, M., and Kraft, G. 1998. Efficiency of thermoluminescent detectors to heavy charged particles. *Nucl. Instrum. Methods. Phys. Res. B* 142:592–598.

22. Verhaegen, F. and Palmans, H. 2001. A systematic Monte Carlo study of secondary electron fluence perturbation in clinical proton beams (70–250 MeV) for cylindrical and spherical ion chambers. *Med. Phys.* 28:2088–95.

23. Palmans, H. 2011. Secondary electron perturbations in Farmer type ion chambers for clinical proton beams. In *Standards, Applications and Quality Assurance in Medical Radiation Dosimetry – Proceedings of an International Symposium*, Vienna, 9–12 November 2010 – Vol. 1, 309–17. Vienna, Austria: International Atomic Energy Agency.

24. Palmans, H. and Verhaegen, F. 2002. Calculation of perturbation correction factors for ionization chambers in clinical proton beams using proton-electron Monte Carlo simulations and analytical model calculations. In *Recent Developments in Accurate Radiation Dosimetry*, eds. J. P. Seuntjens, and P. N. Mobit, 229–45. Madison, WI: Medical Physics Publishing.

25. Zhao, L. and Das. I. J. 2010. Gafchromic EBT film dosimetry in proton beams. *Phys. Med. Biol.* 55:N291–301.

26. Arjomandy, B., Tailor, R., Zhao, L., and Devic, S., 2012. EBT2 film as a depth-dose measurement tool for radiotherapy beams over a wide range of energies and modalities. *Med. Phys.* 39:912–21.

27. Andreo, P., Burns, D. T., Hohlfeld, K., Huq, M. S., Kanai, T., Laitano, F., Smyth, V. G., and Vynckier, S. 2000. *Absorbed Dose Determination in External Beam Radiotherapy – An International Code of Practice for Dosimetry Based on Standards of Absorbed Dose to Water.* IAEA Technical Report Series No. 398, Vienna: International Atomic Energy Agency.

28. Lourenço, A., Shipley, D., Wellock, N., Thomas, R., Bouchard, H., Kacperek, A., Fracchiolla, F., Lorentini, S., Schwarz, M., MacDougall, N., Royle, G., and Palmans, H. 2017. Evaluation of the water-equivalence of plastic materials in low- and high-energy clinical proton beams. *Phys. Med. Biol.* 62:3883–901.

29. Palmans, H., Al-Sulaiti, L., Andreo, P., Shipley, D., Lühr, A., Bassler, N., Martinkovič, J., Dobrovodský, J., Rossomme, S., Thomas, R. A. S., and Kacperek, A. 2013. Fluence correction factors for graphite calorimetry in a low-energy clinical proton beam: I. Analytical and Monte Carlo simulations. *Phys. Med. Biol.* 58:3481–99.

30. Clasie, B., Depauw, N., Fransen, M., Gomà, C., Panahandeh, H. R., Seco, J., Flanz, J. B., and Kooy, H. M. 2012. Golden beam data for proton pencil-beam scanning. *Phys. Med. Biol.* 57:1147–58.

31. Gottschalk, B., Cascio, E. W., Daartz, J., and Wagner, M. S. 2015. On the nuclear halo of a proton pencil beam stopping in water. *Phys. Med. Biol.* 60:5627–54.

32. Grevillot, L., Stock, M., Palmans, H., Osorio, J., Letellier, V., Dreindl, R., Elia, A., Fuchs, H., Carlino, A., and Vatnitsky, S. M. 2018 Implementation of dosimetry equipment and phantoms in practice of light ion beam therapy facility: the MedAustron experience. *Med. Phys.* 45: 352–69.

33. Takayanagi, T., Nihongi, H., Nishiuchi, H., Tadokoro, M., Ito, Y., Nakashima, C., Fujitaka, S., Umezawa, M., Matsuda, K., Sakae, T., and Terunuma, T. 2016. Dual ring multilayer ionization chamber and theory-based correction technique for scanning proton therapy. *Med. Phys.* 43:4150–62.

34. Pedroni, E., Scheib, S., Böhringer, T., Coray, A., Grossmann, M., Lin, S., and Lomax, A. 2005. Experimental characterization and physical modelling of the dose distribution of scanned proton pencil beams. *Phys. Med. Biol.* 50:541–61.

35. Winey, B., and Bussiere, M. 2015. Small field dosimetry: SRS and eyes. In *Principles and Practice of Proton Beam Therapy (AAPM 2015 Summer School)*, eds. I. J. Das, and H. Paganetti, 767–94. Madison, WI: Medical Physics Publishing.

36. Brenner, D. J. and Zaider, M. 1998. Estimating RBEs at clinical doses from microdosimetric spectra. *Med. Phys.* 25:1055–7.

37. Hansen, J. W. and Olsen, K. J. 1985. Theoretical and experimental radiation effectiveness of the free radical dosimeter alanine to irradiation with heavy charged particles. *Radiat. Res.* 104:15–27.

38. Scholze, F., Rabus, H., and Ulm, G. 1998. Mean energy required to produce an electron-hole pair in silicon for photons of energies between 50 and 1500 eV. *J. Appl. Phys.* 84:2926–39.

39. Marinelli, M., Prestopino, G., Verona, C., Verona-Rinati, G., Ciocca, M., Mirandola, A., Mairani, A., Raffaele, L., and Magro, G. 2015 Dosimetric characterization of a microDiamond detector in clinical scanned carbon ion beams. *Med. Phys.* 42:2085–93.

40. Rossomme, S., Hopfgartner, J., Vynckier, S., and Palmans, H. Under response of a PTW-60019 microDiamond detector in the Bragg peak of a 62 MeV/n carbon ion beam. *Phys. Med. Biol.* 61:4551–63.

41. Rossomme, S., Marinelli, M., Verona-Rinati, G., Romano, F., Cirrone, P. A. G., Kacperek, A., Vynckier, S., and Palmans, H. 2017. Response of synthetic diamond detectors in proton, carbon and oxygen ion beams. *Med. Phys.* 44:5445–49.

42. Archambault, L., Polf, J. C., Beaulieu, L., and Beddar, S. 2008. Characterizing the response of miniature scintillation detectors when irradiated with proton beams. *Phys. Med. Biol.* 53:1865–76.

43. Seravalli, E., de Boer, M. R., Geurink, F., Huizenga, J., Kreuger, R., Schippers, J. M., and van Eijk, C. W. 2009. 2D dosimetry in a proton beam with a scintillating GEM detector. *Phys. Med. Biol.* 54:3755–71.

44. Boon, S. N., van Luijk, P., Böhringer, T., Coray, A., Lomax, A., Pedroni, E., Schaffner, B., and Schippers, J. M. 2000. Performance of a fluorescent screen and CCD camera as a two-dimensional dosimetry system for dynamic treatment techniques. *Med. Phys.* 27:2198–208.

45. Archambault, L., Poenisch, F., Sahoo, N., Robertson, D., Lee, A., Gillin, M. T., Mohan, R., and Beddar, S. 2012. Verification of proton range, position, and intensity in IMPT with a 3D liquid scintillator detector system. *Med. Phys.* 39:1239–46.

46. Safai, S., Lin, S., and Pedroni, E. 2004. Development of an inorganic scintillating mixture for proton beam verification dosimetry. *Phys. Med. Biol.* 49:4637–55.

47. Rink, A. 2008. Point-based ionizing radiation dosimetry using radichromic materials and a fibreoptic readout system. *PhD Thesis*, University of Toronto, Canada.

48. Baldock, C., De Deene, Y., Doran, S., Ibbott, G., Jirasek, A., Lepage, M., McAuley, K. B., Oldham, M., and Schreiner, L. J. 2010. Polymer gel dosimetry. *Phys. Med. Biol.* 55:R1–R63.

49. Perles, L. A. Mirkovic, D., Anand, A. Titt, U., and Mohan, R. 2013. LET dependence of the response of EBT2 films in proton dosimetry modeled as a bimolecular chemical reaction. *Phys. Med. Biol.* 58:8477–91.

50. Nichiporov, D., Solberg, K., His, W., Wolanski, M., Mascia, A., Farr, J., and Schreuder, A. 2007. Multichannel detectors for profile measurements in clinical proton fields. *Med. Phys.* 34:2683–90.

51. Vatnitsky, S. M., Miller, D. W., Moyers, M. F., Levy, R. P., Schulte, R. W., Slater, J. D., and Slater, J. M. 1999. Dosimetry techniques for narrow proton beam radiosurgery. *Phys. Med. Biol.* 44:2789–801.

52. Kuess, P., Böhlen, T., Lechner, W., Elia, A., Georg, D., and Palmans, H. 2017 Lateral response heterogeneity of Bragg peak ionization chambers for narrow-beam photon and proton dosimetry. *Phys. Med. Biol.* 62:9189–206.

53. Daartz, J., Engelsman, M., Paganetti, H., Bussière, M. R. 2009. Field size dependence of the output factor in passively scattered proton therapy: influence of range, modulation, air gap, and machine settings. *Med. Phys.* 36:3205–10.

54. Carlino, A., Gouldstone, C., Kragl, G., Traneus, E., Marrale, M., Vatnitsky, S., Stock, M., and Palmans, H. 2018. End-to-end tests using alanine dosimetry in scanned proton beams. *Phys. Med. Biol.* accepted (https://doi.org/10.1088/1361-6560/aaac23).

55. Ableitinger, A., Vatnitsky, S., Herrmann, R., Bassler, N., Palmans, H., Sharpe, P., Ecker, S., Chaudhri, N., Jäkel, O., and Georg, D. 2013. Dosimetry auditing procedure with alanine dosimeters for light ion beam therapy: results of a pilot study. *Radiother. Oncol.* 108:99–106.

56. Cho, J., Summers, P. A., and Ibbott, G. S. 2014. An anthropomorphic spine phantom for proton beam approval in NCI-funded trials. *J. Appl. Clin. Med. Phys.* 15:252–65.

57. Coutrakon, G., Cortese, J., Ghebremedhin, A., Hubbard, J., Johanning, J., Koss, P., Maudsley, G., Slater, C. R., and Zuccarelli, C. 1997. Microdosimetry spectra of the Loma Linda proton beam and relative biological effectiveness comparisons. *Med. Phys.* 24:1499–506.

58. Kellerer, A. M., and Rossi, H. H. 1974. The theory of dual radiation action. In *Current Topics in Radiation Research, Vol. 8,* eds. M. Ebert and A. Howard, 87–158. Amsterdam: North-Holland Publishing Company.

59. Fowler, J. F. 1989. A review: the linear quadratic formula and progress in fractionated radiotherapy. *Br. J. Radiol.* 62:679–94.

60. Moro, D., Chiriotti, S., Conte, V., Colautti, P., and Grosswendt, B. 2015. Lineal energy calibration of a spherical TEPC. *Radiat. Prot. Dosim.* 166(1–4):233–7.

61. ICRU. 1983. *Microdosimetry.* ICRU Report 36. Bethesda, MD: International Commission on Radiation Units and Measurements.

62. Bradley, P. D. 2000. The development of a novel silicon microdosimeter for high LET radiation therapy. *PhD Thesis*, University of Wollongong, Australia.

63. De Nardo, L., Cesari, V., Donà, G., Colautti, P., Conte, V., and Tornielli, G. 2004. Mini TEPCs for radiation therapy. *Radiat. Prot. Dosim.* 108:345–52.

64. Rosenfeld, A. B., Bradley, P. D., Corneluis, I., Kaplan, G. I., Allen, B. J., Flanz J, Goitein, M. Van Meerbeeck, A. Schubert, J. Bailey, J. Takada, Y. Maruhashi, A., and Hayakawa, Y. 2000. New silicon detector for microdosimetry applications in proton therapy. *IEEE Trans. Nucl. Sci.* 47:1386–94.

65. Wroe, A., Rosenfeld, A., and Schulte, R. 2007. Out-of-field dose equivalents delivered by proton therapy of prostate cancer. *Med. Phys.* 34:3449–56 [Erratum in 2008: Med. Phys. 35: 3398.]

66. Agosteo, S., Colautti, P., Fazzi, A., Moro, D., and Pola, A. 2006. A solid state microdosimeter based on a monolithic silicon telescope. *Radiat. Prot. Dosim.* 122:382–6.

67. Wroe, A., Schulte, R., Fazzi, A., Pola, A., Agosteo, S., and Rosenfeld, A. 2009. RBE estimation of proton radiation fields using a ΔE–E telescope. *Med. Phys.* 36:4486–94.

68. Agosteo, S., Fallica, P. G., Fazzi, A., Introini, M. V., Pola, A., and Valvo, G. 2008. A pixelated silicon telescope for solid state microdosimetry. *Radiat. Meas.* 43:585–9.

69. Rollet, S., Angelone, M., Magrin, G., Marinelli, M., Milani, E., Pillon, M., Prestopino, G., Verona, C., and Verona-Rinati, G. 2012. A novel microdosimeter based upon artificial single crystal diamond. *IEEE Trans. Nucl. Sci.* 59:2409–15.

70. Galer, S., Hao, L., Gallop, J., Palmans, H., Kirkby, K., and Nisbet, A. 2011. Design concept for a novel SQUID-based microdosemeter. *Radiat. Prot. Dosim.* 143:427–31.

71. Fathi, K., Galer, S., Kirkby, K. J., Palmans, H., and Nisbet, A. 2017. Coupling Monte Carlo simulations with thermal analysis for correcting microdosimetric spectra from a novel microcalorimeter. *Rad. Phys. Chem.* 140:406–11.

CHAPTER 10

Absolute and Reference Dosimetry

Hugo Palmans

Acoustics and Ionising Radiation

CONTENTS

10.1 INTRODUCTION—REQUIREMENTS FOR TRACEABLE REFERENCE DOSIMETRY

The accuracy needed in dosimetry should be considered in view of the requirement for dose delivered to the target volume, for which, in general, the required relative standard uncertainty levels between 3% and 5% have been quoted in IAEA (International Atomic Energy Agency) Technical Report Series No. 398 (TRS-398) [1]. Reference dosimetry, which is only one step in a chain of procedures leading to dose delivery, should thus be done with relative standard uncertainties well below that, typically around 1%. This chapter, which updates the text on reference dosimetry in the previous edition of this book [2], discusses the primary and reference instruments used for proton beam dosimetry and their uncertainties. Issues particular to scanned beams are emphasized, and the impact of new data published in the International Commission on Radiation Units and Measurements (ICRU) Report 90 [3] is discussed.

10.1.1 Requirements for Primary Standards and Equivalent Instruments

The quantity of interest in radiotherapy is absorbed dose to water. Relevant quantities such as biological response and isoeffective dose are also defined with reference to absorbed dose to water. Given the required uncertainty of dose delivery to the target volume, in line with many industrial standards clinical reference dosimetry would ideally have an uncertainty smaller by an order of magnitude and the uncertainty on primary standards smaller by another other of magnitude. This would require the relative standard uncertainty on primary standards of absorbed dose to water to be lower than 0.1%. This is, however, far from achievable at present, and realistically, we can aim for relative standard uncertainties for primary standards of the order of 0.5% and for reference dosimetry of the order of 1%. These levels are more or less achieved for high-energy photon beam dosimetry using calibrated ionization chambers. For proton beams, no primary standards exist, and the correction factors required for using an ionization chamber with an absorbed dose-to-water calibration coefficient in a photon beam for reference dosimetry in a proton beam result in a relative standard uncertainty of the order of 2% on absorbed dose to water. The most obvious way for reducing this uncertainty is to establish primary standards for proton beam dosimetry.

Ideally, a primary standard measures the quantity absorbed dose directly according to its definition, i.e., energy imparted per unit mass. Calorimeters are the only known devices that can realize such a direct measurement. Given the relation between absorbed dose and fluence, via the stopping power of the medium, it can also be established via a measurement of fluence. Counting the amount of ionization produced or the amount of species produced in radiation-induced chemical reactions can be another route for determining absorbed dose from first principles if accurate values of the production yields of ionization (the mean energy expended per ion pair) or of chemical species (the chemical yields) are known. However, these indirect methods to determine absorbed dose always require interaction-related quantities that are difficult to determine with sufficient accuracy from first principles, and if such methods are used, those interaction quantities are often derived by comparison with calorimeters.

10.1.2 Different Requirements for Scanned and Scattered Beams

For scattered beams with passive beam-shaping elements, such as scatter foils, collimators, range modulators, wedges, or range compensators, a broad scattered beam defined as a reference field has very similar properties as a clinical beam used in a conformal treatment. Absorbed dose to water at a reference point in a field that contains a representative set of beam-shaping elements (e.g., a range modulator and a collimator) is then a relevant quantity for beam monitor calibration, and dose calculations can be based on relative beam models and output factors.

For scanned proton beams, however, dose calculations are based on a superposition of a large number of narrow beam spots, and the dose calculation and optimization are based on the number of particles in each spot. The relevant quantity is either particle number, related to fluence, or its dosimetric equivalent, dose-area-product (DAP). While the direct measurement of particle number appears very attractive since it is relatively straightforward and is directly related to the quantity used by the treatment planning system to define the relative weights of beam spots, it must be realized that, in the dose calculation, the fluence has to be multiplied with the mass stopping power of water (or tissue) that has a considerable uncertainty, given the uncertainty on the stopping power data themselves and the uncertainty on the exact spectrum of charged particles at the measurement depth (primary protons + secondary charged particles produced in nuclear interactions and target fragmentation). As will be explained in Section 10.3.4, using dose-based or DAP based approaches, the experimentally determined DAP has to be divided by the DAP per incident proton (sometimes also called the mean stopping power per unit of incident proton fluence) to derive the number of particles in a spot. The uncertainty on this divider is evenly large as the mass stopping power, but since it is highly correlated with the DAP per incident proton or the mean stopping power per unit of incident proton fluence with which the fluence is multiplied in the dose calculation, this results in a cancelation of most of the uncertainty contribution. This argument favors calibration in terms of DAP rather than particle number.

10.2 PRIMARY INSTRUMENTS FOR DOSIMETRIC QUANTITIES

10.2.1 Absorbed Dose—Calorimeters

Calorimeters provide the most direct method to measure the quantity of interest in reference dosimetry for radiotherapy, absorbed dose to water. The most direct way of measuring this quantity is by use of water calorimeters, but calorimeters using different absorbed media like graphite or the tissue-equivalent plastic A-150 have been used for the same purpose as well. The latter require, in addition to the measurement of dose to the calorimeter medium, a conversion procedure to derive absorbed dose to water.

The basic principle behind calorimetric dosimetry is that all the energy deposited by ionizing radiation will appear as heat if no other changes of internal energy take place in the medium, such as chemical, phase, or lattice energy changes. The energy deposition in calorimeters is quantified by determining the electrical energy dissipation needed to realize the same temperature rise in the medium as the radiation does. This measurement can be realized by a direct substitution of electrical and irradiative energy deposition to keep the medium at a constant temperature, the so-called isothermal operation mode. Alternatively, it can be realized in adiabatic mode when the specific heat capacity, c_{med}, of the medium is known. c_{med} is normally obtained from measuring the temperature rise in the medium due to a known electrical energy dissipation, either in a separate experiment or within the calorimeter setup. From an accurate measurement of the temperature rise, ΔT_{med}, of the medium, absorbed dose to the medium is then derived as

$$D_{med} = c_{med}\Delta T_{med}\frac{1}{1-h}\prod_i k_i \tag{10.1}$$

$\frac{1}{1-h}$ is a correction factor for the heat defect (h is simply called the heat defect and corresponds with the fraction of energy deposited by ionizing radiation that does not appear as heat but is taken away from the medium by changes of its physicochemical state). $\prod_i k_i$ is the product of correction factors for heat transported away or towards the measurement point, field nonuniformity, changed scatter, and attenuation due to the presence of nonmedium equivalent materials.

Typical designs of water and graphite calorimeters that have been used in proton beams are shown in Figure 10.1. The operation of calorimeters in proton beams is very similar as in photon and electron beams, which has been described extensively [4,5], but issues that need to be considered differently in proton beams are heat defect, thermal heat conduction, and dose conversions between different media.

In water calorimeters, the main source of heat defect is the chain of chemical reactions following the radiolysis of water. The reactions following the production of so-called primary species (those present about 10^{-7} s after passage of the ionizing particle) are well known and documented [6,7], and together with values of the chemical yields of primary species and the heats of formation of the six stable chemical species after irradiation [8], the chemical heat defect can be calculated by solving the coupled set of linear differential equations describing all chemical reactions. The production of primary species, however, is linear energy transfer (LET) dependent, and this dependence is not well known. The most

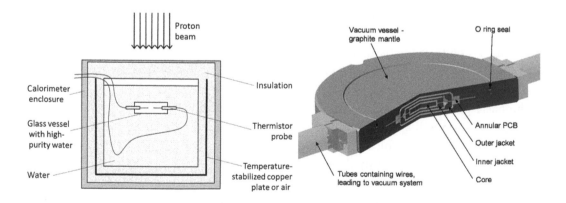

FIGURE 10.1 Left: Schematic representation of a typical sealed water calorimeter design used in scattered and scanned proton beams (showing common features in calorimeters used by Palmans et al. [10], Medin et al. [12], Sarfehnia et al. [24], and Medin [25]). Right: Schematic design of the core, jackets, and mantle of a graphite calorimeter for proton beam dosimetry built at the National Physical Laboratory, UK.

comprehensive information comes from Ross and Klassen [8] and Elliot [9], and based on this, calculations of the chemical heat defect for proton beams has been performed and discussed in various publications [10–14], leading to the following observations:

- For high-energy (low-LET) protons, pure water saturated with a chemically inert gas like argon or nitrogen (the argon or nitrogen system) exhibit a small (<0.1%) initial heat defect, reaching a steady state after a modest irradiation just as it does in photon beams. For high-LET protons, however, a steady increase of the chemical energy in the aqueous system is observed due to a higher production of hydrogen peroxide than what is decomposed, resulting in a nonzero endothermic (positive) heat defect.

- Pure water saturated with hydrogen (the hydrogen system) exhibits a zero heat defect over the entire LET range, which can be explained by an enhanced decomposition of hydrogen peroxide compared to the nitrogen system.

- When initial oxygen concentrations are present, the hydrogen system exhibits an initial exothermic (negative) heat defect that increases until depletion of oxygen, after which the heat defect drops abruptly to zero. This is an attractive system, since this phenomenon offers a way of monitoring when the steady-state zero-heat-defect condition is reached.

- For water with a known quantity of sodium formate as a deliberate impurity saturated with oxygen (the formate system), the exothermic heat defect in a modulated beam is only about half of that in a ^{60}Co beam with the same dose rate. This was explained by the lower chemical yields for certain species at high LET as well as the time structure of the formation of chemical species due to beam modulation.

Two types of experiments can be distinguished that provide support for the earlier theoretical observations:

- Comparison of the heat defect of water with that of a metal for which the heat defect is assumed to be zero (e.g., aluminum or copper). The experiment consists of measuring the thermal heating of a dual water/metal absorber, which forms one thermal body, by totally absorbing the same number of protons in either the water component or the metal component. Using such a setup, it was experimentally confirmed that the endothermic heat defect in pure water increases with LET from a value close to zero for protons to a value of about +4% for 100 keV μm⁻¹ helium ions [15]. A weighted least-squares fit of an exponential function to the data including an uncertainty of 0.3% for the assumption that the heat defect is zero at low-LET [4] gives

$$h=(0.041\pm0.004)\left(e^{-(0.035\pm0.010)\text{LET}}-(1.000\pm0.001)\right) \tag{10.2}$$

where the LET is expressed in keV μm⁻¹. Figure 10.2 shows the experimental data points as well as the fit with the standard uncertainty as a function of LET.

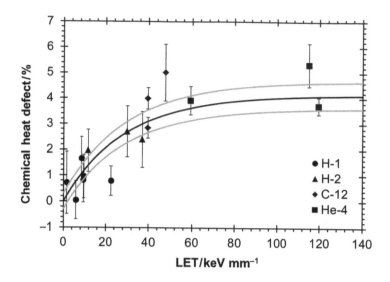

FIGURE 10.2   Chemical heat defect of pure water as a function of LET measured for various ions [15]. The full line represents an exponential fit, and the two gray lines represent the one standard deviation interval based on the uncertainties from the least-square fit parameters and a standard uncertainty of 0.3%, accounting for the assumption that the heat defect is zero at low LET.

- Relative comparison of the heat defect of different aqueous systems by comparing their relative response at the same dose. This way the initial exothermicity of the hydrogen system in the presence of trace oxygen concentrations was demonstrated for protons as well as the relative agreement of the hydrogen system with the nitrogen or argon system after a steady state was reached [10]. Also, the lower heat defect of the formate system in protons as compared to photons in a modulated proton beam was demonstrated experimentally in the same work.

A chemical heat defect can also occur in solid calorimeters. In graphite as well as plastic calorimeters (A-150 tissue equivalent plastic and polystyrene), reaction of the medium with dissolved oxygen may result in an exothermic heat defect. This has been suggested to explain the initial over response of calorimeters made of A-150 and graphite (more than 10% and 2%, respectively), which disappears after sufficient pre-irradiation [16,17]. Another suggested mechanism for a chemical heat defect is the dissociation of polymers, which explains the endothermic heat defect of about 4% for A-150 obtained from similar total absorption experiments as described above using a dual A-150/aluminum absorber [18–20].

In solid calorimeters made of a crystalline or polycrystalline material such as graphite, a physical heat defect is also possible due to the creation and annihilation of interstitial lattice defects. This is generally assumed to be small but may be larger in proton beams than in photon beams due to the higher probability of a sufficient energy transfer to a recoil nucleus. Only at extremely high doses (order 10^9 Gy) received in nuclear reactors, a heat defect has been demonstrated by measuring the release of the Wigner energy when annealing the graphite samples by heating it above 250°C [21]. It is not clear if and how these results would translate to radiotherapeutic dose levels. In a proton total absorption experiment using a graphite/aluminum absorber, an endothermic heat defect of 0.4% with a standard uncertainty of 0.3% was observed [20], indicating that the physical heat defect in graphite must be limited to a few tenths of a percent.

The second issue of concern with calorimeters in proton beams is that of heat conduction, potentially leading away heat from the measurement point or adding heat from the irradiated surroundings of the measurement point. For water calorimeters, where the thermal diffusivity is relatively low, thermal conduction is manageable when the steep gradients are kept a sufficient distance away from the point of measurement. The same criterion as in photon beams could be used so that corrections for heat losses are negligibly small when the distance from the measurement point to steep edges, such as the penumbrae and the Bragg peak, or to the distal edge of the spread-out Bragg peak (SOBP), is at least 3 cm [22]. This cannot be established in narrow low-energy proton beams used, e.g., for treatment of eye melanoma, and corrections have to be applied contributing larger uncertainties to the absorbed dose to water [23]. Also the glass vessel containing high-purity water has to be smaller for low-energy beams, which makes the heat transfer correction factors very sensitive to the positioning of the thermistor within the glass vessel, as shown in Figure 10.3 [23].

In scanned proton beams, an additional complication is that when a pencil beam hits the measurement point, the instantaneous gradients are always substantial, but it has been demonstrated theoretically [11] and confirmed experimentally [24] that if the painting of a target volume takes a time of not more than 2–3 min, the correction for heat conduction and its uncertainty are very similar as for a passively scattered broad proton beam

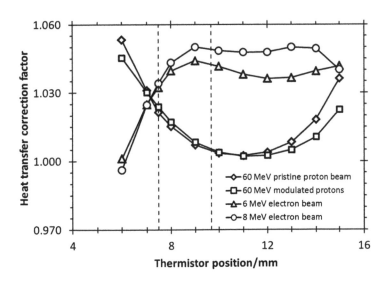

FIGURE 10.3 Heat conduction correction factor in a small water calorimeter developed for low-energy charged particle beams obtained by finite element modeling. The vertical dashed lines indicate the points of measurement in electron (9.68 mm) and proton (7.48 mm) measurements reported by Renaud et al. [23]. (Figure replotted from Renaud et al. [23] with permission of the publisher and authors.)

irradiation of the same duration. Also the excess heat due to the presence of nonwater materials (the glass vessels to contain the high-purity water and the thermistor probes) has been found to cause only minor differences as compared to photon beams [24].

Typical relative standard uncertainties on absorbed dose to water with water calorimeters amount to 0.4%–1% dominated by contributions from the chemical heat defect, heat conduction, and thermistor calibration [10–12,23–25].

In solid calorimeters and in particular graphite calorimeters, the phenomenon of heat conduction is very different given the higher thermal diffusivity. In graphite, the thermal diffusivity is three orders of magnitude higher than in water, meaning that temperature profiles within an irradiated sample redistribute within a time interval much shorter than the irradiation time itself. This is traditionally dealt with by introducing gaps and so-called jackets around the core, a graphite sample with known mass over which the average dose is measured by comparing the temperature increase during irradiation, with the temperature increase resulting from known electrical energy dissipation. In photon beams, advantage is taken from the almost linear attenuation curve; by thermally connecting equal-sized parts of the jackets in front and to the back of the core, the average heating of the jackets is almost equal to that of the core, thus minimizing radiative heat transfer. In proton beams, due to the Bragg peak or the distal edge, this is more difficult to achieve, and the steep gradients need to be either shielded by inducing more thermal barriers or by matching the size of graphite parts beyond the core such that the energy per unit mass of those parts equals dose in the core [26]. For scanned proton beams, it was until recently not clear if this quasi-adiabatic operation mode would be possible, given the rapidly varying and steep instantaneous dose distributions combined with the high thermal diffusivity of graphite. Petrie (2016) studied the heat transfer phenomenon of scanned beams in an analytic way, showing that the heat transfer effects of individual narrow beams can be accurately modeled [27]. Experiments and simulations of fully scanned beam delivery with a high-energy beam showed that in graphite calorimeters the net heat transfer effect in a scanned box field is not substantially different from that in a broad beam uniform dose distribution with the same total irradiation time, as shown in Figure 10.4 [27]. The use of isothermal operation mode has been suggested [2] to overcome the issue of unknown heat transfer effects, but practical experience has revealed that the control system has to respond very fast and probably only an analog implementation would be feasible [27].

Last but not least, for nonwater calorimeters there is the issue of converting dose to the calorimeter medium to the quantity of interest, dose to water. For graphite calorimeters, the conversion from dose-to-graphite to dose-to-water is one of the main uncertainties on the dose-to-water determination in a proton beam [28]. If the charged particle spectra at equivalent depths in water and graphite (related by the continuous slowing down approximation (CSDA) ranges) are identical, then the dose conversion is adequately described by the water-to-graphite mass stopping power ratio for the charged particle spectrum. However, similar to plastic substitutes for water as discussed in Chapter 9, differences in the absorption of primary protons in nonelastic nuclear interactions and differences in the production of secondary charged particles in both target materials are likely to result in unequal charged particle spectra at equivalent depths. Palmans et al. [28] have quantified

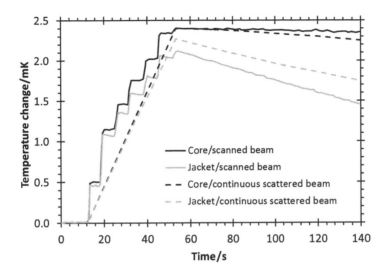

FIGURE 10.4 Comparison of heating of the core and the first jacket with a scanned beam and with scattered beams showing that there is no qualitative difference in the heat loss from the core immediately after irradiation with both modes. (Extracted from Petrie et al. [27] with permission of the author.)

this effect for low-energy proton beams by Monte Carlo simulations and Al Sulaiti et al. [29] by experiment, arriving to the conclusion that the effect is smaller than 1% at typical measurement points in the center of the SOBP. Lourenço et al. [30] showed that for high-energy proton beams the effect can be substantially larger (up to 4% depending on the depth as shown in Figure 10.5), but by a sensible choice of the measurement depth, it can be kept within 2%. Figure 10.5 shows Monte Carlo simulated and experimental fluence correction factors for two proton beam energies. Monte Carlo results are shown for all charged particles but also considering only the proton spectra (primary + secondary) to show that the latter are in better agreement with experiments. This is explained by the short range of the heavier secondary particles; they do not have sufficient energy to penetrate the wall of the ionization chambers, and thus the differences in dose contribution by these heavier charged particles in water and graphite are not picked up by the ionization chamber.

Typical relative standard uncertainties on absorbed dose to water with graphite calorimeters amount to 1%–2% dominated by contributions from the water-to-medium mass stopping power ratio, fluence corrections factors, electrical calibration, and heat transfer [26]. For A-150 calorimeters, in addition, the chemical heat defect contributes substantially to the uncertainty and overall uncertainties are 2%–3% [31].

10.2.2 Fluence—Faraday Cups and Activation Measurements

In a broad proton beam, dose to water at a shallow reference depth, z_{ref}, can, in principle, be derived from the proton fluence at the surface, $\Phi_p(0)$, as

$$D_w(z_{\text{ref}}) = \Phi_p(0)\frac{\text{DAP}_w(z_{\text{ref}})}{n_p(0)}\prod_i k_i \qquad (10.3)$$

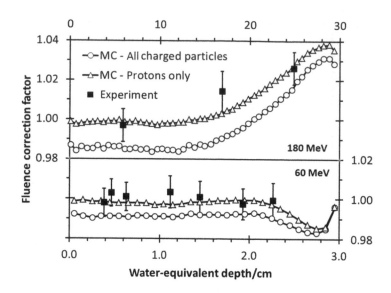

FIGURE 10.5   Fluence correction factors for correcting the conversion of dose to graphite to dose to water using graphite-to-water mass stopping power ratios for two pristine proton beam energies calculated using FLUKA Monte Carlo simulations and determined experimentally using plane-parallel ionization chambers. (Data extracted from Lourenço et al. [30].)

where $\mathrm{DAP}_w(z_{\mathrm{ref}})/n_p(0)$ is the DAP to water of the charged particle spectrum at depth z_{ref} per incident proton incident on the central axis and $\prod_i k_i$ is the product of correction factors for field nonuniformity and beam contamination.

The quantity $\mathrm{DAP}_w(z_{\mathrm{ref}})$, which is unusual and not commonly used in dosimetry protocols, deserves some explanation. It can essentially be calculated from the local charged particle spectrum at the measurement depth as

$$\mathrm{DAP}_w(z_{\mathrm{ref}}) = \sum_i \int_A \int_0^{E_{\max,i}} \Phi_{E,i}(z_{\mathrm{ref}},x,y) \frac{S_w}{\rho}(E,i)\,dE\,dx\,dy \tag{10.4}$$

where $\Phi_{E,i}(z_{\mathrm{ref}},x,y)$ is the charged particle fluence for charged particle type i at depth z_{ref} and lateral coordinates x and y and $S_w/\rho(E,i)$ the mass stopping power of water for particle type i of kinetic energy E. It is essentially defined for a pencil beam, and the area A should be large enough to collect all charged particles for that pencil beam. If the lateral extension of an incident beam is large enough to establish lateral charged particle equilibrium, the ratio $\mathrm{DAP}_w(z_{\mathrm{ref}})/n_p(0)$ can be replaced with the dose to water at the reference depth per unit of incident proton fluence, thanks to the reciprocity principle as explained in ICRU Report 35 [32]. It is obvious that this method relies on accurate values of the proton stopping power for water, which has a relative standard uncertainty of 1%–3% for protons with energies above 1 MeV according to ICRU Reports 49 and 90 [3,33].

Note that the quantity $\mathrm{DAP}_w(z_{\mathrm{ref}})/n_p(0)$ has the dimension of a mass stopping power and is often simply called the mean mass stopping power, but this is a rather misleading nomenclature, since the latter is actually the dose to water divided by the local charged particle fluence:

$$\frac{DAP_w(z_{ref})}{n_p(0)} \ne \frac{\overline{S_w}}{\rho}(z_{ref}) = \frac{\sum_i \int_0^{E_{max,i}} \Phi_{E,i}(z_{ref}) \frac{S_w}{\rho}(E,i)dE}{\sum_i \int_0^{E_{max,i}} \Phi_{E,i}(z_{ref})dE} = \frac{D_w(z_{ref})}{\Phi(z_{ref})} \qquad (10.5)$$

The main instrument in use to measure the incident particle fluence is the Faraday cup, which enables an accurate measurement of the number of protons provided it is well designed. A typical design is shown in Figure 10.6. For broad beams, an additional major uncertainty is due to the determination of the field area. For pencil beams, this uncertainty vanishes when the derived quantity is a laterally integrated dose. A major concern is the influence of electrons generated in the entrance window that reach the collecting electrode (and thus reduce the signal) as well as electrons liberated in and escaping from the collecting electrode (which enhance the signal). Both sources of perturbation to the measurement are usually suppressed by a guard electrode with a negative potential with respect to the electrode and casing sometimes with addition of a magnetic field. The relative standard uncertainty of absorbed dose-to-water determination using a Faraday cup varies between 2% and 3% and is dominated by the collection efficiency of the Faraday cup, the determination of the area, the mass stopping power of water, and the contributions to the absorbed dose to water by secondary charged particles originating from non-elastic interactions.

An alternative method for determining the fluence is to measure the induced activation of a sample [34]. If the number, N, of ^{12}C atoms in a sample is known as well as the cross section, σ, of the $^{12}C(p, pn)^{11}C$ reaction, the measurement of ^{11}C activity, A_0, immediately

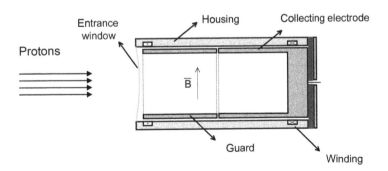

FIGURE 10.6 Schematic diagram of a reference dosimetry-level Faraday cup with internal vacuum. Shown are the collecting electrode, the guard electrode (which is at negative potential with respect to the collecting electrode), the entrance window and the windings creating a magnetic field, B, to suppress the loss of electrons generated in the collecting electrode. (Reproduced from Palmans et al. [13] with permission of the publisher and the authors.)

after an irradiation time τ can be used to derive the proton fluence in the center of the broad beam that hits the sample using the following equation:

$$\Phi_p(0) = \frac{A_0 \, e^{\lambda \frac{\tau}{2}}}{\sigma \, N \, \lambda} \tag{10.6}$$

where λ, is the decay constant of ^{11}C. The activity is usually measured using 4π $\beta\gamma$-coincidence counting, and it is estimated that the fluence can be derived with a relative standard uncertainty of 3%, resulting in an overall relative standard uncertainty of 3%–4% on absorbed dose to water. Advantages of this method are that no determination of the beam area is needed and that it can be used without much loss of accuracy in high dose per pulse beams, such as from synchrotrons or laser-induced beams.

10.2.3 Ionization—Ionization Chambers

Ionization chambers are the backbone for reference dosimetry in radiotherapy, but in principle, they can also be used to derive absorbed dose from first principles. With an ionization chamber, dose to water, $D_{w,Q}$, in a proton beam of beam quality Q is related to the average dose to air over the volume of the air cavity, $\bar{D}_{air,Q}$, via Bragg–Gray cavity theory:

$$D_{w,Q} = \bar{D}_{air,Q} \left(s_{w,air} \right)_Q p_Q \tag{10.7}$$

where $\left(s_{w,air} \right)_Q$ is the water-to-air Bragg–Gray or Spencer–Attix mass stopping power ratio for the charged particle spectrum at the measurement point in water, and p_Q is a correction factor to account for any deviation from the conditions under which Bragg–Gray cavity theory is valid.

The average dose to air can in principle be obtained from the measured ionization in the cavity, M_Q, and knowledge of the mass of air in the cavity $\left(m_{air} = \rho_{air} V_{cav} \right)$ and the mean energy required to produce an ion pair in air, $\left(W_{air} \right)_Q$.

$$\bar{D}_{air,Q} = \frac{M_Q \left(W_{air} / e \right)_Q}{\rho_{air} V_{cav}}, \tag{10.8}$$

where e is the elementary charge. If an accurate estimate of the volume is available independently of a calibration, e.g., from manufacturer's blueprints or measured dimensions (as in primary standards of air kerma), dosimetry using an ionization chamber could be based on first principles using Equations (10.7) and (10.8). It has been shown that this can provide a reliable way of a monitor unit (MU) calibration for a transmission ionization chamber to link with a Monte Carlo based planning system [35].

In practice, however, the volume of commercial ionization chambers is not known with the required precision, and one has to rely on a calibration in a reference beam to estimate the volume or bypass the requirement of knowing the volume. This is the basis of most clinical reference dosimetry, which will be addressed in Section 10.3.

10.2.4 Chemical Yields—Chemical Dosimeters

Chemical dosimeters are based on the knowledge of the amount of a chemical species produced for a given absorbed dose to the medium, mostly in aqueous environment, since the radiolysis of water and the reaction chemistry in water are well known.

The radiation chemical yield, $G(x)$, of a chemical species, x, is defined as

$$G(x) = \frac{n(x)}{\bar{\varepsilon}} \qquad (10.9)$$

where $n(x)$ is the expectation value of the amount of substance of the species x produced, destroyed, or changed in a system by the mean energy imparted $\bar{\varepsilon}$ to the matter of that system. The SI unit is mol J^{-1}, but the radiation chemical yield is more commonly expressed as the number of units of species x produced per 100 eV of energy absorbed. Note that a radiation chemical yield can be negative if an irradiated compound is converted into other chemical species as a result. The molar concentration, c (in mol m^{-3}), of the reaction products can be determined chemically (using chemical probes or scavengers) by photospectroscopy or by fluorescence measurements, depending on the nature of the chemical species. The absorbed dose to the medium can then be determined as

$$D_{med} = \frac{c}{\rho_{med} G} \qquad (10.10)$$

The chemical yields are mostly not known with the required accuracy from first principles to use chemical dosimeters as primary methods of dosimetry. For the ferrous sulfate dosimeter, for example, the chemical yield is determined indirectly by comparing its response with a calorimeter in a ^{60}Co calibration beam. Furthermore, chemical yields are usually strongly dependent on the energy and type of ionizing particles due to the effect of the ionization density on the chemistry within the tracks. The LET dependence of primary species in water, i.e., those formed 10^{-7} s after the passage of ionizing particles, are shown in Figure 10.7 as determined from pulse radiolysis and scavenger experiments.

10.3 REFERENCE DOSIMETRY IN THE CLINIC

As for all forms of external radiotherapy, beam monitor calibration in terms of reference absorbed dose to water is a crucial first step in ensuring clinical delivery of the correct dose to the target region. This section provides a historic overview and discusses state-of-the-art recommendations for the reference dosimetry of clinical proton beams as well as new developments for the reference dosimetry of scanned proton beams.

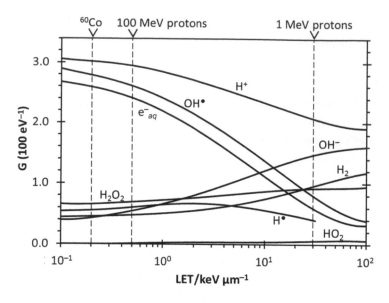

FIGURE 10.7 Chemical yields of primary species 10^{-7} s after the passage of an ionizing particle as a function of LET. (Adapted from Palmans and Vynckier [36] with permission of the publisher and the authors.)

10.3.1 Background—Basic Formalism

To facilitate later the discussion of contributions to differences in dosimetry protocols, the overall expression to derive dose to water from the ionization measurements, obtained from combining Equations (10.7) and (10.8), can be split into three factors as follows:

$$D_{w,Q} = [M_Q] \left[\frac{1}{\rho_{air} V_{cav}} \right] \left[(W_{air}/e)_Q (s_{w,air})_Q \, p_Q \right] \tag{10.11}$$

All protocols for dosimetry of proton beams using ionization chambers and calculated data can be reduced to this factorization in which the second factor, representing an estimate of the ionization chamber volume, is solely related to the calibration conditions, whereas the third factor is solely related to the proton beam.

For example, with the formalism from IAEA TRS-398 [1] using calculated beam quality correction factors, k_Q, the second factor in Equation 10.11 is given by

$$\frac{1}{\rho_{air} V_{cav}} = \frac{N_{D,w,Q_0}}{(W_{air}/e)_{Q_0} (s_{w,air})_{Q_0} \, p_{Q_0}} \tag{10.12}$$

where N_{D,w,Q_0} is the absorbed dose-to-water calibration coefficient in the calibration beam with quality, Q_0, and the quantities in the denominator have the same meaning as quoted earlier but now for the calibration beam quality. The third factor is identical to the one in Equation 10.11, given that the notations used are consistent with those in IAEA TRS-398.

If we take the European Charged Heavy Particle Dosimetry (ECHED) group protocol [37,38] as an example, then the second factor in Equation 10.11 becomes

$$\frac{1}{\rho_{air} V_{cav}} = \frac{N_K (1-g) A_{wall} k_m}{(W_{air}/e)_c} \quad (10.13)$$

where N_K is the air kerma calibration coefficient in a ^{60}Co calibration beam, g is the correction for radiative losses, A_{wall} is the correction factor for absorption and scatter in the chamber wall and buildup cap in ^{60}Co, k_m is the correction factor for the nonair equivalence of the chamber's construction materials and $(W_{air})_c$ is the mean energy required to produce an ion pair in dry air in the calibration beam quality. The third factor would be very similar as in IAEA TRS-398, with the exception that there is no explicit mention of an ionization chamber perturbation correction factor. This is not a de facto difference since in IAEA TRS-398 the assumption is made that p_Q equals unity.

Interesting to note is that there may not be a great difference in uncertainty on the chamber volume as derived by this factor from air kerma or absorbed dose calibrations. However, in the absorbed dose-based protocols, the quantities occurring in the denominator of Equation 10.12 and those in the third factor in Equation 10.11 are the same, except for the difference in beam quality and thus expected to be more strongly correlated than those in the air kerma based approach, where the factors A_{wall} and k_m refer to very different conditions as well as to the buildup cap that is not present with in-phantom measurements. This shows that this factorization is only for illustrative purposes, but that the factors two and three cannot be considered independently.

While it is interesting and necessary to study the influence of the second factor in Equation 10.11 on dosimetry using different dosimetry formalisms, it is not related to the proton beam. Differences in dosimetry related to that factor have been reviewed at length [39] and will not be further discussed. Concerning the third factor in Equation 10.11, each of the constituting quantities deserves separate attention:

- Mean energy required to produce an ion pair in dry air, $(W_{air})_Q$: Although often denoted with a capital "W", sometimes a small "w" is used to clarify that it should be the differential value for the local charged particle spectrum, since protons lose only a fraction of their energy in the ionization chamber cavity (as opposed to photon and electron beams, where the integral value accounts for all energy losses during the complete slowing down of secondary electrons). The W_{air}/e value for protons has been the subject of controversy, since ICRU Report 31 [40] recommended a value of 35.18 J C^{-1}, adopted by the ECHED [37,38], while American Association of Physicists in Medicine (AAPM) Task Group 20 [41] recommended a value of 34.3 J C^{-1}. This discrepancy of 2.6% remained the source of differences in dosimetry recommendations until the publication of ICRU Report 78 [42], which adopted the same value as IAEA TRS-398. There are essentially two ways of measuring this quantity: (i) a simultaneous measurement of the energy loss over an air column and the ionization produced per proton and (ii) by comparing the dose response of an ionization chamber and a calorimeter. The first method is cumbersome and requires a correction for electron losses, which is difficult to determine. The second method has the disadvantage that it does not provide a direct measurement of W_{air}/e, but the pragmatic advantage is that if ionization chamber dosimetry is based on a value derived from calorimetry, it provides consistency with the dosimetry in high-energy photon beams. As discussed in Chapter 9, the W_{air}/e value for protons is energy dependent, but this energy dependence is not very well known. This is mainly of importance for the measurement of depth-dose distributions with ionization chambers.

- Water-to-air mass stopping power ratio, $(s_{w,air})_Q$: The stopping power for protons is governed by the same physics as for electrons and positrons, and the theoretical models for calculations at high energies are based on the same Bethe–Bloch formulae with a series of correction terms. For consistency with photon and electron dosimetry, ICRU Report 49 [33] recommended proton stopping powers using the same values for the mean excitation energy, I, as used in the electron and positron stopping power tables of ICRU Report 37 [43]. Stopping powers for water, graphite, and air have been recently revised based on improved and more abundant experimental data in ICRU Report 90 [3]. An important difference is that compared to electrons, the clinically relevant range of proton energies is at much lower (nonrelativistic) velocities where the density effect is of no importance, but there is a strong energy dependence of the stopping powers. The result is that the water-to-air mass stopping power ratio is fairly constant over the entire clinical energy range. The proton energy dependence of the water-to-air mass stopping power ratio is discussed in Chapter 9, and the standard relative uncertainty is estimated to be 1% [1].

- Ionization chamber perturbation factor, p_Q: The ionization chamber perturbation factor corrects for deviations from Bragg–Gray conditions and can according to IAEA TRS-398 be described as the product of four factors:

    i. a displacement correction factor, p_{dis}, for the deviation of the effective point of measurement from reference point of the ionization chamber;

    ii. a cavity perturbation correction factor, p_{cav}, for the perturbation of the charged particle fluence distribution due to the presence of air cavity;

    iii. a wall perturbation factor, p_{wall}, for the nonwater equivalence of the ionization chamber's wall;

    iv. a central electrode correction factor, p_{cel}, for the presence of the central electrode.

The first one can alternatively be dealt with by positioning the effective point of measurement at the required measurement depth. For proton beams, it is slightly easier to determine an effective point of measurement than for photon beams given the small lateral deflections that protons undergo. A reasonable approximation is thus to regard protons as traveling along straight lines once they enter the ionization chamber geometry and integrate their dose contributions over the cavity volume. For a cylindrical air cavity with radius R_{cyl} in water, it is easy to show that this results in an effective point of measurement that is relative to the center of the cavity positioned a distance $\Delta z_{cyl} = 8R_{cyl}/3\pi \approx 0.85R_{cyl}$

closer to the beam source [44]. For Farmer-type chambers, the higher density of the wall and central electrode materials brings this slightly towards the center of the chamber and closer to the value of $\Delta z_{cyl} \approx 0.75 R_{cyl}$ recommended in IAEA TRS-398 for ion beams [45], but this is not the case for other cylindrical chambers with a thick wall or central electrode for which substantial deviations from this rule may occur. Regarding the other perturbation factors, all the evidence points to corrections of less than 1%, and relative correction factors of 1.005 have been demonstrated both experimentally, by cavity theory and by Monte Carlo simulation for A-150-walled ionization chambers as compared to graphite-walled chambers [46].

Reference dosimetry in clinical beams is usually based on national or international reports providing protocols or codes of practice. This section gives first a historical view followed by a subsection on the most recent recommendation.

10.3.2 History of Reference Dosimetry

The first codes of practice for heavy charged particle dosimetry emerged in the 1980s and early 1990s in AAPM Report 16 produced by the Task Group 20 of the AAPM [41] and ECHED [37,38]. They all had in common that the first recommendation was to use a calorimeter for reference dosimetry. If not available, a Faraday cup could be used as reference instrument or calibrated against a calorimeter or an ionization chamber calibrated in terms of exposure in air or air kerma in a ^{60}Co calibration beam. In AAPM Report 16, the quantity of interest was defined as dose to tissue, and ionization chambers filled with other gasses than air were considered. The third factor of Equation 10.11 thus contained a tissue-to-gas rather than a water-to-air mass stopping power ratio. Also in the original ECHED code of practice, the quantity of interest was defined as dose to tissue, but only air-filled ionization chambers were considered and thus there appeared a tissue-to-air mass stopping power ratio. For both of those two protocols, the second factor in Equation 10.11 was, apart from a few details, similar as what was used in AAPM TG-21 [47] for photon and electron dosimetry, although no consideration was given to the fact that, in the ^{60}Co calibration beam, part of the electron spectrum in the air cavity is not generated in the wall. Both used the stopping power tables of Janni [48]. In the Supplement to the ECHED protocol [38], dose to water was defined as the quantity of interest, and in addition, the stopping powers of ICRU Report 49 [33] were recommended, which was a step in bringing proton dosimetry in better harmony with high-energy photon and electron dosimetry. One of the more substantial differences between the recommendations of the AAPM and the ECHED was, as discussed earlier, the substantial discrepancy of about 2.6% in the W_{air} value for proton beams.

ICRU Report 59 [31] provided a new recommendation that covered both the air kerma and the absorbed dose routes for dosimetry, but in the latter a number of deficiencies of the procedures and data were pointed out [49], including the undocumented derivation of the W_{air} value for proton beams based on published data, the recommendation of using a W_{air} in ambient air rather than dry air, and the omission of a perturbation factor for the ionization chamber in the calibration beam (as in the denominator on the right-hand side of Equation 10.12). IAEA TRS-398 improved on this and provided a recommendation covering ionization chamber dosimetry based on absorbed dose-to-water calibrations for all external beams except neutrons. For protons, the value of W_{air} was obtained from a robust statistical analysis of the available data, and Spencer–Attix water-to-air mass stopping power ratios [50] were used for the first time. Other advantages were better estimates of the uncertainties and better consistency with other external beam modalities regarding the measurement of influence quantities such as recombination and polarity effects. The most recent recommendation in ICRU Report 78 [42] integrally adopts the concepts and data from TRS-398. One of the achievements of ICRU Report 78 was to readdress all the published W_{air} data, correcting a few inconsistencies in earlier used data by ICRU Report 59 and IAEA TRS-398, and to add new data published since 2000. It was decided that for consistency with calorimetry-based data in high-energy photon and electron beams, only the data obtained by calorimetry would be used for a reevaluation [51], resulting in a weighted mean value of $34.15 \pm 0.13\,\mathrm{J\,C^{-1}}$ and a weighted median value of $34.23 \pm 0.13\,\mathrm{J\,C^{-1}}$, both consistent with the value recommended in IAEA TRS-398. A revision of stopping power data combined with new data on perturbation correction factors for photon beams in ICRU Report 90 results in a new estimate of the median value of $34.44 \pm 0.13\,\mathrm{J\,C^{-1}}$.

With the latter three protocols mentioned, the idea of recommending calorimeters in a clinical environment was abandoned, given that calorimeters are too cumbersome and time consuming to operate for routine dosimetry. Nevertheless, the first recommendation of IAEA TRS-398 remains that beam quality correction factors (see Section 10.3.3) are measured using calorimetry for each individual ionization chamber for each individual beam in which they are used, but only a small number of publications have reported beam quality correction factors for proton beams. Beam quality correction factors obtained by comparing ionization chambers with water calorimeters, both in passively scattered proton beams and in scanned proton beams, are compared with calculated data in Figure 10.9 (Section 10.3.3).

10.3.3 Recommendations of IAEA TRS-398 and ICRU Report 78

It is not the aim here to give a comprehensive overview of the code of practice for proton dosimetry in IAEA TRS-398 [1], which has been integrally adopted in ICRU Report 78 [42]. For all details, the reader is referred to those reports. The main steps in the dose determination using ionization chambers calibrated in terms of absorbed dose to water will be briefly outlined, and some important points to pay attention to concerning proton dosimetry will be discussed.

For protons, IAEA TRS-398 only provides recommendations for reference dosimetry at a reference depth in the center of an SOBP, with the exception that the reference depth can be chosen in the plateau region, at a depth of $3\,\mathrm{g\,cm^{-2}}$, for clinical applications with a single-energy proton beam (e.g., for plateau irradiations). The formalism for determination of absorbed dose to water, $D_{w,Q}$, in a proton beam with quality Q is

$$D_{w,Q} = M_Q\, N_{D,w,Q_0}\, k_{Q,Q_0} \tag{10.14}$$

where M_Q is the ionization chamber reading corrected for influence quantities, N_{D,w,Q_0} the absorbed dose-to-water calibration coefficient of the ionization chamber in a calibration beam of quality Q_0, and k_{Q,Q_0} the beam quality correction factor accounting for the use of the calibration coefficient in a different beam quality Q (it thus corrects the reciprocal of the ionization chamber's response in the calibration beam to that in the proton beam).

Due to the limited availability of experimental data for k_{Q,Q_0}, their values are in practice calculated as

$$k_{Q,Q_0} = \frac{(W_{air})_Q (s_{w,air})_Q \, p_Q}{(W_{air})_{Q_0} (s_{w,air})_{Q_0} \, p_{Q_0}}, \tag{10.15}$$

in which all quantities have been defined earlier. The combination of Equations 10.14 and 10.15 leads to the factorization in Equations 10.11 and 10.12 as discussed earlier.

k_{Q,Q_0} values are calculated with Equation 10.15 as a function of the residual range, R_{res}, which serves as the beam quality index and is defined as

$$R_{res} = R_p - z_{ref}, \tag{10.16}$$

where R_p is the practical range, defined as the depth distal to the Bragg peak at which the dose is reduced to 10% of its maximum value and z_{ref} is the reference depth for the measurement of the reference absorbed dose to water. The reference depth is defined at the center of the SOBP, but for the exception of clinical applications with a single-energy proton beam, mentioned earlier, it can also be at a water equivalent depth of $3 \, g \, cm^{-2}$ in the plateau region. R_{res} is related to the most probable energy of the highest-energy proton peak in the spectrum.

Concerning the correction for influence quantities of the ionization chamber reading the corrections for atmospheric conditions (deviations from normal pressure, temperature, and relative humidity at which the calibration coefficient is valid) and electrometer calibration are the same as in all modern protocols for high-energy photon and electron beams and will not be discussed here. The correction for polarity effects and ion recombination are measured in the same way as well, but an important issue to be considered is if a proton beam should be regarded as pulsed or continuous with respect to ion recombination, since this makes a difference in deriving the correction factor from the traditional two-voltage method. IAEA TRS-398 mentions that beams extracted from a synchrotron may have to be regarded as continuous. However, it has been demonstrated that proton beams from a cyclotron, which are inherently pulsed, should be regarded as continuous as well for this purpose given the short time interval between pulses compared with the ion collection time [52]. In high-dose-rate beams like those used in ocular treatments, the error on the ion recombination factor could be up to 2% when applying the two-voltage method with the inappropriate assumption that the beam is pulsed. For scanned proton beams, recent work has demonstrated that recombination can result in large corrections when the beam is pulsed, as is the case when accelerated using a synchrocyclotron [53], while for continuous beams from a cyclotron or long pulse trains from a slow-extraction synchrotron recombination, corrections can be kept sufficiently small [54]. The generic solution is to operate at a higher polarizing voltage but can have the unwanted side effect of a substantial charge multiplication effect that distorts the ideal-model dependence of the ionization charge to the polarizing voltage. Applying the two-voltage method in that condition results in an overestimation of the recombination correction; the two-voltage method should then be applied using two lower voltages as illustrated in Figure 10.8 for a pulsed scanned proton beam.

The calibration coefficient N_{D,w,Q_0} is usually obtained in a ^{60}Co calibration beam at a standards laboratory and in that case IAEA TRS-398 omits the index Q_0 for both the calibration coefficient and the beam quality correction factor in Equation 10.15:

$$D_{w,Q} = M_Q \ N_{D,w} \ k_Q \tag{10.17}$$

k_Q values for most commonly used ionization chambers have been tabulated as a function of R_{res} in IAEA TRS-398 and are for a selection shown in Figure 10.9. It is clear that the variation with beam quality is rather limited. Due to the assumption in IAEA TRS-398 that the perturbation correction factor p_Q is unity for all ionization chambers, the differences between ionization chamber types is solely due to differences in the perturbation correction factors for ^{60}Co. Figure 10.9 also shows k_Q values obtained by comparing ionization chambers with water calorimeters, both in passively scattered proton beams [12,55] and in scanned proton beams [24,25].

Comparing the results for the same chamber types in scattered and scanned beams does not reveal any significant differences, indicating that the same k_Q value can be used for both scattered and scanned beams in spite of the different ways of delivering the beams.

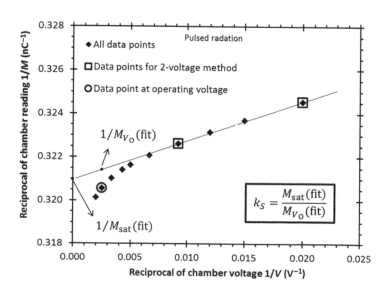

FIGURE 10.8 Illustration of the determination of the recombination correction factor, k_s, in a pulsed pencil beam scanning field in the presence of charge multiplication at the operating voltage.

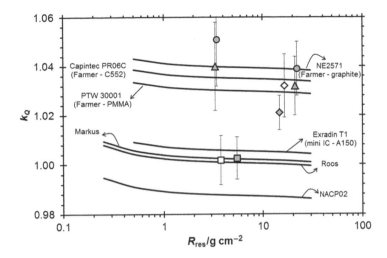

FIGURE 10.9 k_Q values from IAEA TRS-398 (full lines) as a function of the residual range for three Farmer-type ionization chambers, one mini-ionization chamber and three types of plane-parallel ionization chambers. The data points are most of the reported k_Q values measured using water calorimetry (illustrating how scarce such experimental data are); the open symbols represent results in scanned beams, whereas the full symbols represent results in scattered beams. Squares are for the Exradin T1 (Sarfehnia et al. [24]), diamonds are for the NE2571 (Medin et al. [12] and Medin [25]), triangles for the PTW-30001 (Vatnitsky et al. [55]), and circles for the Capintec PR6G (Vatnitsky et al. [55]). The error bars represent standard uncertainties showing that the experimental data are in good agreement with the IAEA TRS-398 data, which themselves have a standard relative uncertainty of 1.7%.

The same conclusion was reached theoretically by Sorriaux et al. [56] based on Monte Carlo simulations. However, as will be discussed in Section 10.3.4, reference dosimetry for scanned beams is mostly performed in the entrance plateau where, depending on the energy, the dose gradient can be substantial and the assumption that $p_{dis} = 1$, which is made in the calculation of k_Q values in IAEA TRS-398, is not valid. The relative displacement correction in beams with energies below 100 MeV can amount to several percent [57]. Vice versa, Monte Carlo calculated k_Q values obtained for single-energy proton beams with the ionization chamber simulated with its center at the reference depth will implicitly include the nonunity displacement correction factor, such that they are not valid for measurements in the SOBP. As it happened in a recent study, Monte Carlo-calculated perturbation k_Q values at shallow depth in a single-energy beam have been published [58], but none have been reported in the SOBP.

Since the k_Q values in IAEA TRS-398 have been calculated with Equation 10.15 using stopping power data for protons in air and water taken from ICRU Report 49 and for ^{60}Co from ICRU Report 37, it could be reasonable to expect that the k_Q values will change when the new stopping power data from ICRU Report 90 are taken into account. However, the value of $(W_{air})_Q$ for proton beams is not independent of the mass stopping power ratios in the formula. Indeed, if the mass stopping power ratios would change, then the evaluation of $(W_{air})_Q$ based on calorimetry data as performed by Jones [51] should be repeated using the new stopping power data. The net result is that the calculated k_Q values are unaffected by the changes in stopping power data, as has been discussed by Andreo et al. [59].

IAEA TRS-398 provides data for both cylindrical and plane-parallel ionization chambers but preference is given to cylindrical ionization chambers as plane-parallel ionization have been reported in the past of having less reliable values of perturbation correction factors in ^{60}Co. Nevertheless, the use of a plane-parallel ionization chamber can be unavoidable, such as for example in an SOBP in a low-energy beam with very limited extension. In that case, if there is any concern about the accuracy of the plane-parallel chamber's tabulated k_Q value, the chamber could be cross-calibrated against a cylindrical ionization chamber in a high-energy proton beam with beam quality Q_{cross}. The cross-calibration coefficient $N_{D,w,Q_{\text{cross}}}$ for the plane-parallel ionization chamber is then obtained as

$$N_{D,w,Q_{\text{cross}}}^{\text{PP}} = \frac{M_{Q_{\text{cross}}}^{\text{cyl}} \; N_{D,w,Q_0}^{\text{cyl}} \; k_{Q_{\text{cross}},Q_0}^{\text{cyl}}}{M_{Q_{\text{cross}}}^{\text{PP}}} \tag{10.18}$$

and absorbed dose to water in the proton beam to be calibrated as

$$D_{w,Q} = M_Q^{\text{PP}} \; N_{D,w,Q_{\text{cross}}}^{\text{PP}} \; k_{Q,Q_{\text{cross}}}^{\text{PP}}. \tag{10.19}$$

$k_{Q,Q_{\text{cross}}}^{\text{PP}}$ can be obtained from the tabulated data in IAEA TRS-398 as

$$k_{Q,Q_{\text{cross}}}^{\text{PP}} = \frac{k_Q^{\text{PP}}}{k_{Q_{\text{cross}}}^{\text{PP}}}, \tag{10.20}$$

The advantage of this approach is that $k_{Q,Q_{\text{cross}}}^{\text{PP}}$ values are close to unity and do not vary much with energy, as can be seen in Figure 10.9 when normalizing the curves at any cross-calibration data point with $R_{\text{res}} > 2\,\text{g cm}^{-2}$. For that beam quality range, the values of $k_{Q,Q_{\text{cross}}}^{\text{PP}}$ are then not more than 0.2% different from unity for any plane-parallel ionization chamber and any clinical proton beam quality.

Currently, the relative standard uncertainty on the reference absorbed dose to water using an ionization chamber calibrated in terms of absorbed dose to water in a ^{60}Co calibration beam or cross calibrated against a cylindrical chamber in a high-energy proton beam is around 2%. This overall uncertainty is dominated by the uncertainty of k_Q values via the uncertainty on the water-to-air mass stopping power ratio, the value of $(W_{\text{air}})_Q$, and the values of perturbation correction factors in both the proton and calibration beam.

10.3.4 Beam Monitor Calibration for Scanned Beams

While scanned proton beams are mentioned in IAEA TRS-398, no dedicated guidance is provided, and reference dosimetry is recommended to be performed in the SOBP as for other proton beams generated by, e.g., double-scattering systems or wobbling systems.

However, the way dose calculations are performed for scanned proton beams requires the monitor chambers to be calibrated either in terms of the number of protons, $n_p(0)$, incident on the phantom surface for the narrow beam which is being scanned or its equivalent in terms of energy deposition, the DAP to water at the reference depth, $\mathrm{DAP}_w(z_{\mathrm{ref}})$. The relation between $n_p(0)$ and $\mathrm{DAP}_w(z_{\mathrm{ref}})$ has already been discussed in Section 10.2.2, establishing the link between Faraday cup measurements and the determination of DAP to water. Since IAEA TRS-398 only provides procedures for the dosimetry of broad fields, either an alternative method has to be used to determine $\mathrm{DAP}_w(z_{\mathrm{ref}})$ or the relation between the absorbed dose to water determined at a point in a broad beam and the quantity of interest has to be established.

For the first method, which provides a direct determination of $\mathrm{DAP}_{w,Q}(z_{\mathrm{ref}})$ for the proton beam with beam quality Q, the reference field is a single static pencil beam. The measurement is performed with a large-area plane-parallel ionization chamber (LAIC) at a shallow depth. Provided the diameter of the collecting electrode is sufficiently large, it can be assumed that all lateral contributions to the energy deposition are integrated (except for the small contributions by neutrons and prompt gammas). In practice, this assumption is not fulfilled, as the size of the collecting electrode of commercially available chambers is not large enough, and the measured value should be corrected for the part of the lateral profile outside of the collecting diameter of the LAIC. This requires either Monte Carlo simulations or measurements with detectors that should be able to cover a very wide range of dose levels (four orders of magnitude). Gottschalk et al. [60] presented a method scanning microionization chambers at multiple offsets from the central axis of a single pencil beam for this purpose.

The use of an LAIC for the determination of $\mathrm{DAP}_w(z_{\mathrm{ref}})$ has been described by Gillin et al. [61]. A formalism, similar as that of IAEA TRS-398, has been introduced to describe the details of this method [62,63]. Using an absorbed DAP to water calibration coefficient N_{DAP,w,Q_0} for the LAIC, $\mathrm{DAP}_{w,Q}(z_{\mathrm{ref}})$ according to this formalism is determined as

$$\mathrm{DAP}_{w,Q}(z_{\mathrm{ref}}) = M_Q \, N_{\mathrm{DAP},w,Q_0} \kappa_{Q,Q_0}. \tag{10.21}$$

The symbol κ_{Q,Q_0} is used here for denoting the beam quality correction factor to emphasize that it corrects a different quantity than the conventional beam quality correction factor k_{Q,Q_0}. κ_{Q,Q_0} is defined as a ratio of DAP to water calibration coefficients in the narrow proton beams with beam qualities Q and the calibration beam with beam quality Q_0:

$$\kappa_{Q,Q_0} = \frac{N_{\mathrm{DAP},w,Q}}{N_{\mathrm{DAP},w,Q_0}} \tag{10.22}$$

At present, the problem with the application of these equations is that calibration coefficients in terms of DAP_w are not provided by Standards laboratories. Additionally, even if such calibrations would become available, then it remains a problem that κ_{Q,Q_0} values are not known for commercially available LAICs. The only option is then to cross-calibrate the LAIC against a conventional ionization chamber [62,63], which can be either a cylindrical or plane-parallel ionization chamber on the central axis of a broad proton beam with cross-calibration beam quality Q_{cross}:

$$N_{DAP,w,Q_{cross}}^{LAIC} = \frac{M_{Q_{cross}}^{REF} \, N_{D,w,Q_0}^{REF} \, k_{Q_{cross},Q_0}^{REF}}{M_{Q_{cross}}^{LAIC}} \times \iint_A OAR(x,y)\,dx\,dy, \tag{10.23}$$

where $OAR(x,y)$ is the off-axis ratio at the reference depth and A the area of the collecting electrode of the LAIC. If the lateral beam profile is uniform over the area of the ionization chamber, then $\iint_A OAR(x,y)\,dx\,dy = A$. DAP to water is then obtained as

$$DAP_{w,Q} = M_Q^{LAIC} \, N_{DAP,w,Q_{cross}}^{LAIC} \, \kappa_{Q,Q_{cross}}^{LAIC}. \tag{10.24}$$

where $\kappa_{Q,Q_{cross}}^{LAIC}$ can be assumed to be unity since both Q and Q_{cross} refer to proton beam qualities.

Apart from the current lack of knowledge of beam quality correction factors for such a DAP formalism, an additional issue is that commercial LAIC may exhibit a response, which is not uniform over the collecting area. This further complicates the Equations (10.21), (10.23), and (10.24) as worked out by Palmans and Vatnitsky [63]. Kuess et al. [64] have investigated the effect of these nonuniformities, concluding that the correction factors are substantial compared to other components contributing to the overall uncertainty of DAP determination.

An alternative, less direct, method that has been proposed by Hartmann et al. [65] is based on the reciprocal relation between DAP and dose as discussed in Section 10.2.2. The reference field for this approach is a single-layer scanned field, of sufficient size to establish lateral charged particle equilibrium at the central axis in a time-averaged sense, a constant number of MUs delivered per spot and equidistant lateral spot spacing in two dimensions. In this field, $D_{w,Q}$ is determined at shallow depth using IAEA TRS-398 (or alternative code of practice) for broad beam dosimetry. DAP to water is then derived as

$$DAP_{w,Q} = D_{w,Q} \, \Delta x \Delta y, \tag{10.25}$$

where Δx and Δy are the constant spot spacings in both orthogonal directions perpendicular to the beam axis.

The dose calculation algorithms in treatment planning systems for scanned proton beams (see Chapter 14) will determine the number of particles for each spot with incident nominal energy E_0, which will then have to be translated to a number of MUs or monitor counts via a calibration function of the form

$$K(E_0) = n_p(0, E_0)/MU \tag{10.26}$$

where $n_p(0, E_0)$ is the number of protons incident on the phantom surface derived from the $DAP_{w,Q}(z_{ref}, E_0)$ determined at the reference depth for the nominal incident energy E_0

using either Equation 10.21, 10.24, or 10.25. Note that z_{ref} may depend on E_0. The derivation of $n_p(0, E_0)$ from $DAP_{w,Q}(z_{ref}, E_0)$ has the form

$$n_p(0, E_0) = \frac{DAP_{w,Q}(z_{ref}, E_0)}{\left[\sum_i \int_A \int_0^{E_{max,i}} \Phi_{E,i}(z_{ref}, x, y) \frac{S}{\rho}(E, i) \, dE \, dx \, dy / n_p(0, E_0) \right]}. \tag{10.27}$$

The expression in square brackets is nothing else than the DAP to water per incident proton for a single spot at the phantom surface, so at first sight, this equation may appear self-evident. But in the common interpretation and use of this expression, the numerator on the right-hand side of the equation is the experimentally determined DAP, while the denominator is the DAP per incident proton known by other means, e.g., a Monte Carlo simulation or a dose calculation engine in a TPS. For a practical, pragmatic implementation, it is important that the dose calculation approach for routine clinical treatment planning uses the same conversion factor or data to calculate doses from a given number of particles per spot so that consistency between the TPS and the monitor calibration using Equation 10.21, 10.24, or 10.25 is ensured. To avoid any confusion with this interpretation, many planning systems will directly use the experimentally determined $DAP_{w,Q}(z_{ref}, E_0)$ values and calculate the calibration curve within the dose calculation engine to ensure that the same $n_p(0, E_0)/DAP_{w,Q}(z_{ref}, E_0)$ ratio is used to establish the calibration curve as is used in the dose calculation. The monitor calibration function could then be written as

$$K(E_0) = \left[\frac{n_p(0, E_0)}{DAP_{w,Q}(z_{ref}, E_0)} \right] \left[\frac{DAP_{w,Q}(z_{ref}, E_0)}{MU} \right], \tag{10.28}$$

where the second factor in square brackets is the experimentally determined DAP to water per MU for a single spot and the first factor in square brackets the $n_p(0, E_0)/DAP_{w,Q}(z_{ref}, E_0)$ ratio used in the dose calculation, which could have, in principle, any arbitrary value. It is only when one aims to compare this ratio with an independent determination of the number of protons using a Faraday cup or activation measurement, as discussed in Section 10.2.2, that its accurate determination is important and thus that accurate values of the charged particle fluence distribution and mass stopping powers in Equation 10.27 are essential.

Figure 10.10 shows, as an example, the beam monitor calibration curve as determined for one of the beam monitors at MedAustron [66] as well as the expression in square brackets in the denominator in Equation 10.27 calculated by Geant4 Monte Carlo simulations using a detailed beam model for the clinical beams at MedAustron [67]. The functional behavior of the beam monitor calibration curve agrees well with the reciprocal of the mass stopping power of nitrogen gas. Small deviations could be due to range shifts and secondary electron perturbations in the monitor chamber as well as uncertainties in the derivation of the number of particles from DAP. The reciprocal of the ratio $n_p(0, E_0)/DAP_{w,Q}(z_{ref}, E_0)$ (simply denoted DAP_w/n in the figure) shows a step at 97.4 MeV, which is the result of switching to a different measurement depth. The discrepancy between DAP_w/n and the mass stopping powers of water

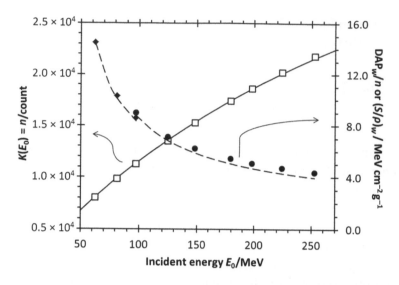

FIGURE 10.10 Example of an experimental beam monitor calibration curve (open squares) and the DAP per incident proton calculated by Monte Carlo (solid symbols) using data from Dreindl et al. [66] and Elia et al. [67]. The beam monitor calibration curve is compared with the reciprocals of the ICRU Report 49 mass stopping power of nitrogen gas (full line—normalized at the 179.2 MeV data point), with which it should agree apart from range shifts and perturbation effects of the monitor chambers. The DAP per incident proton is compared with the ICRU Report 90 mass stopping power of water for monoenergetic protons with the same CSDA range as the residual range at the measurement depth (dashed lines).

from ICRU Report 90, which is growing with increasing incident energy, is due to the contribution of secondary charged particles created in nonelastic nuclear interactions to the DAP.

While the preceding paragraphs show that the beam monitor calibration for scanned beams requires a substantially different approach, the conventional approach of reference dosimetry in the center of the SOBP can still be relevant for scanned beams as well. For beams where range modulation is performed downstream the beam monitor, it can be directly calibrated using reference dosimetry in the SOBP [68]. For scanned beams where the range or energy modulation is performed before the beam monitor chamber, dosimetry in the SOBP is normally performed with the purpose of dose verification. Nevertheless, if it is assumed that the beam monitor calibration function in Equation 10.26 is determined accurately in relative terms as a function of energy but not necessarily in amplitude, an overall multiplicative factor to the calibration function can be derived from reference dosimetry in the SOBP [63].

10.4 DOSIMETRIC INTERCOMPARISON AND REFERENCE DOSIMETRY AUDIT

In the 1980s and 1990s, when the number of proton beam radiotherapy facilities was very limited, dose comparison exercises by the medical physicists were deemed an adequate way of ensuring dosimetric consistency across all participating centers. In the new millennium, the number of proton therapy centers has increased almost exponentially, which has made it impractical to organize dosimetric intercomparisons. Furthermore, the complexity of treatments has increased substantially with the intensity-modulated proton therapy capability of scanned proton beams becoming standard. In this context, dosimetric auditing becomes a more efficient way of ensuring consistency between centers that are, e.g., participating in clinical trials. This section gives an overview of the dosimetric intercomparison exercises performed in the past and the recently developed reference dosimetry audits.

10.4.1 Dosimetric Intercomparisons

In the early 1990s, a few small-scaled dose intercomparison exercises were performed by the proton centers in Clatterbridge (United Kingdom), Orsay (France), Nice (France), Louvain-la-Neuve (Belgium), Tsukuba (Japan), and Cape Town (South Africa) [69–71]. All three of these studies were based on A-150-walled ionization chambers calibrated in terms of air kerma and applying the original ECHED protocol [37]. The first two of those studies were for low-energy proton beams used for ocular tumors, while the third one was performed in a high-energy beam. Overall, these studies concluded that clinically acceptable dosimetric consistency within 3% was achieved for both modulated and unmodulated proton beams using air kerma calibration coefficients determined in the centers where the comparison studies were performed. The study in which the air kerma calibration coefficients from the home countries of the institute were considered [69] didn't observe a significant difference between the two sets of calibration data. Nevertheless, issues in terms of the stability of the cavity volume reported with A-150 walled chambers due to the hygroscopic properties of this wall material may have influenced the standard deviations observed in those studies. These studies also indicated various degrees of agreement between ionization chamber based dosimetry and Faraday cup based dosimetry, ascribed to the potential uncertainties of Faraday cup measurements related to the determination of the beam area and corrections for low-energy proton contamination. This led to the conclusion adopted in later codes of practice not to recommend Faraday cups for reference dosimetry.

By the second half of the 1990s, the number of proton centers had grown considerably and two dosimetry intercomparison of a larger scale were held [72,73]. These revealed larger inconsistencies in the dosimetry performed by different centers around the world, mainly due to the use of different protocols, instrumentation, and data. The earlier of these studies was based on calibrations in terms of exposure or air kerma using the AAPM TG-20 Report [41] and the Supplement to the ECHED protocol [38], and it showed the importance of using a common dosimetry protocol with uniform data to improve dosimetric consistency. When using institution-specific procedures and data, the maximum deviation of reference dose from the average was 3%. When using a common dosimetry protocol, this improved to 1.5%. The second study was performed after the publication of a new protocol in ICRU Report 59, demonstrating that by using this protocol similar consistency could be reached. Since then, no comparably comprehensive dose intercomparison exercises have been performed. This could be partly attributed to the publication of IAEA TRS-398, based entirely on the use of ionization chambers calibrated in terms of absorbed dose to water, which has been implemented by most centers in the world. Also the ICRU in its Report 78 has moved to a consistent recommendation by largely adopting the formalisms and data of

IAEA TRS-398. Another possible reason is that the number of centers has grown to such a level that intercomparison exercises have become impractical to organize.

10.4.2 Reference Dosimetry Audit

A reference dosimetry audit ensures that the traceability of dosimetry to primary standards is adequately and consistently implemented across centers. The first example of such an audit has been described by Moyers et al. [74] and was performed at eight proton therapy centers in North America in collaboration with the Radiological Physics Center (M.D. Anderson Cancer Center, Houston, TX). This audit forms part of the Imaging and Radiation Oncology Core (IROC)'s credentialing for participation in clinical trials with proton therapy and consisted of on-site dosimetry with an Exradin P11 parallel-plate ionization chamber or an Exradin A12 Farmer-type ionization chamber in one of the reference fields of each center. The reference fields were all box fields with a diameter or square field size of at least 10 cm and modulation widths between 6 and 12 cm. Dosimetry was also performed using the cross-calibrated four central ionization chambers of a MatriXX Evolution (IBA Dosimetry, Schwarzenbruck, Germany) array in up to five clinical fields per center. In all cases, the dose to water was expressed per MU and compared with the value of this quantity as reported by the hospital.

The dosimetry by the audit group followed three different procedures as published in

i. ICRU Report 59 using a calibration coefficient in terms of air kerma,

ii. ICRU Report 59 using a calibration coefficient in terms of absorbed dose to water,

iii. ICRU Report 78 using a calibration coefficient in terms of absorbed dose to water (as mentioned earlier, the recommendations in ICRU Report 78 are identical to those in IAEA TRS-398).

Using ICRU Report 78, the mean difference between the audit dose and reported dose was 0.3% with a standard deviation of 1.6%, confirming the adequate traceability and implementation of dosimetric procedures, since most centers currently use ICRU Report 78 for their reference dosimetry. Also using ICRU Report 59 with air kerma calibration coefficients (correcting, however, for the errors in this procedure reported by Medin et al. [49]), a similar level of agreement was found. On the other hand, using ICRU Report 59 with absorbed dose-to-water calibration coefficients, the mean deviation was 3.4%, suggesting that this protocol should not be used in combination with the two others for clinical trials. Overall, the study of Moyers et al. [74] shows that dosimetric auditing is feasible for proton therapy and can aid in ensuring dosimetric consistency for multi-institutional clinical trials.

A European Radiation Dosimetry Group (EURADOS) working group has investigated the suitability of four types of dosimeters (of which the application to protons is discussed in Chapter 9), all calibrated in terms of absorbed dose to water in Co-60, for a mailed reference dosimetry audit in the center of box fields [75]. The study was performed in low-energy beams used for the treatment of eye tumors. Two independent alanine/EPR dosimetry systems exhibited identical behavior, showing an average under response of 5% (varying between 4% and 6%, increasing towards the distal edge) for all modulation widths and ranges and very good reproducibility below 1%. The only disadvantage is the sensitivity of the alanine detector that requires doses of 10 Gy to be delivered for reaching this reproducibility. Optically stimulated luminescent detectors showed a similar but slightly larger under response of 7% with a clear dependence on the residual range. The uncertainty of the system in terms of detector positioning in the reader was not as well established as for alanine systems. Radiophotoluminescent detectors and, even more so, ThermoLuminescent Dosimeters (TLD) exhibit larger corrections and larger dependence on the residual range. This energy dependence needs to be studied in greater detail to make them suitable for audit. One of the motivations of using TLD though is the legal obligation to perform a reference dosimetry audit in Poland prior to start of treatments and the already established use of TLD for dosimetric audit in high-energy photon therapy [75].

10.5 CONCLUSION

This chapter provides an overview of the state-of-the-art of reference dosimetry for clinical proton beams. An overview of the requirements for reference dosimetry and of the primary standards and reference instruments used to establish the chain of traceable reference dosimetry from the standard dosimetry laboratory to the proton therapy center is presented. Current procedures and recommendations for reference dosimetry in the clinic are discussed at great length with a separate section describing the current status of and future needs for accurate reference dosimetry of scanned proton beams. The last section in this chapter reviews dosimetric intercomparisons and reference dosimetry audits that have been performed to ensure consistent reference dosimetry among proton therapy centers worldwide.

REFERENCES

1. Andreo, P., Burns, D. T., Hohlfeld, K., Huq, M. S., Kanai, T., Laitano, F., Smyth, V. G., and Vynckier, S. 2000. *Absorbed Dose Determination in External Beam Radiotherapy – An International Code of Practice for Dosimetry Based on Standards of Absorbed Dose to Water.* IAEA Technical Report Series No. 398, Vienna: International Atomic Energy Agency.
2. Palmans, H. 2011. Dosimetry. In *Proton Therapy Physics*, ed. H. Paganetti, 191–219. London: CRC Press.
3. ICRU. 2016. *Key Data for Ionizing-Radiation Dosimetry: Measurement Standards and Applications.* ICRU Report 90. Oxford: Oxford University Press.
4. Seuntjens, J. and Duane, S. 2009. Photon absorbed dose standards. *Metrologia* 46:S39–58.
5. McEwen, M. R. and DuSautoy, A. R. 2009. Primary standards of absorbed dose for electron beams. *Metrologia* 46:S59–79.
6. Klassen, N. V. and Ross, C. K. 1997. Water calorimetry: the heat defect. *J. Res. Natl. Inst. Stand. Technol.* 102:63–71.
7. Klassen, N. V. and Ross, C. K. 2002. Water calorimetry: a correction to the heat defect calculations. *J. Res. Natl. Inst. Stand. Technol.* 107:171–8.
8. Ross, C. K. and Klassen, N. V. 1996. Water calorimetry for radiation dosimetry. *Phys. Med. Biol.* 41:1–29.
9. Elliot, A. J. 1994. Rate constants and G-values for the simulation of the radiolysis of light water over the range 0–300°C, Technical Report AECL-11073. Chalk River, Ontario, Canada: Atomic Energy of Canada Ltd.

10. Palmans, H., Seuntjens, J., Verhaegen, F., Denis, J. M., Vynckier, S., and Thierens, H. 1996. Water calorimetry and ionization chamber dosimetry in an 85-MeV clinical proton beam. *Med. Phys.* 23:643–50.

11. Sassowsky, M., and Pedroni, E. 2005. On the feasibility of water calorimetry with scanned proton radiation. *Phys. Med. Biol.* 50:5381–400.

12. Medin, J., Ross, C. K., Klassen, N. V., Palmans, H., Grusell, E., and Grindborg, J. E. 2006. Experimental determination of beam quality factors, k_Q, for two types of Farmer chamber in a 10 MV photon and a 175 MeV proton beam. *Phys. Med. Biol.* 51:1503–21.

13. Palmans, H., Kacperek, A., and Jäkel, O. 2009. Hadron dosimetry. In *Clinical Dosimetry Measurements in Radiotherapy*, eds. D. W. O. Rogers and J. E. Cygler, 669–722. Madison, WI: Medical Physics Publishing.

14. Karger, C. P., Jäkel, O., Palmans, H., and Kanai, T. 2010. Dosimetry for ion beam radiotherapy. *Phys. Med. Biol.* 55:R193–234.

15. Brede, H. J., Hecker, O., and Hollnagel, R. 1997. Measurement of the heat defect in water and A-150 plastic for high-energy protons, deuterons and α-particles. *Radiat. Prot. Dosim.* 70:505–8.

16. Bewley, D. K., McCullough, E. C., Page, B. C., and Sakata, S. 1972. Heat defect in tissue-equivalent radiation calorimeters. *Phys. Med. Biol.* 17:95–6.

17. Bewley, D. K. and Page, B. C. 1972. Heat defect in carbon calorimeters for radiation dosimetry. *Phys. Med. Biol.* 17:584–5.

18. Fleming, D. M. and Glass W. A. 1969. Endothermic processes in tissue-equivalent plastic. *Radiat. Res.* 37:316–22.

19. McDonald, J. C. and Goodman, L. J. 1982. Measurements of the thermal defect for A-150 plastic. *Phys. Med. Biol.* 27:229–33.

20. Schulz, R. J., Venkataramanan, N., and Huq, M. S. 1990. The thermal defect of A-150 plastic and graphite for low-energy protons. *Phys. Med. Biol.* 35:1563–74.

21. IAEA. 2000. Stored energy and the thermo-physical properties of graphite. In Irradiation damage in graphite due to fast neutrons in fission and fusion systems, IAEA Technical Document 1154, Vienna, Austria: International Atomic Energy Agency.

22. Palmans, H. 2000. Experimental verification of simulated excess heat effects in the sealed water calorimeter. In *Proceedings of NPL Workshop on Recent Advances in Calorimetric Absorbed Dose Standards*, ed. A. J. Williams, and K. R. Rosser, Centre for Ionising Radiation Metrology report CIRM 42, 74-84. Teddington, United Kingdom: National Physical Laboratory.

23. Renaud, J., Rossomme, S., Sarfehnia, A., Vynckier, S., Palmans, H., Kacperek, A., and Seuntjens, J. 2016. Development and application of a water calorimeter for the absolute dosimetry of short-range particle beams. *Phys. Med. Biol.* 61:6602–19.

24. Sarfehnia, A., Clasie, B., Chung, E., Lu, H. M., Flanz, J., Cascio, E., Engelsman, M., Paganetti, H., and Seuntjens, J. 2010. Direct absorbed dose to water determination based on water calorimetry in scanning proton beam delivery. *Med. Phys.* 37:3541–50.

25. Medin, J. 2010. Implementation of water calorimetry in a 180 MeV scanned pulsed proton beam including an experimental determination of k_Q for a Farmer chamber. *Phys. Med. Biol.* 55:3287–98.

26. Palmans, H., Thomas, R., Simon, M., Duane, S., Kacperek, A., Dusautoy, A., and Verhaegen, F. 2004. A small-body portable graphite calorimeter for dosimetry in low-energy clinical proton beams. *Phys. Med. Biol.* 49:3737–49.

27. Petrie, L. 2016. Characterisation of a graphite calorimeter in scanned proton beams. *PhD Thesis*, University of Surrey, United Kingdom.

28. Palmans, H., Al-Sulaiti, L., Andreo, P., Shipley, D., Lühr, A., Bassler, N., Martinkovič, J., Dobrovodský, J., Rossomme, S., Thomas, R. A. S., and Kacperek, A. 2013. Fluence correction factors for graphite calorimetry in a low-energy clinical proton beam: I. Analytical and Monte Carlo simulations. *Phys. Med. Biol.* 58:3481–99.

29. Al-Sulaiti, L., Shipley, D., Thomas, R., Kacperek, A., Regan, P., and Palmans, H. 2010. Water equivalence of various materials for clinical proton dosimetry by experiment and Monte Carlo simulation. *Nucl. Instrum. Methods Phys A* 619:344–7.

30. Lourenço, A. M., Thomas, R., Bouchard, H., Kacperek, A., Vondracek, V., Royle, G., and Palmans, H. 2016 Experimental and Monte Carlo studies of fluence corrections for graphite calorimetry in low and high-energy clinical proton beams. *Med. Phys.* 43:4122–32.

31. ICRU. 1998. *Clinical Proton Dosimetry Part I: Beam Production, Beam Delivery and Measurement of Absorbed Dose.* ICRU Report 59. Bethesda, MD: International Commission on Radiation Units and Measurements.

32. ICRU. 1984. *Radiation Dosimetry: Electron Beams with Energies between 1 and 50 MeV.* ICRU Report 35. Bethesda, MD: International Commission on Radiation Units and Measurements.

33. ICRU. 1993. *Stopping Powers and Ranges for Protons and Alpha Particles.* ICRU Report 49. Bethesda, MD: International Commission on Radiation Units and Measurements.

34. Nichiporov, D. 2003. Verification of absolute ionization chamber dosimetry in a proton beam using carbon activation measurements. *Med. Phys.* 30:972–8.

35. Paganetti, H. 2006. Monte Carlo calculations for absolute dosimetry to determine machine outputs for proton therapy fields. *Phys. Med. Biol.* 51:2801–12.

36. Palmans, H. and Vynckier, S. 2002. Reference dosimetry for clinical proton beams. In *Recent Developments in Accurate Radiation Dosimetry*, eds. J. P. Seuntjens and P. N. Mobit, 157–94. Madison, WI: Medical Physics Publishing.

37. Vynckier, S., Bonnett, D. E., and Jones. D. T. L. 1991. Code of practice for clinical proton dosimetry. *Radiother. Oncol.* 20:53–63.

38. Vynckier, S., Bonnett, D. E., and Jones. D. T. L. 1994. Supplement to the code of practice for clinical proton dosimetry. *Radiother. Oncol.* 32:174–9.

39. Huq M. S., and Andreo, P. 2004. Advances in the determination of absorbed dose to water in clinical high-energy photon and electron beams using ionization chambers. *Phys. Med. Biol.* 49:R49–104.

40. ICRU. 1979. *Average Energy Required to Produce an Ion Pair.* ICRU Report 31. Washington, DC: International Commission on Radiation Units and Measurements.

41. AAPM. 1986. Protocol for heavy charged-particle therapy beam dosimetry. A report of Task group 20 Radiation Therapy Committee. AAPM Report 16, New York: American Institute of Physics for the American Association of Physicists in Medicine.

42. ICRU. 2008. *Prescribing, Recording, and Reporting Proton-Beam Therapy.* ICRU Report 78. Bethesda, MD: International Commission on Radiation Units and Measurements.

43. ICRU. 1984. *Stopping Powers for Electrons and Positrons.* ICRU Report 37. Bethesda, MD: International Commission on Radiation Units and Measurements.

44. Palmans, H. and Verhaegen, F. 1998. Monte Carlo study of fluence perturbation effects on cavity dose response in clinical proton beams. *Phys. Med. Biol.* 43:65–89.

45. Palmans, H. 2006. Perturbation factors for cylindrical ionization chambers in proton beams. Part I: corrections for gradients. *Phys. Med. Biol.* 51:3483–501.

46. Palmans, H. 2011. Secondary electron perturbations in Farmer type ion chambers for clinical proton beams. In *Standards, Applications and Quality Assurance in Medical Radiation Dosimetry – Proceedings of an International Symposium*, Vienna 9–12 November 2010 – Vol. 1, 309–17. Vienna, Austria: International Atomic Energy Agency.

47. AAPM. 1983. Task group 21: A protocol for the determination of absorbed dose from high-energy photon and electron beams. *Med. Phys.* 10:741–71.

48. Janni, J. F. 1982. Proton range-energy tables 1 keV–10 Gev. *Atom. Data Nucl. Data Tables* 27:147–529.

49. Medin, J., Andreo, P., and Vynckier, S. 2000. Comparison of dosimetry recommendations for clinical proton beams. *Phys. Med. Biol.* 45:3195–211.

50. Medin, J. and Andreo, P. 1997. Monte Carlo calculated stopping-power ratios, water/air, for clinical proton dosimetry (50–250 MeV). *Phys. Med. Biol.* 42:89–105.

51. Jones, D. T. L. 2006. The w-value in air for proton therapy beams. *Rad. Phys. Chem.* 75:541–50.

52. Palmans, H., Thomas, R., and Kacperek, A. 2006. Ion recombination correction in the Clatterbridge Centre of Oncology clinical proton beam. *Phys. Med. Biol.* 51:903–17.

53. Rossomme, S., Horn, J., Brons, S., Jäkel, O., Mairani, A., Ciocca, M., Floquet, V., Romano, F., Rodriguez Garcia, D., Vynckier, S., and Palmans, H. 2017. Ion recombination correction factor in scanned light-ion beams for absolute dose measurement using plane-parallel ionisation chambers. *Phys. Med. Biol.* 62:5365–82.

54. Liszka, M. Stolarczyk, L., Kłodowska, M., Kozera, A., Krzempek, D., Mojżeszek, N., Pędracka, A., Waligórski, M. P. R., and Olko, P. 2018. Ion recombination and polarity correction factors for a plane–parallel ionization chamber in a proton scanning beam. *Med. Phys.* 44:391–401.

55. Vatnitsky, S. M., Siebers, J. V., and Miller, D. W. 1996. k_Q factors for ionization chamber dosimetry in clinical proton beams. *Med. Phys.* 23:25–31.

56. Sorriaux, J., Testa, M., Paganetti, H., Bertrand, D., Lee, J. A., Palmans, H., Vynckier, S., and Sterpin, E. 2017. Consistency in quality correction factors for ionization chamber dosimetry in scanned proton beam therapy. *Med. Phys.* 44:4919–27.

57. Gomà, C., Lorentini, S., Meer, D., and Safai, S. 2014. Proton beam monitor chamber calibration. *Phys. Med. Biol.* 59:4961–71.

58. Gomà, C., Andreo, P., and Sempau, J. 2016. Monte Carlo calculation of beam quality correction factors in proton beams using detailed simulation of ionization chambers. *Phys. Med. Biol.* 61:2389–406.

59. Andreo, P., Wulff, J., Burns, D. T., and Palmans, H. 2013. Consistency in reference radiotherapy dosimetry: resolution of a conundrum when ^{60}Co is the reference quality for charged-particle and photon beams. *Phys. Med. Biol.* 58:6593–621.

60. Gottschalk, B. Cascio, E. W., Daartz, J., and Wagner, M. S. 2015. On the nuclear halo of a proton pencil beam stopping in water. *Phys. Med. Biol.* 60:5627–54.

61. Gillin, M. T., Sahoo, N., Bues, M., Ciangaru, G., Sawakuchi, G., Poenisch, F., Arjomandy, B., Martin, C., Titt, U., Suzuki, K., Smith, A. R., and Ronald Zhu, X. 2010. Commissioning of the discrete spot scanning proton beam delivery system at the University of Texas M.D. Anderson Cancer Center, Proton Therapy Center, Houston. *Med. Phys.* 37:154–63.

62. Palmans, H. and Vatnitsky, S. M. 2015. Dosimetry and beam calibration. In *Principles and Practice of Proton Beam Therapy (AAPM 2015 Summer School)*, eds. I. J. Das and H. Paganetti, 317–51. Madison, WI: Medical Physics Publishing.

63. Palmans, H. and Vatnitsky, S. M. 2016. Beam monitor calibration in scanned light-ion beams. *Med. Phys.* 43:5835–47.

64. Kuess, P., Böhlen, T., Lechner, W., Elia, A., Georg, D., and Palmans, H. 2017 Lateral response heterogeneity of Bragg peak ionization chambers for narrow-beam photon and proton dosimetry. *Phys. Med. Biol.* 62:9189–206.

65. Hartmann, G. H., Jäkel, O., Heeg, P., Karger, C. P., and Krießbach, A. 1999. Determination of water absorbed dose in a carbon ion beam using thimble ionization chambers. *Phys. Med. Biol.* 44:1193–206.

66. Dreindl, R., Osorio, J., Grevillot, L., Böhlen, T., Carlino, A., Kuess, P., Kragl, G., Vatnitsky, S., Stock, M., and Palmans, H. 2017. Application and consistency of PBS beam monitor calibration methods. Abstracts of the 56th Annual meeting of the Particle Therapy Co-operative Group (PTCOG), Chiba, Japan, 8–13 May 2017.

67. Elia, A., Sarrut, D., Carlino, A., Böhlen, T., Fuchs, H., Palmans, H., Stock, M., and Grevillot, L. 2018. A modeling method for non-isocentric proton pencil beam scanning treatment delivery using GATE/Geant4. *Med. Phys.* submitted.

68. Pedroni, E., Scheib, S., Böhringer, T., Coray, A., Grossmann, M., Lin, S., and Lomax, A. 2005. Experimental characterization and physical modelling of the dose distribution of scanned proton pencil beams. *Phys. Med. Biol.* 50:541–61.
69. Kacperek, A., Vynckier, S., Bridier, A., Herault, J., and Bonnet, D. E. 1991. A small scale European proton dosimetry intercomparison. *Proceedings of the Proton Therapy Workshop at PSI* (Feb 28–March 1 1991, Villigen, Switzerland), PSI Bericht no 11176, 76–9.
70. Jones, D. T. L., Kacperek, A., Vynckier, S., Mazal, A., Delacroix, S., and Nauraye, C. 1992. A European proton dosimetry intercomparison. NAC Annual Report NAC/AR/92-01, 61–3.
71. Jones, D. T. L., Schreuder, A. N., Symons, J. E., Vynckier, S., Hayakawa, Y., and Maruhashi, A. 1994. Proton dosimetry intercomparison. NAC Annual Report NAC/AR/94-01, 94–5.
72. Vatnitsky, S., Siebers, J., Miller, D., Moyers, M., Schaefer, M., Jones, D., Vynckier, S., Hayakawa, Y., Delacroix, S., Isacsson, U., Medin, J., Kacperek, A., Lomax, A., Coray, A., Kluge, H., Heese, J., Verhey, L., Daftari, I., Gall, K., Lam, G., Beck, T., and Hartmann, G. 1996. Proton dosimetry intercomparison. *Radiother. Oncol.* 41:169–77.
73. Vatnitsky, S. M., Moyers, M., Miller, D., Abell, G., Slater, J. M., Pedroni, E., Coray, A., Mazal, A., Newhauser, W., Jaekel, O., Heese, J., Fukumura, A., Futami, Y., Verhey, L., Daftari, I., Grusell, E., Molokanov, A., and Bloch, C. 1999. Proton dosimetry intercomparison based on the ICRU report 59 protocol. *Radiother. Oncol.* 51:273–9.
74. Moyers, M. F., Ibbott, G. S., Grant, R. L., Summers, P. A., and Followill, D. S. 2014. Independent dose per monitor unit review of eight U.S.A. proton treatment facilities. *Med. Phys.* 41: 012103.
75. De Saint-Hubert, M., Reniers, B., Stolarczyk, L., De Angelis, C., Knežević, Ž., Kunst, J., Parisi, A., Majer, M., Vanhavere, F., Struelens, L., Harrison, R. M., and Olko, P. 2016. Mailed dosimetry auditing in proton therapy – Eurados intercomparison of passive dosimeter response in proton spot scanning beam. PPRIG workshop 2016. Last accessed online on 5 Oct 2017 at http://www.pprig.co.uk/meetings/docs/20161201-02_pprig_workshop/1-sainthubert.pdf.

IV

Operation

CHAPTER 11

Acceptance Testing and Commissioning

Zuofeng Li, Roelf Slopsema, Stella Flampouri, and Daniel K. Yeung

University of Florida Health Proton Therapy Institute

CONTENTS

11.1 INTRODUCTION

Acceptance testing and commissioning of proton therapy systems must be performed prior to use of the system for clinical patient treatments. In the United States, qualified medical physicists are responsible for these activities [1]. For conventional photon-based radiation therapy, a number of recommendations for the commissioning and periodic quality assurance (QA) of linear accelerator-based radiotherapy has been developed and detailed in several American Association of Physicists in Medicine (AAPM) task group reports [2–4]. While the principles of system acceptance testing, commissioning, and periodic QA program are similar for proton therapy, these specific recommendations nevertheless do not apply well for the practice of proton therapy.

Acceptance testing is designed to ascertain that technical performance parameters and functional capabilities of the system meet the specifications of the system. These specifications are agreed upon in the purchase agreement of the system, for example, in a separate appendix of System Performance Specifications, or embedded in an appendix of Acceptance Testing Procedures (ATPs). Acceptance testing is performed immediately after vender completes its Verification and Validation (V&V) tests, and announces that the system, or, in the case of a multiple-treatment-room system, one or several rooms, is ready for acceptance testing. The time period available for acceptance testing, availability of user physicists to participate in acceptance tests, together with ATP documents, are specified in the purchase agreement. A representative of the purchaser, often the physicist that leads the acceptance tests, signs the ATP documents upon satisfactory completion of these tests. This signature serves as certification that the purchaser accepts the system, as has been tested through acceptance tests and permits release of payment to the vendor, as agreed upon in the purchase agreement. Completion of acceptance tests therefore carries significant technical, clinical, and financial impacts to the purchaser: failure of the system in meeting performance specifications must be brought to the attention of vendor representatives immediately and resolved per the purchase agreement.

The performance of a proton therapy system may vary under different operating conditions, including beam ranges, gantry angles, and field doses/monitor units (MUs) used for beam delivery, due to the extensive use of interpolations and parameters fitting lookup tables for beam tuning and system operation. While it will not be possible to exhaustively test all such combinations of field delivery parameters, the ATP should allow a number of user-selected permutations in these parameters for the key dosimetry and geometric accuracy tests. The physics staff members responsible for performing the acceptance tests may participate in system installation, calibration, and V&V testing processes to gain in-depth understanding of procedures and to facilitate the actual selection of these permutations to be used for subsequent acceptance tests.

Variations in the definitions of beam performance parameters, such as beam range and spread-out Bragg peak (SOBP) widths, depth and lateral beam profile uniformity and symmetry, as well as selection of dosimetry equipment to be used and measurement conditions using these instruments, can significantly impact the interpretations and applications of actual system performance parameters. For example, beam lateral profile uniformity should be measured for the maximum field size available, while SOBP width may be defined to start from proximal 90% or 98% of the depth-dose profiles, and end at distal 90% or 95% depth-dose profiles. Vender-proposed system ATPs should be carefully reviewed to ensure that these definitions and testing procedures meet the user's expectations.

Because of these considerations, a qualified medical physicist should be engaged in the drafting and negotiation of the purchase agreement toward the acquisition of a proton therapy system. Specifically, the physicist will contribute to review and comment on the adequacy and comprehensiveness of proposed system performance specifications; the suitability of the proposed ATP to test and verify these specifications; and the time length available to complete these tests.

Commissioning of proton therapy systems will build on acceptance testing to further explore system performance in a wider set of clinical conditions. System commissioning then continues with measurement, calculation, and validation of computed tomography (CT) scanner Hounsfield unit (HU) to relative stopping power ratio calibration curve, as well as acquisition and validation of proton beam data for patient dose calculation by treatment planning system (TPS). For TPS commissioning, the user performs beam data measurements and collects nondosimetric system parameters that serve as input data for establishment of dose calculation models in the TPS. The established dose calculation model is validated by comparison of calculated dose distributions for various clinically relevant scenarios to measured dose distributions. Dose calculation model fitting parameters may be adjusted iteratively, until acceptable agreement between calculated and measured doses is achieved. TPS commissioning will also attempt to identify weaknesses of TPS dose calculation algorithm in terms of dose calculation accuracy, such that these may be avoided or their effects minimized during clinical patient treatment planning. An end-to-end test, starting with CT imaging of a test phantom, followed by development of a treatment plan, and completed with the image-guided treatment delivery of this treatment plan on the phantom, emulating the expected entire clinical flow, will conclude the technical aspects of system commissioning. In addition, to allow safe and accurate patient treatment using the proton therapy system, commissioning of the system will necessarily include development of a periodic machine QA program, patient-specific QA program, user training, and development of disease-site specific patient simulation, treatment planning, and image-guided treatment delivery procedures. System commissioning therefore may be considered an extended process, with the system released for clinical treatment of patients by disease sites. Adequate discussions should be held with the administrative and clinical teams of the proton center to identify and outline the sequence and expected timeline of such a disease-site-based release of system for clinical operations.

11.2 PREPARATION FOR ACCEPTANCE TESTING

A preliminary radiation survey of areas of the proton facility must be performed as soon as high-energy proton beam is extracted from the accelerator, and when it is transported to the treatment room. The survey ensures the safety of installation personnel, building construction personnel, and user personnel engaged in the facility construction and system installation process. The areas to be surveyed may be limited to the accelerator and beam transport vault, and the treatment room that will be accepted in a facility of multiple treatment rooms. In addition, a survey inside a treatment room for radiation exposure levels due to proton/neutron activation must be performed, especially when construction workers and vendor engineers may share use of the treatment room for completion of room construction and system calibration with beam on at staggered hours. Consider the following scenario: in a multi-room proton facility, acceptance testing of treatment room 1 is performed in an 8-h work shift, while construction workers finish constructions of room 2 in the second shift, and vender engineers perform beam calibration in treatment room 2 in the third shift. As construction workers are in general nonradiation workers, treatment room 2 needs to be surveyed for activation radiation exposure daily before construction shift may start, to ensure compliance of radiation exposure limits in room 2 for members of public during these hours.

Equipment to be used for proton therapy system acceptance testing differs somewhat from what is traditionally used for linear accelerators. Table 11.1 provides a list of such equipment with their intended uses. This equipment, either vendor-provided or facility-purchased, should be validated prior to their use for acceptance testing.

Cylindrical and parallel-plate chambers used in acceptance tests should be evaluated for dosimetric and positional accuracy, and with system leakage/background within specifications. Planar 2D dosimetry systems, including ion chamber arrays, scintillation-plate-based systems, and radiochromic films, together with the software systems used to operate and analyze results of these systems, must be evaluated for accuracy as well, with consideration of the impact of hardware and software resolutions of these systems on measurement results [5–11] (see Chapters 9 and 10 for an overview of dosimetry equipment). Scanning water phantom systems should be evaluated for position accuracy. While most scanning phantoms will be purchased with mounting adapters for standard ion chambers, a mounting adapter for the large diameter parallel plate chamber, to be used for measurement of integrated depth doses (IDDs) of proton pencil beams, may need to be separately acquired.

Acceptance testing of pencil beam scanning delivery technique requires deliveries of SOBPs of various sizes and shapes. As no beam data has been collected for the TPS, yet for creating treatment plans that deliver these SOBPs, they will necessarily be generated based on vender-provided data that is collected or validated in the system V&V process or using

TABLE 11.1 Equipment for Acceptance Testing and Commissioning

Category	Equipment	Used for
Beam dosimetry measurements	Marcus-type parallel-plate ion chamber and electrometer	Absolute dose, depth dose, beam ranges
	Farmer-type ion chamber	Absolute dose calibration
	Large-diameter parallel-plate ion chamber	IDD
	Scanning water tank	Depth dose, beam range, and IDDs
	Multilayer Ion Chamber system (MLIC)	Depth dose, beam ranges
	Scintillation-plate based 2D dosimetry system	Pencil beam spot characteristics, lateral dose profile of SOBP
	2D ion chamber array system	Lateral dose profiles (low resolution) in machine and patient-specific QA measurements
	Radiochromic film, film scanner, and analysis software	Lateral dose profiles (high resolution)
	Calibrated thermometers and barometers	Temperature and pressure corrections
	Plastic or solid water pieces	Absolute dose at depths, lateral profiles at depths
CT scanner	CT density phantom	CT HU—Stopping Power Ration calibration
	CT dose chamber	CT imaging dose measurements
Mechanical and imaging systems	kV meter system	X-ray imaging system calibration and dose measurements
	Planar and cylindrical imaging test phantoms	kV and cylindrical phantoms for X-ray and CBCT imaging system performance tests
	High-Z ball phantoms	Imaging and beam alignments, isocentricity
	Calibrated rulers	Mechanical system tests
Radiation safety	Portable photon and neutron survey meters	X-ray and neutron exposure surveys

vendor-provided treatment plans based on simulated beam data. These treatment plans should be carefully reviewed prior to their use in acceptance testing, both for agreement of beam data used in these plans to those of the system under acceptance testing; and for their ability to deliver uniform SOBPs without excessively larger numbers of spots and layers.

11.3 ACCEPTANCE TESTING

11.3.1 Final Radiation Safety Survey, Photons, and Neutrons

Radiation surveys during acceptance testing serve as a formal demonstration of acceptable exposure levels in all controlled and public areas potentially subject to radiation exposure during system operation (see also Chapter 7). Since acceptance testing may be performed in phases, with one or two treatment rooms of a given proton therapy facility for each phase, radiation safety surveys need to be performed accordingly as well in areas immediately adjacent to the treatment rooms, in addition to the accelerator and beam transport vaults. Radiation surveys of X-ray imaging systems are done at this time as well to ensure the safety of therapists operating these systems. The survey should in addition assure that high-radiation areas within the facility are properly posted with warning signs; that area radiation monitors and beam-on indicating lights function properly; and that available room search alert devices function as intended, as mandated by local radiation protection regulations. A formal report on the results of the radiation safety survey should be produced following completion of the survey for submission to the local regulatory agency.

11.3.2 System Safety Interlocks and Emergency Stops

System emergency stops, such as collision detection emergency stops on the gantry and patient positioning devices, should be tested for correct functioning, as well as for returning to normal operations after the collision condition has been resolved. Power-off emergency stops in a proton therapy system may provide either global or local power-off, with the former shutting off power to the entire system, potentially including the accelerator and beam transport line; and the latter shutting off power to equipment in the treatment room only. Testing of global emergency stops must be done carefully and after detailed discussion on the consequences and recovery procedures to avoid damage to system components. Entry door interlocks that turn off proton beam, or similar interlocks that turn off X-ray beam for imaging, should be tested as well.

11.3.3 Oncology Information System and Treatment Planning System

Oncology Information System (OIS) and TPS are usually installed and should have been accepted prior to acceptance testing of the proton therapy system, even if TPS is not commissioned as yet, as their correct functioning may be required in several acceptance test procedures.

11.3.4 Mechanical Systems and Laser Alignment Systems

All movable mechanical systems of a proton therapy system, such as the gantry, snout, and patient positioning systems (PPS), should be tested for movement accuracy, resolution, limits, isocentricity, and reproducibility. Much of these tests are performed using calibrated levels and rulers in manners similar to what one would use for QA tests of linear accelerators. It should however be noted that, unlike in linear accelerators, proton therapy systems may not provide QA accessories such as front pointer systems for gantry rotation isocenter tests. Users will then need to discuss and collaborate with vendor installation team to fabricate such mechanical devices for these tests, or use alternative radiographic imaging methods, for example, by imaging a metal sphere of appropriate diameter. The test spheres may be inscribed with laser alignment markings for testing laser accuracy. Moyers [12] discussed use of Theodolites for measuring gantry isocentricity, although it should be noted that this technique is time consuming and requires specific expertise and equipment. It is typically used by vendor engineers during gantry installation.

In addition to the gantry and PPS, proton therapy systems are often equipped with telescoping snouts that allow use of additional treatment accessories, such as apertures, range compensators, range shifters, and ridge filters. The snouts should be additionally tested through their range of motion for sagging, with or without maximum weight of accessories installed.

The proton PPSs are generally capable of six degree of freedom (DOF) movements, similar to those now available in higher-end linear accelerators. Some of the proton PPSs are equipped with pressure sensors that automatically correct for tabletop sag under different loading conditions. Some gantry designs do not allow intrinsic mechanical isocentricity of <1 mm radius, and may rely on movement of PPS to achieve such tight isocentricity values based on preconstructed lookup tables similar to the flexmap method [13] used in cone beam CT (CBCT) image reconstruction of linear accelerators. Accuracy of such PPS should be tested with nonzero rotational values in rotation, pitch, and yaw directions, and with distributed weight on the tabletop to approximate clinical scenarios.

11.3.5 Imaging Systems

State-of-the-art proton therapy systems are typically equipped with an orthogonal (or oblique in some cases) pair of kV X-ray tubes/panels supporting 2D planar imaging, from which six DOF corrections in patient alignment may be calculated (see Chapter 20). One of the panel/tube pairs or both may also be used for CBCT acquisition. CT-on-rail or movable CT systems are also used in some installations. Acceptance testing of these systems will be heavily dependent on the specific configuration of the installed imaging system. In principle, however, acceptance testing of imaging systems will necessarily include validation of the system performance and accuracy in image quality, geometric accuracy, radiation exposure characteristics (or patient doses), and speed and accuracy of X-ray or CBCT image registration to reference digitally reconstructed radiographs (DRRs) or CT images.

Appropriate test phantoms should be used for 2D planar X-ray and CBCT image quality evaluations, such as image resolution, low and high contrast, and image geometry accuracy. The measured image quality parameters will be compared to the specifications and serve as baseline values for future periodic QA tests of the imaging systems. The images should be geometrically accurate, as image guidance of proton therapy systems is based on being able to reproduce the projected 3D geometry of known objects on the 2D images for coregistration with DRRs to obtain patient alignment correction values. Distances between known points, lengths, radius, and other similar geometric parameters should be measured, along with source-to-isocenter and source-to-image distances verified to agree with specifications. If the imaging system is designed with movable X-ray tubes and panels, reproducible accuracy of their movements and positioning shall be verified.

X-ray tube performances of both 2D and CBCT imaging systems should be measured at the time of acceptance testing, with the results to be used for guidance in the optimal application of these systems during patient treatment and as baseline values of future QA tests. Where applicable, patient exposure data may need to be submitted to the local regulatory authorities for permit applications. An appropriate kV meter and/or CD dose ion chamber should be used to measure patient doses as well as kV and mA accuracy.

The accuracy of translation and rotation corrections, as well as the speed of calculations, should be tested. These can be done with imaging phantoms of known standard geometry, such as a plastic box phantom embedded with metal markers. The phantom will be set up on the treatment table and aligned to the machine isocenter, with the accuracy of alignment confirmed by the imaging system. The phantom will then be moved away from the isocenter by known distances and degrees using the six DOF PPS. The imaging system will be used to calculate the offsets, with the results compared to the shifts introduced previously. As a second, more clinically relevant step, anthropomorphic phantoms should be used to further test image registration accuracy and calculation speed.

Isocentricity of the imaging system (X-ray isocenter) should be tested, as well as its agreement to the gantry and PPS mechanical isocenter and the proton beam isocenter. For systems installed with an X-ray imaging system that performs proton beam's eye view (BEV) imaging, the central axis of the X-ray beam should be tested to be coincident to that of the proton beam axis and orthogonal to the gantry rotation axis ("colinearity tests").

11.3.6 Beam Parameters

As the available time period for acceptance testing is often limited by the system purchase contract, proton beam performance parameters can only be measured for a subset of available energies, gantry angles, and, in the case of 3DCPT techniques (3D conformal proton therapy delivering SOBPs such as passive scattering, uniform scanning, and wobbling), selected combination of nozzle components for forming the SOBPs during this period. Such selections, therefore, should attempt to allow highly confident sampling of the entire range of available combinations, for example, in the low-, mid-, and high-energy ends of available beam energies. Meeting design specifications at the extreme or border values of the available beam parameters is often more challenging for proton therapy systems. Inclusion of tests at the extremes, for example, measuring spot alignment at the maximum off-axis spot position or verifying the maximum dose rate at the lowest range, should be considered in addition to tests at more typical values. If nozzle hardware is used in designated combinations to form SOBPs of certain spans of beam ranges and SOBP widths, sometimes called "options," representative SOBPs of each option may be selected for measurements. In addition, a number of randomly selected beams at random gantry angles should be measured.

Selection of beam parameter measurements should also allow use of the measured beam data for system commissioning and development of future system periodic QA program. The selected tests would then be repeated periodically during system commissioning for monitoring of system performance stability, and during the daily, monthly, and annual system QA tests.

11.3.6.1 SOBP/3DCPT Techniques: Passive Scattering, Uniform Scanning, and Wobbling

11.3.6.1.1 Range and SOBP Accuracy and Reproducibility; Range/SOBP Resolution; Distal Falloff Values; Depth Profile Uniformity   Measurements of 3DCPT beam range and SOBP width typically use parallel-plate ion chambers. For passive-scattering beams (see Chapter 5), where a continuous beam is available, the ion chamber may be used to scan the beam depth profiles and extract these parameters. For 3DCPT delivered using scanning beam technique (such as uniform scanning or wobbling), point-by-point measurements using parallel-plate chambers are highly inefficient. Multilayer ionization chamber (MLIC) systems, once validated by comparison to parallel-plate chambers, are more suitable for depth profile of such beams.

11.3.6.1.2 Lateral Beam Profile Uniformity and Symmetry, Field Size, Lateral Penumbra   The lateral beam profiles, from which these parameters are extracted, may be measured by use of a small diameter cylindrical ion chamber in a scanning water phantom. Radiochromic films or 2D ion chamber arrays may also be used, especially for 3DCPT produced by scanning beam techniques. It is necessary to process the profiles measured by 2D ion chamber arrays to remove the volume-averaging effect of individual detectors, especially for lateral penumbra measurements. Measured parameters using such systems should also be compared to those of water scans or films, where applicable. Depths of lateral beam profile measurements should be selected to verify specifications in the entrance region, proximal, middle, and distal SOBPs, where applicable. In addition, lateral beam profiles should be measured in the beam distal falloff region to verify beam range uniformity over the treatment fields laterally, as these can be affected by manufacturing variations of beam-scattering devices.

11.3.6.1.3 Dose Monitoring System   While it is not mandatory that the dose monitoring system, or dosimetry system, be calibrated for absolute dose output in Gy/MU for acceptance testing, it is generally helpful to perform a preliminary calibration to allow determination of proper settings of dosimetry systems used during acceptance testing. Absolute dose output calibration of proton beams is typically performed following the International Atomic Energy Agency (IAEA) Technical Report Series 398 [14] (see Chapter 12). The dose monitoring system, consisting of primary and secondary monitor chamber units and their hardware and software controls, are subsequently tested for dose rate accuracy based on this preliminary calibration. Furthermore, the dosimetry system should be tested for output factor dependence on dose rates, output linearity over the entire range of minimum through maximum deliverable MUs (see Chapter 14), resolution of deliverable MUs, as well as output reproducibility over the acceptance testing period. If the system is equipped with automatic temperature and pressure sensors to provide air density corrections for open-to-air monitor chambers, system output variations over the period of testing should be reviewed for accuracy of such corrections. With vendor engineer support, the user may also manually enter arbitrary ambient temperature and pressure values into the system for this test. Systems that require user to enter temperature and pressure values into the system for corrections can be tested in the same manner.

Output factors of a proton therapy system can potentially demonstrate significant gantry angle dependency. Such gantry angle dependency should be measured, for the four prime angles of 0°, 90°, 180°, and 270° for a reference field, and confirmed to be within system specifications.

Tests of dose monitoring system should be performed using a well-qualified or calibrated ion chamber system in water or solid water phantom, with measurement points at the middle of the SOBPs in depths.

11.3.6.1.4 Treatment Accessories   3DCPT techniques use apertures to achieve lateral target dose conformity in the BEV, and range compensators for dose conformity in the distal aspect of target. The apertures and range compensators are calculated in the TPS and exported in the Digital Imaging and Communications in Medicine (DICOM)-RTion plan file. As these accessories are fabricated in commercially available digital milling machines, parameters describing them need to be transcoded into the programming language of the milling machine. The proper functioning of this process, through the physical fabrication of finished products, needs to be verified. In addition, the finished products should be measured for milling accuracy. The process and instruments used for these tests, such as a height gauge to measure thickness of range compensators at various positions, or the plotting of aperture openings to the scale of the physical apertures, serve as the fundamental components of patient-specific QA of these devices. The patient-specific labeling of these devices should be validated as well.

11.3.6.1.5 Partial Field Delivery Recovery   Delivery of a treatment beam may be interrupted for various reasons, such as machine errors or patient movements that require repeat imaging and alignment. The ability to complete a partially delivered treatment by delivering the remaining MUs of the treatment field is therefore a required feature of a proton therapy system. This feature should be available in either stand-alone mode of delivery, where the treatment field may be created and delivered on the treatment delivery console of the system itself, without the oversight of an OIS; or in OIS mode where the dose recording and partial field delivery is controlled by the OIS. Acceptance testing of this feature should include both the functional and dosimetric aspects. The functional aspect of partial field delivery will be limited in the system's ability to record the delivered MU value up to the point of delivery interruption, and the ability to resume delivery of remaining MU with maintenance of all patient- and field data. The dosimetry aspect will evaluate the dosimetric effect of such a process. This may require measurement of both depth and lateral profiles of the total delivered dose distributions, mostly for the uniform scanning and wobbling techniques of 3DCPT delivery, using an MLIC for the depth and film or a 2D ion chamber array for the lateral profiles. The typical passive-scattering delivery technique delivers a treatment field with hundreds of SOBPs. Dosimetric evaluation of partial field delivery for passive-scattering treatment fields is not required in such cases. Significant dosimetric errors, such as a 10% peak or valley in the SOBP or lateral profile of the dose distribution, should be brought to the attention of the vendor and appraised for clinical acceptability.

11.3.6.1.6 Functional End-to-End Test   For acceptance testing, an end-to-end test will be limited to validate the functional performance of the entire patient treatment process. The process will therefore start with CT simulation of a phantom with implanted fiducial markers or similar landmarks suitable for image-guided treatment delivery. CT images of the phantom are imported into the TPS, and a 3DCPT treatment plan was created to deliver a clinically relevant dose to the phantom. The treatment accessories are fabricated, and the treatment plan, segmented structures, and CT images/DRRs are imported into the OIS or the proton therapy system for OIS mode or stand-alone mode delivery. The phantom is aligned on the PPS using image guidance, and treatment is delivered as prescribed using all planned field parameters. After delivery, it is verified that the treatment is correctly recorded in the OIS and that all relevant system logs (e.g., irradiation logs recording the delivered beam parameters and MUs, gantry angles, and PPS parameters) have been properly generated. No dosimetric measurements are required for this test. Note that the end-to-end test as described earlier will be required only if the OIS and TPS are purchased together with the proton therapy system as a turnkey project from the vendor. It would be advisable, however, that this test be performed in the early stage of acceptance testing and commissioning process, even when the systems are purchased separately by the owner, such that any configuration errors or incompatibilities are identified and resolved as soon as possible.

11.3.6.2 Modulated Scanning Beam Techniques

Acceptance testing of modulated scanning beams (see Chapter 6) differs from that of 3DCPT due to the use of highly focused pencil beams scanned laterally in the BEV, and of a number of staggered energies to construct stacked layers of doses to form a 3D dose distribution. The beam pencils may be scanned laterally in a step-by-step fashion (spot scanning), with the beam moved to and held at a prescribed lateral location (spot) for delivery of a predetermined dose/MU. The beam current is turned off upon completion of spot delivery before the beam is moved to the next spot. Variation of doses delivered to each spot (spot weight) allows intensity-modulated proton therapy delivery. The beam may also move continuously in the lateral direction, with motion speed and/or beam current adjusted per treatment plan in the so-called line scanning technique. The beam energy is changed to the next lower one when delivery of the current layer is completed iteratively, until the total planned dose of the field is delivered. Acceptance testing of scanning techniques, therefore, necessarily differs from that of 3DCPT techniques, in the need to verify correct beam spot parameters.

Performance parameters of modulated scanning beams can be heavily dependent on treatment gantry angles, as the beam scanning and focusing magnet controls need to be adjusted for each gantry angle. Acceptance testing of modulated scanning beams, therefore, should attempt to verify that such gantry-angle dependency is within specifications.

A proton therapy system may provide selectable sets of pencil beam sizes (Spot Tune IDs) in a given treatment room. Acceptance testing should be performed for each Spot Tune ID available in the treatment room.

11.3.6.2.1 Range and Range Pullback Accuracy and Reproducibility; Distal falloffs, Range Shifters   For modulated scanning beam techniques, these tests can be performed for either single *unscanned* pencil beams, or for SOBPs constructed from the scanned pencil beams, or for both scenarios. While the former provides greater insight in the properties of a pure pencil beam delivered by the system and may be more convenient for measurement of range pullbacks, it requires the system to operate in a nonclinical mode and can be considered somewhat detached from clinical practice. Use of SOBPs for these measurements, on the other hand, may require construction of a large number of optimized treatment plans before completion of TPS. A combined use of unscanned pencil beams and selected SOBP plans may therefore be a reasonable compromise.

For unscanned beam measurements, a parallel-plate chamber in a scanning water phantom is suitable for high efficiency and accuracy measurements. Measurements of SOBP doses, however, are practically intractable without the use of an MLIC system. The MLIC system should be well qualified by comparison to parallel-plate chamber performance in measurements of unscanned pencil beams, or in a passive-scattering beamline if available, prior to its use in acceptance testing. Of specific concern is the area size of the detectors in the MLIC compared to the beam size. To achieve fluence equilibrium in the measurement, either the detector diameter needs to be larger than the (unscanned) beam size, intercepting all protons and measuring an IDD. Alternatively, the beam is scanned over (the smaller) detector area, achieving proton equilibrium on the measurement axis and measuring a percent depth-dose (PDD).

Range shifters are used to allow treatment of targets located at shallow depths that would have otherwise required delivery of proton energies below the capabilities of the accelerator and/or beam transport system. When purchased from the vendor, the range pullback/reduction value of the device should be measured for compliance to system specifications.

11.3.6.2.2 Spot Size, Position, and Symmetry   These pencil beam parameters are defined at the isocenter, and in general, measured using high-resolution dosimetry instruments, such as radiographic/radiochromic films or a scintillation detector system in air. The beam energies, spot locations, and gantry angles selected for these tests should span the entire available energy range, maximum treatable field size, and the four prime gantry angles as well as selected additional gantry angles (Figure 11.1). As calibration of scanning pencil beams is in general performed at a limited number of gantry angles, the additional gantry angles use during acceptance testing should avoid the ones used in beam calibration.

11.3.6.2.3 Dose Rates for Min, Mid, and Max Energies   Dose rates for modulated scanning beams are typically specified in the time it takes to deliver a given dose, for example, 2 Gy, to a 1 l cubic volume of SOBP dose. It should however be noted that, as discussed in previous section, beam delivery dose rates can also be heavily dependent on the range of SOBP and gantry angle. It is therefore advisable that these tests be run with the use of minimum, medium, and maximum beam ranges, and at different gantry angles. These tests should be performed with dose measured in water or solid water phantom. The time is measured using a stopwatch.

11.3.6.2.4 Dose Monitoring System   Acceptance testing of dose monitoring system of modulated scanning beam technique is mostly identical to that of 3DCPT technique, although in general, dose rates of modulated scanning beam delivery is not adjustable away from

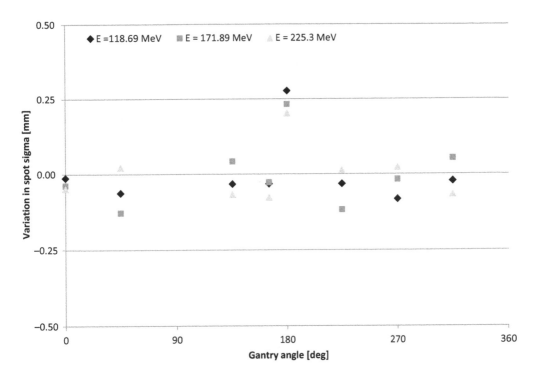

FIGURE 11.1   Measured spot size (sigma of 2D Gaussian) as function of gantry angle for three different beam energies.

their maximum values available from the system. There are therefore no tests of dose rate accuracy or dose rate dependence of output factors required. The remaining tests, including MU dependence on gantry angles, MU linearity, resolution, reproducibility, and temperature and pressure corrections, are performed in the same manner as previously described for 3DCPT.

11.3.6.2.5 Minimum and Maximum MUs Per Spot, Maximum Field Sizes   The minimum and maximum spot MUs available from a modulated scanning beam proton therapy system are the two limitations that must be considered in treatment planning, to assure deliverability of treatment plans. The proton therapy system's ability to consistently deliver spots at min, mid, and max beam energies; at multiple locations up to the maximum field size; and at any gantry angles with minimum and maximum spot MUs must be verified during acceptance testing. These tests, when performed after the dose monitoring system tests, do not necessarily require dosimetry measurements.

The maximum field sizes deliverable from the system are limited by the scanning magnet system, both in their geometric dimensions and their maximum current values. In principle, beams at maximum energy are expected to be more challenging to the scanning magnet system in their demand for high magnet current sustained over the time of beam delivery, while beams with minimum energy and low spot MUs will challenge the dose monitor system in sensitivity and location resolution. These tests are measured in air using either films or scintillation detection systems, using treatment plans with equal spot MUs within each layer spaced a few centimeters apart. These measurements can be used simultaneously with the minimum and maximum MU per spot tests of the previous paragraph, with minimum spot MUs used for the minimum beam energy and maximum spot MUs for the maximum beam energy.

11.3.6.2.6 Partial Delivery Recovery   Partial field delivery from an interrupted modulated scanning beam treatment, in the ideal case, requires identification of the spot location, MU, and beam energy (layer), where the interruption occurred. Availability as well as the presentation of such information should be acquired and reviewed. Recovery from a partially delivered modulated scanning beam field from the specific spot location, with the specific partial spot MU, at the specific layer, may not be possible in all systems. In addition, when an OIS is used for treatment delivery, it is likely that only recovery at the next spot location following interruption is possible. Dosimetric evaluation of interrupted then recovered treatment fields using a selection of SOBPs is therefore necessary. An MLIC and a 2D ion chamber array may be used for these measurements (Figure 11.2).

11.3.6.2.7 Treatment Accessories   Modulated scanning beams may suffer from having larger beam penumbra than 3DCPT beams, if no beam collimating apertures are used [15].

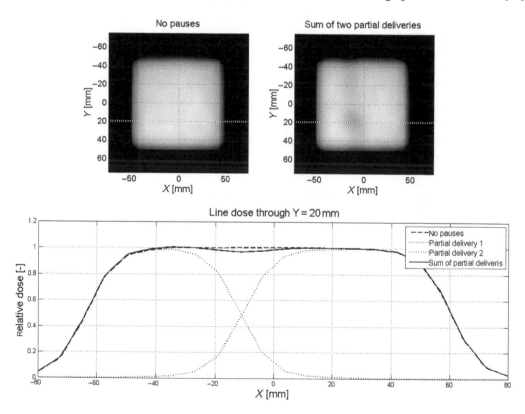

FIGURE 11.2   Two-dimensional dose measurement (using 2D ionization chamber array) of delivery of a PBS uniform-dose cube without interruptions (top left) and delivered as two partial irradiations (top right). Bottom graph shows line dose profile for both cases.

Modulated scanning beam systems may therefore support use of apertures or, alternatively, be equipped with a dynamic collimation system (DCS) that moves collimation blades to trim spot penumbras [16,17]. Additionally, miniridge filters may be used to increase the full-width-at-half-maximum (FWHM) of Bragg peaks for low-energy pencil beams to improve the robustness of treatments, especially for moving targets, and improve delivery efficiency by reducing the total number of layers. These treatment accessories should be tested for their geometric accuracy in agreeing to TPS-calculated values and for the FWHM of Bragg peaks when miniridge filters are used. Acceptance test of apertures will be identical to that of 3DCPT apertures. DCS acceptance tests, on the other hand, will need to be developed in collaboration with the vendor, and, when available, follow practice standards to be developed by professional societies such as the AAPM. In principle though, acceptance tests of DCS should include the collimating blade motion range and accuracy, speed, resolution, and reproducibility; maximum and minimum field sizes; field penumbra values; and potential radiation leakage outside the treatment field. The mechanical tests of DCS should be done for multiple gantry angles to identify any field geometric errors due to gravity. These tests would require delivery of a set of modulated treatment plans, using films or scintillation dosimetry systems for measurements. Activation of the collimation blades, together with its decay characteristics, should be monitored and recorded for use in developing radiation safety procedures at the facility.

11.3.6.2.8 System Stability/Robustness Tests   Compared to the passive-scattering technique, modulated scanning beam technique can be significantly more sensitive to operating condition changes of the accelerator, ion source, beam transport line, treatment nozzle, and treatment room conditions such as noise and vibration levels. A set of tests, consisting of treatment plans calculated to emulate actual patient treatments of different disease sites, such as head and neck, craniospinal irradiation, breast and regional lymph nodes, and prostate with pelvic nodes, may be performed over a period of time to obtain an early indication of system robustness. While no dosimetric measurements are necessary for these tests, the delivery times, system malfunctions such as beam pauses, or other interruptions should be recorded and reviewed with vendor engineers. Any systematic system malfunctions that may significantly disrupt patient treatment delivery in future clinical operations should be resolved, either before completion of acceptance testing or planned to be resolved with an acceptable timeline.

11.3.6.2.9 End-to-End Test   The process of end-to-end testing for modulated scanning beams are mostly identical to that of 3DCPT beams. The user may elect to combine these with system robustness tests.

11.3.6.2.10 Vendor-Specified Composite, Multicriteria Tests   Proton therapy system vendors typically will provide a composite test of modulated scanning beam delivery accuracy (Figure 11.3). These tests may consist of a treatment plan that delivers a pattern similar to what an X-ray test phantom contains. When measured using films or scintillation detection systems, the test will allow users to evaluate spot size, location, spot MU, and other

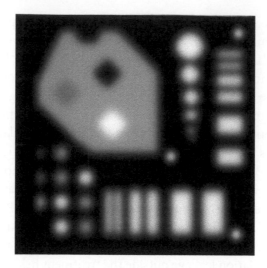

FIGURE 11.3 Measurement (using scintillator detector) of the IBA (Ion Beam Applications, Louvain-La-Neuve, Belgium) test pattern used to verify several properties of a monoenergetic pencil beam.

beam parameters in a composite manner. The measured dose distribution pattern is then compared to the planned pattern using a Gamma Test [18]. Alternatively, dose measurements at depths may be performed using solid water and a 2D ion chamber array for an intensity-modulated treatment plan, with the results analyzed using Gamma Test in the same manner. The beam spot distributions (spot maps) of these composite test plans should be carefully reviewed to understand what their limitations may be, such as testing for a limited field size, beam energies, or specified with Gamma Test analysis parameters and passing criteria that may be too generous for clinical applications.

11.3.6.3 Beam Matching, Room Switching, and Delivery Technique Switching

Multiroom facilities have the opportunity to specify matched beams between rooms for the same technique and room switching times in the purchase agreement of the proton therapy system. In addition, if multiple delivery techniques are available in the same treatment nozzle or the "universal nozzle", the time of technique switching can also be specified. Room-switching and delivery-technique switching times can be tested using a stopwatch, although the start and end times should be reviewed for their clinical implications. Switching the proton beam from one treatment room to another requires performing a hysteresis curing loop for all magnets in the beam transport and gantry systems. The time required to cure the hysteresis is dependent on the previous settings of each magnet. Test of room switching should therefore include several modes of switching: from high-energy treatment delivery in room 1 to low energy in room 2; from low energy in room 1 to high energy in room 2; and switching with similar energies between the two rooms. Note that this test may also include the time required to perform beam tuning in preparation to treatment in room 2. Technique switching of a universal nozzle should be similarly tested for permutations of switching from/to different techniques.

Matched beams in different treatment rooms allow use of the same beam model in the TPS for treatment delivery in different treatment rooms, thus requiring shorter system commissioning time and providing significant patient treatment scheduling advantages. It has been demonstrated that golden beam data can be developed for both passive-scattering [19] and scanning beam techniques [20], based on a limited set of beam quality descriptors. Acceptance test of a matched beam therefore can be significantly shortened as well, with tests for only selected beam parameters for the beam-matched room [21].

11.3.6.4 Use of Vendor Installation Verification and Validation Test Results

Given the complexity of a proton therapy system, it is unlikely that system performance parameters and safety functions can be thoroughly tested during system acceptance testing period. Participation in and a comprehensive understanding of system V&V testing in these areas can significantly increase user confidence in the safety and performance of the system. The qualified physicist responsible for system acceptance testing should be actively engaged in these activities during testing. V&V test procedures and results should be reviewed to help select items to be tested, such as safety interlocks, beam energies, and gantry angles.

11.4 SYSTEM COMMISSIONING

11.4.1 CT to Relative Linear Stopping Power Conversion Calibration

Currently, TPS calculations of the proton energy required to treat a volume within a patient are based on computed tomography images of the patient. The calculations involve conversion of CT HU numbers of each CT voxel to their corresponding relative proton stopping power (RSP) values (see Chapter 14). Commissioning of the CT scanner used for proton therapy therefore includes the measurements and tests necessary for conventional radiotherapy [22] as well as generation and validation of CT HU-to-RSP calibration curve. The process is equivalent to the creation of the HU-to-physical or electron density curves for photon dose calculations, but it is complicated by the absence of guidelines and the lack of phantom materials equivalent to tissues for both X-ray and proton interactions. Common calibration approaches include the stoichiometric method developed by Rutherford et al. [23] and described by Schneider and others [24,25]. Steps include

1. *Setup of standard acquisition and reconstruction protocols for images to be used for proton dose calculation.* Parameters that significantly alter the HU, such as scan energy, should be kept constant for imaging protocols and correspondent calibration curves.

2. *HU measurements under conditions encountered in clinical practice of a heterogeneous phantom containing a set of tissue-equivalent materials of known physical density and composition.* Variable conditions include phantom-related variations such as size, insert position, and phantom position in the bore. Acquisition and reconstruction-related parameters should also be examined within the range of variation during patient imaging.

3. *Use of the stoichiometric method to parameterize the response of the CT scanner based on the data collected in step 2.* The HU of a material of known density and composition for an energy spectrum is given by

$$\text{HU} = \left(\frac{\rho}{\rho \text{H}_2\text{O}} \frac{\sum\left(\frac{\omega_i}{A_i}\left(Z_i \overline{K}^{\text{KN}} + Z_i^{2.86} \overline{K}^{\text{sca}} + Z_i^{4.62} \overline{K}^{\text{ph}} \right) \right)}{\frac{\omega_H}{A_H}\left(\overline{K}^{\text{KN}} + \overline{K}^{\text{sca}} + \overline{K}^{\text{ph}} \right) + \frac{\omega_O}{A_O}\left(8\overline{K}^{\text{KN}} + 8^{2.86} \overline{K}^{\text{sca}} + 8^{4.62} \overline{K}^{\text{ph}} \right)} - 1 \right) \cdot 1000$$

(11.1)

where ρ is the physical density of the material, ω_i is the weight fraction of the ith element, Z_i is its atomic number, and A_i is its atomic weight. The factors $\overline{K}^{\text{KN}}$, $\overline{K}^{\text{sca}}$ and $\overline{K}^{\text{ph}}$, respectively, denote the Klein–Nishina coefficient, the coherent and the binding correction for the incoherent scattering, while the third takes into account the photoelectric absorption. The $\overline{K}$ parameters are determined by fitting the last equation to the measured CT numbers of the previous step. Accurate knowledge of the insert density and elementary composition is critical for the accuracy of CT response parameterization. The vendor-reported values of insert density and composition can be validated by comparison of RSP measurements with calculations, as described in step 4. With the $\overline{K}$ values determined, the HU of any tissue of known physical density and elementary composition can then be predicted with the previous formula. Tissue density and composition tables can be found in International Commission on Radiation Units and Measurements (ICRU) report 46 [26] and International Commission on Radiological Protection (ICRP) reports 23 [27] and 110 [28].

4. *Calculation of the RSP of the same tissues.* The calculation method is based on an approximation of the Bethe–Bloch formula as described by Bichsel [29]:

$$\text{RSP} = \frac{\rho_e^m}{\rho_e^w} \frac{\ln\left(\frac{k(E)}{I_m} \right) - \beta^2}{\ln\left(\frac{k(E)}{I_w} \right) - \beta^2}$$

(11.2)

where $\rho_e = \rho N_A \sum_i \frac{\omega_i Z_i}{A_i}$ is electron density, $k(E) = \frac{2m_e c^2 \beta^2}{1-\beta^2}$ with $\beta^2 = 1 - \left(\frac{m_p c^2}{E + m_p c^2} \right)^2$,

and I is the mean excitation energy of the material. According to ICRU report 49 [30], the mean excitation energy for a compound is calculated from the mean excitation energies of the elements using the Bragg additivity rule: $\ln I = \dfrac{\sum\limits_i \left(\dfrac{\omega_i Z_i}{A_i} \ln I_i \right)}{\sum\limits_i \left(\dfrac{\omega_i Z_i}{A_i} \right)}$

with modified I for various states of compounds.

5. *Generation of calibration curve RSP(HU) based on piecewise linear fit of the tissue points.* Figure 11.4 shows two HU-RSP curves. One is based on measurements of HU and RSP of phantom materials while the other on predicted HU and RSP values of human tissues. The measured curve includes error bars corresponding to RSP measurement uncertainties and HU variability under various clinically relevant conditions. For TPSs that require HU-ρ calibration, the commissioning includes validation of the RSP internal calculations.

HU-RSP calibration curves are considered a major contributor of the range uncertainties of proton therapy [31], and proton centers have performed comprehensive studies to quantify them for their systems [32,33]. Recent advances with the development of dual energy and spectral CT scanners point to better tissue characterization, but clinically significant uncertainty reduction has not yet been shown [34].

11.4.2 Absolute Machine Output Calibrations and Continuous System Performance Monitoring

An accurate system absolute dose calibration, using an ion chamber previously calibrated by a national calibration laboratory or one of its accredited laboratories, should be performed in the initial phase of proton system commissioning, following a suitable calibration protocol, such as the IAEA TRS-398 report [14] (see Chapter 10). The system output should be subsequently measured on a daily basis to assess system stability (see Chapters 12 and 13). Any significant output drift over the course of the commissioning period should be brought to the attention of vendor engineers for resolution.

For 3DCPT technique, the reference field of absolute dose calibration is simply an SOBP, for example, a beam of 15 cm range and 10 cm SOBP width. The depth and lateral profiles

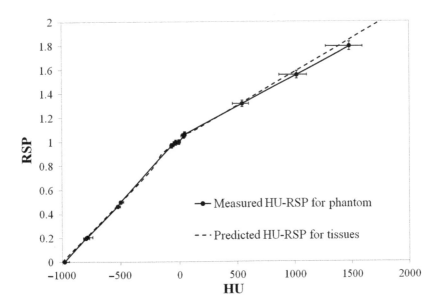

FIGURE 11.4   Measured (phantom) and predicted (tissues) HU-RSP calibration curves.

of the selected field should be verified to be of reproducible uniformity. The calibration will proceed following IAEA TRS-398 report protocol, using a calibrated ionization chamber correct for K_Q as can be looked up in TRS-398 report, and measurement readings corrected for ion recombination, polarity, and temperature and pressure readings.

Determination of absolute dose calibration reference field for the scanning technique is a bit more difficult when the TPS has not been modeled for the pencil beams and cannot produce an SOBP treatment plan for this purpose. The user may perform absolute dose calibration using a single layer of evenly weighted spots, with measurement point in the entrance region of the dose distribution (for example, at 50% of Bragg peak maximum dose depth). Alternatively, the user may manually create an SOBP, containing evenly spaced layers and equally weighted spots within each layer. In either case, the spot and layer spacing, as well as the ratio of spot weights between layers in the SOBP method, will need to be confirmed in depth and lateral profile measurements before absolute dose calibration can be performed.

In addition, a select set of system performance parameters, including beam ranges and SOBP widths; beam spot size, location, and symmetry; mechanical imaging and proton beam isocentricity and agreement; and uniformity of beam depth and lateral profiles, should be measured periodically as well. These measurements should alternate between available beam energies, gantry angles, PPS positions, snout sizes, etc. Demonstrated reproducibility and stability of the system through these measurements will constitute the foundation of the periodic system QA program.

11.4.3 Additional Beam Performance Parameter Measurements Beyond Those of Acceptance Tests

The measurements performed during acceptance testing are only a small subset of the available system operation parameter space, such as gantry angles, PPS translation and rotation angles, snout travel distances and payloads, beam energies, SOBP widths, and field sizes. It is important that additional measurements are performed during the commissioning process to further confirm that the mechanical, imaging, and beam performance parameters consistently meet specifications. Any deficiencies in system performance should be brought to the attention of vendor representatives in search of a resolution.

11.4.4 Data Collection for TPS Modeling

A complete machine model in TPS includes both dosimetric data, or beam data that are necessary for input to the TPS dose calculation algorithm; and nondosimetric parameters that describe the mechanical and imaging system geometry, including X-ray tube locations relative to proton beam, distances of X-ray tubes to isocenter and imaging panels, X-ray image size at isocenter, gantry and PPS, snout movement parameters, etc. The nondosimetric data required for system commissioning are already acquired during acceptance testing or can be acquired from system vendor directly. Parameters that impact treatment delivery accuracy, such as imaging system source-to-axis-distance (SAD) values, should be validated. Much of the dosimetric and nondosimetric data required for a complete TPS modeling are specific to the machine design and delivery techniques of the proton therapy system, and specific to the dose calculation algorithm used in the TPS. TPS vendors usually offer training courses on these topics. The user should participate in these training courses with adequate lead time in the preparation of upcoming system commissioning process.

Many of current proton TPS use proton pencil beam dose calculation algorithm for dose calculations [35] (see Chapter 14). This algorithm calculates, for a given beam energy, the depth dose of a Bragg peak using an analytical equation, corrected for secondary particles. The lateral dose spread characteristics in the system nozzle and in patient are modeled by Gaussian distributions, again corrected for secondary nuclear interaction products.

It should be noted that, while the beam data collected in this section are necessary for TPS beam modeling, they are inadequate for the comprehensive validation of the TPS. It is often more efficient to collect the beam data required for TPS modeling as well as validation at the same time. A complete list of beam data to be measured for both purposes, together with the dosimetry instruments to be used and measurement conditions, should be created prior to beginning of TPS commissioning to allow maximum efficiency of this task.

For 3DCPT techniques, the broad proton beam is calculated by simulated pencil beamlets. SOBPs are formed by stacking Bragg peaks of discrete energies, as produced in the nozzle, and weighted to achieve a flat depth-dose profile. Measured beam data required for these calculations include the following:

- Depth-dose profiles of Bragg (Pristine) peaks for all available options, with small enough separation energies, allow accurate interpolations between the measured depth-dose profiles. These should be measured using a parallel-plate chamber with small electrode separation (≤ 1 mm). The measurements should be performed in an open field of large diameter to minimize the effect of nozzle scattering and ensure proton lateral scattering equilibrium on the measurement axis.

- Lateral dose profiles in air, at several distances upstream and downstream from the isocenter, allow calculation of virtual SAD (VSAD) of the beams. Note that 3DCPT formed using uniform scanning or wobbling beam techniques may have different X-and Y-VSAD values corresponding to the X and Y direction scanning magnet locations in the nozzle.

- Longitudinal dose profiles in air allow calculation of effective SAD (ESAD) for inverse-square-law calculations

- Lateral dose profiles in air of half-blocked fields allow calculation of effective source sizes (Figure 11.5).

- Water-equivalent thicknesses of nozzle components of scatterers and modulation wheel steps, for calculating nozzle lateral scattering Gaussian spread. While these data can be obtained from design drawings of these components from the system vendor, user should verify the accuracy of the design data to rule out significant manufacturing deviations.

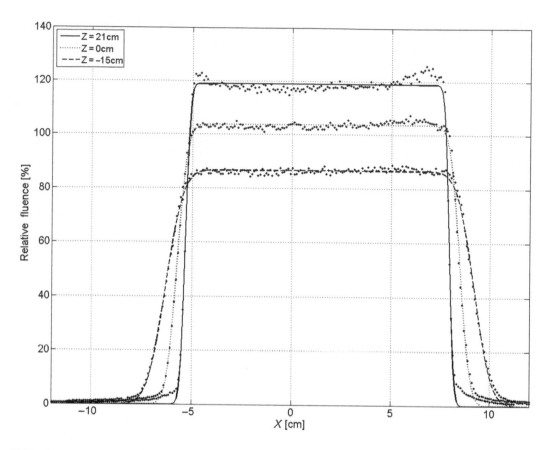

FIGURE 11.5 Measured in-air profiles behind an aperture with an edge at −5 cm and at +7.5 cm from the beam axis, positioned at 20 cm from isocenter. Profiles are measured in the isocenter plane and at 15/21 cm upstream/downstream from isocenter. Treatment planning algorithm determines the "effective source size" by fitting of the lateral falloff with an error function and applying a linear fit to the found sigma as function of position Z.

- Additional nozzle component data, such as modulation wheel step width for systems that use beam current modulation to achieve uniformity of SOBP depth-dose profiles.

- Relative linear stopping power values of treatment accessories of apertures and range compensators

  Scanning pencil beam techniques use magnets to scan a narrow pencil beam across the treatment field, either in a spot-scanning or a line-scanning manner. Pencil beam dose calculation algorithms are common for these techniques as well. Input data for modulated scanning beam modeling are less than what are required for 3DCPT techniques, and include the following:

- IDD curves [36], which are measured using a large-diameter parallel-plate ion chamber, are intended to collect the total (integrated) dose from the pencil beam, for each depth, across its entire lateral spread dimensions. IDD curves of energies with small (several MeV) steps are typically measured to minimize TPS interpolation errors during beam modeling. It should be noted that the large-diameter parallel ion chamber will not be able to collect all large-angle scattered nuclear interaction productions of the beam pencil, or the so-called nuclear halo [37], either produced in the nozzle (ion chambers and other nozzle components) or in patient body. The lateral dose profile of each pencil may be modeled by two superimposed Gaussian distributions, with a smaller sigma Gaussian function of heavier weight modeling the primary proton dose component of the pencil beam; and a second larger sigma Gaussian function of lower weight to model the nuclear halo. Several methods can be used to determine the weight of nuclear halo for such a "double-Gaussian" beam model, including use of Monte Carlo calculations [20,38]. TPS may include Monte Carlo-calculated corrections for nuclear halo for user-provided IDD curves in their beam model module based on user-entered diameter of parallel ion chamber used for IDD measurements. This eliminates the necessary user measurements or calculations of nuclear halo, although the corrections applied should be validated as well. These measurements that are performed with the pencil beam in a static position or "unscanned" therefore require the system to be operated in a service mode.

- Lateral spot profiles in air, at a number of distances away from the isocenter along beam axis, to obtain spot size and symmetry values for TPS beam modeling (Figure 11.6). The spot sigma values are used to calculate the lateral Gaussian spread of IDDs for each depth. Nuclear interaction components of the beam can be measured in air to help define the second Gaussian components. These measurements are typically performed using a validated scintillation detection system as they require

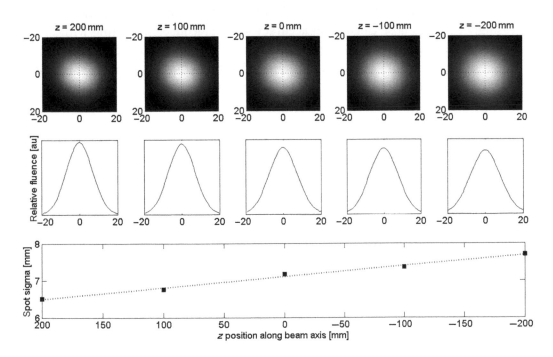

FIGURE 11.6   PBS spot profile measured along the beam axis with a 2D scintillating detector.

high-resolution lateral profile data. Measurements at locations away from isocenter and along beam axis allow evaluation of beam spot characteristics variations away from the isocenter plane for accurate lateral dose spread calculation in water. The pencil beam X- and Y-VSAD values can also be extracted from these measurements.

- The IDDs and lateral in-air spot profiles for each range shifter, if the TPS does not model these. Note that these data need to be measured for validation purpose later.

- Absolute dose output factors with measurement conditions, as discussed in Section 11.4.2.

11.4.5 TPS Dose Calculation Validation

Validation of TPS dose calculations compares absolute and relative dose distributions of various treatment field configurations that are calculated by the TPS vs. measured ones (see Chapter 14). The TPS beam model is established using the data collected in the previous section, as well as adjustable model-fitting parameters required by the specific implementation of the TPS dose calculation algorithm. Errors in beam data used in beam modeling, or incorrect values used for the modeling fitting parameters, can lead to significant differences between TPS-calculated and measured dose distributions.

In addition, TPS dose calculation algorithms may have inherent weaknesses in being able to calculate dose distributions for rare but clinically relevant treatment field configurations. Validation of the TPS should attempt to identify these weaknesses, so that their overall dosimetric impact on patient treatment plans may be minimized, or taken into consideration in the evaluation of such treatment plans.

With these considerations, the process of TPS validation can be divided into steps that increasingly test TPS dose calculation accuracy and inadequacies, starting with tests of open, uncompensated, or nonmodulated fields for 3DCPT and modulated scanning beam techniques, respectively, in homogeneous water phantom, and culminating with tests of blocked, compensated, or modulated fields in heterogeneous anthropomorphic phantoms.

11.4.5.1 Non-modulated Fields in Homogeneous Phantoms; Absolute and Relative Doses

For both 3DCPT and modulated scanning beam techniques, these validation tests use SOBPs (simulated SOBPs for modulated scanning beams) of ranges and SOBP widths that span the deliverable limits of the proton therapy system. As these tests use scanning water phantoms, MLICs, and 2D ion chamber arrays for measurements, they can only be performed for limited gantry angles. These tests should be repeated for each available range shifter and/or miniridge filter of modulated scanning beams, where applicable.

For 3DCPT technique, milled rectangular apertures, of width starting from 3 cm, up to the maximum values available from the system, at 5 cm steps, should be fabricated to shape the treatment fields. Beam range and SOBP widths should span the minimum and maximum limits of each available beam option. Of particular interests are the "full modulation" SOBP fields, where the SOBP widths are equal to or larger than SOBP ranges. Proton pencil beam dose calculation algorithms do not always calculate the entrance doses of such fields with adequate accuracy. Discrepancies between measured and calculated depth and lateral dose profiles should be recorded for future clinical treatment planning considerations. For modulated scanning beam technique, a similar set of SOBP fields should be created, with field sizes, SOBP ranges, and widths spanning system limits.

Depth and lateral dose profiles of these treatment fields should be measured and compared to their TPS counterparts. The comparisons will include field SOBP widths, ranges, and sizes; dose distribution uniformity; lateral penumbra values; distal falloff values; and, for 3DCPT techniques, the proximal lateral profiles including beam entrance regions; proximal, middle, and distal SOBP regions (Figure 11.7). Use of field-shaping apertures will introduce the so-called "slit-scattering" doses or protons that scatter off the apertures

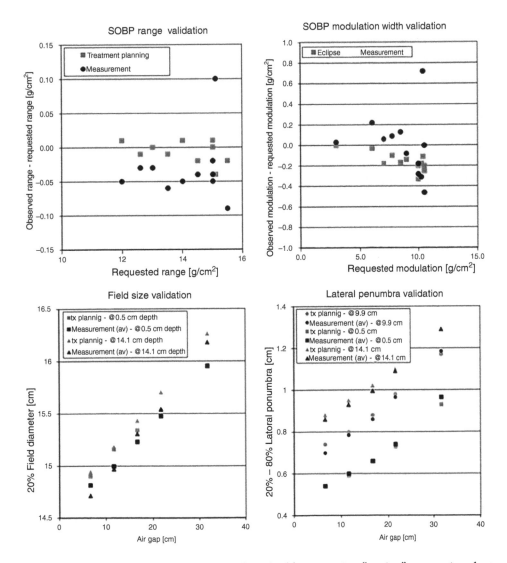

FIGURE 11.7 Select validation measurements for a double-scattering "option" comparing the treatment planning SOBP range (top left), SOBP modulation width (top right), 95% isodose diameter (bottom left), and 80%–20% lateral penumbra (bottom right) to measured values.

and produce "horns" in the shallow depths immediately inside the field edges [39]. Current commercially available proton pencil beam dose calculation algorithms cannot model these horns. The magnitudes of these horns, as well as their depths, should be recorded for future clinical treatment planning considerations (Figure 11.8).

TPS dose calculation for fields with apertures of very small diameters should be evaluated, as well as corners of rectangular fields, to evaluate TPS accuracy of modeling loss of proton lateral scattering in these volumes [40] (Figure 11.9). Additional apertures of smaller than 3 cm width may need to be fabricated to comprehensively evaluate this concern.

For all measurements, the absolute doses for the TPS-calculated MU values, if available, should be measured and compared to TPS-calculated dose values. This is especially important for modulated scanning beams, where significant deviations may indicate inadequate TPS calculation of the nuclear halo effect. If the TPS models nuclear halo using a second Gaussian distribution of the spot lateral profile, the weight of second Gaussian may be adjusted to improve the agreement.

Users should also include tests of TPS dose calculations of

- Beam penumbra variations for 3DCPT technique as a function of varying air gaps

- Depth and lateral dose profiles of modulated scanning beams using range shifters, as a function of varying air gaps. Proton pencil beam dose calculation algorithms may not be able to accurately calculate these dose distributions, especially at shallower depths. The discrepancies, if any, should be recorded and considered in future clinical treatment planning procedures.

- Effects of oblique beam incidence angles. These can be measured in a 3D scanning water tank by introducing a gantry rotation angle in the treatment plans, and the

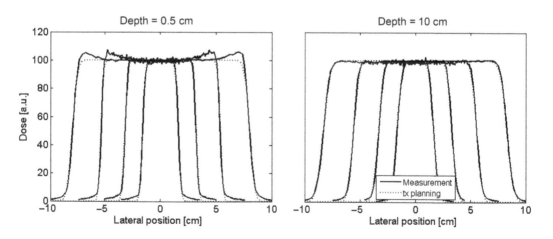

FIGURE 11.8 Lateral profile of a double-scattering SOBP measured behind square apertures of different sizes. Comparing measurement (black, solid) to treatment planning (gray, dash) at 0.5 cm (left) and at 10 cm depth (right) inside a water phantom. Note the "horns" and "tails" in the measured profile at shallow depth, especially for larger aperture size, caused by "slit-scattered" protons.

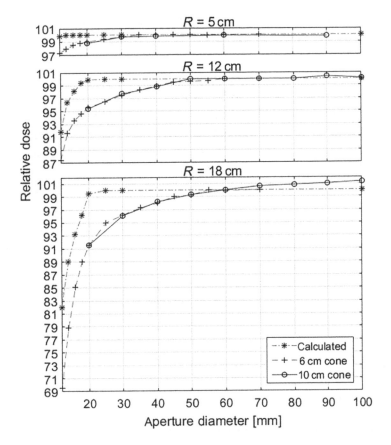

FIGURE 11.9 Variation of dose with aperture size at various depths (upper 5 cm, middle 12 cm, lower 18 cm) for a 2-cm wide SOBP in the Massachusetts General Hospital stereotactic eyeline. (Figure courtesy of Juliane Daartz; for more details see [41])

scanning dosimeter programmed to measure the depth-dose profiles along the beam axis. A combination of oblique beam incidence angle and large air gap for a modulated scanning beam treatment using range shifters may be one of the most challenging scenarios for proton dose calculation algorithms.

11.4.5.2 Modulated Fields in Homogeneous Phantoms; Absolute and Relative Doses

These tests evaluate the dosimetric accuracy of range compensators of 3DCPT technique and beam intensity modulation accuracy of modulated scanning beams. Treatment plans of limited modulations with dose distributions containing stepwise changes in lateral profiles may be used. Lateral dose profiles at multiple depths should be measured, using scanning water phantom for passive-scattering beams, and 2D ion chamber arrays for uniform scanning, wobbling beams, and modulated beams. Subsequently, a set of treatment plans for common tumors to be treated at the proton therapy facility, such as H&N (head-and-neck), lung, breast, CNS (central nervous system), and others, should be calculated and measurements repeated. Comparisons of measured vs. calculated dose distributions and absolute doses should be evaluated for in-field and out-of-field agreements, and supplemented by Gamma Test analysis.

11.4.5.3 Non-modulated Fields in Heterogeneous Slab Phantoms; Absolute and Relative Doses

Measurements of TPS dose calculation accuracy in heterogeneous phantoms can be performed using solid water, bone, and lung approximation slab phantoms. While dose distributions distal to the slab phantoms can be performed by stacking alternative layers of heterogeneous phantom materials over a 2D ion chamber array, film dosimetry is necessary for measurements of dose distributions within the phantoms. These tests should also include absolute dose measurements within the slab phantom, using a solid water phantom slab that accepts an ion chamber. Lateral dose profiles in lung phantom can be measured by placing the film under solid water, bone, and lung phantom slabs, and over a reverse arrangement of these phantom slabs. Dosimetric effect of lateral tissue interface can be evaluated with a piece of bone abutting laterally to a piece of lung, placed over the film. Film dosimetry for depth-dose profiles can be performed in a similar manner by use of a gantry angle approximately 90°, although it may be advisable to avoid having the beam axis parallel to the film surface, to avoid measurement errors due to the non-tissue equivalent RSP of the film. The high linear-energy transfer (LET) of proton beams in the distal falloff region has been reported to cause under response of radiochromic films [42]. Measurements using these films should therefore include an evaluation of the magnitude of such over responses.

11.4.6 Commissioning of Patient Immobilization Devices

All patient immobilization devices that may be traversed by the treatment beams require commissioning measurements [43]. These include all PPS tabletops and inserts, thermoplastic masks, and pillows. Range pullbacks of these devices must be measured and accounted for in treatment planning; or verified that their HU values in simulation CT images, if applicable, will allow range pullback calculations in TPS of adequate accuracy.

11.4.7 Secondary MU Calculation

There does not exist a commercial secondary MU dose calculation system for proton therapy at the present time (see Chapter 13). Several methods have been reported for system-specific 3DCPT secondary MU calculations [44–46]. 3DCPT MU calculations are affected by nozzle components used in delivery of a treatment field, as well as the SOBP ranges and modulation widths, and no universal formalism is currently available. Users of a specific 3DCPT system will therefore need to take the references earlier and develop their own system of secondary MU calculations, such as performing near-exhaustive measurements of SOBPs of various range and width combinations, a process not too dissimilar to what is done for photon radiotherapy, although with significantly increased workload. The task is slightly reduced, however, by the near independence of SOBP output factors relative to larger field sizes. The small-field measurements of Section 11.4.5.1 should be performed to create tables of field-size correction factors, with attention to the range dependence of such correction factors as well.

11.4.8 Peer Review of Output Calibration and End-to-End Tests

Before patient treatment may begin using the newly commissioned proton therapy system, peer review of results of the commissioning tests should be performed. The review will concentrate on absolute dose calibration, CT HU-RLSP calibration, and TPS dose calculations. External experts with experiences in system commissioning may be invited to perform an on-site review. In the United States, the Imaging and Radiation Oncology Core (IROC) offers proton therapy facilities several forms of peer reviews. The IROC provides a remote dosimetry monitoring program, in which a set of thermoluminescence dosimeters (TLDs) can be shipped to the user with instructions of irradiation. A report is issued on the TLD-measured dose for comparison to the user-expected dose. In addition, IROC provides various anthropomorphous phantoms, equipped with implanted fiducial markers for image-guided positioning, as well as TLD and radiochromic film holders, for verification of the entire treatment simulation, planning, and delivery process at the user's facility. These remote dosimetry monitoring reviews provide high confidence in the accuracy of proton therapy treatment at the user's facility and are mandatory for institutions that participate in clinical trials sponsored by the National Cancer Institute.

11.5 SUMMARY

This chapter provides an overview of the acceptance testing and commissioning of a proton therapy system. It must be recognized, however, that the complexity of a multiroom proton therapy system often prevents the continuous and comprehensive completion of system commissioning in a limited time frame. It is likely that system commissioning is completed in stages, for a limited span of beam energies, disease sites, or even gantry rotation angle spans. Development of a comprehensive periodic machine QA program may start with a more detailed daily QA tests in the initial phase of patient treatments and gradually scaled back by moving the daily QA tests to less frequent tests, such as monthly QA tests. Development of patient imaging, simulation, treatment planning, and delivery procedures may likewise be gradual, in step with patient volume ramp-up at the proton facility. Staff training of the system will include vendor-provided training in system operations, internal training of applicable procedures that have been developed at the facility, and mock treatments of anthropomorphic phantoms of various disease sites.

REFERENCES

1. American College of Radiology. *ACR-ASTRO Practice Parameter for the Performance of Proton Beam Radiation Therapy*. Reston, VA: American College of Radiology; 2014.
2. Nath R, Biggs PJ, Bova FJ, Ling CC, Purdy JA, van de Geijn J, Weinhous MS. AAPM code of practice for radiotherapy accelerators: report of AAPM Radiation Therapy Task Group No. 45. *Med Phys*. 1994;21(7):1093–121.
3. Kutcher GJ, Coia L, Gillin M, Hanson WF, Leibel S, Morton RJ, Palta JR, Purdy JA, Reinstein LE, Svensson GK, et al. Comprehensive QA for radiation oncology: report of AAPM Radiation Therapy Committee Task Group 40. *Med Phys*. 1994;21(4):581–618.
4. Klein EE, Hanley J, Bayouth J, Yin FF, Simon W, Dresser S, Serago C, Aguirre F, Ma L, Arjomandy B, Liu C, Sandin C, Holmes T. Task Group 142 report: quality assurance of medical accelerators. *Med Phys*. 2009;36(9):4197–212.

5. International Atomic Energy Agency. The use of plane-parallel ionization chambers in high-energy electron and photon beams. *An International Code of Practice for Dosimetry, Technical Report Series No 381*. Vienna, Austria: International Atomic Energy Agency; 1997.

6. Arjomandy B, Sahoo N, Ding X, Gillin M. Use of a two-dimensional ionization chamber array for proton therapy beam quality assurance. *Med Phys*. 2008;35(9):3889–94.

7. Kirby D, Green S, Palmans H, Hugtenburg R, Wojnecki C, Parker D, LET dependence of GafChromic films and an ion chamber in low-energy proton dosimetry. *Phys Med Biol*. 2010;55(2):417–33.

8. Vatnitsky SM. Radiochromic film dosimetry for clinical proton beams. *Appl Radiat Isot*. 1997;48(5):643–51.

9. Hartmann B, Martisikova M, and J"akel O. Homogeneity of Gafchromic EBT2 film. *Med Phys*. 2010;37(4):1753–6.

10. Mumot M, Mytsin GV, Molokanov AG, Malicki J. The comparison of doses measured by radiochromic films and semiconductor detector in a 175 MeV proton beam. *Phys Med*. 2009;25(3):105–10.

11. Arjomandy B, Tailor R, Anand A, Sahoo N, Gillin M, Prado K, Vicic M. Energy dependence and dose response of Gafchromic EBT2 film over a wide range of photon, electron, and proton beam energies. *Med Phys*. 2010;37(5):1942–7.

12. Moyers MF, Lesyna W. Isocenter characteristics of an external ring proton gantry. *Int J Radiat Oncol Biol Phys*. 2004;60(5):1622–30.

13. Keuschnigg P, Kellner D, Fritscher K, Zechner A, Mayer U, Huber P, Sedlmayer F, Deutschmann H, Steininger P. Nine-degrees-of-freedom flexmap for a cone-beam computed tomography imaging device with independently movable source and detector. *Med Phys*. 2017;44(1):132–142.

14. International Atomic Energy Agency. Absorbed dose determination in external beam radiotherapy. *Technical Reports Series No. 398*. Vienna, Austria: International Atomic Energy Agency; 2000.

15. Safai S, Bortfeld T, Engelsman M. Comparison between the lateral penumbra of a collimated double-scattered beam and uncollimated scanning beam in proton radiotherapy. *Phys Med Biol*. 2008;53(6):1729–50.

16. Hyer DE, Hill PM, Wang D, Smith BR, Flynn RT. Effects of spot size and spot spacing on lateral penumbra reduction when using a dynamic collimation system for spot scanning proton therapy. *Phys Med Biol*. 2014;59(22):N187–96.

17. Hyer DE, Hill PM, Wang D, Smith BR, Flynn RT. A dynamic collimation system for penumbra reduction in spot-scanning proton therapy: proof of concept. *Med Phys*. 2014;41(9):091701.

18. Low DA, Harms WB, Mutic S, Purdy JA. A technique for the quantitative evaluation of dose distributions. *Med Phys*. 1998;25(5):656–61.

19. Slopsema RL, Lin L, Flampouri S, Yeung D, Li Z, McDonough JE, Palta J. Development of a golden beam data set for the commissioning of a proton double-scattering system in a pencil-beam dose calculation algorithm. *Med Phys*. 2014;41(9):091710.

20. Clasie B, Depauw N, Fransen M, Gomà C, Panahandeh HR, Seco J, Flanz JB, Kooy HM. Golden beam data for proton pencil-beam scanning. *Phys Med Biol*. 2012;57(5):1147–58.

21. Langner UW, Eley JG, Dong L, Langen K. Comparison of multi-institutional Varian ProBeam pencil beam scanning proton beam commissioning data. *J Appl Clin Med Phys*. 2017;18(3):96–107.

22. Mutic S, Palta JR, Butker EK, Das IJ, Huq MS, Loo LND, Salter BJ, McCollough CH, Van Dyk J. Quality assurance for computed-tomography simulators and the computed-tomography-simulation process: Report of the AAPM Radiation Therapy Committee Task Group No. 66. *Med Phys*. 2003;30(10):2762–92.

23. Rutherford RA, Pullan BR, Isherwood I. Measurement of effective atomic number and electron density using an EMI scanner. *Neuroradiology*. 1976;11(1):15–21.

24. Schneider U, Pedroni E, Lomax A. The calibration of CT Hounsfield units for radiotherapy treatment planning. *Phys Med Biol.* 1996;41(1):111–24.
25. Schaffner B, Pedroni E. The precision of proton range calculations in proton radiotherapy treatment planning: experimental verification of the relation between CT-HU and proton stopping power. *Phys Med Biol.* 1998;43(6):1579–92.
26. International Commission on Radiation Unites & Measurements. *ICRU Report 46: Photon, Electron, Proton and Neutron Interaction Data for Body Tissues.* Bethesda, MD: International Commission on Radiation Unites & Measurements; 1992.
27. International Commission on Radiological Protection. *ICRP Publication 23: Report of the Task Group on Reference Man.* Ontario: International Commission on Radiological Protection; 1975.
28. International Commission on Radiological Protection. *ICRP Publication 110: Adult Reference Computational Phantoms.* Ontario: International Commission on Radiological Protection; 2009.
29. Bichsel H. Passage of charged particles through matter. In: Billlings BH, Gray DE, editors. *American Institute of Physics Handbook.* 3rd ed. New York: McGraw-Hill; 1972.
30. International Commission on Radiation Unites & Measurements. *ICRU Report 49: Stopping Power and Ranges for Protons and Alpha Particles.* Ontario: International Commission on Radiological Protection; 1984.
31. Paganetti H. Range uncertainties in proton therapy and the role of Monte Carlo simulations. *Phys Med Biol.* 2012;57(11):R99–117.
32. Yang M, Zhu XR, Park PC, Titt U, Mohan R, Virshup G, Clayton JE, Dong L. Comprehensive analysis of proton range uncertainties related to patient stopping-power-ratio estimation using the stoichiometric calibration. *Phys Med Biol.* 2012;57(13):4095–115.
33. Ainsley CG, Yeager CM. Practical considerations in the calibration of CT scanners for proton therapy. *J Appl Clin Med Phys.* 2014;15(3):4721.
34. Li B, Lee HC, Duan X, Shen C, Zhou L, Jia X, Yang M. Comprehensive analysis of proton range uncertainties related to stopping-power-ratio estimation using dual-energy CT imaging. *Phys Med Biol.* 2017;62(17):7056–74.
35. Hong L, Goitein M, Bucciolini M, Comiskey R, Gottschalk B, Rosenthal S, Serago C, Urie M. A pencil beam algorithm for proton dose calculations. *Phys Med Biol.* 1996;41(8):1305–30.
36. Schaffner B. Proton dose calculation based on in-air fluence measurements. *Phys Med Biol.* 2008;53(6):1545–62.
37. Zhang X, Liu W, Li Y, Li X, Quan M, Mohan R, Anand A, Sahoo N, Gillin M, Zhu XR. Parameterization of multiple Bragg curves for scanning proton beams using simultaneous fitting of multiple curves. *Phys Med Biol.* 2011;56(24):7725–35.
38. Lin L, Ainsley CG, Mertens T, De Wilde O, Talla PT, McDonough JE. A novel technique for measuring the low-dose envelope of pencil-beam scanning spot profiles. *Phys Med Biol.* 2013;58(12):N171–80.
39. Van Luijk P, van t' Veld AA, Zelle HD, Schippers JM. Collimator scatter and 2D dosimetry in small proton beams. *Phys Med Biol.* 2001;46(3):653–70.
40. Geng C, Daartz J, Lam-Tin-Cheung K, Bussiere M, Shih HA, Paganetti H, Schuemann J. Limitations of analytical dose calculations for small field proton radiosurgery. *Phys Med Biol.* 2017;62(1):246–57.
41. Daartz J, Engelsman M, Paganetti H, Bussière MR. Field size dependence of the output factor in passively scattered proton therapy: influence of range, modulation, air gap, and machine settings. *Med Phys.* 2009;36(7):3205–10.
42. Arjomandy B, Tailor R, Zhao L, Devic S. EBT2 film as a depth-dose measurement tool for radiotherapy beams over a wide range of energies and modalities. *Med Phys.* 2012;39(2):912–21.
43. Wroe AJ, Bush DA, Slater JD. Immobilization considerations for proton radiation therapy. *Technol Cancer Res Treat.* 2014;13(3):217–26.

44. Kooy HM, Schaefer M, Rosenthal S, Bortfeld T. Monitor unit calculations for range-modulated spread-out Bragg peak fields. *Phys Med Biol.* 2003;48(17):2797–808.
45. Kase Y, Yamashita H, Numano M, Sakama M, Mizota M, Maeda Y, Tameshige Y, Murayama S. A model-based analysis of a simplified beam-specific dose output in proton therapy with a single-ring wobbling system. *Phys Med Biol.* 2015;60(1):359–74.
46. Ferguson S, Chen Y, Ferreira C, Islam M, Keeling VP, Lau A, Ahmad S, Jin H. Comparability of three output prediction models for a compact passively double-scattered proton therapy system. *J Appl Clin Med Phys.* 2017;18(3):108–117.

CHAPTER **12**

Quality Assurance[*]

Juliane Daartz

Massachusetts General Hospital and Harvard Medical School

CONTENTS

12.1 INTRODUCTION

Quality assurance (QA) comprises all measures that ensure that a system functions as it did during acceptance and commissioning. Proton QA categories—machine and patient

[*] Colour figures available online at www.crcpress.com/9781138626508.

specifics—are the same as those of conventional radiotherapy but differ in execution as a consequence of the physics of protons and the delivery systems in particular. Challenges in a proton QA program lie in the absence of standards for QA procedures or acceptance criteria and lack of modality-specific commercially available dosimetry equipment. There is a range of proton delivery technologies in the field, which further complicates the development of unified recommendations. There still are both passive-scattering and beam scanning systems in routine clinical use. Both will be considered in the following sections.

Many institutions design in-house procedures and rely on custom use of standard measurement equipment. While most try to follow the spirit of Task Group (TG) 142 of the American Association of Physicists in Medicine (AAPM) [1], and omit or augment as needed, there is no equivalent publication for proton therapy yet. The report of AAPM task group 224 is in progress and is supposed to fill this void.

Proton therapy is still a niche treatment. Compared to conventional QA procedures and tools, proton therapy QA requires more man hours of complicated procedures utilizing more expensive equipment. It is our responsibility to critically examine our protocols to maximize efficiency. A good resource may be the philosophy of TG 100 [2], which demonstrates some basic methods to understand where to best place resources.

This chapter aims to provide practical considerations of a proton QA program and describes tools and methods most frequently employed in current proton therapy clinics.

12.2 MACHINE QA

This section describes non-patient-specific tests that check technical aspects of beam delivery, such as dosimetric accuracy and mechanical alignment.

12.2.1 Dosimetric Parameters

The clinically important dosimetric parameters as well as their measurement techniques depend on the delivery mode (Figure 12.1). In both passive scattering and pencil beam scanning, a measure of the penetration depth of the beam is monitored. This is usually the distal 90% or 80% point, but, for clinical purposes, can be defined at any stable distinct depth feature. The width of the high-dose region is usually described by the 80%–80% peak width in scanned beams, and 90%–95% or 90%–98% modulation width in passive scattering. The steepness of the distal falloff can be captured by the distal 80%–20% (of peak) depth separation.

In the lateral direction, spot size (characterized by its one sigma width or full-width at half-maximum (FWHM)) and spot position (position of the beam's centroid maximum dose in the isocentric plane) are most important in pencil beam scanning. In passively scattered fields, penumbral width, field width, symmetry, and flatness of the high-dose region are measured. The latter is computed as described in TG 142 [1].

These parameters can vary as a function of gantry angle, beam energy, and nozzle hardware and software configuration, i.e. treatment delivery device equipment settings, needed for a specific treatment beam. Therefore, there is a large set of failure modes for each parameter to fail QA tolerances. Most institutions started operations with an extensive and time-consuming QA program, until sufficient data was gathered to show

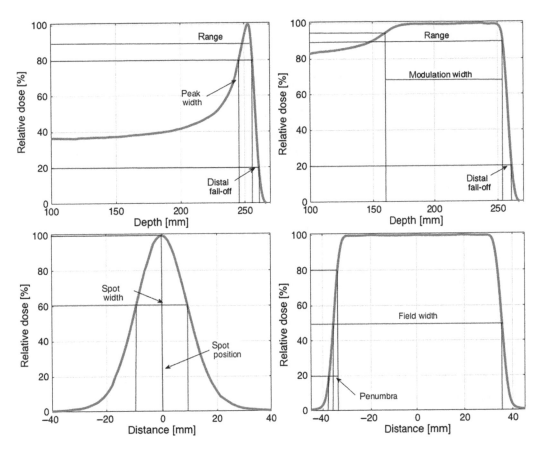

FIGURE 12.1 Top left: Integral depth dose of a pencil beam. For clinical purposes, important parameters are range in water range (in this example, defined at the distal 90% point), peak width (here shown as the distance between the two 80% points) and width of the distal falloff (here the distance between 80% and 20% points). Top right: Central axis depth dose of a double-scattered SOBP. Measured quantities indicated are the distal, modulation width (here shown as the distance between distal 90% and proximal 95% points) and the width of the distal falloff. Bottom left: Profile of a pencil beam (measured at depth in water). Most important quantities are position of the spot center and spot width. Bottom right: Profile of a double-scattered beam collimated by a 70 mm diameter brass aperture. Measured quantities agree with standards in conventional therapy (lateral falloff, symmetry, flatness, 50%–50% width).

most common paths for failure. Procedures are then adapted to increase efficiency and efficacy.

12.2.2 Equipment

12.2.2.1 Depth Doses

Most reference dosimetry is done in water. An essential piece of equipment is therefore a water tank allowing scans in three dimensions. There are devices specifically marketed for proton therapy, featuring modified software and thinner entrance windows, but these are still not fully tailored for proton therapy. Software tools for data acquisition are mostly insufficiently adapted versions of the original designed for photon therapy. Basic data analysis tools exist.

Plane-parallel ionization chambers should be used for depth-dose scans [3], with the center of the inner surface of the window as point of measurement. In large passive-scattering fields, small diameter chambers are the tools of choice to measure only the central axis depth dose in charged particle equilibrium conditions [4]. In pencil beam scanning, large diameter ionization chambers (Bragg Peak Ionization Chamber, PTW, Freiburg, Germany; StingRay, IBA, Dosimetry) are utilized, to capture as much of the primary and secondary dose deposited as possible for an accurate representation of the integral depth dose. It is important to note that no chamber captures the entire beam [5]. Monte Carlo derived correction factors can be used to convert measured data to true integral depth-dose curves [6]. For deeper peaks, a step resolution of 1 mm is sufficient. However, in pencil beam scanning, as the width of the peak region decreases, due to reduced range straggling, it may be necessary to adjust to 0.5 mm steps in the peak region for shallower peaks.

Another possibility for high-resolution depth-dose measurements is utilization of a commercially available, closed water column with a built-in large diameter ionization chamber for scanning in depth only (PEAKFINDER, PTW, Freiburg, Germany). This device may be used at any gantry angle.

A practical issue is the thickness of the front wall of a water tank. It limits the minimum measurable depth and introduces scatter different from water. TRS 398 recommends plastic tank walls between 0.2 and 0.5 cm physical thickness. Its water equivalent thickness should ideally be measured but can be calculated, given the wall thickness and mass density of the plastic.

Although the mechanical accuracy of the available devices is better than 0.5 mm, performing absolute range measurements with an overall accuracy of ±1 mm is challenging and requires very careful setup. Most published procedures for range margins in treatment planning include a 1 mm error to account for machine delivery accuracy [7].

To check for constancy of the depth-dose distribution, the most frequently used devices are multilayer ionization chambers, most commonly with a depth resolution of around 2 mm [8–13]. Their advantage is a vast decrease in measurement time, since the entire length of the depth dose is captured at once. There are both large and small diameter devices available. If carefully commissioned and compared to reference dosimetry in water to ensure differences are well understood—it may even be possible to use these devices to efficiently augment commissioning data. Reference values of single points on a depth-dose curve (such as range, modulation, or peak width) should be obtained in water, by plane-parallel ionization chamber to avoid dose averaging in depth. Figure 12.2 shows the multilayer ionization chamber distributed by IBA, Dosimetry.

Constancy checks of these parameters, however, may be obtained from constancy depth-dose curves described earlier. Even more simplified validation of range constancy can be achieved using the ratio of two single-point dose measurements in plastic phantoms—one in a uniform dose region to normalize, the other located somewhere in the distal falloff region. Many institutions have successfully set up such daily constancy measurements with devices usually used for photon daily QA [14–16]. Ionization chamber measurements

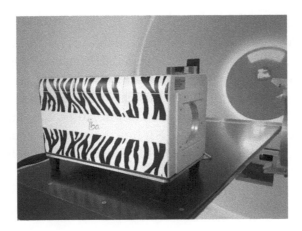

FIGURE 12.2 Multilayer ionization chamber (ZEBRA, IBA, Dosimetry) for measurement of proton depth doses. The device measures up to 33 cm in depth, with 2 mm native resolution. Accuracy of range measurement is reported as ±0.5 mm. The collecting electrode has a diameter of 2.5 cm.

are usually reproducible to within around a percent [17] and therefore are well suited for this task.

For range measurements, multi-layer Faraday cups [18] are in use at some institutions as well. Those devices can be mounted in the treatment head itself and therefore eliminate extra setup time.

Various detectors for dosimetry are described in more detail in Chapter 9.

12.2.2.2 Lateral Profiles

Pencil beam scanning has two clinically important quantities for reference dosimetry: spot position and spot width. If the spot is rather large, it is possible to obtain accurate information on positioning based on commercially available 2D ionization chamber arrays, despite low resolution and volume averaging. For smaller spot sizes, expected in-current use of pencil beam scanning and a wider extent of the measured field, scintillation detectors are used (Lynx, IBA, Dosimetry [19]). The disadvantage of these detectors is their bulkiness and expense. Some authors have reported on the use of strip chambers for this purpose [14,20]. Gafchromic film is an alternative option [21]. Measurements are usually done at isocenter in air, even when used for dosimetric reference. Constancy checks allow a simpler setup, again using commercially available 2D ionization chamber arrays [22,23].

In passive scattering, the important quantities—field flatness, symmetry, and penumbral width—can be derived from scans in a water phantom, using smallest volume thimble ionization chambers, with the caveats of dose averaging over the chamber's extent and long measurement times. Gafchromic film can be used as well. For verification of constancy, 2D ionization chamber arrays are in use [22]. Due to low resolution and dose averaging, these devices are not recommended for exact reference measurements of steep dose gradients, as occurring in the penumbral regions of aperture-collimated proton beams.

12.2.2.3 Absolute Dose

The process for absolute dose calibration is detailed in Chapter 13. Most widely used is the TRS 398 protocol [3]. In summary, a detector is placed at reference depth in water in a large spread-out Bragg peak (SOBP) field (or SOBP-like in the case of scanned beams) and dose measured at various IC voltages to determine chamber correction factors (polarity and recombination). The beam quality correction factor can be determined using tabulated values for the detector utilized. The latter is a function of the residual depth of the point of interest (i.e. water equivalent distance to the distal 10% dose point), which is derived from a depth-dose scan. The absorbed dose is then the product of all correction factors (recombination, polarity, temperature–pressure, electrometer, beam quality), calibration factor as determined by and Accredited Dosimetry Calibration Laboratory (ADCL), and electrometer reading.

The recommended standard for absolute dose measurements according to TRS 398 are graphite wall cylindrical ionization chambers. Plastic-walled detectors provide less stability and more chamber-to-chamber differences. The center of the cavity on its central axis is considered the point of measurement. If the residual range at reference depth is less than $0.5 \, g \, cm^{-2}$ from the practical range, the use of plane parallel chambers is indicated, to avoid dose averaging over a steep dose gradient. Here, the point of measurement is given by the center of the inner front window of the detector. The scanned depth dose used to determine residual depth is acquired with a plane-parallel ionization chamber. The uncertainty in kQ (i.e. the absorbed dose beam quality conversion factor) is 1.7% and 2.1% for cylindrical and plane-parallel chambers, respectively [3]. The Imaging and Radiation Oncology Core (IROC) provides a program for dosimetric intercomparison of proton centers. Approval through the IROC process is, among other criteria, based on annual monitoring of proton beam calibration with TLDs. Summers et al. [24] report that all centers' annual TLD results were within 6% between the years 2011 and 2014.

12.2.3 Nondosimetric Checks

12.2.3.1 Coincidence of Radiation Field and Imaging

Confirming the alignment of imaging and radiation isocenters is essential for accurate patient treatment. The test can be conducted by first aligning a small radio-opaque marker at isocenter using the available clinically applied imaging and setup method (i.e. planar X-rays or cone-beam computed tomography (CBCT)). Gafchromic film is placed downstream of the marker. A single pencil beam or, in passive scattering, a small aperture beam is then used to irradiate the film. The fiducial marker perturbs the beam sufficiently to allow quantification of the distance between the centers of the radiation field and the marker. This can be done using a commercially available film holder with fiducial markers, as shown in Figure 12.3.

At some institutions, a device more commonly applied for photon stereotactic alignment QA is also used for protons [25]—a cone shaped scintillator can measure the intersection of the radiation beam at all gantry angles (XRV-100, Logos Systems Int'l, Scotts Valley, CA).

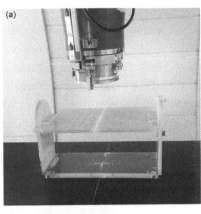

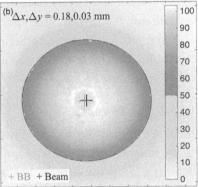

FIGURE 12.3   Left: Setup for verification of coincidence of beam and imaging isocenters in a passively scattered proton beam. The Civco Iso-Align (Civco Radiotherapy, Orange City) holds a rotatable platform with ingrained marks for laser alignment and fiducial markers for X-ray based alignment. For the beam—imaging isocenter coincidence check, the central fiducial is first placed at an imaging isocenter. Gafchromic film is taped to the downstream side of the platform. Right: Image analysis. The distance between the center of the fiducial and center of proton beam is determined using in-house analysis tools. The black circle indicates the 50% isodose line of the proton beam (collimated by a 1 cm diameter brass aperture). The scatter introduced by the fiducial is visible as a hot ring (white).

12.2.3.2 Mechanical Checks

The moving parts include the gantry, patient positioner, and the snout. The latter moves along the beam axis to place field-specific devices such as an aperture, range compensator, or shifter near the patient, thereby minimizing penumbral width. Gantry and couch iso-centricity can be checked with mechanical pointers or using the "Coincidence of Radiation Field and Imaging" procedure at multiple gantry and couch angles [26,27]. General patient positioner translational accuracy and precision can be verified using high-precision mechanical pointers or a simple phantom and the onboard imaging system. Some delivery systems correct for a nonisocentric gantry rotation with an automatic couch positional correction, which must be verified experimentally using the clinically applied workflow. Snout travel should be checked for trueness and positional accuracy.

12.2.3.3 Imaging

Most facilities still use orthogonal X-rays for patient setup, though the use of CBCT is increasing. To verify proper functioning of this equipment, the user is advised to follow standard procedure as defined for conventional therapy. Important documents to consider are the reports from AAPM TG 179 [28], TG 104 [29], and TG 142 [1] that detail measures addressing image quality, accuracy, and precision of the setup procedure. In addition, the connectivity of imaging, dose delivery, and Oncology Information System must be verified.

12.2.3.4 Safety

Most safety checks coincide with the tests done in conventional therapy and are listed in AAPM TG 142 [1]. Measures include checks of mechanical safety features (e.g. collision detection), interlocks (e.g. door, secondary monitor unit (MU) counter), activation levels, and audio-visual systems. Due to the complexity of the treatment process, it is advisable to analyze all procedures leading to and including the treatment itself. TG 100 [2] provides an example for such analysis. Others have published on their experiences using systems-theoretic process Analysis [30] for this purpose. Safety analysis can help in development of new procedures as well as in modification of existing ones.

12.2.3.5 Computed Tomography (CT)

The conversion between Hounsfield units and stopping power ratio is essential in proton radiotherapy. The most common strategy utilized currently is the stoichiometric method as described, for example, by Schneider et al. [31]. In addition to the daily and monthly HU constancy checks typically performed on CTs, it is important to revalidate the HU-SPR curve on an annual basis.

12.2.4 QA Frequency

As of yet, there is no standard when it comes to the frequency of QA measures. An AAPM task group report is underway (TG 224). There are, however, a number of publications detailing QA procedures as performed at various proton therapy centers.

12.2.4.1 Daily

Daily QA verifies constancy of the clinically most important parameters. Ideally, the procedure is efficient and can be handled by radiotherapy technicians, just as is the case in conventional therapy. Unlike conventional therapy, there is no simple and inexpensive commercially available device for daily QA. At some institutions, QA is still performed by physicists, relying on expensive equipment and consuming a lot of time. Multiple authors have reported on in-house developed daily QA routines involving both sparse and higher resolution 2D ionization chamber arrays in combination with plastic phantoms providing beam pullback and/or including radio-opaque markers to incorporate QA of imaging and alignment in the same setup [14–16].

In passive scattering, daily verification must include constancy checks of range and dose, adding symmetry and modulation width are options. Note that large changes in modulation width will be measurable as a deviation in output, i.e. dose per MU, as well. The user might choose to involve all hardware in the nozzle by the daily QA procedure. This can be accomplished without adding time to the daily procedure by rotating through different range and modulation combinations over the course of few days. An example for such a setup is depicted in Figure 12.4. The BeamChecker Plus device (Standard Imaging, Middleton WI), a low-resolution 2D ionization chamber array, is indexed on an in-house built base containing fiducial markers. The base is indexed to the treatment table. QA of the imaging system and patient positioner motion is performed by acquiring orthogonal X-rays of the setup, and noting and executing the shifts as analyzed by the clinically used

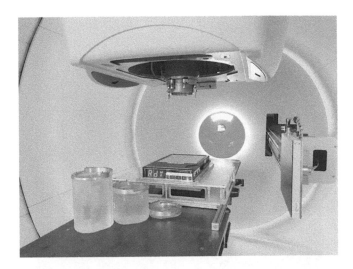

FIGURE 12.4   Daily QA setup for passively scattered beams at the Francis H Burr Proton Therapy Center (Boston, MA). The procedure verifies imaging, setup, patient positioner motion, and dosimetric parameters within 20 min.

alignment software. Dosimetric QA tracks the parameters range, modulation width and dose, as analyzed automatically by in-house written software. Each day one of three different treatment beams is measured, characterized by ranges and modulation widths, and using different range compensator thicknesses to provide the necessary water equivalent thickness. These parameters are chosen such that all nozzle hardware is exercised once over those three fields. The procedure is executed by radiotherapy technicians and takes approximately 20 min. Figure 12.5 shows a 2-month period of data acquired with this procedure for one range-modulation combination.

The parameters to track on a daily basis in pencil beam scanning are range, spot size, spot position, dose, and some authors advocate for a daily measure of range uniformity across the field. Figure 12.6 shows an example for a daily QA setup. A high-resolution 2D

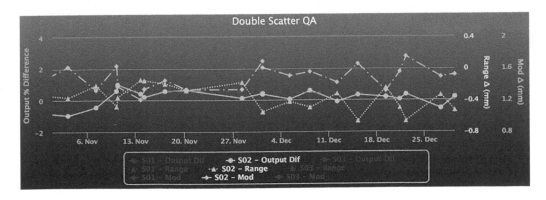

FIGURE 12.5   Example for a 2-month trend review of daily QA tracking proton range, modulation, and dose in a passively scattered beam. The data was acquired using the setup depicted in Figure 12.3.

FIGURE 12.6   Setup for daily QA of the pencil beam scanning system at the Francis H. Burr Proton Therapy Center (Boston, MA). A 2D ionization chamber array (MartriXX, IBA, Dosimetry) is indexed to the treatment table, and a wedge phantom is placed on top.

ionization chamber array (MatriXX, IBA, Dosimetry) is indexed to the table, and a wedge phantom added. The two opposing wedges remove the impact of the positional error of the wedge on range measurement: irradiating along each wedge with beam spots of the same energy allows using the distance of the dose maxima of the two resulting Bragg peaks as surrogate for proton range. All other parameters can be derived from delivery of an appropriately chosen spot map. An example for a spot map for measurement of range constancy with mirrored wedges as well as spot size and position is depicted in Figure 12.7. The number of energies measured differs among institutions between one and all available.

Regardless of treatment modality, mechanical and imaging tests check for constancy of patient positioner motion, laser position, and functionality of image acquisition and analysis. A general check of communication across subsystems is advised.

Across the published literature, most spread can be noted with regard to safety checks. Table 12.1 compiles a list of measured parameters and acceptance thresholds most commonly mentioned in published literature [8,14–16,26,32].

12.2.4.2 Weekly/Monthly
The number of actual weekly checks can be minimal, with many institutions limiting themselves to a weekly trend review of daily QA data.

Monthly QA should verify the same parameters checked on a daily basis, but using different dosimetric equipment, and expanding the range of measurements. Beam quality and mechanical accuracy depend on a multitude of factors such as gantry angle, snout size, or beam energy. Constancy of parameters can be checked as a function of equipment

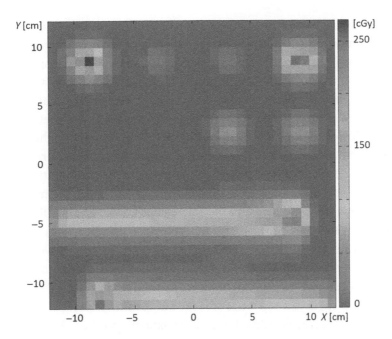

FIGURE 12.7 Example spot map for daily measurement of range, spot size, and spot position constancy in a pencil beam scanning system. (Figure courtesy of B. Clasie, Francis H. Burr Proton Therapy Center.) The data is acquired using the setup shown in Figure 12.5.

TABLE 12.1   Compiled List of Daily Checks as Published by Various Authors [7–9,23–25]

Test Type	Parameter	Acceptance
Dosimetry	Output[a]	2%–3%
	Range[a]	1–2 mm
	Spot width (pencil beam scanning (PBS))[a]	10%–15%
	Spot position (PBS)[a]	1–2 mm
	Symmetry (double scattering (DS))	1%–3%
	Range uniformity (PBS)	0.5 mm
	Modulation width (DS)	3%
Mechanical	PPS translation[a]	1–2 mm
	Laser position[a]	1–2 mm
Imaging	Image acquisition/analysis/communication[a]	Functional
	Proton-X-ray coincidence (uniform scanning (US), DS)	1–2 mm
Safety	Door interlock[a]	
	Beam on indicator[a]	
	AV system[a]	
	Radiation monitor	
	Emergency stops	
	MU interlocks	
	Collision interlocks	
	Beam pause	

[a] Checks are performed by all authors for the applicable delivery mode. Most literature describes additional, equipment-specific checks.

settings in monthly QA to monitor a larger number of clinically relevant scenarios. A practical way to accomplish this rotation through equipment settings is by reserving a weekly time slot and capturing all required parameters once over the course of a month.

For instance, in passive scattering, monthly QA, field symmetry, and flatness can be evaluated in more detail. Measurements can be acquired for clinical treatment beams involving all second scatterers and at multiple gantry angles. If symmetry and flatness are captured with a low-resolution ionization chamber array as part of the daily procedure, checks may be conducted with a high-resolution ionization chamber array for monthlies.

Depth doses are measured for a set of range/modulation combinations to verify constancy. The detector of choice for this task is a multilayer ionization chamber, given the ease of measurement and ability to measure the entire depth dose for all delivery modalities.

Spot size and position can be verified at multiple gantry angles, with either a 2D ionization chamber array or a scintillation detector. The choice of detector involves consideration of beam size. If spots are small, the user might find an ionization chamber array too coarse to capture the data of interest. This can be either just the beam sigma/FWHM, or a gamma pass rate comparing spots to a reference set taken at commissioning.

If a 2D ion chamber array is utilized in daily or patient-specific QA, some facilities check cross-calibration against a single-point reference ionization chamber measurement in a reference cube dose distribution. This also serves the purpose of monthly dose per MU verification with an alternate detector system.

In addition to these extended dosimetry checks, tests should be performed to ensure parameters that are less likely to fail and/or of less dramatic consequences in case of failure. An example for such a test is a check of range uniformity. In pencil beam scanning, it can be performed on a monthly basis (one source recommends this to be done daily [26]) by irradiating a 2D ionization chamber array with a single-energy layer beam. Knowing the water equivalent depth of the ionization chambers and depth-dose profile of the beam, one can translate percent dose difference into mm-range deviation. In passively scattered beams, this check can be included in the flatness/symmetry measurement if the depth of measurement is selected close to the distal falloff. This test will, for instance, show if there are any obstructions (e.g. loose cables in the nozzle) in the path of the beam.

A proton beam–X-ray coincidence measurement (described above) should be done on a monthly basis, ideally rotating through a set of gantry angles. If a machine is used for special treatment procedures (such as radiosurgery), it is advisable to add checks on the day of the procedure to guarantee performance.

According to Ref. [26], snout and couch translation should be monitored on a monthly basis as well as gantry and couch isocentricity. The extent of the mechanical checks is best derived from experience with equipment performance since it differs across vendors.

The quality of imaging for patient setup is essential for both treatment accuracy and minimization of patient exposure. Commonly used for this task is the Leeds phantom, following the procedure laid out in AAPM TG 142.

Safety checks vary according to vendor recommendations and/or user-defined protocols. Arjomandy [26] recommends testing of all emergency stops and interlocks, while TG 142 [1] limits monthly safety tests to laser guard interlocks.

Radiation protection is especially important at proton centers. Nuclear interactions of protons stopping in matter result in greater neutron production as well as activation of beamline components. Activation of various system components may be monitored. Some equipment may be temporarily or even permanently damaged by neutron flux. Shielding and activation are described in more detail in Chapter 7.

Table 12.2 lists an example set of tests and acceptance criteria one might perform on a monthly basis.

12.2.4.3 Annual

Annual dosimetric checks include absolute dose calibration, monitor chamber checks, and depth-dose and lateral profile measurements. Verification of absolute dose is performed as described above according to TRS 398, using an ADCL-calibrated ionization chamber in water. All other dosimetry equipment should be cross-calibrated (if not done on a monthly basis). This includes devices used for daily and monthly procedures.

The monitor chambers are checked for dose linearity and recombination effects (voltage dependence and dose rate dependence). Scans in water can be done as described above for pristine peaks in scanned beams, and for SOBPs in passively scattered fields. This allows verification of range and, if applicable, modulation width. Measuring those parameters for a variety of beam energies and modulation widths is recommended. The data should be compared to values obtained in initial commissioning. If the device is carefully

TABLE 12.2   Example Set of Monthly QA Measures and Acceptance Criteria

Test Type	Parameter	Acceptance
Dosimetry	Range (constancy at various energies, from min to max)	1 mm [24][b]
	Modulation width	2 mm[b]
	Symmetry/flatness (mult. gantry angles)	1%–2%/2%–3% [24,25][b]
	Output (constancy at mult. gantry angles)	1%–2% [24,25][b]
	Spot shape (across entire field size, mult. gantry angles, and energies)	10% of sigma, both directions gamma (2% 2 mm) [24][b]
	Spot position (across entire field size, mult. gantry angles) and energies)	1–2 mm
	Range uniformity across field (PBS)	0.5 mm [24][b]
Mechanical	Imaging-radiation field coincidence (at mult. gantry angles)	1–2 mm [24][b]
	Lightfield-radiation field congruence	2 mm or 1% [24][b]
	Couch translation/rotation	1 mm/1° [24][b]
	Gantry and couch isocentricity[a]	1–2 mm [24][b]
	Snout translation/trueness	1 mm [24][b]
Imaging	Image quality	TG 142 [24][b]
Safety	Emergency stops	Functional [24]
	Interlocks	Functional [24]
	Activation	0.02 mSv/h [24]
	Review vendor PMA	

[a] Some systems utilize a gantry angle-dependent couch correction. This test includes this feature.
[b] Francis H. Burr Proton Therapy Center.

commissioned, some of the depth-dose data may be acquired with a multilayer ionization chamber. When acquiring SOBPs in uniform scanning, these devices are the best option.

Lateral profiles for comparison to baseline data in broad beams are scanned in water as well. When checking symmetry and flatness as a function of gantry angle (at angles different from monthly QA), using a 2D ionization chamber array is most practical.

Spot shapes in scanned beams are most commonly acquired with scintillation screen—camera systems. Validating spot placement and shape across the entire deliverable field size for multiple gantry angles is important, especially if not done on a monthly basis for even a small set of beams. Verification of range uniformity at multiple gantry angles can be part of annual QA as well.

If an institution uses a model or database lookup to determine the dose per monitor unit as a function of range and modulation in passive scattering and uniform scanning beams, the model should be spot-checked in the annual measurements. This usually involves measurement of a rather large set of outputs for varying range/modulation combinations with a thimble ionization chamber in water. Similarly, all factors involved in MU calculation, for instance, range shifter and field size factors, are verified annually.

Range shifters should be checked for damage, and their water equivalent thickness confirmed experimentally.

Finally, an annual end-to-end test is a great way to revalidate the entire chain from CT acquisition to treatment planning system to setup and delivery (Figure 12.8).

Mechanical checks should verify a subset of the commissioning data for all moving parts of the delivery system—the snout, patient positioner, and gantry. High-precision mechanical pointers are the tool of choice for many of these tests [26]. Checks re-establish the baseline for monthly QA, but include measurements at additional angles. This also holds true for all system components involved in patient setup—lasers, light field, X-rays, CBCT, and their coincidence with the treatment beam.

The X-ray systems undergo additional checks to ensure imaging quality (TG179, TG142 [1,28]) as well as any safety features of the imaging system.

Depending on the situation, it might be indicated, or recommended by the vendor, to visually inspect some of the nozzle hardware, such as first scatterers. Inspections likely involve collaboration with the local or vendor's engineering team. The institution should also review the vendor's record of preventive maintenance measures and test procedures.

Similarly, safety checks should be conducted tailored to the specific available equipment following vendor recommendations. For instance, during scanned beam delivery several interlocks are active, whose function may be verified annually. Such interlocks provide online checks of spot position, time spent per spot, and number of protons deliverable per spot. It might not be technically feasible for a user to conduct some or all of the aforementioned tests. In this case, a dialog with the vendor on verification of continued functioning of delivery interlocks should take place.

There is latency to termination of delivery of each spot. It should be verified that this has a negligible effect on a field's delivery by comparing intended versus delivered number of MUs.

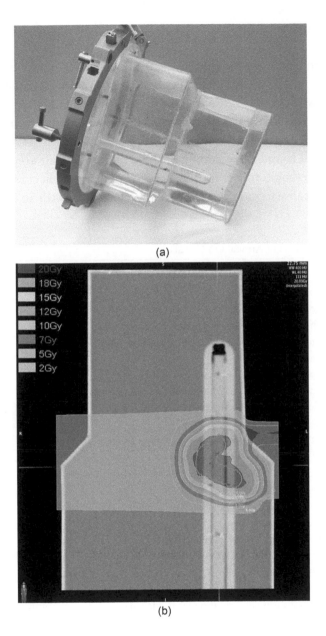

(a)

(b)

FIGURE 12.8 Example for an end-to-end test. Left: This in-house built phantom consists of a thin plastic shell filled with water, containing a slot for a minithimble ionization chamber. The phantom is CT-scanned with a rod containing several fiducial markers. There are several sleeves for the ionization chamber, each positioning the active volume at one of the locations indicated with the fiducials. This allows measurement in various places within the phantom. The phantom itself is designed such that there are no steep density gradients, to avoid uncertainty in depth placement of the chamber. It can be attached to a treatment table with the base of a stereotactic head frame. Right: Coronal view of a CT of the phantom, including isodose lines for a test plan with passively scattered protons. (Figure courtesy of M Bussière, Francis H. Burr Proton Therapy Center.)

The global shutdown is another example for a safety feature that may be verified annually.

Depending on the extent of the monthly activation checks, it may be prudent to gather some data on an annual basis. An example set of annual checks is listed in Table 12.3.

One might ask why QA is necessary given an extensive set of beam delivery interlocks, which are set such that both, safety and quality, are never compromised. There are—at least—two reasons. First, one cannot rely on proper functioning of interlocks alone. For example, it may well happen that in one of the upgrades, which are frequent for proton therapy devices, thresholds are set incorrectly or interlocks are accidentally not activated. Therefore, it is imperative to monitor equipment performance with an independent system.

TABLE 12.3   Example for Set of Annual QA Checks

Test Type	Parameter	
Dosimetry	Output calibration (TRS 398)	2% [24,25][a]
	Range (mult. energies)	1 mm [24][a]
	Depth dose [25] and lateral profile verification	2% [24]
	Output gantry angle dependency (if not done monthly)	2%[a]
	Symmetry/flatness	2% [24]
	Spot position (mult. energies across full field size)	1 mm [24][a]
	Spot check of output calculation model (incl. all factors)	2% [24,25][a]
	Verification of daily and monthly QA equipment	1%/1 mm [24][a]
	Cross-calibration of all field ICs	2% [24,25][a]
	Congruence of X-ray and radiation fields	1 mm [24,25]
	MU IC (test at min and max energy)	
	Linearity	1% [24,25][a]
	Reproducibility	2% [24,25]
	min/max dose/spot	Functional [24][a]
	Delivery latency	Negligible [24,25][a]
	Dose rate and voltage dependency	2%[a]
	Range shifter water equivalent thickness	1 mm [24][a]
Mechanical	Gantry angle accuracy	1° [24][a]
	Gantry isocentricity (mechanical)	2 mm [24]
	Couch translation and rotation (as needed, depending on monthly checks and performance)	1 mm/1° [25]
	Couch sag	1 mm [24]
	Snout rotation accuracy	1° [24]
Imaging	Standard X-ray system checks (imaging dose, system performance)	State regulations, TG 179 [24][a]
Safety	Follow vendor recommendations	[24][a]
	Review vendor annual PMI	
	Interlocks not verified monthly (e.g. aperture, range compensator presence)	Functional
	Emergency stops not verified monthly (e.g. facility shut down)	Functional

Thresholds are stated as difference from baseline.
[a] Francis H. Burr Proton Therapy Center.

Second, obtaining measurements of the clinically relevant parameters provides information on trends in beam characteristics. Trend reviews are important tools for spotting problems early and understanding the system's behavior in general.

12.2.4.4 Vendor Modifications

Many centers are handed over to the vendor at night and/or weekends for maintenance and development. Good communication practices between facility staff and vendor are essential to guarantee that extra physics checks can be performed when needed. This need should be judged by the responsible qualified medical physicist, based on information given by the vendor.

12.3 PATIENT-SPECIFIC QA

12.3.1 Patient-Specific Dosimetry

For irradiation with passively scattered and uniformly scanned beams, there are still institutions measuring the dose per MU with an ionization chamber in water for every single treatment field with or without patient-specific hardware. It is highly recommended to develop a model to calculate output as a function of range and modulation [33] (and additional factors such as field size). If properly commissioned, such a model eliminates the need for patient-specific dosimetry. A secondary, independent MU calculation system has to be employed, analogous to 3D conformal photon therapy. Patient-specific measurement is then limited to verifying proper aperture and range compensator fabrication. The model itself should be revalidated as part of annual QA.

In pencil beam scanning, there are two philosophies to patient-specific QA: actual measurements with an independent dosimetry system vs. log file analysis combined with a second and independent calculation of the dose in-patient.

Measurements most commonly involve a 2D ionization chamber array, which can be placed at variable depths with solid water. More elaborate setups have been reported—using, for instance, an ion chamber array in combination with an in-house built variable water column [34] or an in-house designed solid-state phantom with 24 separate ionization chambers [35]. Commercial solutions are available (DigiPhant PT, IBA, Dosimetry, OCTAVIUS Detector 1500 XDR, PTW Freiburg, Germany). The comparison of dose recomputed in water to measurement can be done with an—ideally 3D [36]—gamma index computation. Passing criteria vary across the published literature. No more than 3% and 3 mm are used for dose and distance agreement, with an expected pass rate of 90% or higher (see Figure 12.9 for an example). In proton pencil beam scanning, the dose for a single beam can vary in all directions; therefore, measurements are commonly done at multiple depths.

Given the resource intensive nature of the measurement process, current development is moving towards more efficient procedures. Several authors have reported on the use of machine log files to determine accuracy of delivery [37,38]. This data is augmented by an independent dose calculation. The latter can be performed with either an analytical method or Monte Carlo simulation [39–41].

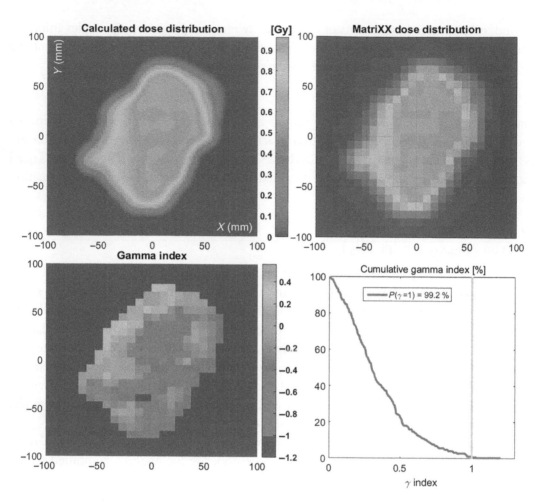

FIGURE 12.9 Example for data analysis in pencil beam scanning patient-specific QA. The dose is recomputed in water (top left) and compared to a measurement (top right, here: MatriXX measurement at depth provided by solid water). A gamma index map is calculated (lower left) and a cumulative gamma index histogram is evaluated. In the procedure used at the Francis H. Burr Proton Therapy Center, 90% of points must pass a 3%/3 mm Gamma index computation. (Figure courtesy of N. Depauw and B. Clasie, Francis H. Burr Proton Therapy Center.)

12.3.2 Patient-Specific Devices

QA of patient-specific devices can be accomplished in a number of ways. For apertures, the simplest method is an overlay of the hardware piece with a paper printout. The thickness of the aperture should be verified as well, since too thin a piece may allow protons to penetrate the block (Figure 12.10). For range compensators, testing can be done by measuring maximum and minimum thickness in addition to one or more arbitrary point thicknesses and qualitative comparison of the fabricated piece with the intended isothickness map. A more comprehensive and automated procedure is possible using modern computer numerical control (CNC) machines. Fabricated hardware can be checked via specific probes or lasers measuring the piece after milling.

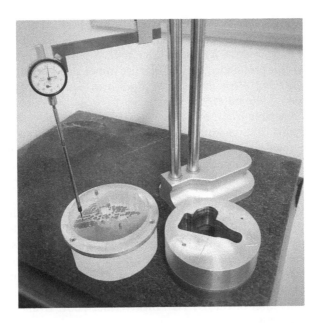

FIGURE 12.10   Sample thickness measurement for a range compensator (left) and brass aperture (right).

ACKNOWLEDGMENTS

Many thanks to Hanne Kooy and Bernie Gottschalk for comments and edits.

REFERENCES

1. Klein EE, Hanley J, Bayouth J, Yin FF, Simon W, Dresser S et al. Task Group 142 report: quality assurance of medical accelerators. *Medical Physics*. 2009;36(9):4197–212.
2. Huq MS, Fraass BA, Dunscombe PB, Gibbons JP, Jr., Ibbott GS, Mundt AJ et al. The report of Task Group 100 of the AAPM: application of risk analysis methods to radiation therapy quality management. *Medical Physics*. 2016;43(7):4209.
3. Andreo P, Burns DT, Hohlfeld K, Huq MS, Kanai T, Laitano F et al., editors. *Absorbed Dose Determination in External Beam Radiotherapy. An International Code of Practice for Dosimetry Based on Standards of Absorbed Dose to Water*. IAEA; Vienna; 2006.
4. Preston WM, Koehler AM. The effects of scattering on small proton beams 1968. Available from: http://users.physics.harvard.edu/~gottschalk/.
5. Gottschalk B, Cascio EW, Daartz J, Wagner MS. On the nuclear halo of a proton pencil beam stopping in water. *Physics in Medicine and Biology*. 2015;60(14):5627–54.
6. Clasie B, Depauw N, Fransen M, Goma C, Panahandeh HR, Seco J et al. Golden beam data for proton pencil-beam scanning. *Physics in Medicine and Biology*. 2012;57(5):1147–58.
7. Paganetti H. Range uncertainties in proton therapy and the role of Monte Carlo simulations. *Physics in Medicine and Biology*. 2012;57(11):R99–117.
8. Actis O, Meer D, Konig S, Weber DC, Mayor A. A comprehensive and efficient daily quality assurance for PBS proton therapy. *Physics in Medicine and Biology*. 2017;62(5):1661–75.
9. Gottschalk B. Multi layer ionization chambers 2007. Available from: https://gray.mgh.harvard.edu/attachments/article/213/24_MultiLayerICs.ppt.
10. Yajima K, Kanai T, Kusano Y, Shimojyu T. Development of a multi-layer ionization chamber for heavy-ion radiotherapy. *Physics in Medicine and Biology*. 2009;54(7):N107–14.

11. Nichiporov D, Solberg K, Hsi W, Wolanski M, Mascia A, Farr J et al. Multichannel detectors for profile measurements in clinical proton fields. *Medical Physics*. 2007;34(7):2683–90.

12. Brusasco C, Voss B, Schardt D, Krämer M, Kraft G. A dosimetry system for fast measurement of 3D depth–dose profiles in charged-particle tumor therapy with scanning techniques. *Nuclear Instruments and Methods in Physics Research Section B: Beam Interactions with Materials and Atoms*. 2000;168(4):578–92.

13. IBA. Dosimetry. Available from: https://www.iba-dosimetry.com/media/1614/rt-ts-e-zebra-omnipro-incline_rev1_0813_a4.pdf.

14. Bizzocchi N, Fracchiolla F, Schwarz M, Algranati C. A fast and reliable method for daily quality assurance in spot scanning proton therapy with a compact and inexpensive phantom. *Medical Dosimetry: Official Journal of the American Association of Medical Dosimetrists*. 2017;42(3):238–46.

15. Ding X, Zheng Y, Zeidan O, Mascia A, Hsi W, Kang Y et al. A novel daily QA system for proton therapy. *Journal of Applied Clinical Medical Physics*. 2013;14(2):4058.

16. Lambert J, Baumer C, Koska B, Ding X. Daily QA in proton therapy using a single commercially available detector. *Journal of Applied Clinical Medical Physics*. 2014;15(6):5005.

17. Fraser D, Parker W, Seuntjens J. Characterization of cylindrical ionization chambers for patient specific IMRT QA. *Journal of Applied Clinical Medical Physics*. 2009;10(4):2923.

18. Gottschalk B. Calibration of the NPTC range verifier 2001. Available from: http://users.physics.harvard.edu/~gottschalk/.

19. Russo S, Mirandola A, Molinelli S, Mastella E, Vai A, Magro G et al. Characterization of a commercial scintillation detector for 2-D dosimetry in scanned proton and carbon ion beams. *Physica Medica: An International Journal Devoted to the Applications of Physics to Medicine and Biology: Official Journal of the Italian Association of Biomedical Physics (AIFB)*. 2017;34:48–54.

20. Pitta G, Lavagno M, Donetti M, La Rosa A, Di Domenico A, Lorentini S, Fracchiola F. Characterization of a quality assurance device for hadron beam 2D profile and position measurements. *International Journal of Particle Therapy*. 2014; 1(3):781.

21. Gambarini G, Regazzoni V, Artuso E, Giove D, Mirandola A, Ciocca M. Measurements of 2D distributions of absorbed dose in protontherapy with Gafchromic EBT3 films. *Applied Radiation and Isotopes: Including Data, Instrumentation and Methods for Use in Agriculture, Industry and Medicine*. 2015;104:192–6.

22. Arjomandy B, Sahoo N, Ding X, Gillin M. Use of a two-dimensional ionization chamber array for proton therapy beam quality assurance. *Medical Physics*. 2008;35(9):3889–94.

23. Lin L, Kang M, Solberg TD, Mertens T, Baeumer C, Ainsley CG et al. Use of a novel two-dimensional ionization chamber array for pencil beam scanning proton therapy beam quality assurance. *Journal of Applied Clinical Medical Physics*. 2015;16(3):5323.

24. Summers PA, Ibbott G, Alvarez P, Moyers M, Lafratta R, Followill D. Compilation of proton therapy data from the approval process for NCI-Funded Clinical Trials 2014. *International Journal of Radiation Oncology, Biology, Physics*. 2014;90(1):S922.

25. Ding X ZJ, Syh J, Petal B, Syh J, Song X, Freund D, Rosen L, Wu H. A novel method for a PBS system spot absolute position check with stereotactic imaging system. *Particle Therapy Cooperative Group*. 2015.

26. Das IJ, Paganetti H. *Principles and Practice of Proton Beam Therapy, AAPM Monograph, 2015 Summer School*. Medical Physics Publishing; Madison, Wisconsin. 2015.

27. Lutz WR, Larsen RD, Bjarngard BE. Beam alignment tests for therapy accelerators. *International Journal of Radiation Oncology, Biology, Physics*. 1981;7(12):1727–31.

28. Bissonnette JP, Balter PA, Dong L, Langen KM, Lovelock DM, Miften M et al. Quality assurance for image-guided radiation therapy utilizing CT-based technologies: a report of the AAPM TG-179. *Medical Physics*. 2012;39(4):1946–63.

29. Yin FF, Wong J, Balter J, Benedict S, Bissonnette JP, Craig T, Dong L, Jaffray D, Jiang S, Kim S, Ma CM, Murphy M, Munro P, Solberg T, Wu QJ. The role of in-room kV X-ray imaging for patient setup and target localization. American Association of Physicists in Medicine; College Park. 2009.

30. Pawlicki T, Samost A, Brown DW, Manger RP, Kim GY, Leveson NG. Application of systems and control theory-based hazard analysis to radiation oncology. *Medical Physics.* 2016;43(3):1514–30.

31. Schneider U, Pedroni E, Lomax A. The calibration of CT Hounsfield units for radiotherapy treatment planning. *Physics in Medicine and Biology.* 1996;41(1):111–24.

32. Arjomandy B, Sahoo N, Zhu XR, Zullo JR, Wu RY, Zhu M et al. An overview of the comprehensive proton therapy machine quality assurance procedures implemented at the University of Texas M. D. Anderson Cancer Center Proton Therapy Center-Houston. *Medical Physics.* 2009;36(6):2269–82.

33. Ferguson S, Chen Y, Ferreira C, Islam M, Keeling VP, Lau A et al. Comparability of three output prediction models for a compact passively double-scattered proton therapy system. *Journal of Applied Clinical Medical Physics.* 2017;18(3):108–17.

34. Trnkova P, Bolsi A, Albertini F, Weber DC, Lomax AJ. Factors influencing the performance of patient specific quality assurance for pencil beam scanning IMPT fields. *Medical Physics.* 2016;43(11):5998.

35. Henkner K, Winter M, Echner G, Ackermann B, Brons S, Horn J et al. A motorized solid-state phantom for patient-specific dose verification in ion beam radiotherapy. *Physics in Medicine and Biology.* 2015;60(18):7151–63.

36. Chang C, Poole KL, Teran AV, Luckman S, Mah D. Three-dimensional gamma criterion for patient-specific quality assurance of spot scanning proton beams. *Journal of Applied Clinical Medical Physics.* 2015;16(5):381–8.

37. Belosi MF, van der Meer R, Garcia de Acilu Laa P, Bolsi A, Weber DC, Lomax AJ. Treatment log files as a tool to identify treatment plan sensitivity to inaccuracies in scanned proton beam delivery. *Radiotherapy and Oncology: Journal of the European Society for Therapeutic Radiology and Oncology.* 2017;125(3):514–9.

38. Zhu XR, Li Y, Mackin D, Li H, Poenisch F, Lee AK et al. Towards effective and efficient patient-specific quality assurance for spot scanning proton therapy. *Cancers.* 2015;7(2):631–47.

39. Fracchiolla F, Lorentini S, Widesott L, Schwarz M. Characterization and validation of a Monte Carlo code for independent dose calculation in proton therapy treatments with pencil beam scanning. *Physics in Medicine and Biology.* 2015;60(21):8601–19.

40. Meier G, Besson R, Nanz A, Safai S, Lomax AJ. Independent dose calculations for commissioning, quality assurance and dose reconstruction of PBS proton therapy. *Physics in Medicine and Biology.* 2015;60(7):2819–36.

41. Shin WG, Testa M, Kim HS, Jeong JH, Lee SB, Kim YJ et al. Independent dose verification system with Monte Carlo simulations using TOPAS for passive scattering proton therapy at the National Cancer Center in Korea. *Physics in Medicine and Biology.* 2017;62(19):7598–616.

CHAPTER 13

Monitor Unit Calibration*

Timothy C. Zhu

University of Pennsylvania

Haibo Lin

New York Proton Center

Jiajian Shen

Mayo Clinic

CONTENTS

---

* Colour figures available online at www.crcpress.com/9781138626508.

13.1 INTRODUCTION

A monitor unit (MU) is a measure of machine output of a proton accelerator. The MUs are measured by monitor chambers built into the treatment head that measure the dose delivered by a beam in a reference condition, e.g., 1 cGy/MU is chosen by some centers for their proton therapy system [1]. This value could be some other artificial values, D_{ref}/MU, in other centers. Some centers use dose per proton, not dose per MU, to define the dose per MU [2]. Here, the MU definition is based on a conversion factor that converts charges generated in the monitor chamber to be defined as 1 MU. The charge generated in the monitor chamber by protons is proportional to the protons passing through the monitor chamber:

$$\text{ion} - \text{pair/proton} = S_{air} \cdot L \cdot \rho_{air} / W_{air}, \tag{13.1}$$

where S_{air} is the mass stopping power in air; L is the gap length of the dose monitor chamber, ρ_{air} is the air density, and W_{air} is the mean energy to generate ion pair in air. Thus, the total charges collected by the monitor chamber can be converted to the number of protons by dividing the charges corrected by conversion factor in Equation 13.1. Using dose per MU is therefore equivalent to the dose per proton [3,4].

The second dose per MU check can be implemented by either analytical (or factor-based) method or Monte Carlo (MC) simulation. In this chapter, we will focus primarily on the analytical method because it is usually more straightforward, easier to implement, and has fast calculation speed. Another advantage of the analytical method is that it usually does not need additional data, since the data for commissioning treatment planning system (TPS) is typically sufficient for commissioning second MU check engine. The MC simulation would provide more accurate dose calculation. But it typically requires an accurate model of the treatment nozzle with slower computational speed. This chapter will focus on the analytical formula, particularly for pencil beam scanning (PBS) beams. For PBS used for intensity modulation proton therapy (IMPT), it is truly necessary to verify the dose distribution in three dimension (3D) unlike that for photon beam based intensity-modulated radiotherapy (IMRT), where generally two dimension (2D) dose profile at a fixed depth is sufficient to verify an IMRT plan, because proton IMPT plans are made of pencil beam pristine peaks modulated in the entire 3D space.

For completeness, we will briefly cover broad proton beams used in conventional proton therapy, where only spread-out Bragg peak (SOBP) beams with nominally flat lateral profile are used. For dose per MU verification of the SOBP beam for a particular range R (or energy), verification of D/MU at one point (e.g., middle of SOBP beam) is sufficient if one can establish the relationship between D/MU there and the range R. Otherwise, it is necessary to verify both D/MU and the range R. One point D/MU check for proton beams is similar to what is currently used for photon beams, where D/MU at one dose prescription point is used. More detailed description for broad proton beams can be found elsewhere [2].

13.2 CATEGORIES OF DOSE-TO-MU FORMALISMS FOR PROTON BEAMS

To perform MU calculations, it is necessary to determine the relationship between the absorbed dose and the measured quantity using fundamental physics quantities. This section establishes some of the basic relationships needed for MU calculations.

13.2.1 Factor-Based Formalism

Factor-based methods determine absorbed dose per MU by using the product of standardized dose ratio measurements. Successive dose ratio factors are multiplied for a chain of geometries, and thus the dose ratio factors, are varied one by one until the geometry of interest is linked back to the reference geometry through an identity equation [2,5]. The ratio of dose at the point of interest to the dose at the reference condition can be expressed as a dose-to-fluence ratio [2]:

$$\frac{D(c,s;z,d)}{D(C_{ref},S_{ref};z_{ref},d_{ref})} = TPR(c,s;z) \cdot PSF(s;d_{ref}) \cdot H_p(c,s;z) \cdot ISF_{air}(z) \qquad (13.2)$$

For simplicity, only the ratios on the central axis are shown, where c is the nozzle setting or multileaf collimator (MLC) setting, or an equivalent square defined at the isocenter plane when a universal nozzle fitted with an MLC is used to shape the field instead of a brass block, s is the field size defined at depth in phantom, z is the source-to-detector distance, and d is the depth in patient (or phantom). The TPR, tissue-phantom ratio, is the depth-dose ratio taking out of the inverse square law and describes the dose variation (range) along the depth direction. For proton beams, the TPR is approximately equal to the percentage depth dose (PDD) divided by $SAD^2/(SSD+d)^2$ because the phantom scatter is negligible, where SAD and SSD are the source-to-axial distance and the source-to-surface distance, respectively. SAD is the distance from the proton virtual source to the isocenter. Since in proton fields PDD is commonly used, we will use PDD for the depth-dose variation in this chapter. PSF, the phantom scatter factor, describes the field size dependence due to proton phantom scattering. Unlike photon beams, this quantity for proton beam is more like a lateral electron disequilibrium factor for an electron beam in that it saturates at a finite lateral field radius r and drops for small field sizes. The proton head scatter factor, H_p, describes the field size dependence due to head scatter inside the accelerator and inverse square factor (ISF)$_{air}$ describes the inverse square dependence determined in air or in a water phantom, respectively. The advantage of fluence-based factor-based formalism is that the contribution from phantom scattering (TPR*PSF) and primary fluence (H_p*ISF$_{air}$) components can be separated.

13.2.2 Pencil Beam Model-Based Formalism

The absorbed dose resulting from proton irradiation is equal to the product of the particle fluence incident onto the patient and linear energy transfer (LET), for a monochromic proton beam. This makes normalization of calculated dose per particle fluence appealing, partially because the fluence is proportional to the dose quantity measured in air and the depth dependence of LET for the clinical proton beam can be determined by the measured PDD in water. The dose in the medium can be calculated as a convolution of a pencil beam kernel, p, which can be calculated from MC simulation, with the incident proton fluence, Φ:

$$D = MU \cdot (D/MU)_{ref} \cdot \frac{\int_x \int_{y'} p(x-x', y-y', z)\Phi(x',y')dx'\,dy'}{\iint_{ref} p(x-x', y-y', z')\Phi_{ref}(x',y')dx'\,dy'}, \qquad (13.3)$$

where $(D/MU)_{ref}$ is the dose per MU at the reference condition. A reference condition is the point where all dosimetrical quantities (e.g., PDD, H$_p$, PSF,...) are normalized to be 1, it can be different from the calibration condition, where $D/MU = 1$.

13.2.3 MC Simulation or Deterministic Algorithms

MC simulations have been used to calculate the absorbed dose per MU [6]. MC simulation platforms have been implemented using either GEANT 4 [7] or MCNPX [8]. This method has been clinically used for dose per MU verification and is considered the gold standard in a heterogeneous medium. This method is now becoming available in some commercial TPS (e.g., Acuros® PT for Varian Eclipse [9]) using Graphics Processing Unit (GPU) to increase dose calculation efficiency. MC simulations are most suitable for relating the absorbed dose to the incident proton fluence directly. We will not include a discussion of D/MU from MC in this chapter.

13.3 FACTOR-BASED DOSE PER MU ALGORITHMS

13.3.1 Passive Scattering and Uniform Scanning Broad Beams

Factor-based MU calculation algorithms are implemented by Massachusetts General Hospital (MGH) [10,11], MD Anderson Cancer Center [12], and others [13,14]. They can be separated to in-air fluence-based and in-water dose-based algorithms, and the accuracy among these models are comparable [15].

13.3.1.1 Fluence-Based Formalism

The fluence-based formalism separates the measurement into in-air (fluence) and in-water (dose) measurements [16]. The basic formalism can be expressed as [17]

$$D = MU \cdot (D/MU)_{ref} \cdot PDD(z) \cdot PSF(s) \cdot H_p \cdot ISF_{air} \cdot POAR(x,y), \qquad (13.4)$$

where $(D/MU)_{ref}$ is the dose per MU for each energy option measured under reference condition (e.g., for Ion Beam Applications (IBA) machines, 10×10 cm^2, SAD = 230 cm, $d = 10$ cm for SOBP or d_{max} for pristine peaks). PSF is the phantom scatter factor that accounts for lateral proton disequilibrium as a function of field size, s. H_p is the proton headscatter factor that accounts for the fluence variation for different blocked fields for the same snout and SSD. OAR is off-axis ratio and ISF is the inverse square factor. PDD = PDD$_b$ * PDD$_f$ is the percentage depth dose for a field with finite circular radius r. In the analytical mode, the calculation is made for the middle of SOBP PDD so that PDD = 1. Alternatively, the energy modulation input file is used to calculate SOBP PDD using the expression:

$$PDD_{SOBP} = \sum_i^n w(R)_i\, PDD_{pris}(R)_i,$$ (13.5)

where $w(R)_i$ is the weight for the pristine PDD with range R. $w(R)$ can also be estimated using the maximum range R and modulation width M using a Cimmino algorithm [18]. For sufficient large field size, e.g., $s > 4$ cm, the phantom scatter factor is PSF(s) = 1.

The ISF$_{air}$ is calculated as

$$ISF_{air} = \left(\frac{SAD - xv}{SSD + z - xv}\right)^2$$ (13.6)

ISF$_{air}$ in Equation 13.6 is consistent with that in Equation 13.2. The term xv is added to represent the virtual source shift (cm) due to the varied incident proton energy (E) and nozzle equivalent thickness for pristine peaks. For SOBP, one usually sets $xv = 0$. The dependence of ISF is instead lumped into the headscatter factor that follows the output factor formula [10]:

$$H_p = (CF/100) \cdot \left(1 + a_1 \cdot r^{a_2}\right) \cdot \left(s_1 + s_2\left(R - R_L\right)\right) \cdot H_0(s)$$ (13.7)

where R is the range, $r = (R - Mm)/Mm$, M is modulation width, and $m = 0.9$ is a conversion factor, CF is a constant to correct for the output change per option, R_L is the minimal range of the option. In this expression for the headscatter factor, the effect of virtual source shift is included as the parameters s_1 and s_2 [10]. The expression $H_0(s)$ is the equivalent square dependence of the headscatter factor and can be determined experimentally. For sufficient large field size, $H_0(s) = 1$.

Uniform scanning delivers uniform intensity proton beams layer by layer along the beam axis by adjusting the proton energy and sweep the beam laterally using scanning magnet system. Due to less beam degrading in the beam delivery system, uniform scanning system generally has the ability to treat larger and deeper target than passive scattering system. In terms of MU calculation, a similar measurement-based approach discussed above has been applied to uniform scanting by different groups [19,20]. The output factor for a uniform scanning system was found as a function of parameters such as the range, modulation, scanning area, field size, and snout position [20]. Considering the inverse square correction for the case that the center of SOBP is not at the isocenter, the author claimed that the excellent agreement between predicted and measured output factors for prostate treatment field has led to the elimination of output measurement for prostate patients at their center [20].

The output of uniform scanning beam increases with the energy of the proton beam for each modulation width and decreases with the modulation width, similar to that reported by Hsi et al. [13] and Kim et al. [21]. The impact of the scanning area on output is unique to a uniform scanning beam. It was also found that the dependence on the scanning area varies little with range and modulation of the proton field, and the output behaves differently for an asymmetric scanning area.

The impact of field size on output factor mainly depends on the depth of the measurement [19,22]. The result indicates that the output is more sensitive to small fields at large depths. In practice, accurate estimation of the effective field size, especially for a small field with irregular field shape, is critical for accurate output calculation. The output also depends on the snout position, which is complicated to model [22]. Therefore, the impact of snout position is generally ignored. However, special attention should be paid when small fields are used for treatment.

13.3.1.2 Dose-Based Formalism

The University of Texas MD Anderson Proton Therapy Center in Houston has developed a conservative measurement-based approach to calculate dose per MU (D/MU), which is following closely the proven method that has been used for external beam photon therapy. The detailed information for this procedure can be found in studies by Sahoo [12]. This approach is similar to the one that has been used at the proton therapy facility at the Loma Linda University Medical Center since 1991 [23]. The dose at a point of interest, D, can be determined as a function of MU as

$$D = MU \cdot ROF \cdot SOBPF \cdot RSF \cdot SOBPOCF \cdot OCR \cdot FSF \cdot ISF \cdot CPSF \qquad (13.8)$$

where ROF is the relative output factor that represents the changes in D/MU relative to the reference calibration condition, such as beam energy and modulation, SOBPF is the SOBP factor that represents the change in D/MU with changes in SOBP width relative to the reference SOBP width used to determine the ROF, RSF is the range shifter factor that accounts for the output change due to the presence of range shifters, SOBPOCF is the SOBP off-center factor for depth of measurements is away from the middle of the SOBP, OCR is the off-center ratio in water to account for the location of the measurement point laterally away from the central axis of the field, FSF is the field size factor that gives the change in D/MU when the field size is different from the reference field size used in ROF measurement, usually 10×10 cm^2, ISF is the inverse square factor, and CPSF is the compensator and patient scatter factor that accounts for the change in D/MU with a compensator relative to the same water-equivalent depth without a compensator. Example data can be found elsewhere [2,12].

13.3.2 Pencil Beam Scanning

Since a proton PBS plan is typically composed of a few thousand spots, small deviation to the singlet spot shape would cause significant dose deviations. Thus, the PBS dose model requires accurate modeling of each single proton spot. The proton spot profile usually has a laterally wide extended low-dose tail that superimposed on the primary Gaussian. The low-dose tails are generated due to the large-angle scattering in the beamline and in medium. To directly measure the low-dose tails, it is reported by many centers that lateral extension to 10^{-4} of the central peak is required [24–28]. However, these measurements are time consuming and challenging. As an approximation, one can use a lookup table to correct the so-called field size effect.

13.4 PENCIL BEAM BASED DOSE PER MU ALGORITHMS

13.4.1 Passively Scattering and Uniform Scanning Broad Beams

A broad beam in proton therapy can be generated by pencil beam kernel convolution to calculate dose using Equation 13.3, where the pencil beam kernel, p, can be expressed as

$$p = \text{ISF}_{\text{air}} \cdot \text{PDD}(z) \cdot \text{LAT}(x,y,z). \tag{13.9}$$

where PDD(z) is the measured depth-dose curve for infinitely large field and infinite SSD, which is proportional to the measured LET for the clinical proton beam, ISF_{air} is the in-air ISF, and LAT(x,y,z) is usually expressed as a sum of two Gaussian functions with a range straggling correction, λ:

$$\text{LAT}(x,y,z) = \frac{\sum_{i=1}^{2} B_i \cdot e^{\frac{x^2+y^2}{\lambda \cdot b_i \cdot r^2}}}{\pi \cdot \overline{r^2} \cdot \lambda \sum_{i=1}^{2} B_i \cdot b_i}, \tag{13.10}$$

where the b_i and B_i, ($i = 1, 2$) are precalculated weighting parameters based on an MC calculation in water. $\overline{r^2}$ is the mean square of radial spread. They are functions of the equivalent proton energy at water surface, E_{eq}, and the fractional depth, z/R_0, R_0 is the mean range of proton with energy E_{eq}. If one ignores the large angular multiple scattering, the lateral spread, LAT, can be expressed as a single Gaussian (SG). One can calculate the radial spread, $\overline{r^2}$, in the medium under Fermi–Eyges small-angle approximation [2].

13.4.2 Pencil Beam Scanning

Two analytical dose algorithms, ray-casting and fluence convolution, have been used for PBS dose calculations [29]. The ray-casting model of a single spot with nominal energy E and an SG lateral profile is shown later:

$$D_{\text{spot}}(x,y,z) = \text{MU} \cdot \text{FSF}(s) \cdot \text{IDD}(E, \text{WET}) \frac{1}{2\pi\sigma_x\sigma_y} \exp\left[-\frac{(x_0-x)^2}{2\sigma_x^2}\right] \exp\left[-\frac{(y_0-y)^2}{2\sigma_y^2}\right], \tag{13.11}$$

Where integral depth-dose (IDD) is the function of spot nominal energy E and water equivalent thickness (WET) at position (x, y, z). The rest of Equation 13.11 is the spot lateral dose distribution LAT(x,y,z) in the medium (e.g., water), which is described by the SG distribution here. However, additional Gaussians (see Equation 13.10) are often necessary to account for the low-dose halo existing over an extended distance of a few centimeters from the spot central peak [24–28]. It is reported that the lateral profile in medium would be a best fit by two Gaussians with the addition of a modified Cauchy–Lorentz distribution [30]. If an SG is used, then it is often necessary to introduce an additional FSF(s) that account for the underestimation of dose at large distances from the spot. When the single-spot dose distribution is explicitly established, the total dose is a summation of the contribution from all spots.

$$D(x,y,z)=\text{ISF}\cdot\text{FSF}(s)\sum_{j=1}^{N}\left[\text{MU}_j\cdot\text{IDD}_j\left(E_j,\text{WET}\right)\cdot\iint\text{LAT}_j\left(x',y',z\right)dx'dy'\right] \quad (13.12)$$

$\text{ISF}=\dfrac{\text{SAD}_{vx}\cdot\text{SAD}_{vy}}{(\text{SSD}_{vx}+z)\cdot(\text{SSD}_{vy}+z)}$, FSF(s) is a field size factor, and LAT$_j$ is the lateral dose distribution that is typically described by one or multiple Gaussians as referenced earlier. The ISF accounts for the diversions between spots. Without this factor, the ray-casting model is valid only for a parallel beam with a small angular divergence. Since the source-to-axis distance for a proton system is usually 2–3 m, the angular divergence of scanning spots is small and spots can be considered parallel [30].

The basic feature of fluence-based models is the convolution of fluence with an elemental pencil beam dose distribution or dose kernel. The fluence-based dose model, shown later, is used in the Eclipse TPS [31,32]

$$D(x,y,z)=\text{FSF}(s)\cdot\sum_{E_k}\sum_{\text{Beamlet }j}\phi_{E_k}\left(x_j,y_j,z\right)D_{E_k}^{\text{Beamlet}}\left(x-x_j,y-y_j,d(z)\right) \quad (13.13)$$

where $\Phi_{E_k}\left(x_j,y_j,z\right)$ is the proton fluence at the position of the beamlet for the kth energy layer, (E_k) is the dose distribution of beamlet (i.e., dose kernel), FSF(s) is a field size factor to correct for underestimation of the dose using an SG, and $d(z)$ is the WET of position z along the beamlet direction. An analytical dose kernel based on MC simulation is used in the Eclipse TPS [31].

The fluence model separates fluence and dose calculation. Its advantage is that the model is not vendor specific since the beamline-related issues (in-air fluence) is separated from general proton physics. However, the model requires an accurate dose kernel, which is not easy to achieve using analytical formulae. The ray-casting model is based on the spot profiles in medium. Its advantage is that all contributions to the beam width such as initial phase space, Multiple Coulomb Scattering (MCS), and nuclear interaction are included.

13.5 BEAM DATA FOR PROTON MU CALCULATIONS

13.5.1 Depth Dose

13.5.1.1 PDD for Broad Beam

For most broad beam application, an SOBP is used in practice. Typically, the dose per MU is determined at the middle of the SOBP, where PDD = 1. However, for doses at other depths in a SOBP, the PDD should be used.

13.5.1.2 IDD for PBS

Both ray-casting model and fluence model need the IDD, in unit of Gy · mm²/MU, as input for dose calculation. IDD is measured for a PBS pristine peak using a large parallel-plate chamber (e.g., 8 cm diameter and sufficient to cover the entire pencil beam lateral area) with a known active area (in mm²) and normalized to the depth at the calibration condition and then multiplied with the Dose/MU at the depth. Since the absolute dose per MU is included in the IDD, it is required to measure the IDD accurately. Any inaccuracy in IDD would be propagated to dose error in the dose calculation. An example of IDD for a synchrotron is shown in Figure 13.1. A similar example for a cyclotron can be found elsewhere [2].

13.5.2 Output Factor

13.5.2.1 Output Factor for Broad Beam

The clinical utilization of proton therapy requires safe and efficient planning and delivery technologies. However, the calculation of output (D/MU) is not supported by proton therapy TPSs, such as Eclipse from Varian Medical Systems. This is due to the complexities

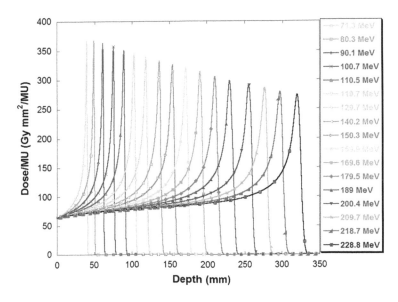

FIGURE 13.1 Dose per MU for integrated pristine peaks of PBS beams from a synchrotron-based Hitachi proton accelerator from 71.3 to 228.8 MeV.

and specifications of beam delivery systems. Historically, output was determined by measurement for each treatment field before treatment. This requires a significant amount of beam time and manpower.

An analytical expression for the depth dose of an SOBP at infinite SAD was derived at MGH [33]. While applying that analytical expression to the surface of SOBP, one can get a relationship between the surface dose and SOBP dose [33]:

$$D_{\text{SOBP,0}} \approx \frac{D_P}{1+0.44r^{0.6}} \tag{13.14}$$

where $r = (R-M)/M$, D_P is the dose in the plateau of SOBP, and $D_{\text{SOBP,0}}$ is the dose at surface of SOBP. The proton MU is proportional to the proton fluence and the surface dose of SOBP. Therefore, a relationship between MU and the SOBP plateau dose could be established (see Equation 13.7), and the proton dose per MU would be calculated analytically.

The IBA double scattering has a total of eight treatment options, each of which is defined by a unique combination of hardware (e.g., scatterer and range modulation wheel), and is limited to a span of beam ranges. MGH applied the earlier analytical formula to each option of their IBA double-scattering system, with the modifications to account for the effective source at finite distance, and effective source position change due to the change of fixed scattered materials [10,11], Equation 13.7. By measuring a few outputs with different range and modulation combinations, the free parameters (a_1, a_2, s) in Equation 13.7 could be fit. The fit has to be performed for each of the energy options. The selection of the output measurements for the model parameter fit should include the range and modulations that cover as large range of r as possible. After the free parameters are fit, the output is just a function of the single variable r. The MGH group can predict outputs within 1.4% (1 standard deviation) of measurements.

13.5.2.2 Output Factor for PBS

For PBS, the output factor is not directly needed for the second MU check, which is different from photon and passive-scattering proton therapy. However, the output factor is directly related to the spot lateral dose distribution. Since a proton PBS plan is composed of thousands of spots, the small inaccurate modeling of the lateral dose distribution for individual spot will accumulate to a significant effect through thousand spots. Figure 13.2 shows that the output could be off by more than 10% for small field because the SG model could not model the low-dose tails. While with a double-Gaussian (DG) model, the second Gaussian is used to model the low-dose tail, which significantly improves the prediction.

Figure 13.2 shows that PBS requires an accurate model of the lateral dose distribution. However, it is a challenge to measure the lateral dose accurately because of the extended low-dose tails. These low-dose tails are generated by two sources: the large-angle scattering in the beamline when the protons pass through the devices in the nozzle and the nuclear halo generated by nonelastic nuclear interaction when protons pass through the medium [35]. The large size Bragg peak chamber is used to measure IDD, so those laterally extended low-dose tails would not be missed. However, it has been reported that even an 8 cm diameter Bragg peak chamber is not big enough for collecting 100% spot dose due to these

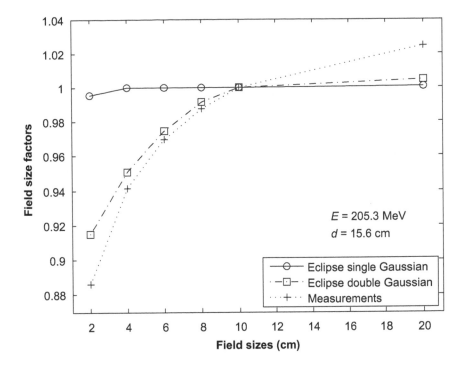

FIGURE 13.2   FSFs (in Equation 13.13) between measurements ("+" symbol) and the TPS models with SG ("circle") and DG ("square") to model spot profiles.

extended low-dose tails [27]. MC simulations are used to correct the finite chamber size effect on the measured IDD.

The ray-casting model needs both the in-air and the in-water profiles as input. The profiles at a few depths from a subset of energies are required to fit the model parameters. The profiles at the other depths and energies would be predicted by the fit parameters. The fluence model needs in air profile as input and the dose kernel, but it does not need the in-medium profile. MCS would be used as a simple approximation to the dose kernel because the principal component of proton beam propagation in medium is due to MCS. The MCS would be represented by an analytical formulae, such as generalized highland approximation [36] and differential Molière method [37,38]. An accurate dose model should also include the nuclear scatter. An analytic approximation based on MC simulations was proposed to account for the dose contributed from nuclear interaction [39].

13.5.3 Primary OAR

13.5.3.1 Broad Beam

The in-air profile of a broad proton beam is typically flat; thus, $POAR_{air} = 1$ is a good assumption for dose per MU calculation at off-axis points.

13.5.3.2 Pencil Beam Scanning

The profile of the PBS spot is very close to a Gaussian shape before it enters into the nozzle. After it enters the nozzle, the protons are scattered by the devices and air in the nozzle. The major effects of the scatterings are to broaden the Gaussian profile, but a small portion due to large-angle scattering can produce a low-dose halo. Although the WET of the devices in the nozzle are manufactured to be as minimum to a few tenth of cm^2/g, their effect to low-dose halo may not be neglected if their distances to isocenter are large. The in-air halo is more significant for the low-energy beam and can extend to beyond 10 cm laterally. It has been reported that a dose deviation up to 10% could be observed if the in-air halo effect is not modeled in the TPS [32].

Among all the devices in the nozzle, the spot profile monitor that is typically positioned at the entrance of the nozzle (about 2–3 m away from the isocenter) is the major resource for the in-air halo effect. The spot profile monitor in new proton machines, such as Hitachi Probeat V5, is moved out of the beamline during patient treatments because it is only needed for beam tuning. Hence, the in-air halo is significantly reduced in these new proton machines. The in-air halo may cause little effect even though it is neglected in modeling. However, no matter how scatter was minimized in the nozzle, the halo is still generated by the range shifter when treating shallow-seated tumors. As the thickness of range shifter (in the order of a few cm) is more than ten times larger than the nozzle WET, the halo produced by range shifter is not negligible (as shown in Figure 13.3a). Hence, correct modeling of the in-air halo becomes more important when a range shifter is used [40].

For high-energy proton beams, the low-dose halo will be produced in the medium due to the nuclear interaction. These low-dose halos are not negligible in the model for accurate dose calculation. As shown in Figure 13.3b, high-energy beam spot without low-dose halo in air produces a significant halo in water.

13.5.4 Inverse Square Law

13.5.4.1 Inverse Square Law for Broad Beams

The ISF for broad beams is determined by the virtual SAD of each beam. Typically, the virtual shift from the nominal SAD of each beam is set to be zero, because the small virtual SAD shift is often applied to the output model (see Equation 13.7).

13.5.4.2 Inverse Square Law for PBS

The ISF for a PBS system due to the nonparallel spots is determined by the virtual SAD of magnets x and y (as shown in Figure 13.4). The virtual SAD could be derived by measuring the output variations with the different SSDs at the broad beam geometry while keeping the same beamline parameters. Figure 13.4 is the ISF for the PBS at Mayo Clinic, which has the virtual SAD 135 and 192.5 cm for magnets x and y, respectively. The beam used for the test has range 20 cm, SOBP 10 cm, and the dose was measured at the center of SOBP (i.e., depth 15 cm).

13.5.5 Generic Beam Data

It is possible to develop generic beam data to be used for PBS, with which appropriate tuning of the proton Gaussian energy spectrum can be matched to a specific PT center [41].

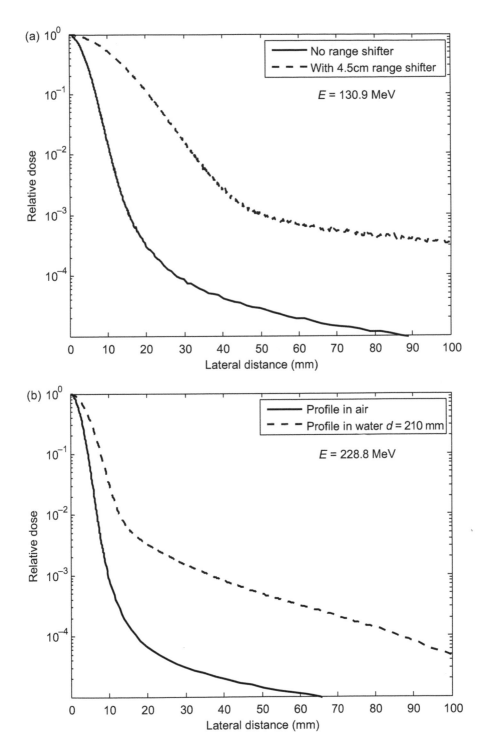

FIGURE 13.3   (a) Measured in-air profile from a pristine 130.9 MeV PBS profile with (dotted line) and without (solid line) range shifter. (b) Comparison between in-air profile and in-water profile at depth of 20 cm from a pristine 228.8 MeV PBS beam.

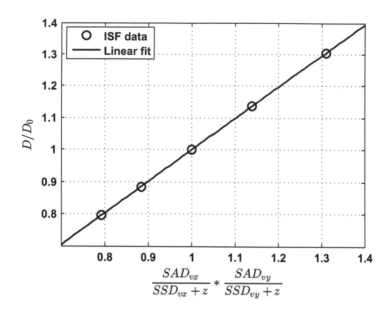

FIGURE 13.4 The ISF defined in Equation 13.13 is shown for the Mayo Clinic proton system. The solid line is the linear fit. The dose was measured for detector at ISO and 10, 20 cm up and down from ISO, respectively. The detector was placed at the center of SOBP for the same beam with range = 20 cm and SOBP = 10 cm (i.e., at depth $z = 15$ cm).

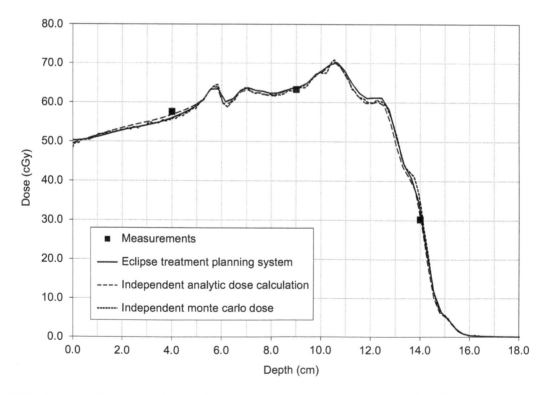

FIGURE 13.5 The measured point doses (square) vs. the dose profiles calculated by TPS, the independent analytic method, and MC for an IMPT head-and-neck field.

13.6 DOSE AND MU VERIFICATION FOR PROTON THERAPY

13.6.1 Broad Beam

Accurate results (better than 3% agreement) were obtained by a number of cancer centers [12,17]. The University of Pennsylvania uses an IBA proton therapy system that is similar to MGH. By applying a linear-quadratic transformation to the nominal r, the MGH model was further improved at University of Pennsylvania [42]. It is reported that the model prediction is within 1% (one standard deviation) while comparing to 1,784 patient-specific fields.

13.6.2 Pencil Beam Scanning

The dose verification by measurements is applied to each clinic treatment plan [43,44]. A 2D array detector is usually used to collect the planar dose distribution. Since the treatment plan is composed of many independent spots, the same plan is delivered multiple times to verify the dose at multiple planes. Hence, PBS patient-specific quality assurance (QA) by measurement alone can be very time consuming. An efficient and effective patient QA program could consist of three components: measurements, independent dose calculation, and analysis of patient-specific delivery log file [45].

Recently, independent dose calculation models, either with analytic methods [30] or MC simulation [46], have been successfully implemented for routine clinical patient-specific QA. These independent dose calculations could reduce the time required for measurements [44]. An example of the patient QA for an IMPT head-and-neck field is shown in Figure 13.5. It includes direct measurements and two dose calculations that were independent of TPS. Since the two independent dose calculations match with the TPS, the selected depth for measurements could be reduced.

Furthermore, the use of a treatment log file from a spot scanning beam (if available) is used for patient-specific QA [47]. The treatment log files record the proton energy, MU, and lateral position of each delivered PBS spot. With an accurate second MU program, the dose delivered to the patient at each treatment fraction would be retrospectively reconstructed. This is helpful to track the delivered dose and would not need any additional efforts if the whole process is automated.

13.7 SUMMARY/CONCLUSIONS

This chapter discusses formalisms used for the dose per MU calculation for broad (uniform) (either passively scattering or uniform scanning) and PBS proton beams. The algorithms are broadly separated into: factor-based and model- (or pencil beam kernel) based methods, with the latter applicable to more general cases, e.g., IMPT. For broad beam, agreements of 1%–3% were generally observed between MU calculation and measurements. For PBS, acceptable agreements of IMPT 3D dose map between second MU calculation and measurements were observed in uniform phantom, although further improvements can be made to improve the modeling of PBS lateral profiles at central and off-axis locations. Novel uses of the MU calculation program, e.g., validating IMPT delivery using log files, are under development to expand the area of its use.

REFERENCES

1. Gillin MT, Sahoo N, Bues M, Ciangaru G, Sawakuchi GO, Poenisch F, Arjomandy B, Martin C, Titt U, Suzuki K, Smith AR, Zhu XR. Commissioning of the discrete spot scanning proton beam delivery system at the University of Texas M.D. Anderson Cancer Center, Proton Therapy Center, Houston. *Med Phys*. 2010;37(1):154–63.
2. Zhu TC, Lin H, Shen J. Chapter 26: monitor unit (MU) calculation. In: Das IJ, Paganetti H, editors. *Principles and Practice of Proton Beam Therapy*. Madison, WS: Medical Physics Publishing; 2015, pp. 231–64.
3. Pedroni E, Bacher R, Blattmann H, Bohringer T, Coray A, Lomax A, Lin S, Munkel G, Scheib S, Schneider U, et al. The 200-MeV proton therapy project at the Paul Scherrer Institution: conceptual design and practical realization. *Med Phys*. 1995;22(1):37–53.
4. Lorin S, Grusell E, Tilly N, Medin J, Kimstrand P, Glimelius B. Reference dosimetry in a scanned pulsed proton beam using ionisation chambers and a Faraday Cup. *Phys Med Biol*. 2008;53(13):3519–29.
5. Zhu TC, Ahnesjo A, Lam KL, Li XA, Ma C-MC, Palta JR, Sharpe MB, Thomadsen B, Tailor RC. Report of AAPM Therapy Physics Committee Task Group 74: In-air output ratio, Sc, for megavoltage photon beams. *Med Phys*. 2009;36(11):5261–91.
6. Koch N, Newhauser WD, Titt U, Gombos D, Coombes K, Starkschall G. Monte Carlo calculations and measurements of absorbed dose per monitor unit for the treatment of uveal melanoma with proton therapy. *Phys Med Biol*. 2008;53(6):1581–94.
7. Paganetti H, Jiang H, Parodi K, Slopsema R, Engelsman M. Clinical implementation of full Monte Carlo dose calculation in proton beam therapy. *Phys Med Biol*. 2008;53(17):4825–53.
8. Stankovskiy A, Kerhoas-Cavata S, Ferrand R, Nauraye C, Demarzi L. Monte Carlo modelling of the treatment line of the Proton Therapy Center in Orsay. *Phys Med Biol*. 2009;54(8):2377–94.
9. Lin L, Huang S, Kang M, Hiltunen P, Vanderstraeten R, Lindberg J, Siljamaki S, Wareing T, Davis I, Barnett A, McGhee J, Simone II CB, Solberg TD, McDonough JE, Ainsley C. A benchmarking method to evaluate the accuracy of a commercial proton Monte Carlo pencil beam scanning treatment system. *J Appl Clin Med Phys*. 2017;18(2):44–9. doi: 10.1002/acm2.12043.
10. Kooy HM, Rosenthal SJ, Engelsman M, Mazal A, Slopsema RL, Paganetti H, Flanz JB. The prediction of output factors for spread-out proton Bragg peak fields in clinical practice. *Phys Med Biol*. 2005;50:5847–56.
11. Kooy HM, Schaefer M, Rosenthal S, Bortfeld T. Monitor unit calculations for range-modulated spread-out Bragg peak fields. *Phys Med Biol*. 2003;48(17):2797–808.
12. Sahoo N, Zhu XR, Arjomandy B, Ciangaru G, Lii M, Amos R, Wu R, Gillin MT. A procedure for calculation of monitor units for passively scattered proton radiotherapy beams. *Med Phys*. 2008;35(11):5088–97.
13. Hsi WC, Schreuder AN, Moyers MF, Allgower CE, Farr JB, Mascia AE. Range and modulation dependencies for proton beam dose per monitor unit calculations. *Med Phys*. 2009;36(2):634–41.
14. Ferguson S, Ahmad S, Jin H. Implementation of output prediction models for a passively double-scattered proton therapy system. *Med Phys*. 2016;43(11):6089–97. doi: 10.1118/1.4965046.
15. Ferguson S, Chen Y, Ferreira C, Islam M, Keeling VP, Lau A, Ahmad S, Jin H. Comparability of three output prediction models for a compact passively double-scattered proton therapy system. *J Appl Clin Med Phys*. 2017;18(3):108–17. doi: 10.1002/acm2.12079.
16. Schaffner B. Proton dose calculation based on in-air fluence measurements. *Phys Med Biol*. 2008;53(6):1545–82.
17. Zhu TC, Li Z, Karunamuni R, Yeung D, Slopsema R. A fluence-based algorithm for MU calculation of proton beams. *Med Phys* 2007;34(6):2403. doi: 10.1118/1.2760657. PubMed PMID: ISI:000247479600371.

18. Zhu TC, Karunamuni R, Slopsema RL. Determination of SOBP PDD for therapeutic proton beams. *Med Phys.* 2008;35(6):2766.

19. Zhao Q, Wu H, Wolanski M, Pack D, Johnstone PA, Das IJ. A sector-integration method for dose/MU calculation in a uniform scanning proton beam. *Phys Med Biol.* 2010;55(3):N87–95. Epub 2010/01/09. doi: 10.1088/0031-9155/55/3/N02. PubMed PMID: 20057011.

20. Zheng Y, Ramirez E, Mascia AE, Ding X, Okoth B, Zeidan O, Hsi WC, Harris B, Schreuder AN, Keole S. Commissioning of output factors for uniform scanning proton beams. *Med Phys.* 2011;38(4):2299–306.

21. Kim DW, Lim YK, Ahn SH, Shin J, Shin D, Yoon M, Lee SB, Kim DY, Park SY. Prediction of output factor, range, and spread-out bragg peak for proton therapy. *Med Dosim.* 2011;36(2):145-52. Epub 2010/07/06. doi: 10.1016/j.meddos.2010.02.006. PubMed PMID: 20599372.

22. Daartz J, Engelsman M, Paganetti J, Bussiere MR. Field size dependence of the output factor in passively scattered proton therapy: influence of range, modulation, air gap, and machine settings. *Med Phys.* 2009;36(7):3205–10.

23. Moyers MF. Proton Therapy. In: van Dyk J, editor. *Modern Technology of Radiation Oncology.* Madison, WI: Medical Physics Publishing; 1999, pp. 863–4.

24. Lin L, Ainsley CG, McDonough JE. Experimental characterization of two-dimensional pencil beam scanning proton spot profiles. *Phys Med Biol.* 2013;57(4):983–97.

25. Lin L, Ainsley CG, Solberg TD, McDonough JE. Experimental characterization of two-dimensional spot profiles for two proton pencil beam scanning nozzles. *Phys Med Biol.* 2014;59(2):493–504.

26. Sawakuchi GO, Mirkovic D, Perles LA, Sahoo N, Zhu XR, Ciangaru G, Suzuki K, Gillin MT, Mohan R, Titt U. An MCNPX Monte Carlo model of a discrete spot scanning proton beam therapy nozzle. *Med Phys.* 2010;37(9):4960–70.

27. Sawakuchi GO, Zhu XR, Poenisch F, Suzuki K, Clangaru G, Titt U, Anand A, Mohan R, Gillin MT, Sahoo N. Experimental characterization of the low-dose envelope of spot scanning proton beams. *Phys Med Biol.* 2010;55(12):3467–78.

28. Schwaab J, Brons S, Fieres J, Parodi K. Experimental characterization of lateral profiles of scanned proton and carbon ion pencil beams for improved beam models in ion therapy treatment planning. *Phys Med Biol.* 2011;56(24):7813–27.

29. Schaffner B, Pedroni E, Lomax A. Dose calculation models for proton treatment planning using a dynamic beam delivery system: an attempt to include density heterogeneity effects in the analytical dose calculation. *Phys Med Biol.* 1999;44(1):27–41.

30. Li Y, Zhu XR, Sahoo N, Anand A, Zhang X. Beyond Gaussians: a study of single-spot modeling for scanning proton dose calulation. *Phys Med Biol.* 2012;57(4):983–77.

31. Ulmer W, Schaffner B. Foundation of an analytical proton beamlet model for inclusion in a general proton dose calculation system. *Rad Phys Chem.* 2011;80(3):378–89.

32. Zhu XR, Poenisch F, Lii M, Sawakuchi GO, Titt U, Bues M, Song X, Zhang X, Li Y, Ciangaru G, Li H, Taylor MB, Suzuki K, Mohan R, Gillin MT, Sahoo N. Commissioning dose computation models for spot scanning proton beams in water for a commercially available treatment planning system. *Med Phys.* 2013;40(4):041723.

33. Bortfeld T, Schlegel W. An analytical approximation of depth-dose distributions for therapeutic proton beams. *Phys Med Biol.* 1996;41(8):1331–9.

34. Shen J, Liu W, Stoker J, Ding X, Anand A, Hu Y, Herman MG, Bues M. An efficient method to determine double Gaussian fluence parameters in the eclipseTM proton pencil beam model. *Med Phys.* 2016;43(12):6544–51.

35. Sawakuchi GO, Titt U, Mirkovic D, Ciangaru G, Zhu XR, Sahoo N, Gillin MT, Mohan R. Monte Carlo investigation of the low-dose envelope from scanned proton pencil beams. *Phys Med Biol.* 2010;55(3):711–21.

36. Gottschalk B, Koehler AM, Schneider RJ, Sisterson JM, Wagner MS. Multiple Coulomb scattering of 160 MeV protons. *Nucl Inst Meth Sec B.* 1993;74(4):467–90.

37. Shen J, Liu W, Anand A, Stoker JB, Ding X, Fatyga M, Herman MG, Bues M. Impact of range shifter material on proton pencil beam spot characteristics. *Med Phys.* 2015;42(3):1335. Epub 2015/03/05. doi: 10.1118/1.4908208. PubMed PMID: 25735288.
38. Gottschalk B. On the scattering power of radiotherapy protons. *Med Phys.* 2010;37(1):352–67.
39. Soukup M, Fippel M, Alber M. A pencil-beam algorithm for intensity modulated proton therapy derived from Monte Carlo simulations. *Phys Med Biol.* 2005;50(21):5089–104.
40. Shen J, Lentz JM, Hu Y, Liu W, Morales DH, Stoker JB, Bues M. Using field size factors to characterize in-air fluence of a proton machine with a range shifter. *Rad Oncol.* 2017;12:52. doi: 10.1186/s13014-017-0783-2.
41. Clasie B, Depauw N, Fransen M, Goma C, Panahandeh HR, Seco J, Flanz JB, Kooy HM. Golden beam data for proton pencil-beam scanning. *Phys Med Biol.* 2012;57(5):1147–58. doi: 10.1088/0031-9155/57/5/1147. PubMed PMID: ISI:000300775600005.
42. Lin L, Shen J, Ainsley CG, Solberg TD, McDonough JE. Implementation of an improved dose-per-MU model for double-scattered proton beams to address interbeamlie modulation width variability. *J Appl Clin Med Phys.* 2014;15(3):297–306.
43. Mackin D, Li Y, Taylor MB, Kerr M, Holmes C, Sahoo N, Poenisch F, Li H, Lii J, Amos R, Wu R, Suzuki K, Gillin MT, Zhu XR, Zhang X. Improving spot-scanning proton therapy patient specific quality assurance with HPlusQA, a second-check dose calculation engine. *Med Phys.* 2013;40(12):121708.
44. Zhu XR, Poenisch F, Song X, Johnson JL, Ciangaru G, Taylor MB, Lii M, Martin C, Arjomandy B, Lee AK, Choi S, Nguyen QN, Gillin MT, Sahoo N. Patient-specific quality assurance for prostate cancer patients receiving spot scanning proton therapy using single-field uniform dose. *Int J Radiat Oncol Biol Phys.* 2011;81(2):552–9.
45. Zhu XR, Li Y, Mackin D, Li H, Poenisch F, Lee AK, Mahajan A, Frank SJ, Gillin MT, Sahoo N, Zhang X. Towards effective and efficient patient-specific quality asurance for spot scanning proton therapy. *Cancer.* 2015;7(2):631–47. doi: 10.3390/cancers7020631.
46. Beltran C, Tseung HWC, Augustine KE, Bues M, Mundy DW, Walsh TJ, Herman MG, Laack NN. Clinical implementation of a proton dose verification system utilizing a GPU accelerated Monte Carlo engine. *Int J Part Ther.* 2016;3(2):312–19. doi: 10.14338/IJPT-10-00011.1.
47. Li H, Sahoo N, Poenisch F, Suzuki K, Li Y, Li X, Zhang X, Lee AK, Gillin MT, Zhu XR. Use of treatment log files in spot scanning proton therapy as part of patient-specific quality assurance. *Med Phys.* 2013;40(2):021703. Epub 2013/02/08. doi: 10.1118/1.4773312. PubMed PMID: 23387726; PMCID: 3555925.

V

Treatment Planning/Delivery

CHAPTER 14

Dose Calculation Algorithms*

Ben Clasie, Harald Paganetti, and Hanne M. Kooy

Massachusetts General Hospital and Harvard Medical School

CONTENTS

---

* Colour figures available online at www.crcpress.com/9781138626508.

14.1 PENCIL BEAM ALGORITHMS

14.1.1 Rationale

The primary dose calculation requirement is to calculate the dose from a radiation field, defined by a geometric relation to the patient and a dosimetric relation to the delivery variables, to a set of points of arbitrary distribution. The set of points often has a metastructure that allows the rapid evaluation of dose in optimization problems. For such problems, the dose algorithm computes dose to a point per unit "intensity" Q, expressed as number of particles or in machine-specific monitor units (MUs), per deliverable proton spot (we use "spot" (or beamlet) to denote a physical beam of protons and "pencil beam" for the mathematical form of transport of protons in medium; note that pencil beam in pencil beam scanning (PBS) refers to spot in our parlance). The dose to a point i from a spot k can be defined as

$$D_i = \sum_j Q_j \times D_{ij}, \tag{14.1}$$

where D_{ij} is the dose per unit spot intensity from spot j to point i and Q_j is the spot intensity. It should be clear that the form offers an immediate conversion from intensity to dose in an optimization problem. The dose calculation only computes the form D_{ij} and intensities are subsequently derived to satisfy treatment constraints. Optimization is the subject of Chapter 19.

We emphasize the dose calculation for PBS delivery systems but include the formalism for spread-out Bragg peak (SOBP) fields. For both, at the dose algorithm core, there is little difference in physics or implementation.

A PBS field is the aggregate of numerous individual narrow proton spots (full width at half maximum (FWHM) ~10 mm). A spot is defined by its location (X,Y) in the isocentric plane reached by magnetic deflection of the proton spot bundle, intensity (Q or MU), and energy (typically specified in the equipment setting for energy but derived from range in water in treatment planning). The relatively large size of a proton spot and the inability of proton accelerators to produce energies at subdermal range requires the use of apertures and range shifters for certain clinical applications and their use is included in the dose calculation formalism. The relative large size of a proton spot does not allow the direct use of the physical spot as a "pencil beam" representation for the mathematical pencil beam to be transported through the patient without loss of resolution and accuracy. Thus, the form D_{ij} is obtained through a *superposition* of mathematical pencil beams into spots, which in turn is summed to represent the entire proton field. The process is described later.

Pencil beam models are the most pragmatic empirical representation for modeling dose transport in medium. The pencil beam model is a convenient representation of a piecewise geometric and physical approximation to the exact model, where each pencil beam allows a sufficiently accurate approximation of all dose depositing processes in the patient by local effects along the axis of the pencil beam. Pencil beam algorithms use the mathematical concept of a set of narrow beams that, as a composite [1], model all the degrees of freedom of the radiation field [2], fill the physical space of the radiation field, and [3] provide a good approximation of modeling the patient as a set of interactions of the pencil beam in a "slab" geometry around the pencil beam axis. Interactions in laterally infinite slab geometry are well understood, and its application in pencil beam models is only limited by the lateral extent of the pencil beam itself. That is, the pencil beam local model is insensitive to heterogeneities lateral to its bounding envelop.

The pencil beam model was first applied in electron dose calculations [1]. The electron field was subdivided in rectangular pencil beams and each pencil beam modeled the electron fluence diffusion in medium in the presence of heterogeneities, as quantified on computed tomography (CT) data. Figure 14.1 shows a simple pseudocode description of a pencil beam model implementation, where line [1] creates an initialized set of dose calculation points, containing the position based on the CT data set and accumulates dose at that position [2], sets up the computation given a decomposition of the field into mathematical pencil beams [3], creates a *trace object* that maintains the geometry and state necessary to resolve the physics interactions as a function of depth [4], places the points in the coordinate system of the pencil beam with the z-axis along the pencil beam axis, and where each trace is to the depth of the next point sorted in depth, now, along the pencil beam axis [5], traces the pencil beam through the volume in increments [6], computes the local *kernel* given the results from the trace and the physics model, and [7] adds dose as a consequence of the *kernel convolution*. The performance of an algorithm implementation is of course of primary significance. The above algorithm scales obviously as $O(P)$, linearly with the number of pencil beams, but the performance of the superposition of the kernel on the points will be very sensitive to the details of its implementation. It is of course the evolution of the kernel and its

```
1       points = new_matrix(sizeof(CT), bounds);
2       for each pencil-beam P in field F {
3           trace = new_trace(P);
4           points_p = place_and_sort_point_in(trace, points);
            while inside(trace, CT) {
5               trace = trace_to_ z_of_ next_p(trace, CT);
6               kernel = compute_kernel(kernel, trace, physics_model);
7               points_p = compute_dose(kernel, points_p);
            }
        }
```

FIGURE 14.1   Pseudocode implementation of a pencil beam algorithm. The algorithm collects the calculation points (points) and sorts the points along the axis of each pencil beam P (points_p) to permit direct stepwise tracing to the depth of each point. At each point, the algorithm computes the extent of the pencil beam and computes the dose to the affected points.

convolution on the dose point geometry that are key to the efficiency of the algorithm implementation.

14.1.2 Physics Model

14.1.2.1 Elastic Scatter in Medium

The physical processes involved in traversal of protons through medium are well understood (see Chapter 2). The primary mode of interaction is through elastic scatter of the proton in the electric field of the atoms in the medium. These numerous scattering events have the statistical consequence that the density of protons lateral to the mean direction approximates a Gaussian distribution. The Gaussian approximation is rigorous, according to the central limit theorem, if every scattering event occurs at a small angle. The protons scattering events, however, have an angular distribution governed, effectively, by the small-angle Rutherford form $\frac{d\sigma}{d\Omega} \approx \frac{1}{\theta^4}$ which means that the lateral distribution will be Gaussian for small angles but will trend toward a scattering tail distribution $\frac{1}{\theta^4}$. A complete multiple Coulomb scattering (MCS) theory was published by Molière in 1947 [2] (see Chapter 2).

We wish to derive a model for the evolution of a beam with initially zero emittance (i.e., zero lateral width and zero angular spread) in medium, whose stopping power is characterized relative to water. The latter is in compliance with the definition of the dose kernels in water, which we apply to the "water-equivalent" representation of the patient. This evolution is largely described by the Gaussian widening of the proton distribution and the involved volume in which those protons will interact.

Fermi–Eyges theory [3] describes the evolution of spatial and angular distributions of particles propagating through matter, with the assumption that the particles undergo many small-angle scattering events. It was initially applied in pencil beam algorithms for electron beams [1], and more recently, although not in the Fermi–Eyges form, to proton beams [4]. The theory is more accurate for protons and heavy ions, because these particles scatter through small angles in interaction with the atoms in the medium.

Fermi–Eyges theory predicts that the lateral spread of a parallel and infinitesimally narrow proton beam as a function of depth in water, z (cm), is Gaussian in shape with width given by [5]

$$\overline{x^2_{MCS}(z)} = \sigma^2_{x,MCS}(z) = \int_0^z (z-z')^2 T(z')dz' \,[\text{cm}^2], \tag{14.2}$$

where $\overline{x^2_{MCS}(z)} = \sum_{i=1}^N x_i^2 / N$ is the lateral variance of the beam (equal to $\sigma^2_{x,MCS}(z)$, the square of the standard deviation of the Gaussian profile, in the limit of many events) and $T(z) \equiv \overline{d\theta^2}/dz$ is the scattering power. Gottschalk [6] gives a parameterization of the scattering power that is accurate almost within experimental errors for protons in water, thin slabs, and high-Z materials. The parameterization, called the Improved Nonlocal Formula, is given by

$$T(z)=T_{dM}=\Big[0.5244+0.1975\ \log_{10}(1-(pv/p_1v_1)^2)+0.2320\log_{10}(pv/\text{MeV})$$

$$-0.0098\log_{10}(pv/\text{MeV})\log_{10}(1-(pv/p_1v_1)^2)\Big]\times\left(\frac{E_s}{pv}\right)^2\frac{1}{L_R},\qquad (14.3)$$

where pv [MeV] is the product of the proton momentum and velocity and is a function of z, p_1v_1 is the initial product of momentum and velocity, $E_s = 15.0$ MeV, and L_R is the radiation length (36.1 cm for water).

Most treatment-planning algorithms, however, use the Highland formula for calculating the spread of proton beams in water, which is accurate within ±5% [4]. The Highland scattering angle is integrated along the beam axis and the result for the lateral standard deviation at z is [7]

$$\sigma_{x,\text{MCS}}^2(z)=\left[1+\frac{1}{9}\log_{10}\left(\frac{z}{L_R}\right)\right]^2\times\left[\int_0^z\left(\frac{14.1\ \text{MeV}}{pv}\times(z-z')\right)^2\frac{1}{L_R}dz'\right],\qquad (14.4)$$

where $\rho = 1$ g/cm^3 is the density of water and the subscript MCS indicates multiple Coulomb scattering. The Highland approximation and Fermi–Eyges theory with the Improved Nonlocal Formula are given in Figure 14.2 for proton beams in water and the two parameterizations agree within 7%.

Each pencil beam evolves in water according to $\sigma_{x,\text{MCS}}(z)$. At z, the total distribution is the convolution of two Gaussian functions, the initial or unperturbed Gaussian beam

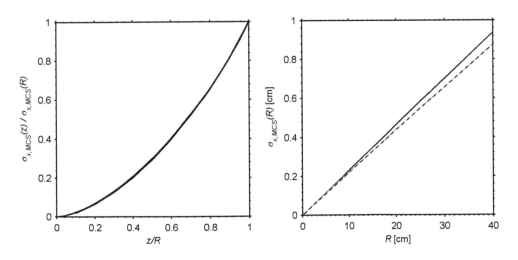

FIGURE 14.2 (left) The normalized Gaussian spread in water vs. normalized depth, z/R, where R is the range of protons to stop in water (cm). This relation is independent of R between 0.1 and 40 cm and is the same for both Equations 14.2 and 14.4 (right). The Gaussian spread at maximum depth (i.e., z = R) for Equation 14.2 (solid line) and Equation 14.4 (dashed line).

shape and the additional spreading from MCS in medium. The beam spread at z for an initially thick beam with zero angular spread is therefore

$$\overline{x_{\text{thick}}^2(z)} = \overline{x_{\text{MCS}}^2(z)} + \overline{x_{\text{thick}}^2(0)}, \tag{14.5}$$

where $\overline{x_{\text{thick}}^2(0)}$ is the initial variance of the thick Gaussian beam. If the initial beam has nonzero angular spread, then

$$\overline{x_{\text{thick}}^2(z)} = \overline{x_{\text{MCS}}^2(z)} + \overline{x_{\text{vac}}^2(z)}, \tag{14.6}$$

where $\overline{x_{\text{xac}}^2(z)} = \overline{x_{\text{thick}}^2(0)} + 2 \cdot \overline{x\theta_{\text{thick}}(0)} \cdot z + \overline{\theta_{\text{thick}}^2(0)} \cdot z^2$ is the spread of the beam in a vacuum at z, $\overline{x\theta_{\text{thick}}(0)} = \sum_{i=1}^{N} x_i \theta_i / N$ is the initial covariance [5], and $\overline{\theta_{\text{thick}}^2(0)} = \sum_{i=1}^{N} \theta_i^2 / N$ is the initial angular variance.

Proton pencil beam dose calculations determine the MCS lateral spread for each pencil beam in a field at each depth, and since a typical field consists of thousands of pencil beams, corresponding to the spatial and energy subdivisions, and hundreds of dose calculation depths, the spread calculation algorithm must be fast and accurate. The use of a lookup table for the results in Figure 14.2 can improve the speed of the dose calculation (see appendices in [4]).

Computational results in water are extended to heterogeneous media by assuming the media is waterlike. That is, the geometric depth z is converted to a water-equivalent depth along the central ray of the pencil beam, and the media is assumed to have the radiation length of water. A general-purpose Monte Carlo simulation should be used to check dose calculations for treatment plans in media that are not waterlike.

14.1.2.2 Large Scattering Events

Dose deposited outside of the central Gaussian region, from large-angle elastic scattering events and secondary particles, is called the beam halo. This leads to extended tails in the spatial and angular distributions of the beam. No correction for this effect is needed for large, uniform fields typical of those produced in scattering systems or uniform scanning systems because there is equilibrium of the lateral scattered dose and the effect is implicitly included in either measurement or model [8]. On the other hand, nonhomogeneous fields are implicit in proton PBS and the beam halo cannot be ignored in the calculation of absolute dose [9–12]. Ignoring the halo in treatment planning can lead to errors of the order of 5%. The halo contribution to the proton spread is characterized by a much larger radial effect. Figure 14.3 shows the buildup of dose measured on the central axis of square fields for a pencil beam of range 25 cm in water. The effect extends up to a distance on the order of 7 cm as shown in Figure 14.3, where full buildup is achieved in a square field of 15 cm.

The halo is not easily described a priori. In practice, however, the halo can be well parameterized as a second Gaussian pencil beam with a fraction f_H relative to the total

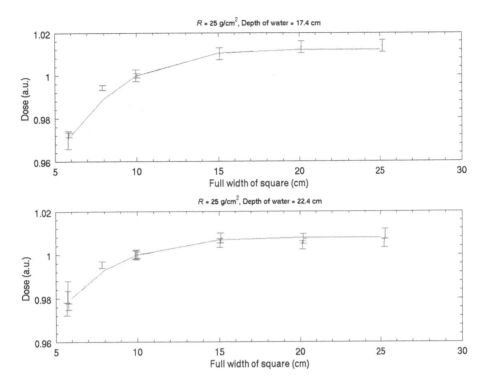

FIGURE 14.3 Observed dose buildup in the center of a square field as a function of field width. The square field is delivered by a set of monoenergetic, 25 cm in water, pencil beams that populate the field area uniformly. The markers are measurements and the lines are based on the model described in the text (Equations 14.7 and 14.8 and Table 14.1). Dose (Y-axis) is normalized to a 10×10 square field central dose (See color online.)

beam. This weight is slowly varying as a function of depth, d, and range, R, and a simple parameterization is adequate to describe dose deposited by the halo, such as

$$1 - f_H(d,R) = D_\infty^M(R,d)/D_\infty^T(R,d) = a_0(R) + a_1(R)t + a_2(R)t^2, \tag{14.7}$$

where t equals d/R, $D_\infty^M(R,d)$ is the MCS depth-dose component integrated over an infinitely broad lateral field, and $D_\infty^T(R,d)$ is the same for the total measured depth-dose. The fraction f_H is derived from measurement and used, subsequently, to decompose measured depth-doses into MCS-depth-doses and halo-depth-doses (Figure 14.4). In our implementation, we parameterize the coefficients in Equation 14.7 for f_H as

$$a_i(R) = b_{0,i} + b_{1,i}R + b_{2,i}R^2, \tag{14.8}$$

where the values for b are given in Table 14.1. This parameterization is derived from measurements of dose buildup in, for example, the center of circular rings of proton pencil beams of increasing radius and as a function of depth and energy. The parameterization allows a decomposition of a measured pristine peak depth-dose in terms of elastic and inelastic dose contributions as a function of energy and depth (see Figure 14.4).

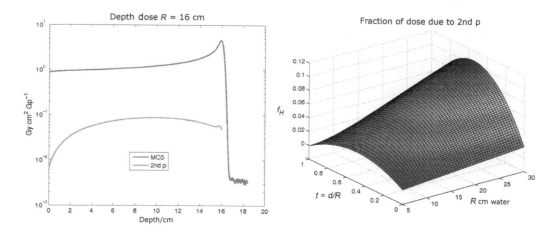

FIGURE 14.4 The total measured depth-dose (left) is decomposed into its component from MCS protons (blue) and from secondary scattered "halo" protons (red). The decomposition (Equation 14.7) can be characterized as a second-order polynomial surface as a function of range-scaled depth $t = d/R$ and range (right) (See color online.)

TABLE 14.1 Fit Parameters for Equation 14.7 to Obtain the Coefficients a_i for the Computation of the Depth-Dependent Fraction f_H in Equation 14.6

	a_0	a_1	a_2
$b_{0,i}$	1.002	2.128e-03	−2.549e-03
$b_{1,i}$	−5.900e-04	−2.044e-02	2.125e-02
$b_{2,i}$	0	3.178e-04	−3.788e-04

These values characterize the beam at the MGH [13].

Finally, the spread of the halo appears to be sufficiently constant (in the beam at the Massachusetts General Hospital (MGH)) as a function of depth as

$$\sigma_H(R) = 6.50 - 0.34R + 0.0078R^2 \tag{14.9}$$

The halo model quantification is sensitive to the details of the measurements and beam conditions. The primary concern, however, is to characterize the energy flow away from our primary dose component characterized by f_H as the redistribution of this secondary scattered energy is homogeneous as compared to the primary energy distribution in the patient.

14.1.3 Dose in Patients

The dose at depth to a point i in patient from a proton pencil beam k (where we use the subscript k to indicate the mathematical concept of a pencil beam instead of the physical spot j) is

$$D_{ik} = \frac{1}{2\pi\sigma^2} \exp\left(-\frac{r_{ik}^2}{2\sigma^2}\right) D_\infty(R_k, \rho_{ik}), \tag{14.10}$$

where ρ_{ik} is the water-equivalent depth along the pencil beam axis, the Z-axis of the pencil beam originating at 0 to the depth (X, Y, Z) of point i, σ is the spread of the pencil beam at depth ρ_{ik}, r_{ik} is the shortest distance from i to the pencil beam k axis and $D_\infty(R_k, \rho_{ik})$ is the broad-beam depth-dose of protons with range (in water) R_k at the radiological depth of point i. The depth-dose has units of [Dose][Area]/[Intensity], which in our practice is Gy (RBE) mm²/Gp (when Q is given in gigaprotons, for example) and where an RBE = 1.1 is assumed per convention.

The above Gaussian form is accurate for in-patient MCS protons and assumes that $D_\infty(R_k, \rho_{ik})$ includes all protons. A physical spot, in reality, has small non-Gaussian components from interactions such as from rare nuclear-scattered protons. Gottschalk et al. [14] enumerate the physical processes contributing to the non-Gaussian components which, when combined, do not have a well-formed equation. The effect, however, can be well approximated for treatment planning purposes by a second Gaussian form, given the small effect of the halo contribution [9]. The decomposed form is

$$D_{ik} = \frac{1}{2\pi\sigma_P^2} \exp\left(-\frac{r_{ik}^2}{2\sigma_P^2}\right) D_\infty^P(R_k, \rho_{ik}) + \frac{1}{2\pi\sigma_H^2} \exp\left(-\frac{r_{ik}^2}{2\sigma_H^2}\right) D_\infty^H(R_k, \rho_{ik}), \quad (14.11)$$

where the primary and halo components are considered separately for the pencil beam k. A practical implementation of the above form, however, must consider the very large range of halo protons and the computational impact (in terms of size) on D_{ik}.

The above form allows a rapid calculation of dose in medium from a uniform broad field of protons of applicable ranges/energies to cover the target volume(s). The set of pencil beams form the base set for the decomposition of the physical spots i. The dose D_{ij} is obtained by convolving the spot intensity profile $Q_j(x,y)$ over this set

$$D_{ij} = \sum_k Q_j F_{kj} D_{ik}, \quad (14.12)$$

where F_{kj} is the fraction of protons from spot j contributed to pencil beam k. Note that this convolution allows for any intensity profile of the spot to accommodate, for example, proton beams collimated by an (small) aperture. For a Gaussian spot profile, F_{kj} is the (normalized) volume of the pencil beam under the profile Q.

14.1.4 Beam Representation

The dose contribution from a specific beam is represented by pencil beam computational variables in Equation 14.10. We have two classes of proton beams: scattered fields (Chapter 5) and PBS fields (Chapter 6).

14.1.4.1 SOBP Beams

Algorithmically, we consider (i) the apertures and range compensators, (ii) the stacking of the pristine peaks, and (iii) some properties of the production system. An SOBP field is produced in a scattering, uniform scanning, or PBS systems (see Chapters 4–6). The use of a scattering system results in a field that appears to emanate from a "virtual" source, while a uniform scanning system has a bivalued source position corresponding to the centers of the deflection magnets. An SOBP field produced in a scattering system includes an inverse square factor in its depth-dose formalism (Equation 14.10). An SOBP field produced in a scanning system, as is the case for a PBS field, implicitly models the inverse square dependency from the set of diverging spots, although the spot itself has no divergence. The source position can be inferred by measuring the increase in width of a collimated field as a function of distance.

A scattering or uniform scanning system produces a source with a size proportional to the thickness of range-shifting material in the beam. The penumbra is quantified by measuring the penumbral edge as a function of distance (from the source) and projecting the penumbral edge at the source position. The projected penumbral width is modeled as a Gaussian spread contribution σ_S from the "virtual" source to the total pencil beam width. The source size in a proton scattering system is significant and on the order of 5 cm. This source size effect is mitigated by placing the source as far away from the patient as possible (and one reason for the large proton gantry diameter; the other being the significant size and mass of the bending magnet) and to place an aperture as close to the patient as possible. The source size, and σ_S, is thus demagnified by the ratio of distances. The source size contribution at the calculation point p is

$$\sigma_s(z_p) = \left(\frac{z_A - z_p}{SAD - z_A} \right) \sigma_s, \qquad (14.13)$$

where z_p is the z-position (along the central axis of the SOBP field) of the point p, z_A is the z-position of the aperture, and SAD is the distance from the isocenter to the source.

SOBP fields use, invariably, a range compensator that shifts the initial range across the field area to that range necessary to "just" place the distal edge of the field beyond the distal target volume surface. The range compensator thus presents various thicknesses of material (typically Lucite) along the lateral extent of the beam. Protons passing through a thickness t will scatter as a consequence and introduce an additional contribution σ_R equal to

$$\sigma_R = L_p \theta_R(t), \qquad (14.14)$$

where $\theta_R(t)$ can be computed by Equation 14.2 (or tabulated as [4]), and L_p is the distance from the range-compensator intersection point to the calculation point p along the ray from the source.

The total spread of a pencil beam at the depth of point p is (including Equation 14.4)

$$\sigma_T^2 = \sigma_S^2 + \sigma_R^2 \sigma_{MCS}^2 \qquad (14.15)$$

Thus, for scattered SOBP fields, produced by set of pristine peaks of energies {R}, the dose to a point p becomes

$$D(p) = \sum_R \frac{1}{2\pi\sigma_T^2} \exp\left(-\frac{r_p^2}{2\sigma_T^2} \right) D_\infty^R(d_p) \left(\frac{SAD - z_p}{SAD} \right)^2, \qquad (14.16)$$

where the inverse square correction accounts for the intrinsic divergence of the field. Dose distributions in treatment planning programs are typically obtained through the application of Equation 14.16. The model holds up well compared to Monte Carlo [15,16] and fails in predictable areas of high-Z materials and distal heterogeneities.

14.1.4.2 SOBP Production Model Effects

There are multiple means of creating SOBP fields. The most common method is to use a rotating modulator wheel with angular segments of widths and thicknesses corresponding to the pullback and weight of individual pristine peaks that comprise an SOBP field (Figure 14.5). The construction of such modulator wheels is an art and, in theory and practice, only achieves a "perfect" uniform SOBP plateau for the design energy.

We demonstrate the production and modeling for a rotating modulator wheel corresponding to the system used by IBA (Ion Beam Applications Ltd, Louvain la Neuve,

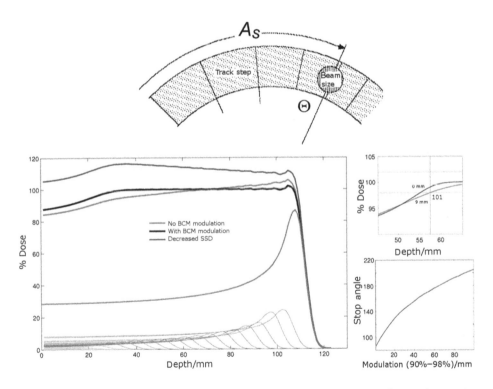

FIGURE 14.5 The SOBP is constructed from the weighted superposition of a single pristine peak, pulled back (shown as the individual pristine peak depth-doses) by successively increasing step thickness on a rotating modulator wheel. The wheel weights are adjusted by modulating the beam current as a function of wheel rotation angle (black curve). In absence of this modulation, the wheel produces the SOBP in red. The SOBP is designed to be flat at the nominal SSD of SAD −30 mm (proximal of the distal edge); other SSD values create a slope in the SOBP. The figure in the top right shows the effect of the source size, where the black line is the SOBP proximal falloff for an infinitesimal spot, while the red line corresponds to the observed softening of the "knee" as a consequence of the source size. The figure in the bottom right shows the stop angle (expressed as an 8-bit value) to achieve the desired modulation. The earlier SOBP has a range (modulation) equal to 110 mm [80] (See color online.)

Belgium) in their double-scattering system design. This design has a set of fixed range-modulator wheels (implemented as up to three concentric tracks on a single modulator wheel), where the thickness of a track step is a combination of a high-Z (lead) and low-Z (carbon/lexan) material, such that the scattering angle remains constant independent of total (water-equivalent) thickness. The beam subsequently passes through one (out of three) scatterer (see Chapter 5) to produce a uniform field up to a 25 cm diameter. The clinical range of energies is between 5 and 30 cm in water and requires the use of eight combinations, dubbed an option, of track and scatterer. Each of the eight options is further subdivided into three suboptions. Thus, a suboption covers about 1 cm of energy interval over which the energy effect on the shape of the SOBP is assumed negligible. The track rotates at 10 Hz and produces a full SOBP dose distribution in patient 10/s.

In our clinical practice, we achieve SOBP plateaus with a ±0.5% uniformity when the isocenter is positioned slightly (~3 cm) from the distal falloff, i.e., SSD = SAD − R + 3 (with SSD being the source-surface distance and SAD the source-axis distance). This requirement is well beyond any theoretical ability to design the proper track thicknesses and angular intervals. The IBA system, however, uses beam current modulation as a function of rotation angle to modulate the track contributions for each pristine peak produced by the track. This modulation as a function of rotation angle allows an increase or decrease of current to correct any design limitations. The track is designed up to a maximum modulation width (corresponding to the maximum track step thickness). The current modulation allows any modulation up to the maximum track thickness by turning the current off before the rotation completes. Thus, modulation is a function of rotation angle, referred to as the "stop angle" in this context. The track has discrete steps, and the finite source size (approximately the track width) causes incomplete contributions of the pristine peak dose when the source stops at a junction (see Figure 14.5). Thus, an SOBP is not a simple superposition of discrete energy pristine peaks.

A model for such a track system specifies, for each track, a base pristine peak depth-dose distribution (defined at a range R equal to the maximum energy of an option), its current modulation distribution, its source size on the track, and its thickness and angular width for each track step (typically on the order of 30 steps). This base pristine peak depth-dose is not trivial to obtain. The scattering system inside of the gantry nozzle presents a complex geometry in which the scattered beam is not easily, if at all, characterized. In addition, the measurement is done while the modulator wheel is rotating. A simple approach has been to measure only the protons that pass through the first, thinnest, step of the track by, essentially, producing an SOBP of zero modulation. The use of this pristine peak, however, fails to produce SOBPs that compare well with measurements. Lu et al. [17] describe an experimental method to measure the pristine peak while the wheel is rotating. This pristine peak does produce an excellent correspondence between calculations and measurements (Figure 14.5). The SOBP is computed as

$$D(d) = Q_0 \sum_i P_0(d - T_i) \int_{\theta_{i-1}}^{\theta_i} \mathrm{BCM}(\theta) A(\sigma_Q, \theta) d\theta, \qquad (14.17)$$

where Q_0 is the total charge delivered in a rotation over the angle θ, P_0 is the SOBP depth-dose distribution at the maximum range defined for a track option, BCM is the (relative) beam current modulation over the rotation, and $A(\sigma_Q, \theta)$ is the fractional area of the beam spot (relative to the total spot area), with spread σ_Q at the rotation angle.

Figure 14.5 shows the result of the model in Equation 14.17 to produce an SOBP at the design SSD and the relevance of the beam current modulation and source size modeling. In clinical practice, the SOBP is specified only in terms of its distal range, typically the 90% (relative to the plateau) range R90, the modulation width from R90 to the proximal falloff, and its dose expressed in MU (the units of the ionization reference chamber). The latter was, by convention, also defined as the 90% value. This, however, leads to a definition of modulation values larger than the range. The latter, in turn, leads (at least in our practice) to issues in beam delivery and field output specification in terms of Gy/MU [8]. Therefore, a more consistent definition is the modulation width between the distal R90 and the proximal 98% falloff. This definition, however, does require an accurate description of this proximal falloff and the implementation of Equation 14.16. Its absence can result in a 5 mm error.

14.1.4.3 Scanning Beams

Proton beams scanned by orthogonal dipole magnets can irradiate arbitrary field areas without the need for a mechanical collimating aperture. Variation of the pencil beam energy allows control of the pristine peak position within the patient and thus also removes the absolute need for a range compensator. Finally, varying the intensity (expressed as total charge) of the pencil beam allows for dose modulation throughout the target volume. Thus, scanned beams have three degrees of control: Energy, Position, and Charge (intensity). The size of the beam tends to be fixed in current scanning system implementation, although there will be an energy dependence. PBS systems are generally assumed to deliver their dose in patient in discrete energy layers as changes in energy may require a mechanical manipulation of the beam and "slow" changes in beamline magnetic systems. Much effort, however, is focused on removing this limitation.

In a PBS field, the "pure" pencil beam emerges from the beamline transport system and traverses one or more foils that separate vacuum from air and one or more ionization chambers required for registering current and position. Thus, the overall "thickness" of space between the beamline and the patient is little and on the order of 1–3 mm. The pencil beam is thus a little perturbed.

The algorithmic representation of the pencil beam is thus as a "cone" of protons with an elliptical spread (expressed in (σ_x, σ_y) in the reference plane of the beam). By and large, the optical focusing of the pencil beam can be ignored [18] and the pencil beam is assumed parallel. The use of a range compensator is not excluded, and thus the spread will include σ_R (Equation 14.15). The effect of including an aperture has not been well studied but is expected to improve the penumbral edge for "large" spot size. An elliptical spot size may rotate as a function of gantry angle, which introduces an implementation requirement. The dose from a set of proton pencil beams {R} is

$$D(p) = \sum_R \frac{1}{2\pi\sigma_T^2} \exp\left(-\frac{r_p^2}{2\sigma_T^2}\right) D_\infty^R(d_p) \tag{14.18}$$

Note the absence of the inverse square correction for the depth-dose as the pencil beam is not diverging. The beam will, however, have an inverse square behavior if one or both axes of the pencil beam pivot around a scanning magnet bending source. This effect manifests itself in the ray tracing of the pencil beam axes through the volume and computationally expresses itself in the computation of r_p in Equation 14.18.

14.1.4.4 Implementation

The dose distribution of proton pencil beams in a uniform medium have the form of Equation 14.18. Lateral heterogeneities can be considered analytically by a ray-casting or ray-tracing model [19], a modified ray-casting model [19], or a superposition of infinitesimally narrow pencil beams [4,10,19]. The ray-casting model calculates the dose distribution based on the water-equivalent path length from the proton source to the point of interest but in the absence of lateral heterogeneities. The modified ray-casting model includes an average density surrounding the ray when computing the water-equivalent path length. While both approaches can be more accurate than the superposition approach, especially when lateral heterogeneities appear near the end of the proton's range, the superposition approach is generally the most accurate [19] and is the approach adopted by modern, analytical proton treatment planning systems. In the superposition approach, the physical spot is decomposed into multiple infinitesimally narrow pencil beams at the location of patient-specific devices (range shifter and/or aperture) with the number of protons assigned to each narrow beam in such a way that their sum is equivalent to the original physical pencil beam at this location. Each narrow beam is transported through the patient-specific devices and the patient. The superposition model, a priori, results in numerous superfluous decompositions due to spot overlap. Instead, the in-medium transport can be considered separately from the physically delivered spots as described in Section 14.1.3. We superimpose the spots, deliverable by the equipment for each field, on top of the pencil beam space (see Figure 14.6). The calculation in patients for Gaussian spots is

$$D(\vec{x}) = \sum_S G_S \tag{14.19a}$$

$$\times \sum_K \left(\oint \frac{1}{2\pi\sigma_O^2(R_S,z)} \exp\left(-\frac{\Delta_{S,K}^2}{2\sigma_O^2(R_S,z)}\right) dA_K\right), \tag{14.19b}$$

$$\times \frac{D_{R_S}^\infty(\rho)}{2\pi\sigma_P^2(R_S,\rho)} \exp\left(-\frac{\Delta_K^2(\vec{x})}{2\sigma_P^2(R_S,\rho)}\right) \tag{14.19c}$$

where the first term (14.19a) is the number of protons G_S (in units of billions, or gigaprotons Gp) in a spot from the set S, the second term (14.19b) is the apportionment of these G_S protons, given the optical spread $\sigma_O^2(R_S,z)$ of the spot, over the set of computational pencil

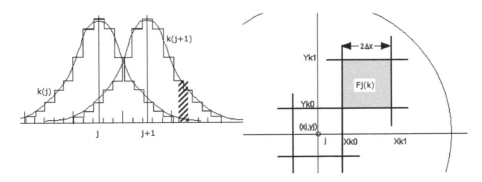

FIGURE 14.6 The transport of spots, physical pencil beams *j*, through the patient first transports "small" bixels (hatch bar in Gaussian, left) through the patient. The bixels have initial 0 spread and thus only spread as a consequence of in-patient scatter. Each spot *j* contributes to *k(j)* bixels proportional to the area of bixel *k* under the spot *j*. The subdivision ensures that the transport in patient is at the highest lateral resolution, limited by the bixel area and the bixel spread at depth, and removes the dependency on spot spread on the calculation resolution and performance.

beams K. This set of computational pencil beams K in Equation 14.19c is defined at the highest resolution necessary to accurately represent the dose in patient. Equation 14.19c models the diffusion of the number of protons, given by the product of 14.19a × 14.19b, in the patient given the scatter spread $\sigma_P(R_S, \rho)$ in the patient due to MCS. In a PBS system, the spot spread $\sigma_0^2(R_S, z)$ in Equation 14.19b is measured in air and is a function of the spot range R_S and position z along the beamlet axis. The parameter $\Delta_{S,K}$ denotes the position of a point in the computational pencil beam area A_K with respect to the spot coordinate system. The final term (14.19c) follows Pedroni et al. [9], where $D_R^\infty(z)$ (in units of Gy cm^2 Gp^{-1}) is the absolute measured depth-dose per proton integrated over an infinite plane at depth z, $\sigma_P(R_S, \rho)$ is the total pencil beam spread at radiological depth ρ caused by MCS in patient (see Hong et al. [4]), and $\Delta_K(\vec{x})$ is the displacement from the calculation point to the K pencil beam axis. Equation 14.19 is a phenomenological description of the distribution in patient of protons delivered by the set of spots S.

The above form, which convolves the spot shape (assumed Gaussian) over the high-resolution set of mathematical pencil beams (k), allows for arbitrary spot shapes to be convolved over this space. For example, a proton beam collimated by an upstream small collimator has a profile that is a convolution of a step function and a Gaussian. This profile (normalized to unity) can be numerically convolved over the k pencil beams to yield the dose in the presence of arbitrary spot shapes.

The use of this set of pencil beams increases the computation performance significantly. The main computational burden in a pencil beam algorithm is the part of the algorithm that, at each step along the pencil beam axis, finds the points that are within the Gaussian envelope (line 7 in Figure 14.1). This search time is proportional to the area (i.e., $O(\sigma^2)$). Thus, tracing the narrowest possible pencil beams improves computational performance. An efficient search algorithm will rely on ensuring that points are efficiently ordered with respect to the pencil beam to ensure that points are indexed as a function of distance along and lateral to the pencil beam axis.

14.1.4.5 Halo Effect Considerations

Nuclear-scattered "halo" protons affect the dose in both SOBP and PBS fields. Their effect in SOBP fields has been shown by Daartz et al. [20] for small-field measurements. The results can be readily modeled by a simple Clarkson integration (see, for example, [21]). Given the location of the pencil beam in the collimated field (Figure 14.7), its halo deficit can be computed for its equivalent circular field as

$$F_H\left(r_{eq}\right)=\frac{1}{2\pi}\sum_T I_k\theta_k F_H\left(r(T_k)\right),\tag{14.20}$$

where $F_H(r)=1-\exp\left(-r^2\big/2\sigma_H^2\right)$ is the halo deficit in a circular field independent of depth as σ_H is largely independent of depth in the Clasie approximation. The local depth-dose becomes

$$D_L(d,r_{eq},R)=D(d,R)\left[\alpha(d,R)+\left(1-\alpha(d,R)\right)F_H\left(r_{eq}\right)\right],\tag{14.21}$$

where α is derived in Clasie et al. Figure 14.7 shows the correction for an SOBP field passing through a 40 mm collimator when the above depth-dose D_L is used for each pencil beam in the field instead of the broad-field depth-dose D.

Consideration of nuclear-scattered halo protons in PBS fields is computationally challenging. The D_{ij} form for Coulomb-scattered protons is sparse due to the limited lateral scatter of those protons; its size is typically on the order of gigabytes. The order of magnitude

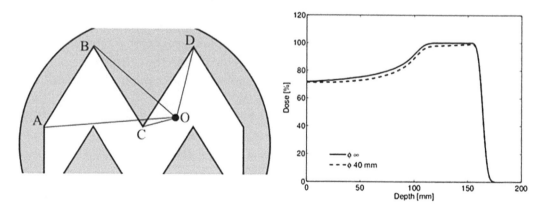

FIGURE 14.7 (a) Left: A zig-zag aperture is shown with a pencil beam position at O. For each triangle (such as OAB), we compute left- or right-handedness ($I_k = \pm1$) to add or subtract the scatter contribution, compute its inner angle θ, and equate the mean chord to the radius of its equivalent circular field $r(T_k)$. The scatter contribution is computed using Equation 14.21 for the angular fraction at O with the radius to the midpoint of the aperture edge segment. (b) Right: Example SOBP ($R = 160\,\mathrm{mm}$, $M = 100\,\mathrm{mm}$) for an aperture opening sufficient for primary and scattered equilibrium and for a 40-mm diameter aperture. Note the depth-dependent effect of the halo because of nuclear scatter buildup and reduction due to lower proton–nucleus cross section as the proton ranges out.

increase of the halo extent would increase the size of the D_{ij} form by a factor of 100. Our approach is to define a new set of "halo" pencil beams whose dimensions are correspondingly larger to compute a H_{im} form that only contains the halo contributions of a spot j to a point i. Note that the index on H is m, not j, as the size of a halo beam is larger than a spot, and hence, multiple spots contribute to a single halo beam. Pencil beams, however, are much smaller compared to a spot. Figure 14.8 shows (for one energy) the subdivision of the isocentric plane in pencil beams, halo beams, and spots. This model gives dose to a point i as

$$D_i = \sum_j Q_j D_{ij} + \sum_m Q_m H_{im}, \qquad (14.22)$$

where $Q_m = \sum_{k \subset m} Q_k$ is the sum of pencil beam charge contained within the halo beam boundary. The pencil beam charge Q_k, in turn, is given (above) as $Q_k = F_{kj} Q_j$. The above decomposition, where pencil beams use the primary depth-dose and halo beam use the halo depth-dose, allows the computation of both matrices in the same time and of the same size. The use of the above form in the optimization loop thus maintains the same performance of the optimization calculation.

14.1.5 A Scatter-Only Monte Carlo

Pencil beam implementations suffer from the assumption that interactions along the central axis are representative within the whole Gaussian envelope. This makes pencil beam

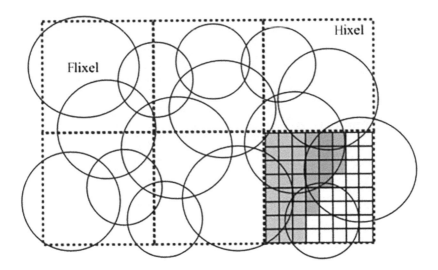

FIGURE 14.8 Halo beams (hixel; large rectangles) and spots (flixel; circular) are superimposed over the pencil beam (bixel; small rectangles) space. The spots contribute protons to the pencil beams (as described in the text). The protons in a halo beam, in turn, are the sum of pencil beam protons contained within the halo beam. Apertures are considered by setting the pencil beam intensity value b to $0 < b < 1$, as shown by the dark/light highlight of the pencil beams in the lower-right bixel. Note that the spot size in a particular plane is not uniform as a consequence of heterogeneities above the plane.

models insensitive to heterogeneities near the pencil beam axis, reduces the lateral sensitivity to that of the Gaussian width, and overemphasizes dosimetric diffusion as a consequence of upstream heterogeneities on the pencil beam axis.

The intrinsic physics pencil beam model for primary and secondary scatter of protons suffices for clinical modeling of the dose distribution. This allows the definition of a simple Monte Carlo implementation, where scatter is modeled on a voxel basis for a proton traversing the voxel. This implementation (pseudo-code in Figure 14.9) uses two random numbers, one to select the scattering mechanism based on Equation 14.7 and one to select the azimuthal angle (0–2π) of scatter along the direction of the incoming proton. The polar angle

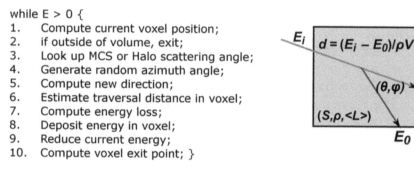

```
while E > 0 {
1.    Compute current voxel position;
2.    if outside of volume, exit;
3.    Look up MCS or Halo scattering angle;
4.    Generate random azimuth angle;
5.    Compute new direction;
6.    Estimate traversal distance in voxel;
7.    Compute energy loss;
8.    Deposit energy in voxel;
9.    Reduce current energy;
10.   Compute voxel exit point; }
```

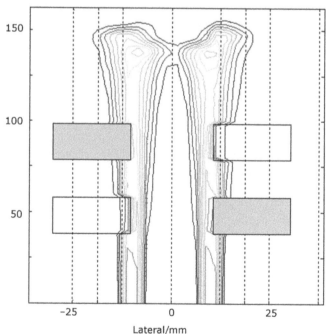

FIGURE 14.9 Pseudocode for the transport of protons through a set of voxels. Only MCS and secondary-scattered protons are considered, each with a mean scattering angle scaled to the voxel dimension and density. The resultant code implementation is well suited to a GPU architecture due to its compactness and lack of secondary particles. The bottom shows two pencil beams traversing a medium with alternating high- and low-density regions.

is the mean scattering angle, MCS, or halo, based on the voxel density. The large number of voxels assures that this approximation converges to the correct overall scatter in the medium.

The Monte Carlo transports individual protons through a volume represented by a rectangular 3D grid of voxels. Each voxel has the relative (to water) stopping power S_v obtained, in practice, from CT Hounsfield units [22]. We score the dose to a voxel as the sum of energy depositions along the individual proton tracks that traverse the voxel and divide by the voxel mass.

A proton, of energy E_S, enters a voxel on one of its faces and exits on another. The model computes the (unscattered) exit point along the incoming proton direction (u,v,w) and the distance L between the entrance and this projected exit point. The mean polar MCS scatter angle in the voxel is $\theta = \theta_0(E_S)\sqrt{S_v}$, while the mean polar halo scatter angle is $\theta = \theta_H(E_S)\sqrt{S_v}$ and the azimuthal angle is randomly uniform between 0 and 2π. The mean scatter angle $\theta_0(E_s)$ is derived and quantified by Gottschalk [6]. The mean scatter angle $\theta_H(E_s)$ is fitted to reproduce the observed spread (Equation 14.9). The fraction of protons that scatter with the halo scatter angle and the azimuthal angle are the random variables. The proton direction is adjusted to (u',v',w'), given the scattering angles and the actual exit point is computed. The energy loss of the proton along the mean voxel track length $\langle L \rangle$ is

$$\Delta_{<\lambda>}(E) = T_W^{\infty}(R,d_p) \times \frac{S_V/\rho_V}{S_W/\rho_W} \times \rho_V \times \langle L \rangle \tag{14.23}$$

The use of a mean track length $\langle L \rangle$ serves solely to reduce computational overhead. Computed pristine Bragg peaks using $\Delta_{\langle L \rangle}(E)$ and their range compared to the original projected range (from PSTAR; National Institute of Standards and Technology) show excellent agreement.

The simplicity of the model offers excellent opportunities for a graphics processing unit (GPU) implementation. In our implementation of the model in CUDA (a parallel computing platform and application programming interface model), on an entry-level nVidia FX1700 card, it achieves a performance of 2,000,000 protons/100 cm^2 in 10 s in 128^3 voxels. This is compared to our analytic pencil beam algorithm implementation, which computes the same distribution in 30 s. Thus, it is clear that at least a simple Monte Carlo, yet improved compared to a pencil beam model, now outperforms significantly traditional algorithmic model implementations.

14.2 MONTE CARLO ALGORITHMS

14.2.1 Monte Carlo Dose Calculation Specific to the Delivery Technique

14.2.1.1 Passive Scattering

A Monte Carlo algorithm starts tracking each particle at a well-defined location (see Chapter 8). The distribution of protons at the treatment head exit or the patient surface is complex in passively scattered because protons exiting the treatment head underwent scattering and interactions events in the beam-shaping devices in the treatment head. This makes a definition of a phase space generally not feasible (see Chapter 8). Thus, for dose calculation in passively scattered proton therapy, the Monte Carlo simulation typically starts at the treatment head entrance with an analytical description of a distribution (in energy, position, etc.) of each particle. Measuring some of the parameters directly can be difficult and one might have to rely on the manufacturer's information. If the planning system prescribes the settings of treatment head components (e.g., the modulator wheel), the phase space can be simulated accordingly. If the planning system prescribes solely range and modulation width, the translation into treatment head settings is typically done by the treatment control software. Consequently, this algorithm needs to be incorporated into the Monte Carlo code [16,23]. Aperture and compensator are also prescribed by the planning systems and can be modeled in the Monte Carlo using the milling machine files. For the modeling of treatment heads in Monte Carlo codes, see Chapter 8.

14.2.1.2 Scanning Beams

To simulate a scanning beam, four-dimensional Monte Carlo techniques can be used to constantly update the magnetic field strength [24–26]. This allows studying beam scanning delivery parameters [25,27,28]. For standard patient-dose calculation, this is typically not necessary. Dose calculations for scanned beam delivery typically do not require a treatment head model but can be based on a phase space at treatment head exit. Other than passive scattering, a beam model can usually be defined for beam scanning at treatment head exit because of the limited amount of scattering material in the beam (see Chapter 8). It has been demonstrated that beam characterization for scanned delivery can be done based on measured depth-dose curves in water alone [29–32]. For example, the beam's energy spread as a function of energy can be deduced by fitting the widths of measured pristine Bragg curves. Secondary particles other than protons generated in the treatment head can typically be neglected [30].

A planning system will provide a matrix of energies, weights, and positions of beamlets, which can be translated into Monte Carlo input settings. One has to be cautious characterizing a beamlet for Monte Carlo by a simple Gaussian fluence distribution because this leads to an overestimation of the fluence in the center of a beamlet due to a halo caused by large-angle protons from nuclear interactions [11,30] (see Chapter 2).

14.2.2 CT Conversion

To recalculate a treatment planning result using Monte Carlo, the patient geometry (typically given as a CT image) and plan information need to be imported into the Monte Carlo code. The CT image can be imported using either a DICOM (Digital Imaging and Communications in Medicine) stream [33] or CT information, based on a planning system specific format [16,34]. Alternatively, the entire plan information can be transferred to the Monte Carlo code using a DICOM-RTion interface. The information provided to the Monte Carlo dose calculation engine will be translated into specific settings in the Monte Carlo code, such as the gantry angle, the patient couch angle, the isocenter position of the CT in the coordinate system of the planning program, the number of voxels and slice dimensions in the CT coordinate system, the size of the air gap between treatment head and patient, and the prescribed dose [16]. This can either be done internally within the

Monte Carlo or using a dedicated program that uses the plan information and translates it into an input file readable by the Monte Carlo code.

Analytical dose calculation algorithms in photon therapy use electron density because the dominant energy loss process is interaction with electrons. Protons lose energy by ionizations, MCS, and nonelastic nuclear reactions. Because each interaction type has a different relationship with the material characteristics obtained from the CT scan [35,36], relative stopping power is being used to define water-equivalent tissue properties in proton therapy [37] (see Section 14.1).

Other than analytical algorithms, Monte Carlo dose calculations are based on tissue (material) compositions and mass densities. Experimental models are being used to determine the correspondence between Hounsfield numbers and human tissues [38]. Stoichiometric calibrations of Hounsfield numbers with mass density and elemental weights allow CT conversion, and several conversion schemes have been published [22,39,40]. A conversion table can be extended to higher Hounsfield units to deal with high-Z implant materials in the patient [41]. A robust division of most soft tissues and skeletal tissues can be done, but soft tissues in the CT number range between 0 and 100 can only be poorly distinguished because CT numbers of soft tissues with different elemental compositions are similar. A relationship between a certain CT number and a combination of materials is not unique, and various fits can lead to a feasible result [40,42]. It is not necessary to consider each Hounsfield number as a separate tissue. Instead, different Hounsfield numbers can share common material properties, i.e., elemental composition and ionization potential [41,43,44]. For better accuracy, independent of the number of tissues, the number of densities is typically the same as the number of gray values (CT numbers) [43]. The accuracy of dose calculations is affected by the ability to precisely define tissues based on CT [38,44]. Not only can the absolute dose vary but also the range might depend on the CT conversion. For head and neck treatments, it was shown that CT conversion schemes can influence the proton beam range in the order of 1–2 mm [42].

Also needed are mean excitation energies for each tissue, which can be interpolated based on the atomic weight of the tissue elements using Bragg's rule. Mean excitation energies for various elements [45] and averaged values for tissues [46,47] are tabulated. Mean excitation energies are subject to uncertainties on the order of 5%–15% for tissues. This is reflected in uncertainties in predicting the proton range [48–50] (see Chapter 17).

Any conversion scheme is valid only for the CT scanner used for the underlying measurements. The CT conversion used by the Monte Carlo needs to be normalized to the departmental CT scanner. This can be done by either doing a stoichiometric calibration or, as an approximation, by simulating relative stopping power values in the Monte Carlo based on an existing CT conversion and then adjust the results according to the planning system conversion curve [16,34,43]. Dual-energy CT provides two different attenuation maps to determine relative electron densities and effective atomic numbers, thus allowing more accurate material composition maps for proton Monte Carlo dose calculation [50–53].

14.2.3 Absolute Dose

Absolute doses are typically reported as cGy per MU (see Chapter 13) and can be obtained by explicitly simulating ionization chamber readings [16,54]. In a segmented ionization chamber, the volume used for absolute dosimetry can be quite small, requiring a large number of histories to be simulated. An alternative method for simulating absolute doses with Monte Carlo is to relate the number of protons at treatment head entrance to the dose in an SOBP in water for a given field specification. With an accurate model of the treatment head, this method is equivalent to a direct MU simulation because instead of relating the dose to the impact of the beam at a given plane in the treatment head (in an ionization chamber), one relates the dose to a specific number of protons at treatment head entrance.

14.2.4 Dose to Water and Dose to Tissue

Dose in radiation therapy is traditionally reported as dose to water. Analytical dose calculation engines calculate dose by modeling physics relative to water (using the relative stopping power). It has been debated whether doses in radiation therapy should be reported as dose to water or dose to tissue [55]. Arguments in favor of using dose to water include the fact that clinical experience is based on dose to water, and that quality assurance and absolute dose measurements are done in water, and the tumor cells in the human body consist mostly of water. Dose constraints in treatment planning are based on our experience with dose to water.

Monte Carlo dose calculation methods are not based on stopping power relative to water but on material properties, which are converted from CT numbers, i.e., material composition, mass density, and ionization potential. Naturally, Monte Carlo dose calculation does result in dose to tissue. To allow a proper comparison between Monte Carlo and analytically generated dose distributions, one has to convert one dose metric to the other.

Dose to water can be higher by ~10%–15% compared to dose to tissue in bony anatomy. For soft tissues, the differences are typically on the order of 1%–2% (Figure 14.10). As the difference in mean dose roughly scales linear with the average CT number, a rough scaling based on the CT numbers can be done based on known relationships for proton energy dependent relative stopping powers and on a nuclear interaction parameterization [56]. Thus, in most cases, it is sufficiently accurate to do a conversion to dose to water retroactively by simply multiplying the dose with energy-independent relative stopping powers.

14.2.5 Impact of Nuclear Interaction Products on Patient Dose Distributions

Protons lose energy via electromagnetic and nuclear interactions. The latter can account to well over 10% of the total dose, specifically in the entrance region of the Bragg curve [57], where secondary protons cause a dose buildup due to predominant forward emission from nuclear interactions. Nuclear interaction cross sections show a maximum at a proton energy of around 20 MeV and decrease sharply if the energy is decreased (see Chapter 8). As a rule of thumb, the average proton energy in the Bragg peak is about 10% of the initial beam energy. The contribution of dose due to nuclear interactions becomes thus negligible close to the Bragg peak position of a pristine Bragg curve because of the decreasing proton fluence and a sharply decreasing cross section. For an SOBP, nuclear interactions still play a role in the peak because dose regions proximal to the Bragg peak contribute. This leads to a tilt of the dose plateau if their contribution is neglected [57]. Figure 14.11 shows the dose

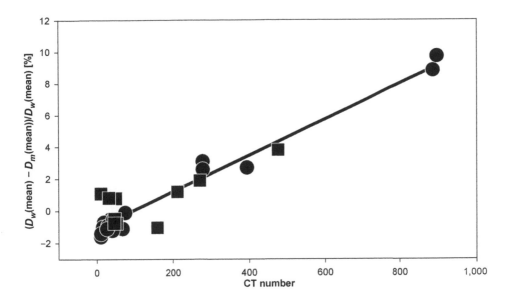

FIGURE 14.10 Percentage difference between dose to water, D_w, and dose to medium, D_m, as a function of the mean Hounsfield unit, CT number, in the volume. Circles show organs at risk and squares refer to target structures. (From Ref. [56], with permission.)

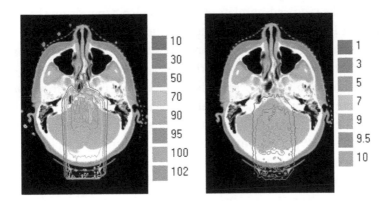

FIGURE 14.11 Dose distribution for one treatment field in a patient treated for a spinal cord astrocytoma. Left: Prescribed dose (in %). Right: Dose due to secondary protons generated in nuclear interactions (in % of the prescribed dose.)

contribution from secondary protons as well as the total dose for a treatment field used to treat a nasopharyngeal carcinoma.

Monte Carlo simulations need to explicitly generate all secondary particles to ensure proper energy balance. While generation of secondary particles is necessary, it might not be necessary (depending on the application) to track them (see Chapter 8). If the projected range of secondary particles is smaller than the region of interest (e.g., the size of a voxel in the patient), it might be sufficient to deposit the energy (dose) locally.

Consideration of nuclear interactions is particularly important for PBS since each pencil is surrounded by a long-range nuclear halo (Chapter 2). The dose distribution is small for each beamlet, but can be significant for a set of beamlets delivering dose to the target volume or in the sharp dose gradient at the distal falloff [11,58]. As Monte Carlo for beam scanning often relies on a beam model (see earlier), proper modeling of the nuclear halo is essential.

14.2.6 Statistical Resolution in a Clinical Setting

Other than with analytical methods, the accuracy of Monte Carlo dose calculation depends on the calculation time (see Chapter 8). While efficiency for many of the applications mentioned in Chapter 8 might not be critical, it becomes important for the clinical use of Monte Carlo, particularly in the context of treatment optimization. The efficiency of Monte Carlo simulations depends on many factors, such as the number of histories or the number of steps each particle takes (see Chapter 8). For dose calculations, the maximum step size is limited by the size of the CT voxels. This is because the physics settings are typically different for different voxels due to their difference in Hounsfield unit and thus material composition or material density.

As the number of histories determines the statistical uncertainty in the dose distribution, it is important to estimate the number of histories needed for a desired uncertainty. It depends on the field parameters, e.g., on the beam energy and treatment volume. If the whole treatment head is being considered (i.e., if the simulation is not based on a phase space distribution), the required number of histories also depends on the efficiency of the treatment head. For PBS, the efficiency of the treatment head is typically in excess of 90%, while for passive scattering, it is typically on the order of 3%–30%. For passive-scattered delivery, a typical number of proton histories for a given patient field is ~25 million at treatment head entrance to reach ~2% statistical accuracy in the target for a single field based on the treatment planning grid resolution (based on a treatment head at the MGH). This number is quite low compared to photon therapy. Because the proton beam's linear energy transfer (LET) is typically higher than that resulting from photon beams, only a few hundred protons need to traverse a voxel for a 2% dose uncertainty. If an analysis is done for the entire plan, i.e., not field specific, the simulation of fewer protons per field may be feasible. A certain statistical precision in the target volume does not guarantee the same precision in organs at risk because of the lower dose and thus potentially fewer particles are involved. On the other hand, the impact of statistical imprecision is less for dose-volume analysis in organs at risk, because the dose distribution is less homogeneous and the dose-volume histograms are shallower [59,60].

In order to allow predicting dose distributions based on fewer histories, it has been suggested that Monte Carlo generated dose distributions can be smoothed [61–63]. This smoothing or denoising is a technique well known in imaging. These methods need to be applied with caution, because regions of low signal are, other than in imaging, not noise but valid information. Some denoising techniques tend to soften dose falloffs, which could have a negative impact when used in proton therapy. Furthermore, other than photon therapy, dose is not directly proportional to particle fluence in proton therapy.

To decrease the statistical uncertainty, one might also be tempted to interpolate the CT grid to a larger grid. Treatment planning systems often present dose distributions on a more course grid than the one provided by the patient's CT scan. When resampling the CT grid, averaging of material compositions is not well defined. Thus, to avoid resampling, the Monte Carlo should operate on the actual CT scan, which is typically in the order of 0.5–5 mm. That of course implies that the Monte Carlo simulations might require to operate on a nonuniform CT grid as often used clinically [43]. Resampling to improve statistical uncertainty can still be done after the dose calculation, where weighting of doses that contribute to a given voxel on a grid can be done accurately based on volume averaging.

14.2.7 Differences between Proton Monte Carlo and Pencil Beam Dose Calculation

Treatment planners are typically aware of dose calculation uncertainties and take these into consideration when prescribing fields, e.g., by avoiding pointing a beam towards a critical structure and by applying safety margins (see Chapters 15–17). Because of such precautions, differences between pencil beam algorithms and Monte Carlo dose calculations turn out to be small when analyzing dose-volume histograms from carefully designed plans. More accurate dose calculations might however lead to a reduction of margins. In general, outcome analysis should be based on the most accurate dose calculation method.

14.2.7.1 Differences in the Predicted Range

The range is a key clinical parameters in proton therapy (see Chapter 4) and uncertainties in range are thus of general concern (see Chapter 17). The range is ideally defined as r_{80}, i.e., as the position of the dose distribution that receives 80% of the prescribed dose in the distal falloff of the proton field (a more accurate abbreviation would be d_{80} (see Chapter 2)). For monoenergetic proton beams, r_{80} is the position where 50% of initial protons have stopped so that r_{80} is independent of the initial proton energy spread. Historically, the range of a clinical proton field is defined as r_{90}, i.e., the position of the 90% dose level in the distal falloff region of a SOBP. The range of a proton field is defined only in water because density variations in the patient will modulate the physical range so that there is generally no clearly defined range of a proton field, rather each small beamlet or pencil has its own range in the patient.

Most analytical algorithms project the range based on the water-equivalent depth in the patient calculated for individual beamlets. This neglects the position of inhomogeneities relative to the Bragg peak depth [64,65]. MCS causes the majority of differences between the analytical and Monte Carlo-based dose calculation. If a proton beam passes through complex heterogeneous geometries, range degradation occurs, which is not well described by analytical dose calculation engines [58,66]. Nevertheless, differences in overall range between analytical algorithms and Monte Carlo in patients are typically small, because both systems have been commissioned using measurements in water. However, locally, significant differences can occur due to tissue interfaces [64,67]. Pencil beam algorithms are less sensitive to complex geometries and density variations, particularly in the presence of lateral heterogeneities [10,19,68]. Figure 14.12 demonstrates the inaccurate consideration of interfaces parallel to the beam path with a pencil beam algorithm due to the treatment of multiple scattering in heterogeneous media. The disequilibrium of scattering events causes protons to be preferentially scattered out of the higher density material into

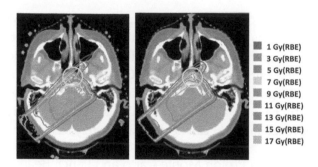

■	1 Gy(RBE)
■	3 Gy(RBE)
■	5 Gy(RBE)
	7 Gy(RBE)
■	9 Gy(RBE)
■	11 Gy(RBE)
■	13 Gy(RBE)
■	15 Gy(RBE)
	17 Gy(RBE)

FIGURE 14.12 Axial views of dose distributions calculated using a commercial planning system based on a pencil beam algorithm (left) and a Monte Carlo system (right). The patient was treated for a spinal cord astrocytoma with three coplanar fields (same as Figure 13.10). The figure only shows one of the fields. The Monte Carlo dose calculation was based on a CT with $176 \times 147 \times 126$ slices with voxel dimensions of $0.932 \times 0.932 \times 2.5/3.75\,\text{mm}^3$ (variable). The yellow circle indicates the dosimetric impact of an interface parallel to the beam path. Doses are in %. (From Ref. [16], with permission.)

the lower density material. Figure 14.13 shows how the range is affected by high-density gradients along the beam direction at lung/tumor, lung/spine, ribs/soft-tissue, and rib/lung interfaces). Note that density gradients that can be tangential to the beam also occur in range compensators. Furthermore, some pencil beam models do not consider aperture edge scattering, which can cause dose errors at patient entrance [69].

In clinical practice, range margins of the order of 3.5% + 1 mm are often assigned to proton fields to cover errors induced by the approximations made in the analytical algorithms (see Chapter 17). While a generic margin is typically applied in treatment planning, range uncertainties depend on the patient geometry. It has been estimated that margins

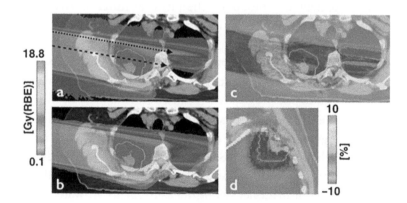

FIGURE 14.13 Example lung patient illustrating the differences in the dose distributions caused by the different calculation algorithms: (a) Monte Carlo simulation (dotted arrow: protons pass a muscle/soft-tissue interface and a bone/lung interface at the distal end of the field; dashed arrow: scattering along the rib and tumor volume in the center of the proton field), (b) analytical calculation, (c) the dose difference between a and b, and (d) the dose difference in coronal view (contour shows target volume shown). (From Ref. [95] with permission.)

can be reduced to 2.7% + 1.2 mm for largely homogeneous patient geometries (such as liver) but should be increased to as high as 6.3% + 1.2 mm for geometries with lateral inhomogeneities (such as head and neck, lung, and breast patients). These numbers assume that tumor coverage is to be maintained for each treatment field, which may be conservative depending on the number of fields prescribed [67]. If Monte Carlo techniques were to be used routinely for dose calculations, these margins could be reduced to 2.4% + 1.2 mm, independent of the complexity of the patients geometry [70]. The remaining uncertainties can in part be tackled by improvements in CT conversion or in vivo range verification (see Chapter 21). These range margins apply to SOBP fields. Beam scanning is typically based on robust optimization to tackle range uncertainties, because each pencil has to be treated individually (see Chapter 19).

14.2.7.2 Differences in the Predicted Dose

While predicting an incorrect range of protons in water might be the main concern when using analytical dose calculation algorithms, there are also expected differences in terms of absolute dose when compared to Monte Carlo. The reason is similar as for range discrepancies, i.e., mainly MCS. Analytical algorithms calculate the spread of the beam for a water-equivalent path length and not the physical distance. An increase in predicted scattering causes an overall broadening of the dose distribution and thus a decrease in dose in the medium- to high-dose region. Consequently, the dose in the target from Monte Carlo simulations is generally lower than the one predicted by analytical algorithms. The underestimation of scattering in analytical algorithms depends on the location of tissues with different densities. It can be seen clearly in dose calculations for lung because protons are scattered in the chest wall, which causes the beam to diverge with distance [71]. The differences in the mean target dose can be up to 4% for head and neck and lung patients or as low as 2% for breast and liver patients (see Figure 14.14) [71,72]. It has been shown that there can be even more significant dose differences between Monte Carlo and analytical methods for small fields (<5 cm in diameter) due to MCS, causing a reduction in dose in the center of the field (see also Chapter 2) [15,20,73]. Small fields lose equilibrium even in the central part of the field resulting in nonflat transverse profiles of the dose distributions.

14.2.8 Clinical Implementation

Monte Carlo simulations have been used to benchmark analytical algorithms (e.g., [10,16]) or to help commissioning planning systems (e.g., [74,75]). Future planning systems will for sure have at least a Monte Carlo option. If proton Monte Carlo dose calculation is not part of a commercial treatment planning system, one might implement an in-house Monte Carlo system in the clinic, which can then be used for routine dose calculation as well as for research purposes. To facilitate data flow between a planning system and the Monte Carlo, the CT information and planning information need to be imported from the planning system, and input files for the Monte Carlo phase space and dose calculation have to be created [76,77]. Ideally, this is done using the DICOM-RTion standard. Within the Monte Carlo, it might be translated into a Monte Carlo specific format, and the calculations can

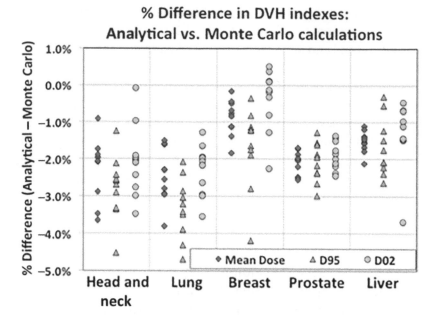

FIGURE 14.14 Differences between analytical and Monte Carlo-based dose calculations in the target volume for five treatment sites. Shown are the mean dose and the maximum dose that covers 95% (2%) of the target volume, D95 (D02). (Adapted from Ref. [72].)

be done on grids identical or different from the CT grid or the grid used for treatment planning.

Besides the radiation field and the patient geometry, other vital information is needed to simulate a patient treatment. These are the CT data, the gantry angle, the patient couch angle, the isocenter position of the CT in the coordinate system of the planning program, the number of voxels and slice dimensions in the CT coordinate system, the size of the air gap between treatment head and patient, and the prescribed dose [16]. Also, couch and gantry rotations have to be applied. The different coordinate systems (treatment control system, treatment head, planning system, and CT system) have to be converted into a common simulation coordinate system.

14.2.9 Accelerated Monte Carlo Dose Calculations

14.2.9.1 Software Acceleration

For proton dose calculation, depending on the beam energy, primary and secondary protons account for roughly 98% of the dose [57]. This includes the energy lost via secondary electrons created by ionizations. The range of most electrons in a clinical proton beam is typically less than 1 mm in water. Thus, explicit tracking of electrons is not necessarily required for dose calculations on a typical CT grid. Efficiency improvements for Monte Carlo dose calculations can be obtained by various variance reduction techniques (see Chapter 8). Furthermore, improvements have been achieved specifically for tracking particles in a voxelized geometry [78]. Particle tracking in a voxel geometry is computationally inefficient because each particle has to stop when a boundary between two different volumes is crossed. Thus, the maximum step size is limited. Such algorithms based on an image segmentation that compromises the regular voxel geometry have been developed to tackle this problem [79,80]. The efficiency (and potential compromise in accuracy) depends on the CT resolution. The speed of the Monte Carlo simulation depends on the grid size, but assuming that a step size above 1 mm is typically not warranted, a larger grid size does not translate into a significant gain.

Some Monte Carlo codes have been specifically designed for fast patient dose calculations only, using approximations to improve computational efficiency. A dedicated Monte Carlo code, VMCpro, was introduced solely optimized for dose calculation in human tissues for proton therapy [81]. A significant speed improvement compared to established Monte Carlo codes was achieved by introducing approximations in the multiple scattering algorithm and using density scaling functions instead of actual material compositions. Furthermore, nuclear interactions were considered explicitly in parameterizations relying on distribution sampling. The agreement with other Monte Carlo codes was excellent. Other codes have been optimized for speed by only tracking primary protons through the treatment head and the patient and treating contributions from nuclear interactions in analytical approximations without the tracking of secondary particles [82]. Also applied have been hybrid methods using spot decomposition techniques, where multiple ray tracings are performed for a single pencil [10,68]. One thus might perform Monte Carlo calculations only on a subset of rays that go through a highly heterogeneous geometry. There is no clear distinction between what one would consider a full-blown Monte Carlo and an analytical algorithm as some fast Monte Carlo codes utilize analytical approaches for certain physics phenomena (see Section 14.1).

Another method is to implement track-repeating algorithms in which precalculated proton tracks and their interactions in material are tabulated after they have been simulated using Monte Carlo [83]. The changes in location, angle, and energy for every transport step and the energy deposition along the track are recorded for all primary and secondary particles and reused in subsequent Monte Carlo calculations [83,84]. During the simulation, at each step there is a track segment loaded from the database that resembles the scenario for the tracked particle.

14.2.9.2 Software Acceleration in Combination with Hardware Acceleration
A GPU is a dedicated hardware designed for accelerated processing of graphics information. It has a much higher number of processing units sharing a common memory space compared to a conventional central processing unit (CPU). A GPU executes a program in groups of parallel threads. Whether a task can be significantly accelerated using GPU hardware depends on the mathematical formalism. Analytical dose calculations are well suited because they can be formulated as vector operations that take advantage of multithreading. The application of proton Monte Carlo dose calculation on GPUs has been reviewed by Jia et al. [85]. The efficiency improvement using GPU codes for Monte Carlo is excellent, allowing dose calculation down to 1% statistical accuracy based on an existing treatment plan to be done in seconds instead of hours.

Particularly, track-repeating algorithms can benefit from the multithreading environment of a GPU [83,86]. In addition to implementing existing algorithms on a GPU, there have been efforts to develop proton Monte Carlo dose calculation algorithms that are tailored to GPU use and are thus very efficient [87–89]. Approximations might be introduced for nuclear interactions because of required data tables that would exceed the memory capability of some GPUs. For instance, the gPMC code [87,90,91] considers proton propagation by a Class II condensed history simulation scheme using the continuous slowing down approximation, while nuclear interactions are modeled using an analytical approach [81]. The code has been compared with CPU-based full Monte Carlo codes for patient dose calculation in seconds and found to be in excellent agreement with a gamma index passing rate of ~99% using a 2%/2 mm gamma index criteria (the gamma index combines features of both dose difference and distance to agreement [92]).

Monte Carlo simulations have been applied to simulate time-dependent patient geometries for studies of breathing motion in proton therapy of the lung (see Section 8.8.1 on Monte Carlo and Chapter 18 on motion). Due to the multitude of simulations necessary to cover the full 4D CT information, GPU-based Monte Carlo has been suggested for such studies [93]. The use of Monte Carlo for these studies is particularly important in proton therapy, because small uncertainties in scattering and range can have a large impact in low-density lung tissue [71].

14.2.10 Monte Carlo-Based Treatment Planning

Calculating dose using Monte Carlo and Monte Carlo treatment planning are two different issues. The discussion earlier deals with the former. If Monte Carlo treatment planning is to be done, it needs to be incorporated into a framework for treatment plan optimization [68,94]. The use of Monte Carlo in an inverse planning framework has stricter requirements in terms of computational efficiency, because the dose has to be calculated more than once in the optimization loop. For example, for intensity-modulated proton therapy, each potentially desired beamlet would have to be precalculated before evoking the optimization algorithm. Each of these would have to be calculated to high statistical precision because the weight of that particular pencil (i.e., its contribution to the final dose distribution) would not be known a priori. If the difference between Monte Carlo and analytical algorithm is small, one might base the optimization on the analytical algorithm and use Monte Carlo simulations for fine-tuning only. The feasibility of this strategy depends on the complexity of the patient geometry, because this determines the difference between the two algorithms.

REFERENCES

1. Hogstrom KR, Mills MD, Almond PR. Electron beam dose calculations. *Phys Med Biol.* 1981;26(3):445–59.
2. Moliere GZ. Theorie der Streuuung schneller geladener Teilchen. I. Einzelstreuung am abgeschirmten Coulomb-Feld. *Z Naturforsch.* 1947;2a:133–45.
3. Eyges L. Multiple scattering with energy loss. *Phys Rev.* 1948;74:1534.
4. Hong L, Goitein M, Bucciolini M, Comiskey R, Gottschalk B, Rosenthal S et al. A pencil beam algorithm for proton dose calculations. *Phys Med Biol.* 1996;41:1305–30.
5. Kinematsu N. Alterntive scattering power for Gaussian beam model of heavy charged particles. *Nucl Instrum Methods Phys Res B.* 2008;266:5056–62.
6. Gottschalk B. On the scattering power of radiotherapy protons. *Med Phys.* 2010;37(1):352–67.

7. Highland VL. Some practical remarks on multiple scattering. *Nucl Instrum Methods.* 1975;129:497–9 (Erratum in Nucl. Instrum Methods 161, 71 (1979)).

8. Kooy H, Schaefer M, Rosenthal S, Bortfeld T. Monitor unit calculations for range-modulated spread-out Bragg peak fields. *Phys Med Biol.* 2003;48:2797–808.

9. Pedroni E, Scheib S, Boehringer T, Coray A, Grossmann M, Lin S et al. Experimental characterization and physical modelling of the dose distribution of scanned proton pencil beams. *Phys Med Biol.* 2005;50:541–61.

10. Soukup M, Fippel M, Alber M. A pencil beam algorithm for intensity modulated proton therapy derived from Monte Carlo simulations. *Phys Med Biol.* 2005;50:5089–104.

11. Sawakuchi GO, Titt U, Mirkovic D, Ciangaru G, Zhu XR, Sahoo N et al. Monte Carlo investigation of the low-dose envelope from scanned proton pencil beams. *Phys Med Biol.* 2010;55(3):711–21.

12. Sawakuchi GO, Zhu XR, Poenisch F, Suzuki K, Ciangaru G, Titt U et al. Experimental characterization of the low-dose envelope of spot scanning proton beams. *Phys Med Biol.* 2010;55(12):3467–78.

13. Clasie B, Depauw N, Fransen M, Goma C, Panahandeh HR, Seco J et al. Golden beam data for proton pencil-beam scanning. *Phys Med Biol.* 2012;57(5):1147–58.

14. Gottschalk B, Cascio EW, Daartz J, Wagner MS. On the nuclear halo of a proton pencil beam stopping in water. *Phys Med Biol.* 2015;60(14):5627–54.

15. Bednarz B, Daartz J, Paganetti H. Dosimetric accuracy of planning and delivering small proton therapy fields. *Phys Med Biol.* 2010;55(24):7425–38.

16. Paganetti H, Jiang H, Parodi K, Slopsema R, Engelsman M. Clinical implementation of full Monte Carlo dose calculation in proton beam therapy. *Phys Med Biol.* 2008;53(17):4825–53.

17. Lu HM, Kooy H. Optimization of current modulation function for proton spread-out Bragg peak fields. *Med Phys.* 2006;33:1281–7.

18. Safai S, Bortfeld T, Engelsman M. Comparison between the lateral penumbra of a collimated double-scattered beam and uncollimated scanning beam in proton radiotherapy. *Phys Med Biol.* 2008;53(6):1729–50.

19. Schaffner B, Pedroni E, Lomax A. Dose calculation models for proton treatment planning using a dynamic beam delivery system: an attempt to include density heterogeneity effects in the analytical dose calculation. *Phys Med Biol.* 1999;44:27–41.

20. Daartz J, Engelsman M, Paganetti H, Bussiere MR. Field size dependence of the output factor in passively scattered proton therapy: influence of range, modulation, air gap, and machine settings. *Med Phys.* 2009;36(7):3205–10.

21. Siddon RL, Dewyngaert JK, Bjarngard BE. Scatter integration with right triangular fields. *Med Phys.* 1985;12(2):229–31.

22. Schneider U, Pedroni E, Lomax A. The calibration of CT Hounsfield units for radiotherapy treatment planing. *Phys Med Biol.* 1996;41:111–24.

23. Paganetti H, Jiang H, Lee S-Y, Kooy H. Accurate Monte Carlo for nozzle design, commissioning, and quality assurance in proton therapy. *Med Phys.* 2004;31:2107–18.

24. Paganetti H. Four-dimensional Monte Carlo simulation of time dependent geometries. *Phys Med Biol.* 2004;49:N75–N81.

25. Paganetti H, Jiang H, Trofimov A. 4D Monte Carlo simulation of proton beam scanning: Modeling of variations in time and space to study the interplay between scanning pattern and time-dependent patient geometry. *Phys Med Biol.* 2005;50:983–90.

26. Shin J, Perl J, Schumann J, Paganetti H, Faddegon BA. A modular method to handle multiple time-dependent quantities in Monte Carlo simulations. *Phys Med Biol.* 2012;57(11):3295–308.

27. Peterson S, Polf J, Ciangaru G, Frank SJ, Bues M, Smith A. Variations in proton scanned beam dose delivery due to uncertainties in magnetic beam steering. *Med Phys.* 2009;36(8):3693–702.

28. Peterson SW, Polf J, Bues M, Ciangaru G, Archambault L, Beddar S et al. Experimental validation of a Monte Carlo proton therapy nozzle model incorporating magnetically steered protons. *Phys Med Biol.* 2009;54(10):3217–29.
29. Kimstrand P, Traneus E, Ahnesjo A, Grusell E, Glimelius B, Tilly N. A beam source model for scanned proton beams. *Phys Med Biol.* 2007;52(11):3151–68.
30. Grassberger C, Lomax A, Paganetti H. Characterizing a proton beam scanning system for Monte Carlo dose calculation in patients. *Phys Med Biol.* 2015;60:633–46.
31. Fracchiolla F, Lorentini S, Widesott L, Schwarz M. Characterization and validation of a Monte Carlo code for independent dose calculation in proton therapy treatments with pencil beam scanning. *Phys Med Biol.* 2015;60(21):8601–19.
32. Grevillot L, Bertrand D, Dessy F, Freud N, Sarrut D. A Monte Carlo pencil beam scanning model for proton treatment plan simulation using GATE/GEANT4. *Phys Med Biol.* 2011;56(16):5203–19.
33. Kimura A, Tanaka S, Aso T, Yoshida H, Kanematsu N, Asai M et al. DICOM interface and visualization tool for Geant4-based dose calculation. *IEEE Nucl Sci Symp Conf Rec.* 2005;2:981–4.
34. Parodi K, Ferrari A, Sommerer F, Paganetti H. Clinical CT-based calculations of dose and positron emitter distributions in proton therapy using the FLUKA Monte Carlo code. *Phys Med Biol.* 2007;52(12):3369–87.
35. Matsufuji N, Tomura H, Futami Y, Yamashita H, Higashi A, Minohara S et al. Relationship between CT number and electron density, scatter angle and nuclear reaction for hadrontherapy treatment planning. *Phys Med Biol.* 1998;43:3261–75.
36. Palmans H, Verhaegen F. Assigning nonelastic nuclear interaction cross sections to Hounsfield units for Monte Carlo treatment planning of proton beams. *Phys Med Biol.* 2005;50:991–1000.
37. Medin J, Andreo P. Monte Carlo calculated stopping-power ratios, water/air, for clinical proton dosimetry (50–250 MeV). *Phys Med Biol.* 1997;42:89–105.
38. Schaffner B, Pedroni E. The precision of proton range calculations in proton radiotherapy treatment planning: experimental verification of the relation between CT-HU and proton stopping power. *Phys Med Biol.* 1998;43:1579–92.
39. du Plessis FCP, Willemse CA, Loetter MG, Goedhals L. The indirect use of CT numbers to establish material properties needed for Monte Carlo calculation of dose distributions in patients. *Med Phys.* 1998;25:1195–201.
40. Schneider W, Bortfeld T, Schlegel W. Correlation between CT numbers and tissue parameters needed for Monte Carlo simulations of clinical dose distributions. *Phys Med Biol.* 2000;45:459–78.
41. Parodi K, Paganetti H, Cascio E, Flanz JB, Bonab AA, Alpert NM et al. PET/CT imaging for treatment verification after proton therapy: a study with plastic phantoms and metallic implants. *Med Phys.* 2007;34(2):419–35.
42. Espana Palomares S, Paganetti H. The impact of uncertainties in the CT conversion algorithm when predicting proton beam ranges in patients from dose and PET-activity distributions. *Phys Med Biol.* 2010;55:7557–72.
43. Jiang H, Paganetti H. Adaptation of GEANT4 to Monte Carlo dose calculations based on CT data. *Med Phys.* 2004;31:2811–8.
44. Jiang H, Seco J, Paganetti H. Effects of Hounsfield number conversions on patient CT based Monte Carlo proton dose calculation. *Med Phys.* 2007;34:1439–49.
45. ICRU. *Stopping Powers and Ranges for Protons and Alpha Particles.* International Commission on Radiation Units and Measurements, Bethesda, MD, 1993; Report No. 49.
46. ICRU. *Tissue Substitutes in Radiation Dosimetry and Measurement.* International Commission on Radiation Units and Measurements, Bethesda, MD, 1989; Report No. 44.
47. ICRU. *Photon, Electron, Proton and Neutron Interaction Data for Body Tissues.* International Commision on Radiation Units and Measurements, Bethesda, MD, 1992; Report No. 46.
48. Andreo P. On the clinical spatial resolution achievable with protons and heavier charged particle radiotherapy beams. *Phys Med Biol.* 2009;54(11):N205–15.

49. Paganetti H. Range uncertainties in proton therapy and the role of Monte Carlo simulations. *Phys Med Biol.* 2012;57:R99–117.
50. Yang M, Virshup G, Clayton J, Zhu XR, Mohan R, Dong L. Theoretical variance analysis of single- and dual-energy computed tomography methods for calculating proton stopping power ratios of biological tissues. *Phys Med Biol.* 2010;55(5):1343–62.
51. Hunemohr N, Paganetti H, Greilich S, Jakel O, Seco J. Tissue decomposition from dual energy CT data for MC based dose calculation in particle therapy. *Med Phys.* 2014;41(6):061714.
52. Hudobivnik N, Schwarz F, Johnson T, Agolli L, Dedes G, Tessonnier T et al. Comparison of proton therapy treatment planning for head tumors with a pencil beam algorithm on dual and single energy CT images. *Med Phys.* 2016;43(1):495.
53. Bar E, Lalonde A, Royle G, Lu HM, Bouchard H. The potential of dual-energy CT to reduce proton beam range uncertainties. *Med Phys.* 2017;44(6):2332–44.
54. Paganetti H. Monte Carlo calculations for absolute dosimetry to determine output factors for proton therapy treatments. *Phys Med Biol.* 2006;51:2801–12.
55. Liu HH, Keall P. D_m rather than D_w should be used in Monte Carlo treatment planning. *Med Phys.* 2002;29:922–4.
56. Paganetti H. Dose to water versus dose to medium in proton beam therapy. *Phys Med Biol.* 2009;54:4399–421.
57. Paganetti H. Nuclear interactions in proton therapy: dose and relative biological effect distributions originating from primary and secondary particles. *Phys Med Biol.* 2002;47:747–64.
58. Sawakuchi GO, Titt U, Mirkovic D, Mohan R. Density heterogeneities and the influence of multiple Coulomb and nuclear scatterings on the Bragg peak distal edge of proton therapy beams. *Phys Med Biol.* 2008;53(17):4605–19.
59. Jiang SB, Pawlicki T, Ma C-M. Removing the effect of statistical uncertainty on dose-volume histograms from Monte Carlo dose calculations. *Phys Med Biol.* 2000;45:2151–62.
60. Keall PJ, Siebers JV, Jeraj R, Mohan R. The effect of dose calculation uncertainty on the evaluation of radiotherapy plans. *Med Phys.* 2000;27(3):478–84.
61. Deasy JO, Wickerhauser M, Picard M. Accelerating Monte Carlo simulations of radiation therapy dose distributions using wavelet threshold de-noising. *Med Phys.* 2002;29:2366–73.
62. Fippel M, Nuesslin F. Smoothing Monte Carlo calculated dose distributions by iterative reduction of noise. *Phys Med Biol.* 2003;48:1289–304.
63. Kawrakow I. On the de-noising of Monte Carlo calculated dose distributions. *Phys Med Biol.* 2002;47:3087–103.
64. Petti PL. Evaluation of a pencil-beam dose calculation technique for charged particle radiotherapy. *Int J Radiat Oncol Biol Phys.* 1996;35:1049–57.
65. Urie M, Goitein M, Wagner M. Compensating for heterogeneities in proton radiation therapy. *Phys Med Biol.* 1984;29(5):553–66.
66. Urie M, Goitein M, Holley WR, Chen GTY. Degradation of the Bragg peak due to inhomogeneities. *Phys Med Biol.* 1986;31:1–15.
67. Schuemann J, Dowdell S, Grassberger C, Min CH, Paganetti H. Site-specific range uncertainties caused by dose calculation algorithms for proton therapy. *Phys Med Biol.* 2014;59(15):4007–31.
68. Soukup M, Alber M. Influence of dose engine accuracy on the optimum dose distribution in intensity-modulated proton therapy treatment plans. *Phys Med Biol.* 2007;52:725–40.
69. Titt U, Zheng Y, Vassiliev ON, Newhauser WD. Monte Carlo investigation of collimator scatter of proton-therapy beams produced using the passive scattering method. *Phys Med Biol.* 2008;53(2):487–504.
70. Paganetti H. Range uncertainties in proton therapy and the role of Monte Carlo simulations. *Phys Med Biol.* 2012;57(11):R99–117.
71. Grassberger C, Daartz J, Dowdell S, Ruggieri T, Sharp G, Paganetti H. Quantification of proton dose calculation accuracy in the lung. *Int J Radiat Oncol Biol Phys.* 2014;89(2):424–30.

72. Schuemann J, Giantsoudi D, Grassberger C, Moteabbed M, Min CH, Paganetti H. Assessing the clinical impact of approximations in analytical dose calculations for proton therapy. *Int J Radiat Oncol Biol Phys*. 2015;92(5):1157–64.

73. Geng C, Daartz J, Lam-Tin-Cheung K, Bussiere M, Shih HA, Paganetti H et al. Limitations of analytical dose calculations for small field proton radiosurgery. *Phys Med Biol*. 2017;62(1):246–57.

74. Koch N, Newhauser W. Virtual commissioning of a treatment planning system for proton therapy of ocular cancers. *Radiat Prot Dosimetry*. 2005;115(1–4):159–63.

75. Newhauser W, Fontenot J, Zheng Y, Polf J, Titt U, Koch N et al. Monte Carlo simulations for configuring and testing an analytical proton dose-calculation algorithm. *Phys Med Biol*. 2007;52(15):4569–84.

76. Verburg JM, Grassberger C, Dowdell S, Schuemann J, Seco J, Paganetti H. Automated Monte Carlo simulation of proton therapy treatment plans. *Technol Cancer Res Treat*. 2015;15:NP35–46.

77. Bauer J, Sommerer F, Mairani A, Unholtz D, Farook R, Handrack J et al. Integration and evaluation of automated Monte Carlo simulations in the clinical practice of scanned proton and carbon ion beam therapy. *Phys Med Biol*. 2014;59(16):4635–59.

78. Schumann J, Paganetti H, Shin J, Faddegon B, Perl J. Efficient voxel navigation for proton therapy dose calculation in TOPAS and Geant4. *Phys Med Biol*. 2012;57(11):3281–93.

79. Sarrut D, Guigues L. Region-oriented CT image representation for reducing computing time of Monte Carlo simulations. *Med Phys*. 2008;35(4):1452–63.

80. Hubert-Tremblay V, Archambault L, Tubic D, Roy R, Beaulieu L. Octree indexing of DICOM images for voxel number reduction and improvement of Monte Carlo simulation computing efficiency. *Med Phys*. 2006;33(8):2819–31.

81. Fippel M, Soukup M. A Monte Carlo dose calculation algorithm for proton therapy. *Med Phys*. 2004;31(8):2263–73.

82. Tourovsky A, Lomax AJ, Schneider U, Pedroni E. Monte Carlo dose calculations for spot scanned proton therapy. *Phys Med Biol*. 2005;50:971–81.

83. Li JS, Shahine B, Fourkal E, Ma CM. A particle track-repeating algorithm for proton beam dose calculation. *Phys Med Biol*. 2005;50(5):1001–10.

84. Yepes PP, Eley JG, Liu A, Mirkovic D, Randeniya S, Titt U et al. Validation of a track repeating algorithm for intensity modulated proton therapy: clinical cases study. *Phys Med Biol*. 2016;61(7):2633–45.

85. Jia X, Ziegenhein P, Jiang SB. GPU-based high-performance computing for radiation therapy. *Phys Med Biol*. 2014;59(4):R151–82.

86. Yepes PP, Mirkovic D, Taddei PJ. A GPU implementation of a track-repeating algorithm for proton radiotherapy dose calculations. *Phys Med Biol*. 2010;55(23):7107–20.

87. Jia X, Schumann J, Paganetti H, Jiang SB. GPU-based fast Monte Carlo dose calculation for proton therapy. *Phys Med Biol*. 2012;57(23):7783–97.

88. Ma J, Beltran C, Seum Wan Chan Tseung H, Herman MG. A GPU-accelerated and Monte Carlo-based intensity modulated proton therapy optimization system. *Med Phys*. 2014;41(12):121707.

89. Li Y, Tian Z, Song T, Wu Z, Liu Y, Jiang S et al. A new approach to integrate GPU-based Monte Carlo simulation into inverse treatment plan optimization for proton therapy. *Phys Med Biol*. 2017;62(1):289–305.

90. Giantsoudi D, Schuemann J, Jia X, Dowdell S, Jiang S, Paganetti H. Validation of a GPU-based Monte Carlo code (gPMC) for proton radiation therapy: clinical cases study. *Phys Med Biol*. 2015;60(6):2257–69.

91. Qin N, Botas P, Giantsoudi D, Schümann J, Tian Z, Jiang SB et al. Recent developments and comprehensive evaluations of a GPU-based Monte Carlo package for proton therapy. *Phys Med Biol*. 2016;61(20):7347–7362.

92. Low DA, Harms WB, Mutic S, Purdy JA. A technique for the quantitative evaluation of dose distributions. *Med Phys.* 1998;25(5):656–61.

93. Botas P, Grassberger C, Sharp G, Paganetti H. Density overwrites of internal tumor volumes in intensity modulated proton therapy plans for mobile lung tumors. *Phys Med Biol.* 2017;63(3):035023.

94. Moravek Z, Rickhey M, Hartmann M, Bogner L. Uncertainty reduction in intensity modulated proton therapy by inverse Monte Carlo treatment planning. *Phys Med Biol.* 2009;54(15):4803–19.

95. Paganetti H, Schuemann J, Mohan R. Dose calculations for proton beam therapy: Monte Carlo. *Principles and Practice of Proton Beam Therapy*: (Eds IJ Das and H Paganetti); Medical Physics Publishing, Madison, WI; 2015:571–94.

CHAPTER 15

Physics of Treatment Planning for Single-Field Uniform Dose*

Martijn Engelsman
Delft University of Technology

Lei Dong
University of Pennsylvania

CONTENTS

---

* Colour figures available online at www.crcpress.com/9781138626508.

15.1 INTRODUCTION

Radiation therapy is a multidisciplinary science. It requires continuous and accurate communication between physicians, clinical physicists, treatment planners, therapists, and nurses. Arguably, the most important step in the radiation treatment of a cancer patient is the act of treatment planning. Treatment planning combines the available clinical information about the patient with the physics aspect of proton therapy and the proton therapy equipment, see Figure 15.1. For optimal treatment plan design, it is important that both physicians and clinical physicists have at least a rudimentary understanding of each other's specialization.

The goal of treatment planning is to design the best possible treatment, given the limitations of the radiation therapy equipment available [1–3]. A good treatment plan ensures the delivery of the desired dose to the tumor while delivering the lowest possible dose to surrounding normal tissues. This requires elaborate tweaking of beam properties, such as beam direction, field shape, and beam weight. Treatment planning allows those responsible for the radiation treatment of a patient, i.e. the radiation oncologist and clinical physicist, to determine the three-dimensional (3D) dose distribution that will be delivered to the patient. With the dosimetric

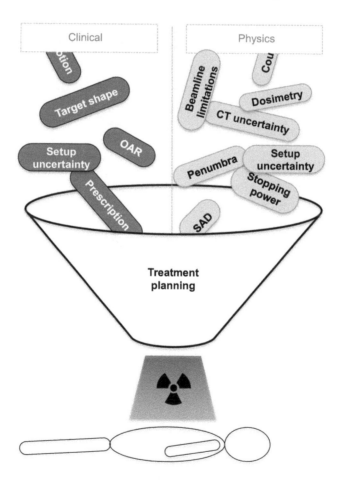

FIGURE 15.1   Treatment planning is the act of combining clinical information and physics information to design the best possible radiation delivery.

consequences of each tweaking visualized before actual treatment delivery, it is possible to design the treatment that best satisfies the wishes of the radiation oncologist, that is deliverable with the equipment available, and that is robust under typical radiotherapy uncertainties.

This chapter will discuss treatment planning of proton radiotherapy for single-field uniform dose (SFUD), in which each proton radiotherapy beam delivers a homogeneous dose to the tumor. SFUD is a fundamental planning concept for proton therapy, which cannot be attained using a single X-ray beam. SFUD proton therapy can be delivered using passively scattered proton therapy (PSPT) as well as pencil beam scanning (PBS). Most of this chapter focused on PSPT. The notable differences between SFUD using PBS and SFUD using PSPT will be discussed in Section 15.4.

Though discussed in detail in Chapters 17 and 18, touching on aspects of the uncertainties in a proton treatment plan cannot be avoided completely as the physicist and the radiation oncologist have to be intimately aware of these uncertainties at the time of treatment plan design.

15.2 PREREQUISITES FOR TREATMENT PLANNING

The prerequisites for proton treatment planning are clinical information and a treatment planning system (TPS).

15.2.1 Clinical Information

Clinical information consists of imaging data and the radiation oncologist's intent, i.e. a dose prescription for the target and dose limits for surrounding healthy tissues (OAR, organs at risk). Even in routine proton therapy treatment planning, an abundance of imaging data, over multiple imaging modalities, is employed by the physician to determine the exact location(s) of cancer within the patient. Without going into detail, the minimum amount of imaging information is the treatment planning computed tomography (CT) scan. On this CT scan, the target and OAR are delineated. Outlining the target in proton therapy follows the guidelines of the International Commission on Radiation Units and Measurements (ICRU) [4–6]. Briefly, the physician delineates the visible tumor (visible on any imaging modality) which is denoted as the gross tumor volume (GTV). Based on clinical experience, the physician can expand the GTV into a clinical target volume (CTV) to account for margin or suspected invisible spread of the cancer.

The physician also provides a prescription, in dose to be delivered to the target and dose constraints for the OAR. In prescribing dose to the target, it is common for the GTV to have a higher prescription dose than the CTV to mitigate the higher tumor burden within GTV and minimize the risk of treatment toxicity when treating the relatively large CTV to a higher dose. A prescription for the target consists of a fraction dose and the total number of treatment fractions. For the target, the goal of treatment planning is to satisfy the prescription as accurately as possible. Overdosing and underdosing (parts of) the target are undesired. Prescriptions for the OAR are an upper limit, and can be expressed in many different ways: a maximum dose to any point, a maximum volume that cannot receive more than a certain dose, the mean dose, etc. For an OAR, the goal is not to exactly meet the constraint but to, if possible, further minimize the dose while still satisfying the dose prescription of the target. Prescriptions for the target and one or more OAR are, however, frequently mutually exclusive. Treatment planning is therefore a balancing act between maximizing the probability of curing the patient (tumor control probability) and minimizing the probability of serious adverse side effects (normal tissue complication probability).

The last input into the design of the treatment plan is an estimate of target motion and the expected setup accuracy of the patient in the treatment room. This information is used to expand the CTV with a safety margin into a planning target volume (PTV), or it can otherwise be taken into account in the design of the treatment field. For a more detailed discussion on the use of the PTV in proton therapy, please see Chapters 17 and 18.

From here on, this chapter will no longer be referring to the "tumor" but to the "target", which can denote either the GTV or the CTV. Also, in this, uncertainties in the target location are directly taken into account in the beam-specific parameters rather than using the geometric PTV as an intermediary step. Because range uncertainties are always along the proton beam direction, PTV is always beam specific for using SFUD planning technique.

15.2.2 Treatment Planning System (TPS)

TPS is a treatment simulation system. The clinical information serves as an input for the TPS. The planning CT scan is a virtual representation of the patient at the moment of treatment delivery. For uncertainties related to the treatment planning CT scan, please see Chapter 17. The TPS also, as much as possible, provides a representation of the capabilities of the treatment delivery system (gantry angles, aperture shapes, available proton energies, beam penumbra, etc., see Figure 15.1).

The treatment planner uses the TPS to define all treatment beam-specific information, such that the treatment prescription is satisfied to the maximum possible extent. The TPS displays the dosimetric result for the planner to review. The dosimetric results can be displayed as dose overlay over patient's anatomy on the CT scan. Statistics can be provided to show the dose-volume histograms for both the targets (GTV, CTV, or PTV) and OARs. If the treatment plan satisfies physician's requirements, the treatment plan will be saved and the planning parameters will be used for actual treatment delivery. The output of a completed treatment plan not only includes dosimetric data such as the prescribed range and modulation width of each treatment field, but also imaging information to be used as reference for accurate patient alignment at the time of treatment, such as digitally reconstructed radiographs.

It is of great importance for the clinical physicist and the physician to understand the limitations of the TPS. Especially, the dose calculation algorithm that is at the heart of the TPS, is an approximation, albeit one that has acceptable accuracy for the vast majority of treatment scenarios. For details on dose calculation algorithms, please see Chapter 14.

15.3 THE TOOLS OF TREATMENT PLANNING

Table 15.1 shows the many attributes under control of the treatment planner during proton therapy planning. For ease of discussion, they have been subdivided into "beam specific" and "treatment plan specific" parameters. Also indicated is what safety margin is affected by which beam specific parameter. The order of discussion of each of these parameters may at times appear somewhat arbitrary. For example, in clinical practice, the treatment planner will first choose the beam direction before deciding on the more detailed aspects such as aperture shape and beam range. The choice of beam angle is, however, easier to understand, with details regarding the choice of range and modulation already explained.

15.3.1 Beam-Specific Choices

15.3.1.1 Lateral Safety Margins

For each beam direction, the aim during treatment planning is to conform the dose closely to the target, both laterally and in the depth direction. In the lateral direction (relative to the beam), this conformality is achieved by using a custom-milled aperture or a multileaf collimator (MLC), see Chapter 5. For simplicity, this chapter will use the word "aperture" to denote either. A target can have a complex 3D shape, and the optimal shape of the aperture can therefore best be determined in the beam's eye view (BEV). The two aspects affecting the lateral safety margin are the penumbra of the proton beam at the depth of the target (dosimetric margin) and the expected uncertainty in the target position due to setup errors and intrafractional tumor motion (setup margin), see Figure 15.2a. The penumbra depends on the proton beamline specifics and varies with range and depth in the patient, see Chapter 4. The dosimetric margin is the distance between the 50% isodose level (the field edge) and the desired isodose coverage (typically, the 95% isodose level). Expected setup errors are part of the clinical input by the physician and clinical physicist. The setup margin can vary from 1 to 2 mm for intracranial radiosurgery treatments, to about 10 mm for prostate treatments without any special internal stabilization method (such as the rectal balloon).

To first order, the shape of the aperture will be a simple geometric expansion of the shape of the target in the BEV, see Figure 15.2b. Just as in photon radiotherapy, the magnitude of

TABLE 15.1 The Many Attributes Under Control of the Treatment Planner in Designing a Passively Scattered Proton Therapy Treatment Plan

Beam Specific	
Range (R)	Distal
Modulation (M)	Proximal
Aperture shape (AP)	Lateral
RC shape	Distal
Smearing	Distal
Air gap/snout extension	Distal
Isocenter location	
Beam direction (gantry angle, couch rotation)	
Patching	
Treatment Plan Specific	
Number of beams	
Relative beam weights	
Beam combinations per treatment fraction ("Fraction Groups")	
Contours for density override (treatment couch; metal artifacts; immobilization devices, etc.)	

Also indicated, for the beam-specific parameters, is the affected safety margin.

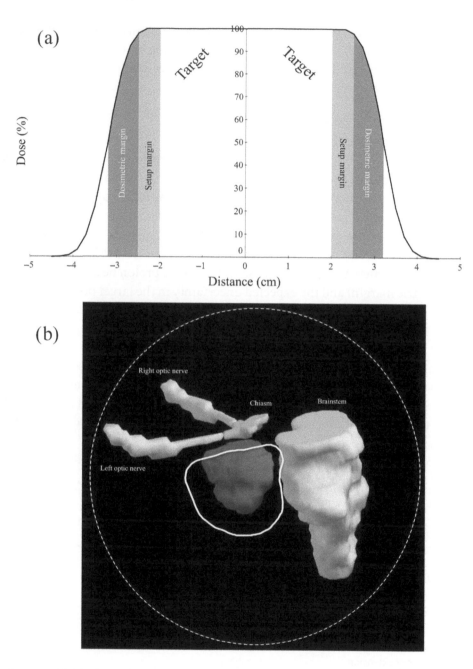

FIGURE 15.2 (a) The lateral margin between aperture edge (50% isodose level) and target edge consists of a dosimetric margin and a setup margin. The central white area indicates the target, which, in this example, needed to be covered with the 95% isodose level assuming a 5-mm setup margin. (b) BEV of a single field for the treatment of an intracranial tumor. The dashed white circle indicates the maximum available aperture size. The dark structure indicates the target while the thick solid white line indicates the aperture shape that was used. A uniform margin of 7 mm was needed to ensure target coverage. To remain within the dose tolerance of the brainstem, a reduced margin was needed. An even smaller margin towards the optical nerves and chiasm was needed to sufficiently spare these OAR.

the lateral safety margin can be based not only on clinical experience but also on a margin recipe [7]. Frequently, the treatment planner has to locally alter the shape of an aperture. Because of multiple Coulomb scattering within the range compensator (RC) and within the patient, a uniform margin between target and aperture edge will typically not result in a uniform margin between the prescription isodose level and the target. This requires local expansion and shrinking of the aperture. Furthermore, dose limits to adjacent OAR may also require the treatment planner to manually alter the aperture shape. In Figure 15.2b, the lateral margin between aperture and target has been reduced in the direction of the brainstem, and even more reduced towards the optic nerves. The achievable shape of an aperture furthermore depends on the limitations of the milling machine (e.g. diameter of the drill) that creates the physical aperture. The TPS has to mimic these limitations accurately. If an MLC is used, the aperture shape is limited by the thickness of the MLC leaves. Dose calculation within the TPS provides feedback to the treatment planner as to the adequacy of the shape of the aperture.

Typically, PSPT beamlines allow motion of the aperture and RC combination along the beam axis by means of snout translation. This allows the treatment planner a choice of air gap, see Figure 15.3, defined as the distance between the downstream side of the RC and the patient skin. The TPS has to model the choice of air gap and the consequences for the shape of the aperture. An increase in air gap increases the penumbra in the patient, which should be modeled by the dose calculation algorithm (Chapter 14). More important, an aperture projection discrepancy may occur in the treatment room if the aperture has been created for a specific air gap that cannot be reproduced at the moment of actual treatment.

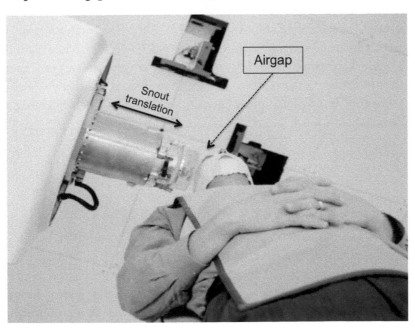

FIGURE 15.3   The snout can be translated along the beam direction to bring the field-specific hardware (aperture and RC) as closely as possible to the patient. Minimizing the air gap reduces the lateral penumbra of the dose distribution.

The patient's shoulder may, for example, not be part of the CT scan and limit snout translation. The increased air gap results in more dose-to-normal tissues, as it leads to an increase in field size and a softening of the lateral penumbra. Especially when an aperture edge is used to exactly control the dose to an OAR, every millimeter is important. At the time of treatment planning, an air gap of a few cm is typically chosen to allow for some wiggle room at the time of treatment. For patient treatments that require abutting fields, an intentional increase in the air gap and a related increase in penumbra can be beneficial as it reduces under- and overdosage at the matchline.

15.3.1.2 Distal and Proximal Safety Margins

The largest benefit of protons for sparing OAR is the finite range of protons. At the time of treatment planning, there may be differences between the estimated range in patient and the required range at the time of treatment delivery. It becomes necessary to use safety margins both distally and proximally to the target. Proximal safety margins can be larger than distal safety margins because of smearing of the RC (see later). The distal safety margin is, however, certainly more important. Keeping the spread-out Bragg peak (SOBP) depth-dose distribution in mind, an error in the range of a proton field can be the difference between 0% and 100% of the beam dose to the target (and to an OAR), while an error in the modulation width has more moderate consequences.

The prescribed range is chosen to ensure distal target coverage with the prescribed dose for that beam direction. Typically, the range of an SOBP is defined by the radiological depth of the distal 90% isodose level. The required range to cover the target is determined by ray tracing the water-equivalent depth (more precisely, the proton stopping power) over the extent of the target in BEV. The rays emanate from the virtual source position, and accumulation of radiological depth will take place from the location where the ray enters the patient geometry to the distal edge of the target. The uncertainty applied to this required range is 3.5% of the range plus an additional millimeter (see Chapter 17), but this may differ slightly between institutions. The prescribed range R, in cm, therefore is

$$R = 1.035 \cdot \max(R_i) + 0.1 \tag{15.1}$$

with R_i the range for each ray i.

If an RC has a minimum thickness (e.g. 1–2 mm, for milling purposes), then this minimum thickness is taken into account for each ray as well. Defining the range by the distal 90% isodose level means that a small part of the target will receive a dose down to 90% rather than 100% of the prescribed dose for that beam direction *if* these range uncertainties are actual (see Chapter 17). The volume of the target receiving a too low dose will be limited as distal conformality in a typical patient scenario is never submillimeter tight. Also, the typical use of multiple beam directions mitigates the magnitude of the possible underdosage.

The prescribed modulation width is chosen to ensure proximal coverage of the target with the prescribed dose. The required modulation width of a beam is also determined by means of ray tracing, but with range uncertainties (3.5% plus 1 mm) applied both at the distal end and at the proximal end:

$$M = \max(1.035 \cdot R_i - 0.965 \cdot P_i) + 0.2 \qquad (15.2)$$

with P_i the water-equivalent depth of the proximal edge of the target along each ray. This modulation width may need to be altered as a function of the amount of smearing applied to the RC (see later). For a treatment plan that consists of multiple beams, one could in principle achieve adequate target coverage even when not all beams individually ensure proximal and distal target coverage. But in clinical practice, such detailed tweaking of a treatment plan is labor intensive and rarely reinforced.

The RC ensures a conformal (tight) distal target coverage. It allows increased sparing of OAR and normal tissues distal to the target by locally pulling back the SOBP as much as possible without affecting target coverage. Interestingly, the proximal side of the target coverage may not be as conformal as the distal surface, because a constant modulation width is needed using the modulation wheel hardware (Chapter 5). The TPS provides a method to determine the 3D shape of the RC, and it can model the presence of the RC and its effect on the beam-specific dose distribution in the patient. As mentioned, the minimum thickness of an RC has to be taken into account in the prescribed range.

A typical method to determine the RC thickness as a function of the BEV position is ray tracing to the distal edge of the target. Along each ray, the thickness of the RC is then determined by

$$RC_i = \max(R_i) - R_i \qquad (15.3)$$

This method of RC design ignores the effect of multiple Coulomb scattering in the patient. However, it is taken into account when dose calculation based on this RC is performed. The result may be that distal dose conformality is inadequate, with location-specific overshoot and undershoot depending on the patient geometry. Worst-case scenario is that an additional "RC volume" has to be manually drawn in the CT scan to which the RC will be designed, thus "faking" the TPS into designing an adequate RC. This is extremely labor intensive, but better methodologies are currently not yet available in commercial TPSs. An improvement would be, for example, to allow the treatment planner to virtually "pull" isodose lines, thereby locally affecting the RC thickness. Best is to have an algorithm that automatically optimizes the RC based on the actual dose calculation algorithm, i.e. a dosimetric rather than a radiologic/geometric determination. The thickness of the RC lateral to the target, but within the aperture circumference, is undefined and typically set to the nearest defined thickness.

An important aspect affecting distal conformality is RC smearing (see, e.g., [6,8]). Smearing is best explained by assuming that the only uncertainties are setup errors (e.g., no range uncertainties, patient density changes). In this case, the RC could be designed to tightly conform the distal falloff of the SOBP to the target, see Figure 15.4a. This is accurate as long as point "p" in the patient is aligned with point "r" in the RC. At the time of delivery

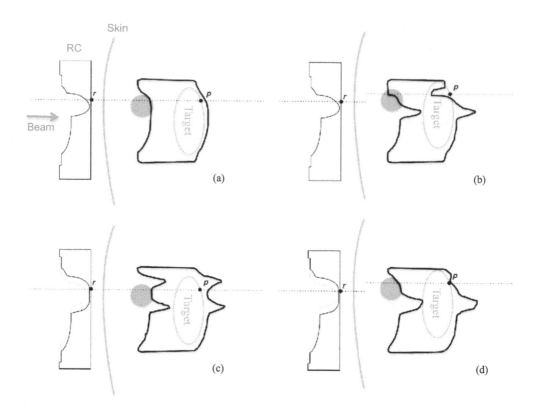

FIGURE 15.4 Schematic diagrams explaining smearing in a two-dimensional geometry. In reality, smearing is applied to a 3D RC. The dark circle indicates a high-density structure. (a) No smearing applied, patient and RC aligned. (b) No smearing applied, patient misaligned. (c) Smearing applied, patient and RC aligned. Also indicated is the shape of the RC prior to smearing. (d) Smearing applied, patient misaligned.

of a treatment fraction, however, the patient may be slightly misaligned with respect to the beam-specific hardware. A typical setup error could be up to a few millimeters, and such a setup error is mimicked in Figure 15.4b. Point "p" is underdosed because protons traveling to this point have to pass through too thick a part of the RC (points "p" and "r" are no longer aligned). The process of smearing is effectively a "hollowing out" of the 3D RC. The thickness at a specific location of the RC *after smearing* is the minimum thickness of the unsmeared RC, as found within a certain distance (i.e. the "smearing radius") from this point. Figure 15.4c shows the dose distribution in the patient for the smeared RC as observed in the treatment planning geometry, i.e. without any patient alignment error. Smearing the RC results in regional overshoot of the proton beam. This overshoot is necessary to ensure target coverage in case of a patient alignment error, see Figure 15.4d. Smearing does not affect the prescribed range, but it may affect the prescribed modulation, see the numerical example later. In fact, the large modulation chosen in Figure 15.4 is necessary to ensure *proximal* target coverage under setup errors and smearing. Please note that no material is taken away at any point on the RC that already has the minimum thickness prior to smearing.

Proper patient alignment minimizes necessary smearing, and the undesired overshoot, and undershoot, of protons into healthy tissue. Although the smearing process does not take into account secondary effects such as altered multiple Coulomb scattering inside the RC and patient, the smearing process is a good strategy to ensure target coverage for setup errors (shifts of the entire patient) with a magnitude up to the smearing radius. Smearing is not a failsafe method to deal with relative changes in the patient density distribution, such as a high-density structure moving to a different upstream location with respect to the tumor, but it may have some merits even in those cases. Smearing has also been suggested as a nonfailsafe method of handling breathing motion for lung cancer (see Section 15.6.2).

As smearing and its application in the TPS are not intuitive, a numerical example follows employing a two-dimensional geometry. Figure 15.5 shows a hypothetical tumor in a, mainly, unit-density phantom. Each voxel is 1 cm × 1 cm, and the dark area in the lower right corner has a density of twice unit density. The proton beam enters from below, lateral target coverage is not taken into consideration, and a 1 cm smearing radius is assumed. Table 15.2 shows the steps in the design of the beam focusing on the rays aimed at positions A–E. The required range to cover the distal edge of the tumor in the case of no uncertainties (D_r) is largest for ray E, i.e. 12 cm. Taking into account range uncertainties of 3.5% + 1 mm (D_u), the prescribed range (R) is 12.52 cm. An RC has to be designed to pull back the distal falloff where necessary (column RC_r). The RC thickness after smearing is shown in column RC_{sm}, leading to a depth of the distal falloff along each ray as according to column $D_{u,sm}$. Especially for rays B and D, smearing increases the proton beam penetration inside the patient. The required modulation along each ray, to ensure proximal target coverage,

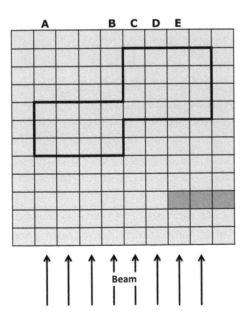

FIGURE 15.5 Stylized unit density 10 × 12 cm² phantom irradiated with a proton beam incident from below. The thick black line indicates the target. The dark gray area indicates a high-density structure.

TABLE 15.2   Steps in Determining, for a Passively Scattered Proton Therapy Field, the Beam-Specific Range, Modulation Width, and Range Compensator Thickness, When Taking into Account Range Uncertainties and Setup Errors, for the Situation as Depicted in Figure 15.5

	Distal, Range					Proximal, Modulation		
	D_r	D_u	RC_r	RC_{sm}	$D_{u,sm}$	P_r	P_u	M_{sm}
A	8.00	8.38	4.14	4.14	8.38	5.00	4.73	3.66
B	8.00	8.38	4.14	1.03	11.49	5.00	4.73	6.76
C	11.00	11.49	1.03	1.03	11.49	7.00	6.66	4.83
D	11.00	11.49	1.03	0.00	12.52	7.00	6.66	5.87
E	12.00	12.52	0.00	0.00	12.52	8.00	7.62	4.90

The underlined numbers designate the prescribed range (R) and modulation width (M).

requires calculation of the water-equivalent depth of the proximal tumor edge (P_r) corrected for the same range uncertainties of 3.5% + 1 mm (P_u). The difference between $D_{u,sm}$ and P_u provides the required local modulation width (M_{sm}), the maximum of which is the prescribed modulation width (M). For this phantom, the largest water-equivalent tumor thickness was 4 cm, but the prescribed modulation taking into account range uncertainties and smearing (setup errors) is almost 7 cm. Counterintuitively, in this example, this modulation is needed to cover the target at its thinnest radiological thickness (ray B).

15.3.1.3 Beam Direction

Now that we have a detailed understanding of how certain beam-specific parameters are chosen and, more importantly, how they affect target coverage and dose to normal tissues, we can discuss the choice of beam directions. Gantries and robotic treatment couches allow the choice of almost any incident beam direction over a 4π solid angle. For isocentric gantries, the beam is always aimed at isocenter, which is, simply put, the point at which the central beam axes for any gantry angle coincide. The choice of isocenter location inside the patient is an important aspect of treatment planning. Typically, it will be near the center of the target. But it is always chosen to be at a location that can be accurately reproduced at the time of patient treatment, based on imaging information provided by the TPS and as acquired by means of imaging on any treatment day. It is important to realize that the isocenter location is "fixed" in the treatment room and one aligns the patient such that the anatomical location of the isocenter as chosen during treatment planning corresponds with the location of the room isocenter. Once the patient is aligned with the room isocenter, the treatment couch allows rotation of the patient around the room isocenter while maintaining alignment between the anatomical and room isocenter.

Many of the considerations in choosing beam directions also play a role in conformal photon radiotherapy, but they are much more important in proton therapy because of the finite range of protons and a relatively high entrance/proximal dose for a given beam direction. An obvious consideration is to geometrically avoid OAR. One tries to prevent beam overlap on the patient skin because skin is a very sensitive OAR, and proton therapy beams do not have the skin-sparing effect of photon beams. The lack of a buildup effect at the patient skin for proton beams on the other hand allows a choice of beam directions through the immobilization devices used without significantly increasing the skin dose. Bulky immobilization devices may, however, require an increase in snout extension, thus increasing the lateral penumbra and the dose-to-normal tissues. Furthermore, immobilization devices may lead to abrupt variations in water-equivalent path length, not only at the edge of the device but also within the device itself, increasing dose-to-normal tissues due to smearing.

In general, beam directions should avoid "skimming" steep density gradients, for example, by preferring angles of beam incidence that are near-perpendicular to the patient skin and by not aiming a beam parallel to the mediastinum–lung interface or diaphragm–lung interface. Beam directions parallel to the auditory canal and the base of the skull should also be avoided as these may challenge the accuracy of the dose calculation algorithm. High-density materials in the patient, such as titanium rods used for stabilizing the spinal column after tumor resection, are avoided as much as possible due to range uncertainties and dose shadowing effects. If it is unavoidable to treat through these high-density materials, one aims to use multiple beam directions to mitigate the possible dosimetric consequences.

To prevent aperture projection errors (see Section 15.3.1.1) the treatment planner needs to be aware of the exact geometry of treatment nozzle and treatment couch, choosing a safe beam direction if necessary. In general, treating obliquely through the treatment couch (less than 45° angles) is avoided. Such a beam direction would increase the penumbra because of the increased snout extension and because of the increase in prescribed range for that beam angle.

A final point of consideration is range variations that occur if the treatment planning CT is an incorrect representation of the patient geometry at the time of treatment delivery. These interfractional variations in the 3D density distribution of the patient make, for example, the following beam directions unfavorable; skimming a patient breast fold, treating through the diaphragm, and treating through an air-filled rectum. Sometimes, it is necessary to bring the patient back to the CT simulator to ensure that the patient model (anatomy) is still intact relative to the beam. If there is a clinically significant dosimetric impact, a new plan may be designed. This is the adaptive radiotherapy process that is frequently used to mitigate systematic or ongoing anatomical changes during treatment course.

15.3.1.4 Patching

Patching is a special planning technique to create conformal avoidance of normal tissues that are next to the treatment target volume. In principle, the choice to do patching is also beam direction related, but it is such a specialized issue in proton therapy that it will be treated independently. Patching is used if no *single* field direction can be chosen that would allow delivery of the required dose to the *entire* target without running the risk of severely overdosing an OAR very close to this target. Patching is typically considered if the target wraps closely around an OAR, with the tolerance dose of the OAR smaller than the prescribed target dose. To deliver the (remaining) dose to such a target with a single field, one would have to, partially or fully, irradiate through the OAR, thereby overdosing it, or one has to aim the beam directly at the OAR while trying to spare it by means of proton range. Because of range uncertainties, this latter solution is not favored unless there is a sufficient gap (e.g. >2 cm water-equivalent path length) between the distal target edge and the OAR. Rather, the choice is to stay off the OAR by means of the much more certain lateral aperture edge.

A typical example case for patching is shown in Figure 15.6. The "through" field treats a part of the target volume from proximal to distal edge. The aperture is designed to cover as

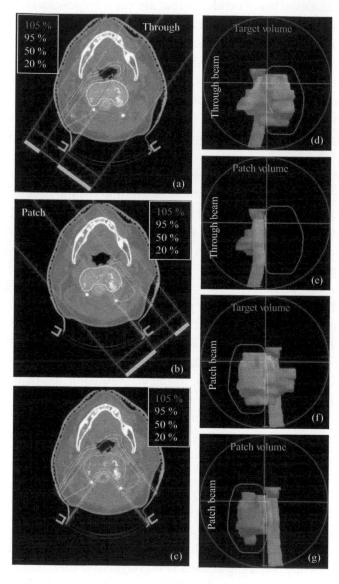

FIGURE 15.6 Example of a patch combination in proton therapy. (a) and (b) show the beam direction and relative dose distribution of the through field and patch field, respectively. The red area indicates the target, and the blue area indicates the spinal cord. (c) The relative dose distribution for the patch combination. (d–g) show the BEV of the patch fields with either the entire target volume or only the patch volume.

much of the target as possible while staying off the OAR. The OARs in this example are the spinal cord and the brainstem. Figure 15.6d–e shows the BEV for the through field with the target volume and patch volume, respectively. The patch volume is that part of the target volume that is not captured within the aperture outline of the through field. The aperture of the "patch" field is subsequently designed to this patch volume. The proton range for the patch field is chosen such that the distal dose falloff matches on the lateral penumbra of the through field along the patchline. Figure 15.6f–g shows the BEV for the fields in the patch combination. Both the through and patch field stay off the OAR by means of aperture edge. The combined dose distribution, as calculated by the TPS, is shown in Figure 15.6c. Unless special measures are taken [9], the distal and lateral SOBP dose gradient are dissimilar enough that dose homogeneity along the patchline is in the order of ±10% even in the absence of any range uncertainties. Furthermore, any subsequent millimeter of overshoot or undershoot of the patch field varies the local dose variation by about another 10%. Some proton therapy institutes will apply smearing to the RC of the patch field, while others do not. In either case, tissues near the patchline can get a considerable over- or underdosage on a per treatment fraction basis depending on the setup error and the uncertainty in the range calculation. For this reason, the total dose delivered by any single patch combination is limited. For targets that need to receive a large (remaining) dose by means of patching, the treatment planner will try to find multiple patch combinations, limiting the use of each patch combination to 3–5 fractions. This substantially averages out the uncertainty in absolute dose along the patchlines. The aim is to have patchlines that do not cross, but sometimes this is unavoidable.

15.3.2 Treatment Plan Specific Choices

As a single proton field can deliver a homogeneous dose to the target, it is possible to deliver the entire prescribed dose with only a single field. For some tumors, e.g. lacrimal gland tumors, this is indeed clinical practice. More frequently, however, multiple beams are used to spread out the dose to normal tissues, just as in photon radiotherapy. A very complex head-and-neck treatment, with for example a large CTV requiring the use of abutting fields and multiple boost regions, can consist of 10–15 unique treatment fields. The fact that each proton field (or patch combination) delivers a homogeneous dose to the target allows the treatment planner great flexibility in choosing individual beam weights. This can help spare certain OAR. Homogenous dose delivery per field also allows treatment of only a limited number of fields for each treatment delivery fraction, instead of treating all fields every day. Many proton therapy centers pursue this strategy of treating fraction groups at a certain cost in biologically corrected dose to the healthy tissues [10]. For this strategy to be safely and effectively applied, one needs to confirm at the time of treatment planning that target coverage is not only achieved by delivery of all fields over the entire treatment course but also for each fraction group.

Once the total dose delivered by a field, and the number of fractions this field will be used are known, the output of the field (dose per monitor unit, MU) can be determined. Current commercially available TPSs are typically unable to calculate the expected number of MUs for a PSPT treatment field. This important characteristic, MU, is therefore at the moment often determined by an in-house developed output prediction model or by individual field measurements (Chapters 12 and 13).

It is possible that photon fields are intermixed into what is mainly a proton therapy treatment plan. The reason would be one of the following: (1) to decrease skin dose, (2) to allow a patient treatment to start even though no proton treatment slots are yet available, and (3) to allow patient treatments to continue in case of unexpected unavailability (down-time) of the proton treatment machine.

15.4 SFUD WITH PBS

In many respects, SFUD treatment planning with PBS is the same as intensity-modulated proton therapy (IMPT) treatment planning. The main difference is that automatic optimization of the spot weights in SFUD is performed on a per-beam basis, ensuring that a homogeneous dose is delivered to the target by every field individually. In IMPT, spot weight optimization is performed over all fields in parallel (simultaneously). Each individual field can deliver a highly inhomogeneous dose distribution to the target, with all fields combined, ensuring a desired (homogeneous) target dose coverage. For details as to the implementation of PBS into the TPS and the use in everyday clinical practice (e.g. choice of spot locations, representation of range shifters, etc.), we refer the reader to Chapter 16. This section is limited to comparing SFUD using PBS and PSPT.

SFUD treatment planning with PBS has the same goal as PSPT: each treatment field ensures a homogeneous dose to the target volume under setup errors and range uncertainties. There are a few notable differences, however, most of which are benefits. The main dosimetric benefit of PBS is that it allows both distal *and* proximal conformality for a single rayline, see Figure 15.7. Although not an accurate description of how spot weights are chosen in PBS, one way of looking at it is that it allows the range and modulation width to be set on a per-proton ray-beam (source to distal target edge) basis. Range uncertainties and

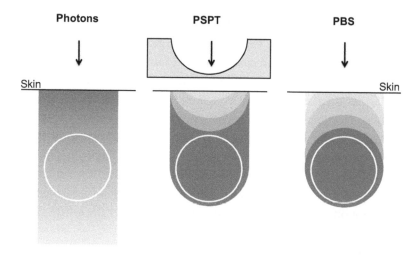

FIGURE 15.7 Schematic diagram showing the lateral conformality of a photon beam, the lateral and distal conformality of a passively scattered proton beam, and the lateral, distal, and proximal conformality of SFUD using PBS. (Figure courtesy of Benjamin Clasie.)

setup uncertainties (smearing), and their effect on the location-specific required range and modulation width will be taken into account for each individual ray beam, thereby sparing normal tissues proximal and distal to the target. The exact choice of range and modulation along any ray beam is, however, limited by the step size between range-energy layers. In other words, without the use of an RC, distal conformality will probably be slightly worse compared to PSPT. Proximal conformality will also not be perfect, even without range uncertainties, but both effects are expected to be clinically insignificant, especially in a multiple-field treatment plan.

Another advantage of PBS may be the possibility to allow for improved dose homogeneity along patchlines under range uncertainties by intentionally softening both the lateral penumbra of the through field and the distal penumbra of the patch field. Just as a wedged field combination in photon therapy is not considered intensity-modulated radiotherapy, use of this patchline smoothing should be considered SFUD instead of IMPT. Newer versions of commercial TPSs can perform robustness optimization using multiple fields simultaneously, which automatically results in a smooth dose gradient along two-beam overlapping region [11]. The planning and optimization using multifield optimization (MFO) will be discussed in Chapter 16.

The main practical benefit of PBS is that design and fabrication of field-specific RC s and apertures is no longer necessary or at least substantially reduced. This improves the workflow considerably and reduces the time between treatment plan completion and first treatment. For example, rather than a laborious manual tweaking of an RC, the dose distribution for a field can be automatically optimized, see Chapter 16.

Adaptive treatment planning in proton therapy using apertures and RCs, i.e. PSPT, is very resource intensive both in manpower and materials. The improved workflow of PBS allows faster treatment adaptation to a changing patient geometry over the course of radiotherapy treatment. Treatment adaptation is more important for proton therapy, because anatomical changes and body contour variations will have a greater dosimetric impact than photon therapy.

Two possible downsides of PBS over PSPT are interplay effects (discussed in Chapter 18) and a relatively large lateral penumbra, especially for shallow ranges (see Chapter 5).

Although it is not the focus of this chapter, it is worth mentioning that a single-field non-uniform dose distribution could be achieved easily by using PBS and IMPT optimization. This is typically called single-field optimization (SFO), which delivers a higher (boost) dose to the GTV while covering the rest of the CTV with a lower dose. A similar dose distribution would require at least two SFUD (PSPT) fields with different aperture sizes to achieve. Simultaneous integrated boost (SIB) treatment technique has been popular recently due to the introduction of IMRT optimization.

15.5 SPECIALIZED TREATMENTS

This section discusses three proton therapy treatments that are a bit less mainstream: eye treatments, proton stereotactic radiosurgery (proton-SRS), and proton stereotactic body radiotherapy (proton-SBRT). In this chapter, SRS denotes a high-precision single fraction (or occasionally double fraction) treatment of a target within the patient's skull. SBRT is used to describe a high-precision treatment of a target lesion within the patient's body over at most a handful of treatment fractions.

15.5.1 Eye Treatments

With a local control rate of well over 95% after 5 years of follow-up (e.g., [12]), treatment of ocular tumors is one of the success stories of proton radiotherapy. Depending on the tumor type, a typical prescription is five fractions of either 10 or 14 Gy each, delivered in five consecutive treatment days. The main OARs are the lens, the macula, and the optic nerve.

Treatment planning is typically based on a recreated geometry rather than on CT-imaging information, see Figure 15.8. Briefly, a model of the eye and tumor is created in the TPS based on ultrasound measurements and orthogonal X-rays. The ultrasound provides information regarding the axial length of the eye, lens thickness, and the thickness ("height") of the tumor. The base of the tumor is reconstructed based on the location of four or more tantalum clips, visible on X-rays, that are sutured near the edge of the tumor during a surgical procedure. These clips also help guide patient setup at the moment of treatment. For a detailed description of the target localization procedure for treatment planning and treatment delivery, please see Refs. [2,13].

The model in the TPS accurately reproduces the location of the OAR for the gaze direction that is preferred for treatment. Apertures are used to tightly conform the dose distribution to the tumor in the lateral direction. If patient setup is based on orthogonal X-rays

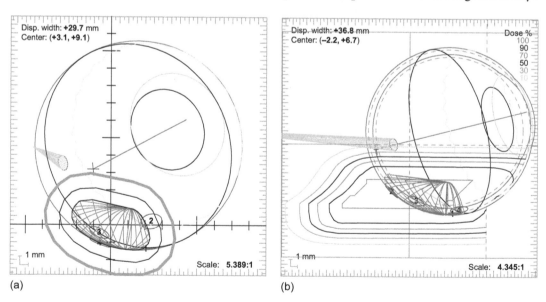

(a) (b)

FIGURE 15.8 Model of the eye as created within EYEPLAN (version 3.05, Martin Sheen, Clatterbridge Centre for Oncology, Bebington, UK). (a) BEV. The thick magenta line indicates the aperture outline. Also indicated are the optic nerve (yellow cone), the lens (green and blue circles), the optic axis (blue line), the macula (magenta cross), and some of the clips (thin magenta ellipses, labeled 1–4). (b) Dose distribution in the vertical beam plane in a slice through the center of the tumor.

of the tantalum clips, the aperture margin is typically 3 mm. Half of this distance (1.5 mm) is to account for the lateral dose gradient between field edge and 90% isodose level. The other 1.5 mm is to take into account target delineation and setup uncertainties. If patient setup is based the beamline light field rather than orthogonal X-rays, the aperture margin is increased by 0.5–1.0 mm because of the additional setup uncertainty. The aperture outline can be locally adjusted to spare OAR. With few exceptions, a typical aperture has an area of less than a few square centimeters. The use of RCs in ocular proton therapy is uncommon. Range and modulation width are chosen to ensure target coverage in the depth direction. Range uncertainties vary between 2.5 and 4.0 mm for anterior to posterior tumors, respectively. An extra mm of range may be added if the location of the tumor brings the proton beam in close proximity to the eye lid. The proximal margin is 3.0 mm unless the target extends all the way towards the surface of the eye.

Every millimeter of the functioning retina can mean an improvement in the quality of life of the patient. Ocular proton beamlines are therefore optimized to have a very sharp lateral falloff to allow maximum sparing of the OAR. The beamline layout of the worldwide available ocular treatment beamlines varies greatly. Depending on the dosimetric characteristics of the beamline, there may be a large effect of aperture shape and size on the output (in cGy/MU) [14]. But in general, the ocular TPS is unable to predict this output, and fields need to be individually calibrated to determine the required MUs to deliver the prescribed dose.

15.5.2 Proton SRS

In proton-SRS, a very high dose of typically 10–20 Gy is delivered in only a single fraction (sporadically: two fractions) to a small tumor. The healthy tissues tolerate the high-fraction dose because of the small irradiated volume. Compared to photon therapy, the characteristic depth-dose distribution of protons can certainly benefit the sparing of healthy tissues, see, e.g., [15].

The main difficulty with proton-SRS treatments is that there is no mitigation of setup and range uncertainties by averaging out the dosimetric consequences over multiple fractions. All dose is delivered in a single fraction, which means that any setup error and range uncertainties can have a maximum negative impact. To ensure target coverage, one would like to apply wide safety margins. The necessity to limit the dose to, and the volume of, irradiated healthy tissues, however, implies a strong need for applying small distal, proximal, and lateral safety margins.

SFUD treatment planning for proton-SRS to a large extent is the same as for normally fractionated treatments, except that there is even more emphasis on high-accuracy treatment delivery. Extra measures are taken in patient immobilization (bite-block or invasive stereotactic frame), target, and OAR delineation by means of additional imaging (e.g. high-resolution magnetic resonance imaging, CT-angiography) and patient position verification (use of fiducial markers), see Refs. [2,6]. Typically, dose is prescribed to the 90% isodose level relative to the dose in the center of the target. This ensures a steep dose falloff (lateral, distal, and proximal) at the cost of increased target dose inhomogeneity. It also makes the minimum dose within the target insensitive to the few-percent dose variations that can be present in the depth direction of the SOBP. The altered prescription isodose level affects the choice of range, modulation width, and target-to-aperture margin only slightly, when compared to standard fractionation in which the aim typically is to ensure coverage of the target with the 95% or even 98% isodose level.

For same-day treatments (CT-scanning, treatment planning, and treatment) one may have insufficient time to create patient-specific hardware and therefore choose to use standard aperture shapes and no RC. The increased convenience for the patient has to be balanced against the increased dose to healthy tissues and risk of complications.

Compared to normal fractionation schedules, it is even more important to choose beam directions that minimize density variations, e.g. avoiding sinuses, especially if the CT scan and patient treatment are not on the same day. It is also important to minimize the dose delivered by beam directions that are challenging for accurate dose calculation, e.g. bony anatomy mixed with air-filled cavities such as the auditory canal. Even in the absence of much tissue inhomogeneity, the dose algorithm has to accurately represent the dose distribution for a single beam, which is challenging for very small fields (see Chapter 14).

Standardized beam arrangements can be used for standard indications such as the treatment of pituitary tumors. The number of treatment fields for a lesion can vary between 2 and 6, with more beams allowing more spreading out of the dose to the healthy tissue and also providing some dose-averaging effect in case one beam inadvertently underdoses the target. The small target sizes make patching (see Section 15.3) an unnecessary technique. If it were beneficial in a rare case to spare an OAR, dose uncertainty on the patchline in combination with the very high fraction dose contraindicates its use.

Designing a treatment plan for a patient with multiple spatially separated intracranial lesions is an especially complicated process, as any overlap between treatment fields for different targets will double the dose to local healthy tissue. The fact that proton beams have no exit dose helps in preventing such beam overlap. It is possible to perform proton-SRS treatments with (SFUD using) PBS as long as the achievable lateral penumbra is sharp enough to provide sufficient sparing of healthy tissues lateral to each field. Especially for small targets, the distance between spot positions also has to be small enough to allow tight lateral, proximal, and distal coverage. For more information on proton-SRS, see [2,16].

15.5.3 Proton-SBRT

Typical treatment sites for proton-SBRT are pancreas, liver, and lung. Just as proton-SRS treatments, proton-SBRT treatments are characterized by a high-dose per fraction. The number of treatment fractions, however, is increased and can vary between three fractions of about 20 Gy and 15 fractions of about 5 Gy. Compared to a single-fraction radiosurgery treatment, the use of multiple fractions will result in some averaging out of the dosimetric consequences of setup errors and range uncertainties. There are, however, a number of additional uncertainties that indicate a need for increased safety margins, see also Chapter 17. Setup accuracy in general is compromised, which indicates a need for increased safety margins. It is harder to align the target laterally with respect to the beam as the target itself may not be visible by means of in-room projection X-ray imaging modalities, and it may move with respect to the more clearly visible bony anatomy. Modern proton therapy systems always equip with cone-beam CT imaging (CBCT) capability, which allows for direct target imaging and target alignment. Nevertheless, the movability of surrounding organs and the target due to, e.g. breathing, heartbeat, and filling of the gastrointestinal tract, can result in large intra- and interfractional range variations. This aside from the "automatic" increase in range uncertainty due to the increased ranges will be needed to treat these deep-seated targets. CBCT imaging at least allows a direct comparison of patient's anatomy with the planning CT, which provides improved alignment quality.

As the number of treatment fractions is still quite limited, it is important for the data used for treatment planning (CT scan) to closely represent the patient density distribution at the time of treatment. For this, one should minimize the time between planning CT-scan acquisition and the first day of treatment, as well as minimize the time of the overall treatment course. In the absence of CBCT capability, one can also acquire more than one CT scan on different days prior to treatment to ensure overall reproducibility of the patient geometry, for example, the level of breathing of the patient.

15.6 PATIENT TREATMENT PLANNING EXAMPLES

Proton radiotherapy nowadays is applied to treat almost any tumor site. This section will, however, be limited to two examples of PSPT only. Section 15.4 provides details as to how these plans would be different if PBS was applied. The C-spine tumor example highlights treatment planning in a geometry without (much) density variation but which requires patching. The lung cancer example highlights the difficulties of proton therapy planning in a time-varying density geometry.

15.6.1 C-Spine Tumor

The patient shown in Figure 15.9a has a tumor within the neck region, wrapped around the spinal cord and the inferior aspect of the brainstem. The challenge for this particular patient and radiotherapy treatment course was to deliver the prescribed dose of 50 Gy to the target in 25 fractions of 2 Gy, while limiting the maximum dose in the spinal cord and brainstem to 30 Gy. This example highlights two other aspects that are especially of concern for proton therapy and that are related to range uncertainties (Chapter 17). Dental fillings resulted in artifacts in the CT. The density around the filling was, however, not overridden as the preferred beam directions for this treatment are lateral to posterior. The large density variations in the mouth (high-density teeth, air cavities, tongue repositioning accuracy) as well as the sensitivity of the oral mucosa to radiation contraindicate use of anterior beam directions. Titanium screws and other hardware are present near the posterior border of the target. Care was taken to ensure that the observed electron density of this hardware was converted into the correct relative proton stopping power. Beam directions for this patient were chosen to minimize the extent of titanium hardware within the BEV, while allowing a proper choice of beam angles for the multiple patch combinations.

The choice as to which beam angles to use for what part of the treatment is driven by the need for multiple patch combinations. Limiting the use of each patch combination to 3–5 fractions, at least two patch combinations are needed to deliver the final 20 Gy on top of 30 Gy with conformal fields. Use of only two patch combinations would, however, allow no

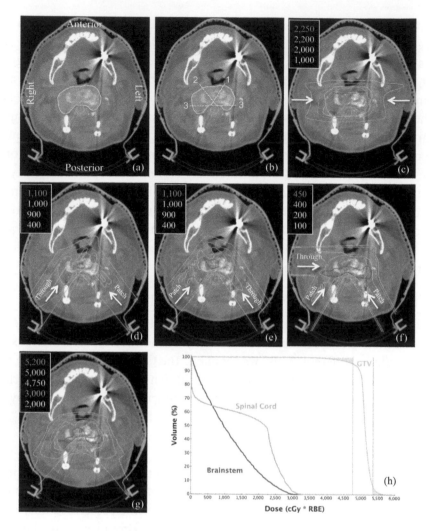

FIGURE 15.9 Treatment planning of a C-spine tumor for a prescribed dose of 50 Gy. (a) Single slice indicating the target (red) and spinal cord (blue). (b) Multiple patchlines (yellow) for multiple patch combinations. (c) Absolute dose delivered by a left-lateral and a right-lateral field. Beam directions are indicated by white arrows. (d) Dose distribution of the first patch combination. (e) Dose distribution of the second patch combination. (f) Dose distribution of the third patch combination, when using the right-lateral through field. (g) Cumulative dose distribution taking into account all treatment fields. (h) Dose-volume histogram of the CTV, spinal cord, and brainstem. The vertical dashed lines indicate 95% and 107% of the prescription dose. The solid red areas indicate under- and overdosing of the CTV.

more doses to be delivered to the spinal cord with any of the patch fields. This then requires each of these fields to have a substantial margin between the aperture edge and the spinal cord, thereby compromising target coverage near the spinal cord. Rather, the choice is to have the conformal fields deliver a reduced dose of, in this case, only 22 Gy, permitting the patch fields to have tighter margins to the spinal cord. This allows treatment of almost the entire target to the prescribed dose, at the cost of finding a third patch combination.

The three patchlines are indicated in Figure 15.9b, and the treatment planner has this in mind when designing the conformal fields and subsequent patch combinations. In other words, the entire treatment is mapped out before the first beam is designed within the TPS.

Treatment planning was started with a right left lateral fields delivering a homogeneous target dose of 22 Gy(RBE). As indicated in Figure 15.9c, these fields did not attempt to spare the spinal cord, and this OAR gets 100% of the dose delivered with these fields. As the subsequent patch fields have intended right-posterior and left-posterior beam directions, the conformal fields were chosen to be right lateral and left lateral to spread out the dose to the healthy tissues.

Figure 15.9d–e shows the first two patch combinations, each delivering 10 Gy(RBE) to the target and using lateral-oblique fields. The isodose lines show some underdosage and overdosage along the patchlines. The third patch combination uses two patch fields and alternating right-lateral and left-lateral through fields to deliver 8 Gy(RBE) to the target. Figure 15.9f shows the patch combination using the right-lateral through field. The cumulative dose distribution for the entire treatment plan is shown in Figure 15.9g. The corresponding dose-volume histograms of the target, spinal cord, and brainstem are shown in Figure 15.9h. The vertical dotted lines indicate 95% and 107% of the prescribed dose, which the ICRU uses as tolerance levels for dose inhomogeneity [4,5]. Underdosage occurs in the region of the target closest to the OAR and is unavoidable as this OAR has to be spared. Overdosage occurs on the patchlines. Over- and underdosage is very limited, as indicated by the red-filled areas in Figure 15.9h. This is despite the close proximity of the OAR to the target and a low dose-constraint for the OAR.

15.6.2 Lung

Proton radiotherapy of a near unit-density tumor that is moving within low-density lung tissue is a major challenge. Variation in the density geometry can occur within the fraction, between fractions, and over the whole course of the treatment. Details regarding these uncertainties are provided in Chapters 17 and 18; this section will be limited to describe a "forward" planning approach. "Forward" in this context means a treatment planning approach that mimics classical conformal photon treatment planning as much as possible, i.e. use of a single CT scan and manual individual beam design.

A number of such approaches for forward planning have been discussed in the literature, e.g. [17–20]. Moyers et al. [17] already point out the difference with classical treatment planning in which the aim is to ensure dose coverage of the CTV by conforming the prescribed dose to the PTV. In their approach, the PTV is only used for determining lateral (aperture) margins, while range, modulation, and RC smearing are chosen on a per-field basis to ensure target coverage under range uncertainties.

Engelsman et al. [18] employed a two-step approach, first choosing parameters of each individual beam to ensure tight dosimetric coverage of the CTV on the mid ex-hale representative CT-scan. This is followed by the application of lateral, distal, and proximal margins on a per-field basis to ensure target coverage under setup errors and range uncertainties. Their simulations show that the magnitude of both the lateral aperture margin and distal smearing margin can be less than the sum of the expected maximal setup error and breathing amplitude. This approach works for the stylized phantom used in their study, as the required range to cover the most distal target edge does not vary over the breathing cycle. For at least some patients, underdosage of the target can occur if motion of the tumor relative to other densities in the patient requires an increase of range, something that the simple approach of smearing the RC cannot provide.

An improved approach is described by Engelsman et al. in which field-specific apertures and RCs are an aggregate of the required aperture and RC for all phases (or at least the most important phases) of the four-dimensional CT scan [19]. This approach is, however, labor intensive and requires automation before it can be routinely applied. Even then it would guarantee tumor coverage over all phases of the planning 4D CT scan, but not necessarily under variations in patient geometry with respect to this planning 4D CT scan.

Kang et al. [20] describe an internal target volume (iGTV) approach (see Ref. [5]) with density override of the volume swept by the moving GTV to ensure target coverage in every phase of the breathing cycle. As they define the iGTV to be the envelope of all GTVs over the breathing cycle, target delineation requires a 4D CT scan. Treatment planning uses only a single CT scan: the density-averaged CT scan. Individual beams can, but do not have to, be designed using the previously mentioned two-step approach. The density override simulates the tumor to be at all possible locations at once and more than ensures target coverage over all phases of the 4D CT scan. It also provides a "buffer" for density variations over the course of the treatment, such as tumor growth or a moderate change in breathing motion. During any treatment delivery, however, the tumor will be only in one location at a time, and the "loss" of density will result in an increased dose to normal tissues distal to the target compared to what is displayed at the time of treatment planning.

In general, and especially for lung tumors, the price of ensuring target coverage under density variations, range uncertainties, setup errors, etc. is an increased dose to surrounding normal tissues. Even the most conservative of the lung tumor treatment planning approaches described [20] does not guarantee target coverage for all patients, and all possible density variations with respect to the treatment planning CT. The aim is therefore to use more than two beam directions such that target regions that may be underdosed by a single beam receive full dose from at least two other beam directions, thus mitigating the possible negative clinical consequences.

Figure 15.10 shows an example of a lung tumor treatment plan, according to the treatment planning protocol as described in Ref. [18]. The fractionation is 70 Gy in 35 fractions of 2 Gy each. Treatment planning is performed on the mid-exhale CT scan and beams are designed in two steps. In step 1, the high-dose region of each beam is separately tightly conformed tightly to the target. Step 2 takes range uncertainties, setup uncertainties, and breathing motion into account by increasing the range, modulation, and smearing distance, and by expanding the apertures. The treatment plan avoids beam directions that are parallel to density gradients. This serves two purposes. First, it minimizes overshoot due to smearing (see Section 15.3.1.2), and second, it reduces the probability of target underdosage should the breathing motion vary from treatment planning CT to fractions of treatment delivery. Figures 15.10a–c show the dose distribution for each of the three individual beam directions after step 1. The overshoot of a few millimeters is deemed acceptable on a multifield

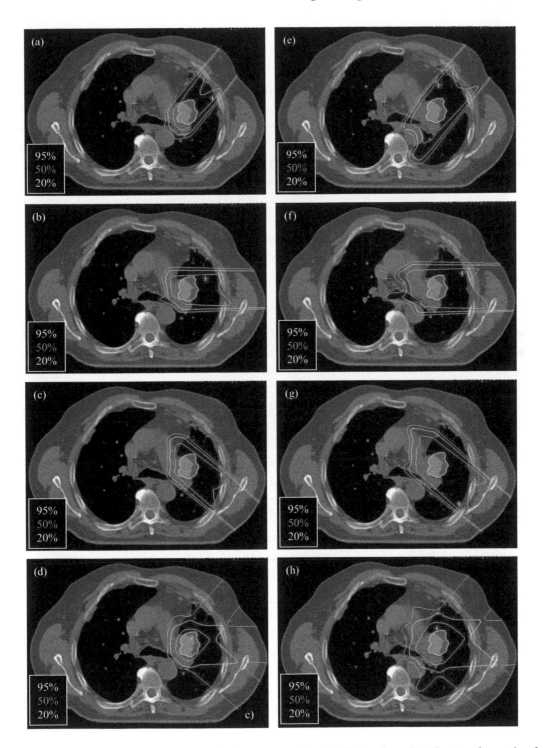

FIGURE 15.10 Treatment planning of a lung tumor. (a–c) Relative dose distribution for each of three individual fields when conforming the high-dose region tightly to the target volume (indicated in red). (d) Cumulative dose distribution for all three fields combined. The right-hand column, (e–h), shows the same information but now for the planning stage it takes into account range uncertainties, breathing motion, and setup errors.

treatment plan, and tweaking this is very time consuming as our TPS does not allow auto-matic dosimetric optimization of the RC. The overshoot is also a consequence of the need to cover the target in more superior or inferior CT slices. The modulation width appears too large for the same reason. The undershoot is only about 5 mm water equivalent but appears especially large because it is in A low-density lung tissue. Figure 15.10d shows the cumulative dose distributions for all beams combined, with each field having an equal weight. Figure 15.10e and f shows similar dose distributions, but after step 2. It is obvious that for any single beam the high-dose region (95% isodose level) does not conform to a geometric expansion of the target. This means that, for proton therapy planning in nonuniform geometries, a PTV-based approach has limited merits. The cumulative dose distribution for step 2 is shown in Figure 15.10h. A similar cumulative dose distribution could have been designed using a PTV-based planning approach. It is, however, important to keep in mind that such an approach would not result in the beam-specific dose distributions (Figure 15.10e–g) that are necessary to guarantee target coverage under range uncertainties, setup errors, and target motion.

15.7 FUTURE PERSPECTIVES OF SFUD

For most tumor sites, SFUD treatment plan design is straightforward using PSPT and beam-specific parameters. PSPT-based SFUD proton therapy is perhaps the most robust planning technique. Field-specific parameters, such as distal and proximal margins, smearing etc. have guaranteed target dose coverage in the presence of expected range uncertainties or setup errors. Nevertheless, PSPT-based SFUD plans have poor conformal-ity, especially near the proximal side of the target.

PBS-based SFUD plans can in principle be designed using a PTV-based optimization technique. Dose distribution within PTV can be optimized to have either uniform dose distribution (SFUD) or nonuniform dose distribution (SFO) using the SIB technique. Unfortunately, a target PTV-based optimization approach ignored the fact that range uncertainties are directional—along the beam directions. Different beam angles may not have the same PTV as in the case of intensity-modulated photon therapy. This is why a single-field beam-specific PTV concept was proposed for PBS SFUD optimization [21]. In such implementations, beam-specific PTV was created for a given beam angle. Ray tracing is used to calculate the distal and proximal surfaces based on expected range uncertainties. Then a lateral smearing is performed to the distal and proximal surfaces to account for setup errors perpendicular to the beam direction. Such a beam-specific PTV is difficult to create manually when the beam angles are not cardinal angles.

While SFUD planning techniques incorporated the expected range uncertainties and setup errors in the beam design process, it is of vital importance to understand the magni-tude of range uncertainties and patient density variations in clinical practice. Unrealistic expectation could result in underdosing the target or overdosing OARs. Excessively con-servative about margins could also result in unnecessary normal tissue toxicities. The treatment planner, physician, and medical physicist have to be continuously aware that the finite range of protons is not only a blessing but also a risk.

Due to its robustness to uncertainties, SFUD proton therapy is likely to continue to be applied for many patients, particularly for cases with large organ motion and anatomical changes. Commercial implementations of MFO and robustness optimization techniques have just started getting into clinic. Newer robustness optimization can bypass PTV concept and directly use CTV to evaluate target coverage [22]. SFUD is a special situation of more complex MFO approach. More advanced treatment planning and optimization techniques will be covered in Chapters 16 and 19.

REFERENCES

1. Meyer JL. *IMRT, IGRT, SBRT: Advances in the Treatment Planning and Delivery of Radiotherapy.* Basel, Switzerland: S. Karger AG; 2007.
2. Delaney TF, Kooy HM. *Proton and Charged Particle Radiotherapy.* Philadelphia, PA: Lippincott, Williams & Wilkins; 2008.
3. Goitein M. *Radiation Oncology: A physicist's-Eye View.* New York: Springer; 2008.
4. ICRU Report 50. Prescribing, recording and reporting photon beam therapy. International Commission on Radiation Units and Measurements; 1993.
5. ICRU Report 62. Prescribing, recording and reporting photon beam therapy. International Commission on Radiation Units and Measurements; 1999.
6. ICRU Report 78. Prescribing, recording, and reporting proton-beam therapy. International Commission on Radiation Units and Measurements; 2007.
7. van Herk M, Remeijer P, Rasch C, Lebesque JV. The probability of correct target dosage: dose-population histograms for deriving treatment margins in radiotherapy. *Int J Radiat Oncol Biol Phys.* 2000;47(4):1121–35.
8. Urie M, Goitein M, Wagner M. Compensating for heterogeneities in proton radiation therapy. *Phys Med Biol.* 1984;29(5):553–66.
9. Li Y, Zhang X, Dong L, Mohan R. A novel patch-field design using an optimized grid filter for passively scattered proton beams. *Phys Med Biol.* 2007;52(12):N265–75.
10. Engelsman M, Delaney TF, Hong TS. Proton radiotherapy: the biological effect of treating alternating subsets of fields for different treatment fractions. *Int J Radiat Oncol Biol Phys.* 2010; 79(2):616–22.
11. Tasson A, Laack NN, Beltran C. Clinical implementation of robust optimization for craniospinal irradiation. *Cancers.* 2018; 10(1):1–7.
12. Egger E, Schalenbourg A, Zografos L, Bercher L, Boehringer T, Chamot L et al. Maximizing local tumor control and survival after proton beam radiotherapy of uveal melanoma. *Int J Radiat Oncol Biol Phys.* 2001;51(1):138–47.
13. Albert D, Miller J, Azar D, Blodi B. *Albert & Jakobiec's Principles & Practice of Ophthalmology.* 3rd ed. Orlando, FL: Saunders Elsevier; 2008.
14. Daartz J, Engelsman M, Paganetti H, Bussière MR. Field size dependence of the output factor in passively scattered proton therapy: influence of range, modulation, air gap, and machine settings. *Med Phys.* 2009;36(7):3205–10.
15. Bolsi A, Fogliata A, Cozzi L. Radiotherapy of small intracranial tumours with different advanced techniques using photon and proton beams: a treatment planning study. *Radiother Oncol.* 2003;68(1):1–14.
16. Chin LS, Regine WF. *Principles and Practice of Stereotactic Radiosurgery.* New York: Springer; 2008.
17. Moyers MF, Miller DW, Bush DA, Slater JD. Methodologies and tools for proton beam design for lung tumors. *Int J Radiat Oncol Biol Phys.* 2001;49(5):1429–38.
18. Engelsman M, Kooy HM. Target volume dose considerations in proton beam treatment planning for lung tumors. *Med Phys.* 2005;32(12):3549–57.
19. Engelsman M, Rietzel E, Kooy HM. Four-dimensional proton treatment planning for lung tumors. *Int J Radiat Oncol Biol Phys.* 2006;64(5):1589–95.

20. Kang Y, Zhang X, Chang JY, Wang H, Wei X, Liao Z et al. 4D proton treatment planning strategy for mobile lung tumors. *Int J Radiat Oncol Biol Phys.* 2007;67(3):906–14.
21. Park PC, Zhu XR, Lee AK, Sahoo N, Melancon AD, Zhang L, Dong L. A beam-specific planning target volume (PTV) design for proton therapy to account for setup and range uncertainties. *Int J Radiat Oncol Biol Phys.* 2012;82(2):e329–36.
22. Cao N, Saini J, Bowen SR, Apisarnthanarax S, Rengan R, Wong TP. CTV-based robustness optimization versus PTV-based conventional optimization for intensity modulated proton therapy planning. *Int J Radiat Oncol Biol Phys.* 2015;93(3):E617–8.

CHAPTER 16

Physics of Treatment Planning Using Scanned Beams*

Anthony Lomax
Paul Scherrer Institute

CONTENTS

---

* Colour figures available online at www.crcpress.com/9781138626508

16.1 INTRODUCTION

It is becoming increasingly clear that the most flexible method of delivering proton (or particle) therapy is by the use of pencil beam scanning (PBS). In this approach, narrow pencil beams of particles are scanned across the target volume in three-dimensions, using deflector magnets in the directions orthogonal to the beam direction, and some form of energy modulation for positioning of the Bragg peak in depth. In its most flexible form, such delivery systems are capable of complete control of the dose delivered by each such pencil beam, resulting in a true fluence modulation in three dimensions from each individual incident field direction [1]. This is the particle therapy equivalent of intensity-modulated radiotherapy (IMRT) with photons, and brings similar (if somewhat more) advantages and potential disadvantages. In this chapter, we will look into both the physics and methods of treatment planning for scanned particle beams, starting with the similarities to conventional IMRT (henceforth referred to here as IMXT) and the main dissimilarities to passive scattering proton therapy. In addition, we will look at different modes of optimizing scanned proton therapy treatments and how possible delivery uncertainties can be dealt with. Finally, a number of case studies will be presented to indicate the potential of these techniques and some of the remaining challenges of treatment planning for PBS proton therapy.

16.2 BASIC PRINCIPLES

16.2.1 There Is No SOBP

Conventional (passive scattering) proton therapy is based very much on the concept of the spread-out Bragg peak (SOBP) (see Chapter 11). This is the basic deliverable depth-dose element, produced by either a continuously rotating range modulator wheel or a ridge filter [2]. As described in more detail in Chapter 5, the SOBP is constructed generally before the beam has been spread laterally, and, anyway, before the laterally scattered beam is subsequently shaped using collimators and compensators. This is an important limitation of this technique. As the SOBP is constructed on the narrow, unbroadened beam, the width of the SOBP along the beam direction is invariable across the field. Put another way, although the length of the SOBP (and in principle its shape) can be varied from field to field, within a single field, the SOBP depth-dose curve is constant across the field.

The consequence of this is shown in Figure 16.1. This shows a single slice of a large Ewing's sarcoma and the calculated dose distribution from a single, passively scattered proton field incident from the posterior aspect. Although this field direction may not be optimal for this particular case, it has been chosen deliberately to show the limitations of the SOBP technique. Indicated in the figure is the "length" of the SOBP that must be

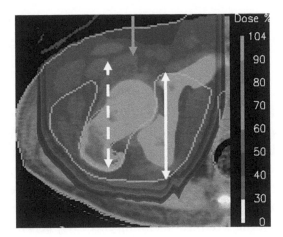

FIGURE 16.1 The dose distribution for a single, passively scattered proton beam incident from the anterior aspect of large and complex Ewings carcinoma. The white double-headed arrows show the minimum SOBP length necessary to cover the whole target, and how, due to the irregularity of the distal edge of the target, such fixed SOBPs can extend well into the normal tissue areas even when the target is quite deep seated.

produced to fully cover the target volume. For this case, the SOBP must be about the same length as the "thickest" portion of the target volume along the beam direction, as indicated by the solid white line in the figure, in this case about 10 cm. However, the target thickness varies extensively across the target, and in this slice, is considerably narrower in the portion directly posterior to the femoral head. Thus, the same SOBP delivered here will be far too wide (as shown by the broken line), with the consequence that the whole femoral head, although proximal to the target volume, will receive the full dose. In addition, at the lateral border of the target, as the distal edge comes closer to the surface and the thickness of the target narrows, the SOBP dose once again extends well beyond the proximal border of the target, in this case, extending right to the surface of the patient.

Once again, it should be pointed out that this is an extremely poor field direction for treating such a target, but nicely illustrates the problem of passive scattering delivery and the use of a field-invariant SOBP depth-dose curve. Indeed, due to this limitation, passive scattering can be directly compared to the delivery of open fields with photons. The depth-dose curve is invariant (a fixed-extent SOBP in the proton case) and the fluence of particles across the field is also uniform. So proton therapy using passive scattering is essentially the direct equivalent of open field therapy with photons.

But this chapter is about beam scanning, so why all the discussion about passive scattering and SOBPs? Because one of the most important differences between beam scanning and passive scattering, particularly when we get to intensity-modulated proton therapy (IMPT), is that the effective depth-dose curve for the proton field can vary across the field. During the delivery, the treatment planning (and the delivery machine) has complete control over the fluence delivered by each Bragg peak delivered to the patient, with the consequence that, in the most general case, Bragg peaks can be distributed in three-dimensions throughout the target volume (see Section 16.2.). Thus, the effective depth-dose curve (resulting from the superposition of all energy/range shifted Bragg peaks along one pencil beam direction) can also be varied. In other words, and as we will see in more detail later, there is no SOBP in PBS.

16.2.2 There Are No Collimators and Compensators

Now let us look at the differences between scanning and passive scattering when conforming the dose-to-target volume. Again, there are two major differences—how the dose is conformed to the distal edge of the target and how the lateral dose conformation is achieved. With PBS, as the Bragg peaks lateral to the beam direction can be freely chosen (and delivered) at the treatment planning stage, the planning process must identify those Bragg peaks that deliver useful dose to the target volume. As we will see later, this preselection process as part of the treatment planning effectively acts like a three-dimensional aperture, simultaneously "cutting" out Bragg peaks that are outside the volume laterally as well as distally and proximally. Thus, for PBS, there is no need for collimators and compensators.

However, in order not to propagate a misunderstanding around this point, this is not the same as saying that collimators and compensators *cannot* be used. Of course, as long as the delivery nozzle supports the mounting of such devices, there is actually no reason why collimators/compensators cannot be used as well with scanning, and particularly for the treatment of superficial tumors, these devices can provide a valuable method for sharpening the lateral penumbra over and above that resulting from the pencil beam size itself [3–5].

16.2.3 Field Design and the Need for Optimization

In the previous sections, we have seen that, for PBS proton therapy, there are no SOBPs, and there is no need for collimators or compensators. Let us now look in detail into field design for pencil beam scanned treatments. The description given here is mainly based on the process used for planning at Paul Scherrer Institute (PSI), but the general principles will be the same for any scanning field.

The main steps of the field design process for scanned proton fields are shown in Figure 16.2. Once a field direction has been defined, the first step is to determine the possible set of all deliverable Bragg peaks within the patient. This set of Bragg peaks will clearly be dependent on the physical characteristics of the delivery machine. For the case shown in the figure, these parameters are as follows. A maximum proton energy of 138 MeV was selected for the field, and the minimum energy is close to zero due to the fact that the energy is modulated at our facility by the insertion of range-shifter plates into the beam directly before the patient (see, e.g., Pedroni et al. [6]). For an energy of 138 MeV, our delivery system has enough such plates to bring the Bragg peaks right to the surface of the patient. In addition, spacing between the Bragg peaks in depth is defined by the thickness of these plates, which are 4.6 mm water equivalent, and pencil beam spacing orthogonal to the beam direction was defined as 5 mm.

Figure 16.2b shows, for one slice of this patient, all the possible Bragg peaks that could be delivered using the parameters defined above. Each red cross is a possible Bragg peak

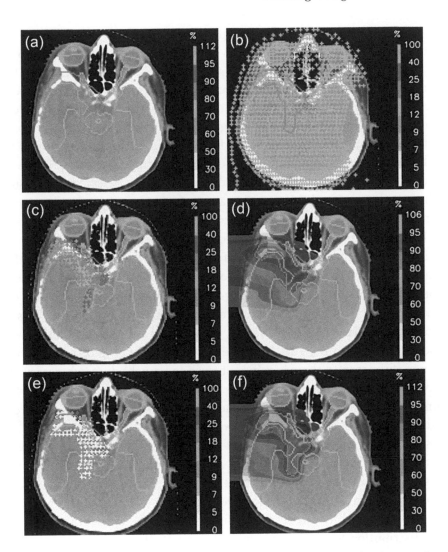

FIGURE 16.2   The main steps in the field design process for an SFUD field for the case shown in (a). (b) All possible Bragg peaks that can be delivered from the selected field direction and with the field geometry parameters as described in the main text. (c) The subset of Bragg peaks automatically selected for subsequent optimization based on their position in relation to the surface of the selected target volume. (d) The initial dose distribution resulting from the preselection process and initial set of Bragg peak fluences ("starting conditions") shown by the colors in (c). (e) Bragg peak fluences after the optimization process, displayed using the same color scale as in (c). (f) The dose distribution resulting from the set of optimized Bragg peak fluences shown in (e).

position, converted from water equivalent range to depth in the patient. This transformation (from a uniform spacing of Bragg peaks in water-equivalent space to an irregular spacing of Bragg peaks within the patient) can clearly be seen from the Bragg peak spacing's in the air of the nasal cavities, as well the irregular shape of the maximum Bragg peak penetration distal to the target volume. That this is irregular in shape is simply due to the non-flat patient surface and different tissue densities within the patient.

Step 2 is to preselect, from all those shown Figure 16.2b, those Bragg peaks that will be most useful for delivering dose to the target volume. As can be seen in Figure 16.2c, which shows the resulting subset of selected Bragg peak positions, Bragg peaks inside the tumor volume, and a little distance outside the target, have been preserved. This is necessary due to the discrete spacing of the Bragg peaks. If only peaks within the target are selected, and as the pencil beams are defined on a 5 mm grid, in the worst case, only the closest peaks to the surface of volume will be selected, and these could be up to 5 mm inside the target volume. This will lead to potential problems of target coverage at the edge of the target. To avoid this, all Bragg peaks up to 5 mm outside have also been selected for the subsequent steps of the calculation. In addition, an initial set of relative weights has been assigned to all Bragg peaks, which we call the "starting conditions."

Figure 16.2d shows the dose distribution resulting from the set of weighted Bragg peaks shown in Figure 16.2c after the application of an analytical dose calculation (outlined later and in Chapter 14). What is immediately clear is that, based on the initial guess for the starting beam weights, the dose in the target is not homogenous. In the middle of the target there is generally too much dose, whilst at the edges, there is too little. This is mostly an effect of the irregular shape of the target, its large size and the substantial overlapping of neighboring pencil beams. That is, where there is overlapping of many pencil beams (in the center of the target), there is sufficient (even too much) dose, whereas at the edges, where inevitably, there are less pencil beams contributing dose to a point, the dose is too low. To improve dose homogeneity across the target, an optimization procedure needs to be applied to find a set of Bragg peak beam weights that satisfies this condition.

The result of this optimization process is shown finally in Figure 16.2e, with the resultant dose distribution in Figure 16.2f. Although the Bragg peaks at the distal and lateral edges generally have higher weights, the resultant internal Bragg peaks (those within the target volume) have very low weights (<5% of the maximum weight in the field indicated by the white crosses in the figure). After the optimization, however, the final dose distribution for this single field (Figure 16.2f) conforms well to the target contour at the 90% dose level (red area) and also provides a much more homogenous dose across the whole target.

16.2.4 Dose Calculations for Scanning and IMPT

To get from Figure 16.2e to 16.2f, a dose calculation is required, the intricacies of which will be covered more in Chapter 13. I will therefore briefly review these from the point of view of PBS.

As with other forms of proton therapy, there are basically three classes of dose calculations that can be used for scanned proton therapy: ray casting, pencil beam (both analytical approaches), and Monte Carlo, which have been nicely reviewed and described by Schaffner et al. [7]. With the ray casting (as still used at our institute), the physical pencil beam incident on the patient is modeled as the smallest element in the dose calculation, with density heterogeneities being dealt with by a simple scaling of the water-equivalent depth of each dose grid calculation point along the field direction (see, e.g., Schaffner et al. [7]). In contrast, for the pencil beam approach, the physical pencil beam is further subdivided into a number of smaller beam elements per physical pencil beam (typically 4–64 depending on the calculation accuracy required), with each beam element being weighted such that the resultant dose envelope approximates to the lateral spread of the physical beam in air. This concept is shown in Figure 16.3a. Such an approach has been described in detail by Soukup et al. [8]. An alternative to this approach was earlier described by Schaffner et al. [7], where, instead of modeling a single physical pencil beam, the total fluence from all applied pencil beams is first calculated and then modeled by the appropriately weighted set of beam elements (see Figure 16.3b). This approach has the advantage of speed in the calculation, as fewer beam elements need to be calculated over the whole field, whereas the first approach (Figure 16.3a) has obvious advantages during the optimization procedure, where the weights of the physical pencil beams (and therefore the total fluence) is changing iteration-by-iteration. In practice, most commercial treatment planning systems (TPSs) for PBS are essentially similar to that shown in Figure 16.3a. Finally, Monte Carlo techniques can also be used [9–11], whereby the physical pencil beams are each represented by many thousands of individual "protons," tracked through the patient geometry, with the number of protons per pencil beam then being proportional to the relative fluence (weight) of that pencil beam. As with other areas of radiotherapy, Monte Carlo calculations provide by far the most accurate results, but at considerable calculational expense. On the other hand, Monte Carlo calculations for scanned fields can, in principle, be considerably faster than

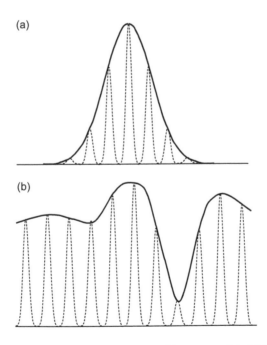

FIGURE 16.3 Analytical calculations using the pencil beam model. (a) The decomposition of an individual Gaussian physical pencil beam into a subset of (Gaussian) beamlets for calculation, showing how a discrete set of such beamlets can be weighted to model the shape of the actual physical pencil beam. (b) The composition of a total fluence (sum of all individual physical pencil beams into a single composite fluence) and its decomposition into beamlets for the calculation (after Schaffner et al. 1999.)

for passive scattering, as the incident pencil beams can be more easily modeled without necessarily transporting particles through the entire treatment nozzle. In addition, the lack of collimators and compensators means that the tracking process can start directly in the patient (see Chapter 10).

16.2.5 The Impact of Secondary Particles

As described in Chapter 3, protons traversing a medium undergo a number of different processes: energy loss, scatter, and interactions with atomic nuclei [12,13]. In the latter process, about 1% of the incident protons are lost per centimeter of traversed matter through interactions with nuclei. These proton losses are not without consequences, however, and although the proton may be "lost" from the primary beam, a spectrum of secondary particles are produced [14]. Although these include heavier, and extremely low-range particles such as deuterium and tritium etc., secondary protons and neutrons are the most predominant.

For the treatment planning of PBS proton therapy, and in particular for IMPT, contributions from secondary protons are the most important. Although the fluence of secondary protons is only a few percent of the fluence of the primary beam, secondary protons have a much wider angular distribution, leading to a long, low-dose tail to the lateral dose profile of the beam—the so-called "halo" effect. The effects of this halo on dosimetry and an analytical model for estimating its effect has previously been published by Pedroni et al. [15]. If this effect is ignored in the dose calculation, it has been found that errors in absolute dose of up 9%–10% can be observed for small fields. This is due to the secondary protons, essentially removing dose from the primary field and into the tissues outside of the irradiated volume.

Figure 16.4 shows this effect. In Figures 16.4a and b, calculated profiles through a single-field uniform dose (SFUD) PBS field (see below) are shown, together with second profiles (shown as closed circles) showing the same field, but this time calculated taking into account the secondary proton halo. There is a clear (and systematic) reduction of about 3% in the dose level throughout the high-dose region, and a small, but just visible, increase in dose in the tails of the profile. Figures 16.4c and d show the same profiles, but this time with the full-dose profile (primary and halo dose) increased in dose globally by 2%. Although there is a slight rounding in the full-dose profile in comparison to the primary dose profile after the global increase, a much improved agreement between the two is found. It is through this global scaling of absolute dose that such effects are currently dealt with at our clinic, although clearly, the best approach is to incorporate the halo effect into the dose calculation during the optimization process, an approach that is now being adopted by most proton treatment planning manufacturers, who have introduced second and even third Gaussians to their analytical dose calculations to model such effects in patient geometries.

16.2.6 The Problem of Superficial Bragg Peaks

Before leaving this section, one additional point needs to be made about field design (and delivery) for PBS proton therapy—the problem of delivering superficial (low-energy) Bragg peaks. This is a problem that occurs surprisingly often, is not trivial, and is,

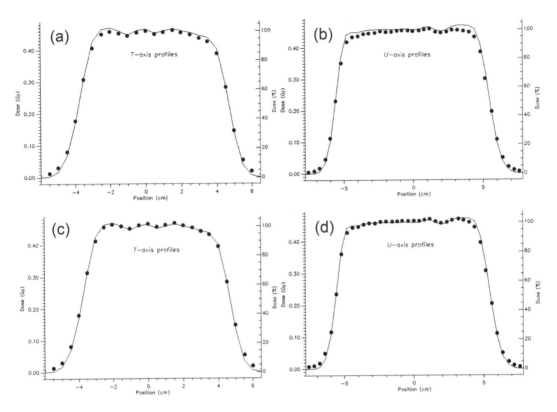

FIGURE 16.4 Effect of the secondary particle "halo" dose associated with scanned proton beams on the absolute dosimetry of homogenous fields. (a and b) Orthogonal dose profiles through a homogenous (SFUD) scanned proton field calculated using primary particle contributions only (solid line) and with the additional effect of the secondary halo dose (crosses). The dose halo essentially removes dose from the primary field and adds dose to the tails of the profile. (c and d) The same profiles, but with the primary and halo dose profile (crosses) globally increased by 2% such as to correct for the effect.

unfortunately, often overlooked. If it is necessary to deliver Bragg peaks close to the surface of the patient, then proton beams of very low energy are required. And here lies the problem. It is extremely difficult to transport low-energy proton beams through a beamline or gantry.

So the next question is, how often do we require such Bragg peaks? After all, when looking at the standard SOBP curve and the relative weights of the Bragg peaks required to produce such a curve, the low-range (and therefore low-energy) Bragg peaks have extremely low weights. Nevertheless, in many circumstances, low-range Bragg peaks can be very important.

Take Figure 16.5a. This shows a slice through a large sarcoma in the pelvis, with the planning target volume (PTV) shown in yellow. Also shown are the Bragg peaks that can be delivered to this tumor from a single posterior beam, assuming a minimum energy of 70 MeV. There is a clear absence of Bragg peaks at the most proximal part of the tumor for all depths within about 3 cm of the patient surface. However, for this slice, this is

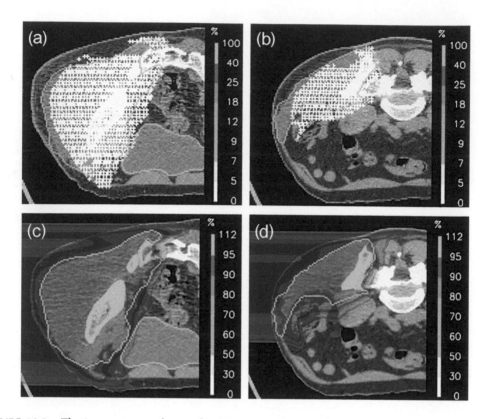

FIGURE 16.5 The importance of superficial Bragg peak positions. (a) An example slice through the PTV of a large chondrosarcoma in the pelvis region. The PTV is shown in green, and the colored crosses show the deliverable Bragg peaks if a minimum deliverable energy of about 70 MeV is assumed (4 cm range). Due to this minimum deliverable limit, there is a 4cm strip of the PTV close to the patient surface, where no Bragg peaks can be applied. (b) A second slice through the same PTV, 5 cm more superior. Here the PTV is extremely narrow and superficial, and with the minimum 4 cm range, this portion of the PTV cannot be sufficiently covered with Bragg peaks to ensure a homogenous coverage of the PTV at this level. (c and d) The resulting dose distributions at the two levels after optimization. Although the coverage at the first level is sufficient, there is a clear problem at the more superior level due to the lack of Bragg peaks in the proximal aspect of the PTV.

not a great problem, as the tumor is thick enough along the field direction that there are enough Bragg peaks stacked from the distal end that the last few energy layers will require extremely low fluences (white crosses in the figure). Thus, the inability to deliver Bragg peaks closer to the patient surface will not be a problem at this level, as they are anyway not required. Figure 16.5b, however, shows another slice for the same case, but 5 cm more superior, again together with the deliverable Bragg peaks. The PTV has a completely different form here, being rather narrow and very superficial. Due to the superficial position of the PTV at this level and the limit on the lowest energy of Bragg peaks that can be delivered, there is a problem in covering the target sufficiently. Indeed, by looking at the resulting dose distribution at this level (Figure 16.5d), it has not been possible to obtain a homogenous dose across the PTV due to the missing Bragg peaks superficially. Consequently, it will be difficult to treat such a target volume from the selected field direction without being able (in the same field) to deliver both high- and extremely low-energy beams of high fluence. For instance, taking Figure 16.4 as an example, to cover the PTV at all depths and levels, it would be necessary to insert a preabsorber to deliver Bragg peaks to the last 4 cm of the target (e.g., for the portion of the field covering the target at the level shown in Figure 16.5b). However, it would be unfortunate to always have the preabsorber in the beam for delivering the more deeply applied Bragg peaks (e.g., the majority at the level shown in Figure 16.5a), as such a device will inevitably degrade the lateral characteristics of the beam.

Ideally, from the delivery machine point of view, one wants a preabsorber that can be automatically inserted into the beam when low penetration Bragg peaks are required, and a TPS that supports this feature. This is exactly what we have implemented on Gantry 2 at PSI, in the form of what we call our mixed pool, and significant improvements in lateral penumbra have been demonstrated for both geometrical and clinical cases [5,16].

16.3 SFUD AND IMPT

16.3.1 SFUD/SFO Planning

In the previous section, the importance of optimization for PBS was described, using the example of a single field. However, as with conventional photon treatment planning and passive scattered proton therapy, it is extremely rare that single-field plans are planned or delivered (the treatment of uveal melanomas being the exception). The reasons for this are twofold. First, additional fields can improve the overall dose homogeneity across the target volume, and second, the "robustness" of the delivered plan can be improved. When one or more, individually optimized and homogenous dose distributions are added together to make a composite plan, we name this a "SFUD" or "single-field optimized" (SFO) plan [17]. An example of such a plan is shown in Figure 16.6. As is clear from the individual fields, the dose across the target for each field is close to uniform, whereas the combination of all the fields improves both the dose homogeneity and dose conformation. The SFUD/SFO

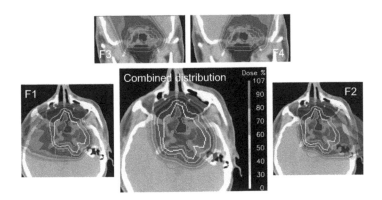

FIGURE 16.6 A first course SFUD plan to a large and complex skull base chordoma, together with the individual field dose distributions making up the total plan. Note that for each field, the dose across the target volume is more or less homogenous.

approach then, although involving an optimization and modulation of the fluence of each individual pencil beam of each field, ensures a smooth dose across the target from each field and can therefore be considered as the scanning equivalent of treating with open fields in photon therapy as well as passively scattered proton therapy.

16.3.2 IMPT/MFO Planning

When one thinks about optimization in radiotherapy, one immediately thinks about IMRT. Although this name is somewhat unfortunate in many ways [1], it has become ubiquitous to describe the process of simultaneously optimizing the cross-field fluences of many, angularly separated photon fields, such as to conform the high dose to the defined target volume, whilst additionally selectively sparing neighboring critical structures. Similarly, if the fluences of all proton pencil beams from multiple fields can also be optimized simultaneously, then we have the proton equivalent to IMRT, typically called IMPT or multiple-field optimization (MFO) [17,18]. And as with IMRT, in IMPT/MFO planning, the optimization can also be driven, not just by the requirement of delivering a therapeutically relevant dose across the tumor but also such that selected critical structures are spared through the definition of dose constraints.

An example IMPT plan with its component individual field dose distributions is shown in Figure 16.7. This shows a complex (and large) skull base chordoma that is close to the brain stem, shown in red. For this plan, the optimization has been asked to cover the PTV as much as possible, whilst sparing the brain stem at about the 60%–70% level, and the resulting individual field dose distributions are highly complex and irregular in form. However, when all these fields are combined into the final plan (shown in the center of the figure), the target is homogenously covered, whilst simultaneously sparing the brain stem. This is of course the power of IMPT. As the optimization is performed simultaneously for all fields, then missing doses from one field can be easily compensated for by the other fields, a possibility that is missing in SFUD planning. However, the individual field distributions also show the potential problem of IMPT planning. The distributions are

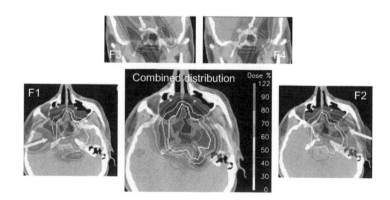

FIGURE 16.7 The second course IMPT plan to the same case as in Figure 8. This time, IMPT has been used to cover the PTV as much as possible, whilst also setting a dose constraint on the brain stem, which partially overlaps with the PTV. The individual field dose distributions are also shown.

extremely irregular and complex, which can have consequences on the robustness of the plan, as we will discuss in more detail later.

16.4 OPTIMIZATION STRATEGIES FOR IMPT

16.4.1 Degeneracy in IMPT Optimization

It will be left to Chapter 16 to deal with optimization theory for proton treatment planning. Nevertheless, a chapter about treatment planning for scanning and IMPT cannot be complete without a discussion about the problem, and potential, of degeneracy in the optimization process. In basic terms, degeneracy simply means that there can be many, sometimes quite different solutions to the specific optimization problem being solved. In general terms, degeneracy will decrease as the number of goals and constraints defined in the optimization process increase. Thus, if the only goal of the optimization process is to achieve a uniform dose to the target volume, the problem will be highly degenerate, whereas if the problem is to achieve a sufficient target coverage, whilst also sparing dose to multiple neighboring critical organs, the degree of degeneracy will rapidly decrease [19–22].

This concept of degeneracy has led some investigators to propose alternative methods of planning and delivering IMPT-type treatments. For example, in Section 16.2.3, the process described for optimizing IMPT plans can be considered to be the most general approach, as it has the largest number of Bragg peaks available to the optimization process. Indeed, it has been previously categorized as 3D-IMPT by Lomax [18], in that the initial Bragg peaks available to the optimization routine are distributed throughout the target volume in three dimensions for each field. However, there is a different alternative technique known as distal edge tracking (DET), which, in some circumstances, can also lead to clinically acceptable and highly conformal treatments.

DET was first proposed by Deasy et al. [23], and in this concept, each field only delivers single Bragg peaks to the distal end of the selected target volume with a homogenous dose across the target being achieved using multiple, angularly spaced fields. Although it has been suggested that the DET approach can be very sensitive to delivery uncertainties [24,25], more recently, the opposite has been shown for certain types of cases [26]. In any case, DET has potentially a number of advantages in that of all the techniques, it uses the smallest number of Bragg peaks/pencil beams and, at least for centrally positioned tumors, it has been shown to minimize the delivered integral dose in comparison to other IMPT approaches [23,27].

16.4.2 When Less Is More: Field Numbers in IMPT Planning

A special consequence of the degeneracy of the IMPT optimization problem is its consequence on the number of fields necessary to achieve a clinically acceptable solution. This has been studied by Stenecker et al. [28] for a simple head-and-neck case. In this work, photon IMXT and IMPT plans were calculated using 3–9 equally spaced fields, with the dose constraints to both parotid glands being successively reduced. The resultant mean doses to these structures were then plotted against the dose inhomogeneity in the PTV.

An example of such a plot for one case is shown in Figure 16.8. For IMXT, as the number of fields increase, the plan also gradually improves. However, there is always an inevitable "playoff" between parotid dose and PTV coverage and, even for the nine field plans, as parotid dose is reduced, PTV dose heterogeneity increases. For the IMPT plans however, the results are quite different. For all IMPT plans, the doses to the parotids are substantially lower than the IMRT plans, for the same (or better) dose homogeneity in the PTV.

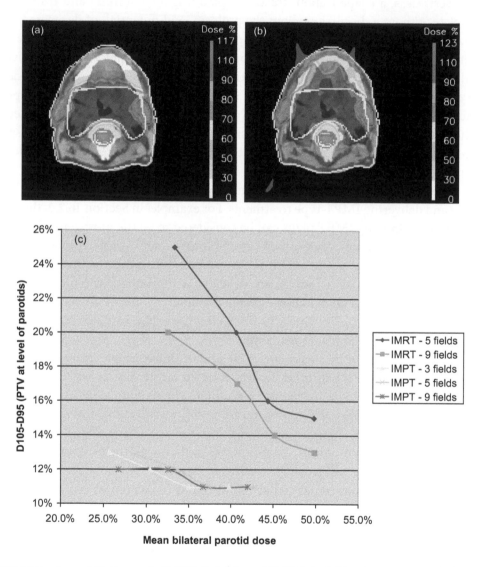

FIGURE 16.8 (a and b) Example IMPT (3 field) and IMXT (9 field) plans to a simple head and neck case. (c) Plots of target dose homogeneity plotted against mean dose to both parotid glands for different IMXT and IMPT plans. Each point on the plot corresponds to one plan, consisting of different numbers of fields (indicated by the lines) and decreasing constraints on the parotid glands. With IMPT, similar (if not better) target dose homogeneity can be achieved for much lower doses to the parotids than for any of the IMXT plans, and little difference is observed between 3, 5, and 7 field IMPT plans.

More interestingly, however, there seems to be little advantage in increasing the number of fields for the IMPT plans.

In summary, it is generally possible to create clinically acceptable, and highly conformal, dose distributions for even complex clinical cases using small numbers of fields, and with fields that are all incident from the same aspect of the patient, the two cases shown in Figures 16.10 and 16.11 being good examples.

16.4.3 The Importance of Starting Conditions in IMPT Optimization

As we have seen, degeneracy is an important issue for IMPT, and in Section 16.7.1, we will look in more detail into how degeneracy can be exploited. But, in daily practice, how do we search the space of possible solutions resulting from the degenerate nature of the problem, or decide which of the many similar solutions we actually want for our final plan (see also Chapter 16)?

Although many sophisticated optimization algorithms are available (Chapter 10), it is still very much the case that optimization algorithms used in commercial and research TPSs are rather simple in concept. Although there are certainly exceptions, most algorithms are gradient based, and based on rather simple dose or dose-volume constraints to the target volume (or volumes) and organs at risk. Given that the problem for the optimizer is then underdefined (or degenerate), such algorithms will inevitably find the nearest solution to the starting conditions that best fits all the constraints. Thus, in many systems, a definition of the starting conditions is a "must" for the user if they want to impose other conditions on the final plan.

Take as an example Figure 16.9. This shows a schematic representation of a simple two-field, parallel opposed plan to a centrally placed target volume with the only constraint being to obtain a homogenous dose across the target volume. Figure 16.9a shows one possible solution for this problem. As a starting condition for both fields, a set of Bragg peaks have been used, which are already weighted such that the resultant dose profile in depth is a mini-SOBP. When these two fields combine, the resulting dose will already be very close to uniform across the target, and the optimizer will perhaps only have to fine-tune Bragg peak weights to achieve the desired result.

In Figure 16.9b, however, all Bragg peaks have the same initial relative weight. If we calculate the resultant depth-dose curve for such a weighting, then this actually gives a linear "wedge" profile in depth, with the maximum dose at the proximal end. As the field from the opposite side has a similar arrangement, when both are combined, the resultant dose across the target volume is once again more-or-less uniform.

So in Figure 16.9, we have two possible IMPT solutions for delivering a uniform dose to the target volume. Are they equivalent? Well yes and no. They are equivalent if one is only interested in target dose uniformity, but not in other aspects. Take for example Figure 16.9c, where we have overlaid the combined dose profiles of the two fields on top of each other and have normalized them such that the dose to the target is the same. In the entrance regions of the two fields, however, there is a clear difference in dose between the two solutions. Solution A (the predefined "SOBP" approach) results in a significantly lower dose in the entrance channels than solution B (the "constant weight" approach). A clinical

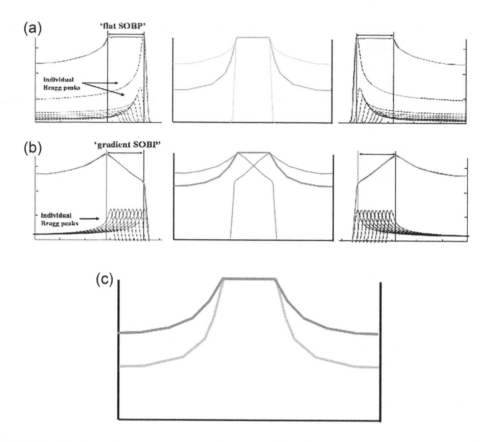

FIGURE 16.9   A schematic representation of two possible solutions to the definition of Bragg peak weights along the beam direction for a simple, parallel opposed field plan. (a) The resulting dose profile along the beam direction as a result of the use of a set of preweighted Bragg peaks delivering a mini-SOBP along the field direction. (b) The dose profile resulting from the same optimization of the same fields, but assuming that the first "guess" for Bragg peak weights along the beam direction is a constant set of weights. (c) The dose profiles from the two solutions superimposed on top of each other.

example of this effect can be found in Albertini et al. [29], and both examples show the importance of understanding the optimization starting conditions in current TPSs for PBS proton therapy.

16.5   CASE STUDIES

Before moving on to other aspects of treatment planning for PBS proton therapy, I would like to briefly present two typical cases that have been treated using PBS proton therapy at PSI, to give the reader a flavor of what can be achieved using the methods outlined in the previous sections.

16.5.1 Case 1. Sacral Chordoma

Figure 16.10 shows the individual series plan and composite treatment for a relapsing sacral chordoma treated at PSI. The PTV had a volume of nearly 1.4 l and was treated to a total of

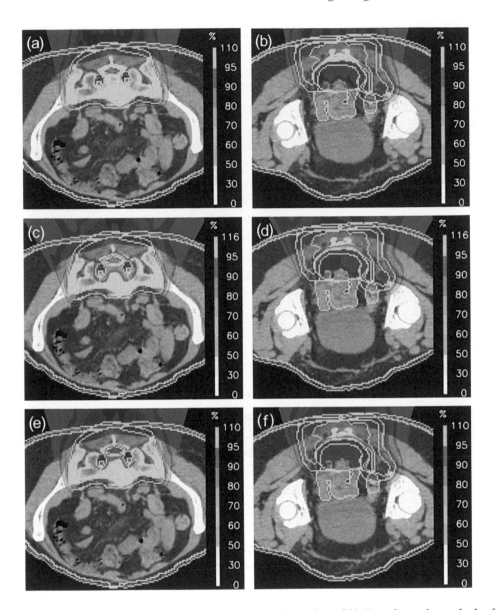

FIGURE 16.10 Case study 1—A relapsing sacral chordoma. (a and b) Two slices through the first series, two-field SFUD plan (0.0–36.0 Gy(RBE)). (c and d) The same slices through the second series two-field IMPT plan (36.0–74.0 Gy(RBE)). (e and f) The composite dose for the complete treatment (74.0 Gy(RBE).)

74 Gy(RBE) using two series. The first series (Figures 16.10a and b) is a two-field SFUD plan using angles of ±15° away from the vertical (posterior), and which was delivered for the first 18 fractions from 0 to 36 Gy(RBE). As with all our plans, a global RBE value of 1.1 has been assumed. Two different slices through this complex case are shown, the first rather more superior and at the level of the cauda equina and the second more inferior and at the level of the rectum. Note the almost total sparing of dose in the abdominal region and pelvic space.

Figures 16.10c and d show the second-series IMPT plan, delivered from 36 to 74 Gy(RBE). This is also a two-field plan, but with the fields now separated from the vertical (posterior) by ±30°. In this case, IMPT has been used to reduce the dose to the cauda equina and nerve roots. The ability to form "donut"-like dose holes around centrally spaced critical structures from an extremely narrow field arrangement is a unique ability of IMPT. Even in this case, the doses to the abdomen, rectum, and pelvic space are extremely low. Finally, Figures 16.10e and f show the composite dose distribution for the whole treatment (to 74 Gy(RBE)).

This case graphically illustrates the power of scanned proton therapy for delivering highly conformal dose distributions to large volumes, whilst minimizing the integral dose to all normal tissues outside of the target volume. It also demonstrates its ability to selectively spare smaller critical structures embedded directly in the target volume, even when small numbers of fields are used. This allows for very large doses to be delivered to large volumes with little in the way of acute side effects to the patient. Indeed, despite the large doses delivered to the large volumes typical for such patients, in our series of 31 patients treated with PBS proton therapy for such indications, only one patient experienced a grade 1 toxicity in the bowel, and no side effects in the bowel of grade 2 or higher have been observed [30].

16.5.2 Case 2. Mucoepidermal Carcinoma of the Parotid

Figure 16.11 shows the second case to be presented here, a simultaneous integrated boost (SIB) IMPT treatment of a mucoepidermal carcinoma of the parotid. Two target volumes were defined for this case, with the low-risk PTV (PTV1) prescribed a total dose of 54Gy(RBE) at 1.8 Gy(RBE) per fraction, and the high-risk volume (PTV2) a dose of 67.5Gy(RBE) at 2.25Gy(RBE) per fraction. Given the very lateral position of both PTVs, this patient has been treated with a three-field arrangement, with all fields coming from the ipsilateral side. The flexibility of 3D IMPT has then been used to deliver a SIB-type treatment, with different fraction doses being delivered to the different PTVs in a total of 30 fractions. As can be seen in the figure, this three-field IMPT approach can deliver well-differentiated dose levels to the two PTVs (the red and blue areas in the figure), ensuring excellent coverage of the two PTVs at their respective 95% dose levels. In addition, due to the narrow angle approach used, the contralateral normal tissues and organs are completely spared.

This is a nice example of the use of IMPT degeneracy to "push the limits" of the optimizer. There is little in the way of angular variation for the fields, and the deliberately nonuniform dose prescription certainly puts more demand on the optimization problem than a simple uniform dose across the whole volume. Nevertheless, using the flexibility of individually modulated Bragg peaks, distributed in three dimensions throughout the PTV volumes, exquisitely conformal dose distributions can be delivered to both PTV volumes in three dimensions, as demonstrated by the sagittal and frontal slices shown on the right. This treatment undoubtedly shows the power of IMPT for complex cases.

PTV 1:
54 Gy(RBE) @
1.8Gy(RBE)

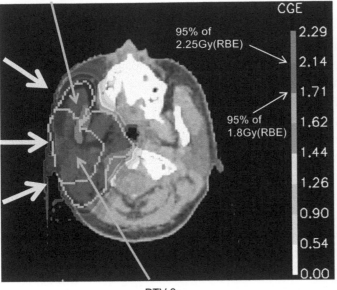

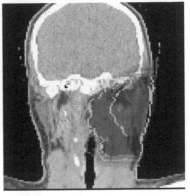

CGE

95% of
2.25Gy(RBE)

2.29

2.14

1.71

1.62

95% of
1.8Gy(RBE)

1.44

1.26

0.90

0.54

0.00

PTV 2:
67.5 Gy(RBE) @
2.25Gy(RBE)

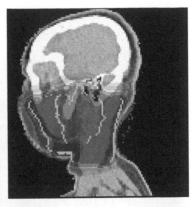

FIGURE 16.11 Case study 2—A SIB IMPT plan to a mucoepidermal carcinoma of the parotid. The low risk (PTV1) and high risk (PTV2) volumes receive fraction doses of 1.8 and 2.25 Gy(RBE), respectively. The blue color wash shows the 95% isodose contour for the 1.8Gy(RBE) plan, whilst the red region shows indicates 95% of 2.25Gy(RBE). Conformation of these dose levels to the two PTVs is excellent in all three dimensions, as indicated by the sagittal and frontal slices shown on the right.

16.6 DEALING WITH UNCERTAINTIES

Having looked in some detail at some of the characteristics and power of PBS proton therapy and IMPT, I move on to additional areas where degeneracy could well be utilized to advantage—to help design plans that are robust to potential delivery errors (see Chapters 14–16).

16.6.1 Magritte's Apple

A famous painting by Belgian artist Rene Magritte is titled "Ceci n'est pas une pomme"—"This is not an apple." Although the image is clearly a very beautiful painting of an apple, Magritte's point was exactly that—it is just a painting of an apple and not a real apple. In many ways, the same can be said of dose distributions calculated (and displayed) by TPSs. Although they are quite accurate representations of the estimated deposited dose, they are exactly that—just representations and estimates. What will actually be delivered to the patient will be, at many different levels, inevitably different due to a multitude of different uncertainties, as discussed in Chapter 15. In this section, we will look at different ways of dealing with such uncertainties and how to analyze and display these when analyzing treatment plans.

16.6.2 To PTV or Not to PTV? That Is the Question

Uncertainty management is nothing new in radiotherapy. In the 1980s, Goitein and coworkers recommended including error estimates as part of the routine planning process, in the form of "maximum" and "minimum" plans, to encapsulate possible variations of the dose distribution about the nominal values [31,32]. Despite this pioneering work, such concepts were slow to be accepted, and uncertainties have instead been typically managed using the concept of the "PTV," as defined in ICRU (International Commission on Radiation Units and Measurements) reports 50 and 62 [33,34]. In this, the PTV is defined as a spatial expansion of the clinical or gross tumor volume, with the margin for the expansion being defined by the likely uncertainties associated with the treatments. This has led to the concept of "margin recipes," where statistical analysis has been applied to the PTV expansion concept to more precisely define (and standardize) margin expansions. The most well known of these is that defined by van Herk et al. [35], which defines the margin based on a separation of the uncertainties into random and systematic components. However, most of this work has been concentrated on photon-type treatments, and little has been published on margin recipes for proton therapy.

But, should there be a difference for PTV definitions and margin recipes between proton and photon therapy? The answer is certainly yes. The problem lies in the additional uncertainty relating to the calculation of the range. Although the effects of positioning errors on the PTV in proton therapy are similar to those for conventional therapy, there is essentially no concept of range uncertainty in photon therapy. However, range uncertainty is almost certainly systematic in nature, and there is no reason to believe that the magnitude of uncertainties in range are the same (or are even correlated) to the positioning uncertainties orthogonal to the beam direction. Thus, as range uncertainties are systematic, and most likely larger than, positional errors, a significantly different margin expansion for the PTV will be required at the distal end of the target than laterally. Consequently, a single PTV expansion for the whole target that is valid for all fields is not necessarily valid for proton therapy. Much more likely is that a field-specific PTV is required, with a different expansion being used along the beam direction to that laterally [36–38].

As an alternative to the PTV concept however, would it not be more elegant to incorporate all delivery uncertainties into the optimization, such that the degeneracy of the optimization problem can be utilized to generate fundamentally "robust" plans? This is an area that is attracting more and more interest in the literature in the form of "robust optimization" and is covered in detail in Chapter 19.

16.6.3 Tools for Evaluating Plan Robustness

As outlined in the previous section, robust planning and optimization is a growing area of research in proton therapy. It is therefore interesting to note that, despite these developments, there is still little agreement on the metric to use for actually evaluating the robustness of a plan. Given the importance put on this aspect by the proton therapy community generally, this is a rather strange omission. As has already been mentioned, such tools were suggested by Goitein in the 1980s [31] but have unfortunately not been generally adopted. However, given the complexity of IMPT plans, it seems to the author that simple tools for evaluating the robustness of treatment plans should be standard tools in any TPS. Without such tools, how can the efficacy of robust planning techniques actually be determined? Or how will it be possible, in the degenerate world of IMPT planning, to differentiate between two IMPT solutions with very similar resulting dose distributions but for which robustness may be very different (i.e., between a 3D and DET-type solution).

In the last few years, we have tried to follow on from the work of Goitein et al. [31] and develop tools for representing robustness to the treatment planner using "worse-case" distributions [24,39]. More recently, we have developed the concept of "error-bar" distributions, which essentially display as a three-dimensional distribution the width of the two-sided error bar associated with the nominal dose at every point of the three-dimensional distribution. The full details of this approach can be found in Albertini et al. [40], and just one example will be shown here.

Figures 16.12a and b show the nominal dose distributions of two plans to a skull base case; one is an SFUD plan and the other an IMPT plan. Field directions for the two cases

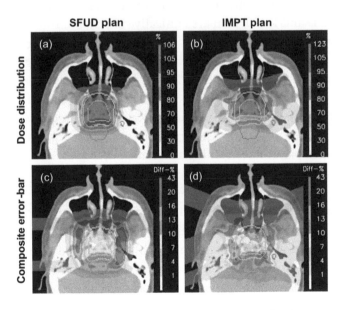

FIGURE 16.12 Example application of "error-bar" distributions to display potential dose errors for proton treatment plans. (a) A three-field (1 lateral and 2 superior lateral obliques) SFUD plan to a skull base chordoma. (b) A four-field (right and left lateral anterior and posterior oblique) IMPT plan to the same case with a strict dose constraint on the brain stem. (c and d) "Composite" error-bar distributions for the two plans, which combine random and systematic errors into a single "error-bar" distribution. The potential variation of dose within the CTV (inner yellow contour) for the IMPT is clearly much higher than that for the SFUD plan, whereas the opposite is true for the brain stem (spared in the optimization process for the IMPT plan.)

are the same. Figures 16.12c and d show the resultant "Error-bar" distributions, here the so-called "composite" distributions that combine the possible random effects of positional spatial errors with the systematic effects of range errors into a single error distribution. In these figures, the colors now show the possible variation in dose at each point about the nominal value and within a certain confidence limit (in this case, the 85% confidence limit). In a similar way to dose distributions, cumulative error-bar-volume histograms (EVH) can also be calculated for any delineated structure, which then indicate the amount of dose variance that can be anticipated in any structure. These can be interpreted like normal tissue dose-volume histograms (DVHs), in that the more the curves are towards the bottom left-hand corner of the plot, the less dose variance there will be in that structure and the more robust is the plan for this structure. The corresponding EVH plots for the two plans shown in Figure 16.12 are shown in Figure 16.13.

Although the two dose distributions are very similar (as may be expected as the selected field directions are identical), there are some subtle, but maybe significant differences in the dose variance distributions. Directly at the border of the high-dose region with the spinal cord, the dose variance of the IMPT plan is actually somewhat smaller than that of the SFUD plan, indicating that the IMPT plan could, in this case, be a little more robust than the SFUD plan when considering dose to the spinal cord. This is confirmed by the EVHs, where the IMPT reduces the volume of the spinal cord that could experience large dose variances. Although it could be argued that the differences here are insignificant, this is a case where the SFUD and IMPT plan are clinically similar, but the variance distributions clearly different, indicating that there could well be differences in the robustness of the two plans to delivery errors. In other words, evaluating robustness using tools such as those suggested here provide an additional criterion by which the planner can more accurately navigate through the degenerate world of SFUD and IMPT plans. Happily, the method shown here is just one of a number robustness metrics now being developed and reported [41–45], showing that, 30 years after the first work of Goitein in this important area, the error bar in treatment planning is making a return.

16.6.4 Relative Biological Effectiveness

The final "uncertainty" to be discussed is that of the biological effect of protons. This is comprehensively discussed in detail in Chapter 22, and only the consequences of RBE, or rather its uncertainties, for treatment planning will be discussed here.

The RBE of protons is a widely discussed topic that is not without its controversies. However, what is clear is that, as the physical processes of the interactions of protons with tissue are very different to those of photons, the biological effect of protons, at least at the micro- and nanoscale, must also be different [46–48]. What is also clear is that such differences will also vary as a function of the position of interaction in the Bragg peak curve, due to the rapidly increasing energy deposition, and therefore ionization density, as the proton energy approaches zero. So, at first sight at least, it seems obvious that a variable RBE, with the RBE increasing towards the end of range of the proton field, should be included into the treatment planning process. On the other hand, humans, organs, cells, and even DNA molecules are extremely complex systems, where predictions of biological reactions

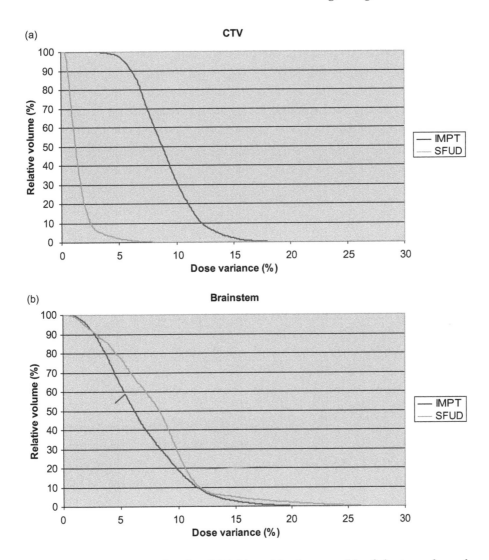

FIGURE 16.13 Example EVHs for the CTV (a) and brain stem (b) of the two plans shown in Figure 16.12. As with conventional DVHs for critical structures, EVHs that are more towards the bottom left hand corner of the plot are more robust (less dose variation at a given volume) than EVH curves towards the top right. The EVH plots clearly show that within the CTV, the SFUD plan is far more robust than the IMPT plan, whereas in the brain stem, the IMPT plan is a little more robust than the SFUD approach.

to well-defined physical interactions are extremely difficult to understand and model. As such, the actual *clinical* consequences of such well-known and expected *physical* differences are still very *unclear*, as confirmed by the large variation in RBE values measured for proton beams of similar characteristics, but measured with different cell lines and biological end points. The problem becomes orders of magnitude more complicated when scaling such in vitro data to the organ/tumor scale, where additional factors such as overall radiation sensitivity, local repair mechanisms, organ architecture, and other complex *in vivo* processes and patient-specific parameters will all play a role in determining "biological" response.

So, although including biological effect into the treatment planning process would seem eminently sensible, in practice, the inherent complexity of the problem necessarily challenges the assumption that RBE at any given point can be represented by a single, relative figure. Nevertheless, much is being published on incorporating RBE into the plan optimization process, and such models are even slowly appearing in planning systems (see, e.g., [49]). Although this may be a welcome development, one must be aware of the uncertainties associated with the models, and also of the potential consequences of over- or underestimating RBE effects in the planning process. For instance, if it is expected that the biological effect at the distal end of a proton field is increased, but this increase is *over*-estimated, then the consequence could be a biological *under*dosage of the distal end of the tumor, potentially reducing the chances of tumor control.

So, how should varying biological effect be incorporated into the planning process? My answer would be—with caution and pragmatically. As already described by Paganetti [50], an enhanced RBE at the distal end of a field is similar to a (one-sided) range uncertainty and could be included into the planning process in a similar way, either by avoiding single fields "ranging" out on critical structures and/or by applying multiple field directions such as to "wash out" any possible end-of-range biological effects. These have been approaches taken by ourselves at PSI, and of the many patients treated at our facility, we have as yet seen no clinical evidence of enhanced side effects (or potential tumor recurrences) that could be attributed to a varying biological effect as opposed to normal dose responses and/ or patient-specific risk factors.

As a surrogate for biological effect, however, a more well-defined and verifiable approach is to use the calculable and measurable surrogate of biological response known as the linear energy transfer (LET). LET provides a spatially varying function that varies in the same direction as RBE (i.e., increased LET likely results in an increased biological effect), without pretending to predict an actual biological effect (see, for example, Figure 16.14). As will be described more in Section 16.7.2, the use of such a parameter in the optimization in the form of "LET painting" may provide the best biologically relevant parameter for inclusion in the optimization process, such that the degenerate nature of IMPT optimization can be used to "move" areas of high LET away from critical structures (if desired), whilst preserving the same target coverage and conformity as a standard plan.

16.7 FUTURE DIRECTIONS FOR TREATMENT PLANNING OF PBS PROTON THERAPY

PBS proton therapy is a rapidly developing technology, and advances in treatment planning must reflect and, in some cases, drive such developments. In this final section, some future directions of treatment planning for PBS proton therapy will be briefly reviewed and discussed.

16.7.1 Utilizing Degeneracy

One of the most promising areas for future developments in treatment planning for PBS proton therapy is in the exploitation of planning degeneracy, the concept of which was discussed in Section 16.4.1. To recall, degeneracy implies that the problem we are trying to

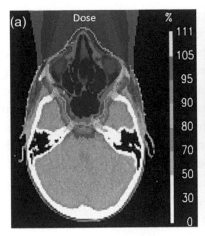
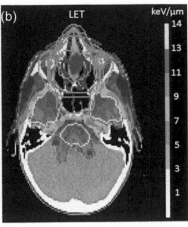

FIGURE 16.14 The physical dose distribution for a three-field plan applied to a nasopharynx tumor (a) and the LET distribution for the same plan (b). This has been calculated using an analytical model developed in-house at PSI, and shows the expected higher LET values in the normal tissues at the distal end of the fields. As the fields have all been planned to come in from the anterior aspect of the patient, the higher LET values in this example are "splashed" in the posterior normal tissues. However, also note that the higher LET values (>5 keV/μm) correspond with regions of very low dose in the physical distribution.

solve clinically is rather too simple for the number of variables available (i.e., beam angles and pencil beams), and that consequently, there are many different sets of pencil beam weights, positions, and angles that could all provide clinically acceptable, and similar, dose distributions. Although if not fully understood, this can be problematic, it also indicates that these additional, and underutilized, degrees of freedom resulting from the flexibility of PBS proton therapy are there to be exploited. Indeed, in the first publication on IMPT in 1999 [18], such potential was already discussed. In that publication, four alternative approaches to IMPT were described, namely 3D, 2D, 2.5D, and DET showing how the degeneracy of the problem could somehow be manipulated.

Searching the solution space provided by degeneracy can be quite interesting, particularly when only a limited number of field directions are available or desired (see, e.g., Figures 16.10 and 16.11). For instance, even for a SFUD field, do we need to have Bragg peaks distributed completely over the whole target volume? Take Figure 16.2e for instance. After the optimization, the majority of Bragg peaks within the target have very low relative weights (the white crosses in Figure 16.2e). Do we need to deliver all these, or are many redundant?

In a study we performed a few years ago, it was found that by building a "spot-reduction" option into the optimization algorithm [51,52], the number of delivered Bragg peaks could be reduced by up to 85%. When employing the same approach for IMPT (multiple-field) optimization, the number of Bragg peaks could be reduced even more, as would be expected perhaps when comparing the 3D-IMPT and DET approaches. Interestingly however, when applying the spot-reduction approach to a simple cylindrical target volume, it was even found that DET is not necessarily the optimum approach for reducing Bragg peak numbers. Figure 16.15 shows the Bragg peak positions resulting from the spot-reduction approach for this case. As can be seen, only distal Bragg peaks in the central portion of the sphere are actually needed to ensure a homogenous dose across the target when planning using five IMPT fields. The more lateral distal peaks have been successfully removed. For this solution, 20% fewer Bragg peaks are required than for the DET solution! More sophisticated methods for reducing the number of pencil beams delivered have recently been published by van de Water et al. [53], also demonstrating the striking potential of such methods for significantly reducing the number of pencil beams required per field and plan.

Although in these examples, no clinical advantage of pencil beam reduction has been demonstrated, reducing the numbers of delivered Bragg peaks per field could have some indirect advantages from the point-of-view of delivery. As the average fluence per delivered pencil beam increases as the number of applied pencil beam reduces, this could allow one to deliver the treatment using higher beam intensities. This in turn could result in significant reductions in treatment time.

An alternative approach for exploiting degeneracy is to explore different strategies for the placement of Bragg peaks, as shown in Figure 16.16. Typically, pencil beams are placed on a rectilinear grid orthogonal to the beam direction, as shown schematically in Figure 16.16a (see also 16.2b). As an alternative however, one can imagine selectively

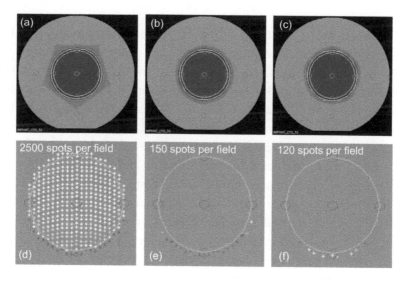

FIGURE 16.15 (a–c) Dose distributions for five field plans to a cylindrical target volume. (a) 3D-IMPT, (b) DET, (c) using the spot-reduction algorithm explained in the main text. (d–f) The corresponding Bragg peak positions and weights for the posterior field of each plan only, with the colors representing the relative weight of the individual Bragg peaks. The corresponding number of nonzero-weighted Bragg peaks for each approach is also shown. Using a spot-reduction scheme directly in the optimizer, a substantial reduction in the number of Bragg peaks required for dose coverage can be achieved, and this approach even finds a solution where less Bragg peaks than the DET approach are required (120 peaks per field as opposed to 150 peaks per field for the DET approach.)

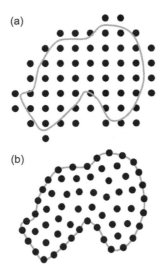

FIGURE 16.16   A schematic comparison of spot placement on a rectilinear and regular grid (a) and the gain that could be achieved by "optimizing" Bragg peak/pencil beams such that placement is performed first directly on the target surface, and then pencil beams/Bragg peaks are "filled-in" in the internal region (b). As demonstrated in the work of Meier et al. [54], such an approach can significantly improve penumbra around the target volume.

placing Bragg peaks directly on the contours of the target volume and then "filling in" the remaining Bragg peaks internally to the target as required (in this case on an irregular grid, see, e.g., Figure 16.16b). Such an approach has recently been shown to improve dose falloff outside the target volume [54] and could potentially lead to a more efficient delivery. As such, irregular and contour-based Bragg peak positioning will certainly be a requirement, and perhaps will even be standard, in future active scanning delivery and TPSs. Such thoughts on optimizing pencil beam placement in PBS proton fields has been further investigated recently in the work of van de Water et al. [53] who have essentially included pencil beam positioning as a variable in the optimization process, with very interesting results.

16.7.2 LET Painting

In the previous section, we discussed how degeneracy in the planning process could be exploited from a more or less geometric aspect. However, there are certainly also clinical and biological "drivers" for treatment plans for which degeneracy could also be a useful characteristic. One of these is the issue of biological effectiveness of protons (see Section 16.6.4).

As discussed earlier, the differential biological effectiveness of protons is a controversial, and complex, subject. Nevertheless, there are some fundamental principles that can be relied on and therefore could be used as additional information (or constraints) in the planning and optimization process. One of these is LET and the knowledge that, at least over the range of LET typically expected in clinical proton plans, an increased LET is a potential predictor of enhanced biological effect (EBE).

So, can information about the LET, as a surrogate for RBE, be used to better exploit degeneracy? This is certainly an area that has been explored in the literature, and that seems to show some promise. Taking a pleasingly "pragmatic" approach in line with that suggested in Section 16.6.4, Unkelbach et al. [55] have included an LET model into the optimization process to use the degeneracy of IMPT to "steer" regions of high LET out of designated critical structures. In their work, they have shown how a clinically equivalent physical dose distribution to a conventionally optimized IMPT plan can be constructed, but such that no regions of high LET are deposited in the brain stem or optic nerve. This approach makes no more assumption than the one stated above (i.e., that a higher LET *may* lead to an EBE), but nevertheless provides the clinician with an additional level of confidence that any potential biological differences in these organs resulting from the use of protons will be minimized. This is a neat approach, also proposed by others (see, e.g., An et al. [56]). Other authors have, however, attempted to go further with LET as a parameter in the optimization process, moving towards "LET painting."

The basic principle of LET painting is to use planning degeneracy to concentrate high LET regions in the tumor (thus potentially increasing the biological effectiveness of the treatment) or even to deliver higher LET to localized regions of the tumor which are expected to have a higher tumor cell density or radiation resistance. An interesting example of this is shown in the work of Farger et al. [57]. By manipulating the target volume for a prostate case, and delivering wedge, rather than homogenous, dose distributions to the tumor from two, parallel opposed fields, they show that the average LET to the tumor can be increased in comparison to a standard (SFUD)-type approach. However, this is an interesting but also a paradoxical example. Although the LET is undeniably increased in the tumor, this has not been achieved without a substantial increase on the physical and biological effective dose to all lateral lying normal tissues, including the femoral head. Indeed, the increase of dose to the normal tissues appears to be of a much higher magnitude (tens of percent) than the gain in biological dose in the tumor (a few percent). In attempting to increase LET across the tumor uniformly then, the solution shown by Farger et al. is similar to that shown in Figure 16.9b, and also discussed by Albertini et al. [29]. With the field configuration used, the only way to deliver a homogenous LET to the tumor is to cover it with Bragg peaks all of a more or less consistent weight (remember, the highest LET is always in the Bragg peak). In a one-dimensional sense, this corresponds to the weighting in Figure 16.9b which, as we have seen, has the consequence of substantially increasing the entrance dose. This example not only shows the potential of LET painting but also that it may not come without a cost. Nevertheless, I hope that this example shows that there is a potential in LET painting, but that to be best exploited, we must not focus just on one aspect (tumor dose) but must look at the whole picture.

16.7.3 Towards (Daily) Adaptive Planning

As I have attempted to indicate in this chapter, treatment planning for PBS proton therapy is a sophisticated, accurate, and flexible tool for personalizing every single treatment delivered to a patient. However, it is not only a "personalization" of the treatment, it is also a very precise tool for accurately "driving" the delivery machine such that the applied dose can be conformed very precisely to the defined tumor volume, and just as precisely "carved out" of neighboring critical structures. As such, a TPS-driven PBS proton therapy facility is indeed a very sharp knife.

So, in many ways, it is a shame that, with such predictive and effective power of modern TPSs, their accuracy and effectiveness will be ultimately limited by the timescale of therapy. For instance, a radiotherapy course with protons, as in conventional radiotherapy, typically lasts many weeks. However, the treatment planning process is still predominantly performed (with all its precision and accuracy) on imaging data sets of the patient taken, in many cases, weeks before the treatment day. So, are we fooling ourselves when evaluating (and showing at conferences and in papers) beautiful, conformal dose distributions, which are calculated based on information of the patient's tumor and anatomical constitution that is essentially historical?

Well, the answer is yes and no. "No," because the clinical success of radio- and proton therapy shows that, as we improve our delivery and associated planning capabilities, outcomes improve. "No" also because we have, and are developing, tools for understanding the effects of uncertainties resulting from changes in the tumor and patient over time (e.g., margins and robust analysis/optimization). And finally "no" because there are few other areas of medicine that have such accurate tools to predict the treatment "agent" (in our case, deposited energy) in three-dimensions throughout the patient. On the other hand, however, the answer has to be "yes," simply because for a large number of patients, the tumor and surrounding anatomy will change over the course of therapy. And for proton therapy, this can be one of the largest sources of uncertainty that we have in our treatments.

Anatomical changes during treatment and their effects on proton therapy treatments can be substantial [58–62]. Although many of these changes are over a slow timescale (days), and thus fractionation and margins can help mitigate the effects, there is no doubt that, if plans could be changed (adapted) based on such changes, then the precision, and hopefully effectiveness, of treatments could be substantially improved. This is the idea of adaptive therapy, first proposed by Yan in 1997 [63].

The idea is simple, and in many ways obvious. Instead of imaging once at some point, and then basing the whole, fractionated treatment on this "snapshot" of the patient's anatomy at some time, weeks in advance of the treatment day, why not image more regularly, and then replan the treatment based on these regular, and presumably more accurate, representations of the tumor and patient anatomy? In other words, why not move proton therapy (and radiotherapy generally) from a *planned* treatment, to a regular *intervention*, with the latter process becoming an *imaging-adapt-deliver* paradigm, potentially performed on a daily basis?

The idea is fantastically attractive, and much is in place in modern TPSs to support regular or daily adaptive therapy. Advanced in-room imaging is at last being introduced as standard for proton therapy systems, in the form of cone-beam CT systems or in-room diagnostic CTs. In addition, the time to optimize and calculate plans is coming down as computer power increases. For instance, even for Monte Carlo calculations, the use of graphical processing units has brought dose calculations down to a few seconds [64–66]. When such technology is also applied to analytical calculations, it is expected that single dose calculations can be reduced to a few milliseconds, and full plan optimizations to a few seconds. Thus, in principle, there is little in the way, at least from the technological point of view, of a move towards daily adaptive therapy. So why are we not there yet?

I think this is for two main reasons. First, the problem of plan quality assurance (QA) and second the uncertainties associated with deformable registration. So before finishing this section, let us just briefly look at these.

A treatment plan, and the resulting information that needs to be transmitted to the treatment machine, are both rather complex, containing much information. For the treatment plan, this is in the form of a three-dimensional map of (predicted) dose in relation to the patient geometry, whereas the machine data contains a wealth of geometric, technical, and positional data, fully representing every nuance of the geometry of the fields to be applied. For PBS proton therapy, this necessarily needs to represent the many thousands of individually weighted and modulated pencil beams that make up an individual field of the plan. One of the issues of daily adapted therapy then, is how to check that the "intervention of the day," a necessarily automatically calculated and perhaps newly optimized treatment for the patient, really reflects the wishes of the clinical staff. For example, is target coverage sufficient, or, perhaps more importantly, will sparing of neighboring critical structures be sufficient. Although many succinct and efficient tools for evaluating plans are already available (e.g., DVHs, biological models, etc.), it is well known that such checks are still very time consuming, and could be a major bottleneck in the daily adaptive therapy process. In addition, the traditional methods of checking the validity of the machine data generated by the daily plan (e.g., plan-specific dosimetric verifications) will have to be rethought, as such measurements will clearly be impossible in a daily interventional approach. This is not to say that these problems are unsurmountable. The recent introduction of daily adaptive approaches for photon therapy using online magnetic resonance (MR) guidance [67] shows that novel solutions to these problems are being introduced. As such, it is only a matter of time before similar solutions appear for proton therapy.

The second issue is a little more subtle. Although it seems obvious to calculate a new intervention on a daily basis if all the technology and QA is in place, this leads to an additional problem that can add to the overall uncertainty of the treatment. By definition, the "image-adapt-deliver" paradigm requires a new image (or images) of the patient each day. Assuming that the tumor and/or anatomy has changed (which is, after all, what we are trying to "catch" by moving to an adaptive regime), then the dose distribution resulting from the newly calculated intervention will necessarily be calculated on a somewhat different anatomy to that of the day before (or after). Thus, an inherent part of an adaptive approach is the ability to accumulate dose on some form of *reference* geometry, such that the total dose delivered to the tumor and critical organs can be estimated at the end of treatment. As such, methods for registering doses calculated on these different geometries onto a single reference are obligatory, and for most anatomical areas, these need to be deformable in nature.

And there lies the problem. Deformable registration is an inherently ill-defined problem for many anatomical regions and, depending on the algorithm and parameters used, quite different results can be obtained. As such, any dose accumulation as a result of a daily interventional approach in RT will be associated with a new type of uncertainty that is simply not there for the "single-plan-fits-all" paradigm of traditional radiotherapy. The big question therefore for adaptive therapy is whether these new uncertainties will be larger or smaller than those of an unadapted plan being delivered to anatomical geometries that we know are changing over time.

To this author, it is unconceivable that this would be the case, and as such, adaptive therapy, maybe on a daily basis, *will* be the future of PBS proton therapy, and the tools for the design, and QA, of daily interventions are the next major steps in treatment planning development.

16.8 SUMMARY

PBS proton therapy is currently experiencing what can only be called a boom period. Particularly in Europe and the United States, a number of hospital-based sites are nearing completion, are in a late stage of planning, or are currently being seriously discussed. The reasons for this are clear. As we have tried to outline in this chapter, scanning provides the most flexible method for delivering proton therapy, either for achieving improved dose conformation per field or, perhaps more importantly, allowing for IMPT. The latter approach truly allows the treatment planner to fully exploit the advantageous characteristics of protons in ways that are not possible with photon-based techniques or with passive scattering (see, e.g., the case studies in Section 16.5).

However, as we have also tried to explain here, this flexibility does not come without consequences. In particular, great attention must be paid to the characteristics of the delivery device to get the most out of scanning. Clearly, as the maximum lateral penumbra that can be achieved is determined on the scanned pencil beam size, it is imperative that the delivery machine minimizes this as much as possible. In addition, and as described in detail in Section 16.2.6, much attention must also be paid to the problem of the delivery of superficial Bragg peaks. Then in treatment planning, given the degeneracy of the problem, there is a great potential for either exploiting this to the maximum or ignoring this and leaving the user frustrated with suboptimal plans. Whilst it is tempting to think that existing IMRT optimization methods will suffice for scanned protons, the additional degrees of freedom available to the optimizer together with the characteristics of protons mean that, already in the planning system, special care needs to be paid to these factors, for instance, by the use of multiple-criteria planning approaches (as described elsewhere in this book) and/or the ability for the treatment planner to set the starting conditions to "drive" the result of the optimization in a desired direction (as described in Section 16.4.4 of this chapter). Finally, although plan "robustness" is a sadly underevaluated characteristic of any radiotherapy treatment, it is certainly true that, due to their finite range, this aspect is more important for proton therapy than it is for conventional therapies. Although much of this aspect can be gained through experience and good training of staff, it is also an aspect that should be more closely incorporated into the planning and quality assurance aspects of proton therapy. For this, tools must be provided by both the treatment planning and delivery machine manufacturers by which the consequences of delivery uncertainties can be estimated at the time of planning, and their magnitudes determined during treatment. For this, uncertainty analysis tools should be provided in the TPSs and advanced imaging and verification tools provided at the treatment machine. Ultimately however, uncertainties can only be substantially reduced by moving to a daily adaptive therapy regime. As such, there is much interesting and important technological research and development still to be done in the field of scanned proton therapy.

REFERENCES

1. Webb S, Lomax AJ. There is no IMRT? *Phys Med Biol.* 2001; 46: L7.
2. Koehler AM, Schneider RJ, Sisterson JM. Range modulators for protons and heavy ions. *Nucl Instrum Methods.* 1975; 131: 437–440.
3. Daartz J, Bangert M, Bussiere MR, Engelsman M, Kooy HM. Characterization of a mini-multileaf collimator in a proton beamline. *Med Phys.* 2009; 36: 1886–1894.
4. Hyer DE, Hill PM, Wang D, Smith BR, Flynn RT. A dynamic collimation system for penumbra reduction in spot-scanning proton therapy: Proof of concept. *Med Phys.* 2014; 41: 091701.
5. Winterhalter C, Lomax AJ, Oxley D, Weber DC, Safai S, A comprehensive study of lateral fall-off (penumbra) optimisation for pencil beam scanning (PBS) proton therapy. *Phys Med Biol.* Oct 2017.
6. Pedroni E, Bacher E, Blattmann H, Boehringer T, Coray A, Lomax AJ, et al. The 200 MeV proton therapy project at PSI: Conceptual design and practical realization. *Med Phys.* 1995; 22: 37–53.
7. Schaffner B, Pedroni E, Lomax AJ. Dose calculation models for proton treatment planning using a dynamic beam delivery system: an attempt to include density heterogeneity effects in the analytical dose calculation. *Phys Med Biol.* 1999; 44: 27–42.
8. Soukup M, Fippel M, Alber M. A pencil beam algorithm for intensity modulated proton therapy derived from Monte Carlo simulations. *Phys Med Biol.* 2005; 50: 5089–5104.
9. Tourovsky A., Lomax AJ, Schneider U, Pedroni E. Monte Carlo dose calculations for spot scanned proton therapy. *Phys Med Biol.* 2005; 50: 971–981.
10. Fippel M, Soukup M. A Monte Carlo dose calculation algorithm for proton therapy. *Med Phys.* 2004; 31: 2263–2273.
11. Jiang H, Paganetti H. Adaptation of GEANT4 to Monte Carlo dose calculations based on CT data. *Med Phys.* 2004; 31: 2811–2818.
12. Goitein M. *Radiation Oncology: A Physicist's-Eye View.* New York: Springer Science+Business Media; 2008.
13. Lomax AJ. Charged particle therapy: The physics of interaction. *Cancer J.* 2009; 15: 285–291.
14. Paganetti H. Nuclear interactions in proton therapy: dose and relative biological effect distributions originating from primary and secondary particles. *Phys Med Biol.* 2002; 47: 747–764.
15. Pedroni E, Scheib S, Boehringer T, Coray A, Lin S, Lomax AJ. Experimental characterisation and theoretical modelling of the dose distribution of scanned proton beams: the need to include a nuclear interaction beam halo model to control absolute dose directly from treatment planning. *Phys Med Biol.* 2005; 50: 541–561.
16. Wilk-Kohlbrecher A, Albertini F, Bolsi A, Fachouri N, Oxley D, Perrin R, Romokos V, Safai S, Weber DC, Lomax AJ. Lateral penumbra improvement for PBS proton therapy using Bragg peak specific pre-absorption, PTCOG 56, Yokohama, Japan, May 2017.
17. Lomax AJ. Intensity Modulated Proton Therapy. In: Delaney T, Kooy H, editors. *Proton and Charged Particle Radiotherapy.* Boston: Lippincott, Williams and Wilkins; 2008.
18. Lomax AJ. 1999 Intensity modulated methods for proton therapy. *Phys Med Biol.* 1999; 44: 185–205.
19. Alber M, Meedt G, Nüsslin F. On the degeneracy of the IMRT optimisation problem. *Med Phys.* 2002; 29: 2584–2589.

20. Lacer J, Deasy J, Bortfeld T, Solberg TD, Promberger C. Absence of multiple local minima effects in intensity modulated optimisation with dose-volume constraints. *Phys Med Biol.* 2003; 48: 183–210.

21. Llacer J, Agazaryan N, Solberg T, Promberger C. Degeneracy, frequency response and filtering in IMRT optimization. *Phys Med Biol.* 2004; 49: 2853–2880.

22. Webb S. The physical basis of IMRT and inverse planning. *Br J Radiol.* 2003; 76: 678–689.

23. Deasy JO, Shephard DM, Mackie TR. Distal edge tracking: A proposed delivery method for conformal proton therapy using intensity modulation. In: Leavitt DD, Starkschall GS editors. *Proceedings of the XIIth ICCR.* Madison: Medical Physics Publishing, 1997, p. 406–409.

24. Lomax AJ. Intensity modulated proton therapy and its sensitivity to treatment uncertainties 2: the potential effects of inter-fraction and inter-field motions. *Phys Med Biol.* 2008; 53: 1043–1056.

25. Lomax AJ. Intensity modulated proton therapy and its sensitivity to treatment uncertainties 1: the potential effects of calculational uncertainties. *Phys Med Biol.* 2008; 53: 1027–1042.

26. Albertini F, Hug EB, Lomax AJ. The influence of the optimization starting conditions on the robustness of intensity-modulated proton therapy plans. *Phys Med Biol.* 2010; 55: 2863–2878.

27. Oelfke U, Bortfeld T. Intensity modulated radiotherapy with charged particle beams: studies of inverse treatment planning for rotation therapy. *Med Phys.* 2000; 27: 1246–1257.

28. Stenecker M, Lomax AJ, Schneider U. Intensity modulated photon and proton therapy fort he treatment of head and neck tumors. *Radiother Oncol.* 2006; 80: 263–267.

29. Albertini F, Lomax AJ, Hug EB. In regard to Trofimov et a.l: Radiotherapy treatment of early-stage prostate cancer with IMRT and protons: a treatment comparison. *Int J Radiat Oncol Biol Phys.* 2007; 69: 1333–1334.

30. Schneider RA, Vitolo V, Albertini F, Koch T, Ares C, Lomax AJ, Goitein G, Hug EB. Small bowel toxicity after high dose spot scanning-based proton beam therapy for paraspinal/retroperitoneal neoplasms Strahlenther. *Onkol.* 2013; 189: 1020–1025.

31. Goitein M. Calculation of uncertainty in the dose delivered in radiation therapy. *Med Phys.* 1985; 12: 608–612.

32. Urie M, Goitein M, Doppke K, Kutcher G, LoSasso T, Mohan R, et al. The role of uncertainty analysis in treatment planning. *Int Radiat Oncol Biol Phys.* 1991; 47: 1121–1135.

33. ICRU50 International Commission on Radiation Units and Measurements, Prescribing, Recording, and Reporting Photon Beam Therapy, ICRU report 50, International Commission on Radiation Units and Measurements, Washington, 1993.

34. ICRU62 International Commission on Radiation Units and Measurements, Prescribing, Recording, and Reporting Photon Beam Therapy, Supplement to ICRU Report No. 50, ICRU report 62, International Commission on Radiation Units and Measurements, Washington, 1999.

35. van Herk M, Reneijer P, Rasch C, Lebesque JV. The probability of correct target dosage:dose-population histograms for deriving treatment margins in radiotherapy. *Int J Radiat Oncol Biol Phys.* 2000; 47: 1121–1135.

36. Bert C and Rietzel E. 4D treatment planning for scanned ion beams. *Radiat Oncol.* 2007; 2: 24.

37. Graeff C, Durante M and Bert C. Motion mitigation in intensity modulated particle therapy by internal target volumes covering range changes. *Med Phys.* 2012; 3910.

38. Knopf A-C, Boye D, Lomax AJ, Mori S. Adequate margin definition for scanned particle therapy in the incidence of intrafractional motion. *Phys Med Biol.* 2013; 58: 6079–6094.

39. Lomax AJ, Pedroni E, Rutz HP, Goitein G. The clinical potential of intensity modulated proton therapy. *Z Med Phys.* 2004; 14: 147–152.

40. Albertini F, Hug EB and Lomax AJ. Is it necessary to plan with safety margins for actively scanned proton therapy? *Phys Med Biol.* 2011; 56: 4399–4413.

41. Pflugfelder D, Wilkens JJ, Oelfke U. Worst case optimization: A method to account for uncertainties in the optimization of intensity modulated proton therapy. *Phys Med Biol.* 2008; 53: 1689–1700.

42. Fredriksson A. A characterization of robust radiation therapy treatment planning methods-from expected value to worst case optimization. *Med Phy.* 2012; 39: 5169–5181.
43. Lowe, M et al. Incorporating the effect of fractionation in the evaluation of proton plan robustness to setup errors. *Phys Med Biol.* 2016; 61: 413–429.
44. Perko Z, van der Voort SR, van de Water S, Hartmann CM, Hoogeman M, Lathouwers D. Fast and accurate sensitivity analysis of IMPT treatment plans using polynomial chaos expansion. *Phys Med Biol.* 2016; 61: 4646–4664.
45. Malyapa R, Lowe M, Bolsi A, Lomax AJ, Weber DC, Albertini F. Evaluation of robustness to set-up and range uncertainties for head and neck patients treated with pencil beam scanning proton therapy. *Int J Radiat Oncol Biol Phys.* 2016; 95: 154–162.
46. Tommasino F and Durante M, Proton radiobiology. *Cancers.* 2015; 7: 353–381.
47. Grosse N, Fontant AO, Hug EB, Lomax AJ, Coray A, Augsburger M, Paganetti H, Sartori AA, Pruschy M. Deficiency in homologous recombination renders mammalian cells more sensitive to proton versus photon irradiation. *Int J Radiation Oncol Biol Phys.* 2014; 88: 175–181.
48. Fontana AO, Augsburger M, Grosse N, Guckenberger M, Lomax AJ, Sartori AA, Pruschy MP. Differential DNA repair pathway choice in cancer cells after proton- and photon-irradiation Radiother. *Oncol.* 2015; 116: 374–380.
49. Tseung HSWC, Ma J, Kreofsky CR, Ma DJ, Beltran C. Clinically applicable Monte Carlo-based biological dose optimization for the treatment of head and neck cancers with spot-scanning proton therapy. *Int J Radiat Oncol Biol Phys.* 2016; 95: 1535–1543.
50. Paganetti H. Range uncertainties in proton therapy and the role of Monte Carlo simulations. *Phys Med Biol.* 2012; 57: 99–117.
51. Bosshardt M., Lomax AJ. Optimising spot numbers for IMPT. ESTRO Physics Congress, Lisbon, Portugal, September 2005.
52. Albertini F, Gaignat S, Bosshardt M, Lomax AJ. Planning and Optimizing Treatment Plans for Actively Scanned Proton Therapy. In: Censor Y, Jiang M, Wang G, editors. *Biomedical Mathematics: Promising Directions in Imaging, Therapy Planning and Inverse Problems.* Madison: Medical Physics Publishing. 2010. p.1–18.
53. Van de Water S, Kooy HM, Heijmen BJ, Hoogemann MS. Shortening delivery times of intensity modulated proton therapy by reducing proton energy layers during treatment plan optimization. *Int J Radiat Oncol Biol Phys.* 2015; 92: 460.
54. Meier G, Leiser D, Besson R, Mayor A, Safai S, Weber DC, Lomax AJ. Contour scanning for penumbra improvement in pencil beam scanned proton therapy. *Phys Med Biol.* 2017; 62: 2398–2416.
55. Unkelbach J, Botas P, Giantsoudi D, Gorissen BL, Paganetti H. Reoptimisation of intensity modulated proton therapy plans based on linear energy transfer. *Int J Radiat Oncol Biol Phys.* 2016; 96: 1097–1106.
56. An Y, Shan J, Patel SH, Wong W, Schild SE, Ding X, Bues M, Liu W. Robust intensity-modulated proton therapy to reduce high linear energy transfer in organs at risk. *Med Phys.* 2017 doi:10.1002/mp.12610
57. Farger M, Toma-Dasu I, Kirk M, Dolney D, Diffenderfer ES, Vapiwala N, Carabe A. Linear energy transfer painting with proton therapy: A means of reducing radiation doses with equivalent clinical effectiveness. *Int J Radiat Oncol Biol Phys.* 2015; 91: 1057–1064.
58. Wang Y, Efstathiou JA, Sharp GC, Lu HM, Ciernik IF, Trofimov AV. Evaluation of the dosimetric impact of interfractional anatomical variations on prostate proton therapy using daily in-room CT images. *Med Phys.* 2011; 38: 4623–4633.
59. Albertini F, Bolsi A, Lomax AJ, Rutz HP, Timmerman B, Goitein G. Sensitivity of intensity modulated proton therapy plans to changes in patient weight. *Radiother Oncol.* 2008; 86: 187–194.

60. Müller BS, Duma MN, Kampfer S, Nill S, Oelfke U, Geinitz H, Wilkens JJ. Impact of interfractional changes in head and neck cancer patients on the delivered dose in intensity modulated radiotherapy with protons and photons. *Phys Med.* 2015; 31: 266–272.

61. van de Schoot AJ, de Boer P, Crama KF, Visser J, Stalpers LJ, Rasch CR, Bel A. Dosimetric advantages of proton therapy compared with photon therapy using an adaptive strategy in cervical cancer. *Acta Oncol.* 2016; 55: 892–829.

62. Placidi L, Bolsi A, Lomax AJ, Schneider R, Malyapa R, Weber DC, Albertini F. The effect of anatomical changes on pencil beam scanned proton dose distributions for cranial and extra cranial tumors. *Int J Radiat Oncol Biol Phys.* 2017; 97: 616–623.

63. Yan D, Vinci F, Wong J, Martinez A. Adaptive radiation therapy. *Phys Med Biol.* 1997; 42: 123–132.

64. Qin N, Botas P, Giantsoudi D, Schuemann J, Tian Z, Jiang SB, Paganetti H, Jia X. Recent developments and comprehensive evaluations of a GPU-based Monte Carlo package for proton therapy. *Phys Med Biol.* 2016; 61: 7347–7362.

65. Ma J, Beltran C, Tseung HSWC. Herman MG A GPU-accelerated and Monte Carlo-based intensity modulated proton therapy optimization system. *Med Phys.* 2014; 41: 121707.

66. Jia X, Schümann J, Paganetti H, Jiang SB. GPU-based fast Monte Carlo dose calculation for proton therapy. *Phys Med Biol.* 2012; 57: 7783–7797.

67. Bohoudi O, Bruynzeel AME, Senan S, Cuijpers JP, Slotman BJ, Lagerwaard FJ, Palacios MA. Fast and robust online adaptive planning in stereotactic MR-guided adaptive radiation therapy (SMART) for pancreatic cancer. *Radiother Oncol.* 2017; 12: S0167–8140(17)32498–2.

CHAPTER 17

Precision and Uncertainties in Planning and Delivery*

Daniel K. Yeung

University of Florida Health Proton Therapy Institute

Jatinder R. Palta

Virginia Commonwealth University

CONTENTS

---

* Colour figures available online at www.crcpress.com/9781138626508.

17.1 INTRODUCTION

Proton therapy allows for conformal dose distributions with sharp dose falloff for complex target volumes and unprecedented lower doses in normal tissue as compared to state-of-the-art conventional radiotherapy. Unlike three-dimensional (3D) conformal radiotherapy, the precision and accuracy of both the treatment planning and delivery of proton therapy are greatly influenced by random and systematic uncertainties associated with the delineation of volumes of interest in 3D imaging, imaging artifacts, tissue heterogeneities, patient immobilization and setup, interfractional and intrafractional patient and organ motion, physiological changes, and treatment delivery. Furthermore, the locations, shapes, and sizes of diseased tissue can change significantly because of daily positioning uncertainties and anatomical changes during the course of radiation treatments. Transient intrafractional changes, such as rectal and bladder filling status, in the treatment of prostate cancer can also introduce uncertainties in dose delivery. Because of these changes, the 3D computed tomography (3DCT) images used for radiation treatment planning do not necessarily correspond to the actual position of the anatomy at the delivery time of each treatment fraction or even to the mean treatment position. Therefore, the traditional assumption that the anatomy discerned from 3DCT images acquired for planning purposes is applicable for every fraction is treated with suspicion in proton therapy.

The published literature concerning acceptable planning and delivery precision and accuracy in proton therapy is sparse. However, the dose–response curve in radiation therapy can be quite steep, and evidence suggests that a 7%–10% change in the dose to the target volume may result in a significant change in tumor-control probability [1]. Similarly, such a dose alteration may also result in a sharp change in the incidence and severity of radiation-induced morbidity. Surveying the evidence on effective and excessive dose levels, Herring and Compton [2] concluded that a therapeutic system should be capable of delivering a dose to the tumor volume within 5% of the dose prescribed. The International Commission on Radiation Units and Measurements (ICRU) Report 24 lists several studies in support of this conclusion. Since the finite range of protons makes therapy more susceptible to tissue-density uncertainties than photon therapy, achieving the aforementioned dose accuracy in proton therapy is a challenge. In conventional photon therapy, intrafractional variations in the shape, size, and position of anatomical structures; tissue heterogeneities; uncertainties in the conversion of CT numbers to relative electron densities; imaging artifacts; and beam-delivery uncertainties smear dose distribution in a patient [3–5]. On the other hand, similar uncertainties in proton therapy can result in significantly compromised target coverage and normal-tissue sparing, which limit its full potential. This chapter describes precisions and uncertainties associated with proton therapy planning and delivery and currently understood mitigation strategies.

17.2 RANGE UNCERTAINTIES IN CLINICAL PROTON BEAMS

Proton beams are used in radiation therapy because of their physical characteristics of energy loss as they penetrate into matter. Specifically, protons have a finite depth of penetration into material; the magnitude of this depth depends on their energy and on the stopping power of the irradiated material. In addition, protons exhibit a Bragg peak with negligible dose at the end of their range. Finally, the dose from a proton beam falls off sharply, both laterally and distally. Figure 17.1 shows the relationship between the proton beam energy and its maximum penetration in water. However, clinical proton beams are not monoenergetic and do not exhibit the same relationship as shown in Figure 17.1. Clinical proton beams have energy and angular spread, which is a result of energy losses and scattering in beam-modifying devices, dosimetric equipment in the beamline, and the air gap.

Therefore, the factors that contribute to range uncertainties in clinical proton beams include inherent uncertainties in linear stopping power, in the formation of broad clinical proton beams (laterally and in-depth), and in the determination of radiological thicknesses of bolus/compensator materials and accessories.

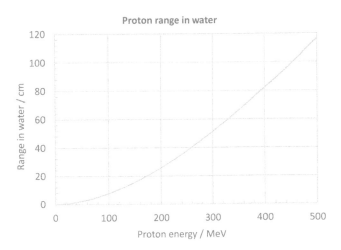

FIGURE 17.1   Proton range in water as a function of energy.

17.2.1 Inherent Uncertainties in Linear Stopping Power

Andreo [6] demonstrated that the range of I-values for water and tissue-equivalent materials stated in ICRU reports 37, 49, and 73 [7–9] for the collision stopping power formulae, namely 67, 75, and 80 eV, yield a spread of the Bragg peak's depth as well as a spread of up to 3 mm for a 122 MeV proton beam (Figure 17.2). He also found that the uncertainty in the Bragg peak is energy dependent due to other energy-loss competing-interaction mechanisms. Although accurate dose-distribution depth measurements in water can be used when empirical dose calculation models are developed, the energy dependence of range uncertainties causes substantial limitations. In the case of *in vivo* human tissues, where distribution measurements are not feasible, a spread of the Bragg peak's depth due to the various soft-tissue compositions is of the same magnitude or more than that of water. Thus, the inherent uncertainty in computing the linear stopping power of water- and tissue-equivalent material can result in a concomitant uncertainty of ±1.5%–2.0% in the range calculation of a clinical proton beam.

17.2.2 Uncertainties in the Formation of Broad Clinical Proton Beams (Laterally and in Depth)

A broad proton beam can be formed by passive scattering or by dynamic scanning of a pencil beam, both laterally and in depth. In passive scattering, the placement of scattering material in the beam provides a near-uniform dose within the field; a variable thickness propeller that rotates in the beam gives uniform dose in depth (see Chapter 5). In beam scanning, lateral and in-depth uniform dose distribution is achieved by scanning a pencil

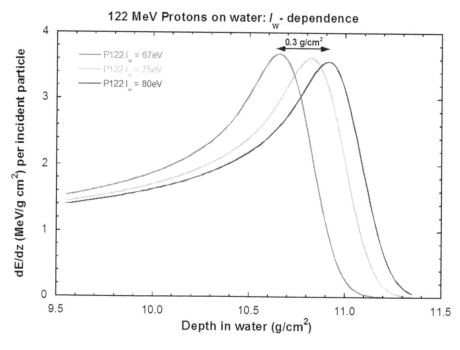

FIGURE 17.2 Variation of the depth of Bragg peak for the I-values of water 67, 75, and 80 eV for a 122 MeV proton beam. (Reproduced with permission from [6].)

beam or a spot laterally and in depth (by changing its energy; see Chapter 6). The lateral positions and weights of each pencil beam or spot of a particular energy level determine the lateral distribution for proton energy, and weighting the pencil beams or spots at each position within the field determines the in-depth distribution. In both methods, the range of protons is changed either by inserting absorbers into the beam path or by changing the beam energy upstream. Typically, proton beam energy uncertainty introduces a systematic range of uncertainty of ±0.6–1.0 mm; however, because of the physical introduction of beam-modifying devices in the beamline, the reproducibility of the range is ±1.0 mm for passively scattered proton beams.

17.2.3 Uncertainties in the Determination of Radiological Thicknesses of Bolus/Compensator Materials and Accessories

Passively scattered proton beams utilize physical range compensators to achieve distal conformance of the dose distribution to the target volume. The thickness profile of the compensator is calculated based on the difference between the distal ranges of the given ray to that of the global maximum for the beam. Based on the stopping power of the compensator material used, the water-equivalent thicknesses necessary for the range pullbacks are converted to physical thicknesses. The compensator is milled out of acrylic, wax, or other low-impedance (Z) materials to minimize additional scatter-increased penumbra from the compensator. The nonuniformity of stock compensator materials introduces small uncertainties in their relative stopping power (RSP), which in turn affects the range of protons. Sharp gradients in the compensator-thickness profile can also induce fluence perturbation as protons are preferentially scattered away from the thicker parts towards adjacent thinner areas. This scattering can result in hot and cold spots of 10%–20% in dose near such gradients and minor range degradation. Furthermore, nonuniformities in the thickness and composition of the accessories (e.g., immobilization devices, tabletops, and head holders) can also increase range uncertainties. Range uncertainties introduced by both compensators and accessories are generally systematic and ±1.0 mm.

17.3 UNCERTAINTIES IN TREATMENT PLANNING

CT data are considered the gold standard for computing dose distributions in a proton treatment planning system. Planar dose distribution displays, dose-volume histograms (DVHs), and dose-volume indices derived from 3D dose computation are used to make treatment decisions based on a balance among target volume and critical normal tissues doses. It is generally assumed that the dose distribution computed on a static 3D imaging data set at the beginning of the treatment course represents the dose distribution actually delivered to the patient during the whole course of radiotherapy. Such dose distributions for cohorts of patients are also used to determine associations between dose and dose–volume indices and dose–response models versus observed tumor and normal tissue responses. However, dose distributions computed on a static 3D imaging data set at the beginning of the course of radiotherapy may differ significantly. Some contributing factors are described later.

17.3.1 CT Number and Stopping Power Ratios

Calculating the proton range and energy deposition in treatment planning requires the determination of the RSPs of the human tissues within the body. This is often based on a scanner-specific calibration curve that maps CT numbers (relative attenuation with respect to water) of tissues to RSPs. The most commonly used method to establish this mapping is the stoichiometric calibration proposed by Schneider [10]. In this method, the calibration curve is determined based on theoretical CT numbers and RSPs of human body tissues, which are calculated according to the known elemental compositions of human tissues and the CT-specific parameters determined from the measurement of tissue substitutes.

Using CT scans of phantoms with various tissue substitutes of known chemical composition, the response of the CT is parameterized as a function of the material's composition. The mean attenuation coefficient of a given material can be described by the formula [11]:

$$\bar{\mu} = \grave{O}_e N_A \left(Z \bar{K}^{KN} + Z^{2.86} \bar{K}^{sca} + Z^{4.62} \bar{K}^{ph} \right), \tag{17.1}$$

where N_A is Avogadro's number; $\grave{O}_e$ and Z are the electron density and effective atomic number of the material, respectively; and $\bar{K}^{KN}$, $\bar{K}^{sca}$, and $\bar{K}^{ph}$ are the scanner-dependent parameters that account for the Klein–Nishina coefficient, coherent and incoherent scattering, and the photoelectric effect. These parameters are determined experimentally using a least-squares fit to the measured CT numbers of the tissue substitutes. The CT-to-RSP calibration curve for real biological tissues is then determined by calculating the theoretical CT numbers and RSPs as follows:

a. Use Equation 17.1 with the fitted parameters to calculate the theoretical CT numbers of the biological tissues using chemical compositions and effective densities taken from the literature [12–14].

b. Calculate the theoretical RSPs using the approximated Bethe–Bloch equation

$$RSP = \varrho_{e.\,rel} \frac{\ln \dfrac{2m_e c^2 \beta^2}{\left(1-\beta^2\right)I_{m.x}} - \beta^2}{\ln \dfrac{2m_e c^2 \beta^2}{\left(1-\beta^2\right)I_{m.w}} - \beta^2}, \tag{17.2}$$

where $m_e c^2$ is the rest mass energy of the electron, β is the relative velocity of the particle to the speed of light, and $I_{m.x}$ and $I_{m.w}$ are the mean ionization energies of the material and water, respectively. There are uncertainties in determining the CT to RSP calibration, which result in uncertainties in the proton range. For example, patient size and beam hardening [15], changes in the photon energy spectra [16], and the choice of the reconstruction algorithm [17] are known factors that affect the measured CT number.

The uncertainty of the real tissue composition, tissue variations in patient population [18–20], and the uncertainties in the mean ionization energies (I-values) of tissue and water [6,21] all contribute and limit the accuracy in determining RSPs. In addition, there is no perfect one-to-one correspondence between CT numbers and RSP of human tissues [21]. Yang et al. [21] comprehensively analyzed proton range uncertainties related to patient stopping power ratio estimation using stoichiometric calibration. Both systematic and statistical uncertainties were estimated with tissues classified as one of three categories: low-density (lung), soft, and high-density (bone) tissues. The combined uncertainty for different treatment sites was determined to be fairly consistent at approximately 3.0%–3.4%. They also noted that the uncertainties of the I-values for tissue and water were likely correlated (since water is a major constituent of soft tissues); a 10% variation in I-values was found to result in only 0.2% difference in RSP.

Doolan et al. [20] performed an intercomparison of several RSP estimation models. They found that extracting the published I-values from Bishsel [22], Janni [23], and ICRU [14], and using them in conjunction with Schneider resulted in larger errors in the approximation of RSP calculations, as each set of I-values were adjusted to match with the original calculation models. Thus, they proposed a new method using actual stopping power measurements of tissue substitutes to optimize the set of I-values such that the theoretical RSPs calculated would closely match the measurements. This approach reduced the mean errors from −0.53% to +0.11%.

In principle, proton CT scanners could directly provide a 3D map of the patient's RSPs. While research in this area has been robust [24–32], no commercially viable systems have emerged. Schneider et al. and Doolan et al. [18,19] have suggested calibrating X-ray CT data sets using proton radiographic images, but this work remains in a preliminary stage. In clinical practice, a margin of 2.5%–3.5% of the beam range is added to account for the proton beam range uncertainty, with this conversion from Hounsfield units (HU) to RSP contributing ±0.5% to the uncertainty and uncertainties in the tissue I-values contributing a further ±1.5% (both based on 1.5 standard deviations) [33].

Dual-energy CT (DECT) offers an alternative method and potential improvement in determining RSP. Hunemohr et al. [34] demonstrated a two-step method of DECT-based RSP prediction. Using a proprietary algorithm, they calculated the relative electron density and an effective atomic number image (referred to as a Rho/Z map). The effective atomic number was then used to determine the I-value through an empirical relation established by Yang et al. [35]. Measurements based on homogenous tissue substitutes indicated a mean absolute deviation of 0.6% in comparison with water-equivalent path lengths measured in a carbon-ion beamline. Mohler et al. [36] pointed out that Yang's I-calibration curve relied on unmixed tabulated tissues of well-defined elemental composition; therefore, it may not accurately account for real human tissues. Also, a gap in the effective atomic number between the soft tissue and bony tissues make it difficult to model the volume-averaging effect for multiple tissues in CT. As an improvement, Mohler proposed using a new parameter, photon absorption cross section, to determine the RSP.

Computer tomography artifacts, contrast agents, and metal implants can introduce significant range uncertainties. Indeed, contrast-free CT and artifact suppression reconstruction algorithms should be used. Residual artifacts should be contoured with their HU overwritten to match the RSP of the underlying tissues. Beam angles should be chosen to avoid or minimize the path length through the implant. The specifications from the manufacturer should be consulted to determine the dimension and estimate the RSP (with direct measurements if possible) of the implant. Such area should be contoured and over-written with HU to match the expected RSP value.

17.3.2 Uncertainties in Dose-Computation Algorithms

Analytic dose algorithms (ADCs), commonly known as pencil beam algorithms [37–39], are widely used for treatment planning in proton therapy. Clinical implementation of ADCs is often optimized for speed rather than accuracy to achieve a reasonable calculation time. Notably, ADCs have limitations with clinical implications (see also Chapter 14).

17.3.2.1 Range Uncertainties in Analytic Algorithms

To account for the range uncertainties related to patient stopping power ratio estimation, a generic range margin is typically added in proton treatment planning. Based on the assumption of 2% uncertainty in determining the HU and an additional 1% from the conversion of CT to water-equivalent densities in tissue, a margin recipe of 3.5% of the prescribed range +1 mm was introduced at the Harvard Cyclotron Laboratory. The 1-mm value accounts for uncertainties in patient setup, the design of the range compensator, and determining the skin surface. Although similar recipes (with varying specific values) have been widely adopted in various proton centers, the formulation does not fully include uncertainties related to dose calculation. ADCs typically project the range based on the water-equivalent depth in the patient to estimate the lateral spread at a given depth, ignoring the relative position of inhomogeneities to the Bragg peak. Comparative studies using Monte Carlo simulations [40,41] have shown that, for complex geometries and density variations (e.g., at bone–soft tissue interfaces), ADCs often fail to correctly predict the effects of range degradations due to their limited ability to model multiple-Coulomb scattering (MCS). Range degradation can result in underestimation of doses to critical structures distal to the target, especially in low-density regions [42].

Schuemann et al. [43] assessed the range uncertainties caused by the dose algorithm in several disease sites (liver, prostate, breast, medulloblastoma, lung, and head and neck). These clinical sites feature various degrees of inhomogeneity, with the liver and prostate representing a relatively homogeneous patient geometry. The lung and head and neck, on the other hand, present significant lateral inhomogeneities. The study recommended the adoption of site-specific range margins with the values given in Table 17.1. The margins were derived from a fit of the root-mean-square deviation (RMSD) in percent of the prescribed range for R90 calculated with ADC versus Monte Carlo (Figure 17.3). The total margins listed in the table included estimation of the nondose calculation-related uncertainties of 2.4% + 1.2 mm [33]. For relatively homogenous sites (liver and prostate), ADCs sufficiently predict the range; the uncertainty from dose calculation is estimated to be ~1.4% (R90). Adding the other uncertainties (2.4% + 1.2 mm) in quadrature, it may be possible to reduce the generic margins to 2.8% + 1.2 mm. For heterogeneous sites (lung, breast, and head and neck), the range margin for R90 associated with dose calculation is estimated to be ~5.8%. This corresponds to a recommended margin of 6.3% + 1.2 mm,

TABLE 17.1 Recommended Site-Specific Range Margins If an Analytical Dose Calculation Is Used and Coverage Is Warranted for Every Field

Site	Dose Calc. Only (%)	Total (mm)
Liver + prostate	1.4	2.8% + 1.2
Whole brain	2.0	3.1% + 1.2
Lung + breast	5.3	5.9% + 1.2
Head and neck	6.0	6.5% + 1.2
Lung + breast + head and neck	5.8	6.3% + 1.2

Source: Data from [43].
All margins are based on RMSDR90.

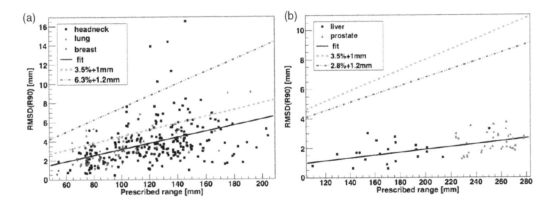

FIGURE 17.3 Linear fit to sites (a) with and (b) without significant lateral inhomogeneities as indicated in the legend together with the current range margins (3.5% + 1 mm) and the recommended range margins. (Reproduced with permission from [43].)

which is significantly larger than the generic margin of 3.5% + 1 mm that is commonly used. Notably, using MC for treatment planning can significantly reduce the margins for all of these sites to 2.4% + 1.2 mm [33]. For a field with a 10-cm range, this translates to 3.7 mm of spared normal tissue. The above results observed from proton fields with passive scattering will also apply to scan beams when the single-field uniform dose technique is used. However, when intensity-modulated proton therapy (IMPT) is used to deliver non-homogenous fields, the range uncertainty must be defined on a pencil-by-pencil basis.

17.3.2.2 Dose Uncertainties in Analytic Algorithms

Schuemann et al. [44] also assessed the clinical impact of approximations in analytic dose calculations for five treatment sites (head and neck, lung, breast, prostate, and liver). Dose plans calculated with ADCs were compared with Monte Carlo simulations [45] and evaluated using DVH analysis, γ-index analysis, and estimates of tumor control probabilities (TCP). ADCs were found to overestimate the tumor dose by 1%–2% on average. The mean dose to the target volume as well as D95, D50, and D02 (dose to 95%, 50%, and 2% volume, respectively) predictions were within 5% of the delivered dose. The γ-index passing rate using 3%/3 mm criterion for target volumes was above 96%. TCP differences were 2%–2.5% for liver and breast cancer patients; and up to 6%, 6.5%, and 11% for prostate, head and neck, and lung cancer patients, respectively. Figure 17.4 shows a head and neck case, where ADC resulted in significant dose discrepancies. ADCs overestimated the TCP by 6.5% compared to Monte Carlo. Streaks of overdose and underdose are clearly visible distal to the interface of the nasal septum and air cavities (Figure 17.4c). The limited ability of ADCs to model the effects of MCS at high-density gradients, especially when such interfaces are parallel to the beam direction, resulted in regions of the clinical target volume (CTV) receiving doses 10% lower than predicted. Interestingly, the target region has a deceiving γ-index passing rate of 97.5%, revealing a potential flaw of the metric. The overall study concluded that algorithms based on ADC could lead to underdosage of the target by as much as 5% and differences in TCP of up to 11%. Approximations in dose calculation can lead to geometric miss from overestimating the range or predicting unrealistic dose homogeneity caused by underestimation or overestimation of scattering effects in tissue. For clinical sites with complex geometries, more advanced methods for dose calculation, such as Monte Carlo simulations, are often necessary.

The shortcomings of ADCs in lung treatments warrant further emphasis. Taylor et al. [46] compared analytic and Monte Carlo-based algorithms for proton dose calculation using measurements of an anthropomorphic lung phantom as a benchmark. The phantom simulated a left thorax with a centrally embedded target and a sinusoidal motion of 2-cm amplitude to mimic the free breathing pattern. Dose plans from five institutions were compared to dose delivered to the phantoms using TLD measurements and film dosimetry in planes through the target. ADCs overestimated the dose to the target center by an average of 7.2%, whereas Monte Carlo calculations were within 1.6% of measurements. The overall passing rate for ADCs was 61% despite a relatively lax acceptance criterion (±7% point dose agreement and > 85% of pixels passing a 7.5%/5 mm gamma criterion on planar analysis). There was no significant difference in the pass rate between irradiations, while the phantom was moving (internal target-volume approach) or static (breath-hold approach) or between using passively scattered protons versus scanning beams. The two most common causes of failure were low dose to the target center and poor agreement in the dose profiles (both in shape and magnitude) in the distal third of the treatment fields

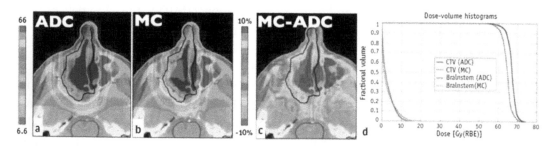

FIGURE 17.4   An example of a head-and-neck case with Rx to CTV 66 Gy(RBE). (a and b) Dose distributions for ADC and MC, (c) difference (MC—ADC), (d) DVH for CTV and brain stem. (Reproduced with permission from [44].)

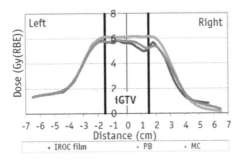
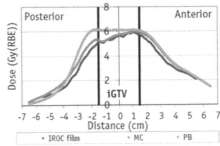

FIGURE 17.5   Dose profile through the center of the PTV in the left-right and posterior-anterior directions. ADC predicts a uniform profile that differs significantly from MC and physical measurements. (Reproduced with permission from [46].)

(Figure 17.5). Schuemann et al. [44] also reported overestimation of target dose to the lung by ADCs. Similarly, Grassberger et al. [47] found that ADCs overestimated both the dose and range of the proton beam in the lung. For lung treatments, analytic algorithms failed to account for lateral scatter and range degradations; thus, planning with Monte Carlo is the preferred choice.

17.3.3 Impact of Tissue-Density Heterogeneities

The influence of and compensation for heterogeneities on dose distribution is far more critical for protons than for photons. Heterogeneities alter the penetration and lateral scattering of protons in the patient. The dosimetric impact of these two effects in the presence of heterogeneities in proton beams, relative to what occurs in a homogeneous medium, can potentially be substantial because of the sharp dose falloff characteristics of protons. When designing treatment beams, one must account for the presence of heterogeneities proximal to or within the target volume by not only calculating their influence on target volume coverage but also compensating for range modifications. The influence of heterogeneities varies by clinical situation. Three potential scenarios are possible: (i) the heterogeneity extends through the entire proton beam. (ii) the heterogeneity intercepts part of the proton beam in the lateral direction, or (iii) the heterogeneity is small and complexly structured in the proton beam.

17.3.3.1 Bulk Heterogeneities Intersecting the Full Beam

A proton beam's energy loss in a section of material of a particular areal density (gm/cm^2) is similar for all materials, with the exception of highly hydrogenous substances (in which $Z/A > 0.5$) and high-Z elements (in which $Z/A < 0.5$). Protons lose energy in a medium primarily through electromagnetic interactions with atomic electrons. Since the mass of protons is large compared with the mass of electrons, they lose only a small fraction of their energy and are deflected very little in each interaction. Although the probability of nuclear interactions increases with the energy of protons, their impact in the therapeutic energy range is minimal except for the shape of the Bragg peak. Nuclear interactions essentially decrease the intensity of protons in the beam by producing secondary particles. These particles may be important from the biological point of view because of their higher relative biological effectiveness (RBE) values, but their impact on physical dose distribution is negligible. Therefore, interposing a material composed of a substance other than that of the surrounding medium primarily increases or decreases the beam's range but does not affect the shape of the depth dose in the region distal to the heterogeneity. The change in a beam's range (ΔR) in such a situation (measured in units of length and not medium-equivalent density) is altered by an amount given by

$$\Delta R = \frac{t\left(\rho_{eq}^{medium} - \rho_{eq}^{slab}\right)}{\rho_{water}} \quad (17.3)$$

In this equation, t is the physical thickness of the interposed slab, ρ_{eq}^{medium} is the water-equivalent density of the interposed slab, and ρ_{eq}^{slab} is the water-equivalent density of the surrounding medium. The water equivalent density is estimated by comparing the mass stopping power of the material in question with the mass stopping power of water at the energy of therapeutic interest. Alternatively, it can be obtained by measuring the change in residual range in water of protons passing through a water tank with and without a physical thickness, t of the interposed slab. These relationships hold equally when the interposed slab replaces the entire surrounding medium. The interposed slab also affects the beam's penumbra, since penumbras are largely caused by upstream multiple scattering, which is dependent on the chemical composition of the interposed material; however, this phenomenon has little effect on tissue-equivalent materials and can be ignored.

17.3.3.2 Bulk Heterogeneities Partially Intersecting the Beam

The influence of bulk heterogeneity that partially intersects the proton beam is primarily limited to the interface between the two media. The beam penetration is altered in the shadow of the heterogeneity and decreases if the water-equivalent density of the interposed heterogeneity is greater than that of the medium. Equation 17.3 calculates the magnitude of this decrease. The penetration is the same in the region unshadowed by the heterogeneity. However, the difference between the MCS of protons and materials of various densities results in a perturbation of the dose at the interface (i.e., hot and cold spots at the interface). When any material of a density different from that of the surrounding medium is interposed in the beam cross section, the beam penetration is altered in the shadow of the material (just as for the case of a fully intersecting heterogeneity) and is unchanged in the region unshadowed by the heterogeneity. However, at the interface region, the difference between the MCS in the two adjacent materials creates a hot spot on the low-density side and a cold spot on the high-density side. Consequently, dose perturbation in air at the interface can be as high as 50% in the tissue–air interface [48]. If one side of the interface is not air, but rather the interface is between two materials of different scattering powers, then the dose perturbation is much reduced; in the case of a bone–tissue interface, from ±50% to approximately ±9% [48]. Notably, the magnitude of hot and cold spots diminishes with depth in the phantom due to the increasing angular distribution of the protons. Figure 17.6 illustrates the impact of introducing a compensator with sharp edges

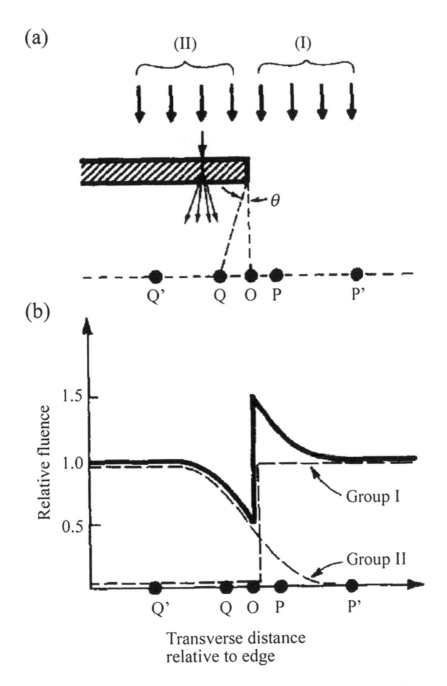

FIGURE 17.6   Illustrates edge-scattering effect. (ICRU Report 78, reproduced with permission.)

in a 150-MeV proton beam. The difference in MCS between a tissue-equivalent compensator material and air creates hot and cold spots of almost 10% at a midrange depth in water.

This is probably the worst-case scenario of dose perturbation at an interface in clinical proton beams. The dose perturbation in the presence of bulk heterogeneity that partially intersects the beam within a patient is often much smaller, especially if one side of the interface is not air but rather between two materials of different scattering powers. For example, in the case of a bone, the tissue interface is much smaller than for that air–tissue interface.

17.3.3.3 Small But Complexly Structured Heterogeneities Intersecting the Beam

In most clinical situations, the patient presents a complex pattern of heterogeneities. The most extreme scenarios are found in the region at the base of skull where protons may be directed along extended bone surfaces or in a complex bone–tissue–air structure (e.g., the petrous ridge or the paranasal sinuses). These complex heterogeneities create range perturbations and MCS-induced dose nonuniformities. Urie et al. [49] studied the influence of these types of heterogeneities on proton beams and concluded that MCS is the main cause of Bragg peak degradation. They concluded that Bragg peak degradation cannot be predicted by simply using the stopping powers of the materials composing the heterogeneities. They also suggested that Bragg peak degradation can be not only diminished by increasing the angular divergence of the beam but also at the expense of widening the lateral falloff. More recently, Sawakuchi et al. [40] carried out systematic Monte Carlo simulation studies to understand this phenomenon. Their Monte Carlo simulation data (Figure 17.7) confirmed the findings of Urie et al. and revealed a trend of increasing distal falloff width with increasing complexity of heterogeneities. They concluded that MCS is the primary cause of Bragg peak degradation, nuclear scattering contributes approximately 5% to the distal falloff of the Bragg peak, and the energy spectra of the proton fluence downstream of various heterogeneity volumes are well correlated with Bragg peak

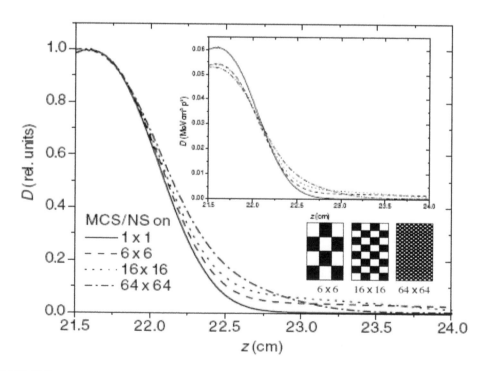

FIGURE 17.7 Degradation of a 230 MeV proton beam traversing different density heterogeneities. Black areas shown in the inset represent compact bone. (Reproduced with permission from [40].)

distal falloff widths. Most treatment planning algorithms that use analytical models cannot explicitly account for this effect; thus, there is a potential uncertainty in distal edge degradation of ±1.0 mm.

17.3.4 Uncertainties in RBE of Protons

The RBE is defined as the ratio of doses necessary to reach the same level of effect when comparing one type of ionizing radiation relative to another (see Chapter 22). Current clinical practice for proton therapy adopts an RBE of 1.1 relative to high-energy photons. Based on extensive review and analysis of published *in vitro* and *in vivo* cell survival data, Paganetti [50] found that the average RBE was ~1.15 at the center of a typical spread-out Bragg peak (SOBP) for a dose of 2 Gy per fraction. The dependency on linear energy transfer (LET) was investigated by observing the RBE values over a range of LET: the entrance region of an SOBP up to the SOBP center (LET 0–3 keV/μm), the downstream half of an SOBP (LET 3–6 keV/μm), the distal edge region (LET 6–9 keV/μm), and the distal falloff (LET 9–15 keV/μm). The average RBE for these regions was ~1.1, ~1.21, ~1.35, and ~1.7, respectively, yielding a value of ~1.15 at the center of the SOBP. The RBE for scanning beams (sharper distal falloff) was also higher compared to passive scattering beam.

Although these averages over many cell lines with a wide range of $(\alpha/\beta)_x$ ratios might not truly represent clinically relevant tissues, there is a significant increase in RBE over depth [51] in clinical proton fields. The increase in RBE as a function of depth also results in a shift in the falloff, increasing the effective range up to 4 mm depending on the dose and $(\alpha/\beta)_x$ [52–54]. From a theoretical basis, one would expect RBE to increase as the $(\alpha/\beta)_x$ decreases; however, the analyzed data [50] do not fully support this theory, with the exception of low-LET data. Previous studies have indicated that RBE increases with decreasing $(\alpha/\beta)_x$ are significant only at low $(\alpha/\beta)_x$ values (below about 5 Gy) [55,56].

The majority of experiments have little or no data below 2 Gy. Interpretation of data is further complicated by potential limitations of the linear quadratic model in this region. The dose dependency of the RBE is, therefore, difficult to assess clinically relevant doses at ~2 Gy per fraction. Most experiments predict an increase of RBE as the dose decreases, particularly at higher LET and/or lower $(\alpha/\beta)_x$, which is in general agreement with theoretical considerations [53].

By definition, RBE depends on the biological endpoint. The most clinically relevant endpoints are TCP and normal tissue complication probability (NTCP). While TCP can be based on cell survival studies, direct response of human tumors to radiation can also be measured. A study of tumor growth delay of human hypopharyngeal squamous cell carcinoma cells in mice indicated an RBE between 1.1 and 1.2 at ~20 Gy for a 23 MeV proton beam relative to 6 MV photons [57]. A study on the recurrence of mouse mammary carcinoma in mice resulted in an RBE of ~1.1 (at ~50 Gy relative to cobalt-60 [58]).

For NTCP, endpoints such as early effects in erythema and late effects in lung fibrosis, lung function, or spinal cord injury are more clinically relevant. The multitude of effects complicates the definition of a variable RBE. Because the slope of the NTCP dose–response curve is steeper than the TCP dose–response curve, RBE uncertainties could be even more significant. Multiple factors at the cellular level affect NTCP, including reactive oxygen species, DNA strand breaks, foci formation, repair proteins and gene expressions, chromosome aberrations, mutations, and apoptosis [50]. The variety of endpoints and the large spread of results do not currently allow a comprehensive analysis of clinical RBE. The most promising endpoint to relate RBE to NTCP might be the induction of inflammatory molecules, such as cytokines related to pneumonitis and fibrosis, after irradiation. The concept of RBE is also complicated by the fact that dose distributions to organ at risks with proton therapy are typically more heterogeneous compared to photon therapy. Doses to normal tissue within a given organ can vary and have a large range of values, yet most dose constraints are defined based on mean dose. The evaluation of RBE for NTCP may require a voxel-by-voxel analysis and the concept of equivalent uniform biological dose.

17.4 UNCERTAINTIES IN TREATMENT DELIVERY

While protons allow for control of dose deposition along and laterally across the beam, the potential of dose uncertainties in proton therapy can be quite large if adequate attention is not paid to both the treatment planning and delivery processes. In particular, one must consider patient selection. Proton therapy is an ideal option for patients who may experience problems with dose to normal tissue and/or organs at risk that surround the target volume (e.g., spinal cord irradiation in pediatric patients and skull-based tumors that are close to sensitive normal tissues, such as the brain stem and optical chiasm). On the other hand, proton therapy is not optimal if there are large heterogeneities in the beam's path, great potential for uncertainties in patient positioning and intrafractional internal-organ motion, or a significant possibility of physiological changes throughout treatment. Therefore, all potential candidates for proton therapy must be carefully screened and evaluated using relevant imaging studies that establish the clinical appropriateness of this treatment modality. The precision and accuracy of proton therapy delivery is influenced by beam-delivery techniques, positioning, immobilization, localization, and interfractional and intrafractional motion, as described later.

17.4.1 Patient Alignment, Setup, and Localization

The accuracy of beam placement relative to the patient (i.e., relative to the target volumes and the organs at risk) is far more important in proton therapy than in photon therapy. The sharp dose falloff in proton beams is a double-edged sword. A small positioning inaccuracy can result in no dose or too much dose at the point of interest. On the other hand, a small beam-placement inaccuracy in photon beam therapy has very little effect; it simply smears the dose distribution. Therefore, the full advantage of protons can only be achieved if there is good registration between the real or virtual compensator and any heterogeneity within the patient. Placement accuracies of 1–2 mm or even less are essential in proton therapy; indeed, the emphasis on the beam-placement accuracy has less to do with target-volume conformation (the target-volume definition by itself has the largest uncertainty) than with the conformal avoidance of nearby critical structures.

Achieving the desired placement accuracy of 1–2 mm requires excellent immobilization of the patient and accurate localization of the patient relative to the treatment equipment (see also Chapter 20). The latter is usually accomplished by the localization of bony landmarks or implanted fiducials, as seen in diagnostic quality orthogonal–planar radiographs. Selecting an appropriate surrogate for the target volume on a planar radiograph is essential. Timmerman and Xing [59] document that in some disease sites bony landmarks may not serve as a good surrogate for the target; thus, the only possible solution is volumetric imaging. At the present time, most proton therapy systems do not include onboard volumetric imaging for more accurate localization. Hence, strategies for quantifying residual uncertainties in target localization for each disease site need to be developed and implemented in each clinic.

Multimodality volumetric imaging is essential in all advanced radiotherapy techniques, especially in proton therapy. Developing a realistic patient model is far more important in proton therapy than in conventional radiation therapy. A quality volumetric CT image set without contrast is required for treatment planning and accurate characterization of tissue densities and heterogeneities. A volumetric cine magnetic resonance imaging (MRI) set is needed for better definition of soft-tissue anatomy and to create a patient-specific motion model. Some clinical situations, such as the lung, may require a positron emission tomography (PET) image for more accurate delineation of CTVs. Each imaging modality has its own limitation (e.g., high-Z material artifacts in CT, image distortion in MRI, and poor resolution in PET). As a result, multimodality registration is bound to have residual uncertainties. The magnitude of these uncertainties is reported to be 5–6 mm and is technique and disease-site dependent [60]; consequently, each clinic should independently quantify these uncertainties.

17.4.2 Relative Motion of Internal Structures with Respect to the Target Volume

Internal structure motion with respect to the target volume can occur both over the course of the entire therapy (interfractional motion) and during the delivery of a single fraction (intrafractional motion) (see Chapter 18). Interfractional movement is attributed to changes in bowel or bladder filling, tumor shape and size changes, or changes in the patient's weight on a day-to-day or week-to-week basis. These changes are problematic for proton therapy because variable position/shape of critical structures can affect the penetration of protons passing through those structures, thereby increasing the uncertainty in dose delivered to the target volume. Intrafractional motion, which can occur on a range of timescales, can be problematic as well. Motion caused by the beating of the heart is periodic in nature, with a cycle time of approximately 1 s; motion caused by respiration is periodic, with a cycle time of approximately 4 s; motion caused by peristalsis is aperiodic and can take place over timescales of up to 1 min. Of these motions, respiration is probably most important, as it can produce quite large displacements and can affect the dose to the tumors in the abdomen as well as in the thorax region. Even with patient immobilization and motion management techniques, some degree of residual motion remains, which must be appropriately taken into account in planning the proton treatment.

17.5 MANAGEMENT OF UNCERTAINTIES IN PLANNING AND DELIVERY

Dose distributions in a proton plan can be significantly degraded due to uncertainties in range, setup errors, and changes in anatomy due to weight gain or loss, or changes in air cavities during treatment. For passive scattering, the range and setup errors are addressed by adding proximal and distal margins to the target, modifying aperture, and smearing in the compensator design. IMPT can provide highly conformal and homogenous dose to tumors with complex geometries while sparing adjacent organs at risk. Dose from individual beams, however, can be rather inhomogeneous, and the desirable dose plan is contingent on an optimal combination of dose contributions from beams coming from different directions. As a result, an optimized IMPT plan can be rather sensitive to range and setup errors, leading to underdosing of the tumor not only at the boundary but also in the interior regions.

Setup errors introduce spatial shifts and translocations of heterogeneities along the beam path and result in misalignment of dose maps from individual beams relative to each other. These errors, as well as range overshoots, can create a dose sum that exceeds the tolerance of critical structures and compromises target coverage. In photon therapy, setup errors and uncertainties in tumor location are handled by dose coverage to a planning target volume (PTV) defined by added margins around the CTV. This assumes an approximate static dose cloud in which the CTV will remain adequately covered as it gets shifted. For proton therapy, the static dose approximation is invalid, as the dose distribution can be significantly perturbed.

17.5.1 Minimization of Uncertainties

As we have shown, inherent and patient-specific planning and delivery uncertainties have far more dire consequences in proton therapy than in photon therapy. Consequently, one must seek to understand not only the sources of uncertainty and reduce them whenever possible but also the magnitude and implications of the inevitable residual uncertainties. The mere process of identifying the sources and magnitudes of uncertainties can help establish safe clinical practices. In particular, overall uncertainty estimates tend to be local; in other words, a local clinical environment that includes patient immobilization, imaging protocol, delineation strategy and protocol, treatment planning strategy, treatment localization, plan evaluation, and treatment delivery equipment and technique has far greater influence on the overall uncertainty in proton therapy planning and delivery than inherent uncertainties in the physics of proton therapy. Thus, each proton therapy clinic should estimate the sources and magnitude of residual uncertainties in each step of the proton therapy process and develop mitigation strategies. Table 17.2 illustrates an example of such an uncertainty analysis for non-moving clinical targets and summarizes the source and magnitude of each uncertainty. These uncertainties can only be mitigated with better physics data; otherwise, the only solution is to account for them in the margin recipes. However, other systematic uncertainties in the proton range that are attributed to the reproducibility of the delivery system, CT calibration uncertainties, compensators, and accessories, among other elements, can be substantially minimized through rigorous

TABLE 17.2   Summary of Estimated Uncertainties in Treatment Planning and Delivery with Proton Therapy

Source of Uncertainty	Uncertainty Before Mitigation	Mitigation Strategy	Uncertainty After Mitigation
[a]Inherent range uncertainty (pristine Bragg peak)	± 1.3 mm	None	± 1–3 mm
[a]Inherent range uncertainty (SOBP)	±.6–1.0 mm	None	±.6–1.0 mm
Range reproducibility	±1.0 mm	Rigorous QA	±.5 mm
Compensator	±1.0 mm	Rigorous QA of compensator material	±.5 mm
Accessories (tabletop, immobilization jig etc.)	±1.0 mm	Rigorous QA of all accessories	±.5 mm
CT	±3.5% of range	Site-specific imaging protocols	±1%–2.0% of range
Patient setup	±1.5 mm	Rigorous patient selection criteria	±1.0 mm
Intrafractional patient motion	Variable	Rigorous patient selection criteria	±1.0 mm
Compensator position relative to patient	Variable	Rigorous patient selection criteria	±1.0 mm
Range uncertainty (straggling) due to complex heterogeneities	± 1 mm	Rigorous patient selection criteria	±.5 mm
CT artifacts	Variable	Rigorous patient selection criteria	± 1.0 mm
Range computation in water in a treatment planning system (TPS)	Variable	Rigorous patient selection criteria and image edits	±.5 mm
Range computation in tissue of known composition and density in a TPS	±.5 mm	None	±.5 mm
Multimodality image registration	±1 mm	Better dose computation algorithms	±.5 mm
Treatment delivery (target coverage uncertainty)	±1–3 mm	Site specific image registration protocols	±1–2 mm
Treatment delivery (dosimetric uncertainty)	±1–3 mm	Rigorous site-specific delivery technique selection	±1 mm
	±1%–3.0%	Rigorous QA	±1.0%

[a] Inherent uncertainty in the particle range determination caused by uncertainty of stopping powers and its basic components, notably the mean excitation energy or I-value of a substance.

quality assurance (QA) (see Chapter 12). Finally, patient-specific uncertainties can only be minimized by enforcing rigorous patient selection criteria and better clinical protocols. Patients who have poor clinical dispositions and significant heterogeneities or implants in the treatment area should not be usually considered for treatment with protons. If proton therapy is clinically warranted for such a patient, then appropriate margins should be selected based on anticipated uncertainties in the planning and deliver of proton therapy.

It is helpful to have a reasonable estimate of the magnitude and impact of each piecewise uncertainty before developing a strategy to mitigate it. Each uncertainty and its consequences can be reduced by understanding the underlying causes. Although uncertainties can be minimized with improved knowledge and more sophisticated tools and processes, they cannot be eliminated altogether. Therefore, residual uncertainties must be managed.

They must be incorporated into planning, plan evaluation, and optimization processes. Many techniques have been developed to mitigate these uncertainties; however, not all techniques are extensible to proton therapy, and alternative strategies must be explored. Indeed, the need to reduce uncertainties and manage residual uncertainties is greater for protons than for photons.

17.5.1.1 Immobilization and Image Guidance

Patient immobilization is much more crucial in proton therapy because any positioning error can result in significant dose perturbation. Protons are relatively insensitive to motion along the beam direction; however, the dose conformation in a plane perpendicular to the beam is sensitive to lateral motion, thus necessitating good immobilization. Immobilization devices with sharp density gradients and nonuniform thicknesses in the path of the proton beam can perturb the planned dose distribution due to interfractional misalignment and should be avoided. This is especially true when physical compensators are used to compensate for heterogeneities.

Although image-guided setup using orthogonal kV planar X-ray images has been used in proton therapy for many years, image guidance using bony anatomy on orthogonal planar images may not be adequate for target localization, as small differences in the amount of soft tissue in the beam path between the treatment position and the position for planning imaging cannot be accurately quantified with this approach. Therefore, image-guidance techniques, such as onboard cone-beam CTs and in-room CT-on-rails for volumetric imaging, are essential for proton therapy. See Chapter 20 for a more in-depth discussion.

17.5.1.2 Implanted Fiducials

Implanted fiducials have been used in various clinical sites to improve target localization in image-guided radiation therapy. Most commercially available fiducial markers are made of high-Z materials that cause artifacts in CT scans and can significantly perturb dose because of the steep dose gradient in particle beams, especially for IMPT [61–64]. Fiducial markers should, therefore, possess the following desirable features: visibility in the required imaging modality (CT, MRI, X-ray), absence of (or with minimal) artifacts, low perturbation of the target dose, and stability after insertion. A thorough study by Habermehl et al. [65] included Visicoils (IBA (Ion Beam Applications, Louvain-La-Neuve, Belgium): gold coils with diameters 1.1, 0.75, 0.5, and 0.35 mm), Gold Anchor (Naslund Medical AB, gold, 0.27 mm diameter), Beammarks (Beampoint AB: nitinol, 1.2 mm diameter), and BiomarC (Carbon Medical Technologies: Zirconium oxide covered with pyrolytic carbon, 1 mm diameter). All markers demonstrated good visibility. Carbon-coated thin gold markers produced the least artifacts and minor dose perturbation. Gold markers >0.5 mm are not recommended, except for use laterally or distally to the treatment field. For markers with high-Z material and thickness greater than 0.5 mm, the water-equivalent path length and dosimetric impact of the markers should be carefully evaluated.

Fiducials are most commonly used for proton therapy of prostate cancer. The prostate usually exhibits interfractional motion relative to the bony anatomy due to the daily variation in rectal and bladder filling [66]. The use of fiducials provides superior verification of the prostate position. The best alignment is achieved with 3–4 gold seeds implanted transrectally under ultrasound guidance near the apex, middle, and base of the prostate [67]. Fiducial migration generally is minimal, although loss of misplaced seeds in the bladder and through the urethra is not uncommon. A 2007 MRI-based study [68] revealed a random interfractional deformation uncertainty of 1.5 mm and an average reduction in prostate volume of 0.5% per fraction during treatment. CT imaging studies [69,70] have suggested that translational and rotational movements of the prostate can be treated as a rigid body transformation. Image-guided proton therapy with fiducials can effectively determine if PTV margins are sufficient to overcome the adverse effects of interfractional and intrafractional organ motion for individual patients to ensure dosimetric coverage [70,71]. Olsen et al. [70] found that the first 10 fractions can reliably predict if the margins are adequate for the entire course of treatment. Fiducials markers are also essential for target localization of clinical sites that exhibit significant interfractional and intrafractional motion, such as the lung, pancreas, and liver [72,73].

17.5.1.3 Adaptive Replanning

Any interfractional or intrafractional physiological and/or anatomical changes have far greater potential of geographical miss of the tumor and overdosing of critical structures in proton therapy compared to photon therapy. Essentially, anatomical perturbations in photon beam therapy smear the dose distributions, while any perturbations in normal tissue anatomy with respect to the proton beam direction contribute substantially to uncertainties in dose delivered to a patient. These issues can only be mitigated in proton therapy by frequent imaging and adaptive replanning either online or offline. Early experience shows that adaptive replanning is much more frequent and necessary for patients treated with protons compared to patients treated with photons. Accurate adaptive planning also requires robust deformable registration between treatment planning imaging and online imaging.

17.5.1.4 Monte Carlo Dose Calculations

One of the sources of uncertainty in proton dose calculations is the approximation and assumption of semiempirical analytical dose calculation formalism algorithms used clinically. Monte Carlo techniques remain the gold standard for dose computation; however, they are plagued by long computation times, especially for proton therapy, where energy as an added dimension increases the central processing unit time by an order of magnitude. Furthermore, computation resource requirements can be many times greater for IMPT. Although Monte Carlo techniques are currently limited to benchmarking for protons, increased computer power, including parallel processing clusters and graphics processing units, will hopefully allow these techniques to be more widely embedded into clinics.

17.5.2 Incorporation of Residual Uncertainties

While some uncertainties can be mitigated, one must account for residual uncertainties in treatment planning and delivery to ensure adequate coverage of the target with the prescribed dose and sparing of normal tissues. As described earlier, the consequences of uncertainties are far greater with protons than with photons. Therefore, appropriate measures must be taken in proton therapy to minimize and account for uncertainties. One cannot assume that the treatment plan reflects exactly what the patient will receive. The following subsections describe approaches to account for uncertainties and evaluate dose distributions in the presence of uncertainties.

17.5.2.1 Margins

The recipes for margins to account for all types of uncertainties in photon treatments are well described and understood [48,74–76]. In photon beam therapy, the expansion of the CTV to PTV with appropriate margins ensures that there is a high probability (95% or higher) that the clinical target will receive the prescribed dose. The defined PTV is analogous to the shell of an egg while the CTV is the yoke. If the prescribed dose covers the PTV, the coverage of clinical target is assured.

However, this concept of PTV is not appropriate for protons. The major uncertainties for protons are (i) the range uncertainty, which depends on the direction of the beam and size of the clinical target, and (ii) the anatomic variations in the path of protons that can considerably perturb dose distributions in the clinical target and normal tissues. Therefore, for proton treatment planning, beam-specific proximal and distal margins are assigned to the CTV for each beam individually. These margins are essentially independent of the change in position of the anatomy along the beam direction. Moyers et al. [77] and Schuemann et al. [43] have provided the most comprehensive margin recipes for protons. The lateral margins assigned for beams in proton therapy are analogous to the lateral CTV to PTV margins in photon therapy. However, additional margins for the lateral spreading (penumbra) of the beam, which is depth dependent for protons, must also be included.

17.5.2.2 Proton Plan Evaluation in the Face of Uncertainties

Uncertainties in proton therapy are often a result of the complex interplay of a variety of error sources. Historically, estimating and reporting uncertainties has been at best implicit. Experienced physicians recognize that "what you see in the treatment plan is not what you get in the patient." Currently available radiotherapy treatment planning systems cannot explicitly show the consequences of uncertainties on displayed plans. As a result, physicians make mental assessments of the magnitude of the known uncertainties and their dosimetric consequences. These assessments are acceptable if the physician is able to discern all potential uncertainties and their consequences. Unfortunately, these assessments become exceedingly difficult when greater geometric accuracy in dose delivery is warranted. Two novel methods to estimate the uncertainty limit associated with a particular treatment were originally described almost 30 years ago [78,79] but never implemented into treatment planning systems. Goitein [78] proposed three separate dose calculations to set higher and lower doses at any point using extreme values for a few parameters, whereas Leong [79] introduced a convolution method for blurring the planned dose with a normal distribution of spatial displacement to investigate the effects of random geometrical treatment uncertainties. More recently, Jin et al. [3,80] showed that for uncertainties that are uncorrelated, one can predict the standard deviation of a planned dose distribution.

Thus, with a comprehensive knowledge of spatial and dosimetric uncertainties in planning and delivery, one can compute and display confidence-weighted dose distribution (with a preset confidence interval), confidence-weighted DVHs, and dose-uncertainty-volume histograms. ICRU Report 78 has also described several other approaches for presenting uncertainty in proton therapy and provides specific recommendations for reporting proton therapy uncertainties.

17.5.2.3 Robust Optimization

Lomax and Pflugfelder [81–83] have stressed the need to assess the robustness of IMPT plans and suggested methods to reduce the adverse dosimetric effects of potential errors. Optimization typically involves the tradeoff of multiple and often conflicting objectives (see Chapter 19). The objectives may include one or more target volumes and several organs at risk. Kraan et al. [84] performed a detailed analysis of dose uncertainties in IMPT for oropharyngeal cancer in the presence of anatomical, range, and setup errors. An in-house system with criteria optimization of pencil beam energies, directions, and weights was used to create retrospective plans for ten patients. The prescription dose was 54 Gy for elective neck levels and 66 Gy for the primary tumor using a simultaneous integrated boost over 30 fractions. Each patient also had a repeat CT during the course of treatment. Dose plans were recalculated and evaluated in 3,700 simulations of setup, range, and anatomical uncertainties. The acceptance criteria were that 90% of cases met the requirement of 98% of the target volume, receiving 95% or higher of the prescribed dose. Range and setup errors were assumed to follow a normal distribution. Dose plans were acceptable for range errors up to 4% and setup errors up to 1.5 mm ($1\,\sigma$ values) when each uncertainty was considered individually. Dose plans were unacceptable when the uncertainties were combined. For anatomical changes, adaptive planning using the repeat CTs was necessary. The study also showed that robustness was not improved by using more beams (e.g., 5 or 7 beams versus the three-field plan).

Robust optimization aims to achieve an effective treatment plan that is protected against uncertainties. Unkelbach et al. [85] provided an example based on stochastic programming. Range and setup errors were incorporated using a probabilistic approach in the optimization of an intensity-modulated treatment plan. An objective function was used to optimize the treatment plan, the dose distribution of which depended on a set of random variables that parameterized the uncertainty. In a discrete implementation, each combination of possible realizations of range and setup errors represented a scenario that yielded a certain value in the objective function. The overall objective was a weighted sum of objectives for the individual scenarios, the weighting of which was given by the incident probability of the corresponding scenario. By optimizing the expected value of the objective function, an optimal tradeoff was achieved to yield a dose plan that was "immune" to range and setup uncertainties.

As pointed out by the authors, a treatment plan with steep dose gradients in the beam direction is sensitive to range errors, whereas one with steep lateral dose gradients is vulnerable to setup errors. The more robust plan is obtained by redistributing dose among different beam directions to remove unfavorable dose gradients that are sensitive to errors.

The resulting plan quality is often not as good as the nominal optimized dose plan that does not account for range and setup errors (albeit unrealistic). The tradeoff, however, aims at retaining the best possible dose distribution while making it robust.

Liu et al. [86] compared the robustness of head-and-neck IMPT plans using PTV-based conventional optimization methods with those using CTV-based "worst-case" robust optimization. Setup errors were modeled with six cases of isocenter shifts in x ($\pm 2\,mm$), y ($\pm 3\,mm$), and z ($\pm 3\,mm$), the magnitudes of which matched the margins used for PTV expansion. Two cases were used to represent range uncertainties ($\pm 3.5\%$). For the worst-case optimization, the objective function was evaluated for the "worst-case" dose distribution during each iteration. For voxels within the CTV, this was defined by the minimum dose (worst target coverage) in each voxel among the nine cases (eight cases modeling the uncertainties plus the nominal case). For voxels outside the CTV, the maximum dose was used (worst case for normal tissues).

Both PTV-based and CTV-based optimized plans were used to determine the dose distributions in 21 scenarios, representing the possible combinations of setup and range uncertainties. The RMSD per voxel was used to quantify the robustness of dose in a voxel. For the 21 dose distributions, the RMSD-volume histograms (RVH) were calculated for each of the two optimization methods. The area under the RVH curve (AUC) provides a numerical index indicating the plan robustness similar to the way that equivalent uniform dose summarizes a DVH. The smaller AUC value represents a better plan robustness. Conventional DVH analysis was also used to evaluate target coverage, homogeneity, and normal tissue sparing. The worst-case scenario robust optimization was found to provide a statistically significant improvement in target coverage over PTV-based plans in the presence of uncertainties. The robustly optimized plans also reduced the dose to the organs at risk compared with the PTV-based plans in the worst-case scenarios.

Since robustness to uncertainty often comes at the expense of plan quality, it is essential that the uncertainties considered are realistic. Fractionated treatment deliveries are known to lessen the effect of uncertainties resulting from random setup errors [87]. Existing robust optimization approaches, however, have neglected such effects. Lowe et al. [88] proposed to use "fractionation incorporated robust optimization" and demonstrated that it could reduce sensitivity to uncertainty compared to conventionally optimized plans and a reduced integral dose compared to robustly optimized plans. Another study that included fractionation dependence can be found in Oden et al. [89]. In summary, robust optimization to account for uncertainties is essential for IMPT planning. Algorithms for criteria and robust optimization were recently made available in commercial treatment planning systems. Active developments are also underway for the introduction of biological optimization to account for biological effects, such as LET and RBE [89–91].

17.5.2.4 Management of Uncertainties in RBE

The assumption of an RBE value of 1.1 for protons in current clinical practice is consistent with the average value observed from experimental data. However, RBE increases with

LET and is significantly higher, especially at the distal falloff towards the end of range. The higher RBE also increases the biological effective range. To take advantage of the sharp distal falloff, proton beams are often arranged to create a sharp dose gradient to spare critical structures distal to the target. However, because of the higher RBE and uncertainties, one should avoid multiple beams having the distal falloff within critical structures. The higher RBE would result in a dose level (not accounted for in planning) that could exceed normal tissue tolerance. Giantsoudi et al. [92] investigated incidences of brain stem necrosis in a cohort of patients with medulloblastoma receiving proton therapy. The area of injury did coincide with a region with higher LET, therefore receiving a higher biological dose. However, all cases in the cohort had similar beam arrangements and LET distributions, and the majority of patients (115 out of 119) were not injured. The variations in tissue sensitivity among patients were likely a contributing factor for the few cases resulting in injury. Because RBE also depends on $(\alpha/\beta)_x$ which for most tissues is not very well known, it would be difficult to clinically evaluate proton plans based directly on biological equivalent dose. Unkelbach et al. [93] proposed a technique to reoptimize IMPT plans to avoid high values of LET in critical structures located near the target volume. The IMPT plan was first optimized based on physical dose. The LET optimization step was performed based on objective functions evaluated for the product of LET and physical dose (LET×D). The latter represents, to first approximation, a measure of the additional biological dose that is caused by high LET. This represents a pragmatic approach to circumvent the lack of precise knowledge of the $(\alpha/\beta)_x$ and RBE values for accurate determination of biological dose (see Chapters 19 and 22).

17.6 SUMMARY

Accurate dose calculation methods, improvements in positioning and motion management techniques, robust optimization, and robust evaluation can make the delivered dose distributions more similar to planned dose distributions and improve confidence in their accuracy. Furthermore, improved understanding of RBE variability and its clinical relevance, and the development of reliable RBE predictive models, should allow us to better correlate dose distributions with treatment response and lead to improved and more consistent clinical outcomes. A reduction in uncertainties will also enable us to reliably intercompare competing treatment modalities in terms of their clinical and cost effectiveness.

The rationale for the use of proton beams in radiotherapy, which is based on their ability to provide uniform dose to the target while sparing the surrounding tissue, is very compelling. This rationale is simply a consequence of the physical characteristics of energy loss by protons as they penetrate into matter. Protons have a finite depth of penetration in material; the magnitude of this penetration depends on the protons' energy and the density of the irradiated material. This distinct advantage can turn into a double-edged sword if adequate consideration is not given to the potential sources of uncertainties in proton therapy. One must recognize that uncertainties are an inevitable part of the planning and delivery of radiotherapy. In this way, protons are no different from radiotherapy delivered with photons; however, the impact of these uncertainties is much more profound in proton therapy. Patient-positioning errors, patient motion, misalignment of beam modifiers, changes in tissue characteristics, and dosimetric errors in planning and delivery can result in a delivered dose that is very different from the planned dose. Therefore, we recommend the following procedures for each patient considered for treatment with proton therapy:

- Analyze potential sources of uncertainty by evaluating the patient's 3D/4D imaging data and clinical disposition.

- Minimize the sources of uncertainties.

- Document the magnitude of the residual uncertainties.

- Develop strategies for the analysis, quantification, and display of residual uncertainties.

- Implement a QA program that ensures that treatment can be given with a confidence level that is established for the patient. For example, a 95% confidence level will give a 95% assurance that the patient receives the prescribed dose distribution.

REFERENCES

1. ICRU. Report 24: Determination of Absorbed Dose in Patients Irradiated by Beams of X or Gamma Rays in Radiotherapy Procedures. *Journal of the International Commission on Radiation Units and Measurements*, Volume os13, Issue 1, 15 September 1976.
2. Herring D. The degree of precision required in the radiation dose delivered in cancer radiotherapy. *Br J Radiol*. 1971; 5(10): 51–8.
3. Jin H, Palta J, Suh TS, Kim S. A generalized a priori dose uncertainty model of IMRT delivery. *Med Phys*. 2008; 35(3): 982–96.
4. Yan D, Xu B, Lockman D, Kota K, Brabbins DS, Wong J, et al. The influence of interpatient and intrapatient rectum variation on external beam treatment of prostate cancer. *Int J Radiat Oncol Biol Phys*. 2001; 51(4): 1111–9.
5. Yan D, Lockman D. Organ/patient geometric variation in external beam radiotherapy and its effects. *Med Phys*. 2001; 28(4): 593–602.
6. Andreo P. On the clinical spatial resolution achievable with protons and heavier charged particle radiotherapy beams. *Phys Med Biol*. 2009; 54(11): N205–15.
7. Berger MJ, Inokuti M, Anderson HH, Bichsel H, Dennis JA, Powers D, Seltzer SM, Turner JE. Report 37: Stopping Powers for Electrons and Positrons. *Journal of the International Commission on Radiation Units and Measurements*, Volume os19, Issue 2, 1 December 1984.
8. Berger MJ, Inokuti M, Andersen HH, Bichsel H, Powers D, Seltzer SM, Thwaites D, Watt DE. Report 49: Stopping Powers and Ranges for Protons and Alpha Particles. *Journal of the International Commission on Radiation Units and Measurements*, Volume os25, Issue 2, 15 May 1993,
9. ICRU. Report 74: Patient Dosimetry for X Rays Used in Medical Imaging. *Journal of the International Commission on Radiation Units and Measurements*, Volume 5, Issue 2, 1 bDec 2005, Pages 1–113.
10. Schneider U, Pedroni E, Lomax A. The calibration of CT Hounsfield units for radiotherapy treatment planning. *Phys Med Biol*. 1996; 41(1): 111–24.
11. Schneider W, Bortfeld T, Schlegel W. Correlation between CT numbers and tissue parameters needed for Monte Carlo simulations of clinical dose distributions. *Phys Med Biol*. 2000; 45(2): 459–78.

12. Woodward HW, White DR. The composition of body tissues. *Br J Radiol.* 1986; 59: 1209–18.

13. White DW, Woodard H; Hammond S. Average soft-tissue and bone models for use in radiation dosimetry. *Br J Radiol.* 1987; 60: 907–13.

14. ICRU. *Tissue Substitutes in Radiation Dosimetry and Measurement* (Report 44). Bethesda, MD: International Commission on Radiation Units and Measurements, 1989.

15. Schaffner B, Pedroni E. The precision of proton range calculations in proton radiotherapy treatment planning: Experimental verification of the relation between CT-HU and proton stopping power. *Phys Med Biol.* 1998; 43(6): 1579–92.

16. Qi ZY, Huang SM, Deng XW. Calibration of CT values used for radiation treatment planning and its impact factors. *Ai Zheng.* 2006; 25(1): 110–4.

17. Kanematsu N, Matsufuji N, Kohno R, Minohara S, Kanai T. A CT calibration method based on the polybinary tissue model for radiotherapy treatment planning. *Phys Med Biol.* 2003; 48(8): 1053–64.

18. Schneider U, Pemler P, Besserer J, Pedroni E, Lomax A, Kaser-Hotz B. Patient specific optimization of the relation between CT-hounsfield units and proton stopping power with proton radiography. *Med Phys.* 2005; 32(1): 195–9.

19. Doolan PJ, Testa M, Sharp G, Bentefour EH, Royle G, Lu HM. Patient-specific stopping power calibration for proton therapy planning based on single-detector proton radiography. *Phys Med Biol.* 2015; 60(5): 1901–17.

20. Doolan PJ, Collins-Fekete CA, Dias MF, Ruggieri TA, D'Souza D, Seco J. Inter-comparison of relative stopping power estimation models for proton therapy. *Phys Med Biol.* 2016; 61(22): 8085–104.

21. Yang M, Zhu XR, Park PC, Titt U, Mohan R, Virshup G, et al. Comprehensive analysis of proton range uncertainties related to patient stopping-power-ratio estimation using the stoichiometric calibration. *Phys Med Biol.* 2012; 57(13): 4095–115.

22. Bichsel H. Passage of charged particles through matter. In: Gray E, editor. *American Institute of Physics Handbook.* New York, NY: McGraw-Hill, 1972.

23. Janni J. Energy loss, range, path length, time-of-flight, straggling, multiple scattering, and nuclear interaction probability: In Two Parts. Part 1. For 63 Compounds Part 2. For Elements $1 \leq Z \leq 92$. *Atomic Data and Nuclear Data Tables*, Volume 27, Issues 4–5, July–September 1982, Pages 341–529.

24. Hanson KM, Bradbury JN, Cannon TM, Hutson RL, Laubacher DB, Macek RJ, et al. Computed tomography using proton energy loss. *Phys Med Biol.* 1981; 26(6): 965–83.

25. Schneider U, Pedroni E. Proton radiography as a tool for quality control in proton therapy. *Med Phys.* 1995; 22(4): 353–63.

26. Zygmanski P, Gall KP, Rabin MS, Rosenthal SJ. The measurement of proton stopping power using proton-cone-beam computed tomography. *Phys Med Biol.* 2000; 45(2): 511–28.

27. Sadrozinski H. Issues in proton computed tomography *Nucl Instrum Methods Phys Res.* 2003; 511: 275–81.

28. Schulte RW, Bashkirov V, Klock MC, Li T, Wroe AJ, Evseev I, et al. Density resolution of proton computed tomography. *Med Phys.* 2005; 32(4): 1035–46.

29. Schulte R. Conceptual design of a proton computed tomography system for applications in proton radiation therapy *IEEE Trans Nucl Sci.* 2004; 51: 866–72.

30. Hurley RF, Schulte RW, Bashkirov VA, Wroe AJ, Ghebremedhin A, Sadrozinski HF, et al. Water-equivalent path length calibration of a prototype proton CT scanner. *Med Phys.* 2012; 39(5): 2438–46.

31. Testa M, Verburg JM, Rose M, Min CH, Tang S, Bentefour el H, et al. Proton radiography and proton computed tomography based on time-resolved dose measurements. *Phys Med Biol.* 2013; 58(22): 8215–33.

32. Esposito M, Anaxagoras T, Evans PM, Green S, Manolopoulos S, Nieto Camero J, Parker D, Poludniowski G, Price T, Waltham C, Allinson N. CMOS Active Pixel Sensors as energy-range detectors for proton Computed Tomography. *Journal of Instrumentation.* 2015; 10. C06001.

33. Paganetti H. Range uncertainties in proton therapy and the role of Monte Carlo simulations. *Phys Med Biol.* 2012; 57(11): R99–117.
34. Hunemohr N, Krauss B, Tremmel C, Ackermann B, Jakel O, Greilich S. Experimental verification of ion stopping power prediction from dual energy CT data in tissue surrogates. *Phys Med Biol.* 2014; 59(1): 83–96.
35. Yang M, Virshup G, Clayton J, Zhu XR, Mohan R, Dong L. Theoretical variance analysis of single- and dual-energy computed tomography methods for calculating proton stopping power ratios of biological tissues. *Phys Med Biol.* 2010; 55(5): 1343–62.
36. Mohler C, Wohlfahrt P, Richter C, Greilich S. Range prediction for tissue mixtures based on dual-energy CT. *Phys Med Biol.* 2016; 61(11): N268–75.
37. Hong L, Goitein M, Bucciolini M, Comiskey R, Gottschalk B, Rosenthal S, et al. A pencil beam algorithm for proton dose calculations. *Phys Med Biol.* 1996; 41(8): 1305–30.
38. Szymanowski H, Mazal A, Nauraye C, Biensan S, Ferrand R, Murillo MC, et al. Experimental determination and verification of the parameters used in a proton pencil beam algorithm. *Med Phys.* 2001; 28(6): 975–87.
39. Schaffner B. Proton dose calculation based on in-air fluence measurements. *Phys Med Biol.* 2008; 53(6): 1545–62.
40. Sawakuchi GO, Titt U, Mirkovic D, Mohan R. Density heterogeneities and the influence of multiple Coulomb and nuclear scatterings on the Bragg peak distal edge of proton therapy beams. *Phys Med Biol.* 2008; 53(17): 4605–19.
41. Paganetti H, Jiang H, Parodi K, Slopsema R, Engelsman M. Clinical implementation of full Monte Carlo dose calculation in proton beam therapy. *Phys Med Biol.* 2008; 53(17): 4825–53.
42. Yamashita T, Akagi T, Aso T, Kimura A, Sasaki T. Effect of inhomogeneity in a patient's body on the accuracy of the pencil beam algorithm in comparison to Monte Carlo. *Phys Med Biol.* 2012; 57(22): 7673–88.
43. Schuemann J, Dowdell S, Grassberger C, Min CH, Paganetti H. Site-specific range uncertainties caused by dose calculation algorithms for proton therapy. *Phys Med Biol.* 2014; 59(15): 4007–31.
44. Schuemann J, Giantsoudi D, Grassberger C, Moteabbed M, Min CH, Paganetti H. Assessing the clinical impact of approximations in analytical dose calculations for proton therapy. *Int J Radiat Oncol Biol Phys.* 2015; 92(5): 1157–64.
45. Paganetti H. Monte Carlo simulations will change the way we treat patients with proton beams today. *Br J Radiol.* 2014; 87(1040): 20140293.
46. Taylor PA, Kry SF, Followill DS. Pencil beam algorithms are unsuitable for proton dose calculations in lung. *Int J Radiat Oncol Biol Phys.* 2017; 99(3): 750–6.
47. Grassberger C, Daartz J, Dowdell S, Ruggieri T, Sharp G, Paganetti H. Quantification of proton dose calculation accuracy in the lung. *Int J Radiat Oncol Biol Phys.* 2014; 89(2): 424–30.
48. ICRU. *Prescribing, Recording, and Reporting Proton Beam Therapy*, ICRU Report 78. Bethesda, MD: Journal of the ICRU, 2007.
49. Urie M, Goitein M, Holley WR, Chen GT. Degradation of the Bragg peak due to inhomogeneities. *Phys Med Biol.* 1986; 31(1): 1–15.
50. Paganetti H. Relative biological effectiveness (RBE) values for proton beam therapy. Variations as a function of biological endpoint, dose, and linear energy transfer. *Phys Med Biol.* 2014; 59(22): R419–72.
51. Wouters BG, Lam GK, Oelfke U, Gardey K, Durand RE, Skarsgard LD. Measurements of relative biological effectiveness of the 70 MeV proton beam at TRIUMF using Chinese hamster V79 cells and the high-precision cell sorter assay. *Radiat Res.* 1996; 146(2): 159–70.
52. Robertson JB, Williams JR, Schmidt RA, Little JB, Flynn DF, Suit HD. Radiobiological studies of a high-energy modulated proton beam utilizing cultured mammalian cells. *Cancer.* 1975; 35(6): 1664–77.

53. Carabe A, Moteabbed M, Depauw N, Schuemann J, Paganetti H. Range uncertainty in proton therapy due to variable biological effectiveness. *Phys Med Biol*. 2012; 57(5): 1159–72.
54. Matsumoto Y, Matsuura T, Wada M, Egashira Y, Nishio T, Furusawa Y. Enhanced radiobiological effects at the distal end of a clinical proton beam: in vitro study. *J Radiat Res*. 2014; 55(4): 816–22.
55. Gerweck LE, Kozin SV. Relative biological effectiveness of proton beams in clinical therapy. *Radiother Oncol*. 1999; 50(2): 135–42.
56. Paganetti H, Goitein M. Radiobiological significance of beamline dependent proton energy distributions in a spread-out Bragg peak. *Med Phys*. 2000; 27(5): 1119–26.
57. Zlobinskaya O, Siebenwirth C, Greubel C, Hable V, Hertenberger R, Humble N, et al. The effects of ultra-high dose rate proton irradiation on growth delay in the treatment of human tumor xenografts in nude mice. *Radiat Res*. 2014; 181(2):177–83.
58. Urano M, Verhey LJ, Goitein M, Tepper JE, Suit HD, Mendiondo O, et al. Relative biological effectiveness of modulated proton beams in various murine tissues. *Int J Radiat Oncol Biol Phys*. 1984; 10(4): 509–14.
59. Timmerman RX, Xing L. *Image-Guided And Adaptive Radiation Therapy*. Philadelphia, PA: Lippincott Williams & Wilkins, 2009.
60. Brock KK. Deformable Registration Accuracy C. Results of a multi-institution deformable registration accuracy study (MIDRAS). *Int J Radiat Oncol Biol Phys*. 2010; 76(2): 583–96.
61. Jakel O, Reiss P. The influence of metal artefacts on the range of ion beams. *Phys Med Biol*. 2007; 52(3): 635–44.
62. Herrmann R, Carl J, Jakel O, Bassler N, Petersen JB. Investigation of the dosimetric impact of a Ni-Ti fiducial marker in carbon ion and proton beams. *Acta Oncol*. 2010; 49(7): 1160–4.
63. Newhauser WD, Koch NC, Fontenot JD, Rosenthal SJ, Gombos DS, Fitzek MM, et al. Dosimetric impact of tantalum markers used in the treatment of uveal melanoma with proton beam therapy. *Phys Med Biol*. 2007; 52(13): 3979–90.
64. Cheung J, Kudchadker RJ, Zhu XR, Lee AK, Newhauser WD. Dose perturbations and image artifacts caused by carbon-coated ceramic and stainless steel fiducials used in proton therapy for prostate cancer. *Phys Med Biol*. 2010; 55(23): 7135–47.
65. Habermehl D, Henkner K, Ecker S, Jakel O, Debus J, Combs SE. Evaluation of different fiducial markers for image-guided radiotherapy and particle therapy. *J Radiat Res*. 2013; 54(Suppl 1): i61–8.
66. Antolak JA, Rosen, II, Childress CH, Zagars GK, Pollack A. Prostate target volume variations during a course of radiotherapy. *Int J Radiat Oncol Biol Phys*. 1998; 42(3): 661–72.
67. Kudchadker RJ, Lee AK, Yu ZH, Johnson JL, Zhang L, Zhang Y, et al. Effectiveness of using fewer implanted fiducial markers for prostate target alignment. *Int J Radiat Oncol Biol Phys*. 2009; 74(4): 1283–9.
68. Nichol AM, Brock KK, Lockwood GA, Moseley DJ, Rosewall T, Warde PR, et al. A magnetic resonance imaging study of prostate deformation relative to implanted gold fiducial markers. *Int J Radiat Oncol Biol Phys*. 2007; 67(1): 48–56.
69. Deurloo KE, Steenbakkers RJ, Zijp LJ, de Bois JA, Nowak PJ, Rasch CR, et al. Quantification of shape variation of prostate and seminal vesicles during external beam radiotherapy. *Int J Radiat Oncol Biol Phys*. 2005; 61(1): 228–38.
70. Olsen JR, Noel CE, Baker K, Santanam L, Michalski JM, Parikh PJ. Practical method of adaptive radiotherapy for prostate cancer using real-time electromagnetic tracking. *Int J Radiat Oncol Biol Phys*. 2012; 82(5): 1903–11.
71. Zhang X, Dong L, Lee AK, Cox JD, Kuban DA, Zhu RX, et al. Effect of anatomic motion on proton therapy dose distributions in prostate cancer treatment. *Int J Radiat Oncol Biol Phys*. 2007; 67(2): 620–9.

72. Ohta K, Shimohira M, Sasaki S, Iwata H, Nishikawa H, Ogino H, et al. Transarterial fiducial marker placement for image-guided proton therapy for malignant liver tumors. *Cardiovasc Intervent Radiol.* 2015; 38(5): 1288–93.

73. Kubiak T. Particle therapy of moving targets-the strategies for tumour motion monitoring and moving targets irradiation. *Br J Radiol.* 2016; 89(1066): 20150275.

74. ICRU. *ICRU Report 62: Prescribing, Recording and Reporting Photon Beam Therapy* (supplement to ICRU report 50). Bethesda, MD: International Commission on Radiation Units and Measurements, 1999.

75. ICRU. *ICRU Report 50: Prescribing, Recording, and Reporting Photon Beam Therapy.* Bethesda, MD: International Commission on Radiation Units and Measurements, 1993.

76. van Herk M, Remeijer P, Rasch C, Lebesque JV. The probability of correct target dosage: dose-population histograms for deriving treatment margins in radiotherapy. *Int J Radiat Oncol Biol Phys.* 2000; 47(4): 1121–35.

77. Moyers MF, Miller DW, Bush DA, Slater JD. Methodologies and tools for proton beam design for lung tumors. *Int J Radiat Oncol Biol Phys.* 2001; 49(5): 1429–38.

78. Goitein M. Calculation of the uncertainty in the dose delivered during radiation therapy. *Med Phys.* 1985; 12(5): 608–12.

79. Leong J. Implementation of random positioning error in computerised radiation treatment planning systems as a result of fractionation. *Phys Med Biol.* 1987; 32(3): 327–34.

80. Jin H, Palta JR, Kim YH, Kim S. Application of a novel dose-uncertainty model for dose-uncertainty analysis in prostate intensity-modulated radiotherapy. *Int J Radiat Oncol Biol Phys.* 2010; 78(3): 920–8.

81. Lomax AJ, Boehringer T, Coray A, Egger E, Goitein G, Grossmann M, et al. Intensity modulated proton therapy: A clinical example. *Med Phys.* 2001; 28(3): 317–24.

82. Pflugfelder D, Wilkens JJ, Szymanowski H, Oelfke U. Quantifying lateral tissue heterogeneities in hadron therapy. *Med Phys.* 2007; 34(4): 1506–13.

83. Lomax AJ. Intensity modulated proton therapy and its sensitivity to treatment uncertainties 1: The potential effects of calculational uncertainties. *Phys Med Biol.* 2008; 53(4): 1027–42.

84. Kraan AC, van de Water S, Teguh DN, Al-Mamgani A, Madden T, Kooy HM, et al. Dose uncertainties in IMPT for oropharyngeal cancer in the presence of anatomical, range, and setup errors. *Int J Radiat Oncol Biol Phys.* 2013; 87(5): 888–96.

85. Unkelbach J, Bortfeld T, Martin BC, Soukup M. Reducing the sensitivity of IMPT treatment plans to setup errors and range uncertainties via probabilistic treatment planning. *Med Phys.* 2009; 36(1): 149–63.

86. Liu W, Frank SJ, Li X, Li Y, Park PC, Dong L, et al. Effectiveness of robust optimization in intensity-modulated proton therapy planning for head and neck cancers. *Med Phys.* 2013; 40(5): 051711.

87. Lowe M, Albertini F, Aitkenhead A, Lomax AJ, MacKay RI. Incorporating the effect of fractionation in the evaluation of proton plan robustness to setup errors. *Phys Med Biol.* 2016; 61(1): 413–29.

88. Lowe M, Aitkenhead A, Albertini F, Lomax AJ, MacKay RI. A robust optimisation approach accounting for the effect of fractionation on setup uncertainties. *Phys Med Biol.* 2017; 62(20): 8178–96.

89. Oden J, Eriksson K, Toma-Dasu I. Incorporation of relative biological effectiveness uncertainties into proton plan robustness evaluation. *Acta Oncol.* 2017; 56(6): 769–78.

90. Giantsoudi D, Grassberger C, Craft D, Niemierko A, Trofimov A, Paganetti H. Linear energy transfer-guided optimization in intensity modulated proton therapy: Feasibility study and clinical potential. *Int J Radiat Oncol Biol Phys.* 2013; 87(1): 216–22.

91. Giantsoudi D, Sethi RV, Yeap BY, Eaton BR, Ebb DH, Caruso PA, et al. Incidence of CNS injury for a cohort of 111 patients treated with proton therapy for medulloblastoma: LET and RBE associations for areas of injury. *Int J Radiat Oncol Biol Phys.* 2016; 95(1): 287–96.

92. Giantsoudi D, Adams J, MacDonald SM, Paganetti H. Proton treatment techniques for posterior fossa tumors: Consequences for linear energy transfer and dose-volume parameters for the brainstem and organs at risk. *Int J Radiat Oncol Biol Phys*. 2017; 97(2): 401–10.

93. Unkelbach J, Botas P, Giantsoudi D, Gorissen BL, Paganetti H. Reoptimization of intensity modulated proton therapy plans based on linear energy transfer. *Int J Radiat Oncol Biol Phys*. 2016; 96(5): 1097–106.

Precision and Uncertainties for Moving Targets*

Christoph Bert

Universitätsklinikum Erlangen

Martijn Engelsman

Delft University of Technology

Antje C. Knopf

University of Groningen

CONTENTS

* Colour figures available online at www.crcpress.com/9781138626508

18.1 INTRODUCTION

Time-dependent variation in the target location and the patient density distribution can occur on different timescales; i.e., over the entire treatment course (weight loss/gain, treatment response of the target), between subsequent treatment fractions (interfractional motion like setup variations or changes in cavity fillings), and within a treatment fraction (intrafractional motion due to heartbeat, breathing, or relaxation). Some examples of density variations are shown in Figure 18.1. All of these possible geometrical and anatomical changes will be briefly discussed in the first section of this chapter "Motion from a Clinical Perspective." In the section "Magnitude of the Dosimetric Effect of Target Motion" the dosimetric consequences of motion, especially for treatments delivered with scanned proton beams, will be elaborated. In the section "Dealing with Motion," we will subsequently discuss how to effectively deal with each type of motion. In the section "Quality Assurance for Moving Tumors" methods to analyze the sensitivity of treatments towards density

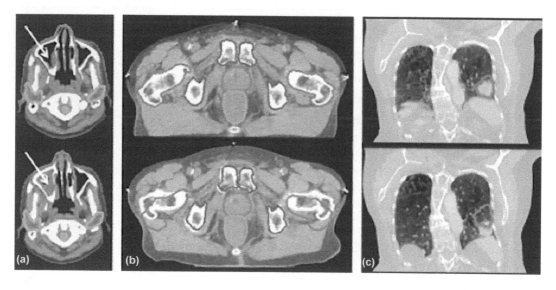

FIGURE 18.1 Motion and density variation as a function of time. (a) Two CT slices acquired a few weeks apart as an example of density variation over the course of the treatment, as indicated by the white arrow. (Courtesy of Francesca Albertini, Paul Scherrer Institute.) (b) Two CT slices for a prostate cancer patient on different treatment days as an example of interfractional variation. The variation in femoral head rotation affects the dose delivered by the typically applied lateral beam directions. (Courtesy of Lei Dong, MD Anderson Cancer Treatment Center, Dallas, TX.) (c) Two breathing phases for a lung cancer as an example of intrafractional variation. The red contours indicate the target position when in the exhale phase.

variations, motion, and interplay effects are elaborated. The chapter will conclude with an outlook on future short- and long-term perspectives on the treatment of moving targets. Recently, the Particle Therapy Co-Operative Group (PTCOG) thoracic and lymphoma subcommittee published consensus guidelines for implementing pencil beam scanning (PBS) Proton Therapy for Thoracic Malignancies that review the current clinical status [1].

18.2 MOTION FROM A CLINICAL PERSPECTIVE

18.2.1 Over the Course of Treatment

"Motion" describes gradual, but systematic, variations in the patient-density distribution. These can occur either due to a redistribution of density within the patient or due to the addition or disappearance of matter. Examples of such variations are tumor shrinkage and growth [2], weight gain or loss, increase or decrease in lung density, and systematic variation in bowel and rectal filling. Redistribution of density includes changes in the motion pattern such as breathing trajectory or changes in the baseline [3]. McDermott et al. [4] reported on the use of portal imaging to detect anatomy changes in head and neck, prostate, and lung cancer. Barker et al. [5] reported on the use of an integrated computed tomography (CT)-linear accelerator system to track tumor shrinkage, edema, and overall weight loss.

18.2.2 Interfractional Motion

Interfractional density variations can be as large as the gradual variations over the entire treatment course, but they have a more random character. Examples are variations in the amplitude of breathing, in the average tumor position over the breathing cycle, and daily variation in rectal, bowel, and bladder filling. For prostate tumors, daily variation in the rotation of the femoral heads and corresponding variation in the patient exterior surface may affect the proton range within the patient. Langen and Jones [6] provide an extensive overview of interfractional position variation.

18.2.3 Intrafractional Motion

Much data has been published on the extent of intrafractional motion. Langen and Jones [6] provide an overview of many of these studies. Without attempting to be complete, we will discuss motion parameters for a few tumor sites.

For lung tumors, motion is typically largest for those tumors close to the diaphragm and typically largest in the superior–inferior direction. Although peak-to-peak motion of 30 mm can be observed, the typical motion is less than 10 mm (65% of patients in the study of Sonke et al. [3]). Hysteresis in the motion trajectory can be observed [7] as well as a baseline drift over the duration of the treatment fraction [8]. Intrafractional variations in proton beam penetration can be especially large for tumors near the diaphragm or heart. As the liver is relatively close to the diaphragm, peak-to-peak motion with an average of 17 mm has been observed for these tumors under normal breathing conditions [9]. Also for cardiac sarcomas, the heartbeat causes both tumor excursion and, depending on the choice of beam angles, a large variation (up to several centimeters) in water-equivalent depth of the tumor. Intrafractional motion of the prostate is mainly a consequence of moving gas and feces and is typically less than a few millimeter (1 standard deviation) [10], but the effects on proton penetration can be quite severe. Intrafractional effects also depend on the specific positioning and immobilization devices that are used [11].

18.3 MAGNITUDE OF THE DOSIMETRIC EFFECT OF TARGET MOTION

In photon radiotherapy, great effort is made by means of image-guided adaptive radio-therapy (e.g., [12]), to ensure translational and rotational alignment of the target with the treatment fields. Density variations are less of a concern. They have only a minor effect on the dose distribution of the patient, as the photon depth-dose distribution is rather shallow and insensitive to density variations. For a proton beam, such (motion-induced) density variations have a large dosimetric effect, as it can mean the difference, for example, between 100% and 0% of the prescribed dose for the distal target edge (see Figure 18.2).

Interplay effects in intensity modulated radiotherapy (IMRT) have limited dosimetric consequences if a sufficient number of treatment fractions are delivered (e.g., [13]). In pho-ton therapy, one can therefore estimate the additional safety margin needed for treating moving tumors by a rather straightforward blurring of the dose distribution inside the patient (e.g., [14–16]).

In proton therapy, the finite range and steep dose gradients make it much more com-plex to deliver an accurate treatment for moving tumors [17,18]. Safety margins for pro-ton treatments should consider changes of water-equivalent path length in addition [19]. Due to the finite range, density variations over time as well as the interplay effect can have severe dosimetric consequences that cannot be addressed solely by safety margins. The next two sections describe dosimetric consequences of motion and highlight their magnitude.

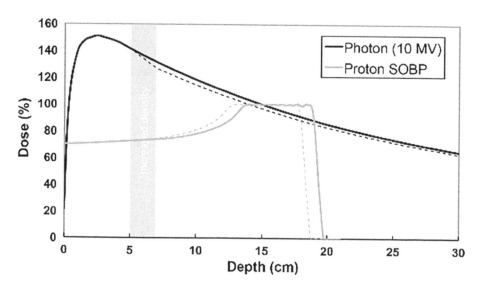

FIGURE 18.2   Central axis percent depth-dose (PDD) curves for a 10-MV photon beam and a pro-ton spread-out Bragg peak. For each modality, the thin dashed lines are the PDD for the situation of an increased density being present at a depth from 5 to 7 cm.

18.3.1 Density Variations

In the case of single-field uniform dose (SFUD) proton therapy, density variations (without interplay effect) affect the dose near the proximal and distal edge of the target (see Figure 18.3). For intensity-modulated proton therapy (IMPT), these density variations may furthermore have dosimetric consequences within the target volume because of mismatches of the possibly very steep dose gradients of the individual fields.

Engelsman et al. [20] used a simplified phantom geometry and beam setup to analyze the effect of setup errors and intrafractional lung tumor motion and concluded that significant dosimetric disturbances can occur if density variations are not proactively taken into account in the design of the treatment plan. Kang et al. [21] analyzed a variety of forward planning approaches for lung tumors in real patient geometries under intrafractional motion and showed that none of the planning approaches accurately predicted the actually delivered dose distribution in the target (planning target volume (PTV)). Mori et al. [17] analyzed the variation in radiological path length over the breathing cycle for a set of 11 lung cancer patients. Figure 18.4 shows an example of range variation for a single beam direction aimed at a lung tumor. Hui et al. [22] analyzed the effect of interfractional motion and anatomic changes for lung cancer patients by means of weekly 4DCT scans and dose recalculation at the end-inhale and end-exhale phases and indicated that adaptive replanning should be considered in selected patients.

In current clinical practice, the changing density distribution is taken into consideration at the time of treatment plan design based on the experience of the dosimetrist or medical physicist. This approach leads to acceptable dose distributions, but several studies showed that 4D treatment plan assessment and/or adaptive planning approaches

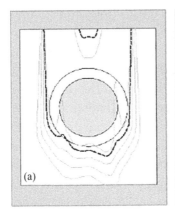
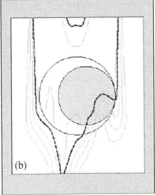

FIGURE 18.3 Phantom simulating a unit density tumor (gray circle) in low-density lung tissue (white). The thin black circle indicates a 10-mm 3D geometric expansion of the tumor into the PTV (see Chapters 10 and 11 for the applicability of the PTV for proton therapy). (a) The dashed lines are isodose lines (50%, 80%, 90%, 95%, 100%) for a single passively scattered proton field, incident from the top and designed to cover the PTV. (b) Isodose lines for a 10-mm displacement of the tumor with respect to the treatment beam. The tumor remains within the PTV but is underdosed, whereas the dose to certain parts of the lung is increased.

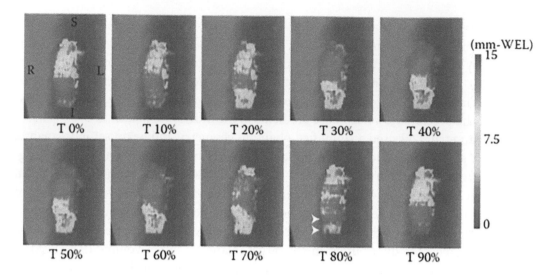

FIGURE 18.4 Range fluctuation for a posterior to anterior beam direction aimed at a tumor moving in lung. The range variation is expressed in water-equivalent length and superimposed on 4D CT images. (From Mori et al. [17].)

could further improve dose coverage of the target volume and reduce dose to organ at risk (OAR).

18.3.2 Interplay

The interplay effect describes the detrimental effect on the dose distribution for moving targets if delivering a (dosimetric) fraction of the intended dose distribution is not near-instantaneous, as for example in PBS. These interplay effects can occur due to any motion not mitigated by motion reduction (e.g., gating, breath-hold) and for any remaining motion within a gating window. SFUD proton therapy, even if delivered by means of fast spinning range modulator wheels and not by stationary ridge filters, does not suffer from interplay effects because the motion of the beam is much faster (e.g., periods of 100 ms) than target motion. Interplay effects can, however, be substantial in the case of layer-stacking [23,24] and, more generally, in the case of PBS. This section will not discuss layer stacking but will focus on interplay effects for PBS only. Figure 18.5 shows an example of the possible severity of the interplay effect for a single-field irradiation of a moving lung tumor.

A PBS dose distribution is typically delivered (and treatment planned) by grouping the necessary spot positions in range layers (isoenergy layers). Although dose delivery within a range layer can be near instantaneous, the time needed to switch between layers can give rise to interplay effects, since the target motion phase changes between individual layers.

Interplay effects for PBS have long been recognized to be a potential issue for proton radiotherapy [18,25–31] and are one of the reasons why PBS is currently restrainedly used in a clinical setting to treat patients with moving tumors. Interplay affects the dose at the tumor edge and within the tumor (e.g., [18]). Exactly how interplay affects the dose distribution depends on the target motion characteristics as well as the characteristics of the dose delivery system, such as the choice of scan pattern and the scanning speed.

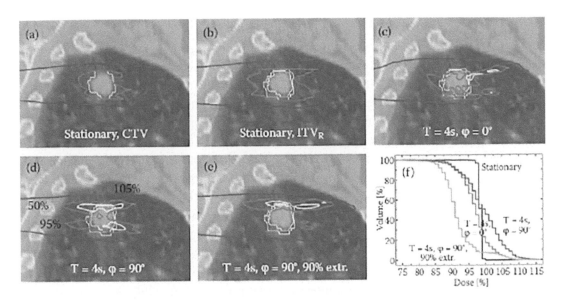

FIGURE 18.5 Isodose distributions for a single PBS irradiation of a moving lung tumor showing the interplay effect between dose delivery and tumor motion. The blue, red, and yellow lines indicate the 50%, 95%, and 105% isodose levels, respectively. (a and b) Stationary dose distributions for treatment plans to the clinical target volume (CTV) and to an internal target volume accounting for respiratory motion only (ITVr [117]). (c) Dose distribution for a breathing period of $T = 4\,s$ with the beam starting at the inhale breathing phase ($\varphi = 0°$). The interplay pattern is sensitive to changes in initial breathing phase, (c) versus (d), and to changes in the scan speed, (d) versus (e), simulated in this case by changing the extraction rate from the synchrotron to 90% of the initial value. (f) Corresponding DVHs of the target. (Reprinted from Bert et al. [28].)

Nowadays, the effects of motion for actual patient geometries have been shown in several studies [28,32–35]. They conclude that especially for small fraction numbers (i.e., hypofractionation), beam arrangements perpendicular to the main motion direction, small spot sizes, and motion amplitudes above 5 mm treatment of moving targets with scanned particle beams require motion mitigation strategies such as rescanning, gating, or tracking. For robustly optimized treatment plans on the other hand, setup and range uncertainties, breathing motion, and interplay effects have only limited impact on target coverage, dose homogeneity, and OAR dose parameters.

18.4 DEALING WITH MOTION

There are a variety of strategies available to mitigate the effects of motion in proton radiotherapy. The best approach to deal with motion may differ for different tumor sites and is likely a combination of some of the strategies described later. Although some of these strategies act only at the time of treatment delivery, all of them have to be taken into account already at the time of treatment plan design. Gating, for example, controls the patient and target position inside the treatment room, but the treatment plan has to be designed to the correct phase(s) of a 4DCT scan. Residual uncertainty in the target position (e.g., within the gating window) also has to be addressed already during treatment planning.

18.4.1 The Importance of Motion Monitoring

Most of the strategies discussed in the following sections, e.g., beam gating and especially beam tracking, depend heavily on accurate assessment of the time-dependent target position and patient geometry. Although motion monitoring is not unique to proton therapy, it should be clear that the importance of motion monitoring, when compared to photon radiotherapy, is much more important. We will therefore provide a brief overview of the various motion monitoring strategies.

Motion monitoring can be as simple as observing if the patient indeed holds his breath during beam-on, for example, in deep-inspiration breath-hold (e.g., [36]). It can also be as technologically advanced as real-time image-guided radiotherapy by means of integrating an magnetic resonance imaging (MRI) scanner into the treatment room (e.g., [37]), although this has not yet been attempted for proton therapy. Other strategies include the 4D tracking of implanted markers (e.g., [38]) and motion phase/amplitude monitoring by means of external markers. This latter strategy relies on assuming or measuring a correlation between external markers and internal target position (e.g., [39]). A hybrid solution that does not require ionizing radiation for measuring the position of implanted fiducials is described by Shirato et al. [38], but that can provide comparable accuracy, is model-based prediction of internal target motion. In these approaches, a model is trained by parallel observation of internal motion (e.g., based on implanted fiducials) and an external motion surrogate, such as the height of the chest wall. During treatment delivery, the model then predicts the internal target position based on a frequent assessment of the external surrogate position. In case of longer irradiations the model might be updated or checked by additional measurements of the internal motion. Clinically, such a model approach is used in the Cyberknife Synchrony [40].

Accurate motion monitoring is not only limited to the treatment room but also should obviously take place during the treatment preparation phase; e.g., motion assessment during 4DCT scanning. Ideally, the same motion monitoring system is used both at the treatment preparation phase and during treatment delivery. A more comprehensive overview about the importance of imaging and image guidance in advanced particle therapy from a clinical point of view can be found in [41].

18.4.2 General Treatment Planning Considerations

During treatment planning one can (try to) address uncertainties in the actually delivered dose distribution as a consequence of a time-varying geometry. A simple approach is to choose beam directions that minimize the effect of motion. For example, one should choose not to use beam directions that skim the heart or that treat through the diaphragm. The resulting variation in local proton range affects the required distal safety margin as well as the probability to spare OAR distal to the target. In choosing beam directions, it may also be valuable to increase the number of fields in the treatment plan/treatment fraction, as this averages out the dose uncertainties of single beams. Knopf et al. [32] investigated the importance of the number and direction of fields in a treatment plan. They propose the use of multiple-field plans for which field directions are chosen as much as possible parallel to the main motion direction.

With beam directions chosen, it is still a challenge to design a plan that ensures target coverage despite (remaining) motion. Chapter 15 provides a few approaches for the case of SFUD delivered by means of passive-scattering proton therapy (PSPT). Typically, in PSPT one performs forward (or manual) treatment planning and chooses beam parameters (range, modulation width, and range compensator shape) that either ensure target coverage in every separate breathing phase or that will likely ensure target coverage when the dose is accumulated over the motion cycle. SFUD using PBS is similar, but next to providing target coverage under density variation, it furthermore requires analysis of the interplay effect e.g., by multiple 4D dose calculations with varying motion parameters.

Bert et al. [30] suggested, for PBS and limited target motion, to optimize the spot size (Bragg peak shape) and the distance between spots (energy layers). They show that increasing the overlap between pencil beams provides a sort of "intrinsic" rescanning. As more pencil beams contribute to the dose at each position in the target, no additional rescanning may be needed. Within the range layer, decreasing the distance between spots or increasing the spot size increases the pencil beam overlap. In the latter case, the lateral beam penumbra may, however, be negatively affected. In the direction parallel to the beam, one can decrease the distance between range layers and potentially also the shape of the Bragg peak by, for example, incorporating a small ripple filter to broaden the peak. The effectiveness of the strategy of increasing the overlap of spots is, however, reduced at very high-dose gradients, where some spots will deliver no or very little dose.

IMPT, by definition, adds the ability to optimize the dose distribution over multiple beams angles all at once. In case of static tumors and no uncertainties, the automatic optimization of IMPT plans allows the use of pencil beams that are most beneficial in geometrically sparing OAR. With range and setup uncertainties and, more important for this section, a variable density geometry, robust optimization (Chapter 19) will furthermore allow the use of those pencil beams that suffer least from density variations.

Regardless of the use of SFUD or IMPT, one should not expect the treatment plan on the static CT scan to appear perfectly conformal to the target. Target motion and other density variations add to the necessary distal overshoot and proximal undershoot already needed to ensure target coverage for range uncertainties and setup errors.

Choosing the number and direction of beams wisely, and robust optimization, can help mitigate the effects of both (remaining) intrafractional motion (e.g., for lung cancer) and interfractional density variations (e.g., for prostate cancer). Increasing pencil beam overlap reduces the effect of intrafractional motion.

18.4.3 Robust Optimization/4D Optimization
Plan robustness concerns the sensitivity of a treatment plan to potential disturbing effects [42]. Several factors can contribute to the deterioration of proton treatment plans. Setup and range errors can result in an underdosage of the target or an overdosage of nearby healthy structures. Machine uncertainties can result in deviations between the planed and delivered dose distribution. And in the case of moving targets, interplay effects can occur. To enable a safe treatment of targets in the thorax, treatment plans are required to be robust against all these uncertainties.

Robust optimization is used to incorporate uncertainties such as patient positioning in the treatment planning process. Four-dimensional robust optimization additionally incorporates tumor motion and its consequences into the optimization process. More details about treatment plan optimization are provided in Chapter 19. Four-dimensional robust optimization can be seen as a motion mitigation option. It makes specific use of the motion, and it is interpreted as an additional degree of freedom, which can, e.g., help to avoid dose to OAR as indicated in Figure 18.6 [43].

Four-dimensional robust optimization has been investigated in photon beam therapy for a number of years [44–49]. In scanned particle therapy, however, the methodology has been addressed only recently and by a few research teams. Graeff et al. [50] developed and tested in phantom experiments a 4D robust optimization method, in which the target is virtually split into multiple sectors. Each of those sectors is linked to a motion phase of the target volume and targeted by a dedicated raster field. This approach is currently restricted to the temporal resolution of the originating 4D imaging study (typically 8–10 phases through a single breathing cycle). As an alternative, the same group worked on a type of "temporal tracking", in which each pencil beam of the field is assigned to a particular motion phase [43,51]. Recently, a method for 4D robust optimization based on the union of the gross tumor volume (GTV) in all motion phases was proposed [52]. Motion is dealt by so-called uncertainty scenarios. As in robust optimization for stationary targets the proposed optimization technique does not require synchronization with the patient's motion during treatment delivery. The latter approaches are based on the 4DCT and thus limited in their temporal resolution and assuming that the 4DCT represents the patient's moving anatomy also during treatment delivery. For irregular breathing patterns, this will not necessarily be the case. Bernatowicz et al. [53] described an alternative 4D optimization method for PBS proton therapy, which uses a 4D dose calculation based on a deforming

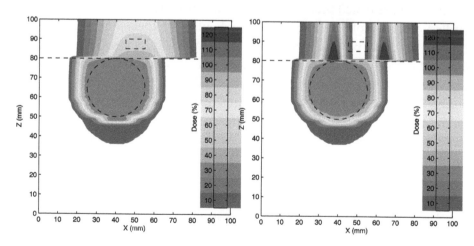

FIGURE 18.6 Simulated irradiation of a spherical target volume (dashed circular line) with an anterior quadratic OAR. In stationary conditions (left), the target volume is covered with therapeutical dose level but the OAR is irradiated with >80% of the prescribed dose. In the presence of organ motion (right), the motion of the target can be compensated by tracking, and the OAR will be spared if 4D-optimized treatment plans are used. Full details are reported by Eley et al. [43].

calculation grid. Using such a calculation, the optimization approach is not necessarily bound to discrete motion phases and can incorporate irregular motion, e.g., from 4DCT-MRI [54], and detailed delivery dynamics of the treatment machine directly into the optimization process. As such, and in analogy to "direct" optimization techniques in IMRT or volumetric intensity modulated arc therapy (VMAT) [55–57], this approach can be considered as a "direct" 4D optimization approach, which potentially could provide additional flexibility to the 4D optimization process. Four-dimensional robust optimization is currently finding its way into the clinic at scanning proton facilities and its clinical feasibility will be shown in the coming years.

The differences between the planned and delivered dose distributions can be assessed before treatment by evaluating the robustness of the treatment plan. Previous studies often analyzed and reported the robustness of plans towards individual uncertainties. For example, as mentioned earlier, several authors assessed specifically the impact of interplay effects on dose distributions [28,32,33,35,58,59]. Inoue et al. [34] developed a robustness evaluation tool considering the impact of setup and range uncertainties, breathing motion, and interplay. Similarly, Dowdell et al. [60] studied the influence of motion effects and setup uncertainties in fractionated lung IMPT and Tang et al. [61] analyzed the effect of intrafractional prostate motion based on Calypso data.

Besides in silico studies, the development of dynamic, anthropomorphic, thorax phantoms [62–64] and the execution of 4D dose verifications experiments will help to realistically determine the robustness of scanned proton treatments to moving targets. Richter et al. showed the feasibility of 4D dose reconstructions in phantom settings, which served as a validation of the corresponding 4D treatment planning code [65]. The infrastructure was then used for 4D dose reconstruction in hepatocellular carcinoma (HCC) treatments with a scanned ion beam [66] and also in preclinical studies addressing radiation therapy of atrial fibrillation [67].

Even in the most complex setting of the studies mentioned earlier, not all combinations of all uncertainties are incorporated. However, the studies showed that taking into account possibly occurring uncertainties in combination can lead to significantly different clinical decision concerning the application of motion mitigation techniques than looking at the effects individually. It is thus recommended to combine as many as reasonably possible of the clinically present uncertainties in a simulation, which should ideally be performed for the decision making of each patient's treatment. The latter is only feasible with highly integrated plan robustness modules into clinical treatment planning systems that can be validated, e.g., by dose calculations against measurements.

18.4.4 Breath-Hold and Mechanical Motion Suppression

A logical method to prevent unwanted effects of motion is to reduce the motion itself. Reduction of tumor motion is not unique to proton therapy, but the potential gain is larger than in photon radiotherapy as it may improve reproducibility in both target position and patient density distribution.

A passive way of suppressing motion is the use of an abdominal compression [68]. The additional required equipment in the treatment area might however limit the possible incident beam directions.

An active approach towards the reduction of tumor motion is breath-hold (e.g., [69]). The quasi-stationary geometry during breath-hold should be represented by a planning CT acquired under identical breath-hold conditions. During a treatment delivery in breath-hold, the patient is asked to maintain the correct breath-hold target position for a prolonged time period of up to 20 s or as long as the patient can comfortably manage. Often, an audio or visual feedback is used for support or a spirometer controls the air flow. An in-depth discussion of the rationale and technical implementation of deep inspiration breath-hold-based photon and proton treatments of upper abdominal tumors was published by Boda-Heggemann et al. [70]. An extreme form of breath-hold is the treatment during apnea under controlled short anesthesia ("general intubation anesthesia under oxygen insufflations in apnea"), as utilized at the Rinecker Proton Therapy Center [71].

18.4.5 Gating

A technique frequently used in photon radiotherapy for motion mitigation and also implemented at the first particle therapy centers is beam gating (e.g., [72,73]). For gating, the patient breathes freely with radiation delivery inhibited if the target is not within a certain distance of the intended position, i.e., the gating window. The treatment beam is triggered on and off automatically by a motion monitoring system. Residual motion within the gating window requires mitigation for interplay effects in case of PBS as described earlier (increased spot overlap) or by combination with other mitigation schemes such as rescanning [27].

Gating offers two clinical advantages, namely clinically acceptable levels of dose conformation and sparing of OARs as illustrated in Figure 18.7. It does this by minimizing the magnitude of intrafractional organ motion and beam field size, respectively.

The gating technique faces several technical challenges. For the success of gating, it is crucial to minimize time latency. For synchrotrons, specific extraction protocols such as radio frequency (RF)-knock-out extraction had to be developed to enable gating [74,75]. Furthermore, both gating and breath-hold increase the overall time to deliver a treatment fraction, which may be a challenge for a sick patient.

Most treatment techniques based on gating rely on an external surrogate for determination of the gating window using the phase of the respiratory cycle, i.e., "phase-based gating." Due to interfractional positional variation, however, the tumor/organ position is not always the same in corresponding respiratory phases. Thus, phase-based gating involves the risk to systematically miss the target, especially for a small gating window. In addition, external surrogates always involve the risk of systematic mismatch between surrogate and internal tumor motion. Ideally, gating should be based on imaging of the internal motion using the amplitude of the target, i.e., its position. The target is then irradiated within a specific position defined during treatment planning ("amplitude-based gating"). Internal motion can either be determined in the images directly [76,77], using fiducials with the risk of influencing the dose distribution of particle beams [78,79], or by correlation models to reduce the imaging dose of frequent fluoroscopy [80].

Abdominal compression, gating, breath-hold, and anesthesia reduce (the effect of) intrafractional motion. These techniques may, however, reduce intrafractional motion at

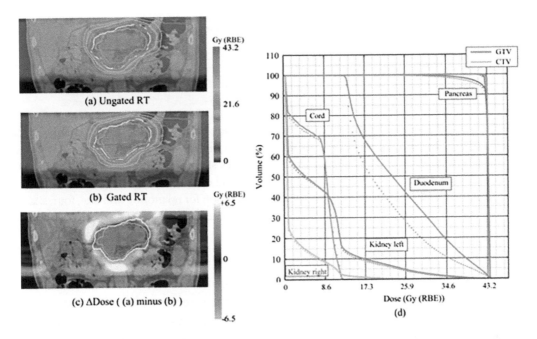

FIGURE 18.7 Accumulated carbon-ion passive scattering beam dose distributions in pancreatic treatment for (a) ungated and (b) gated treatment, and (c) dose distribution differences. Large positive dose differences (over 10%) were observed mainly on the inferior aspect, resulting from the fact that gated treatment irradiates only during exhalation phases. (d) DVH for the dose distributions in (a) (solid line) and (b) (dotted line). While yellow and green lines show the PTV, CTV, and GTV contours, respectively. The gating technique prevented dose degradation of the target in this case and reduced excessive dose to the duodenum (Reproduced with permission from Mori et al. [118].)

the cost of increasing interfractional variations, i.e., variation in the daily "frozen" target position. Rietzel and Bert [81] stated that the reproducibility of the target position between beam-on periods should be in the order of 3–5 mm to allow accurate delivery of (intensity-modulated) particle therapy.

18.4.6 Rescanning

Rescanning, also called repainting, is a strategy to mitigate interplay effects by controlling the timing of beam delivery with respect to the motion cycle. Rather than delivering all dose continuously in one pass, dose delivery to an appropriately sized internal target volume is broken up into several cycles, thereby smoothing the delivered dose distribution. This decreases the risk of large localized dose discrepancies at the cost of accepting smaller dose discrepancies to a larger volume. To first order, multiple rescans within a treatment fraction has the same dose-averaging effect as described by Bortfeld et al. [13] for interplay effects in multiple-fraction IMRT treatments, i.e., the standard deviation in dose to a point in the target is expected to reduce by the square root of the number of rescans. Figure 18.8, however, shows that more rescans is not automatically better. The effectiveness of the chosen rescanning strategy depends as much on the target motion characteristics as on the timing and dosimetric characteristics of treatment delivery. There are many variables to be

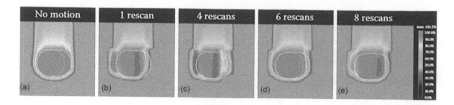

FIGURE 18.8 Four-dimensional treatment planning study of a tumor moving in a unit-density phantom (gray). The thin dashed line indicates the target outline. The solid white line indicates a proton range adjusted internal target volume. The figures show the calculated dose distributions versus the number of rescans, assuming the dosimetric and dose delivery timing characteristics of Gantry 2 of the Paul Scherrer Institute. (a–e) Dose distribution for no motion and one, four, six, and eight rescans, respectively. (Courtesy of Dirk Boye, Paul Scherrer Institute.)

adjusted as part of a rescanning strategy (see later), and the optimal rescanning solution is best determined at the time of treatment planning [81].

The only parameter that is typically fixed is that the direction of slowest rescanning is the depth direction. Anything else (e.g., the scan path or the scanning speed with a layer) can be a subject of optimization, and many of these variables have not yet been extensively researched in clinically realistic density geometries. Where possible, in this section, we will follow the (re) scanning nomenclature as used by Zenklusen et al. [28,31].

Within a range layer, the dose can be delivered by means of discrete spot scanning, continuous scanning, or continuous line scanning (see Chapters 6 and 16). The choice of which method to use affects the time structure of dose delivery within a layer and thereby the interplay effect. Indirectly, for example, if the aim is to limit the total treatment time, this can also affect the number of rescans that can be performed.

There are two approaches for rescanning within a range layer. In isolayered rescanning, there is a maximum to the dose delivered per spot that is small in absolute dose (particle numbers). This means that spots with a higher weight will be rescanned more often than lower-weighted spots, and that the beam path varies for each rescan since the number of remaining beam positions within a layer changes. In scaled rescanning, each spot is rescanned equally often. This can lead to arbitrarily low spot weights, the application of which may be limited by the technical limitations of the treatment delivery system.

There are also several possibilities for rescanning the dose distribution to the entire target volume. In volumetric rescanning, the 3D dose distribution for a single field is delivered one after another, with the absolute dose delivered per scan reduced proportionally to the number of rescans. As in scaled rescanning, this may lead to too low spot weights. In nonvolumetric rescanning (also known as slice-by-slice rescanning), all dose is delivered within an energy layer by means of rescanning, before switching to the next energy layer and repeating this process. It is also feasible to only rescan the deepest layers, which typically have the highest spot weights. Because of the proton depth-dose distribution, this will result in intrinsic rescanning of the more proximal layers as well [81].

The time structure of rescanning is the final parameter that can be varied. Furukawa et al. [27] suggest the use of phase-controlled rescanning (PCR). PCR aims at temporally adjusting the number of rescans to the periodical breathing cycle. A similar approach was proposed by Seco et al. [29] who investigated five different rescanning strategies by means of a simulation study on a homogeneous phantom aiming to deliver a homogeneous dose over 30 fractions. The most effective method was breath-sampled rescanning, in which the layer scan start times of each rescan are distributed evenly throughout the breathing cycle.

Zenklusen et al. [31] analyzed various rescanning strategies. Their extensive simulations showed that rescanning needs to achieve a dose inhomogeneity of less than 1.5% (root-mean-square) to comply with the ICRU recommendations of a target dose between 95% and 107%. For motion up to 5 mm amplitude, their simulations indicated that, when using spot scanning, it is best to have the number of rescans of a spot be proportional with the spot weight (i.e., "isolayered rescanning"). Furthermore, for motion amplitudes above 5 mm, rescanning has to be combined with a strategy to control the motion amplitude, such as breath-hold or beam gating, as also confirmed by Zhang et al. [82]. Bernatowicz at al. [83] compared layered and volumetric rescanning for different scanning speeds of proton beams and concluded that layered rescanning is the method of choice for slow scanning systems. Analysis of dose homogeneity showed that layered rescanning leads to a smoother decrease in dose inhomogeneity as a function of the number of rescans than volumetric rescanning.

For limited motion, rescanning is an effective method to reduce interplay effects in PBS. More research is needed to determine the optimal beam scanning parameters for specific tumor sites and motion characteristics. For rescanning to be most effectively applied to patient treatments, commercial treatment planning systems need to allow optimization of all parameters, taking into account the limitations of the various treatment delivery systems and the patient motion characteristics, i.e., 4D treatment planning.

It is important to realize that rescanning addresses the consequences of interplay effect only (i.e., one of the effects of intrafractional motion). The amplitude of target motion and the corresponding density variation, either intrafractional or interfractional, still has to be addressed in the treatment planning phase, by using adequate safety margins [19] and any of the approaches mentioned in section "General Treatment Planning Considerations."

18.4.7 Adaptive Radiotherapy

Gradual density variations that occur during the treatment course may not have a clinically relevant dosimetric effect on a timescale of one treatment fraction to the next, but over the course of a few fractions, they can have just a devastating effect on the dose to the tumor as, for example, a sizable setup error. The practice of repeat imaging is at the moment not routine in the proton therapy community. Perhaps justifiably so, good clinical results have been obtained with proton therapy without taking this into account. On the other hand, proton therapy has historically mainly been applied for tumor locations that suffer minimally from density variations (e.g., eye and brain tumors) and long-term follow-up and data for other tumor sites (e.g., pancreas, prostate, lung) may not yet be sufficient to show an effect.

Simply tracking the patient weight or the more sophisticated use of patient localization images (either portal images or cone-beam CT (CBCT) [4]) can serve as the basis for a decision protocol to indicate a need for replanning, or at least to assess continued adequacy of the existing treatment by recalculating the dose distribution on the new patient geometry.

Variation in patient geometry over the course of treatment can be addressed via 4D robust optimization as discussed earlier and in Chapter 19. An alternative is adaptive radiotherapy treatment approaches. PBS allows for adaptive radiotherapy or even online changes of the treatment plan, since no patient-specific hardware is required. Adaptive schemes that take into account patient-specific changes of the target area are thus applicable and have the potential to increase the precision of proton beam therapy by minimizing margins.

Adaptive treatment schemes might also be the method of choice to deal with some motion scenarios. For "slowly" moving cervical tumors, for example, the feasibility and advantage of image-guided adaptive proton therapy using a plan-library based plan-of-the-day approach has been shown [84]. Similar approaches are also imaginable for choosing a plan-of-the day for a specific breathing motion characteristic predominate during a specific fraction.

18.5 QUALITY ASSURANCE FOR MOVING TUMORS

Proton therapy in general, and IMPT specifically, provides benefits for many patients, but the clinical staff, i.e., radiation oncologists and clinical physicists, have to be intricately aware of the uncertainties mentioned in this and the previous chapter. The tighter we try to control and conform the dose by means of new and evolving technology and treatment modalities, the more susceptible we may be to "misses" due to remaining uncertainties that have not been properly taken into account in the treatment plan design.

It is hard to draw a single conclusion as the best strategy to deal with time-varying density variations. There are many treatment delivery parameters that can be optimized, see section "Dealing with Motion," and one can combine approaches such as gating and rescanning. The best strategy may also depend on beamline performance, something that is gradually improving over time. In any case, it is important not to simply apply proton therapy to any tumor site without at least performing a rudimentary sensitivity analysis to density variations, motion, and interplay effects.

18.5.1 Treatment Planning Quality Assurance (QA)

The effectiveness of any proton therapy delivery technique in dealing with target motion is, judging by the available literature, typically validated in silico using homogeneous or highly stylized phantoms. This simplification helps in assessing the merits of the various techniques but does not exclude the possibility of severe unexpected dose variations for any actual patient characteristics. It is impossible to underemphasize the need for a priori assessment of the accuracy of a proton therapy treatment plan, especially for moving tumors, by means of using a 4D treatment planning simulation platform that incorporates 4DCT, the patient's breathing trace, the time-resolved dose delivery characteristics of the beamline, and that takes any technique-specific parameters into account [1,81,85,86].

In general, an estimate of target dose and normal tissue dose can be obtained by recalculating the 3D dose distribution for a variety of possible treatment scenarios (e.g., [14]).

In photon therapy, the shape of the dose distribution to first order can be considered invariant under density variations and setup errors. This makes a probabilistic approach towards the likelihood of "a good treatment plan" as suggested by, e.g., van Herk et al. [14] quite feasible. Rather than recalculating the dose distribution for every possible setup error and density variation, one only has to shift it. Interplay effects are proven to be small, given a sufficient number of treatment fractions and/or beam portals [13]. But even in photon radiotherapy, clinical use of such dose-error simulation platforms is very limited. Typically, the appropriateness of a treatment plan (and of a treatment) is judged by the radiation oncologist and the clinical physicist, based on dose-volume histograms (DVHs) and axial dose distributions for a single static patient geometry: the treatment planning CT scan.

The application of proton therapy to tumor sites with a variable density geometry (e.g., lung and liver instead of intracranial treatments) in combination with the advance of IMPT is recent developments and warrants caution. Interestingly, the development and application of dose-error simulation platforms in proton therapy has not kept up with these new developments.

A similar probabilistic approach as described earlier [14] may be feasible for proton therapy. Many variables can, however, have a significant degrading effect on the cumulative dose; Hounsfield unit to proton stopping power conversion, setup errors, intrafractional and interfractional density variation, variation in the motion characteristics, timing of treatment delivery with respect to this motion, etc. A probabilistic plan evaluation approach may therefore require a nearly prohibitive number of recalculations of the dose distribution, making a priori plan evaluation an immense numerical challenge. Some of the variables mentioned may have only a limited effect on the cumulative dose to the tumor, but this needs to be proven comprehensively. Chapter 19 (Section Robust Optimization) provides a framework for taking (some of these uncertainties) into account intrinsically in the treatment plan design. Furthermore, Mori and Chen [87] describe a tool that allows a priori selection of beam directions with reduced sensitivity to range variations over the motion cycle, thereby increasing the confidence that what is treatment planned will also be delivered to the patient.

18.5.2 Treatment Delivery Quality Assurance (QA)

Proton radiotherapy, like any other means of radiation therapy, requires a quality management program that assesses and ensures the continued accuracy of treatment delivery. This means system-wide QA as well as patient-specific QA. Meier et al. [88] showed that QA based on independently reconstructed dose distributions is favorable to pencil beam by pencil beam comparisons for the detection of delivery uncertainties and the estimation of their effects. As in photon beam therapy, where independent algorithms are frequently used for IMRT treatment QA, a tight machine QA program is required for checking the patient's treatment plans.

Aside from the general dosimetric QA (see Chapter 12), system-wide QA for moving tumors has to validate the motion-effect mitigation technique applied. General testing of the system under reference conditions provides a great amount of confidence. Such tests can be quite elaborate. The minimal solution typically involves dosimetric measurements in a (solid) water phantom positioned on a motion platform using a known motion trajectory. But more elaborate motion phantoms should be developed to mimic the clinical situation more closely. Validation of each step in the radiation therapy chain individually may not be able to anticipate all circumstances that may occur during clinical operation. For completeness, an end-to-end test should be designed and executed, simulating an actual treatment as closely as possible.

Compared to static phantom geometries, phantoms developed for validating treatments in moving geometries face a much higher complexity. A moving phantom requires, in addition to aforementioned criteria, a realistic tumor motion and relative motion of the target with respect to modeled bony anatomy. Some ingenious and complex phantoms have been manufactured and reported in the literature, reproducing the dynamics commonly present in the irradiation of abdominal or thoracic tumors. Phantoms produced by Kashani et al. [89], Vinogradskiy et al. [90], Serban et al. [91], and Perrin et al. [62] included a deformable lung, even more closely mirroring realistic patient anatomy. Such a phantom allows end-to-end testing from 4DCT scanning, to deformable registration, to (robust) plan optimization, to in-room patient positioning, and to dose delivery.

True a priori patient-specific verification is very hard (if not impossible), as the time-varying density geometry cannot be mimicked. The next best approach is to limit the duration of undesired dose delivery. In vivo dosimetry (see Chapter 21), for example, can provide an early warning as to the inadequateness of a delivered treatment fraction and allows the clinical team to take corrective action. Another approach, that also only allows correction after delivery of at least some dose, may be to perform closed-loop dose accumulation. Using the treatment fraction-specific 3D/4D patient imaging data or a model of such data [92], the online measured motion characteristics and a detailed log of all machine parameters as a function of time, one may be able to recalculate a best estimate of the actually delivered dose, as shown without daily imaging of data by Richter et al. [66,67]. Patient-specific QA and adaptive therapy are thus closely interrelated.

18.6 FUTURE PERSPECTIVES

The finite range of protons not only allows for substantial sparing of OAR but also requires great caution in mitigating the detrimental dosimetric effects of time-varying density variations. Figure 18.9 attempts to visualize the susceptibility of a number of proton and photon treatment modalities to these variations. Photon radiotherapy dose distributions suffer relatively little from these uncertainties, or they can often be taken into account adequately by means of a straightforward margin approach. The term "density variations" in this graph denotes range and setup uncertainties, as well as density variations over time (intrafractional, interfractional, and gradual changes over the entire treatment course). For the proton modalities, PSPT is the only one that does not suffer from interplay effects. IMPT is the most sensitive treatment modality as, by definition, multiple inhomogeneous dose distributions have to be matched accurately (in 3D) to ensure the intended, and typically homogeneous, target dose coverage.

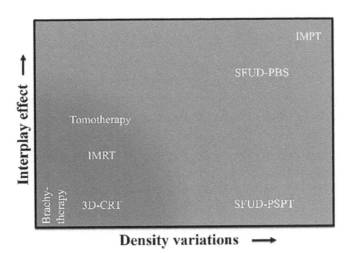

FIGURE 18.9 Sensitivity of the proton modalities, and some photon modalities, to density variations and interplay effect. Blue means less sensitive, and red means more sensitive. The abbreviation 3D-CRT denotes three-dimensional conformal (photon) radiotherapy.

18.6.1 Short Term

At the moment, motion in proton radiotherapy is a problem that has not yet been completely solved. For PSPT, one may be able to choose a practical solution, realizing that the approach used may not ensure target coverage for all possible patients and motion characteristics. However, most of the new proton therapy centers deliver scanned proton therapy. For these centers, a transition from research to clinical application is required in the next years [86].

Recently, the first particle centers equipped with scanning started treating moving tumors. At the MD Anderson Cancer Center in Houston and at Scripps Proton Therapy Center in San Diego, a subset of lung patients are treated with active scanning [93–97]. The same holds true for the University of Pennsylvania in Philadelphia. The proton radiotherapy group in Sapporo will soon start online X-ray image-guided scanned proton treatment [79]. Researchers at Heidelberg Ion-Beam Therapy Centre have treated liver patients with scanned carbon beams [98] and the group at the National Institute of Radiological Sciences started gated scanned treatment for lung and liver [77,99,100]. These are only a few examples for the rapid clinical development towards scanned particle treatments for moving targets. Many new proton facilities are solely equipped with scanning and thus will have to use PBS SFUD or IMPT plans for static as well as moving indications.

The first scanning proton facilities will be using 4D robust optimization tools available in commercial treatment planning software soon. When choosing between 4D robust optimization and adaptive treatment regimes, 4D robust optimization is the more straightforward choice, not requiring any changes in the accelerator infrastructure but potentially in the treatment control system. However, with more imaging capabilities coming available at proton facilities, a shift towards adaptive treatment protocols might happen.

With more and more proton centers aiming to treat moving targets, in the coming years, it will be important to define standards for motion monitoring, to establish 4D treatment planning guidelines including commercial solutions of those and to develop 4D QA tools.

Imaging is crucial when treating moving targets as emphasized earlier. For an ideal consideration of motion, the patient geometry (target volume and OAR) should be monitored prior to and on each day of fractionated treatment delivery. Imaging capabilities during radiotherapy are more advanced in photon therapy than in particle therapy. However, image guidance in proton therapy is slowly beginning to catch up with conventional therapy. In-room imaging in the form of in-room CT and CBCT is becoming available [101]. CBCT-based proton dose calculations have been investigated [102,103], and the utilization of 4D CBCT data for dose calculation has been proposed [104]. As more in-room image-guidance options have become available for particle therapy in the coming years, there will be a greater opportunity to exploit the dosimetric advantages particle therapy has to offer potentially, even by exploiting adaptive treatment strategies.

For motion evaluation before treatment, 4DCT imaging is still standard. Recently, it has been shown that prospective 4DCT reconstruction is superior to retrospective reconstruction resulting in less artifacts. Dou et al. [105] have proposed a promising implementation that was recently applied clinically. To capture motion variations and drift effects, the 4D MR may play a bigger role in the future [54].

18.6.2 Long Term

In recent years, there has been a trend towards MRI-guided radiotherapy. MR offers exquisite soft-tissue contrast and is highly versatile, capable of imaging a wide variety of structures. The ability to visualize lymph nodes, for example, enables the use of stereotactic boost to individual nodes. In the context of moving targets, MRI offers the ability to perform 4D imaging, implemented as a repeated acquisition of 3D image volumes. Currently achieved temporal resolutions are in the order of 2–3 s for full 4D acquisitions and 250 ms for single slices [106]. Further development is needed to obtain real-time 4D data acquisition. However, there has been recent progress on retrospectively reconstructing 4D-MRI data sets from multislice acquisitions [107] and MRI-based 4DCT generation [108]. It is also believed that online MRI can provide a feedback loop to enable gating in clinical routine [109,110]. With respect to beam tracking, feasibility studies on phantom geometries have been performed [106]. All of these developments were reported for photon beam therapy in MR-linacs; however, since initial studies on the effect of the magnetic field on particle beams have been reported [111,112], a swift transition to proton therapy seems reasonable.

Then, also beam tracking with particle beams might have a chance for clinical transition. It is the most sophisticated and technically challenging approach to deal with motion. In beam tracking, the aim is to adjust the position of the beam to the time-varying position of the target in three dimensions. Figure 18.10 shows a possible implementation of beam tracking and some dosimetric results obtained with a nonclinical prototype system [113,114].

There are three prerequisites to successful application of beam tracking in proton radiotherapy. First, a highly accurate treatment delivery system that allows fast position and energy switching. Saito et al. [115] estimate that the scanning system should be able to

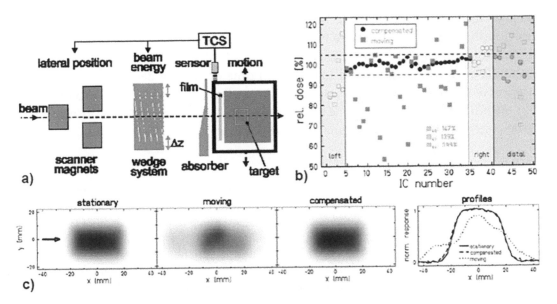

FIGURE 18.10   (a) Experimental motion tracking system as designed at the GSI Helmholtzzentrum für Schwerionenforschung facility in Darmstadt, Germany. The scanner magnets allow fast variation of the pencil beam in the direction perpendicular to the central beam axis (Δx, Δy). Required range adaptation is performed by a fast wedge system (Δz). (b) Results of ionization chamber measurements positioned in the target volume normalized to measurements with a stationary target. (c) Radiographic film responses and corresponding dose profiles along the line indicated by the arrow in the panel labeled "stationary." (Based on Bert et al. [113] and Bert et al., [119]. With permission.)

adapt to an updated target position within a timescale comparable or even shorter than the irradiation time of a single spot. Updating the spot position in the depth direction on such a short timescale is a challenge. An intermediate solution that is easier to translate into clinical use is lateral tracking only. Then the scanner magnets compensate transversal motion, leaving the challenging range compensation to be addressed by other motion mitigation options such as margins and rescanning.

The second prerequisite is accurate real-time monitoring of the target position. Van de Water et al. [116] found that the positioning error (root-mean square) should be less than 1 mm to ensure a sufficiently homogeneous target dose (D5-D95 of less than 5%). This is in agreement to the report of Rietzel and Bert [81] who expressed that the millimeter precision that is achievable with state-of-the-art fluoroscopic motion monitoring systems should be sufficient for beam tracking.

The third prerequisite is near-instantaneous range correction estimates. As this third dimension, the target depth, cannot (yet) be measured online in real time, the current approach to account for time-dependent range variations is precalculation of the range correction for each motion phase and scan position [85]. This requires information available at the time of treatment planning as well as a fast method to determine the required 3D correction in spot position on the fly.

It is unlikely that tracking will become a widely used clinical approach to address target motion in current proton facilities, since the existing technical infrastructure is not designed for beam tracking, especially with respect to fast range variations. The weakest link limiting clinical application of beam tracking is accurate range correction estimates. The accuracy of precalculated range corrections is limited by the representativeness of the 4DCT scan used with respect to the actual time-dependent density geometry at the moment of treatment delivery. To employ tracking to its full potential new imaging capabilities, online 3D imaging has to be introduced in proton therapy facilities. As mentioned earlier, integrated MR imaging could bridge this gap.

REFERENCES

1. Chang JY, Zhang X, Knopf A, Li H, Mori S, Dong L, et al. Consensus guidelines for implementing pencil-beam scanning proton therapy for thoracic malignancies on behalf of the PTCOG thoracic and lymphoma subcommittee. *Int J Radiat Oncol Biol Phys.* 2017; 99(1): 41–50.
2. Britton KR, Starkschall G, Tucker SL, Pan T, Nelson C, Chang JY, et al. Assessment of gross tumor volume regression and motion changes during radiotherapy for non-small-cell lung cancer as measured by four-dimensional computed tomography. *Int J Radiat Oncol Biol Phys.* 2007; 68(4): 1036–46.
3. Sonke JJ, Lebesque J, van HM. Variability of four-dimensional computed tomography patient models. *Int J Radiat Oncol Biol Phys.* 2008; 70(2): 590–8.
4. McDermott LN, Wendling M, Sonke JJ, van Herk M, Mijnheer BJ. Anatomy changes in radiotherapy detected using portal imaging. *Radiother Oncol.* 2006; 79(2): 211–7.
5. Barker JL, Jr., Garden AS, Ang KK, O'Daniel JC, Wang H, Court LE, et al. Quantification of volumetric and geometric changes occurring during fractionated radiotherapy for head-and-neck cancer using an integrated CT/linear accelerator system. *Int J Radiat Oncol Biol Phys.* 2004; 59(4): 960–70.
6. Langen KM, Jones DTL. Organ motion and its management. *Int J Radiat Oncol Biol Phys.* 2001; 50(1): 265–78.
7. Seppenwoolde Y, Shirato H, Kitamura K, Shimizu S, van Herk M, Lebesque JV, et al. Precise and real-time measurement of 3D tumor motion in lung due to breathing and heartbeat, measured during radiotherapy. *Int J Radiat Oncol Biol Phys.* 2002; 53(4): 822–34.
8. Sonke JJ, Rossi M, Wolthaus J, van Herk M, Damen E, Belderbos J. Frameless stereotactic body radiotherapy for lung cancer using four-dimensional cone beam CT guidance. *Int J Radiat Oncol Biol Phys.* 2009; 74(2): 567–74.
9. Balter JM, Ten Haken RK, Lawrence TS, Lam KL, Robertson JM. Uncertainties in CT-based radiation therapy treatment planning associated with patient breathing. *Int J Radiat Oncol Biol Phys.* 1996; 36(1): 167–74.
10. Smitsmans MH, Pos FJ, de Bois J, Heemsbergen WD, Sonke JJ, Lebesque JV, et al. The influence of a dietary protocol on cone beam CT-guided radiotherapy for prostate cancer patients. *Int J Radiat Oncol Biol Phys.* 2008; 71(4): 1279–86.
11. Li W, Purdie TG, Taremi M, Fung S, Brade A, Cho BCJ, et al. Effect of immobilization and performance status on intrafraction motion for stereotactic lung radiotherapy: Analysis of 133 patients. *Int J Radiat Oncol Biol Phys.* 2011; 81(5): 1568–75.
12. Dawson LA, Jaffray DA. Advances in image-guided radiation therapy. *J Clin Oncol.* 2007; 25(8): 938–46.
13. Bortfeld T, Jokivarsi K, Goitein M, Kung J, Jiang SB. Effects of intra-fraction motion on IMRT dose delivery: statistical analysis and simulation. *Phys Med Biol.* 2002; 47(13): 2203–20.
14. van Herk M, Remeijer P, Rasch C, Lebesque JV. The probability of correct target dosage: dose-population histograms for deriving treatment margins in radiotherapy. *Int J Radiat Oncol Biol Phys.* 2000; 47(4): 1121–35.

15. Killoran JH, Kooy HM, Gladstone DJ, Welte FJ, Beard CJ. A numerical simulation of organ motion and daily setup uncertainties: Implications for radiation therapy. *Int J Radiat Oncol Biol Phys*. 1997; 37(1): 213–21.
16. Engelsman M, Sharp GC, Bortfeld T, Onimaru R, Shirato H. How much margin reduction is possible through gating or breath hold? *Phys Med Biol*. 2005; 50(3): 477–90.
17. Mori S, Wolfgang J, Lu HM, Schneider R, Choi NC, Chen GT. Quantitative assessment of range fluctuations in charged particle lung irradiation. *Int J Radiat Oncol Biol Phys*. 2008; 70(1): 253–61.
18. Lambert J, Suchowerska N, McKenzie DR, Jackson M. Intrafractional motion during proton beam scanning. *Phys Med Biol*. 2005; 50(20): 4853–62.
19. Knopf AC, Boye D, Lomax A, Mori S. Adequate margin definition for scanned particle therapy in the incidence of intrafractional motion. *Phys Med Biol*. 2013; 58(17): 6079–94.
20. Engelsman M, Kooy HM. Target volume dose considerations in proton beam treatment planning for lung tumors. *Med Phys*. 2005; 32(12): 3549–57.
21. Kang Y, Zhang X, Chang JY, Wang H, Wei X, Liao Z, et al. 4D Proton treatment planning strategy for mobile lung tumors. *Int J Radiat Oncol Biol Phys*. 2007; 67(3): 906–14.
22. Hui Z, Zhang X, Starkschall G, Li Y, Mohan R, Komaki R, et al. Effects of interfractional motion and anatomic changes on proton therapy dose distribution in lung cancer. *Int J Radiat Oncol Biol Phys*. 2008; 72(5): 1385–95.
23. Farr JB, Mascia AE, Hsi WC, Allgower CE, Jesseph F, Schreuder AN, et al. Clinical characterization of a proton beam continuous uniform scanning system with dose layer stacking. *Med Phys*. 2008; 35(11): 4945–54.
24. Fujitaka S, Takayanagi T, Fujimoto R, Fujii Y, Nishiuchi H, Ebina F, et al. Reduction of the number of stacking layers in proton uniform scanning. *Phys Med Biol*. 2009; 54(10): 3101–11.
25. Phillips MH, Pedroni E, Blattmann H, Boehringer T, Coray A, Scheib S. Effects of respiratory motion on dose uniformity with a charged particle scanning method. *Phys Med Biol*. 1992; 37(1): 223–33.
26. Seiler PG, Blattmann H, Kirsch S, Muench RK, Schilling C. A novel tracking technique for the continuous precise measurement of tumour positions in conformal radiotherapy. *Phys Med Biol*. 2000; 45(9): 103–10.
27. Furukawa T, Inaniwa T, Sato S, Tomitani T, Minohara S, Noda K, et al. Design study of a raster scanning system for moving target irradiation in heavy-ion radiotherapy. *Med Phys*. 2007; 34(3): 1085–97.
28. Bert C, Grözinger SO, Rietzel E. Quantification of interplay effects of scanned particle beams and moving targets. *Phys Med Biol*. 2008; 53(9): 2253–65.
29. Seco J, Robertson D, Trofimov A, Paganetti H. Breathing interplay effects during proton beam scanning: Simulation and statistical analysis. *Phys Med Biol*. 2009; 54(14): N283–94.
30. Bert C, Gemmel A, Saito N, Rietzel E. Gated irradiation with scanned particle beams. *Int J Radiat Oncol Biol Phys*. 2009; 73(4): 1270–5.
31. Zenklusen SM, Pedroni E, Meer D. A study on repainting strategies for treating moderately moving targets with proton pencil beam scanning at the new Gantry 2 at PSI. *Phys Med Biol*. 2010; 55(17): 5103–21.
32. Knopf AC, Hong TS, Lomax A. Scanned proton radiotherapy for mobile targets-the effectiveness of re-scanning in the context of different treatment planning approaches and for different motion characteristics. *Phys Med Biol*. 2011; 56(22): 7257–71.
33. Grassberger C, Dowdell S, Lomax A, Sharp G, Shackleford J, Choi N, et al. Motion interplay as a function of patient parameters and spot size in spot scanning proton therapy for lung cancer. *Int J Radiat Oncol Biol Phys*. 2013; 86(2): 380–6.
34. Inoue T, Widder J, van Dijk LV, Takegawa H, Koizumi M, Takashina M, et al. Limited impact of setup and range uncertainties, breathing motion, and interplay effects in robustly optimized intensity modulated proton therapy for stage III non-small cell lung cancer. *Int J Radiat Oncol Biol Phys*. 2016; 96(3): 661–9.

35. Kang M, Huang S, Solberg TD, Mayer R, Thomas A, Teo BK, et al. A study of the beam-specific interplay effect in proton pencil beam scanning delivery in lung cancer. *Acta Oncol* (Stockholm, Sweden). 2017; 56(4): 531–40.
36. Mageras GS, Yorke E. Deep inspiration breath hold and respiratory gating strategies for reducing organ motion in radiation treatment. *Semin Radiat Oncol*. 2004; 14(1): 65–75.
37. Lagendijk JJ, Raaymakers BW, Raaijmakers AJ, Overweg J, Brown KJ, Kerkhof EM, et al. MRI/linac integration. *Radiother Oncol*. 2008; 86(1): 25–9.
38. Shirato H, Shimizu S, Shimizu T, Nishioka T, Miyasaka K. Real-time tumour-tracking radiotherapy. *Lancet* (London, England). 1999; 353(9161): 1331–2.
39. Berbeco RI, Nishioka S, Shirato H, Chen GT, Jiang SB. Residual motion of lung tumours in gated radiotherapy with external respiratory surrogates. *Phys Med Biol*. 2005; 50(16): 3655–67.
40. Nioutsikou E, Seppenwoolde Y, Symonds-Tayler JR, Heijmen B, Evans P, Webb S. Dosimetric investigation of lung tumor motion compensation with a robotic respiratory tracking system: an experimental study. *Med Phys*. 2008; 35(4): 1232–40.
41. Mori S, Zenklusen S, Knopf AC. Current status and future prospects of multi-dimensional image-guided particle therapy. *Radiol Phys Technol*. 2013; 6(2): 249–72.
42. McGowan SE, Albertini F, Thomas SJ, Lomax AJ. Defining robustness protocols: a method to include and evaluate robustness in clinical plans. *Phys Med Biol*. 2015; 60(7): 2671–84.
43. Eley JG, Newhauser WD, Luchtenborg R, Graeff C, Bert C. 4D optimization of scanned ion beam tracking therapy for moving tumors. *Phys Med Biol*. 2014; 59(13): 3431–52.
44. Keall PJ. 4-dimensional computed tomography imaging and treatment planning. *Semin Radiat Oncol*. 2004; 14(1): 81–90.
45. Schlaefer A, Fisseler J, Dieterich S, Shiomi H, Cleary K, Schweikard A. Feasibility of four-dimensional conformal planning for robotic radiosurgery. *Med Phys*. 2005; 32(12): 3786–92.
46. Ma Y, Lee L, Keshet O, Keall P, Xing L. Four-dimensional inverse treatment planning with inclusion of implanted fiducials in IMRT segmented fields. *Med Phys*. 2009; 36(6): 2215–21.
47. Falk M, Munck af Rosenschold P, Keall P, Cattell H, Cho BC, Poulsen P, et al. Real-time dynamic MLC tracking for inversely optimized arc radiotherapy. *Radiother Oncol*. 2010; 94(2): 218–23.
48. Li X, Wang X, Li Y, Zhang X. A 4D IMRT planning method using deformable image registration to improve normal tissue sparing with contemporary delivery techniques. *Radiat Oncol* (London, England). 2011; 6: 83.
49. Trofimov A, Rietzel E, Lu HM, Martin B, Jiang S, Chen GT, et al. Temporo-spatial IMRT optimization: Concepts, implementation and initial results. *Phys Med Biol*. 2005; 50(12): 2779–98.
50. Graeff C, Luchtenborg R, Eley JG, Durante M, Bert C. A 4D-optimization concept for scanned ion beam therapy. *Radiother Oncol*. 2013; 109(3): 419–24.
51. Graeff C. Motion mitigation in scanned ion beam therapy through 4D-optimization. *Phys Med*. 2014; 30(5): 570–7.
52. Liu W, Schild SE, Chang JY, Liao Z, Chang YH, Wen Z, et al. Exploratory study of 4D versus 3D robust optimization in intensity modulated proton therapy for lung cancer. *Int J Radiat Oncol Biol Phys*. 2016; 95(1): 523–33.
53. Bernatowicz K, Zhang Y, Perrin R, Weber DC, Lomax AJ. Advanced treatment planning using direct 4D optimisation for pencil-beam scanned particle therapy. *Phys Med Biol*. 2017; 62(16): 6595–609.
54. Boye D, Lomax T, Knopf A. Mapping motion from 4D-MRI to 3D-CT for use in 4D dose calculations: a technical feasibility study. *Med Phys*. 2013; 40(6): 061702.
55. Shepard DM, Earl MA, Li XA, Naqvi S, Yu C. Direct aperture optimization: A turnkey solution for step-and-shoot IMRT. *Med Phys*. 2002; 29(6): 1007–18.
56. Broderick M, Leech M, Coffey M. Direct aperture optimization as a means of reducing the complexity of intensity modulated radiation therapy plans. *Radiat Oncol* (London, England). 2009; 4: 8.

57. Papp D, Unkelbach J. Direct leaf trajectory optimization for volumetric modulated arc therapy planning with sliding window delivery. *Med Phys*. 2014; 41(1): 011701.

58. Wölfelschneider J, Friedrich T, Lüchtenborg R, Zink K, Scholz M, Dong L, et al. Impact of fractionation and number of fields on dose homogeneity for intra-fractionally moving lung tumors using scanned carbon ion treatment. *Radiother Oncol*. 2016; 118(3): 498–503.

59. Jakobi A, Perrin R, Knopf A, Richter C. Feasibility of proton pencil beam scanning treatment of free-breathing lung cancer patients. *Acta Oncol* (Stockholm, Sweden). 2017: 1–8.

60. Dowdell S, Grassberger C, Sharp G, Paganetti H. Fractionated lung IMPT treatments: Sensitivity to setup uncertainties and motion effects based on single-field homogeneity. *Technol Cancer Res Treat*. 2016; 15(5): 689–96.

61. Tang S, Deville C, McDonough J, Tochner Z, Wang KK, Vapiwala N, et al. Effect of intrafraction prostate motion on proton pencil beam scanning delivery: A quantitative assessment. *Int J Radiat Oncol Biol Phys*. 2013; 87(2): 375–82.

62. Perrin RL, Zakova M, Peroni M, Bernatowicz K, Bikis C, Knopf AK, et al. An anthropomorphic breathing phantom of the thorax for testing new motion mitigation techniques for pencil beam scanning proton therapy. *Phys Med Biol*. 2017; 62(6): 2486–504.

63. Haas OC, Mills JA, Land I, Mulholl P, Menary P, Crichton R, et al. IGRT/ART phantom with programmable independent rib cage and tumor motion. *Med Phys*. 2014; 41(2): 022106.

64. Steidl P, Richter D, Schuy C, Schubert E, Haberer T, Durante M, et al. A breathing thorax phantom with independently programmable 6D tumour motion for dosimetric measurements in radiation therapy. *Phys Med Biol*. 2012; 57(8): 2235–50.

65. Richter D, Schwarzkopf A, Trautmann J, Kramer M, Durante M, Jakel O, et al. Upgrade and benchmarking of a 4D treatment planning system for scanned ion beam therapy. *Med Phys*. 2013; 40(5): 051722.

66. Richter D, Saito N, Chaudhri N, Hartig M, Ellerbrock M, Jakel O, et al. Four-dimensional patient dose reconstruction for scanned ion beam therapy of moving liver tumors. *Int J Radiat Oncol Biol Phys*. 2014; 89(1): 175–81.

67. Richter D, Lehmann HI, Eichhorn A, Constantinescu AM, Kaderka R, Prall M, et al. ECG-based 4D-dose reconstruction of cardiac arrhythmia ablation with carbon ion beams: application in a porcine model. *Phys Med Biol*. 2017; 62(17): 6869–83.

68. Negoro Y, Nagata Y, Aoki T, Mizowaki T, Araki N, Takayama K, et al. The effectiveness of an immobilization device in conformal radiotherapy for lung tumor: reduction of respiratory tumor movement and evaluation of the daily setup accuracy. *Int J Radiat Oncol Biol Phys*. 2001; 50(4): 889–98.

69. Wong JW, Sharpe MB, Jaffray DA, Kini VR, Robertson JM, Stromberg JS, et al. The use of active breathing control (ABC) to reduce margin for breathing motion. *Int J Radiat Oncol Biol Phys*. 1999; 44(4): 911–9.

70. Boda-Heggemann J, Knopf AC, Simeonova-Chergou A, Wertz H, Stieler F, Jahnke A, et al. Deep inspiration breath hold-based radiation therapy: A clinical review. *Int J Radiat Oncol Biol Phys*. 2016; 94(3): 478–92.

71. Bachtiary B, Hillbrand M, Rinecker H, Skalsky C. Krebstherapie mit protonen – 2015 Protonentherapie – Anerkannt Und Oft Besser. Munich, Germany: Rinecker Proton Therapy Center, 2015.

72. Minohara S, Kanai T, Endo M, Noda K, Kanazawa M. Respiratory gated irradiation system for heavy-ion radiotherapy. *Int J Radiat Oncol Biol Phys*. 2000; 47(4): 1097–103.

73. Lu HM, Brett R, Sharp G, Safai S, Jiang S, Flanz J, et al. A respiratory-gated treatment system for proton therapy. *Med Phys*. 2007; 34(8): 3273–8.

74. Noda K, Kanazawa M, Itano A, Takada E, Torikoshi M, Araki N, et al. Slow beam extraction by a transverse RF field with AM and FM. *Nucl Instrum Methods Phys Res A*. 1996; 374: 269–77.

75. Iwata Y, Kadowaki T, Uchiyama H, Fujimoto T, Takada E, Shirai T, et al. Multiple-energy operation with extended flattops at HIMAC. *Nucl Instrum Meth A*. 2010; 624(1): 33–8.

76. Matsuura T, Miyamoto N, Shimizu S, Fujii Y, Umezawa M, Takao S, et al. Integration of a real-time tumor monitoring system into gated proton spot-scanning beam therapy: An initial phantom study using patient tumor trajectory data. *Med Phys.* 2013; 40(7): 071729.

77. Mori S, Karube M, Shirai T, Tajiri M, Takekoshi T, Miki K, et al. Carbon-ion pencil beam scanning treatment with gated markerless tumor tracking: An analysis of positional accuracy. *Int J Radiat Oncol Biol Phys.* 2016; 95(1): 258–66.

78. Shirato H, Harada T, Harabayashi T, Hida K, Endo H, Kitamura K, et al. Feasibility of insertion/implantation of 2.0-mm-diameter gold internal fiducial markers for precise setup and real-time tumor tracking in radiotherapy. *Int J Radiat Oncol Biol Phys.* 2003; 56(1): 240–7.

79. Shimizu S, Miyamoto N, Matsuura T, Fujii Y, Umezawa M, Umegaki K, et al. A proton beam therapy system dedicated to spot-scanning increases accuracy with moving tumors by real-time imaging and gating and reduces equipment size. *PloS one.* 2014; 9(4): e94971.

80. Schweikard A, Shiomi H, Adler J. Respiration tracking in radiosurgery without fiducials. *Int J Med Robot.* 2005; 1(2): 19–27.

81. Rietzel E, Bert C. Respiratory motion management in particle therapy. *Med Phys.* 2010; 37(2): 449–60.

82. Zhang Y, Knopf AC, Weber DC, Lomax AJ. Improving 4D plan quality for PBS-based liver tumour treatments by combining online image guided beam gating with rescanning. *Phys Med Biol.* 2015; 60(20): 8141–59.

83. Bernatowicz K, Lomax AJ, Knopf A. Comparative study of layered and volumetric rescanning for different scanning speeds of proton beam in liver patients. *Phys Med Biol.* 2013; 58(22): 7905–20.

84. van de Schoot AJ, de Boer P, Crama KF, Visser J, Stalpers LJA, Rasch CRN, et al. Dosimetric advantages of proton therapy compared with photon therapy using an adaptive strategy in cervical cancer. *Acta Oncol.* 2016; 55(7): 892–9.

85. Bert C, Rietzel E. 4D treatment planning for scanned ion beams. *Radiation Oncol.* 2007; 2: 24.

86. Knopf AC, Stutzer K, Richter C, Rucinski A, da Silva J, Phillips J, et al. Required transition from research to clinical application: Report on the 4D treatment planning workshops 2014 and 2015. *Phys Med.* 2016; 32(7): 874–82.

87. Mori S, Chen GT. Quantification and visualization of charged particle range variations. *Int J Radiat Oncol Biol Phys.* 2008; 72(1): 268–77.

88. Meier G, Besson R, Nanz A, Safai S, Lomax AJ. Independent dose calculations for commissioning, quality assurance and dose reconstruction of PBS proton therapy. *Phys Med Biol.* 2015; 60(7): 2819–36.

89. Kashani R, Lam K, Litzenberg D, Balter J. Technical note: a deformable phantom for dynamic modeling in radiation therapy. *Med Phys.* 2007; 34(1): 199–201.

90. Vinogradskiy YY, Balter P, Followill DS, Alvarez PE, White RA, Starkschall G. Verification of four-dimensional photon dose calculations. *Med Phys.* 2009; 36(8): 3438–47.

91. Serban M, Heath E, Stroian G, Collins DL, Seuntjens J. A deformable phantom for 4D radiotherapy verification: Design and image registration evaluation. *Med Phys.* 2008; 35(3): 1094–102.

92. Wolfelschneider J, Seregni M, Fassi A, Ziegler M, Baroni G, Fietkau R, et al. Examination of a deformable motion model for respiratory movements and 4D dose calculations using different driving surrogates. *Med Phys.* 2017; 44(6): 2066–76.

93. Chang JY, Cox JD. Improving radiation conformality in the treatment of non–small-cell lung cancer. *Semin Radiat Oncol.* 2010; 20(3): 171–7.

94. Liu H, Chang JY. Proton therapy in clinical practice. *Chin J Cancer.* 2011; 30(5): 315–26.

95. Kardar L, Li Y, Li X, Li H, Cao W, Chang JY, et al. Evaluation and mitigation of the interplay effects of intensity modulated proton therapy for lung cancer in a clinical setting. *Pract Radiat Oncol.* 2014; 4(6): e259–68.

96. Li Y, Kardar L, Li X, Li H, Cao W, Chang JY, et al. On the interplay effects with proton scanning beams in stage III lung cancer. *Med Phys*. 2014; 41(2): 021721.

97. Liu W, Liao Z, Schild SE, Liu Z, Li H, Li Y, et al. Impact of respiratory motion on worst-case scenario optimized intensity modulated proton therapy for lung cancers. *Pract Radiat Oncol*. 2015; 5(2): e77–86.

98. Habermehl D, Debus J, Ganten T, Ganten MK, Bauer J, Brecht IC, et al. Hypofractionated carbon ion therapy delivered with scanned ion beams for patients with hepatocellular carcinoma—feasibility and clinical response. *Radiat Oncol* (London, England). 2013; 8: 59.

99. Mori S, Inaniwa T, Furukawa T, Takahashi W, Nakajima M, Shirai T, et al. Amplitude-based gated phase-controlled rescanning in carbon-ion scanning beam treatment planning under irregular breathing conditions using lung and liver 4DCTs. *J Radiat Res*. 2014; 55(5): 948–58.

100. Takahashi W, Mori S, Nakajima M, Yamamoto N, Inaniwa T, Furukawa T, et al. Carbon-ion scanning lung treatment planning with respiratory-gated phase-controlled rescanning: simulation study using 4-dimensional CT data. *Radiat Oncol* (London, England). 2014; 9: 238.

101. Veiga C, Janssens G, Teng CL, Baudier T, Hotoiu L, McClelland JR, et al. First clinical investigation of cone beam computed tomography and deformable registration for adaptive proton therapy for lung cancer. *Int J Radiat Oncol Biol Phys*. 2016; 95(1): 549–59.

102. Veiga C, Alshaikhi J, Amos R, Lourenço AM, Modat M, Ourselin S, et al. Cone-beam computed tomography and deformable registration-based "dose of the day" calculations for adaptive proton therapy. *Int J Part Ther*. 2015; 2(2): 404–14.

103. Kurz C, Dedes G, Resch A, Reiner M, Ganswindt U, Nijhuis R, et al. Comparing cone-beam CT intensity correction methods for dose recalculation in adaptive intensity-modulated photon and proton therapy for head and neck cancer. *Acta Oncol* (Stockholm, Sweden). 2015; 54(9): 1651–7.

104. Cai W, Dhou S, Cifter F, Myronakis M, Hurwitz MH, Williams CL, et al. 4D cone beam CT-based dose assessment for SBRT lung cancer treatment. *Phys Med Biol*. 2016; 61(2): 554–68.

105. Dou TH, Thomas DH, O'Connell DP, Lamb JM, Lee P, Low DA. A method for assessing ground-truth accuracy of the 5DCT technique. *Int J Radiat Oncol Biol Phys*. 2015; 93(4): 925–33.

106. Yun J, Wachowicz K, Mackenzie M, Rathee S, Robinson D, Fallone BG. First demonstration of intrafractional tumor-tracked irradiation using 2D phantom MR images on a prototype linac-MR. *Med Phys*. 2013; 40(5): 051718.

107. Paganelli C, Summers P, Bellomi M, Baroni G, Riboldi M. Liver 4DMRI: A retrospective image-based sorting method. *Med Phys*. 2015; 42(8): 4814–21.

108. Bernatowicz K, Peroni M, Perrin R, Weber DC, Lomax A. Four-dimensional dose reconstruction for scanned proton therapy using liver 4DCT-MRI. *Int J Radiat Oncol Biol Phys*. 2016; 95(1): 216–23.

109. Glitzner M, Crijns SP, de Senneville BD, Kontaxis C, Prins FM, Lagendijk JJ, et al. On-line MR imaging for dose validation of abdominal radiotherapy. *Phys Med Biol*. 2015; 60(22): 8869–83.

110. Stam MK, van Vulpen M, Barendrecht MM, Zonnenberg BA, Crijns SP, Lagendijk JJ, et al. Dosimetric feasibility of MRI-guided external beam radiotherapy of the kidney. *Phys Med Biol*. 2013; 58(14): 4933–41.

111. Hartman J, Kontaxis C, Bol GH, Frank SJ, Lagendijk JJ, van Vulpen M, et al. Dosimetric feasibility of intensity modulated proton therapy in a transverse magnetic field of 1.5 T. *Phys Med Biol*. 2015; 60(15): 5955–69.

112. Oborn BM, Dowdell S, Metcalfe PE, Crozier S, Mohan R, Keall PJ. Future of medical physics: Real-time MRI-guided proton therapy. *Med Phys*. 2017; 44(8): e77–90.

113. Bert C, Gemmel A, Saito N, Chaudhri N, Schardt D, Durante M, et al. Dosimetric precision of an ion beam tracking system. *Radiat Oncol*. 2010; 5(1): 61.

114. Grözinger SO, Bert C, Haberer T, Kraft G, Rietzel E. Motion compensation with a scanned ion beam: a technical feasibility study. *Radiat Oncol*. 2008; 3(34).

115. Saito N, Bert C, Chaudhri N, Gemmel A, Schardt D, Rietzel E. Speed and accuracy of a beam tracking system for treatment of moving targets with scanned ion beams. *Phys Med Biol.* 2009; 54(16): 4849–62.
116. van de Water S, Kreuger R, Zenklusen S, Hug E, Lomax AJ. Tumour tracking with scanned proton beams: Assessing the accuracy and practicalities. *Phys Med Biol.* 2009; 54(21): 6549–63.
117. ICRU. *Prescribing, Recording and Reporting Photon Beam Therapy*, ICRU Report 62. Bethesda, Md: International Commission on Radiation Units and Measurements, 1999.
118. Mori S, Yanagi T, Hara R, Sharp GC, Asakura H, Kumagai M, et al. Comparison of respiratory-gated and respiratory-ungated planning in scattered carbon ion beam treatment of the pancreas using four-dimensional computed tomography. *Int J Radiat Oncol Biol Phys.* 2010; 76(1): 303–12.
119. Bert C, Saito N, Schmidt A, Chaudhri N, Schardt D, Rietzel E. Target motion tracking with a scanned particle beam. *Med Phys.* 2007; 34(12): 4768–71.

Treatment Planning Optimization*

Alexei V. Trofimov and David Craft
Massachusetts General Hospital and Harvard Medical School

Jan Unkelbach
UniversitätsSpital Zürich

CONTENTS

---

* Colour figures available online at www.crcpress.com/9781138626508

19.1 OPTIMIZATION OF PLANNING FOR SCATTERED PROTON FIELDS

A spread-out Bragg peak (SOBP) is the foundation of forward treatment planning for 3D-conformal proton therapy with uniform scattered fields. It is used to achieve a longitudinal conformality of the required dose to the target. In his seminal paper on therapeutic use of protons, Dr. Robert Wilson recognized the need for optimization of proton dose distribution for clinical treatments, by pointing out that Bragg peaks ought to be spread out uniformly to cover large tumor volumes. In his assessment, the depth-dose shaping could be "easily accomplished by interposing a rotating wheel of various thickness" in the beam path [1], the method of modulation that is now widely used for proton therapy (see Chapter 5). By choosing the word "easily," Dr. Wilson, perhaps, anticipated that the problem of optimization of modulation of SOBP would be relatively easy compared to the optimization problems yet to arise in the new field of proton therapy.

19.1.1 Optimization of SOBP Longitudinal Profiles

To create a clinically relevant SOBP of the desired depth-dose profile in a passive beam scattering system, a variety of components must operate in conjunction to produce the desired beam parameters. Koehler et al. [2] described one of the earliest examples of design of uniform dose or "flat" SOBPs using computer-based optimization. Based on the input values of range and modulation width, the code written in Fortran IV iteratively searched for the set of amplitudes of shifted pristine peaks and spacings between them (in other words, relative width, and thickness of the wheel steps), which realized the desired SOBP (see Figure 19.1a). Notably, because the shape of the Bragg peak curve varies with the beam energy, the weights of individual peaks in the SOBP need to be optimized separately for different ranges in tissue, to avoid sloping in the SOBP, as shown in Figure 19.1b,c.

In the early days of proton therapy, the wide variety of clinically required combinations of range and SOBP modulation required a large number of premanufactured wheels, with separate wheels required for shallow and deep tumors, one wheel for a close set of modulation width (smallest steps of the propeller could be added or removed to allow for some variation in the total modulation width). A more flexible modern solution employs a beam current modulation system, with a limited number of wheel tracks (see Chapter 5). The pulled-back Bragg peaks can be individually controlled to produce uniform dose profiles for a large range of treatment depths using only a small number of modulator wheels [3–5].

In principle, by temporally optimizing the beam current during the modulation cycle, one can create SOBPs with arbitrary depth-dose profiles. This includes "intensity-modulated" fields, according to the common definition, namely, dose distribution, that are inhomogeneous by design. Notably, the beam current modulation literally means "intensity modulation" of the beam (in a cyclical pattern synchronized with the wheel rotation), regardless of whether the resulting distribution is inhomogeneous intensity-modulated proton therapy (IMPT) or conventional "3D-conformal." An example of inhomogeneous dose achievable with range modulation is the SOBP, including a simultaneous integrated dose boost delivered to a subsection of the target, as in Figure 19.1d. It should be noted though that this technique allows for dose profile modulation only in the longitudinal (depth) direction, whereas the dose within the field remains uniform laterally.

19.1.2 Forward Planning with SOBP Fields

Procedures of forward planning for 3D-conformal proton therapy have been described in detail by Bussière and Adams [6], as well as in Chapter 15. Figure 19.2 illustrates how a "manual optimization" of a treatment plan might be undertaken. The search for a satisfactory solution does involve iterative adjustment; however, it is rather subjective (i.e., depends on the planner's training, habit, judgment etc.) and is not systematic (e.g., iterations do not always lead toward a more preferable solution). Thus, the process cannot be termed optimization in a strictly mathematical sense.

First, the irradiation directions are selected, and the range and SOBP modulation width are chosen as necessary to cover the target. Range compensators are designed to

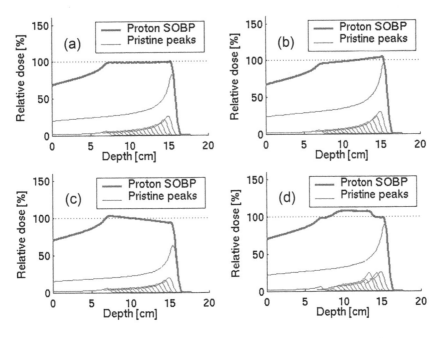

FIGURE 19.1 Depth-dose profile of an SOBP and constituent pristine peaks: optimization of pristine peak weights leads to (a) uniform SOBP dose, while variation in the pristine peak dose profile may introduce a (b) raising or (c) falling slope in SOBP. In principle, arbitrary profiles of the peak dose can be achieved by optimization, i.e., (d) a profile with an integrated dose boost of 10% to the middle part of the SOBP.

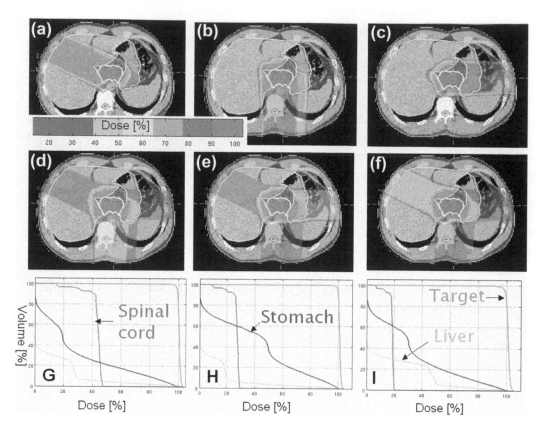

FIGURE 19.2   Forward planning and manual optimization for a case of retroperitoneal tumor. Dose distributions from three beam directions are shown in figures (a–c). These can be combined with various weights, leading to a variety of clinically acceptable dose distributions, e.g., with doses shown in (d–f), and the DVH in (g–i), respectively.

conform the dose to the distal aspect of the target, and accommodations are made to prevent underdosing of the target in case of misalignment of treatment field and tissue heterogeneities, e.g., using the technique of compensator expansion, or "smearing" [7] (also see Chapter 5). Once these steps are completed, a forward calculation is performed to determine the dose from the given field, based on the assumed fluence distribution of the proton beam. The task of the planner is then to iteratively adjust the relative contributions, or "weights," of multiple beams and to combine their doses so that the resulting distribution suits a particular set of requirements. For example, in the case illustrated in Figure 19.2, irradiating the spinal cord up to the tolerance (Figure 19.2d,g) may be considered acceptable in a certain situation, as this configuration minimizes the integral dose, and the main irradiation direction is least affected by internal motion (e.g., of liver with respiration for the right-anterior beam) or variations in the stomach and bowel filling (for the left beam lateral). In other situations, such as repeat treatments, cord tolerance may be reduced, and other directions may have to be used. In those cases, the clinically optimal balance, between irradiation of various structures, need to be selected (compare, e.g., Figure 19.2h vs. i).

19.2 PENCIL BEAM BASED PROTON THERAPY AS AN OPTIMIZATION PROBLEM

Pencil beam scanning (PBS) delivery allows one to achieve highest conformality of proton dose distributions to the target volume and best sparing of healthy tissue. Unlike 3D-conformal treatments, in which each SOBP field delivers a uniform (within a few per cent) dose to the whole target volume, individual scanned fields may be arbitrarily nonuniform (such as, classic "IMPT" [8], integrated boost) or fairly uniform (also known as "single-field uniform dose" (SFUD)) dose distributions (e.g., see Figure 19.3). Similar to intensity-modulated radiation therapy (IMRT) with photons, these multiple field contributions combine to produce the desired therapeutic dose distribution, which may be shaped to conform to the clinical prescription. An important difference from photon IMRT is that the Bragg peak of the proton depth-dose distribution introduces an additional degree of freedom in modulation of the dose in depth along the beam axis, in addition to the modulation in the transverse plane, which is available with both IMRT and PBS. Despite this difference, IMRT and PBS planning are very similar regarding the mathematical formulation of the treatment planning problem.

To take full advantage of the possibility to sculpt the dose in depth, PBS treatments use narrow proton pencil beams, which can be scanned across the transverse plane while changing energy and intensity to control the dose to a point. The most common and

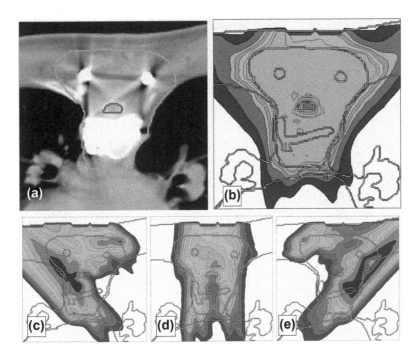

FIGURE 19.3   IMPT plan for a paraspinal tumor: (a) CT scan showing outlines of the tumor and the spinal cord, (b) dose distribution from a 3D IMPT treatment plan, using three beam directions. Dose contributions from individual beams are shown for (c) right-posterior oblique, (d) posterior, and (e) left-posterior oblique fields. (The conventional IMPT plan did not include any consideration of delivery uncertainties.)

versatile PBS-planning technique is the 3D-modulation method, in which individually weighted Bragg peak "spots" are placed throughout the target volume [8]. The examples in this chapter employ 3D-modulation; however, most optimization methods described later are equally applicable to other techniques, such as SFUD treatments or the distal edge tracking (DET). SFUD planning is a special case of 3D-modulation PBS-planning, in which an additional constraint is imposed on dose inhomogeneity within each field, with the goal to deliver a uniform dose to the target from every individual field direction. In the DET method, Bragg peaks are placed only at the distal surface of the tumor [9].

19.2.1 Setup of the PBS Optimization Problem

To apply general optimization methods to radiation therapy planning, technical limitations and treatment goals need to be formulated mathematically as objectives and constraints. For that purpose, the patient image data are partitioned into volumes of interest (VOI), which could include targets, critical organs at risk of undesired side effects (OAR), and other tissue volumes. VOIs are further divided into basic geometric elements called voxels.

The total dose distribution from the PBS field delivered with a scanned beam can be calculated as the sum of contributions from "static" pencil beams fixed at various positions along the scan path. The dose from individual pencil beams to various voxels of interest can be represented in the form of a dose influence matrix D_{ij}, where i is the voxel index and j is the beam index. The total dose to any voxel is then calculated as

$$d_i = \sum_j x_j \cdot D_{ij}, \tag{19.1}$$

where x_j is the relative "weight" of the beam j, which is proportional to the total number of protons delivered at a given spot, which is the position of Bragg peak. The weights x_j are the optimization variables that need to be determined in treatment planning. Due to the large number (thousands or tens of thousands) of such pencil beams involved, PBS treatment planning requires mathematical optimization methods [10,11]. The output of the plan optimization is a set of beam weight distributions, often called intensity or fluence maps. Unlike in IMRT, where a single two-dimensional fluence map characterizes a field, for PBS, many beam energies may be used to irradiate the target from the same direction, and optimization will yield separate maps for every energy setting.

Dosimetric or other planning objectives may be defined for volumes or individual voxels. The planning objectives and their priorities can be expressed in the objective function (OF). The term "optimization," in the context of treatment planning, typically signifies the search for a set of plan parameters which minimize the value of the OF, subject to a set of constraints that have to be fulfilled.

A widely used OF that aims at minimizing the volume, within a given OAR n, that exceeds the maximum tolerance dose D^{max} is given by the quadratic penalty function:

$$f_n(d) = \frac{1}{N_n} \sum_{i=1}^{N_n} \left(d_i - D^{max} \right)_+^2 \tag{19.2}$$

Similarly, one can define a quadratic function that aims to reduce volumes of the tumor, which receive less than the minimum dose D^{min}. OFs may also include the generalized equivalent uniform dose (gEUD) [12]:

$$f_n(d) = \left[\frac{1}{N_n} \sum_{i=1}^{N_n} (d_i)^p \right]^{1/p} \tag{19.3}$$

where N_n is the number of voxels in the VOI n, and p is an organ-specific parameter.

In addition, there may be constraints on the dose in a VOI that have to be fulfilled to make the treatment plan acceptable. For example, one can request that the dose in every voxel belonging to the tumor should be between a minimum dose D^{min} and a maximum dose D^{max}. This would result in the hard constraint

$$D^{min} \le d_i \le D^{max} \quad \forall i \in VOI. \tag{19.4}$$

In clinical situations, treatment objectives often directly conflict each other: for example, a target may not be completely irradiated to the prescribed level if a dose-sensitive critical structure is immediately adjacent to it. In this case, hard dosimetric constraints have to be used with care, and it is often necessary to reformulate a constraint as an objective. For example, it may be necessary to minimize the dose to an OAR that exceeds the tolerance dose through a quadratic objective, rather than enforcing the dose to be below the maximum dose in every voxel through a constraint. Such an objective is often referred to as a "soft constraint" in the medical physics community. Multicriteria optimization (MCO) methods, discussed in Section 19.3, address such inherent treatment planning contradictions.

Thus, one can formulate the general PBS dose optimization problem as follows:

Minimize (with respect to the beam weights x):

$$\sum_n \alpha_n f_n(d)$$

subject to the constraints:

$$d_i = \sum_j x_j \cdot D_{ij}$$

$$l_m \le C_m(d) \le u_m$$

$$x_j \ge 0. \tag{19.5}$$

In the earlier formulation, the different objectives f_n are multiplied by respective weighting factors α_n and added together to form a single composite objective. By selecting and adjusting the weighting factors, the treatment planner can prioritize different objectives, and control the trade-off between them. The approach of a weighted sum of objectives is pursued in most current treatment planning systems. (An alternative to this standard approach is MCO.) The functions C_m denote a general constraint function, and l_m and u_m are upper and lower bounds. An example of a simple constraint function is the minimum or maximum dose constraint mentioned earlier. Alternatively, constraints on equivalent uniform dose (EUD) can be imposed [12].

A number of additional parameters often need to be specified before optimization of beam weights is performed. These include, for example, the choice of the algorithm for placement of Bragg peaks [8] as well as the volume used for placement (which may be larger than the target), the spacing of peaks and layers in depth [13], and the size of the pencil beam used for delivery of therapy [14]. These additional treatment parameters or hyper-parameters affect the outcome of optimization; however, they are not determined through an optimization algorithm in the mathematical sense, and are instead chosen based on experience, planning studies, and physical or theoretical considerations.

19.2.2 Solving the Optimization Problem

IMPT represents a textbook example of a large-scale optimization problem, especially if convex objectives and constraints are used. The variables, the beam weights x, are continuous, i.e., can take any nonnegative value. The objectives can typically be formulated in closed form as a function of the optimization variables and also the gradient of the objective can be calculated analytically. Therefore, a large variety of algorithms can be applied. Those can be categorized into constrained and unconstrained methods. In the case of unconstrained methods, no dosimetric hard constraints are applied, i.e., all treatment goals are formulated as objectives. The only constraints that always need to be fulfilled to yield a physically meaningful plan are the variable-bound constraints $x_j \geq 0$. However, those can be treated through relatively simple methods like gradient projection methods.

Constrained optimization for PBS is still challenging due to the large number of variables (10^3–10^5) and the large number of voxels (10^5–10^7). If only linear objectives and constraints are used, the linear programming framework can be applied. Linear programming is used in the treatment planning system Astroid (http://astroidplanning.com). Most treatment planning systems, however, use nonlinear objectives and constraints. These systems typically build on quasi-Newton methods such as the limited-memory Broyden–Fletcher–Goldfarb–Shanno (BFGS) algorithm. Different approaches for handling constraints are being used, including sequential quadratic programming (RayStation, Raysearch Laboratories) and augmented Lagrangian methods (Pinnacle, Philips Healthcare).

Optimization problems may further be classified as convex or nonconvex. In a convex optimization problem, all of the constraints as well as the minimized OF are convex functions. For example, linear functions, and therefore, linear programming problems are convex. The feasible region (i.e., the set of all spot-weight vectors x that fulfill the constraints) is then also convex, being the intersection of convex constraint functions. With convex objectives and a convex feasible region, a local optimal solution is also a global optimal solution. Thus, optimization would either yield a globally optimal solution or demonstrate that there is no feasible solution. All the objectives and constraints described earlier (e.g., quadratic function, EUD) are convex.

Conversely, a nonconvex optimization problem is any problem where the objective is nonconvex or nonconvex constraint functions give rise to a nonconvex feasible region. In this setting, multiple local optimal solutions are possible and, in practice, considering the large number of variables in PBS, it is typically not possible to guarantee that an algorithm used to solve the optimization problem indeed converges to a globally optimal solution. The most common nonconvex constraints that are used in practice are dose-volume constraints, which can be conveniently defined to specify the desired shape of the dose-volume histogram (DVH) directly [15], for example: "the fraction of the volume of a specific OAR irradiated to 40 Gy is not to exceed 30%."

19.3 MULTICRITERIA OPTIMIZATION

Optimization theory is built up around the single criterion optimization problem, where there is one objective and other problem considerations are included as constraints. In radiotherapy, the main objective—to cover the target with the prescription dose—is in direct conflict with the other objectives of keeping the dose to the healthy organs to a minimum. If biological response models such as tumor control probability (TCP) and normal tissue complication probability (NTCP) were reliable, one might be able to solve radiotherapy optimization well in a single criterion mode: maximize TCP subject to the NTCPs of the relevant OARs being below acceptable levels. But even in this setting, depending on the patient-specific trade-off (for the treatment plan under consideration, how much gain in TCP is there if you allow NTCP for some organ to increase by some amount), were there a tool to easily explore other options, a physician might choose a different plan than the plan returned from the single criterion optimization.

Presently, the standard commercial systems available for treatment optimization still attempt to solve the radiotherapy optimization problem with a single-criterion approach, and this leads to a lengthy optimization iteration cycle, where treatment planners try to find the set of weights and function parameters that give a plan that best matches the physician's goals for treatment. The problem is it is very difficult to guess those weights and function parameters to get a good plan, and as the number of organs to consider increases, this task becomes increasingly more difficult. Several groups worked to bring MCO into routine clinical usage [16–22].

There are two main approaches to MCO for radiotherapy treatment planning: prioritized optimization and the Pareto surface (PS) approach. Later, we describe the two approaches, show how they are related and discuss their pros and cons.

19.3.1 Prioritized Optimization

Prioritized optimization, or lexicographic ordering, as it is sometimes called in the literature, is a natural approach for dealing with multiple objectives when the objectives can be ranked in terms of importance [23,24]. Letting f_1 denote the highest priority objective, f_2

the second highest, etc., prioritized optimization solves the following sequent of optimization problems for k priority levels:

(1) *minimize* $f_1(x)$ *subject to* $x \in X$;

(2) *minimize* $f_2(x)$ *subject to* $x \in X$, *and* $f_1(x) \leq f_1^* \cdot (1+\varepsilon)$;

...

(k) *minimize* $f_k(x)$ *subject to* $x \in X$, *and, for all* $i < k$, $f_i(x) \leq f_i^* \cdot (1+\varepsilon)$, (19.6)

where x is a vector of the decision variables, X is a constraint set that represents constraints on the beamlet fluences (upper and lower bounds), and is also used to denote hard dosimetric constraints, such as voxel dose, organ mean dose, or EUD that must be met by every considered solution. f_1^* is the optimal objective value from the first optimization, i.e., the fluence values in the case of PBS, and ε is a small positive slip factor. Multiplication by $(1+\varepsilon)$ allows a small degradation in the value of the first optimization, thus hopefully permitting the second priority objective to achieve a good value, and so forth. The result of the final optimization is the single result of the prioritized optimization approach. The choice of ε (and whether it is the same for each step) and the priority ordering of the objectives will influence the final result.

19.3.2 Pareto Surface Approach

The PS approach does not prioritize the objectives, but instead treats every objective equally. Unlike prioritized optimization, the PS approach yields not a single plan, but a set of optimal plans that trade-off the objectives in a variety of ways. Given a set of objectives and constraints, a plan is considered Pareto-optimal if it is feasible, and there does not exist another feasible plan, which is strictly better with respect to one or more objectives, and is at least as good for the rest. Assuming that the objectives are chosen correctly, Pareto-optimal plans are the plans of interest to planners and doctors. The set of all Pareto-optimal plans comprises the PS.

The PS-based MCO problem can be formulated as

$$\text{minimize} \{f_1(x), f_2(x), \ldots, f_N(x)\} \text{ subject to } x \in X, \quad (19.7)$$

where X is used, as before, to represent all beamlet and dose constraints, and N is the number of OFs. The algorithmic decisions to be made for this approach are (i) how to compute a reasonable set of diverse Pareto optimal plans and (ii) how to present the resulting information to the decision makers. Radiotherapy seems to be one of the first fields, if not the first, to fully address the question of populating PSs for $N \geq 3$. Two main types of strategies populating the PS have been put forward for the radiotherapy problem: weighted sum methods and constraint methods.

Weighted sum methods are based on combining all the objectives into a weighted sum and solving the resulting scalar optimization problem. By solving the problem for a variety of weights, a variety of different Pareto-optimal plans are found. If the underlying objectives and constraint set are convex, every Pareto-optimal point can be found by some weighted sum. Several publications describe methods to choose the weights appropriately, to produce a small set of plans that covers the PS sufficiently well [18,21,25]. These methods intrinsically take into account convex combinations of calculated PS points when evaluating the goodness of a set of Pareto plans. All of these methods get bogged down when the number of objectives is large (e.g., over 8). Fortunately, on a practical level, even as few as N+1 PS plans are often sufficient to determine good treatment plans [26,27].

Constraint methods use the OFs as constraints (like in prioritized optimization), and by varying the constraint levels, different Pareto-optimal solutions are found. The state of the art of constraint-based methods is the improved normalized normal constraint (NNC) method [28]. The main deficit of constraint-based methods is that error measures, which give the quality of the PS approximation, are not a natural part of the algorithm or output, as they are in the methods of Craft et al. [18] and Rennen et al. [25]. Weighted sum and constraint methods are graphically depicted for 2D PSs in Figure 19.4.

19.3.3 Navigation of the Pareto Surface

The final task in a PS-based approach to treatment planning is to allow the user to select a plan from the PS. Since the PS is represented by a finite set of Pareto-optimal treatment plans, there are two natural approaches to plan selection. The easiest way is simply to allow the treatment planner to select one of the computed Pareto-optimal treatment plans. In the case of PBS, where treatment plans can be weighted and combined to form other valid treatment plans, it makes sense to allow users to smoothly transition between the computed solutions. When navigating across convex combinations of the database plans, either forcing Pareto-optimality or not, the standard method is to present N sliders, one for each objective, and the underlying algorithmic task is to determine how to move in the objective space in response to a slider movement [21,29].

An alternative to presenting the users with N sliders is to allow them to select two of the N objectives and then display a 2D trade-off for those two objectives. For the other N-2 objectives, the user can impose upper bounds, influencing the 2D trade-off surface being evaluated. The benefit of this method is that it allows the user to visualize a 2D slice of PS, which may yield intuition into the problem at hand. Figure 19.5 shows what this might

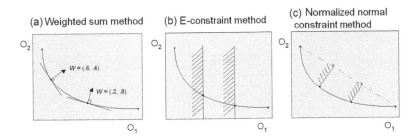

FIGURE 19.4   Methods to compute a database of PS points: (a) weighted sum, (b) e-constraint and (c) NNC method.

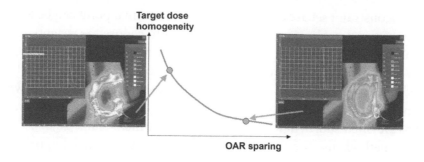

FIGURE 19.5 Illustration of two Pareto-optimal plans, showing trade-offs in OAR sparing vs. target dose homogeneity.

look like for examining the trade-off between sparing the lung and controlling hot spots within a target.

19.3.4 Comparing Prioritized Optimization and Pareto Surface-Based MCO

Prioritized optimization and PS MCO are compared graphically in Figure 19.6. It is important to note that both methods rely on optimization with hard constraints. In the prioritized approach, this is obvious since objectives move into the constraint section. In the PS method, constraints are important in the problem formulation, to restrict the domain of the PS to a useful one. For example, it makes sense to put an absolute lower bound on target doses normally, even if a user is interested in exploring some underdosing of the tumor to improve OAR sparing (otherwise, anchor plans for OAR will be "all 0" dose plans, which are not helpful for planning). Similarly, a hard upper maximum dose on all voxels is useful. Therefore, MCO methods in general are best used when a constrained solver is at hand. Solvers implemented in RayStation (RaySearch Laboratories), Pinnacle (Phillips), Monaco (Electa), UMPlan/UMOpt (University of Michigan), and Astroid (http://astroidplanning. com) are examples of solvers that allow true hard constraints (as opposed to those that handle constraints approximately by using a penalty function with a high weight).

The advantage of the prioritized approach is that it is a programmable procedure that results in a single Pareto-optimal plan, but the disadvantage is, there is only one plan presented to the

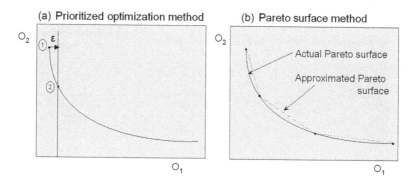

FIGURE 19.6 Illustration of approaches to MCO (a) prioritized optimization and (b) PS method.

user at the end of the process. PS methods on the other hand present all optimal options to the user, but might be considered overwhelming for routine planning since the user would have to decide on selecting a single plan manually from a large number of options on PS. However, plan selection in standard cases may be fast, even with many options, since sliding with navigation sliders is much more efficient than the reoptimization iteration loop. Notably, since the navigation process is user-driven, it is not as reproducible as the prioritized approach.

The effect of clinical introduction of MCO in proton therapy has been investigated, e.g., by Kamran et al. [30], who found that physicians showed preference for multicriteria plans compared to the conventional trial-and-error approaches, because the former generally achieved improved sparing of OAR without loss of target coverage. MCO plans more readily met the physicians' approval, and, thus, planning process required a shorter time overall.

19.4 ROBUST OPTIMIZATION METHODS FOR PBS

Uncertainties in proton therapy have been addressed in detail in Chapters 17 and 18 and, e.g., in publications by Lomax [31,32]. From the delivery point of view, an optimal plan needs to be "robust," that is, designed in such a way that slight deviations from the plan due to various uncertainties during treatment delivery will not affect the quality of treatment outcome. In other words, a robust treatment plan will deliver a clinically acceptable dose distribution as long as the deviations from the planned do not exceed the assumed levels.

Heuristics to mitigate the effects of uncertainty in patient setup and proton range have been developed for different treatment modalities. These methods include compensator smearing in passively scattered proton therapy or selection of favorable beam angles (see Chapter 15). Here, we discuss robust optimization strategies specifically for IMPT.

19.4.1 IMPT Dose in the Presence of Uncertainties

Doses delivered from different directions in IMPT are typically inhomogeneous and require the use of a number of proton energies. For this reason, variations in the target setup and penetration depth during delivery can lead to misalignment and mismatch of doses from individual fields, and, consequently, alter the combined dose distribution.

To satisfy the requirement of dose conformity to the target, steep dose gradients are often delivered at the target border. Such steep dose gradients in the dose contributions of individual beams make IMPT plans yet more sensitive to both range and setup errors. In particular, dose gradients in the beam direction make the treatment plan vulnerable to range errors, since an error in the range of the proton beams corresponds to a relative shift of these dose contributions longitudinally inside the patient. As a consequence, the dose within the target may not add up to a homogeneous dose as desired. Hot and cold spots may arise. Moreover, dose may be shifted into critical organs. Generally, the more conformal the combined IMPT dose is, the more complex the fluence maps per field are, and the more sensitive the plans are to the delivery uncertainties.

As an illustration, consider a conventional IMPT plan for a case of paraspinal tumor, shown in Figure 19.3. The target entirely surrounds the spinal cord, which is to be spared. An IMPT dose distribution was optimized using a quadratic OF, thus aiming at a homogeneous target dose. As is characteristic of IMPT, the homogeneous dose distribution in

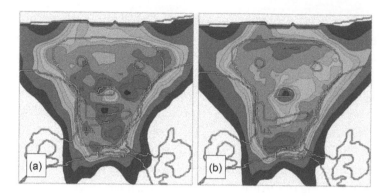

FIGURE 19.7   Estimated dose distribution from the plan in Figure 19.3, assuming (a) a 5 mm range overshoot of all pencil beams and (b) a systematic 3.5 mm setup error (posterior shift.)

the target is achieved through a superposition of highly inhomogeneous contributions delivered from three beam directions.

The dose distribution that results from a range overshoot of all pencil beams in this plan (i.e. protons penetrate further into the patient than anticipated during planning) would lead to a higher dose to the spinal cord, as shown in Figure 19.7a. Sensitivity of the same plan to setup errors is illustrated in Figure 19.7b, which shows the dose distribution resulting from a 3.5 mm setup error posteriorly (upwards in the picture). This shift has no impact on the dose contribution of the posterior beam. However, the oblique beams hit the patient surface at a different point. For a posterior shift, the dose contributions of the oblique beams are effectively shifted apart, which results in cold spots around the spinal cord.

From this illustration, it is evident that, unlike in conventional X-ray therapy, plan degradation in the presence of range and setup uncertainties in IMPT cannot be prevented, to a satisfying degree, with safety margins. Expanding the irradiated area around the target with margins could potentially reduce the dose shortfall at the edge of the target in the presence of an error. However, the general problem of misaligning the dose contributions of different fields, which leads to dose uncertainties in all of the target volume, cannot be solved through margins. This problem instead relates to steep dose gradient in the dose contributions of individual fields. The fundamental limitations of safety margins in IMPT can further be interpreted via the following consideration. In general, the planning target volume (PTV) concept is based on the assumption that, as long as the clinical target volume (CTV) moves within the PTV, and the PTV is irradiated to the prescription dose, then the CTV will receive the prescribed dose. While this underlying assumption is approximately valid in photon therapy, it breaks down for protons.

19.4.2 Robust Optimization Strategies

To address the shortcomings of the PTV concept in IMPT, robust treatment planning methods have been developed, which directly incorporate uncertainty into the IMPT optimization problem [33–42]. Most literature on robust IMPT optimization focuses on range and setup uncertainty. In parts, these methods have previously been investigated to incorporate setup errors and interfraction organ motion into IMRT planning. However, only the necessity to address fundamental limitations of the PTV concept in IMPT has led to the first implementation of robust optimization methods in commercial planning systems. Robust optimization is now available in several commercial planning systems for treatment planning in clinical practice.

Several approaches to robust planning are distinguished in the literature, which will be discussed later. All approaches have in common that they require a model of the uncertainty to be accounted for. It is typically assumed that the fluence map x of an IMPT plan can be delivered with high accuracy. However, the dose distribution in the patient that this fluence map results in is uncertain. Mathematically, this can be described as uncertainty in the dose-influence matrix D_{ij}. In particular, geometric uncertainty such as range and setup errors is modeled as uncertainty in the dose-influence matrix.

Most methods assume that uncertainty can be sufficiently well described by a discrete set of error scenarios that may occur. For example, a simple model of range uncertainty, where it is assumed that all pencil beams simultaneously overshoot or undershoot, could be described by three scenarios:

1. The nominal scenario (i.e., no error)

2. Range overshoot by a specified amount

3. Range undershoot by a specified amount.

Range overshoot or undershoot can be modeled by downscaling or upscaling the Hounsfield units in the planning CT. A simple model of a systematic setup error consists of seven scenarios, containing the nominal scenario plus a specified patient shift in superior, inferior, anterior, posterior, left, and right directions. If we enumerate the error scenario by an index s, the treatment may result in different dose distributions d_s, which are given by

$$d_i^s = \sum_j D_{ij}^s x_j$$

where D_{ij}^s is the dose-influence matrix for scenario s. In the literature, three approaches to incorporate the set of dose distributions into IMPT optimization can be distinguished:

• The probabilistic approach

• The minimax approach

• Treatment planning based on a worst-case dose distribution

19.4.2.1 The Probabilistic Approach
In the probabilistic or stochastic programming approach [33,35], the scenarios s are associated with importance weights p_s. The importance weights can also be interpreted as the

probability for a given error scenario to occur. Treatment plan optimization is performed by optimizing the expected value of the OF:

$$\underset{x}{\text{minimize}} \ \sum_s p_s f\left(d^s\right)$$

$$\text{subject to } d_i^s = \sum_j D_{ij}^s x_j$$

$$x_j \geq 0 \tag{19.8}$$

This composite OF can be interpreted in a multicriteria view: The composite objective is a sum of objectives for every possible error scenario weighted with the probability of that error to occur. The general goal is to find a treatment plan that is good for all possible errors, but larger weights are assigned to those scenarios that are likely to occur, and lower weights to large errors that are less likely to happen.

For a pure quadratic OF that minimizes the deviation from a prescribed dose D^{pres},

$$f(d) = \sum_i \left(d_i - D^{pres}\right)^2 \tag{19.9}$$

the expected value of the OF is

$$\sum_s p_s f\left(d_i^s\right) = \sum_i \left[\left(\sum_s p_s d_i^s - D^{pres}\right)^2 + \sum_s p_s\left(d_i^s - \sum_s p_s d_i^s\right)^2\right] \tag{19.10}$$

which is the sum of two terms: the first term is the quadratic difference of the expected dose $\sum_s p_s d_i^s$ and the prescribed dose, and the second term is the variance of the dose.

Hence, minimizing the expected value of the quadratic OF aims at bringing the expected dose close to the prescribed dose in every voxel; and simultaneously minimizes the variance of the dose in every voxel, such that the expected dose is approximately realized even if an error occurs.

19.4.2.2 The Minimax Approach

The minimax approach [36] optimizes the treatment plan for the worst error scenario that may occur. Mathematically, this can be expressed as minimizing (with respect to the pencil beam intensities x) the maximum of the OF (taken over the error scenarios s).

$$\underset{x}{\text{minimize}} \ \underset{s}{\max} \ f\left(d^s\right)$$

$$\text{subject to } d_i^s = \sum_j D_{ij}^s x_j$$

$$x_j \geq 0 \tag{19.11}$$

Hence, the minimax aims to determine the treatment plan that is as good as possible for the most unfavorable error that may occur. Unlike the probabilistic approach, the minimax approach does not assign importance weights p_s to the error scenarios.

Solving the minimax problem requires constrained optimization techniques. In such methods, each scenario would typically be associated with a Lagrange multiplier. To relate the minimax approach to the probabilistic approach, one can interpret the Lagrange multipliers as importance weights. This means that the treatment plan obtained by the minimax approach corresponds to the result of the probabilistic approach for a specific set of importance weights.

19.4.2.3 Optimization of the Worst-Case Dose Distribution

There are several different flavors of the minimax approach. In its pure form as defined earlier, the maximum of the composite OF over the error scenarios is considered. In practice, the composite OF is typically a weighted sum of objectives of different structures. One variation of the minimax approach is to consider the maximum with respect to the error scenarios over each objective separately. This approach has been referred to as the objective-wise worst-case method as opposed to the composite worst-case method and can formally be written as

$$\underset{x}{\text{minimize}} \sum_k \max_s \left(f_k\!\left(d^s\right) \right)$$

$$\text{subject to } d_i^s = \sum_j D_{ij}^s x_j$$

$$x_j \geq 0 \tag{19.12}$$

where k is the index over the objectives associated with different structures. An advantage of the objective-wise worst-case method is that it is easily compatible with the MCO methods [37].

In many cases, the objectives associated with a structure represent a sum over the contributions of individual voxels in that structure. This is, for example, the case for the widely used quadratic penalty function. In this case, one can consider the maximum over the error scenario of the objective contribution in every individual voxel. The treatment plan optimization problem can then be written as

$$\underset{x}{\text{minimize}} \sum_i \max_s \left(f_i\!\left(d_i^s\right) \right)$$

$$\text{subject to } d_i^s = \sum_j D_{ij}^s x_j$$

$$x_j \geq 0 \qquad\qquad (19.13)$$

In this case, $\max_s \left(f_i\left(d_i^s\right)\right)$ can be interpreted as the worst-case dose that voxel i may receive. If f_i is a quadratic penalty function for tumor underdose, $\max_s \left(f_i\left(d_i^s\right)\right)$ is the minimum dose that this voxel may receive for any of the error scenarios considered. Analogously, if f_i is a quadratic penalty function for overdosage of on OAR, $\max_s \left(f_i\left(d_i^s\right)\right)$ is the maximum dose to that voxel. Therefore, the distribution of $\max_s \left(f_i\left(d_i^s\right)\right)$ can be interpreted as a worst-case dose distribution. This distribution is unphysical because every voxel is considered independently. Although in one voxel, the worst case may correspond to a patient shift anteriorly, the worst case in another voxel may correspond to a patient shift posteriorly. Hence, the worst-case dose distribution cannot be realized. However, it can be considered as a lower bound for the quality of a treatment plan. The idea to perform IMPT optimization based on the worst-case dose distribution was originally suggested by Pflugfelder et al. [34].

19.4.3 Examples of Robust Optimization

Incorporating uncertainty in IMPT optimization yields increasingly robust treatment plans. Consider two treatment plans: a conventional plan optimized without accounting for uncertainty and a plan optimized for range and setup uncertainty using the probabilistic approach, i.e., the setup and range uncertainties modeled with a Gaussian distribution. Figure 19.8 shows the DVHs corresponding to dose distributions calculated for range and setup errors randomly sampled from these Gaussian distributions. For the conventional plan, target coverage is strongly degraded in many cases and the dose to the spinal cord can be very high for some scenarios. The variation in the DVHs of the robust plan is greatly reduced, ensuring better target coverage and lower spinal cord doses.

To gain some insight into how this robustness is achieved, let us consider the dose contributions of individual beams. Figure 19.9 compares four treatment plans: the conventional plan, a plan optimized for range uncertainty only, a plan optimized for setup uncertainty

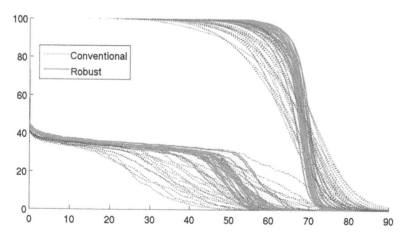

FIGURE 19.8 DVH comparison between a conventional and a robust IMPT plan. DVHs for the CTV and the spinal cord are shown for randomly sampled range and setup errors.

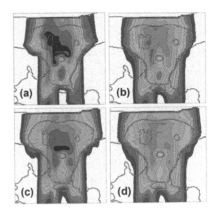

FIGURE 19.9 For the case illustrated in Figure 19.3, dose contributions from the posterior beam from four differently optimized plans: (a) conventional IMPT, robust IMPT incorporating (b) range uncertainty only, (c) setup uncertainty only, and (d) considering both range and setup uncertainty.

only, and a plan incorporating both types of errors. The conventional plan is characterized by steep dose gradients both in beam direction and laterally—especially around the spinal cord. The plan optimized for range uncertainty shows reduced dose gradients in beam direction and avoids placing a steep distal falloff of a Bragg peak in front of the spinal cord. The lateral falloff is used instead of the distal falloff to shape the dose distribution around the spinal cord. The plan optimized for setup errors only shows reduced dose gradients in the lateral direction, but it does not avoid placing a distal Bragg peak falloff in front of the critical structure, and therefore does not provide robustness against range errors per se. The plan optimized for both range and setup errors shows reduced dose gradients both longitudinally, in the beam direction, and laterally.

In summary, robustness is achieved through a redistribution of dose contributions among the beam directions and through avoiding unfavorable dose gradients. For our sample paraspinal case, the price of robustness is a higher dose to the spinal cord for the nominal case. In a conventional plan, the steep distal Bragg peak falloff is utilized, which allows for optimal sparing of the spinal cord. If range errors are to be accounted for, the shallower lateral falloff is used, leading to a shallower dose gradient between tumor and spinal cord for the nominal case. Publications by Pflugfelder et al. [34] and Unkelbach et al. [35] provide a more detailed analysis for this treatment geometry.

19.4.4 Commercial Implementations and State of Research

The fundamental limitations of safety margins in IMPT led to the first implementations of robust optimization in commercial planning systems. Interestingly, the currently available implementations cover all three approaches to robust planning described above. Raystation (Raysearch Laboratories) was the first planning system to offer this functionality and implements the minimax approach. Pinnacle (Philips Healthcare) has an implementation of the probabilistic approach, and Eclipse (Varian) utilizes the worst-case dose distribution. Hence, robust IMPT optimization is now available not only in a research environment but also for clinical use.

Since robust planning techniques for IMPT were first investigated, the question whether one method is generally superior has been discussed. The probabilistic approach optimizes the average plan quality. It is possible that a plan does not achieve the desired dose quality for the worst scenario, for example in situations where a large number of scenarios are modeled. On the other hand, the minimax approach optimizes the plan for the worst case only and has no incentive to improve plan quality for more likely scenarios. It was shown that some methods do yield undesirable results in specific situations [36], but there is no comprehensive evidence that one method is generally superior. To first approximation, all methods achieve the features of robust plans described above if used adequately.

Most works have used a small set of discrete range and setup errors to model uncertainty or sampled errors from a Gaussian distribution. More recently, researchers developed improved methods to quantify dose uncertainty. Bangert et al. [43] developed analytical methods to calculate expectation value and variance of the dose distribution for Gaussian range and setup errors without relying on sampling. Perko et al. [44,45] used a technique called polynomial chaos expansion to develop a parameterized model of the dose distribution as a function of range and setup errors.

19.5 ACCOUNTING FOR BIOLOGICAL EFFECTS IN PBS OPTIMIZATION

Treatment planning for proton therapy usually employs a constant relative biological effectiveness (RBE) factor of 1.1 for the conversion of physical dose d_i to "biological" dose (see Chapter 22). Under the assumption of a constant RBE, treatment plan optimization can be performed based on the physical dose alone, as described in the preceding sections of this chapter. In other words, the physical dose is the only measure that is needed to characterize the radiation field and to assess the quality of the treatment plan. However, this may be oversimplifying, and a second quantity may be needed to characterize the radiation field and its radiobiological effectiveness. In vitro cell survival experiments suggest that the amount of radiation-induced cell kill increases with higher linear energy transfer (LET), and consequently at the end of range of the proton beams [46]. Hence, LET is often used as a second physical quantity to characterize a treatment plan.

Several models have been developed to model RBE as a function of physical dose, LET, and tissue-specific parameters (see Chapter 22) [47–50]. These models have in common that, for a fixed physical dose and tissue parameter, they essentially model an approximately linear increase of RBE with LET. For the purpose of including variable RBE effects into IMPT planning, it is illustrative to consider the following, simplified RBE model. We consider an exponential cell kill model

$$S = \exp(-\alpha d) \tag{19.14}$$

where S denotes the surviving fraction of cells after delivering dose d and α is the radiosensitivity parameter. To first approximation, it is assumed that α increases linearly with dose-averaged LET, which we denote by L:

$$\alpha = \alpha_0 (1+cL) \tag{19.15}$$

In analogy to the biologically effective dose model, the total biological dose b can be defined as

$$b = -\frac{\log(S)}{\alpha_0} = (1+cL)d = d + cLd \tag{19.16}$$

Hence, the product of LET and dose, scaled by a parameter c, can be interpreted as the additional biological dose due to the LET effect, which is added to the physical dose to obtain the total biological dose b. Alternatively, $(1 + cL)$ can be interpreted as the RBE, so that b represents the RBE-weighted dose.

For IMPT planning, the biological dose model is extended to multiple pencil beams. Let $D_{ij}x_j$ denote the physical dose that pencil beam j delivers to voxel i. Here, D_{ij} denotes the dose contribution of pencil beam j to voxel i for unit fluence, and x_j denotes the fluence of pencil beam j. The total cell survival in voxel i is given by

$$S_i = \prod_j \exp\left(-\alpha_0 \left(1+cL_{ij}\right)D_{ij}x_j\right) \tag{19.17}$$

where L_{ij} is the dose-averaged LET of pencil beam j in voxel i. The RBE-weighted dose is thus given by

$$b_i = -\frac{\log(S_i)}{\alpha_0} = \sum_j \left(1+cL_{ij}\right)D_{ij}x_j = d_i + c\sum_j L_{ij}D_{ij}x_j \tag{19.18}$$

where d_i is physical dose in voxel i. By defining the dose-averaged LET over all pencil beam contributions as

$$\bar{L}_i = \frac{\sum_j L_{ij}D_{ij}x_j}{d_i} \tag{19.19}$$

the biological extra dose due to the LET effect is given by the product of physical dose and dose-averaged LET, $\bar{L}_i d_i$, multiplied by a constant c.

The parameter c is uncertain and may depend on tissue type and the dose level. One approach to select a plausible value for c consists in reproducing the clinically used RBE of 1.1 for a certain radiation field. For example, c can be selected such that the RBE in the center of an SOBP is 1.1. For example, to obtain an RBE of 1.1 in the center of an SOBP with a range of 10 cm and a modulation of 5 cm, c is set to $c = 0.04$ μm/keV. Figure 19.10 illustrates the simplified RBE model. Figure 19.10a shows the physical dose for 10^9 protons for a single pencil beam, delivering approximately 4.5 Gy at the Bragg peak. Figure 19.10b shows the additional biological dose $c\bar{L}_i d_i$ in Gy for $c = 0.04$ μm/keV. The sum of $c\bar{L}_i d_i$ and d_i yields the RBE-weighted dose in Figure 19.10c. The RBE (Figure 19.10d) is the ratio of RBE-weighted dose and physical dose, which becomes ill defined where the dose goes to zero. For this simple model, RBE essentially represents dose-averaged LET. For $c = 0.04$ μm/keV, $c\bar{L}_i d_i$ adds another 1.5 Gy to the physical dose near the Bragg peak, resulting in an RBE of approximately 1.3. In the falloff region distal to the Bragg peak, the RBE increases to values around 1.5–1.6.

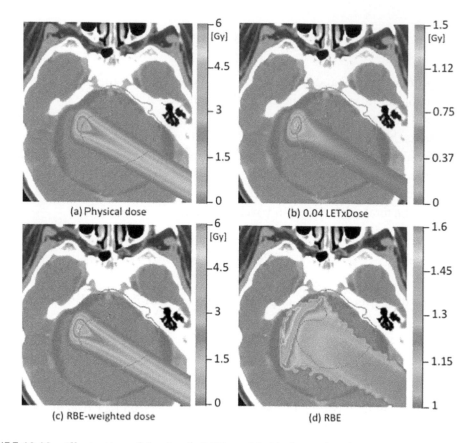

FIGURE 19.10   Illustration of the simple RBE model. (a) shows the physical dose distribution of a single pencil beam with 10^9 protons; (b) shows the corresponding additional biological dose due to high LET, i.e., LETxDose scaled with $c = 0.04$ µm/keV; (c) shows the total biological dose, i.e., the sum of (a) and (b); (d) shows the RBE, which is the ratio of RBE-weighted dose and physical dose.

The product of LET and physical dose, which we refer to as LETxDose, has several advantages that make it suitable for consideration in IMPT planning:

- LETxDose is a physical quantity that is well defined just like physical dose. It does not depend on uncertain biological parameters.

- LETxDose, when scaled with a constant c, represents a first-order approximation of the biological extra dose that results from high LET. This is an advantage over LET itself.

- Given dose coefficients D_{ij} and LET coefficients L_{ij}, LETxDose is a linear function of the optimization variables x_j, the pencil beam fluence. Hence, the same mathematical optimization techniques as in physical dose optimization can be applied.

The use of the distribution of LETxDose for treatment plan evaluation is demonstrated for a patient treated for an atypical meningioma shown in Figure 19.11. The PTV (blue contour) overlaps with the optic nerve (black contour) and the brain stem (green contour). Figure 19.11a shows the dose distribution of an IMPT treatment plan consisting of three beams, where 50 Gy physical dose is prescribed to the target volume. Additional objectives to limit the maximum dose to OARs to 50 Gy, conformity, and for minimizing brain stem gEUD and integral dose were used. Figure 19.11b shows the distribution of LETxDose, scaled with $c = 0.04$ um/keV. This reveals that high values of LETxDose are observed in the brain stem and the optic structures. The reason for this is that conventional IMPT planning preferentially uses pencil beams whose Bragg peaks fall at the distal edge of the target volume. This is because these pencil beams deliver dose "for free" to the tumor while traversing the target volume. As a consequence, high LETxDose is observed at the edge of the target volume while LETxDose is comparatively small in the center of the gross tumor volume (GTV) (red contour)—an undesired situation.

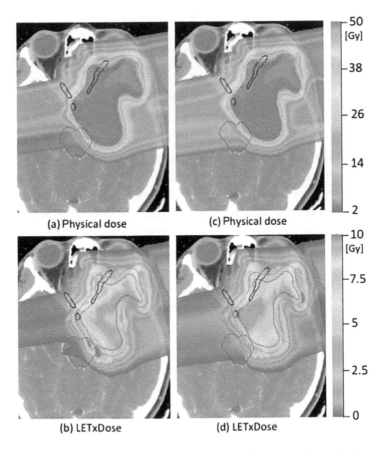

FIGURE 19.11 Illustration of LET-based treatment plan evaluation and reoptimization for an atypical meningioma patient. The contours show the GTV (red), PTV (blue), brain stem (green), optic nerves, and chiasm (black). (a) shows the physical dose delivered to the target volume for a conventional IMPT plan optimized based on physical dose; (b) shows the corresponding LETxDose distribution scaled with $c = 0.04$ μm/keV; (c) physical dose distribution after LET-based reoptimization, indicating minor changes in the physical dose distribution; (d) LETxDose distribution after LET-based reoptimization, showing that high LETxDose is avoided in the brain stem and optic structures.

602 ■ Proton Therapy Physics

One approach to incorporate LET-dependent RBE into IMPT planning is to perform treatment plan optimization based on RBE-weighted dose. For the simplified RBE model above, and equivalently for other RBE models, the treatment planning problem would then be written as

$$\underset{x}{\text{minimize}} \ f(b)$$

$$\text{subject to} \ \ b_i = d_i + cL_i d_i$$

$$d_i = \sum_j D_{ij} x_j$$

$$L_i = \frac{1}{d_i} \sum_j L_{ij} D_{ij} x_j$$

$$x_j \geq 0 \tag{19.20}$$

This means that the OF, and possibly constraint functions, are evaluated for RBE-weighted dose rather than physical dose. Such an approach has been investigated by Wilkens et al. [51], in a modified form. However, this approach typically raises concerns due to uncertainty in the RBE model, in this case the parameter c. The difficulty can be illustrated for the goal of target coverage. Let us assume that a quadratic penalty function for underdose is evaluated for RBE-weighted dose instead of physical dose. Then, IMPT planning will typically yield a treatment plan that lowers the physical dose in the target volume in regions of LET, based on the assumption that RBE is larger than 1.1. If the RBE model overestimates the RBE increase with LET, these areas would potentially be underdosed.

To address the problem of uncertain RBE models, approaches to LET-based IMPT planning have been investigated. The idea is that treatment planning is in parts based on physical dose to remain consistent with the current practice of dose prescription and dose reporting. However, in addition to satisfying planning goals for physical dose, the distribution of LET is controlled. This can have primarily two goals:

1. In the situation that critical normal tissues are located within or near the target volume, it may be desired to avoid high LET in these regions [52–54].

2. In a situation where treatment planning aims at dose escalation in parts of the target volume, high LET in these regions may be desired [55,56].

In analogy to the term dose painting, this has sometimes been called LET painting. A method to avoid high LET in critical structures within or near the target volume has been suggested [54]. In this approach, an initial treatment plan is created based on physical dose alone, which reflects current practice. Subsequently, the LETxDose distribution is evaluated. If this raises concerns as for the example shown in Figure 19.11, the

LETxDose is reoptimized in a second step. The second optimization step can be performed using prioritized optimization is described earlier in this chapter. To that end, the objectives used for physical dose optimization are constrained to values close to their optimal value. This means that the degradation of the treatment plan in terms of the physical dose distribution is tightly constrained. The OF is evaluated for LETxDose. The result of the reoptimization step is shown in Figure 19.11 (right panel) for the atypical meningioma example. In this case, target coverage and conformity of the physical dose distribution were constrained not to deteriorate. A 3% increase in the brain stem gEUD and the mean dose to normal brain was allowed. The OF is a quadratic penalty function evaluated for LETxDose, which penalizes LETxDose values in the optic structures and the brain stem that exceed the minimum LETxDose value observed in the target volume. It is observed that high LETxDose in OARs can be avoided when accepting a minor degradation in the physical dose distribution. Further examples and details can be found in reference [54]. Prioritized optimization is an appealing methodology for LETxDose based reoptimization of IMPT because shaping the physical dose distribution is typically considered more important that the LETxDose distribution. However, PS navigation methods could similarly be applied.

Several authors have investigated the possibility of elevating LET in the target volume, for example in hypoxic subvolumes [55,56]. It has been shown that it is often possible to increase LET in the GTV compared to conventional IMPT plans that are optimized based on physical dose alone. However, these modifications are typically associated with more substantial degradation of the physical dose distribution. The treatment plan that minimizes the healthy tissue dose in the beam entrance regions preferentially uses Bragg peaks placed at the distal edge of the target. In contrast, high LET in the GTV is achieved if as many protons as possible stop within the GTV. In terms of physical dose, this corresponds to an inefficient use of protons, which conflicts with the goal of minimizing healthy tissue dose in the entrance region [54]. A demonstration that, for the same normal tissue entrance dose, LET escalation in the target volume achieves significantly higher RBE-weighted dose than simply allowing the physical dose to increase is yet to be provided.

19.6 OTHER APPLICATIONS OF MATHEMATICAL OPTIMIZATION IN PROTON THERAPY

19.6.1 Management of Intrafractional Motion with "4D Optimization"

Precision of therapy delivery can be affected not only by the changes in setup and patient anatomy between treatment fractions but also by the intrafractional motion of the target, which could be due to respiration, peristalsis, or organ settling due to gravity (see Chapter 18 for more details). If no action is taken, there is always a risk that parts of the target may move outside of the treatment field, resulting in a loss of dose coverage. Even in cases where treatment planning margins are generous enough to cover the full amplitude of motion, intrafractional motion would degrade dose gradients, and increase irradiation of surrounding healthy tissues. An important difference from photon therapy is that, in particle therapy, due to the limited range, the use of margin expansions, such as internal target volumes, requires explicit consideration of possible changes in radiological depth to target, as these are often affected by organ motion [57]. Additionally, like X-rays, in dynamically delivered intensity-modulated therapy, certain patterns of superposition of motion of the target and the scanned beam, or so-called "motion interplay," can have a severe impact on the delivered dose [58].

Numerous approaches have been put forward, which aims to mitigate the impact of intrafractional motion: these include recommendations for selection of planning image set, compensator expansion, internal margins [59,60], delivery methods employing beam gating [61], field rescanning [62,63], and target tracking [64,65]. Some of the challenges to clinical implementation of 4D planning have been described in collaborative workshop reports [66–68].

In this section, we review efforts to incorporate intrafractional motion into the optimization of beam weights in IMPT. Many of these approaches have been inspired by insights originally developed in the context of IMRT with photons. Although the methodology can be often transferred to IMPT, special consideration must be given to the specific challenges mentioned earlier, i.e. interplay effects and sensitivity to changes in radiological path length.

Methods to incorporate intrafractional motion in plan optimization require a characterization of the geometrical variation of the patient's anatomy. For respiratory motion, this can be obtained from respiratory-correlated CT (often called 4DCT), which provides the geometry of the patient in several phases of the breathing cycle [69] or surrogate marker data. Accurate evaluation of the actual dose distribution delivered to a moving target requires first calculating instantaneous dose to all phases of the 4DCT. Figure 19.12 illustrates variation in the proton dose distribution delivered to a changing anatomy, throughout the respiratory cycle. Such instantaneous doses can then be mapped onto a reference anatomical set, by using the correspondence established between the voxels of different CT

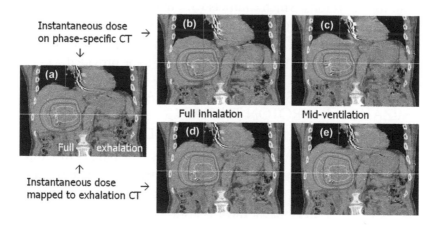

FIGURE 19.12 Dosimetric evaluation of a treatment plan for a tumor in the liver, using respiratory-correlated CT: Estimated instantaneous dose delivered during (a) the full exhalation, (b) full inhalation, and (c) midventilation phases of the respiratory cycle. To estimate the total dose, contributions from various instances of the anatomy have to be mapped onto the reference CT set, e.g., for (d) full inhalation dose (dose "B" mapped onto the full exhalation CT "A"), and (e) midventilation (dose "C" mapped onto CT "A"). (Images courtesy of Dr. S. Mori (NIRS).)

sets, obtained through elastic image registration. The mapped dose can be subsequently added along with contributions from all instances of variable anatomy, to yield the dose accumulated throughout the respiratory cycle [70,71].

In a simple approach to include motion, it is assumed that motion is sufficiently well described by the reconstructed phases of a 4DCT and that the dose delivered to a voxel is obtained by summation of the dose contributions from all phases:

$$d_i = \sum_s p_s d_i^s = \sum_s \left[p_s \sum_j x_j D_{ij}^s \right] = \sum_j x_j \left[\sum_s p_s D_{ij}^s \right] \quad (19.21)$$

Here, s is an index to the instance of geometry, and the voxel index i refers to an anatomical voxel defined in the reference phase. D_{ij}^s is the dose influence matrix for phase s. Its calculation requires elastic registration of the CT of phase s with the reference phase. The parameters p_s are probabilities that the patient is in phase s and are referred to as the "motion probability distribution function (PDF)," which can be estimated from a recorded breathing signal.

Treatment planning can be performed by optimizing the beam weights x_j based on objectives and constraints evaluated with the cumulative dose from Equation 19.21 [72].

The general idea is that, rather than passively let the motion deteriorate the original plan, one should anticipate it, and, in fact, actively engage it in shaping the desired dose distribution. The resulting treatment plan would deliver an inhomogeneous dose distribution to a static geometry. However, the inhomogeneities are designed such that, after accumulating dose over the whole breathing cycle, the desired dose distribution is obtained. Because one of the most manifest effects of motion on the dose is the smoothing or washout of gradients both within the target and at its borders, the logical way to counter this effect is dose boosting at the edges of the target, in what is termed "edge enhancement" [73].

The exact pattern of optimum inhomogeneity enhancement is determined by the form of motion PDF. Generally, the effect of motion on the dose may be approximated as convolution of the dose with the PDF; thus, the desired motion-compensated plan can be roughly approximated with the inverse process: deconvolution. This is however constrained by the requirement that the fluences delivered at all pencil beam spots are physical, thus, if negative values arise from deconvolution or during optimization, those need to be reset to zero (or the minimum allowed, if the beam cannot be completely turned off, e.g., in a continuous scan).

Since the PDF does not depend on time, the use of probabilistic planning does not require complex technical delivery modifications to ensure synchronization of the beam with the motion cycle, and thus delivery of such fields can be relatively easily implemented in practice. However, PDF-based optimization methods rely on the reproducibility of target motion patterns during delivery, and sufficient sampling of the motion PDF. When motion deviates from the expectation, a significant dosimetric deviation may occur.

Through the use of robust planning techniques, the dosimetric outcome of a treatment plan based on a known motion PDF can be made less vulnerable to variations in the breathing pattern. This approach thus aims at finding a treatment plan that yields an acceptable cumulative dose distribution even if the actual breathing pattern during treatment differs from the estimated motion PDF assumed for treatment plan optimization. Chan et al. [74,75] investigated robust optimization for respiratory motion by modeling the variability in the breathing motion via uncertainties in the motion PDF parameters p_s.

Instead of optimizing a single fluence map that is delivered without synchronization with the breathing pattern, one could also optimize a separate fluence map x_j^s for specific phases of motion. The objectives and constraints for treatment plan optimization are formulated in terms of the cumulative dose, with the distinction that the fluence map is different in every phase [72]. The delivery of such treatment plans would require a synchronization of the layer delivery with the breathing motion, which has become more feasible with development of fast PBS. Several studies proposed phase-synchronized plans for particle therapy [76–78], assuming the availability of reliable control system for both targeting and delivery.

In the effort to increase robustness of 4D solutions to irregularities and uncertainties of motion, and building upon the 4D dose calculation via deforming calculation grid [79], an optimization technique was proposed, which incorporated both irregular motion and delivery dynamics of the treatment machine [80].

Others explored the possibilities for circumventing the requirement of perfect motion-delivery synchronization by combining robust optimization and 4D treatment planning. Liu et al. [81] demonstrated that a single, non-phase-specific plan could be constructed, which made the target dose distribution more resilient to uncertainties and simultaneously mitigated the interplay effect, while improving conformity compared to conventional 3D planning.

19.6.2 Spot Grid and Scan Path Optimization

In PBS, beam spots are typically placed on a 3D grid throughout the entire tumor region. As a result of weight optimization, a large number of beam spots will be assigned zero (or negligible, below the minimum deliverable by the given system) weight in the optimization of the treatment plan. Nevertheless, in a naïve implementation of spot scanning, the beam would be steered in a zigzag pattern over the entire grid, including the spot positions that correspond to zero weight. Early studies, such as Kang et al. [82], formulated the approaches to minimize the delivery time by optimizing scan path to avoid regions with zero-weight spots: the problem corresponds to a traveling salesman problem, and simulated annealing has been applied to solve the problem. More recently, the focus shifted to identifying more optimum irregular placement patterns, both in terms of efficient selection of beam energies for layer irradiation as well as optimized clustering and culling of spots within a given layer to improve delivery efficiency [83–85].

Further studies probed the hypothesis that since various scanning sequences may be affected by motion and uncertainties differently, a more favorable delivery sequence might be identified through optimization, taking into consideration characteristics of motion [86].

19.6.3 Beam Angle Optimization

The automated beam selection problem has been extensively studied in photon-based radiation therapy with a number of distinct approaches proposed [87–90]. However, its indication in proton treatment planning has remained debatable. While early reports have demonstrated the importance of selection of beam angle approaches, e.g., for base of skull tumors [91] and a number of *in silico* treatment planning studies reported on potentially improved dose distributions achieved by using nonstandard angles, any stated benefits must be considered in the context of delivery uncertainties. The reason being that highly conformally designed doses delivered with plans optimized under idealized conditions may deteriorate dramatically due to delivery errors [92]. Recent proposals tend to tackle beam selection as a part of robust optimization problem. As an example, Cao et al. [93] adapted a worst-case optimization approach to optimize treatment plans, in which the beam angles were optimized for achieving not only dosimetric benefits but also robustness against uncertainties.

19.6.4 Optimization for Rotational Proton Therapy

The potential for improved sparing of the normal lung tissue with rotational approach to proton delivery was first described over two decades ago [94]; however, technical implementation was hindered by, to name just a few, the difficulties of sufficiently quick adjustment of the beam range (for both scattered-field and PBS-based delivery), distal confirmation and longitudinal modulation of the dose (whether by changing the modulator devices or the time structure of the beam current modulation), or synchronization of continuous delivery with the gantry rotation.

With the development, refinement, and broader implementation of fast PBS, proton arc delivery attracted more interest recently: in particular, its advantage in lung treatment was revisited in terms of not only reducing the tissue dose but also maintaining target conformity due to improved robustness to range uncertainties [95]. Sánchez-Parcerisa et al. [96,97] investigated rotational proton modulated arc therapy solutions using fast-range switching and dynamic beam current modulation in a multileaf collimation system. Ding et al. [98,99] formulated spot-scanning proton-arc (SPArc) approach based on robust optimization. SPArc integrates beam spot resampling with energy layer reorganization and redistribution among delivery angles to achieve a more robust and delivery-efficient plans for continuous arc delivery.

REFERENCES

1. Wilson R. 1946. Radiological use of fast protons. *Radiology.* 47:487.
2. Koehler AM, Schneider RJ, Sisterson JM. 1977. Flattening of proton dose distributions for large-field radiotherapy. *Med Phys.* 4:297–301.
3. Lu HM, Kooy H. 2006. Optimization of current modulation function for proton spread-out Bragg peak fields. *Med Phys.* 33:1281–7.
4. Lu HM, Brett R, Engelsman M, Slopsema R, Kooy H, Flanz J. 2007. Sensitivities in the production of spread-out Bragg peak dose distributions by passive scattering with beam current modulation. *Med Phys.* 34:3844–53.
5. Engelsman M, Lu HM, Herrup D, Bussiere M, Kooy HM. 2009. Commissioning a passive-scattering proton therapy nozzle for accurate SOBP delivery. *Med Phys.* 36:2172–80.
6. Bussière MR, Adams AA. 2003. Treatment planning for conformal proton radiation therapy. *Technol Cans Res Treat.* 2:389–399.

7. Urie M, Goitein M, Wagner M. 1984. Compensating for heterogeneities in proton radiation therapy. *Phys Med Biol*. 29:553–66.

8. Lomax A. 1999. Intensity modulation methods for proton radiotherapy. *Phys Med Biol*. 44:185–205.

9. Deasy JO, Shepard DM, Mackie TR. 1997. Distal edge tracking: A proposed delivery method for conformal proton therapy using intensity modulation. In *Proceedings of XII-th ICCR*, eds. D.D. Leavitt, and G. Starkshall, 406–409. Salt Lake City, Utah: Medical Physics Publishing.

10. Oelfke U, Bortfeld T. 2001. Inverse planning for photon and proton beams. *Med Dosim*. 26:113–24.

11. Nill S, Bortfeld T, Oelfke U. 2004. Inverse planning of intensity modulated proton therapy. *Z Med Phys*. 14:35–40.

12. Thieke C, Bortfeld T, Niemierko A, Nill S. 2003. From physical dose constraints to equivalent uniform dose constraints in inverse radiotherapy planning. *Med Phys*. 30:2332–9.

13. Kang JH, Wilkens JJ, Oelfke U. 2008. Non-uniform depth scanning for proton therapy systems employing active energy variation. *Phys Med Biol*. 53:N149–55.

14. Trofimov A, Bortfeld T. 2003. Optimization of beam parameters and treatment planning for intensity modulated proton therapy. *Technol Cancer Res Treat*. 2:437–44.

15. Bortfeld T. 1999. Optimized planning using physical objectives and constraints. *Semin Radiat Oncol*. 9:20–34.

16. Lahanas M, Schreibmann E, Baltas D. 2003. Multiobjective inverse planning for intensity modulated radiotherapy with constraint-free gradient-based optimization algorithms. *Phys Med Biol*. 48:2843–71.

17. Romeijn E, Dempsey J, Li J. 2004. A unifying framework for multi-criteria fluence map optimization models. *Phys Med Biol*. 49:1991–2013.

18. Craft D, Halabi T, Bortfeld T. 2005. Exploration of tradeoffs in intensity-modulated radiotherapy. *Phys Med Biol*. 50:5857–68.

19. Hong TS, Craft DL, Carlsson F, Bortfeld TR. 2008. Multicriteria optimization in intensity-modulated radiation therapy treatment planning for locally advanced cancer of the pancreatic head. *Int J Radiat Oncol Biol Phys*. 72:1208–14.

20. Thieke C, Küfer KH, Monz M, et al. 2007. A new concept for interactive radiotherapy planning with multicriteria optimization: first clinical evaluation. *Radiother Oncol*. 85:292–8.

21. Monz M, Küfer KH, Bortfeld TR, Thieke C. 2008. Pareto navigation: algorithmic foundation of interactive multi-criteria IMRT planning. *Phys Med Biol*. 53:985–98.

22. Clark VH, Chen Y, Wilkens J, Alaly JR, Zakaryan K, Deasy JO. 2008. IMRT treatment planning for prostate cancer using prioritized prescription optimization and mean-tail-dose functions. *Linear Algebra Appl*. 428:1345–1364.

23. Ehrgott M. 2005. *Multicriteria Optimization*. Berlin: Springer.

24. Jee KW, McShan DL, Fraass BA. 2007. Lexicographic ordering: Intuitive multicriteria optimization for IMRT. *Phys Med Biol*. 52:1845–61.

25. Rennen G, van Dam ER, den Hertog D. 2009. Enhancement of Sandwich Algorithms for Approximating Higher Dimensional Convex Pareto Sets. Discussion Paper 2009–52, Tilburg University, Center for Economic Research.

26. Craft D, Bortfeld T. 2008. How many plans are needed in an IMRT multi-objective plan database? *Phys Med Biol*. 53:2785–96.

27. Craft D. 2010. Calculating and controlling the error of discrete representations of Pareto surfaces in convex multi-criteria optimization. *Phys Med*. 26: 184–91.

28. Messac A, Mattson CA. 2004. Normal constraint method with guarantee of even representation of complete Pareto frontier. *J AIAA*. 42:2101–2111.

29. Craft D, Monz M. 2010. Simultaneous navigation of multiple Pareto surfaces, with an application to multicriteria IMRT planning with multiple beam angle configurations. *Med Phys*. 37:736–41.

30. Kamran SC, Mueller BS, Paetzold P, et al. 2016. Multi-criteria optimization achieves superior normal tissue sparing in a planning study of intensity-modulated radiation therapy for RTOG 1308-eligible non-small cell lung cancer patients. *Radiother Oncol.* 118:515–20.

31. Lomax AJ. 2008. Intensity modulated proton therapy and its sensitivity to treatment uncertainties 1: The potential effects of calculational uncertainties. *Phys Med Biol.* 53:1027–42.

32. Lomax, AJ. 2008. Intensity modulated proton therapy and its sensitivity to treatment uncertainties 2: The potential effects of inter-fraction and inter-field motions. *Phys Med Biol.* 53:1043–56.

33. Unkelbach J, Chan TC, Bortfeld T. 2007. Accounting for range uncertainties in the optimization of intensity modulated proton therapy. *Phys Med Biol.* 52:2755–73.

34. Pflugfelder D, Wilkens JJ, Oelfke U. 2008. Worst case optimization: a method to account for uncertainties in the optimization of intensity modulated proton therapy. *Phys Med Biol.* 53:1689–700.

35. Unkelbach J, Martin BC, Soukup M, Bortfeld T. 2009. Reducing the sensitivity of IMPT treatment plans to setup errors and range uncertainties via probabilistic treatment planning. *Med Phys.* 36:149–63.

36. Fredriksson A, Forsgren A, Hardemark B. 2011. Minimax optimization for handling range and setup uncertainties in proton therapy. *Med Phys.* 38:1672–84.

37. Chen W, Unkelbach J, Trofimov A, Madden T, Kooy H, Bortfeld T, Craft D. 2012. Including robustness in multi-criteria optimization for intensity-modulated proton therapy. *Phys Med Biol.* 57:591–608.

38. Liu W, Zhang X, Li Y, Mohan R. 2012. Robust optimization of intensity modulated proton therapy. *Med Phys.* 39:1079–91.

39. Liu W, Frank SJ, Li X, et al. 2013 Effectiveness of robust optimization in intensity-modulated proton therapy planning for head and neck cancers. *Med Phys.* 40:051711.

40. Liu W, Mohan R, Park P, et al. 2014. Dosimetric benefits of robust treatment planning for intensity modulated proton therapy for base-of-skull cancers. *Pract Radiat Oncol.* 4: 384–91.

41. Fredriksson A, Bokrantz R. 2014. A critical evaluation of worst case optimization methods for robust intensity-modulated proton therapy planning. *Med Phys.* 41:081701.

42. Li H, Zhang X, Park P, et al. 2015. Robust optimization in intensity-modulated proton therapy to account for anatomy changes in lung cancer patients. *Radiother Oncol.* 114:367–72.

43. Bangert M, Hennig P, Oelfke U. 2013. Analytical probabilistic modeling for radiation therapy treatment planning. *Phys Med Biol.* 58:5401–19.

44. Perko Z, van der Voort SR, van de Water S, Hartman CM, Hoogeman M, Lathouwers D. 2016. Fast and accurate sensitivity analysis of IMPT treatment plans using Polynomial Chaos Expansion. *Phys Med Biol.* 61:4646–64.

45. van der Voort S., van de Water S, Perkó Z, Heijmen B, Lathouwers D, Hoogeman M. 2016. Robustness recipes for minimax robust optimization in intensity modulated proton therapy for oropharyngeal cancer patients. *Int J Radiat Oncol Biol Phys.* 95:163–70.

46. Paganetti H. 2014. Relative biological effectiveness (RBE) values for proton beam therapy. Variations as a function of biological endpoint, dose, and linear energy transfer. *Phys Med Biol.* 59:R419–72.

47. Carabe A, Moteabbed M, Depauw N, Schuemann J, Paganetti H. 2012. Range uncertainty in proton therapy due to variable biological effectiveness. *Phys Med Biol.* 57:1159–72.

48. McNamara AL, Schuemann J, Paganetti H. 2015.A phenomenological relative biological effectiveness (RBE) model for proton therapy based on all published in vitro cell survival data. *Phys Med Biol.* 60:8399–416.

49. Wedenberg M, Lind BK, Hardemark B. 2013. A model for the relative biological effectiveness of protons: the tissue specific parameter alpha/beta of photons is a predictor for the sensitivity to LET changes. *Acta Oncol.* 52:580–8.

50. Wilkens JJ, Oelfke U. 2004 A phenomenological model for the relative biological effectiveness in therapeutic proton beams. *Phys Med Biol.* 49:2811–25.

51. Wilkens JJ, Oelfke U. 2005. Optimization of radiobiological effects in intensity modulated proton therapy. *Med Phys.* 32:455–65.

52. Giantsoudi D, Grassberger C, Craft D, Niemierko A, Trofimov A, Paganetti H. 2013 Linear energy transfer-guided optimization in intensity modulated proton therapy: feasibility study and clinical potential. *Int J Radiat Oncol Biol Phys.* 87:216–22.

53. Grassberger C, Trofimov A, Lomax A, Paganetti H. 2011. Variations in linear energy transfer within clinical proton therapy fields and the potential for biological treatment planning. *Int J Radiat Oncol Biol Phys.* 80:1559–1566.

54. Unkelbach J, et al. 2016. Reoptimization of intensity modulated proton therapy plans based on linear energy transfer. *Int J Radiat Oncol Biol Phys.* 96:1097–1106.

55. Fager M, et al. Linear energy transfer painting with proton therapy: a means of reducing radiation doses with equivalent clinical effectiveness. *Int J Radiat Oncol Biol Phys.* 2015. 91(5): 1057–64.

56. Bassler N, et al. 2014. LET-painting increases tumour control probability in hypoxic tumours. *Acta Oncol.* 53(1): 25–32.

57. Mori S, Chen GT. 2008. Quantification and visualization of charged particle range variations. *Int J Radiat Oncol Biol Phys.* 72:268–77.

58. Rietzel E, Bert C. 2010. Respiratory motion management in particle therapy. *Med Phys.* 37:449–60.

59. Engelsman M, Rietzel E, Kooy HM. 2006. Four-dimensional proton treatment planning for lung tumors. *Int J Radiat Oncol Biol Phys.* 64:1589–95.

60. Kang Y, Zhang X, Chang JY, Wang H, Wei X, Liao Z, et al. 2007. 4D Proton treatment planning strategy for mobile lung tumors. *Int J Radiat Oncol Biol Phys.* 67:906–14.

61. Lu HM, Brett R, Sharp G, Safai S, Jiang S, Flanz J, Kooy H. 2007. A respiratory-gated treatment system for proton therapy. *Med Phys.* 34:3273–8.

62. Seco J, Robertson D, Trofimov A, Paganetti H. 2009. Breathing interplay effects during proton beam scanning: simulation and statistical analysis. *Phys Med Biol.* 54:N283–94.

63. Zhang Y, Huth I, Wegner M, Weber DC, Lomax AJ. 2016 An evaluation of rescanning technique for liver tumour treatments using a commercial PBS proton therapy system. *Radiother Oncol.* 121:281–287.

64. van de Water S, Kreuger R, Zenklusen S, Hug E, Lomax AJ. 2009. Tumour tracking with scanned proton beams: assessing the accuracy and practicalities. *Phys Med Biol.* 54:6549–63.

65. Bert C, Gemmel A, Saito N, et al. 2010. Dosimetric precision of an ion beam tracking system. *Radiat Oncol.* 5:61.

66. Knopf A, Nill S, Yohannes I, et al. 2014. Challenges of radiotherapy: Report on the 4D treatment planning workshop. *Phys Med.* 30: 809–15.

67. Knopf A, Bert C, Heath E, et al. 2010. Special report: Workshop on 4D-treatment planning in actively scanned particle therapy-Recommendations, technical challenges, and future research directions. *Med Phys.* 37:4608–4614.

68. Knopf AC, Stützer K, Richter C, et al. 2016. Required transition from research to clinical application: Report on the 4D treatment planning workshops 2014 and 2015. *Phys Med.* 32:874–82.

69. Keall P. 2004. 4-dimensional computed tomography imaging and treatment planning. *Semin Radiat Oncol.* 14:81–90.

70. Bert C, Rietzel E. 2007. 4D treatment planning for scanned ion beams. *Radiat Oncol.* 2:24.

71. Zhang X, Zhao KL, Guerrero TM, et al. 2008. Four-dimensional computed tomography-based treatment planning for intensity-modulated radiation therapy and proton therapy for distal esophageal cancer. *Int J Radiat Oncol Biol Phys.* 72:278–87.

72. Trofimov A, Rietzel E, Lu HM, et al. 2005. Temporo-spatial IMRT optimization: concepts, implementation and initial results. *Phys Med Biol.* 50:2779–98.

73. Chan TC, Tsitsiklis JN, Bortfeld T. 2010. Optimal margin and edge-enhanced intensity maps in the presence of motion and uncertainty. *Phys Med Biol.* 55:515–33.

74. Chan TC, Bortfeld T, Tsitsiklis JN. 2006. A robust approach to IMRT optimization. *Phys Med Biol.* 51:2567–83.
75. Bortfeld T, Chan TCY, Trofimov A, Tsitsiklis JN. 2008. Robust management of motion uncertainty in intensity-modulated radiation therapy. *Oper Res.* 56:1461–1473.
76. Graeff C, Luechtenborg R, Eley JG, et al. 2013. A 4D-optimization concept for scanned ion beam therapy. *Radiother Oncol.* 109:419–424.
77. Graeff C. 2014. Motion mitigation in scanned ion beam therapy through 4D-optimization. *Phys Med-Eur J Med Phys.* 30:570–577.
78. Richter D, Schwarzkopf A, Trautmann J, Krämer M, Durante M, Jäkel O, Bert C. 2013. Upgrade and benchmarking of a 4D treatment planning system for scanned ion beam therapy. *Med Phys.* 40:051722.
79. Van de Water S, Kreuger R, Zenklusen S, Hug E., Lomax AJ. 2009. Tumour tracking with scanned proton beams: Assessing the accuracy and practicalities. *Phys Med Biol.* 54:6549.
80. Bernatowicz K, Zhang Y, Perrin R, Weber DC, Lomax AJ. 2017 Advanced treatment planning using direct 4D optimisation for pencil-beam scanned particle therapy. *Phys Med Biol.* 62:6595–6609.
81. Liu W, Schild SE, Chang JY, et al. 2016. Exploratory study of 4D versus 3D robust optimization in intensity modulated proton therapy for lung cancer. *Int J Radiat Oncol Biol Phys.* 95:523–33
82. Kang JH, Wilkens JJ, Oelfke U. 2007. Demonstration of scan path optimization in proton therapy. *Med Phys.* 34:3457–64.
83. Morávek Z, Rickhey M, Hartmann M, Bogner L. 2009 Uncertainty reduction in intensity modulated proton therapy by inverse Monte Carlo treatment planning. *Phys Med Biol.* 54:4803–1.
84. van de Water S, Kooy HM, Heijmen BJ, Hoogeman MS. 2015. Shortening delivery times of intensity modulated proton therapy by reducing proton energy layers during treatment plan optimization. *Int J Radiat Oncol Biol Phys.* 92:460–8.
85. van de Water S, Kraan AC, Breedveld S, et al. 2013. Improved efficiency of multi-criteria IMPT treatment planning using iterative resampling of randomly placed pencil beams. *Phys Med Biol.* 58:6969–8.
86. Li H, Zhu XR, Zhang X. 2015. Reducing dose uncertainty for spot-scanning proton beam therapy of moving tumors by optimizing the spot delivery sequence. *Int J Radiat Oncol Biol Phys.* 93:547–56.
87. Bortfeld T, Schlegel W. 2003. Optimization of beam orientations in radiation therapy—some theoretical considerations. *Phys Med Biol.* 38:291–304.
88. Rowbottom CG, Webb S, Oldham M. 1999. Beam-orientation customization using an artificial neural network. *Phys Med Biol.* 44:2251–2262.
89. Li Y, Yao J, Yao D. 2004. Automatic beam angle selection in IMRT planning using genetic algorithm. *Phys Med Biol.* 49:1915–1932.
90. Wang X, Zhang X, Dong L, Liu H, Gillin M, Ahamad A, Ang K, Mohan R. 2005. Effectiveness of noncoplanar IMRT planning using a parallelized multiresolution beam angle optimization method for paranasal sinus carcinoma. *Int J Radiat Oncol Biol Phys.* 63:594–60.
91. Jäkel O., Debus J. 2000. Selection of beam angles for radiotherapy of skull base tumours using charged particles. *Phys Med Biol.* 45:1229–1241.
92. Trofimov A, Unkelbach J, DeLaney TF, Bortfeld T. 2012. Visualization of a variety of possible dosimetric outcomes in radiation therapy using dose-volume histogram bands. *Pract Radiat Oncol.* 2:164–71.
93. Cao W, Lim GJ, Lee A, Li Y, Liu W, Ronald Zhu X, Zhang X. 2012. Uncertainty incorporated beam angle optimization for IMPT treatment planning. *Med Phys.* 39:5248–56.
94. Sandison GA, Papiez E, Bloch C, Morphis J. 1997. Phantom assessment of lung dose from proton arc therapy. *Int J Radiat Oncol Biol Phys.* 38:891–897.

95. Seco J, Gu G, Marcelos T, Kooy H, Willers H. 2013. Proton arc reduces range uncertainty effects and improves conformality compared with photon volumetric modulated arc therapy in stereotactic body radiation therapy for non-small cell lung cancer. *Int J Radiat Oncol Biol Phys*. 87:188–94.

96. Sánchez-Parcerisa D, Pourbaix JC, Ainsley CG, Dolney D, Carabe A. 2014 Fast range switching of passively scattered proton beams using a modulation wheel and dynamic beam current modulation. *Phys Med Biol*. 59:N19–26.

97. Sanchez-Parcerisa D, Kirk M, Fager M, Burgdorf B, Stowe M, Solberg T, Carabe A. 2016. Range optimization for mono- and bi-energetic proton modulated arc therapy with pencil beam scanning. *Phys Med Biol*. 61:N565–N574.

98. Ding X, Li X, Zhang JM, Kabolizadeh P, Stevens C, Yan D. 2016. Spot-scanning proton arc (SPArc) therapy: The first robust and delivery-efficient spot-scanning proton arc therapy. *Int J Radiat Oncol Biol Phys*. 96:1107–1116.

99. Ding X, Li X, Qin A, et al. 2017. Have we reached proton beam therapy dosimetric limitations?— A novel robust, delivery-efficient and continuous spot-scanning proton arc (SPArc) therapy is to improve the dosimetric outcome in treating prostate cancer. *Acta Oncol*. 3:1–3.

VI

Imaging

CHAPTER 20

Proton Image Guidance*

Jon Kruse

Mayo Clinic

CONTENTS

20.1 INTRODUCTION

X-ray based external beam radiotherapy has advanced dramatically over the last 20 years, with innovations including inverse planning, intensity modulation, and image guidance. Of these, daily online image analysis and patient positioning before treatment has arguably done the most to reduce radiotherapy toxicity related to normal tissue dose. Reliable localization of the target has allowed practitioners to reduce treatment margins without sacrificing target coverage. Additionally, precise and accurate localization is required to exploit the steep dose gradients that are made possible with intensity-modulated radio-therapy. Daily pretreatment localization became commonplace with the advent of digital flat panel X-ray imagers [1]. These devices, electronic portal imaging devices (EPIDs), were

---

* Colour figures available online at www.crcpress.com/9781138626508.

initially mounted opposite the megavoltage X-ray source on treatment linear accelerators and delivered high-quality images for immediate analysis with relatively low imaging dose [2,3]. In the next generation of image guidance development, a diagnostic X-ray tube and an additional flat panel imager were mounted on X-ray linac gantries, orthogonal to the treatment beam direction. Owing to the enhanced photoelectric attenuation of kilovoltage (kV) X-rays by bony anatomy and to the increased interaction probability of low-energy X-rays, these kV imagers generated much higher image quality at lower imaging dose than EPIDs [4,5]. Finally, high-speed acquisition of digital radiographs coupled with gantry rotation led to the development of cone beam computed tomography (CBCT) and volumetric soft tissue localization before treatment [6].

Many essential elements of the image guidance process such as digitally reconstructed radiographs (DRRs) [7] and computerized analysis of orthogonal radiographs [8] were developed for proton therapy at the Harvard Cyclotron Laboratory. However, even though proton dose distributions are particularly sensitive to setup deviations [9,10], the advancement of image guidance in proton therapy lagged behind X-ray clinics in the early 21st century. This is due partially to the relatively small market share of proton therapy equipment and proportionally limited development dollars invested by equipment manufacturers, but there are also physical constraints within some proton therapy installations that limit the application of standard radiographic localization equipment. As the number of proton therapy centers has increased rapidly, however, the sophistication of the image guidance systems in these centers has followed and closed the technological gap with X-ray therapy.

When comparing image guidance for X-ray and proton radiotherapy, it is important to recognize important differences on how these two modalities are affected by deviations in patient position and anatomy. Figure 20.1 shows a single-field proton plan treating a circular target volume. In Panel A, the field is properly aligned to the phantom and the dose distribution conforms very tightly to the target. In Panel B, the beam is misaligned

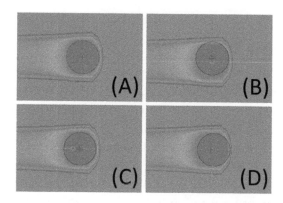

FIGURE 20.1   Panel A shows a conformal proton plan treating a circular target. In Panel B, the phantom is misaligned laterally, and a lateral portion of the target is untreated. In Panel C, the phantom is misaligned longitudinally, but the target coverage is unaffected. In Panel D, the target is properly aligned to the beam, but it has moved closer to the phantom surface, resulting in a loss of coverage to the distal target.

to the phantom with a lateral offset, and the dose distribution misses the target in the same way an X-ray field would. In Panel C, the beam is misaligned to the phantom with a longitudinal offset, but the target coverage is nearly identical to Panel A with no setup error. Finally, in Panel D the target is aligned properly to the field, but it has moved internally relative to the phantom surface, and the distal portion of the target is missed. These simple diagrams illustrate an important property of proton treatment plans. X-ray dose distributions are only minimally affected by changes in the patient contour or heterogeneities, and therefore placing the target in the correct position relative to isocenter generally guarantees adequate target coverage. Proton dose distributions, however, are substantially modified by any changes in patient anatomy. Simply placing the target at isocenter does not guarantee adequate dosimetric coverage; practitioners must also be concerned with the location of the target within the patient and any changes in anatomy between the target and the patient surface.

20.2 RADIOGRAPHIC LOCALIZATION

20.2.1 Geometry

The most common modality for image guidance in either X-ray or proton external beam radiotherapy is planar radiography. Typically, in X-ray therapy there is a single flat panel imager mounted opposite to the treatment X-ray source and a second kV tube and flat panel mounted orthogonal to the treatment beam direction. A primary advantage of this arrangement in X-ray radiotherapy is that the treatment beam itself may be used to image the patient. Naturally, there is no radiation exiting the patient during proton therapy to form an image, but the beam's eye view (BEV) of a patient is often the preferred imaging angle for judging the localization of a patient before treatment, especially in treatments with an aperture or multileaf collimator shaping the field. For this reason, a number of proton therapy systems have kV X-ray tubes mounted in the beam nozzle [11,12]. These tubes are driven into the beam path for pretreatment imaging and retracted before the proton beam is delivered. While there is no distinct disadvantage to this arrangement in a passively scattered proton beam system, most new facilities deliver actively scanned proton beams. A primary measure of the quality of a scanned proton beam therapy system is the size of the proton pencil beam at isocenter—smaller beam spots naturally allow for more conformal dose distributions. Mounting an X-ray tube in the proton path, downstream of the beamline vacuum window, increases the thickness of the air column the beam must traverse before reaching the patient and increases the size of the proton beam at the patient, especially for low energies. An example of this nozzle design is pictured in Figure 20.2 [12].

Newer spot scanning systems tend to mount the BEV X-ray tube under vacuum upstream of the beam window, or the BEV is relinquished altogether and a pair of X-ray tubes are mounted in air with their beam axes oriented 45° away from the proton beam axis [13]. While these designs help minimize growth of the beam spot through scattering in air, X-ray tube cooling under vacuum presents an additional technological challenge, and X-ray tubes mounted at angles offset from the treatment beam direction may require additional gantry rotations to image the patient at cardinal angles. A novel equipment design in Gantry 2 at Paul Scherrer Institute in Switzerland has an X-ray tube located

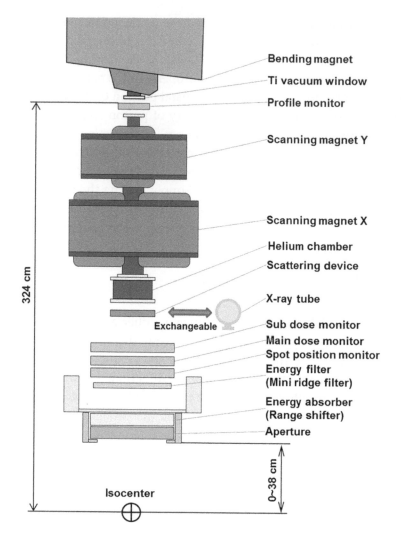

Bending magnet
Ti vacuum window
Profile monitor
Scanning magnet Y
Scanning magnet X
Helium chamber
Scattering device
X-ray tube
Exchangeable
Sub dose monitor
Main dose monitor
Spot position monitor
Energy filter
(Mini ridge filter)
Energy absorber
(Range shifter)
Aperture
324 cm
0–38 cm
Isocenter

FIGURE 20.2 A scanning beam proton nozzle design features an X-ray tube that can be positioned in the beam path.

behind the last gantry bending magnet. The return yoke of the magnet has an aperture to allow X-rays to pass through and create a 20×25 cm^2 BEV image at isocenter [14].

Not all radiographic localization systems are gantry mounted, however. A number of systems have been developed for X-ray radiotherapy with kV sources and image detectors affixed to the treatment room, rather than the gantry. One example of this is the BrainLAB ExacTrac system, with X-ray tubes mounted under the floor and the image receptors suspended from the ceiling. This geometry allowed for development of a third party commercial image guidance system without altering the structure of the X-ray linac gantry. The other advantage of this geometry is that the patient anatomy at isocenter can be imaged at any point in the treatment process, independent of gantry or couch angle. A similar configuration can be seen in CyberKnife X-ray treatment rooms, with a pair of X-ray tubes hanging from the ceiling and imaging panels either on, or under, the floor. In this case, the geometry is required by the fact that a CyberKnife radiotherapy source is mounted on a robotic arm and there is no gantry to support the IGRT apparatus.

Because of the massive size and cost of a proton therapy gantry, many proton treatment rooms are built with either a single-fixed beam port, multiple-fixed beam ports, or a half gantry [15]. Each of these applications requires an IGRT installation without gantry-mounted imagers. Attaching X-ray tubes and panels to the floor, ceiling, and walls of the room remains a common approach. An example can be seen in Figure 20.3, showing a treatment room at Northwestern Medicine Chicago Proton Center. A single treatment nozzle pivots between two fixed locations on one wall. One X-ray tube/imager pair is mounted on the wall orthogonal to the beam direction. The second X-ray tube is mounted under the floor and pointed towards a panel hanging from the ceiling.

Other facilities, such as the fixed beam rooms at the Heidelberg Ion Therapy Center [16] and the half gantry treatment rooms at St Jude Children's Research Hospital [17] each feature kV C-arm imagers mounted on robotic arms. While more mechanically complex than fixed geometry systems, robotic C-arms offer the ability to image at any tube angle without the support of a full treatment gantry. Robotic C-arms may also be used to produce CBCT images of the patient—a feature that will be discussed in Section 20.4.

20.2.2 Matching Techniques

In most applications, localization radiographs are compared to reference DRRs from the planning system to determine setup corrections, which are automatically transferred to the patient couch. The simplest systems rely on human operators to adjust alignment of

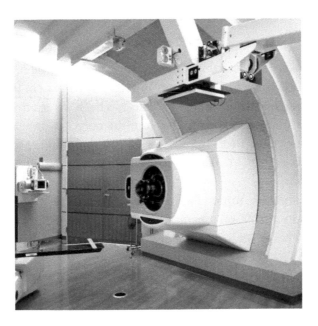

FIGURE 20.3   The fixed beam treatment room at Northwestern Medicine Chicago Proton Center has a nozzle that pivots between two beam angles. The IGRT equipment is mounted on the walls, floor, and ceiling. Image courtesy of Mark Pankuch.

radiographs overlaid on the reference images. Out-of-plane rotations are difficult to judge and apply manually, so human-guided matching algorithms generally provide only translational corrections or perhaps translations and a single rotational axis.

Modern image guidance systems feature automated analysis tools that can extract full 6 degree of freedom correction vectors from a pair of planar radiographs. There are a number of techniques for determining all six translational and rotational deviations from a pair of radiographs, but most modern approaches adjust the position and rotation of the reference CT scan and generate DRRs until a the best match to the acquired radiographs is found [18]. An important feature of an automated registration algorithm is the ability to manually indicate a region of interest where pertinent anatomy is located in the radiographs. Human patients are not rigid bodies, and many radiographically evident features tend not to correlate well with location of the target tissue. Cervical spine and skull base are typically good surrogates for target location in head and neck treatments, for instance, while mandible and clavicles may not be. Users should be able to guide the algorithm to focus on pertinent anatomy or to "paint out" features which are to be ignored.

These automated registration techniques are a natural complement to 6 degree of freedom robotic patient positioners that are found in many proton treatment rooms. However, best practices dictate a limit on the magnitude of both translational and rotational shifts that should be applied as a result of image guidance. Translational shifts of more than 3 cm away from initial positioning should trigger a careful pretreatment review of the setup and image analysis to ensure that the users have not aligned to an incorrect vertebral body, for example. Similarly, users may wish to consider limiting pitch and roll corrections to no more than 5° to avoid changes in the patient anatomy or involuntary patient motion as a result of the applied correction.

20.2.3 Fluoroscopic Image Guidance

Modern flat panel imagers are capable of capturing images in cine mode with frame rates of 30 Hz or higher to allow for fluoroscopic evaluation of the patient in treatment position. Pretreatment fluoroscopic examination of patients has been shown to be useful in tailoring respiratory gating parameters [19] or to measure the correlation between tumor motion and external surrogates [20]. These techniques can be applied immediately before proton therapy irradiation to validate an external gating signal, which will be used to control the proton beam.

In addition to examining tumor motion before treatment, fluoroscopy can be used to track tumor motion during treatment and gate the treatment beam on and off based on tumor location. This capability was developed for X-ray radiotherapy at Hokkaido University [21]. Four sets of X-ray tubes and TV imagers were mounted in the treatment room about isocenter of a megavoltage linac. Fiducial markers are placed in the tumor and the markers are automatically identified and tracked by the fluoroscopy system. When the markers are within the allowed volume, the treatment beam is enabled.

More recently, Hokkaido University has installed a proton therapy system and applied a similar approach to respiratory gating of the proton beam. The proton treatment facility at Hokkaido features a compact 360° gantry with a pair of X-ray imagers mounted at

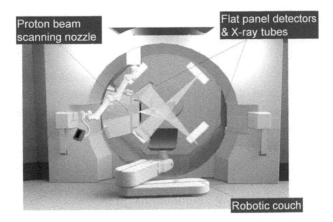

FIGURE 20.4   The compact gantry with fluoroscopic gating capability at Hokkaido University. (Adapted from Shimizu et al. [13] with permission.)

45° relative to the proton beam as shown in Figure 20.4. Fiducial markers are detected and recorded every 0.033 s, and the locations of the markers are used to gate the proton beam on or off [13]. Simulations of this approach for treatment of lung tumors showed that enabling the proton beam gate when markers were within 2 mm of planned position represents a viable trade-off between dosimetric degradation due to respiratory motion and increased treatment time [22].

Imaging dose to the patient can be substantial under fluoroscopic guidance and must be carefully considered. Five minutes of fluoroscopy, for example, may result in an imaging dose exceeding 100 mGy [23]. The tracking systems at Hokkaido have been designed to minimize imaging dose by tracking with a single imager when possible, and reducing the fluoroscopic frame rate during treatment [24].

20.3 FIDUCIAL MARKERS

The primary downside to radiographic localization is that the images usually depict only bony anatomy, which may not be a good surrogate for location of pertinent soft tissues, either target or critical organs. Radiographic fiducial markers were probably first applied to image-guided radiotherapy in proton treatments of cranial lesions [8]. The anatomical features of the skull are reliably correlated to the location of target tissues in the brain, but in this case, the fiducials were used to locate discrete points that were used to calculate both translational and rotational correction vectors. Around the same time, groups proposed implanting markers in soft tissue in or around the target volume to make it radiographically evident [25,26]. With the advent of amorphous silicon flat panel imagers, however, daily online localization of fiducial markers became a widespread practice for a large number of treatment sites.

Prostate was among the first treatment sites to see application of image guidance via implanted fiducial markers [27,28]. Later, fiducial markers were used for pretreatment localization of pancreatic tumors [29] or to track respiratory motion of liver tumors [30,31] and pancreatic volumes [32]. Finally, fiducial markers have been placed in the lung and tracked to generate a respiratory gating signal [21,33]. In addition to markers placed with the intent of image guidance, titanium surgical clips in the tumor bed have been used to localize target volumes in the prostatic bed or breast lumpectomy cavity [34].

For most treatment sites, the position of markers within soft tissue has been shown to be relatively stable over the course of treatment [35,36]. To be sure, though, at least three markers are usually placed, when possible. This way, if one marker does migrate it is obviously displaced relative to the others and can be ignored. Ideally the markers are placed in the tumor volume, but in sites such as liver or lung, there are concerns with spreading tumor cells via the marker implantation [37,38]. It has been shown, however, that the closer a fiducial can be placed to a liver tumor, the stronger its position correlates with the target tissue [39].

In addition to these concerns, in proton therapy, there are a few additional issues related to fiducial markers that must be considered. Larger, high-density fiducial markers will clearly be more visible in radiographs of pelvic or abdominal anatomy, but large markers have been shown to generate excess Hounsfield unit (HU) artifact in treatment planning CT scans. This could be a problem for proton dose distributions that are particularly sensitive to the CT numbers of the tissues that the protons traverse [40]. In addition to perturbations to the dose calculation, excessively large, dense fiducial markers may also cause localized dose deficits immediately downstream of the marker. The localized dosimetric shadow of radiopaque fiducial markers has been studied extensively, usually through Monte Carlo calculations [41,42] or film measurements [40,43]. The dosimetric effect of fiducial markers has been shown to depend on the size of the markers, the material of their composition, their orientation with respect to the beam direction, and their position in the target with respect to end of range. Markers near the proximal region of the target interact with high-energy protons and the difference in Coulomb scattering in the marker compared to the surrounding tissue causes a modest dose deficit. Markers near the end of the proton path, however, may be dense enough to actually stop the protons that strike them and cause a much more severe underdosage downstream [42]. Figure 20.5 shows a Monte Carlo calculation of a single proton field planned to deliver uniform dose. The dosimetric shadows are much more pronounced for fiducials oriented in the beam direction and near the end of proton range. Localized dose deficits of up to 85% are observed for the largest gold markers, situated near the end of the beam path and oriented parallel to the beam direction. These effects can be minimized by using smaller markers made of stainless steel or titanium rather than gold and by using more than a single treatment field. Matsuura et al. [44] estimated the change in tumor control probability (TCP) which could be attributed to a dose shadow from fiducial markers in prostate patients. They found that TCP was not affected by gold markers with 1.5 mm diameter as long as two fields are used. However, if only two fields are used in conjunction with a 2 mm diameter gold marker, a TCP reduction of more than 3% can be expected.

20.4 VOLUMETRIC LOCALIZATION

Radiographic planar imaging is a viable technique for pretreatment localization if either bony anatomy or implanted fiducials are good indicators of the location of both target tissue and nearby critical structures. A cohort of prostate cancer patients received a planning

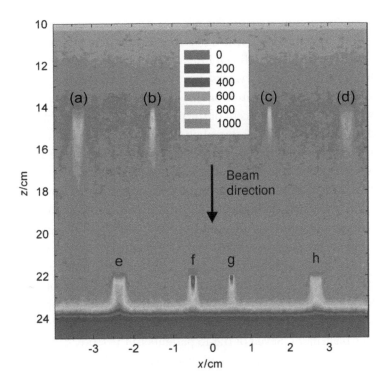

FIGURE 20.5   Monte Carlo calculations depicting the impact of marker size, orientation, and depth of gold markers on a single-field proton dose distribution. (From Newhauser et al. [42] with permission.)

CT and three follow-up CT scans to examine the constancy of the correlation between marker location and the surface of the prostate and seminal vesicles. While the prostate showed little deformation over the course of treatment, the position of the seminal vesicles was less correlated to the marker position [45]. In another study, patients with fiducial markers placed in their prostates received magnetic resonance imaging (MRI) scans at the time of their treatment planning CT and at a randomly assigned treatment fraction. The fiducial markers, prostate, bladder, and rectum, were contoured on each of the MRI images. It was found that fiducial marker location was not a good surrogate for the prostate surface in the anterior–posterior or craniocaudal directions [46]. Additionally, in some radiotherapy treatment sites, sparing of a nearby critical organ is at least as important as adequately treating the tumor volume, although fiducial markers are rarely placed in normal tissues, and so these organs are not visualized and localized relative to the high-dose regions of the treatment plan. Finally, fiducial marker placement could potentially seed tumor cells along the needle path [38], and marker placement in the lung has been shown to carry a high risk of pneumothorax [47]. For these reasons, direct volumetric imaging of soft tissues may be a preferable modality for pretreatment localization.

An additional benefit of volumetric localization is that the image data acquired in the treatment room each day may also be useful for adaptive radiotherapy protocols. Proton therapy dose distributions are particularly sensitive to changes in patient anatomy—weight gain or loss, or changes to tumor size or density over a course of treatment. Lung treatments especially have been shown to be sensitive to changes in patient anatomy [48]. In a phase II trial of proton therapy for nonsmall cell lung cancer, all patients received a four dimensional CT (4DCT) resimulation in weeks 3 or 4 of their treatment. The treatment plans were recalculated on these resimulation scans, and the tumor coverage and normal tissue sparing were evaluated. Twenty percent of the patients had their treatment plans modified as a result of these follow-up calculations [49]. If daily localization images could be used for dose calculation, plan quality could be evaluated or modified on a more frequent basis and without the need for a separate CT simulation appointment.

An alternative use of volumetric imaging for patient setup has been proposed by Cheung et al. [50], in which the patient's treatment plan is calculated on a CT data set acquired immediately before treatment and optimal isocenter placement is determined dosimetrically. The technique was studied for 15 lung cancer patients who received weekly CT scans. The initial dose calculation was performed with the treatment isocenter aligned geometrically to the patient anatomy, as would be done in traditional IGRT protocols. Additional calculations were performed with the isocenter shifted up to 8 mm, 2 mm at a time. Target coverage and mean dose to the ipsilateral lung were analyzed to determine optimal isocenter placement. Seven of the 15 patients in the study did not meet dosimetric constraints when the treatment plan was aligned anatomically, but all 15 were able to meet target coverage constraints when the plan was aligned dosimetrically.

20.4.1 Cone Beam CT

Volumetric image guidance became possible with the development of CBCT localization systems for X-ray linear accelerators in the early 2000. CBCT systems [6] have been shown to offer superior accuracy and allow for correspondingly reduced treatment margins [51]. Ten to 15 years later, CBCT systems began to appear in proton therapy treatment rooms. The most straightforward implementation of CBCT for proton therapy involves the use of an X-ray tube and flat panel imager mounted on a 360° capable gantry [52]. The obvious advantage of this geometry is that the patient can be imaged at a treatment isocenter. However, as discussed in section 2 of this chapter, many proton treatment rooms are built without a full 360° gantry or perhaps without a gantry at all and therefore lack the infrastructure required for a traditional CBCT acquisition.

Robotic arms fitted with radiographic C-arms have been developed to acquire CBCTs in a number of treatment rooms without full gantries. The proton therapy system at St Jude Children's Research Hospital in Memphis Tennessee features 190° gantries and a ceiling mounted robotic C-arm [17]. The robotic C-arm is mechanically decoupled from the proton gantry and therefore allows for imaging at three different locations, as shown in Figure 20.6. Patients receiving cranial or head and neck treatments are able to receive CBCT scans at the treatment isocenter. Because of tight clearances between the patient and the proton nozzle, however, most other patients must be moved away from treatment isocenter for CBCT acquisition. The robotic positioner moves the patient 100 cm away from treatment isocenter for CBCT acquisition and then returns them to treatment isocenter with an added correction vector derived from the CBCT registration.

FIGURE 20.6   A robotic C-arm allows for image acquisition at treatment isocenter (Position 0), 27 cm from treatment isocenter along the axis perpendicular to gantry rotation (Position 1), and 100 cm from treatment isocenter along the gantry rotation axis (Position 2). (From Hua et al. [17]; with permission.)

A similar robotic CBCT system has been installed at Centro Nazionale de Adroterapia Oncologica (CNAO) in Italy. The treatment rooms are outfitted with fixed beam nozzles and a robotic patient positioner. A second robotic arm mounted on the floor is equipped with an X-ray tube and flat panel imager to allow for planar radiographic or CBCT imaging [53]. The imaging isocenter is several meters from treatment isocenter, and the robotic patient positioner moves the patient between the treatment and imaging locations with submillimeter accuracy [54]. The imaging system rotates 220° around the patient to acquire projection radiographs which are used to reconstruct a $30 \times 30 \times 15$ cm volumetric image. Because of regulations limiting the speed of robotic arms near humans, the process of moving between isocenters and acquiring a CBCT with this system adds approximately 15 min to the treatment time.

A novel design for CBCT acquisition incorporates the X-ray tube and flat panel detector into a ring that is mounted on the robotic couch. The ImagingRing, under development by medPhoton in Austria, translates longitudinally up and down the treatment couch top and both the X-ray tube and detector rotate individually by more than 470° around the patient. By rotating the tube and detector independently and dynamically collimating the X-ray tube, the system can reconstruct 3D images that are not centered on the rotation axis of the ring [55]. Independent rotation may allow for a larger diameter field of view, while helical acquisition can produce a longer image than standard CBCT systems.

The most obvious drawback to using CBCT images for adaptive replanning is diminished HU accuracy when compared with standard helical CT. In proton treatment planning, the CT HU is used to determine tissue composition, and subsequently, the relative stopping power of those tissues [56]. Under ideal conditions with a state-of-the-art diagnostic CT scanner, there are a number of uncertainties in the conversion of HU to stopping power such as ambiguity in excitation energies [57], degeneracy in HU for human tissues [58], beam hardening in the CT scanner [58], or CT artifact [59]. These uncertainties were originally estimated to translate to a range uncertainty of 3.5% + 1 mm [60], a value that is typically still used clinically.

Range uncertainties arising from image artifacts and HU uncertainties are substantially worse when performed on CBCT images, compared with helical scanners. The primary difference between helical and CBCT images is the increased scatter component in

CBCT imaging [61], and the size of the patient can impact the magnitude of this effect [62]. Field-of-view limitations and ring artifacts [63] may also limit the ability to perform accurate dose calculations on CBCT images. Li et al. [64] showed that without scatter correction, photon dose calculation errors arising from scatter in CBCT images may be greater than 10%, although they proposed scatter correction methods to reduce the dose differences to less than 1%. Proton dose calculations are considerably more sensitive to artifacts and HU variations, because of the finite range of the beam and steep dose gradients in the beam direction. Therefore, use of CBCT images for adaptive radiotherapy clearly requires some strategy for dealing with the shortcomings of this modality.

One approach to more reliably calculate proton dose distributions on CBCT images is to deform the treatment planning CT onto the CBCT image acquired at treatment delivery, creating a virtual CT. Landry et al. [65] built a deformable phantom resembling human head and neck anatomy and tested the ability of a deformable image registration to create a daily virtual CT. Intensity-modulated proton fields were planned on the reference CT of the phantom and were recalculated on virtual CT images derived from deformed CBCTs. They were able to achieve water-equivalent thickness differences of less than 2% between the reference and virtual CT images, which is comparable to the accepted range accuracy of approximately 3.5% for proton dose calculations on helical scans. They suggest, however, that while this technique could illustrate clinically meaningful variations in proton dose distributions, any adaptive planning should be performed on a new helical CT scan, and not the virtual image.

Deformable image registration between a reference CT and CBCT can be challenging, especially in treatment sites where air pockets may appear or change size from day to day. For that reason, Park et al. [66] evaluated two techniques for scatter correction in CBCT as a means to improve the accuracy of proton dose calculations directly on CBCT data. The first approach, which was originally developed to perform two-dimensional exit dosimetry with an EPID, assumes a uniform distribution of scatter across the imaging plane [67]. Projections were corrected assuming a uniform scatter distribution and then used to reconstruct a CBCT image set. Differences in absolute HU between the CBCT and reference CT images were corrected by shifting all HU in the CBCT to match soft tissue values on the periphery of each image [62]. The second approach used an a priori scatter correction in which the reference CT was deformably registered to a CBCT. Beam projections from the deformed CT were used to estimate scatter maps which could be subtracted from the raw projections and used to create a scatter-free CBCT [68]. While the uniform scatter correction was found to be inadequate for proton dose calculation, the authors were encouraged by the potential of the a priori method to derive CBCT images that were suitable for proton dose calculations. In a subsequent study, Kim et al. performed a priori scatter correction on a set of CBCT images acquired from patients undergoing X-ray radiotherapy for head and neck cancer [69]. The water-equivalent path lengths to the distal edges of the target volumes were calculated on the reference CTs and compared to weekly scatter corrected CBCT images. They were able to identify clinically relevant changes in target depth which were attributed to variations in patient anatomy.

Besides HU accuracy, the other major challenge in using CBCT for proton therapy is related to respiratory motion. Because CBCT acquisition requires rotating an unshrouded gantry or C-arm around the patient, image acquisition typically takes 30 s for a half-fan scan or 60 s for a full 360° scan. This acquisition time is long compared to a typical respiratory period, and so the CBCT image of a mobile tumor will appear blurred. If a tumor that moves under respiratory action is to be treated with a scattered proton beam, typically an internal target volume (ITV) is created from a four-dimensional planning CT to define the full extent of tumor motion [70]. Because, to first order, the scattered beam irradiates the entire treatment volume uniformly during the delivery, the target receives the intended dose as long as it stays within the ITV. The blurred image of the tumor in a CBCT is likely a viable target to be compared to an ITV contour in volumetric localization techniques.

In spot scanning proton therapy, however, the dose deposition from the proton beam is time dependent, and so whenever a mobile target is treated, the interplay effect is a concern (see Chapter 18). This phenomenon describes the potential overtreatment or undertreatment of portions of a target because it moves relative to the scanning proton beam [71,72]. A number of approaches have been proposed to mitigate the interplay effect, including rescanning or 4D treatment plan optimization, that may be compatible with a time-averaged slow acquisition CBCT alignment [73]. Others, however, such as respiratory gating or breath-held treatment, only enable beam delivery during a portion of the respiratory cycle and should not be aligned with imaging that is averaged over the full respiratory cycle. Guckenberger et al. [74] developed a technique for CBCT guidance of moving targets in the liver, where the liver and diaphragm dome are visible in the time-averaged CBCT images. While the interface between liver and lung is blurred, the end-inhale and end-exhale positions can be identified and compared to contours from corresponding phases of a 4DCT planning scan. Blessing et al. [75] have demonstrated a technique for breath-held CBCT acquisition within 15 s using simultaneous kV/MV CBCT acquisition in X-ray therapy. This approach could potentially be applied to breath-held proton treatments in gantries with dual-kV imaging systems. If a treatment is to be gated on a portion of the respiratory cycle, however, the patient should be aligned using only image data that were acquired within the respiratory gate. Four-dimensional CBCT has been achieved experimentally by sorting individual projections according to phases of the respiratory cycle and then reconstructing volumetric images from each of those phase-grouped projection sets [76,77]. A single phase of a 4D CBCT image, at full exhale for example, could be used for online guidance of a scanning beam treatment meant to be gated at that same point in the respiratory cycle. Long acquisition and reconstruction times, however, have so far prevented the widespread clinical application of 4D CBCT for either X-ray or proton radiotherapy guidance.

20.4.2 CT on Rails

An alternative to CBCT for volumetric localization is CT on rails. These devices are diagnostic-quality helical CT scanners in the treatment room that move on rails over a stationary patient during image acquisition, rather than the CT couch moving through the scanner. The first CT on rails was developed in the 1990s in Japan to guide stereotactic radiotherapy on an X-ray linac [78]. The geometry of that CT on rails has been implemented in other X-ray therapy treatment rooms since. Typically, the CT scanner is mounted opposite the therapy linac, and the two devices share a common treatment couch that rotates 180° between treatment and imaging isocenters. The main sources of geometric uncertainty in one of these systems were mechanical positioning of the couch within the CT (0.5 mm) and in alignment of contours to the CT images (0.4 mm). Mechanical precision of this system was improved by incorporating radio-opaque markers to correlate the linac and CT isocenters [79].

CT on rails has now been incorporated into the design of several modern proton therapy treatment rooms, such as at the Universitäts Protonen Therapie Dresden (UPTD) in Dresden, Germany, Gantry 2 at the Paul Scherrer Institut (PSI) in Switzerland [14], ATreP in Trento, Italy [80], and at the Mayo Clinic proton therapy centers in both Rochester, MN and Phoenix, AZ. Figure 20.7 shows the CT on rails installation in Gantry 1 at the Mayo Clinic in Rochester. The distance between treatment and imaging isocenters is approximately 1.5 m, and the patient can be moved from one to the other in approximately 60 s. The bore of the CT scanner is 80 cm in diameter to accommodate large patients or immobilization devices, and the range of the CT scanner allows for more than 1 m of the treatment couch to be imaged.

The major drawback of CT on rails compared to radiographic localization or some CBCT systems is that the patient must be moved between treatment and imaging isocenters, with potentially negative impact on both accuracy and efficiency. The advantages of CT on rails address many of the shortcomings of CBCT addressed previously. Because the CT is a diagnostic helical scanner, the images are of higher quality, with larger field of view, and can be acquired relatively quickly once the patient is positioned within the bore. The improved image quality from CT on rails is notable for its impact on the ability to localize anatomy, but the most important factor is improved HU fidelity. Without the excessive scatter and artifact seen with CBCT, localization images from CT on rails can be used directly for proton dose calculation to either verify the dose distribution on the patient's anatomy of the day or used within adaptive protocols to modify the treatment plan in the presence of anatomical changes. The other important advantage of CT on rails compared to CBCT is rapid image acquisition. Patients who are to be treated with a breath-held technique to mitigate the interplay effect can usually hold their breath for a typical helical CT acquisition time much more easily than for a CBCT. Similarly, a CT on rails can be equipped with a respiratory tracking device to generate a 4DCT. A single phase of the 4DCT can be used to localize a patient before respiratory-gated therapy.

Figure 20.8 illustrates the value of being able to accurately calculate proton dose distributions directly on CT on rail localization images. Panel A shows a treatment plan for a

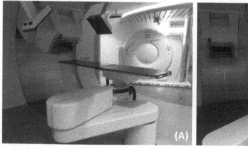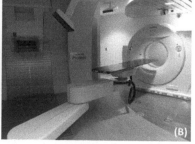

FIGURE 20.7 Gantry Room 1 at the Mayo Clinic in Rochester, MN features a robotic patient positioner that can move the patient between treatment isocenter (Panel A) and a CT on rails (Panel B).

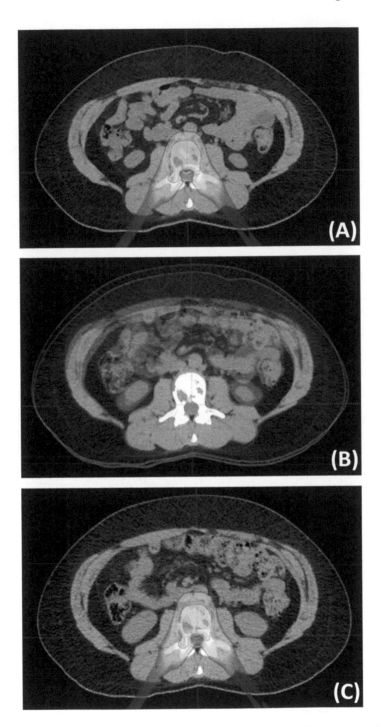

FIGURE 20.8   Panel A shows a treatment plan for hypofractionated treatment of a lumbar vertebral body that spares the nerves of the cauda equina. Panel B is the treatment planning CT blended with a pretreatment localization CT. The images are aligned at the vertebral body, but the external contours are slightly different. Panel C shows the treatment plan recalculated on the localization scan. The change in the external contour causes a potential overdose of the nerves.

patient who received hypofractionated proton radiotherapy to her lumbar spine. The entire vertebral body was to be covered by the prescription dose, but dose to the nerves in the cauda equina was of particular concern. The plan was robustly optimized to account for expected geometric and range uncertainties, and the dose wash in panel A shows that the interior of the vertebral body is kept below threshold dose. The patient presented for her first treatment fraction and was localized via CT on rails. The vertebral body was easily localized in the correct location, which could have been performed via planar radiography as well. However, panel B in Figure 20.8 shows the planning CT blended with the localization CT acquired in the treatment room. With the vertebral body registered between the two images, the blended view shows a 2 mm deviation between posterior external contours in the planning and localization CTs, a difference that exceeded expectations at the time of planning. Before delivering the first treatment fraction, the localization CT was transferred to a treatment planning station just outside the treatment room, and the plan was recalculated within minutes of acquisition. Panel C shows the dose distribution calculated on the localization CT, and in this calculation, the dose in the center of the vertebral body exceeds the nerve threshold dose. The patient was not treated on this day, but the localization CT was able to generate a new treatment plan which was more robust to variations in the external contour. This subtle deviation in anatomical presentation would not have been visible with planar radiographic localization, and CBCT may have lacked the HU fidelity required to identify the dosimetric impact of 2 mm difference in target depth.

20.5 OPTICAL LOCALIZATION

All of the image guidance techniques discussed to this point have used X-ray interactions in the patient to form an image. Alternatively, a number of systems for localization or monitoring patient position have been developed using optical or infrared imaging. Advantages of the optical localization systems include the ability to image the patient with no ionizing radiation, high spatial resolution, and high-frequency continuous imaging [81].

20.5.1 Marker Tracking

The first optical localization system for external beam radiotherapy used a pair of charge-coupled device (CCD) cameras to track the position of an array of infrared light-emitting diodes attached to a custom bite block. This system was used at the University of Florida to localize patients for X-ray cranial radiosurgery with better than 0.5 mm accuracy [82]. A similar system marketed as ExacTrac (BrainLab, Germany) is able to localize and track a nonrigid array of passive infrared reflecting spheres, allowing the system to be used for treatment sites on the thorax or abdomen. Daily repeatability in the placement of the spheres on the patient's body was shown to impact the targeting accuracy of the optical system [83]. The accuracy of an optical localization system for extracranial targets will also be impacted by the diminished correlation between external marker location and internal target tissue.

In particle therapy applications, the treatment rooms at CNAO in Italy use both radiographic localization techniques and an optical tracking system [53]. Patients have infrared reflective spheres affixed to their thermoplastic immobilization hardware. They are brought into the room in their immobilization equipment, and after initial localization via lasers, the patient position is refined by positioning the reflective spheres. The patient position is further refined by planar radiography or CBCT. Radiographic refinements to the optical localization averaged less than a millimeter for head and neck patients and were over a millimeter for pelvis treatments [84]. This difference, again, is attributable to the decreased correspondence between external markers and internal anatomy for extracranial treatments. The authors note that the complementary nature of the optical and radiographic localization systems provides an additional layer of quality assurance to their treatment process.

The high-frequency nature of optical monitoring makes it an appealing source of a respiratory signal for applications in which the beam delivery will be gated or will track a mobile tumor. The RPM system from Varian Medical Systems uses a single infrared camera to track a reflective box on the patient's abdomen [85]. The respiratory trace can be used to gate a beam delivery or to sort CT data by phase to produce a 4DCT scan. An experimental system that tracks reflective spheres has been integrated with the control system of the carbon ion delivery system at Holmholtzzentrum für Schwerionenforshung (GSI) in Darmstadt, Germany. Experiments with a breathing phantom have demonstrated promise for this system to correlate external reflector position with internal anatomy and someday actively track a moving target with a scanning ion beam [86].

20.5.2 Surface Imaging

In addition to tracking discrete markers affixed to the patient, there are systems that can generate, localize, and monitor a 3D surface map of the patient. A logical application of this technology is the setup of treatment sites in which the external surface of the patient correlates well with the target tissues. Breast cancer is one candidate, especially given the variability of the shape and location of the breast surface relative to bony anatomy. In an early test of this process, cohort of patients receiving accelerated partial breast irradiation (APBI) with X-rays and electrons were set up with orthogonal port films. After port film exposure, the patient's surface was monitored with a surface imaging system (AlignRT; Vision RT, London, UK), and deviations between the surface and a reference image were recorded. Adjustments to the patient position based on surface imaging were not performed, but the investigators report that the surface imaging would have provided improved congruence of the breast surface, although the correlation with target volume for APBI was not established [87]. In a subsequent study, a cohort of APBI patients had surgical clips which could radiographically identify the location of the target volume within the breast. Patients were setup using kV radiographs of bony anatomy and with 3D surface imaging. Additional postshift radiographs were obtained, and the positional deviation of the surgical clips indicated the residual error for each setup modality. Radiographic imaging of bony anatomy was found to localize the target cavity to within 5 mm, but the surface imaging residual error was only 2 mm [88]. Peng et al. [89] studied a surface imaging system for intensity-modulated radiation therapy of intracranial targets. Patients were initially set up with a stereotactic bite block with reflective markers. CBCT imaging validated correlation of surface imaging to internal target anatomy, and patient position was monitored throughout treatment with both the optical bite block and the surface imaging system. They found that the combination of volumetric radiographic imaging from the CBCT and real-time monitoring of intrafractional motion with the surface imaging system was a favorable approach for frameless radiosurgery [90].

Surface-based localization has been implemented for proton therapy as well; Massachusetts General Hospital reported on a trial in which postmastectomy patients received spot scanning proton therapy to their chestwall after localization with a surface imaging system. This treatment site was found to be well suited to surface imaging because radiographically evident bony anatomy is often far away from the superficial target tissue. In addition to increased accuracy compared to radiographic localization, the surface imaging system was more efficient as well and led to reduced treatment times and imaging dose [90]. This treatment approach has since been incorporated into routine clinical practice at that center [91].

Besides localizing and monitoring patient anatomy, surface imaging systems are also used to validate the operation of proton therapy equipment. Most modern proton therapy systems position the patient on a 6 degree of freedom couch top mounted to an industrial robot, and in treatment rooms with half gantries or fixed beam angles, the robotic positioner may perform several isocentric rotations within a single treatment fraction. Compared to the simple mechanical pedestal of a linac couch, isocentric rotations on a robotic positioner are especially complicated. The position of treatment isocenter relative to the couch top will vary with each patient setup, and rotating about an arbitrary point requires the coordination of up to six individual motors. In addition, the mass and location of the center of gravity of the load on the couch top must be measured so that the vector between the couch surface and isocenter can be compensated for sag of the couch under load. Despite this complexity, robotic positioners have been shown to isocentrically rotate patients with errors of less than 1 mm [92]. Each center should validate the mechanical and operational integrity of its own system, however, especially in the first few years of operation.

At Mayo Clinic in Rochester, MN, patients are treated on a half gantry and their initial localization is typically performed radiographically with the patient's body perpendicular to the axis of gantry rotation. Most patients are then rotated 90° for treatment, and subsequently rotated 180° in the other direction for bilateral treatments. Initially, the fidelity of these isocentric rotations was validated radiographically, acquiring a new set of X-ray images at each couch angle. The drawbacks to this conservative approach include increased X-ray dose to the patients, loss of efficiency, and the inability to observe anatomy in both imagers for some combinations of couch and gantry angles. The treatment rooms are also equipped with an array of three surface imaging cameras, and so surface imaging has been studied as a replacement for radiographic validation of couch rotations. The intrinsic accuracy of the surface imaging array has been shown to be comparable to the radiographic system for a rigid phantom. However, when an image of the patient surface is used to verify mechanical operation of the robotic couch, the couch behavior is convolved with other unrelated factors such as patient motion. To study the sensitivity of the surface imaging system in detecting nonisocentric couch rotations, volunteers were placed on the couch and rotated isocentrically 20 times. At each couch rotation, the measured deviation from reference position was recorded. Additional rotations were performed with planned mechanical shifts of 2, 4, 6, and 8 mm, simulating suboptimal robot performance. The measured deviations were histogrammed and compared to the purely isocentric rotations. This study was performed for a simulated cranial treatment in which the volunteer was immobilized in a thermoplastic mask, and a thoracic treatment in which only a knee cushion was used. The data for these two measurement sets are shown in Figure 20.9. For each case, the surface imaging system reported a mechanical runout that was within 2 mm of the actual value. It was determined that by setting an action threshold of 4 mm the surface imaging system would identify any gross mechanical deviations with a very low false-positive rate.

20.6 IMAGING OUTSIDE THE TREATMENT ROOM

Proton therapy centers are extraordinarily expensive, but one way to mitigate the high cost of a facility is to treat as many patients as possible, minimizing the cost per patient treated. This is typically done by operating with expanded treatment hours compared to an X-ray center, but maximizing the efficiency of proton treatments can also support this objective. To that end, a number of proton therapy facilities perform anesthesia induction, immobilization, or even localization outside the treatment room. At both Mayo Clinic facilities, for example, pediatric patients may begin their daily treatment process in a dedicated anesthesia suite. There are two induction rooms in the center of the suite, where the patients are placed in their immobilization devices and anesthesia is started. The treatment couch top is on a specialized gurney, and a compact anesthesia system is mounted at the foot of the couch top. Patients who will require extensive pretreatment manipulation within their immobilization equipment, such as craniospinal irradiations, move to one of two

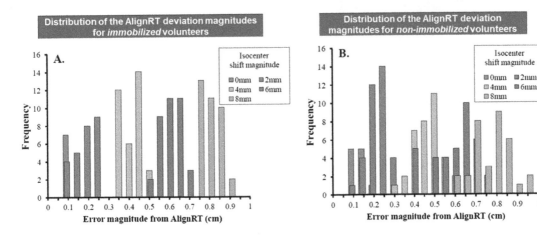

FIGURE 20.9 Volunteers were rotated isocentrically on a robotic couch, and after each rotation, the deviation from isocenter was measured with a surface imaging system. The experiment was repeated with simulated mechanical errors (isocenter shifts) of 2, 4, 6, and 8 mm. Panel A shows the measured deviations for volunteers immobilized in a thermoplastic mask, while Panel B shows measurements for nonimmobilized volunteers.

immobilization rooms. A robotic arm automatically locates the gurney in this room and lifts the couch top off of the gurney. The patient is imaged with a pair of orthogonal radiographic panels, and an anatomic pose is compared to treatment planning DRRs, as shown in Figure 20.10. When the patient is immobilized correctly, the robot places the treatment couch back on the gurney and the patient is transported to the treatment room. The robotic positioner in the gantry room will lift the patient from the gurney and move them to isocenter for final localization and treatment. After treatment, the robotic positioner places the patient back on the gurney so that they may return to one of four recovery rooms in the anesthesia suite. While a craniospinal patient with three treatment isocenters may be on the treatment couch top for 90 min or more, they occupy a treatment room for only

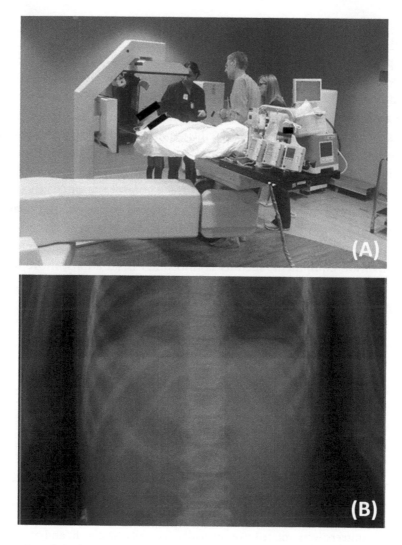

FIGURE 20.10   Panel A shows an anesthetized patient positioned under a pair of radiographic panels to assess patient pose. Deviations in the patient immobilization, such as spinal curvature shown in Panel B, can be corrected in this room before moving to the treatment room for final localization and treatment.

a portion of that time. By performing anesthesia and radiographic immobilization outside the treatment room, a facility may maximize the number of complex treatments such as craniospinal irradiations for which proton treatment is ideally indicated but otherwise tie up an inordinate share of treatment room time.

This concept is further expanded at centers such as PSI in Switzerland or Westdeutsches Protonentherapiezentrum Essen (WPE) in Germany. In these facilities, the patient is immobilized and imaged outside the treatment room, but the isocenter of the imaging equipment is registered to a treatment isocenter in the gantry room. At PSI, patients are immobilized, and a motorized transporter moves the patient and treatment couch to a CT scanner. A pair of orthogonal topograms are acquired in the CT scanner and registered to images from the treatment planning CT scan. X, Y, and Z displacements are calculated and applied to the patient position. The transporter moves the immobilized patient approximately 20 m to the treatment room where the couch is docked, and the patient is treated with no further imaging. To verify systematic integrity, a subset of treatments end in the CT scanner where posttreatment topograms are acquired and analyzed [93]. Fava et al. [94] performed a Monte Carlo study of proton facility throughput and found that as many as 45% more patients could be treated in a single room center by using remote positioning. The advantage of remote positioning was found to decrease as the number of treatment rooms sharing the proton beam increased or the speed of the transporters decreased [94]. At WPE, the trolley system is intended not only to increase efficiency within the treatment room but to facilitate localization with modalities such as CT or MRI elsewhere in the center.

20.7 OCULAR TREATMENTS

In general, the disease sites and treatment processes described here are not substantially different than those that would be described for image guidance of X-ray radiotherapy. Some proton therapy centers, however, have dedicated beamlines for treating orbital tumors or uveal melanomas. The beam delivery systems are typically single-scattered beam lines developed to maximize the sharpness of the beam's lateral penumbra and limit dose to critical structures nearby. To target the tumor in the eye while sparing structures such as the optic nerve or retina, it is necessary not only to localize the bony anatomy around the eye but also to optimize the gaze angle—the rotation of the globe within the orbit. When these treatments were planned at the Harvard Cyclotron Laboratory, the gaze angle was set by asking the patient to look at a point light source. Tantalum markers sewn onto the eye were imaged radiographically before treatment to verify the position of the eye relative to the beam axis and the gaze angle. During treatment delivery occasional X-ray exposures verified that patient position and gaze angle were maintained [95]. Later ocular treatment systems continuously verified gaze angle by monitoring the pupil with a video image. At Massachusetts General Hospital, the location of the pupil is marked on a video screen after the eye position has been validated radiographically. A member of the treatment team monitors the video image of the eye and manually pauses the beam if the location of the pupil deviates. Later, automated systems were developed to identify and track the pupil in video images of the eye. These systems automatically pause beam delivery in response to eye motion [96,97].

20.8 CLINICAL APPLICATION

This chapter has focused primarily on imaging systems which can be used to precisely localize the patient relative to treatment isocenter and maximize the potential of proton therapy. Successful application of image guidance equipment requires more than just the imaging hardware, however. The systems must be integrated with carefully designed clinical protocols which match the treatment planning objectives to achievable accuracy of the image guidance and delivery system. The mechanical limit of any of the imaging systems described here allows a rigid phantom to be localized with submillimeter precision, but that performance is unlikely to be met on a human subject. Intrafractional motion, tumor variations, or changes in patient immobilization all represent potential sources of dosimetric degradation of a patient plan. The geometric and range uncertainties associated with a given treatment site must be estimated so that a treatment plan can be developed that reliably delivers the prescription dose to the target and spares normal anatomy. Often, the geometric uncertainties for a given treatment site can be determined from previous image guidance experience. Trofimov et al. [98] collected a set of serial CT scans for a cohort of patients receiving intensity modulated radiation therapy (IMRT) for prostate cancer. Proton plans were developed for these patients and recalculated on the subsequent CT images. They found that daily proton dose distributions were affected by femur position and soft tissue deformation, even if the prostate is properly aligned to the treatment isocenter. The magnitude of these dosimetric variations validated the size of their planning target volume (PTV) expansions for the treatment of prostate with protons [98].

Rather than PTV expansions about the clinical target volume (CTV), some proton centers will account for expected positional and range errors by performing robust optimization of the individual beam spot weights [99] (see Chapters 16, 18, and 19). The expected errors are programmed into the optimization process and the result should be a plan that reliably treats the target within those clinical variations. It is critical that the treatment chart for each patient includes a description of PTV margins or robust optimization parameters as well as pertinent anatomy so that therapists performing the daily image guidance know exactly how to localize the patient. Some treatment sites such as head and neck feature a large number of prominent anatomical features such as vertebral bodies, mandible, and skull base that do not move in a collectively rigid fashion. The therapists must know for each patient which features are to be localized and which can be given less consideration. Additionally, they should know the PTV margins or robustness parameters so that they know when a patient is localized well enough to meet the dosimetric intentions of the plan or when more manipulation or imaging is required.

REFERENCES

1. Antonuk LE, Boudry J, el-Mohri Y, et al. Large area, flat panel, amorphous silicon imagers. *Proc SPIE.* 1995; 2432: 216–27.
2. Antonuk LE. Electronic portal imaging devices: A review and historical perspective of contemporary technologies and research. *Phys Med Biol.* 2002; 47: 31–65.
3. Herman MG, Balter JM, Jaffray DA, et al. Clinical use of electronic portal imaging: Report of AAPM radiation therapy committee task group 58. *Med Phys.* 2001; 28: 712–37.

4. Fox TH, Elder ES, Crocker IR, Davis LW, Landry JC, Johnstone PA. Clinical implementation and efficiency of kilovoltage image-guided radiation therapy.

5. Pisani L, Lockman D, Jaffray D, Yan D, Martinez A, Wong J. Setup error in radiotherapy: online correction using electronic kilovoltage and megavoltage radiographs. *Int J Radiat Oncol Biol Phys.* 2000; 47: 825–39.

6. Jaffray DA, Siewerdsen JH, Wong JW, Martinez AA. Flat-panel cone-beam computed tomography for image-guided radiation therapy. *Int J Radiat Oncol Biol Phys.* 2002; 53: 1337–49.

7. Goitein M, Abrams M, Rowell D, Pollari H, Wiles J. Multi-dimensional treatment planning: II. Beam's eye-view, back projection, and projection through CT sections. *Int J Radiat Oncol Biol Phys.* 1983; 9: 789–97.

8. Gall KP, Verhey LJ, Wagner M. Computer-assisted positioning of radiotherapy patients using implanted radiopaque fiducials. *Med Phys* 1993; 20: 1153–9.

9. Liebl J, Paganetti H, Zhu M, Winey BA. The influence of patient positioning uncertainties in proton radiotherapy on proton range and dose distributions. *Med Phys.* 2014; 41: 091711.

10. Park PC, Cheung JP, Zhu XR, et al. Statistical assessment of proton treatment plans under setup and range uncertainties. *Int J Radiat Oncol Biol Phys.* 2013; 86: 1007–13.

11. Wang N, Ghebremedhin A, Patyal B. Commissioning of a proton gantry equipped with dual x-ray imagers and a robotic patient positioner, and evaluation of the accuracy of single-beam image registration for this system. *Med Phys.* 2015; 42: 2979–91.

12. Gillin MT, Sahoo N, Bues M, et al. Commissioning of the discrete spot scanning proton beam delivery system at the University of Texas M.D. Anderson Cancer Center, Proton Therapy Center, Houston. *Med Phys* 2010; 37: 154–63.

13. Shimizu S, Miyamoto N, Matsuura T, et al. A proton beam therapy system dedicated to spot-scanning increases accuracy with moving tumors by real-time imaging and gating and reduces equipment size. *PLoS One.* 2014; 9.

14. Safai S, Bula C, Meer D, Pedroni E. Improving the precision and performance of proton pencil beam scanning. *Transl Cancer Res.* 2012; 1: 196–206.

15. Devicienti S, Strigari L, D'Andrea M, Benassi M, Dimiccoli V, Portaluri M. Patient positioning in the proton radiotherapy era. *J Exp Clin Cancer Res.* 2010; 29: 47.

16. Haberer T, Debus J, Eickhoff H, Jakel O, Schulz-Ertner D, Weber U. The Heidelberg Ion Therapy Center. *Radioth Oncol.* 2004; 73(Suppl 2): S186–90.

17. Hua C, Yao W, Kidani T, et al. A robotic C-arm conebeam CT system for image-guided proton therapy: Design and performance. *Br J Radiol.* 2017: 20170266.

18. Murphy MJ. An automatic six-degree-of-freedom image registration algorithm for image-guided frameless stereotaxic radiosurgery. *Med Phys.* 1997; 24: 857–66.

19. Vedam SS, Keall PJ, Kini VR, Mohan R. Determining parameters for respiration-gated radiotherapy. *Med Phys.* 2001; 28: 2139–46.

20. Hoisak JD, Sixel KE, Tirona R, Cheung PC, Pignol JP. Correlation of lung tumor motion with external surrogate indicators of respiration. *Int J Radiat Oncol Biol Phys.* 2004; 60: 1298–306.

21. Shirato H, Shimizu S, Kunieda T, et al. Physical aspects of a real-time tumor-tracking system for gated radiotherapy. *Int J Radiat Oncol Biol Phys.* 2000; 48: 1187–95.

22. Kanehira T, Matsuura T, Takao S, et al. Impact of real-time image gating on spot scanning proton therapy for lung tumors: A simulation study. *Int J Radiat Oncol Biol Phys.* 2017; 97: 173–81.

23. Murphy MJ, Balter J, Balter S, et al. The management of imaging dose during image-guided radiotherapy: Report of the AAPM Task Group 75. *Med Phys.* 2007; 34: 4041–63.

24. Shirato H, Oita M, Fujita K, Watanabe Y, Miyasaka K. Feasibility of synchronization of real-time tumor-tracking radiotherapy and intensity-modulated radiotherapy from viewpoint of excessive dose from fluoroscopy. *Int J Radiat Oncol Biol Phys.* 2004; 60: 335–41.

25. Balter JM, Lam KL, Sandler HM, Littles JF, Bree RL, Ten Haken RK. Automated localization of the prostate at the time of treatment using implanted radiopaque markers: technical feasibility. *Int J Radiat Oncol Biol Phys.* 1995; 33: 1281–6.

26. Balter JM, Sandler HM, Lam K, Bree RL, Lichter AS, ten Haken RK. Measurement of prostate movement over the course of routine radiotherapy using implanted markers. *Int J Radiat Oncol Biol Phys.* 1995; 31: 113–8.

27. Herman M, Pisansky TM, Kruse JJ, Prisciandaro JI, Davis BJ, King BF. Technical aspects of daily on-line positioning of the prostate for three-dimensional conformal radiotherapy using an electronic portal imaging device. *Int J Radiat Oncol Biol Phys.* 2003; 57: 1131–40.

28. Schallenkamp JM, Herman MG, Kruse JJ, Pisansky TM. Prostate position relative to pelvic bony anatomy based on intraprostatic gold markers and electronic portal imaging. *Int J Radiat Oncol Biol Phys.* 2005; 63: 800–11.

29. Varadarajulu S, Trevino JM, Shen S, Jacob R. The use of endoscopic ultrasound-guided gold markers in image-guided radiation therapy of pancreatic cancers: A case series. *Endoscopy.* 2010; 42: 423–5.

30. Berbeco RI, Hacker F, Ionascu D, Mamon HJ. Clinical feasibility of using an EPID in CINE mode for image-guided verification of stereotactic body radiotherapy. *Int J Radiat Oncol Biol Phys.* 2007; 69: 258–66.

31. Kitamura K, Shirato H, Shimizu S, et al. Registration accuracy and possible migration of internal fiducial gold marker implanted in prostate and liver treated with real-time tumor-tracking radiation therapy (RTRT). *Radiother Oncol.* 2002; 62: 275–81.

32. Knybel L, Cvek J, Otahal B, et al. The analysis of respiration-induced pancreatic tumor motion based on reference measurement. Radiat Oncol 2014; 9: 192.

33. Willoughby TR, Forbes AR, Buchholz D, et al. Evaluation of an infrared camera and X-ray system using implanted fiducials in patients with lung tumors for gated radiation therapy. *Int J Radiat Oncol Biol Phys.* 2006; 66: 568–75.

34. Weed DW, Yan D, Martinez AA, Vicini FA, Wilkinson TJ, Wong J. The validity of surgical clips as a radiographic surrogate for the lumpectomy cavity in image-guided accelerated partial breast irradiation. *Int J Radiat Oncol Biol Phys.* 2004; 60: 484–92.

35. Kupelian PA, Willoughby TR, Meeks SL, et al. Intraprostatic fiducials for localization of the prostate gland: monitoring intermarker distances during radiation therapy to test for marker stability. *Int J Radiat Oncol Biol Phys.* 2005; 62: 1291–6.

36. Pouliot J, Aubin M, Langen KM, et al. (Non)-migration of radiopaque markers used for on-line localization of the prostate with an electronic portal imaging device. *Int J Radiat Oncol Biol Phys.* 2003; 56: 862–6.

37. Chang S, Kim SH, Lim HK, et al. Needle tract implantation after percutaneous interventional procedures in hepatocellular carcinomas: lessons learned from a 10-year experience. *Korean J Radiol.* 2008; 9: 268–74.

38. Patel Z, Retrouvey M, Vingan H, Williams S. Tumor track seeding: A new complication of fiducial marker insertion. *Radiol Case Rep.* 2014; 9: 928.

39. Seppenwoolde Y, Wunderink W, Wunderink-van Veen SR, Storchi P, Mendez Romero A, Heijmen BJ. Treatment precision of image-guided liver SBRT using implanted fiducial markers depends on marker-tumour distance. *Phys Med Biol.* 2011; 56: 5445–68.

40. Habermehl D, Henkner K, Ecker S, Jakel O, Debus J, Combs SE. Evaluation of different fiducial markers for image-guided radiotherapy and particle therapy. *J Radiat Res.* 2013; 54(Suppl 1): i61–8.

41. Giebeler A, Fontenot J, Balter P, Ciangaru G, Zhu R, Newhauser W. Dose perturbations from implanted helical gold markers in proton therapy of prostate cancer. *J Appl Clin Med Phys.* 2009; 10: 2875.

42. Newhauser W, Fontenot J, Koch N, et al. Monte Carlo simulations of the dosimetric impact of radiopaque fiducial markers for proton radiotherapy of the prostate. *Phys Med Biol.* 2007; 52: 2937–52.

43. Cheung J, Kudchadker RJ, Zhu XR, Lee AK, Newhauser WD. Dose perturbations and image artifacts caused by carbon-coated ceramic and stainless steel fiducials used in proton therapy for prostate cancer. *Phys Med Biol.* 2010; 55: 7135–47.

44. Matsuura T, Maeda K, Sutherland K, Takayanagi T, Shimizu S, Takao S, Miyamoto N, Nihongi H, Toramatsu C, Nagamine Y, Fujimoto R, Suzuki R, Ishikawa M, Umegaki K, Shirato H. Biological effect of dose distortion by fiducial markers in spot-scanning proton therapy with a limited number of fields: a simulation study. *Med Phys.* 2012; 39(9): 5584–91. doi:10.1118/1.4745558.

45. van der Wielen GJ, Mutanga TF, Incrocci L, et al. Deformation of prostate and seminal vesicles relative to intraprostatic fiducial markers. *Int J Radiat Oncol Biol Phys.* 2008; 72: 1604–11.e3.

46. Nichol AM, Brock KK, Lockwood GA, et al. A magnetic resonance imaging study of prostate deformation relative to implanted gold fiducial markers. *Int J Radiat Oncol Biol Phys.* 2007; 67: 48–56.

47. Bhagat N, Fidelman N, Durack JC, et al. Complications associated with the percutaneous insertion of fiducial markers in the thorax. *Cardiovasc Intervent Radiol* 2010; 33: 1186–91.

48. Berman AT, St. James S, Rengan R. Proton beam therapy for non-small cell lung cancer: Current clinical evidence and future directions. *Cancers.* 2015; 7: 1178–90.

49. Koay EJ, Lege D, Mohan R, Komaki R, Cox JD, Chang JY. Adaptive/nonadaptive proton radiation planning and outcomes in a phase II trial for locally advanced non-small cell lung cancer. *Int J Radiat Oncol Biol Phys.* 2012; 84: 1093–100.

50. Cheung JP, Park PC, Court LE, et al. A novel dose-based positioning method for CT image-guided proton therapy. *Med Phys.* 2013; 40: 051714.

51. Grills IS, Hugo G, Kestin LL, et al. Image-guided radiotherapy via daily online cone-beam CT substantially reduces margin requirements for stereotactic lung radiotherapy. *Int J Radiat Oncol Biol Phys.* 2008; 70: 1045–56.

52. Veiga C, Janssens G, Teng CL, et al. First Clinical Investigation of Cone Beam Computed Tomography and Deformable Registration for Adaptive Proton Therapy for Lung Cancer. *Int J Radiat Oncol Biol Phys.* 2016; 95: 549–59.

53. Fattori G, Riboldi M, Pella A, et al. Image guided particle therapy in CNAO room 2: implementation and clinical validation. *Phys Med.* 2015; 31: 9–15.

54. Pella A, Riboldi M, Tagaste B, et al. Commissioning and quality assurance of an integrated system for patient positioning and setup verification in particle therapy. *Technol Cancer Res Treat.* 2014; 13: 303–14.

55. Rit S, Clackdoyle R, Keuschnigg P, Steininger P. Filtered-backprojection reconstruction for a cone-beam computed tomography scanner with independent source and detector rotations. *Med Phys.* 2016; 43: 2344.

56. Schneider U, Pedroni E, Lomax A. The calibration of CT Hounsfield units for radiotherapy treatment planning. *Phys Med Biol.* 1996; 41(1): 111–24.

57. Andreo P. On the clinical spatial resolution achievable with protons and heavier charged particle radiotherapy beams. *Phys Med Biol.* 2009; (54): N205–N215

58. Yang M, Virshup G, Clayton J, Zhu XR, Mohan R, Dong L. Theoretical variance analysis of single- and dual-energy computed tomography methods for calculating proton stopping power ratios of biological tissues. 2010.

59. Newhauser WD, Giebeler A, Langen KM, Mirkovic D, Mohan R. Can megavoltage computed tomography reduce proton range uncertainties in treatment plans for patients with large metal implants? *Phys Med Biol.* 2008; 53: 2327–44.

60. Goitein M. Calculation of the uncertainty in the dose delivered during radiation therapy. *Med Phys* 1985; 12: 608–12.

61. Yoo S, Yin FF. Dosimetric feasibility of cone-beam CT-based treatment planning compared to CT-based treatment planning. *Int J Radiat Oncol Biol Phys.* 2006; 66: 1553–61. Epub 2006 Oct 23.

62. Richter A, Hu Q, Steglich D, et al. Investigation of the usability of conebeam CT data sets for dose calculation. *Radiat Oncol.* 2008; 3: 42. doi:10.1186/748-717X-3-42.

63. Schulze R, Heil U, Gross D, et al. Artefacts in CBCT: A review. *Dentomaxillofac Radiol.* 2011; 40: 265–73. doi:10.1259/dmfr/30642039.

64. Li J, Yao W, Xiao Y, Yu Y. Feasibility of improving cone-beam CT number consistency using a scatter correction algorithm. *J Appl Clin Med Phys.* 2013; 14: 4346. doi:10.1120/jacmp. v14i6.4346.

65. Landry G, Dedes G, Zollner C, et al. Phantom based evaluation of CT to CBCT image registration for proton therapy dose recalculation. *Phys Med Biol.* 2015; 60: 595–613. doi:10.1088/0031-9155/60/2/595. Epub 2014 Dec 30.

66. Park YK, Sharp GC, Phillips J, Winey BA. Proton dose calculation on scatter-corrected CBCT image: Feasibility study for adaptive proton therapy. *Med Phys.* 2015; 42: 4449–59.

67. Boellaard R, van Herk M, Uiterwaal H, Mijnheer B. Two-dimensional exit dosimetry using a liquid-filled electronic portal imaging device and a convolution model. *Radiother Oncol.* 1997; 44: 149–57.

68. Niu T, Sun M, Star-Lack J, Gao H, Fan Q, Zhu L. Shading correction for on-board cone-beam CT in radiation therapy using planning MDCT images. *Med Phys.* 2010; 37: 5395–406.

69. Kim J, Park YK, Sharp G, Busse P, Winey B. Water equivalent path length calculations using scatter-corrected head and neck CBCT images to evaluate patients for adaptive proton therapy. *Phys Med Biol.* 2017; 62: 59–72.

70. Kang Y, Zhang X, Chang JY, et al. 4D Proton treatment planning strategy for mobile lung tumors. *Int J Radiat Oncol Biol Phys.* 2007; 67: 906–14.

71. Bert C, Grozinger SO, Rietzel E. Quantification of interplay effects of scanned particle beams and moving targets. *Phys Med Biol.* 2008; 53: 2253–65. doi:10.1088/0031-9155/53/9/003. Epub 2008 Apr 9.

72. Kraus KM, Heath E, Oelfke U. Dosimetric consequences of tumour motion due to respiration for a scanned proton beam. *Phys Med Biol.* 2011; 56: 6563–81. doi:10.1088/0031-9155/56/20/003. Epub 2011 Sep 21.

73. Knopf A, Bert C, Heath E, et al. Special report: Workshop on 4D-treatment planning in actively scanned particle therapy—recommendations, technical challenges, and future research directions. *Med Phys.* 2010; 37: 4608–14.

74. Guckenberger M, Sweeney RA, Wilbert J, et al. Image-guided radiotherapy for liver cancer using respiratory-correlated computed tomography and cone-beam computed tomography. *Int J Radiat Oncol Biol Phys.* 2008; 71: 297–304. doi:10.1016/j.ijrobp.2008.01.005.

75. Blessing M, Stsepankou D, Wertz H, et al. Breath-hold target localization with simultaneous kilovoltage/megavoltage cone-beam computed tomography and fast reconstruction. *Int J Radiat Oncol Biol Phys.* 2010; 78: 1219–26. doi:10.016/j.ijrobp.2010.01.030. Epub May 28.

76. Sonke JJ, Zijp L, Remeijer P, van Herk M. Respiratory correlated cone beam CT. *Med Phys.* 2005; 32: 1176–86.

77. Dietrich L, Jetter S, Tucking T, Nill S, Oelfke U. Linac-integrated 4D cone beam CT: first experimental results. *Phys Med Biol.* 2006; 51: 2939–52. Epub 006 May 24.

78. Uematsu M, Fukui T, Shioda A, et al. A dual computed tomography linear accelerator unit for stereotactic radiation therapy: A new approach without cranially fixated stereotactic frames. *Int J Radiat Oncol Biol Phys.* 1996; 35: 587–92.

79. Court L, Rosen I, Mohan R, Dong L. Evaluation of mechanical precision and alignment uncertainties for an integrated CT/LINAC system. *Med Phys.* 2003; 30: 1198–210.

80. Amelio D, Cianchetti M, Rombi B, et al. P13.02proton Therapy for Brain and Skull Base Tumors at the New Opening Trento Facility. *Neuro Oncol.* 2014; 16: ii65–6. doi:10.1093/neuonc/nou174.248.

81. Meeks SL, Tome WA, Willoughby TR, et al. Optically guided patient positioning techniques. *Semin Radiat Oncol.* 2005; 15: 192–201.

82. Bova FJ, Buatti JM, Friedman WA, Mendenhall WM, Yang CC, Liu C. The University of Florida frameless high-precision stereotactic radiotherapy system. *Int J Radiat Oncol Biol Phys.* 1997; 38: 875–82.

83. Yan H, Yin FF, Kim JH. A phantom study on the positioning accuracy of the Novalis Body system. *Med Phys.* 2003; 30: 3052–60.

84. Desplanques M, Tagaste B, Fontana G, et al. A comparative study between the imaging system and the optical tracking system in proton therapy at CNAO. *J Radiat Res.* 2013; 54: i129–35.

85. Kubo HD, Len PM, Minohara S, Mostafavi H. Breathing-synchronized radiotherapy program at the University of California Davis Cancer Center. *Med Phys.* 2000; 27: 346–53.

86. Fattori G, Saito N, Seregni M, et al. Commissioning of an integrated platform for time-resolved treatment delivery in scanned ion beam therapy by means of optical motion monitoring. *Technol Cancer Res Treat.* 2014; 13: 517–28. doi:10.7785/tcrtexpress.2013.600275. Epub 2013 Dec 17.

87. Bert C, Metheany KG, Doppke KP, Taghian AG, Powell SN, Chen GT. Clinical experience with a 3D surface patient setup system for alignment of partial-breast irradiation patients. *Int J Radiat Oncol Biol Phys.* 2006; 64: 1265–74.

88. Chang AJ, Zhao H, Wahab SH, et al. Video surface image guidance for external beam partial breast irradiation. *Pract Radiat Oncol.* 2012; 2: 97–105. doi:10.1016/j.prro.2011.06.013. Epub Jul 30.

89. Peng JL, Kahler D, Li JG, et al. Characterization of a real-time surface image-guided stereotactic positioning system. *Med Phys.* 2010; 37: 5421–33.

90. Batin E, Depauw N, MacDonald S, Lu HM. Can surface imaging improve the patient setup for proton postmastectomy chest wall irradiation? *Pract Radiat Oncol.* 2016; 6: e235–e41. doi:10.1016/j.prro.2016.02.001. Epub Feb 11.

91. Depauw N, Batin E, Daartz J, et al. A novel approach to postmastectomy radiation therapy using scanned proton beams. *Int J Radiat Oncol Biol Phys.* 2015; 91: 427–34. doi: 10.1016/j.ijrobp.2014.10.039.

92. Nairz O, Winter M, Heeg P, Jakel O. Accuracy of robotic patient positioners used in ion beam therapy. *Radiat Oncol.* 2013; 8: 124. doi:10.1186/748-717X-8-124.

93. Bolsi A, Lomax AJ, Pedroni E, Goitein G, Hug E. Experiences at the Paul Scherrer Institute with a remote patient positioning procedure for high-throughput proton radiation therapy. *Int J Radiat Oncol Biol Phys.* 2008; 71: 1581–90.

94. Fava G, Widesott L, Fellin F, et al. In-gantry or remote patient positioning? Monte Carlo simulations for proton therapy centers of different sizes. *Radiother Oncol.* 2012; 103: 18–24. doi:10.1016/j.radonc.2011.11.004. Epub Nov 25.

95. Goitein M, Miller T. Planning proton therapy of the eye. *Med Phys.* 1983; 10: 275–83.

96. Shin D, Yoo SH, Moon SH, Yoon M, Lee SB, Park SY. Eye tracking and gating system for proton therapy of orbital tumors. *Med Phys.* 2012; 39: 4265–73. doi:10.1118/1.4729708.

97. Via R, Fassi A, Fattori G, et al. Optical eye tracking system for real-time noninvasive tumor localization in external beam radiotherapy. *Med Phys.* 2015; 42: 2194–202. doi:10.1118/1.4915921.

98. Trofimov A, Nguyen PL, Efstathiou JA, et al. Interfractional variations in the setup of pelvic bony anatomy and soft tissue, and their implications on the delivery of proton therapy for localized prostate cancer. *Int J Radiat Oncol Biol Phys.* 2011; 80: 928–37. doi:10.1016/j.ijrobp.2010.08.006. Epub Oct 13.

99. Unkelbach J, Bortfeld T, Martin BC, Soukup M. Reducing the sensitivity of IMPT treatment plans to setup errors and range uncertainties via probabilistic treatment planning. *Med Phys.* 2009; 36: 149–63.

CHAPTER 21

In Vivo Treatment Verification*

Katia Parodi

Ludwig-Maximilians-Universität München (LMU Munich)

CONTENTS

---

* Colour figures available online at www.crcpress.com/9781138626508

21.1 INTRODUCTION: THE NEED OF IN VIVO RANGE OR DOSE VERIFICATION IN PARTICLE THERAPY

Because of increased physical selectivity, proton therapy is more sensitive than conventional radiation modalities to changes of the actual treatment situation with respect to the planned one. In particular, the finite "beam range" in tissue is strongly influenced by the radiological path length, determining the position of the Bragg peak (see Chapter 2) and thus the precise localization of the intended dose delivery. Therefore, uncertainties in the knowledge of the in vivo beam range in the patient is one of the major concerns in proton therapy, hampering full clinical exploitation of the dosimetric advantages in clinical practice.

Major sources of range (and thus dose delivery) errors during fractionated proton therapy are discussed in Chapters 14, 17, and 18. To account for all possible sources of random and systematic errors in clinical practice, cautious safety margins are added to the tumor volume when designing the treatment plan. Moreover, treatment strategies tend to rely on the more controllable but less conformal lateral penumbra of the beam, rather than placing the sharper distal dose falloff in front of radiosensitive structures (Chapters 15 and 16).

Reduction of margins and reliable application of doses tightly sculpted to the tumor target is a major goal of modern radiotherapy to avoid excess toxicity and promote dose escalation for increased tumor control. Conventional and novel methods of image-guided radiotherapy (IGRT) are thus increasingly well established in photon therapy for evaluation of patient geometry and tumor position directly at the treatment site. While corresponding IGRT solutions of orthogonal X-ray radiographies were already pioneered in the early phase of ion therapy since the 1980s [1], volumetric X-ray image guidance solutions only recently entered particle therapy (see Chapter 20). Additional methods are being investigated to provide both anatomical and stopping power ratio (SPR) information before proton irradiation [2]. But regardless of their successful clinical translation, they all lack information on proton interaction in tissue during treatment. Hence, techniques assessing beam range and ideally delivered dose in vivo and non-invasively would be highly beneficial. For optimal quality of patient care, such methods should enable an independent validation of the entire therapy chain from treatment planning to delivery. Also, they should enable prompt identification and quantification of unexpected deviations between planned and actual dose delivery for promoting adaptive strategies towards dose-guided radiation therapy (DGRT) and, ideally, inhibiting improper delivery during treatment ("real-time monitoring"). This promises to play an important role in the increasingly considered hypofractionated regimens, where less or even no subsequent fractions are available for compensation of errors.

21.2 THE CHALLENGE OF IN VIVO TREATMENT VERIFICATION AND DGRT IN PROTON THERAPY COMPARED TO PHOTON THERAPY

The physical properties of penetrating megavoltage photon beams enable detection of the radiation traversing the patient simultaneously to therapeutic treatment. This is typically achieved by using planar detectors such as electronic portal imaging devices (EPIDs) [3]. Unfortunately, the different properties of charged particles with respect to neutral photons prohibit the usage of EPID dosimetry for in vivo quality assurance and DGRT of proton therapy, because the primary therapeutic protons are completely stopped in the patient. To overcome this intrinsic limitation, some approaches under investigation aim at utilizing low-dose irradiation fields or single pencil beams of higher beam energies than for therapy, to reach special implanted dosimeters in specific anatomical sites [4] or to completely traverse the patient (when sufficiently high beam energies are available) for pretreatment dose and range probing [5,6], respectively. Although these approaches enable valuable quality assurance checks and refined SPR assessment before treatment, they do not convey information generated by the actual therapeutic dose delivery and are thus not suitable for true DGRT or real-time monitoring. This chapter will review other techniques currently under development or already under clinical evaluation for exploitation of surrogate signals induced by the therapeutic dose delivery, which can be detected during or after proton irradiation for in vivo non-invasive verification of the actual treatment.

21.3 SECONDARY EMISSIONS AND PHYSIOLOGICAL CHANGES CORRELATED TO THE DELIVERED TREATMENT IN PROTON THERAPY

Secondary emissions usable for DGRT or even real-time monitoring of proton therapy can be distinguished into their direct link to measurable physical or physiological processes. The former physical emissions can be produced as a result of nuclear fragmentation reactions between the incoming protons and the target nuclei of the penetrated tissue, or of thermoacoustic effects induced by the energy deposition process, naturally enhanced at the Bragg peak. In the first case, due to intrinsic physical differences between the nuclear and electromagnetic interactions underlying nuclear fragmentation and dose deposition, respectively, the surrogate emission signal can only be correlated to but not directly matching the delivered dose. In the second case, the resulting acoustic signal carries a more direct correlation to the energy deposition process, but is more affected by interaction in the traversed tissue prior to detection. Nevertheless, in both scenarios valuable information can be inferred from the comparison of the measured signal with an expectation, e.g., based on a previous measurement (reproducibility check) or on an analytical or Monte Carlo (MC) calculation, taking into account the planned treatment, the dynamic beam delivery, and the chosen detection strategy (accuracy check). The corresponding range monitoring techniques at different levels of maturity feature positron emission tomography (PET), prompt gamma detection, and so-called "ionoacoustics," as addressed in Sections 21.4–21.6, respectively. In addition to physical emissions, proton irradiation can induce physiological changes, like fatty tissue replacement of bone marrow, which can manifest in magnetic resonance imaging (MRI) scans. These physiological processes can also enable range monitoring at a different (typically slower) timescale than physical emissions, as addressed in Section 21.7. A summary of the different possible techniques with related timescales is given in Figure 21.1.

21.4 PET IMAGING
21.4.1 The Production of Irradiation-Induced β⁺-Activity

Nuclear interactions between the impinging proton beam and the target nuclei of the traversed tissue may yield positron-emitting nuclei (cf. clinically relevant cross sections

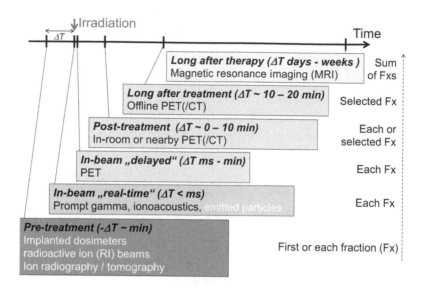

FIGURE 21.1 Schematic representation of different imaging approaches for in vivo dose or range verification on different timescales and delays ΔT with respect to the time of irradiation (upper axis). White characters refer to methods not (further) addressed in the text, because either limited to heavy ion applications (radioactive ion (RI) imaging, emitted particles) or not directly induced by the therapeutic irradiation (implanted dosimeters, ion radiography/tomography). The right arrow suggests possible clinical implementations for monitoring first, individual, or selected fractions (FXs), up to the integral fractionated treatment course (cf. text).

in Figure 21.2), thus opening the possibility of PET-based in vivo treatment verification. Nuclear fragmentation (cf. schematic representation in Figure 21.3) is a complex two-stage process, resulting in the fast (within ca. 10^{-22} s) production of excited prefragments, which eventually approach the final state via nucleon evaporation and photon emission in about 10^{-21} to 10^{-16} s [9], with the latter process being relevant for the prompt gamma imaging techniques described in Section 21.5.

Among all possible reaction yields, neutron-deficient nuclei likely to undergo β^+-decay are produced. In living tissue, typical positron emitters produced include light isotopes such as ^{11}C, ^{15}O, ^{13}N, with half-lives $T_{1/2}$ of about 20, 2, and 10 min, respectively. These β^+-active target fragments are formed all along the beam path as long as the proton energy is above the nuclear reaction threshold, mostly located at 10–20 MeV/u. This corresponds to ca. 1–4 mm residual proton range in tissue. Due to the interaction kinematics, activated target recoils remain approximately at the production place. Thus, according to the typically weak energy dependence of reaction cross sections for most of the therapeutically relevant proton energies in the most abundant tissue constituents (cf. Figure 21.2), the characteristic activation track in homogeneous media exhibits a rather constant or slowly rising slope, dropping to zero few millimeters before the Bragg peak (Figure 21.4a). Additional contributions of secondary radiation other than protons (e.g., neutrons) to tissue activation can be assumed negligible in proton therapy [11].

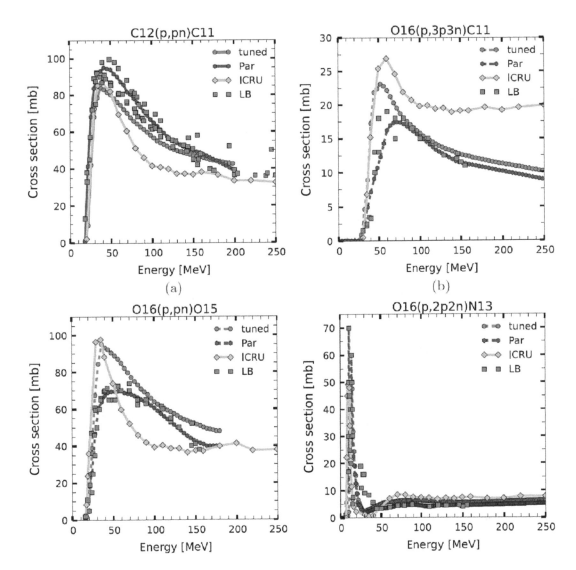

FIGURE 21.2   Comparison of cross sections based on different compilations and theoretical values as well as tuned on experimentally determined positron emitter yields (see details in Ref. [7]) for main proton induced reactions on carbon and oxygen producing ^{11}C, ^{15}O, and ^{13}N, respectively. (Published in Ref. [7]. ©Institute of Physics and Engineering in Medicine. Reproduced by permission of IOP Publishing. All rights reserved.)

21.4.2 The Imaging Process

The transient pattern of irradiation-induced β^+-activation can be imaged via well-established PET techniques, exploiting coincident detection of the annihilation photon pairs resulting from the β^+-decay of the produced positron-emitting isotopes. The first step in the image formation process is the β^+-decay of the irradiation-induced positron emitter $A(Z,N)$ with mass number A and neutron number N at a random time depending on the isotope half-life:

$$A(Z,N) \rightarrow A(Z-1, N+1) + e^+ + \nu_e \tag{21.1}$$

Proton

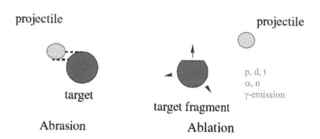

FIGURE 21.3   Schematic representation of peripheral nuclear collisions, adapted from Ref. [8], illustrating different reaction products such as β⁺-active target fragments and evaporation of nucleons and light fragments.

In this radioactive transformation, one proton is transformed into one neutron, resulting in the emission of a positron and a neutrino with a continuous energy spectrum. While the neutrino escapes without interaction, the positron is slowed down typically within few millimeters in the medium, losing energy in Coulomb inelastic collisions with the atomic electrons and suffering several angular deflections. Once almost at rest, the positron either annihilates as a free particle with an electron of the medium into two photons or it captures an electron to form an unstable bound state $e^- - e^+$, so-called positronium. Although the two different atomic states of the positronium can lead to different annihilation processes into two (para-positronium) or three (ortho-positronium) photons, the three γ-emissions are in practice negligible. Thus, all detectable radiation can be attributed to the two annihilation γ-quanta which, according to momentum and energy conservation laws, are emitted in opposite directions and carry an energy of 511 keV each, equal to the positron and electron rest mass. Deviations from perfect colinearity can occur because of the residual energy of the electron–positron system, resulting into emission angles that

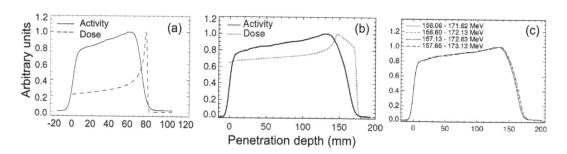

FIGURE 21.4   Depth distributions of calculated dose (dashed or dotted line) and in-beam measured β⁺-activity (solid line) for monoenergetic (a) and spread-out Bragg peak (b, SOBP, built by 11 energies in the 156.06–171.62 MeV interval) irradiation of polymethyl methacrylate (PMMA, $C_5H_8O_2$) targets [10]. The resulting range resolving power is shown in panel (c) for activity depth distributions separately measured after changing the nominal SOBP plan in energy steps corresponding to incremental variations of less than 1 mm in depth [10].

approximately follow a Gaussian distribution centered at 180° with ~0.3° full-width at half-maximum (FWHM) for typical residual energy values of about 10 eV.

The 511 keV photon pairs are energetic enough to penetrate the tissue surrounding the place of annihilation and eventually escape from the patient. Therefore, the PET signal can be acquired by opposite detector pairs surrounding the patient and operated in coincidence (Figure 21.5). Due to the relatively high photon energy, dense high-Z inorganic scintillators are usually employed for stopping the penetrating photon radiation and, at the same time, for enabling fast timing performances of coincident detection typically within a few nanoseconds time window. Commonly used PET detectors are based on bismuth germanate (BGO), gadolinium orthosilicate (GSO), or lutetium oxyorthosilicate (LSO) scintillator crystals coupled to photomultipliers, though more recent solutions combine latest generation scintillators such as lutetium fine silicate (LFS) with state-of-the-art silicon photomultipliers [12].

The detected coincidences can be attributed to radiation, either emerging from a single annihilation event ("true" or "scattered") or belonging to independent emissions accidentally occurring close in time ("random"). Under assumption of perfect colinearity of the annihilation photon pairs, "true" coincidences refer to emission events originating along the line of response (LOR) connecting the two opposed crystals fired in coincidence. Differently, "scattered" coincidences are generated by events, where at least one photon has experienced a scattering process and thus a change with respect to the initial direction. This results in a mismatch between the detected LOR and the original annihilation place. A similar mismatch obviously applies to random coincidences due to the different origins of the independent annihilation events accidentally detected in the same coincidence window.

For a reliable reconstruction of the β^+-activity distribution underlying the measured signal, the amount of "true" coincidences has to be disentangled from the entire collected data. This can be achieved via proper corrections for random [13] and scattered [14] coincidences. In this process, the properties of the used detector system are crucial. In particular, energy resolution is helpful for discrimination of scattered events, while short decay constants are essential for good coincidence timing and related random suppression. The latter has also been the main argument for the rapid establishment of lutetium-based

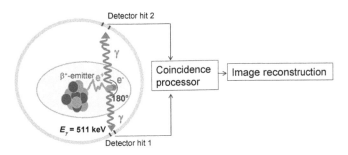

FIGURE 21.5 Schematics of the PET imaging process in case of a full-ring detector surrounding the activity source, depicting all steps of β^+-decay, positron emission, moderation in the medium, annihilation, and coincident detection of the resulting γ-pairs for image reconstruction.

scintillators like LSO or LFS, whose timing properties (and light output) outperform the more traditional detector systems based on BGO. This is at the expense of a minor amount of intrinsic radioactivity, which is responsible for a small background (especially of random coincidences) in the measurable signal. This detector radioactivity is of no concern in standard nuclear medicine imaging, but it may become more relevant at the typical low activity levels of ion therapy monitoring. However, recent studies suggest that, in the latter scenario, image quality is mostly constrained by statistical limitations of low amount of counts, rather than the relatively high lutetium-related random background [15].

The quantitative amount of decays from the activity source can be recovered from the estimated number of "true" coincidences by taking into account the total photon attenuation along each LOR as well as the quantum detection efficiency of the hit crystal pairs. The three-dimensional (3D) β^+-activity distribution can be finally recovered from the measured projections using tomographic reconstruction algorithms, typically approximating the position of β^+-decay with the annihilation emission place and including a normalization for the space-dependent geometrical probability of detection from each emission position. Iterative methods using the maximum-likelihood expectation maximization [16] or the ordered subset expectation maximization [17] algorithms in general provide the most reliable solutions. Due to typical detector granularity of 4–6 mm as well as intrinsic limitations coming from the nonperfect colinearity of the annihilation photons and the finite positron range, spatial resolutions of ca. 4–5 mm FWHM are generally attainable in the center of the scanner field of view (FOV). Indeed, such coarse resolution introduces a blurring of the imaged map of activation, but it does not alter the center of gravity of the distributions. Thus, it does not prevent localizing with millimeter accuracy the distal activation falloff, correlated with the primary proton beam range, as long as sufficient counting statistics is detected (cf. Section 21.3.1 and Figure 21.4). This fact was experimentally demonstrated in phantom studies [18]. Moreover, spatial resolution improvements are already possible with current detector technologies when refining the data processing, e.g., by incorporating the actual position-dependent point-spread function in the reconstruction process [19]. When available, additional information on arrival time difference (time of flight, TOF) of the photons in the crystals fired in coincidence can help reduce the uncertainty of the reconstructed emission position along the LOR. According to the 3×10^{10} cm/s speed of light c and the time resolutions τ of 350–600 ps achievable with the latest generation PET scanners, the uncertainty in the localization of the emission event can be reduced to ca. ± 5–9 cm ($c\cdot\tau/2$). Thus, current TOF information cannot dramatically improve spatial resolution, but it can offer increased signal-to-noise ratio in the reconstructed images. Moreover, attempts to push time resolutions of PET scanners below 100 ps are ongoing, with the final goal of enabling direct TOF-based reconstruction for real-time visualization of the created activity [20].

In comparison to conventional tracer imaging in diagnostic nuclear medicine, a major challenge for PET-based range monitoring is the extremely low counting statistics. The irradiation-induced activity signal approximately amounts to 0.5–15 kBq/Gy/cm³ depending on the time course of irradiation and involved anatomical site [21–22]. For typical fraction doses of 0.5–2 Gy, this corresponds to activity densities, which can be far below those of ≈10–100 kBq/cm³ reached in standard PET imaging. Moreover, once positron emitters are formed in living body by external irradiation, they undergo a complex hot chemistry and can be partly spread out by diffusion or even carried away ("washed out") from the location of activity production by different physiological processes (e.g., perfusion) occurring at different timescales. As a result, the irradiation-induced activity tends to disappear more quickly in living tissues than in inorganic matter, thus following not only a physical but also a so-called "biological" decay with varying half-lives from 2 s up to 10,000 s according to investigations in dead and living animals [23,24].

21.4.3 Clinical Implementation

Different strategies can be implemented clinically for imaging the irradiation-induced β⁺-activity, exploiting different time windows of the characteristic activity buildup and decay during and after therapeutic beam application (Figure 21.6). In situ detection with the patient directly in treatment position can be performed during ("in-beam") or immediately after ("in-room onboard") end of irradiation by means of customized systems fully integrated in the dose delivery environment [25,26]. Alternatively, the patient can be moved to a remote installation for imaging, starting from 2 up to 20 min after therapeutic beam application using conventional nuclear medicine PET scanners located inside ("in-room") or outside ("offline") the treatment room [22].

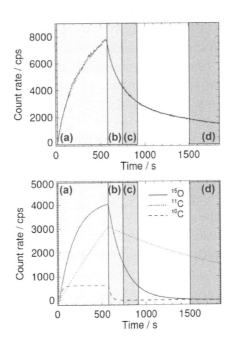

FIGURE 21.6 Different time windows for in-beam (a), in-room onboard (b), in-room (c), and offline (d) detection of β⁺-activity buildup and decay during and after irradiation. (Adapted from Refs. [10,18].) The top panel depicts the measured (solid line) coincidence rate detected for ~10 min proton irradiation of a PMMA target, compared to a calculation (dashed) using the estimated contributions of the different isotopes separately shown in the bottom panel.

Online implementations (in-beam and in-room onboard) better preserve correlation between measurable signal and delivered dose by detecting the major activity contribution from short-lived emitters such as ^{15}O and by minimizing signal degradation from positioning uncertainties and biological washout. However, they are also very demanding, since they require development of dedicated detector instrumentation to avoid interference with the beam as well as to enable flexible patient positioning. Typical geometrical configuration feature planar [26] or tomographic [25] dual-head systems. However, also "openPET" concepts of dual-rings with a gap in between or of a single ring with slanted or axially shifted detector arrangements have been proposed, to maximize geometrical coverage, still preserving an open space for the beam and treatment couch [27–30]. In-beam implementations additionally offer the possibility to measure the time-resolved activity formation during irradiation. However, they also require customized data acquisition systems for synchronization with the beam delivery and for rejection or at least suppression of undesired background, e.g., due to prompt-gamma emission, during the "beam-on" time [31]. In fact, the latter radiation background during beam extraction typically limits meaningful in-beam PET detection only to the pauses of pulsed beam delivery, thus reducing the amount of measurable decays from the produced isotopes in dependence of the accelerator duty cycle and dose rate. This is especially crucial for conventional in-beam PET data acquisition at efficient accelerator systems such as continuous-wave cyclotrons and dedicated clinical synchrotrons [32]. Nevertheless, first solutions for measuring a reasonable signal during proton beam delivery have been reported [33–34]. Depending on the amount of usable data accumulated during dose delivery, in-beam solutions may still require prolongation of the acquisition for few minutes after the end of irradiation for sufficient counting statistics, similar to the in-room onboard solutions. This is obviously at the expense of patient throughput in the treatment room. In terms of imaging performances, the space-variant detector response of dual-head configurations or unconventional dual- and single-ring openPET designs, in combination with the generally low counting statistics of irradiation-induced activation, pose challenging issues for trustful image reconstruction. This requires careful design of the detector geometry and cautious interpretation of the reconstructed activity distributions. In particular, first dual-head installations were either restricted to central planar imaging [26] or suffered from severe degradation of the imaging performances far from the central FOV plane in tomographic reconstruction [25,27]. However, TOF can help mitigate limited angle reconstruction artifacts, with major benefits expected from the aforementioned investigations of ultrafast detectors with timing resolution below 100–200 ps.

Postradiation in-room and offline imaging can rely on commercially available full-ring tomographs, but require accurate replication and fixation of the treatment position. Especially for in-room instrumentation, this can be achieved using modern robotic positioning systems. Availability of computed tomography (CT) imaging in combined PET/CT scanners can considerably help coregistration at the expense of additional radiation exposure of the patient. For in-room instrumentation, moderate efforts are required for integration of the scanner in the treatment environment and for optimization of the workflow to minimize occupation of the treatment room after irradiation. Offline imaging outside of the treatment room is less demanding in terms of integration efforts, but requires longer measuring times for accumulation of sufficient counting statistics as in the in-beam or in-room acquisition during or shortly after treatment (cf. Figure 21.6). Moreover, offline imaging can only enable detection of the integral fraction dose delivery, without the capability to resolve the contribution of the first applied treatment field. This restricts in vivo range verification to nonopposing beam portals for multifield irradiation. Individual monitoring of the first delivered treatment field is instead possible with in-room instrumentation, provided that the clinical staff accepts modifying the patient position for transfer to/from the PET scanner before applying the remaining fields. Obviously, minimization of the time elapsed between irradiation and imaging is crucial to keep degradation of the measurable signal at an acceptable level, taking into account the typical half-lives of 2 and 20 min for the main ^{15}O and ^{11}C products, as well as the typical timescale of 2 s up to 10,000 s for the different components of biological washout [24]. This would ideally require delays of a couple of minutes at most, which appears feasible for in-room installations. Larger delays may be unavoidable for offline implementations. The loss of activity in the time elapsed between irradiation and imaging consequently affects the choice of the acquisition time window, which has to be selected as a trade-off between counting statistics, patient throughput in the treatment room (in-room imaging) and patient comfort in prolonged acquisitions (offline imaging). This typically translates into measuring times of approximately 5 min for in-room imaging up to 30 min for offline imaging. In terms of tomographic imaging, conventional full-ring scanners typically provide better performances than the above-mentioned in situ installations, but may suffer from shortcomings of postradiation applications in terms of reduced counting statistics and increased washout, depending on the imaging delay and acquisition time window. Furthermore, reconstruction algorithms of commercial scanners are typically optimized for imaging high levels of localized activity concentrations as typically administered in standard tracer imaging and may thus suffer limitations in the imaging of low-level, extended activity distributions as induced by therapeutic irradiation [15].

The irradiation-induced activity within the patient represents only a "surrogate" signal correlated to but not directly matching the delivered dose (cf. Figure 21.4). Ideal clinical exploitation of PET imaging would require a solution to the challenging ill-posed inverse problem directly relating the measured activity to the actual dose delivery. Despite promising attempts to establish such relationship [35,36], a straightforward solution is not yet available for application to low statistics and washout-affected patient data. Nevertheless, useful clinical information on the correct delivery of the intended treatment can be obtained from comparing the measured PET images with an expected β^+-activation pattern. This can be either the reference activity measured at the first day of treatment [37] or a detailed calculation taking into account the patient-specific treatment plan, the fraction-specific time course of irradiation and PET acquisition, as well as the detector-dependent imaging performances [38,39]. The former approach only enables a consistency check over the course of fractionated therapy, whereas the latter may allow direct in vivo validation of the planned beam range in the patient. To this end, the computational model must carefully take into account both tissue elemental composition and different positron emitter production channels of relevance for the considered acquisition time window. Preclinical studies in known homogeneous media highlighted the benefit of directly using available experimental cross sections folded with the energy-dependent proton fluence rather than relying on the predictions of general-purpose nuclear models [18,40]. Moreover, they highlighted changes of the distal activity falloff up to a couple of millimeters depending on the acquisition time window, due to the signal contribution from different isotopes such as ^{11}C, ^{10}C, and ^{15}O, as well as the different shapes and energy thresholds of the reaction cross sections [18]. First computations of proton-induced activation in clinical cases were implemented using either MC simulation methods [11,39] or analytical convolution of the planned dose distribution with reaction-dependent filter functions in [39,41]. Clearly, accuracy of the computational methods strongly depends on the used cross section data and tissue elemental composition, shown to introduce 1 mm uncertainty in PET-based range verification when extracted from CT images [42]. To overcome the former issue, cross section data can be tuned to reproduce positron emitter yields disentangled from temporal analysis of activity decay in irradiated phantoms of known compositions [7,43]. The latter issue of tissue segmentation could be improved using dual-energy CT [44], which is currently under investigation for clinical integration in proton therapy.

Besides physical production and decay of positron emitters, computational models must also account for complex biological washout effects, considerably affecting magnitude and spatial distribution of measured activity especially in postradiation imaging. A first solution described in Ref. [39] was based on the idea [24] to decompose biological processes (C_{bio}) into fast, medium, and slow components affecting the physical activity (A_{phys}) decay in the final measurement (A_{meas}) as

$$A_{\text{meas}}(t) = C_{\text{bio}}(t) A_{\text{phys}}(t)$$

$$C_{\text{bio}}(t) = \left[M_f \exp(-\lambda_{\text{bio}f}t) + M_m \exp(-\lambda_{\text{bio}m}t) + M_s \exp(-\lambda_{\text{bio}s}t) \right]; \quad \sum M = 1$$

(21.2)

A generic CT segmentation was then proposed to identify different tissues (e.g., brain, muscle, fat) for assignment of biological M and λ coefficients based on literature data for implanted radioactive ions in animal tissue [24] and time analysis of activity decay in regions of interest set in different tissues [39]. This modeling already provided encouraging results with 5%–30% quantitative agreement between calculation and prediction especially for head-and-neck tumor locations [39]. However, other investigations reported more severe deviations in extracranial anatomical sites, especially in the pelvic region [45], thus demanding further model refinements. In this context, more recent results showed that data extracted from animal experiments with implanted radioactive beams are likely not directly applicable to proton therapy [46]. As an interesting exploitation of washout processes, Grogg et al. [47] suggested a dynamic model to reconstruct maps of ^{15}O produced in tissue. By observing changes of biological clearance rates over the treatment course, this approach promises both information on ^{15}O production maps, for direct comparison to

MC predictions and geometric verification of the applied dose, as well as functional information of washout rates, as potential indicators of treatment response.

In addition to range verification, PET imaging can also confirm lateral position of the irradiation field and detect unpredictable deviations between treatment fractions [37,48] or between planned and actual treatment [48,49]. Their possible causes are minor misalignments or local anatomical and/or physiological changes of the patient, introducing density modifications in the beam path. Although modifications of patient position and/or morphology can also be deduced from in-room 2D or 3D kilovoltage X-ray imaging, PET-based monitoring does not deliver additional radiation exposure to the patient (when using the planning CT for attenuation correction and anatomical coregistration). Furthermore, in-beam implementations may detect transient modifications occurring during treatment, which could be missed by pre- or posttreatment imaging.

When discrepancies between measurement and expectation are detected, a PET-guided quantification of the most likely applied dose corresponding to the experimental observation can be performed. This requires careful inspection of the PET images possibly supported by computer-aided tools for interactive or ideally automated assessment of the underlying reason of mismatch (e.g., mispositioning), in combination with access to the treatment planning system for recalculation of dose delivery. Recent studies showed different automated analysis tools for identification of critical deviations, even incorporating unavoidable uncertainties of the PET verification process for a more robust decision support system [50–53]. Eventually, improvement of the PET imaging performances and better understanding of current uncertainties in the knowledge of tissue stoichiometry, reaction cross sections, and washout parameters might enable direct dose reconstruction based on the promising preclinical approaches proposed in [36,41].

21.4.4 Worldwide Installations and Clinical Experience

The original idea of using PET for monitoring ion irradiation dates back to the end of the 1960s in connection with the ion therapy program at the Lawrence Berkeley Laboratory (LBL), USA [54]. Following the encouraging results at LBL, the method was further pursued for heavy ions at the new carbon ion therapy facilities established at the Heavy Ion Medical Accelerator at Chiba, Japan [55], and the Gesellschaft für Schwerionenforschung Darmstadt (GSI), Germany [56]. For protons, detectability and usability of the irradiation-induced β^+-activation was first investigated at the Brookhaven National Laboratory in the late 1970s [23]. Later on in the 1990s, feasibility phantom studies were carried out using commercial offline PET scanners by different groups in the Netherlands and Canada [57,58]. Thorough investigations of activity detection during and after proton irradiation of homogeneous and inhomogeneous targets were performed with an in-beam limited-resolution BGO-based prototype at the Michigan State University, USA [59], and with the dedicated in-beam PET installation at GSI Darmstadt [18,40].

For clinical application with passively scattered proton beams, first trials were performed with commercial full-ring PET scanners few minutes after treatment, thus suffering from coregistration issues between treatment and imaging positions, and lacking a calculation modeling for comparison to the measured images [22,60]. The first attempt to overcome coregistration issues of offline imaging by using a combined PET/CT scanner, along with the establishment of a detailed modeling of the expected activation pattern including biological washout, was realized at Massachusetts General Hospital (MGH), USA [39]. This study highlighted the importance of using the same treatment immobilization device at the imaging site despite availability of the additional CT for coregistration to the planning CT. Also, it indicated the possibility to achieve good spatial correlation and quantitative agreement between measured and calculated activity distributions when taking metabolic processes into account according to Equation 21.2. In particular, for head-and-neck patients, the beam range could be verified within 1–2 mm in favorable locations of well-coregistered low-perfused bony structures, for the considered cases of single or multiple angulated treatment fields. However, low spine and eye sites indicated the need for better fixation and coregistration methods. Following these promising results, a further study extended the population of patients to a total of 23 subjects (including the pilot trial) receiving at least one postradiation PET/CT acquisition after treatment [61]. Both studies identified challenges especially for extracranial tumor sites, due to major limitations from biological washout, breathing motion, coregistration issues, and lack of HU-tissue correlativity for reliable extraction of elemental tissue composition to be used in the activation modeling.

Following the MGH experience, other proton therapy facilities addressed the potential of offline PET/CT not only for in vivo range verification and confirmation of the treated volume [62] but also for assessment of information on prostate motion and patient position variability during daily proton beam delivery to help establish patient-specific planning target volume margins [63].

To overcome limitations of offline imaging, a BGO-based dual-head planar camera, adapted from a system originally designed for small animal imaging, was integrated into the proton gantry at the National Cancer Center of Kashiwa, Japan, for routine verification of treatment delivery with respect to the reference activation at the first fraction [37]. The system was also investigated for patient position verification via radioactive markers, alternatively to X-ray 2D IGRT [64].

As a compromise between more demanding in-beam/onboard solutions and suboptimal offline approaches, a prototype mobile full-ring neuroPET scanner based on cesium iodide (CsI)-scintillators was clinically investigated at MGH for in-room PET acquisitions shortly after end of irradiation [65]. Such a mobile system is rolled in the treatment room and the immobilized patient is moved in the scanner FOV using a robotic table few minutes after completion of the dose delivery. The first clinical study for nine patients [66,67] showed that the considerable reduction of time elapsed between irradiation and imaging enables a short-length PET scan of 5 min to yield similar results as a 20 min scan [67]. For eight patients treated with a single field, average range differences under 5 mm (<3 mm for six out of eight patients) with root-mean-square deviations of 4–11 mm were observed between measurements and MC simulations based on planning CT. A major limitation of the PET system alone was coregistration errors of ca. 2 mm. Hence, a second-generation prototype, based on state-of-the-art PET detectors and integrating CT imaging for attenuation correction, coregistration to the planning CT and/or assessment of anatomical changes, was developed and is currently being evaluated in a clinical protocol. However, also this system exhibits a small bore opening of application limited to cranial or pediatric indications.

While all the earlier reported clinical experiences were limited to passively scattered proton beam delivery, monitoring of pencil beam scanning (PBS) delivery was first pursued at the Heidelberg Ion Beam Therapy Center in Germany, using a commercial latest generation PET/CT scanner (with TOF and point-spread-function reconstruction) installed outside of the treatment rooms. Compared to previous offline implementations, a dedicated shuttle solution was realized to share the same tabletop with immobilization devices between the robotic treatment couch and the PET/CT. Data reported for proton treatments indicated absolute range agreement between (planning) CT-based MC simulations and measurements of 4.2 ± 2.2 mm (mean ± std for ten patients with primary glioma [68]) and −2.7 ± 4.9 mm (from five patients with tumor in the head [52]). In all cases, reproducibility typically better than 1 mm between treatment sessions was observed, with only few reported higher deviations up to 2 mm. Recently, a study including MC calculations on properly calibrated CT images of the PET/CT scanner was reported for tumor indications in brain, head and neck, sarcoma, and spine [48]. The analysis showed feasibility of detecting range shifts up to ±3 mm from both PET measurements and simulations, found well correlated (typically within 1.8 mm) to anatomical changes derived from CT scans, in agreement with dose data. Although confirming the promising sensitivity of PET to detect interfractional range variations, this study acknowledged limitations of offline imaging due to insufficient accuracy of patient-specific biological washout modeling and low signal. Better results are expected from a dedicated dual-head scanner, based on state-of-the-art LFS scintillators coupled to multipixel photon counters, just entering clinical testing at the Centro Nazionale di Adroterapia Oncologica (CNAO) in Italy. It enables dynamic (with few seconds time resolution to accumulate sufficient statistics) reconstruction of applied treatment, as shown in the first clinical investigation with PBS proton irradiation [69]. Examples of installations and results for different implementations are shown in Figures 21.7 and 21.8, respectively.

21.4.5 Ongoing Developments and Outlook

In addition to clinical studies with new-generation dedicated in-beam PET detectors [69], several efforts on improved TOF include scintillator detectors or alternative technologies like resistive plate channels [70] and even low-density plastic scintillators [71]. The final goal is to enable direct, event-by-event reconstruction of the activity measured during patient irradiation, with minimal degradation of image quality in case of limited angle geometry [27]. Regardless of the detector technology, data acquisition is also crucial for in-beam application at high duty cycle machines like continuous-wave cyclotrons, to obtain usable events despite the huge "beam-on" radiation background for optimal patient throughput. Nevertheless, the PET signal is intrinsically "delayed," making IGRT and DGRT difficult to achieve in "real-time." This motivates research on other viable approaches of in vivo imaging or possibilities to exploit irradiation-induced short-lived positron emitters like ^{12}N for so-called "beam-on PET" to provide real-time feedback [72].

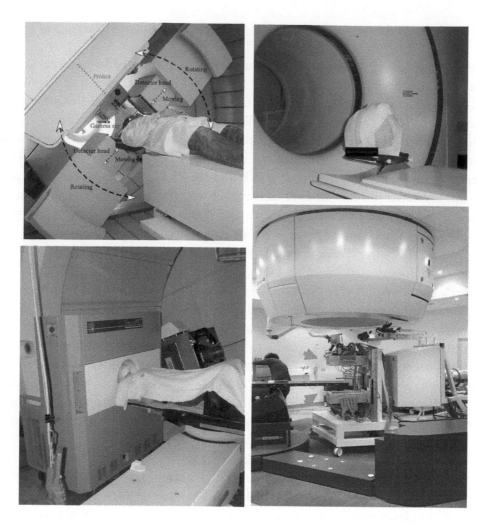

FIGURE 21.7   Installations used for clinical in vivo PET verification of proton therapy: dual-head in-room onboard camera integrated into the beam gantry at the National Cancer Center of Kashiwa, Japan (top left, [37] with permission), offline PET/CT (top right, [39]) and movable in-room PET scanner (bottom left, [66]) at MGH Boston, and new in-beam dual-head scanner (bottom right, see arrows) at CNAO (courtesy of Giuseppina Bisogni, INFN Pisa and CNAO).

21.5 PROMPT GAMMA

21.5.1 The Production of Irradiation-Induced Prompt Gamma

Prompt gamma-based treatment verification aims at exploiting the energetic γ emissions from excited nuclear states typically occurring in the final de-excitation phase of (pre)fragments formed in nuclear interactions between the therapeutic proton beam and the irradiated tissue (Figure 21.2). This is also the process responsible for the major component of undesired "beam-on" radiation background of in-beam PET acquisitions [31]. According to the typical energy dependence of reaction cross sections, the prompt gamma yield increases towards the end of the beam path, until the energy of the primary protons and of the secondary hadronic radiation drops below the nuclear reaction threshold, resulting in

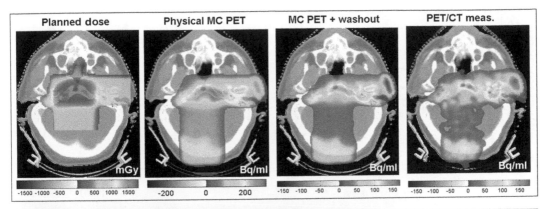

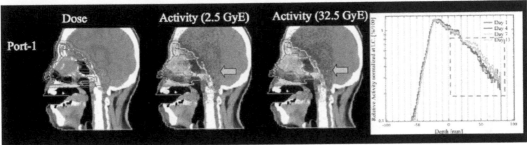

FIGURE 21.8 Clinical implementations of in vivo PET-based verification of proton therapy. (Adapted from Refs. [37,39] with permission.) The top row refers to offline PET/CT imaging long after irradiation at MGH ([39], cf. Figure 21.9), where a mismatch of the PET/CT measurement ("PET/CT Meas") with the "Planned Dose" is evident. Instead, good agreement is obtained for the PET MC modeling when taking biological washout into account ("MC PET + washout"), in contrast to physical activation alone ("Physical MC PET"). The rainbow color bars give absolute dose or activity values, while the gray scale refers to arbitrary rescaled CT values. The bottom row refers to in-room onboard imaging immediately after irradiation at NCC Kashiwa ([37], cf. Figure 21.9). Here, the planar activity measurements at different days (cf. quoted accumulated dose on top) revealed inconsistency, with a trend to activation at larger depth (cf. activity profiles in right inset). This finding prompted the detection of a serious anatomical modification of the patient (over 50 cm³ tumor shrinkage on a new CT), thus triggering plan adaptation.

a distal falloff correlated with the primary beam range (Figure 21.9). De-excitation prompt gamma are typically emitted in a short time (<1 ns) after nuclear interaction and isotropically in space. The prompt gamma energy spectrum can be very broad up to several McV or even tens of MeV. The shape exhibits an almost exponentially decreasing continuum superimposed onto several emission lines corresponding to transitions between discrete energy levels of the excited nuclei. Since de-excitation emission can also occur in flight, the width of the spectral lines can be affected by Doppler broadening.

21.5.2 The Detection Approach

Energetic prompt gammas can escape from the patient after production, with lower (factor of about 5 [75]) attenuation compared to 511 keV annihilation photons. Therefore, detection

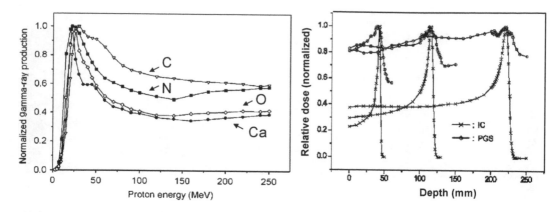

FIGURE 21.9 *Left*: Relative γ-ray production as function of energy for proton interaction on C (open triangles), N (filled squares), O (open diamond), and Ca (filled circles). (Adapted from Ref. [73], ©Institute of Physics and Engineering in Medicine. Reproduced by permission of IOP Publishing. All rights reserved.) *Right*: Correlation between depth-dose profiles measured with an IC and corresponding prompt gamma scans (PGS) along proton beam penetration in water. (Adapted from Ref. [74] with permission.) This pioneering detection of right-angled prompt gammas was achieved in Korea with a mechanically collimated and neutron-shielded CsI scintillator.

of this penetrating radiation preserving spatial information on the place of emission may be exploited for "real-time" verification of the treated area and, in particular of the beam range [74]. In addition, spectroscopic identification of the characteristic emission lines can be used to determine tissue composition, as suggested in Ref. [73]. However, as discussed in the following, technologies well established in nuclear medicine for single photon imaging are not applicable, thus requiring the development of novel detector concepts aiming to exploit different information (spatial location of emission, energy, TOF) of prompt gammas, or their conversion into secondary electrons.

To recover the spatial distribution of prompt gamma emissions for correlation to the beam range and delivered dose, the most straightforward detection approach is similar to standard single photon emission imaging. It requires a photon detector and a collimation system to select a particular angular direction from the isotropic gamma emission. However, following considerations rule out applicability of conventional photon detection systems used in diagnostic nuclear medicine imaging. First of all, the emission energy of the radiation to be detected is not known a priori. Moreover, the high energies in the MeV range challenge efficient collimation, requiring much more bulky material for mechanical solutions similar to gamma (Anger) cameras. Therefore, first solutions aimed only at recovering the signal emitted from the last portion of beam range in tissue, detected at 90° from the beam direction with a pinhole or a single-slit knife-edge-shaped collimator combined with a position-sensitive gamma detector, e.g. a scintillation crystal readout by photomultipliers [76–78]. The collimator can be located at adjustable distances (e.g., 15 cm) from the patient, depending on available space and desired trade-off between detection efficiency, spatial resolution, and FOV. Limitation to one-dimensional detection was deemed a viable solution for monitoring the distal range of individual pencil beams with a relatively simple and cost-effective instrumentation [78]. A straightforward extension to an array of scintillators behind a multislit or multislat collimator could enable recovery of the prompt gamma signal along the entire beam penetration depth, to provide additional information on the entrance in the patient of relevance in case of anatomical changes [79–81]. Depending on the time resolution of the employed scintillator(s), TOF can be additionally used to suppress the considerable background from the slower neutrons [82]. Besides spatial information (given by the collimator) and temporal information (for background suppression), the photon energy can also be exploited to identify the original nuclear emission and thus enable prompt gamma spectroscopy. In combination with a priori knowledge of the discrete prompt gamma ray emissions along proton pencil beams in different tissue elements, spectroscopic information can be used to simultaneously determine the beam range and oxygen and carbon concentration of the irradiated matter [83]. In the implementation by Verburg et al [84], additional reduction of background from Compton scattering and neutron-induced gamma rays was achieved with an active shield consisting of four optically separated BGO crystals, surrounding the primary lanthanum bromide ($LaBr_3$) detector of excellent energy and timing resolution.

As alternative to mechanical collimation, electronic collimation can be achieved by exploiting the known kinematics of Compton scattering. Along with elimination of the massive collimator, this approach promises 3D imaging capabilities [85]. However, conventional Compton camera setups are not applicable since the penetrating scattered photon is typically not completely stopped in the subsequent absorber. Therefore, alternative designs are being explored, which exploit multiple scattering processes until final interaction (ideally complete absorption) of the energy-degraded prompt gamma in the last component of the multistage detector assembly [86]. Directional information is recovered from intersection of the characteristic Compton cones (in which scattering at a certain angle took place) with the emission plane. In cases where the track of the scattered Compton electron can be reconstructed, the photon source position can be obtained from the intersection of Compton arcs, thus relaxing the requirement of full scattered-photon absorption [87]. Background suppression can be achieved via shielding and/or proper data acquisition strategies such as event triggering, pulse-shape discrimination, and TOF [88], capable to separate the signal induced by the different radiation components.

In the latter case of timing applications to background suppression, simulation studies suggested the need to adapt the TOF selection window to account for the lower propagation speed of protons, generating a few ns time shift of the ideal window for γ-emissions from deeper beam penetration depths [82]. As a reverse of this concept, the different photon arrival times were proposed to provide information on beam penetration depth, so-called prompt gamma timing [89]. Here no mechanical collimation is required, and very fast detectors can be used to measure the time spectrum of prompt gammas emitted at a certain backward angle with respect to the proton beam direction after a reliable start signal. Different momenta of the distribution can be then analyzed to enable reconstruction of the proton beam penetration depth [90]. Alternatively, the integral of the prompt gamma peak in the time spectrum can be monitored, exploiting the correlation between beam penetration depth and detectable prompt gamma count rates for a well-controlled intensity of the beam delivery [91]. Similar to TOF background suppression, reliability of prompt gamma timing techniques is higher for smaller bunch widths, making such time-based approaches especially suitable for ns-wide bunches delivered by cyclotron accelerators.

Conversion of prompt gammas into electrons of easier tracking was also suggested [92].

21.5.3 Clinical Implementation

Similar to β^+-activation, also the irradiation-induced prompt gamma production only represents a "surrogate" signal correlated to but not directly matching the delivered dose (cf. Figure 21.9). Therefore, unless a straightforward relationship between prompt-gamma emission yield and dose deposition can be established, this technique has to rely on the comparison of the measured signal with an expectation, e.g., deduced from detailed MC calculations [93,94] or analytical approaches [95,96]. Due to the manifold reaction channels leading to nuclear excitation and subsequent gamma emission, the computational problem is even more demanding than for PET. Hence, accuracy of available nuclear models and cross sections, and possibility to incorporate experimentally obtained data is crucial to the reliability of the calculation engines. Clinical applications will thus require extensive experimental validation of prompt gamma yields and their detection depending on beam properties, irradiated tissue, and selected detection approach.

The final choice of detector system and clinical implementation strategy will also largely depend on the specific beam delivery and time structure. Passively shaped scattered broad beams spread therapeutic proton beam intensities over large (several cm^3) volumes, most likely requiring time-integrated (even time-resolved, if properly synchronized with the beam modulation devices) 3D reconstruction of prompt-gamma emissions for drawing conclusions on the correctness of the entire (energy-resolved) dose delivery. Differently, if efficient detection per incident pencil beam can be guaranteed during PBS irradiation, at least for the most energetic layers typically carrying the majority of the dose, even 1D prompt gamma detection might offer a tool for "real-time" monitoring of the delivered beam range. Here, spatial information of the impinging pencil beam direction could be obtained from upstream beam monitors or by placing a fast position-sensitive detector (e.g., a hodoscope) in front of the patient. This information could be complemented by computation of the most likely path of protons traveling in the body, similar to computational approaches developed for image reconstruction in proton radiography or tomography [97]. Artistic views of possible clinical implementations originally depicted for application to the less scattering ^{12}C ion beams are shown in Figure 21.10 for two detector solutions.

In all cases of time-resolved and "real-time" acquisition, fast comparison with the expectation might open the possibility of prompt interruption of the irradiation if relevant range deviations are detected, besides an integral control of the daily delivery before the next treatment fraction. Alternatively, prompt-gamma imaging of a subset of pencil beams could be evaluated as a range probe before application of the entire treatment [75]. Since the fast γ-emission is insensitive to washout processes, dose deconvolution ideas initially proposed for PET monitoring can find application in prompt gamma imaging, opening the perspective of in vivo dose verification [99].

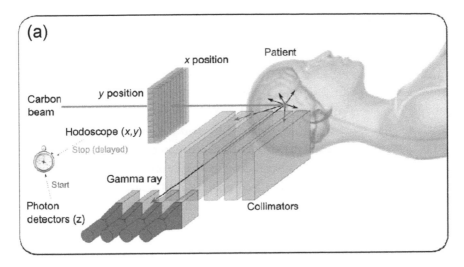

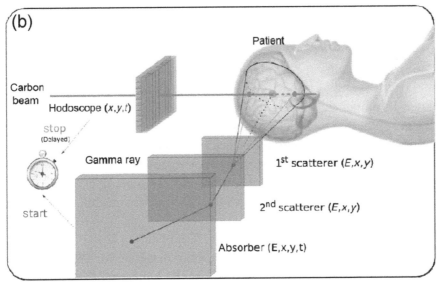

FIGURE 21.10   Possible clinical implementations of real-time prompt gamma imaging of scanned pencil-like ion beams (in this example, carbon) using either a multicollimated multidetector prompt gamma camera (top, a) or a multistage Compton camera (bottom, b). The upstream hodoscope detector serves to tag the ions in time and space prior to entrance in the patient, while TOF is used to separate the γ signal from neutron background (cf. text). (Adapted from Ref. [98]).

21.5.4 Worldwide Investigations and First Clinical Experience

Following the first theoretical proposition of using prompt gamma for treatment verification by Stichelbaut and Jongen [100], first experimental investigations were limited to the detection of collimated (mostly right-angled) prompt gammas, scanned along the proton beam penetration depth in homogeneous targets for 1D assessment of spatial correlation with the dose delivery and beam range [74]. For water, such scans were successfully compared to ionization chamber (IC) measurements of depth dose to illustrate the promising correlation with the Bragg-peak location (Figure 21.9). In addition to further efforts of the

Korean National Cancer Center team in testing alternative detector concepts such as pin-hole or Compton cameras [101], several groups undertook many computational and experimental investigations to identify suitable detection systems [90]. In particular, Polf et al. put forward the innovative idea to provide spectroscopic insight on tissue composition from prompt gamma energy spectra [73,102]. This idea of prompt gamma spectroscopy and active shielding was further elaborated at MGH, resulting in the first promising collimated setup of [84], combining excellent energy and time resolution, of which a second-generation prototype was recently realized and is close to enter clinical testing.

For Compton camera imaging, several detector concepts have been proposed, featuring different multistage designs with either a combination of scattering layer(s) and absorber, with or without secondary electron tracking or multiple scattering layers only. Proposed detector technologies span from scintillators (LaBr$_3$, LSO, BGO) to semiconductors (Si, Cadmium Zinc Telluride, CZT) as well as gaseous detectors (time projection chambers) and thereof combination [90]. The system closer to clinical testing at the University of Maryland consists of four pixelated CZT detection stages. In preclinical testing at the Roberts Proton Therapy Center (Philadelphia, USA), 2D prompt gamma distributions in water could be successfully reconstructed, and corresponding 1D profiles proved detectability of 3-mm range shifts at a dose of 400 cGy [102].

The more recent prompt gamma timing [89] and peak integral [91] techniques were first demonstrated in homogeneous and, for the former one, heterogeneous phantoms [103]. This experience indicated the need of dedicated beam monitors for reliable start information of TOF measurements, of multiple detectors for increased sensitivity and recovery of target position, as well as of detection systems of very good timing resolution and high data throughput [90]. Gamma Electron Vertex Imaging is also being experimentally investigated using a 1-mm Be plate converter and a tracking hodoscope, but is still challenged by sensitivity issues [104].

The first system reaching maturity for clinical application is the knife-edge slit camera developed by the company IBA [78]. Originally designed for PBS, it was first clinically deployed for passively scattered treatment at the University Proton Therapy Center Dresden, Germany [105]. With background suppression based on open and close collimator measurements, the data indicated ±2 mm interfractional range variations, consistent with dose calculations based on three in-room control CTs of the same patient. More recently, a similar system was deployed in Philadelphia to monitor six fractions of PBS treatment of a brain tumor, proving the ability to monitor the range of individual or even better aggregated spots in nine energy layers (Figure 21.11), with averaged interfractional changes from −0.8 to 1.7 mm observed for the three fields over three weeks [106]. When aggregating the prompt gamma signal from neighboring spots to increase counting statistics, a range shift retrieval precision of 2 mm was achieved (cf. Figure 21.11), however, with major limitations attributed to the limited accuracy of the trolley-mounted prototype [106].

The edge detection ability of relevance for range monitoring primarily depends on counting statistics [106,107]. All prompt gamma cameras reported so far show detection efficiencies from 10^{-5} (collimated cameras) to 10^{-4} (Compton cameras) [90], well below the few percent detection efficiency of PET. Hence, despite the estimated superior (up to

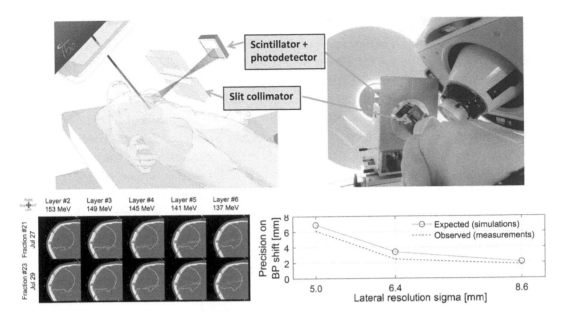

FIGURE 21.11 Schematic view (top left) and realization (top right) of the single-slit knife-edge collimated camera, along with its first clinical application to PBS (bottom left), comparing planned (red) and measured (green) Bragg peak depth with respect to the planning target volume (narrow yellow line) for different energy layers and fractions, using a 7 mm aggregation kernel. Expected and observed shift retrieval precisions as a function of the lateral spatial resolution (defined as the sum in quadrature of the spot sigma and the sigma of the aggregation kernel) are depicted in the bottom right panel. (Adapted from Refs. [78,105,106] with permission.) (See color online).

a factor of 80) photon yield with respect to a typical in-room PET implementation (5 min acquisition starting 2 min after a 30-s long irradiation) [75], the signal-to-noise available in prompt-gamma measurements critically depends on the achievable detection efficiency for reliable clinical application.

21.5.5 Ongoing Developments and Outlook

Along with improvements of computational models and, for tomographic approaches, image reconstruction, several tools are being developed for automated evaluation of prompt gamma distributions, particularly 1D profiles, adopted from strategies originally proposed for PET or adapted to the specific prompt gamma falloff shape [78,79]. First concepts of hybrid systems combining prompt gamma detection with 3D PET imaging were also proposed [108].

21.6 IONOACOUSTICS

When protons penetrate a medium, the energy deposition mainly due to electronic energy losses causes a localized heating generating thermal expansion and related thermoacoustic emissions, similar to those produced through local absorption of laser light in optoacoustics [109]. Owing to the fast (~ns) stopping process of proton beams and resulting secondary electrons, including the very fast (~ps, [110]) conversion of radiation into heat, such local heating can be assumed as instantaneous and related to the temporal beam shape. For the typical (sub)ms length of ion pulses and individual spot delivery in PBS, heat diffusion processes are negligible, and the energy deposition can be considered adiabatic. Moreover, for proton beam pulses shorter than the transit time of the sound wave across the Bragg peak width (~5 µs at 70 MeV [111]), the energy deposition process also fulfills the condition of so-called stress confinement [112], maximizing the amplitude of acoustic emission. In this context the relevant temporal structure refers to the modulating envelope of the beam pulses in the ~µs timescale, since much shorter timescales arising from the accelerator radiofrequency (e.g., ~10 ns microstructure period for isochronous cyclotrons) are not relevant to the acoustic signal formation at clinical proton beam energies.

Under the aforementioned conditions of thermal and stress confinement, the heating function becomes proportional to the first time derivative of the heating rate and can be factorized into a temporal and spatial component, resulting in the simplified solution of the general thermoacoustic equation for the acoustic pressure $p(\mathbf{r},t)$:

$$p(\mathbf{r},t)=\frac{\beta}{4\pi v_s \kappa \rho C_V}\frac{\partial}{\partial t}\int_{A(t)}\frac{Q(\mathbf{r}')}{R}dA' \tag{21.3}$$

where $Q(\mathbf{r}')$ denotes the energy deposition density represented by the Bragg curve, integrated for $R=|\mathbf{r}-\mathbf{r}'|$, $A(t)$ is the surface on which $R=v_s t$ for the speed of sound v_s in the propagating medium, β is the thermal coefficient of volumetric expansion, k the medium isothermal compressibility, ρ the mass density, and C_V the specific heat capacity at constant volume [113].

Hence, the acoustic wave amplitude depends on both the deposited energy and temporal shape of the excitation pulse. In particular, Refs. [114–116] suggested distinguishing between the signal originating before the Bragg peak, resulting in a cylindrical wave emitted laterally, and that originating at the Bragg peak, approximately corresponding to a spherical wave (Figure 21.12). The latter was postulated to be of higher interest for range verification based on TOF techniques, when placing single detectors in axial or near-axial configuration distal to the Bragg peak [117].

In terms of detector choice, it has to be considered that not only the amplitude of the acoustic wave but also its frequency spectrum strongly depends on the temporal scale of the heating process, related to the beam temporal profile and sharpness of the Bragg peak (converted to time through division of the Bragg peak width by v_s). In particular, reported mean values span from few MHz for preclinical beam energies of 20 MeV investigated in a wide range of temporal pulses (from 8.0 ns to 4.3 µs at 3 ns rise time [117]) down to tens of kHz for clinical energies above 140 MeV and pulse lengths of few microseconds [111,118]. Hence, ionoacoustic detection has to rely on low-frequency, broadband hydrophones placed in acoustic contact with the patient, similar to the first clinical reported observation of acoustic signatures in Ref. [119]. Interestingly, similar to PET and prompt gamma imaging, resolvability of the Bragg peak position (i.e., edge detection) with TOF was found not affected by the imaging spatial resolution, which is related to the signal frequency in conventional ultrasound imaging [111]. Hence, the lower frequency spectrum

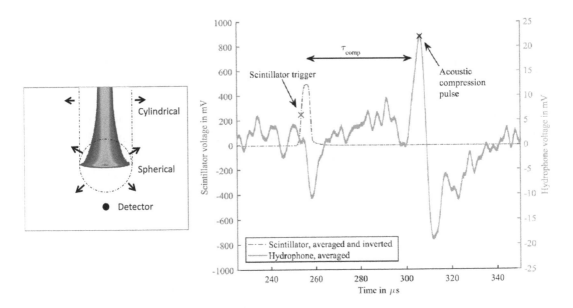

FIGURE 21.12 *Left:* Typical setup for ionoacoustic measurements in water [116], showing the two components of the acoustic wave generated by a proton pencil beam. *Right:* Waveform of scintillator and hydrophone (in axial configuration distal to the Bragg peak) measured in water for a 220 MeV proton beam accelerated from a synchrocyclotron [111]. ©Institute of Physics and Engineering in Medicine. Reproduced by permission of IOP Publishing. All rights reserved.

in the kHz region does not prevent the prospect of obtaining millimeter-accurate range monitoring, and can also help reducing signal attenuation. In fact, the measured signal strength is the major factor influencing the achievable resolution of ionoacoustics for range monitoring, depending on the initial acoustic emission (related to the spatial and temporal characteristics of the energy deposition), the signal attenuation in tissue and the sensitivity of the transducer.

Whereas new-generation compact synchrocyclotrons already exhibit suitable pulsing of few µs for ionoacoustic monitoring during treatment delivery [111], conventional isochronous cyclotrons require special beam modulation into pulses well matched to optimize thermoacoustic emissions [118]. In this latter case of modified beam sources, Refs. [117,118] foresaw clinical workflows entailing a pretreatment range verification of few "diagnostic spots" (possibly combined with ultrasound US/CT markers) delivered with artificially created proton pulses of reduced temporal length and high current, followed by delivery of the conventional treatment.

Although further speculations on future clinical workflows can still be considered too premature, the commonly perceived strategy already envisioned in the early clinical attempts [119] should combine multiple broadband transducers placed in contact to the patient, ideally with robotic or autoadhesive means. The method could greatly benefit from additional prior tissue knowledge, derived either from pretreatment X-ray CT images (similar to proposed aberration corrections in US image guidance, [120]) or from coregistered US images. With extremely fast data processing, it might become possible to visualize both the Bragg peak position and tissue anatomy in quasi real-time, thus promising an appealing means for checking the beam delivery on a spot-by-spot basis and enable image-guided tumor motion compensation. Hence, the method might prove highly beneficial for those sonic accessible critical tumor indications currently challenged by interfractional organ motion, such as prostate, breast, liver, and pancreas [117].

Back in the 1970s, Sulak et al. [121] reported a first thorough experimental characterization of acoustic signals generated by protons in water for beam diagnostic purposes. The first clinical study during passively scattered proton treatment of a hepatic tumor followed in the 1990s [119]. More recently, several experimental and theoretical investigations demonstrated the use of ionoacoustics for range verification in homogeneous water phantoms with the more favorable PBS delivery, mostly under idealized conditions [115,117,122,123]. Extension to ionoacoustic tomography with first-time coregistration to optoacoustic and US imaging was reported for an ex vivo mouse experiment [124]. Shortly afterwards, Patch et al. reported the first intrinsically coregistered ionoacoustic and ultrasonic acquisition with the same US transducer array for high dose rate irradiation (~2 Gy/pulse at the Bragg peak) of a water tank and a gelatine phantom with a cavity left empty or filled with olive oil [125]. First ionoacoutic measurements in water at a clinical artificially pulsed (18 μs rise time) isochronous cyclotron and a clinical synchrocyclotron were reported by Jones et al. [118] and Leharck et al. [111], showing the potential of millimeter range resolution for doses of a few Gy, as typically carried by the most energetic pencil beams in clinical treatment. Preliminary simulation studies in realistic patient geometries (Figure 21.13) support the possibility to maintain millimeter range accuracy in favorable anatomical sites, when using multiple detectors or detector arrays along with proper triangulation techniques [126], thus encouraging ongoing developments toward a potentially precise and cost-effective modality.

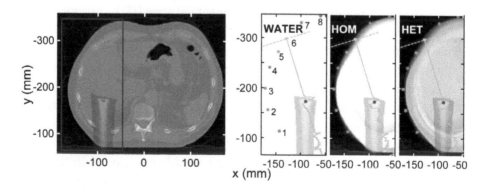

FIGURE 21.13 Simulation of a single PBS spot dose deposition to a liver tumor (left) and corresponding representative acoustic wave arriving on a given detector on the patient surface (right), calculated for three different scenarios of water, homogeneous, and heterogeneous tissue (from the left to the right, respectively). (Reproduced from Ref. [126], ©Institute of Physics and Engineering in Medicine. Reproduced by permission of IOP Publishing. All rights reserved.)

21.7 MAGNETIC RESONANCE IMAGING

Radiation can also induce physiological changes correlated to the beam range, as shown in follow-up MRI investigations of patients receiving craniospinal proton irradiation [127]. Such hyperintense T1-weighted signal was attributed to radiation-induced fatty replacement of bone marrow, detectable as early as 10 days after therapy and persisting up to 21 months after treatment. Visual comparison with treatment plans indicated a clear correlation of the sharply delineated bone marrow changes with the distal dose falloff for posterior–anterior fields, directly placing the characteristic proton Bragg peaks only in the posterior part of the vertebral body.

Although such post treatment information cannot enable patient-specific IGRT nor DGRT for correction of the already entirely delivered therapeutic dose, it can still provide essential clinical information for improvement of population-based treatment margins in specific indications. Hence, in a second study [128], the same researchers tried quantifying the relationship between radiation dose and MRI signal intensity (SI) on a population-based approach, analyzing the MR signal correlation with the more controllable lateral penumbra of the proton dose in the irradiated sacrum. Using the so established general dose-SI relationship, the distal falloff of the "actually delivered" dose was deduced from MRI images and compared to the treatment plan to estimate proton range delivery errors in the lumbar spine (Figure 21.14). Initial findings indicated a tendency to overshoot of about 1.9 mm (95% confidence interval, 0.8–3.1 mm), i.e., well within clinical margins [128]. The magnitude of this overshoot was judged in the same order of magnitude of the inherent uncertainties of the method. These latter uncertainties were mainly attributed to the patient-specific validity of a general dose-SI relationship deduced in a different anatomical region than the lumbar spine, the usage of heuristic rules for determination of data to be reliably used in the analysis, and coregistration between MRI and planning CT data.

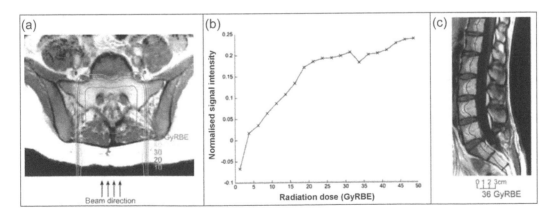

FIGURE 21.14 Clinical workflow and results of MRI-based in vivo range verification. The less controllable lateral penumbra of the beam in the sacrum (left, a) is used to determine a dose-SI curve (middle, b). From the analysis of more patient data a general dose-SI curve is obtained and used to calculate the 50% distal isodose deduced from the MRI scan (red lines in the right panel, c) for comparison with the planned one (blue lines), showing in this example a generalized beam overpenetration in the lumbar spine. (Adapted from Ref. [128], with permission.)

Since then, along with further attempts to perform a quantitative exploitation of the MRI signal for in vivo range determination in craniospinal irradiation, additional studies have been performed to identify other potential mechanisms for application to different tumor sites. Impairment of the liver function can become visible as hypointense areas on MRI scans when using special contrast agents. A retrospective study investigated MRI data for five patients with intrahepatic carcinoma metastatic to the liver, treated by 3D conformal proton stereotactic body radiotherapy using two cross-fired fields, with radiation dose of either 44 or 55 Gy in five fractions [129]. Contrast-enhanced T1-weighted gradient echo images were acquired at different scanners within 11–25 weeks after end of treatment, starting 20 min after injection of the hepatocyte-directed contrast agent. A statistically significant correlation was observed between radiation dose and MRI SI. For more quantitative analysis, the dose-SI relationship had to be established, similar to the previously reported work, using the region that includes two superior/inferior penumbrae of the two employed proton treatment fields. The comparison indicated a trend of undershoot with respect to the planning scenario, with mean difference of -2.18 ± 4.89 mm for anterior–posterior and posterior–anterior beams, while -3.90 ± 5.87 mm for lateral beams. A possible multistep model for interpretation of the signal formation mechanism was also proposed [130]. However, the authors acknowledged several limitations of the study due to uncertainties of nonrigid registration, coarse (2.0–3.5 mm slice thickness) resolution of MRI images, possible shrinkage of the dysfunctional volume induced by irradiation several weeks after end of treatment and possible limitations of the used SI (related to uptake of the contrast agent by the hepatocytes) as surrogate for the irradiation-induced physiological change. Future studies might benefit from MRI sequences, enabling earlier detection of the irradiation-induced physiological changes and from growing interest in MRI integration in proton therapy [131].

21.8 CONCLUSION AND OUTLOOK

This chapter has reviewed the main methods currently under experimental or clinical investigation for in vivo range (dose) verification in proton therapy. All methods hold great promise to complement each other and enable proton treatment verification at different timescales (cf. Figure 21.1) and in different anatomical sites. Indeed, despite the discussed limitations of "delayed" emission and biological washout, PET still represents the most mature technique readily available for clinical implementation with relatively moderate efforts. While ongoing developments toward next-generation ultrafast TOF-PET detectors might open the perspectives of real-time monitoring, the latter possibility has been recently demonstrated with a first prototype of prompt gamma camera with a single-slit knife-edge collimator. Despite the current limitation to 1D projection of the prompt gamma signal, promising results for both passively scattered and actively scanned clinical treatments were reported, confirming the potential of the method for in vivo range verification and DGRT. For certain tumor indications, ionoacoustics offers an intriguing idea for real-time in-vivo range monitoring, combining the advantage of compact and cost-effective instrumentation with the unique potential of quasi-simultaneous morphological imaging with conventional

US. Such possibilities for fraction-specific imaging of physical emissions could be complemented by MRI at regular time intervals during the treatment course, to optimize treatment margins for reduction of toxicity and safe dose escalation studies.

REFERENCES

1. Verhey LJ, Goitein M, McNulty P, Munzenrider JE, Suit HD. Precise positioning of patients for radiation therapy. *Int J Radiat Oncol Biol Phys* 1982; 8: 289–94.
2. Landry G, Dedes G, Pinto M, Parodi K, Imaging and particle therapy: current status and future perspectives, in *Advances in Particle Therapy: a Multidisciplinary Approach* edited by Manjit Dosanjh and Jacques Bernier, CRC Press, Boca Raton, 2018.
3. Antonuk LE. Electronic portal imaging devices: a review and historical perspective of contemporary technologies and research. *Phys Med Biol* 2002; 47: R31–65.
4. Hoesl M, Deepak S, Moteabbed M, Jassens G, Orban J, Park YK, Parodi K, Bentefour EH, Lu HM. Clinical commissioning of an in vivo range verification system for prostate cancer treatment with anterior and anterior oblique proton beams. *Phys Med Biol* 2016; 61: 3049.
5. Koehler AM. Proton radiography. *Science* 1968; 160: 303–4.
6. Mumot M, Algranati C, Hartmann M et al. Proton range verification using a range probe: definition of concept and initial analysis. *Phys Med Biol* 2010; 55(16): 4771.
7. Bauer J, Unholtz D, Kurz C, Parodi K. An experimental approach to improve the Monte Carlo modelling of offline PET/CT-imaging of positron emitters induced by scanned proton beams. *Phys Med Biol* 2013; 58: 5193–213.
8. Schardt D, Elsässer T, Schultz-Ertner D. Heavy-ion tumor therapy: physical and radiobiological benefits. *Rev Mod Phys* 2010; 82: 383–425.
9. Hüfner J. Heavy fragments produced in proton-nucleus and nucleus-nucleus collisions at relativistic energies. *Phys. Rep.* 1985; 125: 129.
10. Parodi K. On the feasibility of dose quantification with in-beam PET data in radiotherapy with 12C and proton beams. *PhD Thesis*, Dresden University of Technology; 2004, in Forschungszentrum Rossendorf Wiss-Techn-Ber FZR-415 (2004).
11. Parodi K, Ferrari A, Sommerer F, Paganetti H. Clinical CT-based calculations of dose and positron emitter distributions in proton therapy using the FLUKA Monte Carlo code. *Phys Med Biol* 2007; 52: 3369–87.
12. Ferrero V et al. The INSIDE project: in-beam PET scanner system features and characterization. *JINST* 2017; 12: C03051.
13. Brasse D, Kinahan PE, Lartizien C, Comtat C, Casey M, Michel C. Correction methods for random coincidences in fully 3D whole-body PET: impact on data and image quality. *J Nucl Med* 2005; 46: 859–67.
14. Ollinger JM. Model-based scatter correction for fully 3D PET Phys Med Biol 1996; 41: 153–176.
15. Kurz C, Bauer J, Conti M, Guérin L, Eriksson L, Parodi K. Investigating the limits of PET/CT imaging at very low true count rates and high random fractions in ion-beam therapy monitoring. *Med Phys* 2015; 42: 3979–91.
16. Shepp LA, Vardi Y. Maximum likelihood reconstruction for emission tomography. *IEEE Trans Med Imaging* 1982; 1: 113–22.
17. Hudson HM, Larkin RS. Accelerated image reconstruction using ordered subsets of projection data. *IEEE Trans Med Imaging* 1994; 13: 601–9.
18. Parodi K, Pönisch F, Enghardt W. Experimental study on the feasibility of in-beam PET for accurate monitoring of proton therapy. *IEEE Trans Nucl Sci* 2005; 52: 778–86.
19. Alessio AM, Stearns CW, Tong S, Ross SG, Kohlmyer S, Ganin A, Kinahan PE. Application and evaluation of a measured spatially variant system model for PET image reconstruction. *IEEE Trans Med Imaging* 2010; 29: 938–949.

20. Crespo P, Shakirin G, Fiedler F, Enghardt W, Wagner A, Direct time-of-flight for quantitative, real-time in-beam PET: a concept and feasibility study. *Phys Med Biol* 2007; 52: 6795–6811.
21. Parodi K, Paganetti H, Cascio E, Flanz J, Bonab A, Alpert N, Lohmann K, Bortfeld T, PET/CT imaging for treatment verification after proton therapy: a study with plastic phantoms and metallic implants. *Med Phys* 2007; 34: 419–35.
22. Vynckier S, Derreumaux S, Richard F et al. Is it possible to verify directly a proton-treatment plan using positron emission tomography? *Radiother Oncol* 1993; 26: 275–7.
23. Bennett GW, Archambeau JO, Archambeau BE, Meltzer JI, Wingate CL. Visualization and transport of positron emission from proton activation in vivo. *Science* 1978; 200: 1151–3.
24. Mizuno H, Tomitami T, Kanazawa M et al. Washout measurements of radioisotopes implanted by radioactive beams in the rabbit. *Phys Med Biol* 2003; 48: 2269–81.
25. Enghardt W, Crespo P, Fiedler F et al. Charged hadron tumour therapy monitoring by means of PET. *Nucl Instrum Methods A* 2004; 525: 284–8.
26. Nishio T, Ogino T, Nomura K et al. Dose-volume delivery guided proton therapy using beam on-line PET system, *Med Phys* 2006; 33: 4190–7.
27. Crespo P, Shakirin G, Enghardt W. On the detector arrangement for in-beam PET for hadron therapy monitoring. *Phys Med Biol* 2006; 51: 2143–63.
28. Yamaya T, Yoshida E, Inaniwa T, Sato S, Nakajima Y, Wakizaka H, Kokuryo D, Tsuji A, Mitsuhashi T, Kawai H, Tashima H, Nishikido F, Inadama N, Murayama H, Haneishi H, Suga M, Kinouchi S. Development of a small prototype for a proof-of-concept of OpenPET imaging. *Phys Med Biol* 2011; 56: 1123–37.
29. Tashima H, Yamaya T, Yoshida E, Kinouchi S, Watanabe M, Tanaka E. A single-ring OpenPET enabling PET imaging during radiotherapy. *Phys Med Biol* 2012; 57: 4705–18.
30. Tashima H, Yoshida E, Inadama N, Nishikido F, Nakajima Y, Wakizaka H, Shinaji T, Nitta M, Kinouchi S, Suga M, Haneishi H, Inaniwa T, Yamaya T. Development of a small single-ring OpenPET prototype with a novel transformable architecture. *Phys Med Biol* 2016; 61: 1795–809.
31. Parodi K, Crespo P, Eickhoff H, Haberer T, Pawelke J, Schardt D, Enghardt W. Random coincidences during in-beam PET measurements at microbunched therapeutic ion beams. *Nucl Instrum Methods Phys Res A* 2005; 545: 446–58.
32. Parodi K, Bortfeld T, Haberer T. Comparison between in-beam and offline PET imaging of proton and carbon ion therapeutic irradiation at synchrotron- and cyclotron-based facilities. *Int J Rad Oncol Biol Phys* 2008; 71: 945–56.
33. Sportelli G, Belcari N, Camarlinghi N, Cirrone GA, Cuttone G, Ferretti S, Kraan A, Ortuño JE, Romano F, Santos A, Straub K, Tramontana A, Del Guerra A, Rosso V. First full-beam PET acquisitions in proton therapy with a modular dual-head dedicated system. *Phys Med Biol* 2014; 59: 43–60.
34. Helmbrecht S, Enghardt W, Fiedler F, Iltzsche M, Pausch G, Tintori C, Kormoll T. In-beam PET at clinical proton beams with pile-up rejection. *Z Med Phys* 2017; 27: 202–217.
35. Fourkal E, Fan J, Veltchev I. Absolute dose reconstruction in proton therapy using PET imaging modality: feasibility study. *Phys Med Biol* 2009; 54(11): N217–28.
36. Remmele S, Hesser J, Paganetti H, Bortfeld T. A deconvolution approach for PET-based dose reconstruction in proton radiotherapy. *Phys Med Biol* 2011; 56 :7601–19.
37. Nishio T, Miyatake A, Ogino T, Nakagawa K, Saijo N, Esumi H. The development and clinical use of a beam ON-LINE PET system mounted on a rotating gantry port in proton therapy. *Int J Radiat Oncol Biol Phys* 2010; 76(1):277–86.
38. Pönisch F, Parodi K, Hasch BG, Enghardt W. The description of positron emitter production and PET imaging during carbon ion therapy. *Phys Med Biol* 2004; 49: 5217–32.
39. Parodi K, Paganetti H, Shih HA et al. Patient study on in-vivo verification of beam delivery and range using PET/CT imaging after proton therapy. *Int J Rad Oncol Biol Phys* 2007; 68: 920–34.

40. Parodi K, Enghardt W, Haberer T. In-beam PET measurements of b+-radioactivity induced by proton beams. *Phys Med Biol* 2002; 47: 21–36.
41. Parodi K, Bortfeld T. A filtering approach based on Gaussian-powerlaw convolutions for local PET verification of proton radiotherapy. *Phys Med Biol* 2006; 51: 1991–09.
42. España S, Paganetti H. The impact of uncertainties in the CT conversion algorithm when predicting proton beam ranges in patients from dose and PET-activity distributions. *Phys Med Biol* 2010; 55(24):7557–71.
43. Miyatake A, Nishio T, Ogino T. Development of activity pencil beam algorithm using measured distribution data of positron emitter nuclei generated by proton irradiation of targets containing 12C, 16O, and 40Ca nuclei in preparation of clinical application. *Med Phys* 2011; 38: 5818–5829.
44. Berndt B, Landry G, Schwarz F, Tessonnier T, Kamp F, Dedes G, Thieke C, Würl M, Kurz C, Ganswindt U, Verhaegen F, Debus J, Belka C, Sommer W, Reiser M, Bauer J, Parodi K. Application of single- and dual-energy CT brain tissue segmentation to PET monitoring of proton therapy. *Phys Med Biol* 2017; 62: 2427–2448.
45. Knopf A, Parodi K, Bortfeld T, Shih HA, Paganetti H, Systematic analysis of biological and physical limitations of proton beam range verification with offline PET/CT scans. *Phys Med Biol* 2009; 54: 4477–4495.
46. Ammar C, Frey K, Bauer J, Melzig C, Chiblak S, Hildebrandt M, Unholtz D, Kurz C, Brons S, Debus J, Abdollahi A, Parodi K. Comparing the biological washout of β+-activity induced in mice brain after 12C-ion and proton irradiation. *Phys Med Biol* 2014; 59: 7229–44.
47. Grogg K, Alpert NM, Zhu X, Min CH, Testa M, Winey B, Normandin MD, Shih HA, Paganetti H, Bortfeld T, El Fakhri G. Mapping 15O production rate for proton therapy verification. *Int J Radiat Oncol Biol Phys.* 2015; 92: 453–459.
48. Handrack J, Tessonnier T, Chen W, Liebl J, Debus J, Bauer J, Parodi K. Sensitivity of post treatment positron-emission-tomography/computed-tomography to detect interfractional range variations in scanned ion beam therapy. *Acta Oncol.* 2017; 1–8 doi: 10.1080/0284186X.2017.1348628. [Epub ahead of print].
49. Enghardt W, Parodi K, Crespo P, Fiedler F, Pawelke J, Pönisch F. Dose quantification from in-beam positron emission tomography. *Radiother Oncol* 2004; 73: S96–8.
50. Helmbrecht S, Santiago A, Enghardt W, Kuess P, Fiedler F. On the feasibility of automatic detection of range deviations from in-beam PET data. *Phys Med Biol* 2012; 57: 1387–1397.
51. Kuess P, Helmbrecht S, Fiedler F, Birkfellner W, Enghardt W, Hopfgartner J, Georg D. Automated evaluation of setup errors in carbon ion therapy using PET: feasibility study. *Med Phys* 2013; 40: 121718.
52. Frey K, Unholtz D, Bauer J, Debus J, Min CH, Bortfeld T, Paganetti H, Parodi K. Automation and uncertainty analysis of a method for in vivo range verification in particle therapy. *Phys Med Biol* 2014; 59: 5903–5919.
53. Chen W, Bauer J, Kurz C, Tessonnier T, Handrack J, Haberer T, Debus J, Parodi K. A dedicated software application for treatment verification with off-line PET/CT imaging at the Heidelberg Ion Beam Therapy Center. *J Phys: Conf Ser* 2018; 777: 012021.
54. Maccabee HD, Madhvanath U, Raju MR. Tissue activation studies with alpha-particle beams. *Phys Med Biol* 1969; 14: 213–224.
55. Tomitani T, Kanazawa M, Yoshikawa K et al. Effect of target fragmentation on the imaging of autoactivation of heavy ions. *J Jpn Soc Ther Radiol Oncol* 1997; 9(S2): 79–82.
56. Enghardt W, Crespo P, Fiedler F et al, Charged hadron tumour therapy monitoring by means of PET. *Nucl Instrum Methods A* 2004; 525: 284–8.
57. Paans AMJ, Schippers JM, Proton therapy in combination with PET as monitor: a feasibility study. *IEEE Trans Nucl Sci* 1993; 40: 1041–4.
58. Oelfke U, Lam GKY, Atkins MS, Proton dose monitoring with PET: quantitative studies in Lucite. *Phys Med Biol* 1996; 41: 177–96.

59. Litzenberg DW, Roberts DA, Lee MY et al, On-line monitoring of radiotherapy beams: experimental results with proton beams. *Med Phys* 1999; 26: 992–1006.

60. Hishikawa Y, Kagawa K, Murakami M et al, Usefulness of positron-emission tomographic images after proton therapy. *Int J Radiat Oncol Biol Phys* 2002; 53: 1388–91.

61. Knopf AC, Parodi K, Paganetti H, Bortfeld T, Daartz J, Engelsman M, Liebsch N, Shih H. Accuracy of proton beam range verification using post-treatment positron emission tomography/computed tomography as function of treatment site. *Int J Radiat Oncol Biol Phys* 2011; 79: 297–304.

62. Nishio T, Miyatake A, Inoue K, Gomi-Miyagishi T, Kohno R, Kameoka S, Nakagawa K, Ogino T, Experimental verification of proton beam monitoring in a human body by use of activity image of positron-emitting nuclei generated by nuclear fragmentation reaction. *Radiol Phys Technol* 2008; 1: 44–54.

63. Hsi WC, Indelicato DJ, Vargas C, Duvvuri S, Li Z, Palta J. In vivo verification of proton beam path by using post-treatment PET/CT imaging, *Med Phys* 2009; 36: 4136–46.

64. Yamaguchi S, Ishikawa M, Bengua G, Sutherland K, Nishio T, Tanabe S, Miyamoto N, Suzuki R, Shirato H. A feasibility study of a molecular-based patient setup verification method using a parallel-plane PET system. *Phys Med Biol* 2011; 56(4):965–77.

65. España S, Zhu X, Daartz J, Liebsch N, El Fakhri G, Bortfeld T, Paganetti H. Feasibility of in-room pet imaging for in vivo proton beam range verification. *Med Phys* 2010; 37: 3180.

66. Zhu X, España S, Daartz J, Liebsch N, Ouyang J, Paganetti H, Bortfeld TR, El Fakhri G. Monitoring proton radiation therapy with in room PET imaging. *Phys Med Biol* 2011; 56: 4041–4057.

67. Min CH, Zhu X, Winey BA, Grogg K, Testa M, El Fakhri G, Bortfeld TR, Paganetti H, Shih HA. Clinical application of in-room positron emission tomography for in vivo treatment monitoring in proton radiation therapy. 2013; 86: 183–9.

68. Nischwitz SP, Bauer J, Welzel T, Rief H, Jäkel O, Haberer T, Frey K, Debus J, Parodi K, Combs SE, Rieken S. Clinical implementation and range evaluation of in vivo PET dosimetry for particle irradiation in patients with primary glioma. *Radiother Oncol* 2015; 115: 179–85.

69. Bisogni MG, Attili A, Battistoni G, Belcari N, Camarlinghi N, Cerello P, Coli S, Del Guerra A, Ferrari A, Ferrero V, Fiorina E, Giraudo G, Kostara E, Morrocchi M, Pennazio F, Peroni C, Piliero MA, Pirrone G, Rivetti A, Rolo MD, Rosso V, Sala P, Sportelli G, Wheadon R. INSIDE in-beam positron emission tomography system for particle range monitoring in hadrontherapy. *J Med Imag* 2017; 4: 011005.

70. Blanco A, Chepela V, Ferreira-Marquesa R, Fonte P, Lopesa MI, Peskove V, Policarpo A. Perspectives for positron emission tomography with RPCs. *Nucl Instrum Methods Phys Res A* 2003; 508: 88–93.

71. Moskal P, Alfs D, Bednarski T, Białas P, Czerwiński E, Gajos A, Gorgol M, Jasińska B, Kamińska D, Kapłon Ł, Krzemień W, Korcyl G, Kowalski P, Kozik T, Kubicz E, Mohammed M, Niedźwiecki Sz, Pałka M, Pawlik-Niedźwiecka M, Raczyński L, Rudy Z, Rundel O, Sharma NG, Silarski M, Słomski A, Strzelecki A, Wieczorek A, Wiślicki W, Zgardzińska B, Zieliński M. J-PET: a novel TOF-PET scanner based on plastic scintillators. *Radiother Oncol* 2016; 118: S76.

72. Buitenhuis HJT, Diblen F, Brzezinski KW, Brandenburg S, Dendooven P. Beam-on imaging of short-lived positron emitters during proton therapy. *Phys Med Biol* 2017; 62: 4654–72.

73. Polf J, Peterson S, Ciangaru G, Gillin M, Beddar S. Prompt gamma-ray emission from biological tissues during proton irradiation: a preliminary study. *Phys Med Biol*. 2009; 54: 731–743.

74. Min CH, Kim CH, Youn MY, Kim JW, Prompt gamma measurements for locating the dose falloff region in the proton therapy. *App Phys Lett* 2006; 89: 1835171–3.

75. Moteabbed M, Espana S, Paganetti H. Monte Carlo patient study on the comparison of prompt gamma and PET imaging for range verification in proton therapy. *Phys Med Biol* 2011; 56: 1063–83.

76. Kim D, Yim H, Kim JW. Pinhole camera measurements of prompt gamma rays for detection of beam range variation in proton therapy. *J Korean Phys Soc* 2009; 55: 1673–6.

77. Bom V, Joulaeizadeh L, Beekman F. Real-time prompt gamma monitoring in spot-scanning proton therapy using imaging through a knife-edge-shaped slit. *Phys Med Biol* 2012; 57: 297–308.

78. Smeets J, Roellinghoff F, Prieels D, Stichelbaut F, Benilov A, Busca P, Fiorini C, Peloso R, Basilavecchia M, Frizzi T, Dehaes JC, Dubus A. Prompt gamma imaging with a slit camera for real-time range control in proton therapy. *Phys Med Biol* 2012; 57: 3371–405.

79. Schmid S, Landry G, Thieke C, Verhaegen F, Ganswindt U, Belka C, Parodi K, Dedes G. Monte Carlo study on the sensitivity of prompt gamma imaging to proton range variations due to interfractional changes in prostate cancer patients. *Phys Med Biol* 2015; 60: 9329–47.

80. Min CH, Lee HR, Kim CH, Lee SB. Development of array-type prompt gamma measurement system for in vivo range verification in proton therapy. *Med Phys* 2012; 39: 2100–7.

81. Pinto M, Dauvergne D, Freud N, Krimmer J, Letang JM, Ray C, Roellinghoff F, Testa E. Design optimisation of a TOF-based collimated camera prototype for online hadrontherapy monitoring. *Phys Med Biol* 2014; 59: 7653–74.

82. Biegun AK, Seravalli E, Lopes PC, Rinaldi I, Pinto M, Oxley DC, Dendooven P, Verhaegen F, Parodi K, Crespo P, Schaart DR. Time-of-flight neutron rejection to improve prompt gamma imaging for proton range verification: a simulation study. *Phys Med Biol* 2012; 57: 6429–44.

83. Verburg JM, Seco J. Proton range verification through prompt gamma-ray spectroscopy. *Phys Med Biol* 2014; 59: 7089–106.

84. Verburg JM, Riley K, Bortfeld T, Seco J. Energy- and time-resolved detection of prompt gamma-rays for proton range verification. *Phys Med Biol* 2013; 58: L37–49.

85. Mackin D, Peterson S, Beddar S, Polf J. Evaluation of a stochastic reconstruction algorithm for use in Compton camera imaging and beam range verification from secondary gamma emission during proton therapy. *Phys Med Biol* 2012; 57: 3537–3553.

86. Peterson SW, Robertson D, Polf J, Optimizing a three-stage Compton camera for measuring prompt gamma rays emitted during proton radiotherapy. *Phys Med Biol* 2010; 55: 6841–6856.

87. Thirolf PG, Aldawood S, Böhmer M, Bortfeldt J, Castelhano I, Dedes G, Fiedler F, Gernhäuser R, Golnik C, Helmbrecht S, Hueso-Gonzalez F, v.d., Kolff H, Kormoll T, Lang C, Liprandi S, Lutter R, Marinsek T, Maier L, Pausch G, Petzoldt J, Römer K, Schaart D, Parodi K. Development of a Compton Camera Prototype for Medical Imaging. *EPJ Web Conf* 2016; 117: 05005.

88. Testa M. Physical measurements for ion range verification in charged particle therapy. *PhD Thesis*, University Claude Bernard Lyon 1, Lyon, France, 2010 (https://twiki.ific.uv.es/twiki/pub/Main/IrisGroup/PhD_Manuscript_Mauro_Testa.pdf).

89. Golnik C, Hueso-Gonzalez F, Muller A, Dendooven P, Enghardt W, Fiedler F, Kormoll T, Roemer K, Petzoldt J, Wagner A, Pausch G. Range assessment in particle therapy based on prompt gamma-ray timing measurements. *Phys Med Biol* 2014; 59: 5399–422.

90. Krimmer J, Dauvergne D, Létang JM, Testa É. Prompt-gamma monitoring in hadrontherapy: a review. *Nucl Instrum Methods A* 2018; 878: 58–73.

91. Krimmer J, Angellier G, Balleyguier L, Dauvergne D, Freud N, Herault J, Letang JM, Mathez H, Pinto M, Testa E, Zoccarato Y. A cost-effective monitoring technique in particle therapy via uncollimated prompt gamma peak integration. *Appl Phys Lett* 2017; 110: 154102 1–5.

92. Kim CH, Park JH, Seo H, Lee HR, Erratum: Gamma electron vertex imaging and application to beam range verification in proton therapy. *Med Phys* 2012; 39: 6523–6524.

93. Polf JC, Peterson S, McCleskey M, Roeder BT, Spiridon A, Beddar S, Trache L. Measurement and calculation of characteristic prompt gamma ray spectra emitted during proton irradiation. *Phys Med Biol* 2009; 54: N519–27.

94. Kraan AC. Range verification methods in particle therapy: underlying physics and Monte Carlo modelling. *Front Oncol* 2015; 5.

95. Sterpin E, Janssens G, Smeets J, Stappen FV, Prieels D, Priegnitz M, Perali I, Vynckier S. Analytical computation of prompt gamma ray emission and detection for proton range verification. *Phys Med Biol* 2015; 60: 4915–46.
96. Huisman BFB, Létang JM, Testa É, Sarrut D. Accelerated prompt gamma estimation for clinical proton therapy simulations. *Phys Med Biol* 2016; 61: 7725–7743.
97. Williams DC. The most likely path of an energetic charged particle through a uniform medium Phys. *Med. Biol.* 2004; 49: 2899–2911.
98. Testa M. Physical measurements for ion range verification in charged particle therapy. *PhD Thesis*, University Claude Bernard Lyon 1, Lyon, France, 2010 (https://twiki.ific.uv.es/twiki/pub/Main/IrisGroup/PhD_Manuscript_Mauro_Testa.pdf).
99. Schumann A, Priegnitz M, Schoene S, Enghardt W, Rohling H, Fiedler F. From prompt gamma distribution to dose: a novel approach combining an evolutionary algorithm and filtering based on gaussian-powerlaw convolutions. *Phys Med Biol* 2016; 61: 6919–34.
100. Stichelbaut F, Jongen Y. Verification of the proton beams position in the patient by the detection of prompt gamma-rays emission. *39th Meeting of the Particle Therapy Co-Operative Group* (San Francisco, CA), 2003.
101. Kim D, Yim H, Kim JW. Pinhole camera measurements of prompt gamma rays for detection of beam range variation in proton therapy. *J Korean Phys Soc* 2009; 55: 1673–6.
102. Polf JC, Avery S, Mackin DS, Beddar S. Imaging of prompt gamma rays emitted during delivery of clinical proton beams with a Compton camera: feasibility studies for range verification. *Phys Med Biol* 2015; 60: 7085–99.
103. Hueso-González F, Enghardt W, Fiedler F, Golnik C, Janssens G, Petzoldt J, Prieels D, Priegnitz M, Römer KE, Smeets J, Vander Stappen F, Wagner A, Pausch G. First test of the prompt gamma ray timing method with heterogeneous targets at a clinical proton therapy facility. *Phys Med Biol* 2015; 60: 6247–72.
104. Lee HR, Kim SH, Park JH, Jung WG, Lim H, Kim CH. Prototype system for proton beam range measurement based on gamma electron vertex imaging. *Nucl Instrum Methods Phys Res A* 2017; 857: 82–97.
105. Richter C, Pausch G, Barczyk S, Priegnitz M, Keitz I, Thiele J, Smeets J, Stappen FV, Bombelli L, Fiorini C, Hotoiu L, Perali I, Prieels D, Enghardt W, Baumann M. First clinical application of a prompt gamma based in vivo proton range verification system. *Radiother Oncol* 2016; 118: 232–7.
106. Xie Y, Bentefour EH, Janssens G, Smeets J, Vander Stappen F, Hotoiu L, Yin L, Dolney D, Avery S, O'Grady F, Prieels D, McDonough J, Solberg TD, Lustig RA, Lin A, Teo BK. Prompt gamma imaging for in vivo range verification of pencil beam scanning proton therapy. *Int J Radiat Oncol Biol Phys* 2017; 99: 210–8.
107. Roellinghoff E, Benilov A, Dauvergne D, Dedes G, Freud N, Janssens G, Krimmer J, Létang JM, Pinto M, Prieels D, Ray C, Smeets J, Stichelbaut F, Testa E. Real-time proton beam range monitoring by means of prompt-gamma detection with a collimated camera. *Phys Med Biol* 2014; 59: 1327–38.
108. Parodi K. On- and off-line monitoring of ion beam treatment. *Nucl Instrum Methods Phys Res A* 2016; 809: 113–9.
109. Wang LHV, Hu S. Photoacoustictomography:Invivoimagingfrom organelles to organs. *Science* 2012; 335: 1458.
110. Toulemonde M, Surdutovich E, Solov'yov A. Temperature and pressure spikes in ion-beam cancer therapy. *Phys Rev E* 2009; 80: 031913.
111. Lehrack S, Assmann W, Bertrand D, Henrotin S, Herault J, Heymans V, Vander Stappen F, Thirolf P, Vidal M, Van de Walle J, Parodi K. Submillimeter ionoacoustic range determination for protons in water at a clinical synchrocyclotron. *Phys Med Biol* 2017; 62: L20–L30.
112. Wang LV, Wu H-I, *Biomedical Optics: Principles and Imaging*, John Wiley & Sons, Hoboken, NJ, 2007.

113. Parodi K, Assmann W. Ionoacoustics: a new direct method for range verification. *Mod Phys Lett* 2015; A30: 1540025.
114. Albul VI, Bychkov VB, Vasil'ev SS, Gusev KE, Demidov VS, Demidova EV, Krasnov NK, Kurchanov AF, Luk'yashin VE, Sokolov AY. Acoustic field generated by a beam of protons stopping in a water medium. *Acoust Phys* 2005; 51: 33–37.
115. Jones KC, Witztum A, Sehgal CM, Avery S. Proton beam characterization by proton-induced acoustic emission: simulation studies. *Phys Med Biol* 2014; 59: 6549–6563.
116. Jones KC, Vander Stappen F, Sehgal CM, Avery S. Acoustic time-of-flight for proton range verification in water. *Med Phys* 2016; 43: 5213–24.
117. Assmann W, Kellnberger S, Reinhardt S, Lehrack S, Edlich A, Thirolf PG, Moser M, Dollinger G, Omar M, Ntziachristos V, Parodi K. Ionoacoustic characterization of the proton Bragg peak with submillimeter accuracy. *Med Phys* 2015; 42: 567–574.
118. Jones KC, Vander Stappen F, Bawiec CR, Janssens G, Lewin PA, Prieels D, Solberg TD, Sehgal CM, Avery S. Experimental observation of acoustic emissions generated by a pulsed proton beam from a hospital-based clinical cyclotron. *Med Phys* 2015; 42: 7090–7.
119. Hayakawa Y, Tada J, Arai N, Hosono K, Sato M, Wagai T, Tsuji H, Tsujii H. Acoustic pulse generated in a patient during treatment by pulsed proton radiation beam. *Radiat Oncol Invest* 1995; 3: 42–5.
120. Fontanarosa D, Pesente S, Pascoli F, Ermacora D, Abu Rumeileh I, Verhaegen F. A speed of sound aberration correction algorithm for curvilinear ultrasound transducers in ultrasound-based image-guided radiotherapy. *Phys Med Biol* 2013; 58: 1341–60.
121. Sulak L, Armstrong T, Baranger H, Bregman M, Levi M, Mael D, Strait J, Bowen T, Pifer AE, Polakos PA, Bradner H, Parvulescu A, Jones WV, Learned J. Experimental studies of the acoustic signature of proton-beams traversing fluid media. *Nucl Instrum Methods* 1979; 161: 203–217.
122. Alsanea F, Moskvin V, Stantz KM. Feasibility of RACT for 3D dose measurement and range verification in a water phantom. *Med Phys* 2015; 42: 937–46.
123. Jones KC, Seghal CM, Avery S. How proton pulse characteristics influence protoacoustic determination of proton-beam range: simulation studies. *Phys Med Biol* 2016; 61: 2213–42.
124. Kellnberger S, Assmann W, Lehrack S, Reinhardt S, Thirolf P, Queiros D, Sergiadis G, Dollinger G, Parodi K, Ntziachristos V. Ionoacoustic tomography of the proton Bragg peak in combination with ultrasound and optoacoustic imaging. *Sci Rep* 2016; 6: 29305.
125. Patch SK, Kireeff Covo M, Jackson A, Qadadha YM, Campbell KS, Albright RA, Bloemhard P, Donoghue AP, Siero CR, Gimpel TL, Small SM, Ninemire BF, Johnson MB, Phair L. Thermoacoustic range verification using a clinical ultrasound array provides perfectly co-registered overlay of the Bragg peak onto an ultrasound image. *Phys Med Biol* 2016; 61: 5621–38.
126. Jones KC, Nie W, Chu JCH, Turian JV, Kassaee A, Sehgal CM, Avery S. Acoustic-based proton range verification in heterogeneous tissue: simulation studies. *Phys Med Biol Phys Med Biol* 2018; 63: 025018.
127. Krejcarek SC, Grant PE, Henson JW, Tarbell N, Yock T Physiological and radiographic evidence of the distal edge of the proton beam in craniospinal irradiation. *Int J Rad Biol Phys* 2007; 68: 646–9.
128. Gensheimer MF, Yock TI, Liebsch NJ, Sharp GC, Paganetti H, Madan N, Grant PE, Bortfeld T. In vivo proton beam range verification using spine MRI changes. *Int J Radiat Oncol Biol Phys* 2010; 78(1): 268–75.
129. Yuan Y, Andronesi OC, Bortfeld TR, Richter C, Wolf R, Guimaraes AR, Hong TS, Seco J. Feasibility study of in vivo MRI based dosimetric verification of proton end-of-range for liver cancer patients. *Radiother Oncol.* 2013; 106: 378–82.
130. Richter C, Seco J, Hong TS, Duda DG, Bortfeld T. Radiation-induced changes in hepatocyte-specific Gd-EOB-DTPA enhanced MRI: potential mechanism. *Med Hypotheses.* 2014; 83: 477–81.
131. Oborn BM, Dowdell S, Metcalfe PE, Crozier S, Mohan R, Keall PJ. Future of medical physics: Real-time MRI-guided proton therapy. *Med Phys* 2017; 44: e77–e90.

VII

Biological Effects

CHAPTER **22**

The Physics of Proton Biology*

Harald Paganetti

Massachusetts General Hospital and Harvard Medical School

CONTENTS

---

* Colour figures available online at www.crcpress.com/9781138626508

22.1 INTRODUCTION

Radiation interacts with tissue atoms and causes molecules to go to excited states, leading to vibrations that produce heat (more than 95% of the energy in radiation therapy is transferred into heat). Nevertheless, the most important impact of radiation is energy deposition events causing ionizations. Ionizations are predominantly responsible for damage that leads to chemical reactions threatening the DNA via highly reactive radicals (indirect effects). These have to occur within a few nanometers of the DNA, i.e., within the diffusion distance of the radicals. In addition, radiation can cause lethal damage from direct deposition of energy (direct effects). For low linear energy transfer (LET) radiation, effects are caused mainly via δ-electrons creating free radicals. The LET is the energy transferred to the absorbing medium per unit track length of a particle. Figure 22.1 illustrates the different interaction channels. While all molecules in the cell are affected by radiation, the damage to the DNA molecules is decisive for mutation induction, carcinogenic transformation, and killing of most cell types. The term cell death is usually used for loss of reproductive capacity, although for some cell types, apoptosis or programmed cell death may also be initiated by damage to the cell membrane. DNA damage is an alteration in the chemical structure of DNA. The most relevant damage is a double-strand break (DSB). Damaged DNA undergoes transformations that invoke different repair mechanisms where a vast majority of single-strand breaks (SSBs) are typically repaired. Only a small portion of the initial biochemical damage leads to a lasting effect.

Clustering of radiation damage is largely responsible for the effectiveness of high-LET radiation. There are several spatial levels of damage such as clustering of DNA damage at the level of a few base pairs, clustering of DNA damage at the level of chromosomes, and the spatial distribution of damage across cells. It is not so much the type of DSBs, but the spatial distribution of DNA damage as well as the properties of the intrinsic cell repair mechanisms that determine a biological effect.

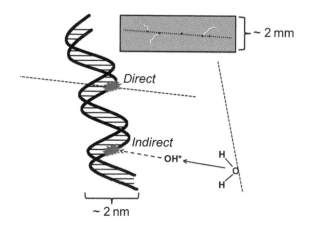

FIGURE 22.1   Direct and indirect (via diffusion of free radicals) radiation effects on the DNA. The dashed lines show two potential particle paths leading to the two interaction types. The inset illustrates the size of a proton track with ionization events and δ-electrons (white tracks.)

The track structure, defined as a track core causing interactions of the primary particle and a surrounding halo generated by δ-electrons, differs amongst different ionizing radiations and LET. For a typical proton radiation field, ~70% of the energy lost is transferred to secondary electrons (δ-rays), ~25% is needed to overcome their binding energy, and the residual 5% produces neutral excited species [1]. Each proton track is associated with δ-rays with a wide distribution of electron energies and a maximum energy of roughly 500 keV. An electron with an energy of 100 keV will cause ~500 energy deposition events with energies larger than 10 eV in a 6 nm^2 target but less than 10 with energies larger than 150 eV [2]. The energy of these electrons, and thus their range, is typically much lower than the electrons produced in photon beams.

The track structure affects not only the type of damage but also the capacity of the cell to repair it [3,4]. Cluster of strand breaks that are more concentrated in space and the associated damage renders damage less amenable to repair, or rather, to competent repair [5]. Furthermore, clustered strand breaks increase the likelihood of misjoining, which can cause chromosome aberrations. For example, at a given LET, protons are in fact more effective than heavier ions [6–8]. This is because protons have smaller track radii (as defined by the range of δ-electrons) in a certain LET range, implying higher ionization density in the submicrometer range, ultimately leading to more complex damage [9]. Radiation is more effective per unit dose when the energy deposition is more concentrated in space.

Even if photons and protons might cause the same number of DNA DSBs per unit dose, the distribution of DSBs among cells and among the individual chromosomes or parts of a chromosome may differ. Table 5.1 shows the approximate number of events in a mammalian cell in different radiation fields after a dose of 1 Gy [10,11]. The number of ionizations per cell and the initial yield of DSBs shows little variation with ionization density, but the number of residual breaks after cellular repair substantially differs. The same number of initial DSBs gives rise to a larger number of chromosome aberrations following high-LET vs. low-LET irradiation. Furthermore, higher LET radiation increases the frequency of complex aberrations (Table 22.1).

The LET is a macroscopic parameter that does not describe energy deposition in a biological target (e.g., a cell nucleus) but rather the energy deposition per path length of a

TABLE 22.1 Average Yield of Damage in a Single Mammalian Cell after 1 Gy Delivered by Photons (Low LET) or Low-Energy α Particles (High LET) (Nucleus Diameter: 8 m; Energy Deposition per Ionization: 25 eV)

Radiation	Photons	α Particles
Tracks in nucleus	1,000	2
Ionizations in nucleus	10^5	10^5
Ionizations in DNA	1,500	1,500
DNA single-strand break	700–1,000	300–600
DNA DSBs (initially)	18–60	70
DNA DSBs (after 8 h)	6	30
Chromosome aberrations	0.3	2.5
Complex aberrations	10%	45%
Lethal lesions	0.2–0.8	1.3–3.9
Cells inactivated	10%–50%	70%–95%

Source: Adopted after Refs. [10,11].

particle. For the dose-averaged LET (LET_d), the LET is averaged in such a way that the contribution of each particle is weighted by the dose it deposits. This concept is applicable for proton beams where the number of particle tracks crossing a subcellular structure is quite large for high-LET ion beams, the number of tracks per subcellular target may be much smaller and the track averaged LET might become more meaningful.

In addition to differences in single track structure, the energy deposited per incident proton is significantly higher than the energy deposited per incident photon. Consequently, for the same dose, the number of protons crossing a region of interest is typically lower than the corresponding number of photons. This, in turn, causes energy depositions to be more heterogeneous in proton radiation fields compared to photon radiation. In general, the spatial distribution of damage might be random for low-LET radiation but follows more closely specific particle tracks at high LET values [12].

22.2 DOSE–RESPONSE MODELS

Assuming a particle crossing a sensitive area in tissue, e.g., a proton crossing a cell nucleus, the mean number of effective hits per dose is defined by Poisson statistics. The probability of finding a nucleus with ω lethal events if an average of Ω lethal events per nucleus is produced in the whole population is

$$P(\omega\Omega)=\frac{\Omega^{\omega}}{\omega!}\exp(-\Omega)$$

(22.1)

Since only cells without lethal events will by definition survive, the survival probability as a function of absorbed dose, D, with N_0 being the initial number of cells and N being the number of unaffected cells, is given as

$$N/N_0(D)=P(0,\Omega)=\exp(-\Omega(D))$$

(22.2)

The response of cells to dose is typically visualized as a dose–response curve. Cell survival curves represent the number of cells that have lost the ability for unlimited proliferation. The assumption that there are individual targets requiring a single hit to be inactivated would result in a straight line on the logarithmic dose–response plot. A linear-quadratic parameterization is the simplest mathematical formulation to fit most survival curves [13,14]:

$$N/N_0(D)=\exp(-\alpha D-\beta G D^2)$$

(22.3)

The linear-quadratic model describes the surviving fraction of cells as a function of dose, using α (Gy^{-1}) and β (Gy^{-2}) to characterize the intrinsic radiation sensitivity of cells. In its general form, it includes a dose protraction factor, G, to account for the temporal pattern of radiation delivery [15]. For a short delivery time and for one fraction delivery, G equals 1. Accordingly, a biological effect can be defined as (with the dose D delivered at once or in n fractions of d):

$$E(D)=\alpha D+\beta D^2 =n(\alpha d+\beta d^2)$$

(22.4)

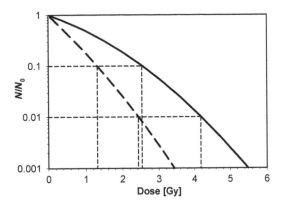

FIGURE 22.2   Example of two dose–response curves. The solid line might resemble the response after photon irradiation and the long dashed line might be caused by very low energy proton irradiation. The RBE at 10% survival would be ~2.55/1.3=1.96, while at 1% survival it would be ~4.15/2.4=1.73 (indicated by short dashed lines.)

Figure 22.2 shows two dose–response curves. The curvature, or shoulder, of a survival curve can be interpreted based on damage repair capacity or damage induction mechanisms, i.e., by the α/β ratio. For a given endpoint, dose–response curves from proton radiation are typically steeper than the ones from photon radiation.

The linear-quadratic formalism might not be valid below ~1 Gy (e.g., due to hypersensitivity or adaptive response [16,17]), which could affect dose–response estimations for organs at risk typically receiving less than 1 Gy for a treatment using 2 Gy per fraction to the target. The linear-quadratic formalism is also questionable for doses higher than ~10 Gy, which might affect its predictive power for hypofractionated treatments [18,19]. A more general description using a number of sensitive sites in a cell nucleus predicts an exponential asymptote for high doses [20]. Furthermore, when fitting α and β to a response curves, one has to keep in mind that the result may depend on the dose range considered [17,21].

While the linear-quadratic function is the most common in radiation therapy, dose-response curves can also be described with other mathematical equations, e.g., the multitarget/single-hit equation with the parameters D_0 (intrinsic radiosensitivity) and n. The parameter n is the extrapolation number representing the ability to accumulate and repair sublethal damage:

$$\frac{N}{N_0}=1-\left(1-\exp\left[\frac{-D}{D_0}\right]\right)^{n} \qquad (22.5)$$

The multitarget formalism assumes that from a collection of targets in the cell nucleus, n have to be hit at least k times each to inactivate the cell (k equals 1 in the equation above).

Note that the dose–response formalisms consider dose to individual cells, which does not necessarily reflect the response of an entire organ (see Chapter 23). Also, at high doses, damage to vascular structures might impact tumor control not reflected by simple cell kill considerations.

22.3 THE RELATIVE BIOLOGICAL EFFECTIVENESS (RBE) OF PROTONS

22.3.1 The RBE at Therapeutic Doses

The proton relative biological effectiveness (RBE) is the ratio of the absorbed doses that produce the same biological effect or endpoint X between a reference radiation (e.g., ^{60}Co γ-rays or 6 MV X-rays) and a proton irradiation (Equation 22.6).

$$\text{RBE}\left(\text{Dose}, \text{EndpointX}, \text{ proton beam properties}\right) = \frac{\text{Dose}_{\text{reference}}\left(\text{EndpointX}\right)}{\text{Dose}_{\text{protons}}\left(\text{EndpointX}\right)} \quad (22.6)$$

The dose in Equation 22.6 refers to the dose per fraction. In addition to dose, the proton RBE varies with proton energy (thus LET) and depends on other factors such as the intrinsic radiosensitivity of the tissue and nature of the biological endpoint (molecular, cellular). Doses in proton therapy are specified as Gy(RBE) to reflect that the dose was multiplied with an RBE value [22].

The definition of RBE is most meaningful in regions or tissues that receive a uniform absorbed dose. In regions that receive a nonuniform dose of radiation, the RBE must be computed for each (uniformly irradiated) region of tissue within the larger tissue region of interest. In patients, RBE values are assigned on a CT voxel-by-voxel basis as a dose-modifying factor, considering the dose in each voxel independently. Organ effect modeling can then be based on the RBE-weighted dose distribution (see Chapter 23). In organs at risk exposed to inhomogeneous dose distributions, the determination of RBE values is further complicated by the interplay between dose–volume structural and functional effects and inhomogeneities in the dose distribution.

Assuming that the linear quadratic equation (Equation 22.3) is a valid approximation, one can deduce a relationship of RBE with the α_x and β_x values for the reference photon radiation, the α_p and β_p of the proton radiation and the proton dose per fraction, D_p (Equation 22.7).

$$RBE\left(D_p, \alpha_x, \beta_x, \alpha_p, \beta_p\right) = \frac{\sqrt{\alpha_x^2 + 4\beta_x D_p\left(\alpha_p + \beta_p D_p\right)} - \alpha_x}{2\beta_x D_p} \quad (22.7)$$

Using the asymptotic values of RBE at doses of 0 and ∞ Gy, one can also define the relationship using RBE_{max} and RBE_{min}, respectively [23]:

$$RBE_{max}\left(LET_d, \alpha_p, \alpha_x\right) = \frac{\alpha_p\left(LET_d\right)}{\alpha_x} \quad RBE_{min}\left(LET_d, \beta_p, \beta_x\right) = \sqrt{\frac{\beta_p\left(LET_d\right)}{\beta_x}} \quad (22.8)$$

$$RBE\left(LET_d, D_p, \left[\alpha/\beta\right]_x\right) = \frac{1}{2D_p}\left(\sqrt{\left[\alpha/\beta\right]_x^2 + 4\left[\alpha/\beta\right]_x RBE_{max}D_p + 4RBE_{min}^2 D_p^2} - \left[\alpha/\beta\right]_x\right)$$

$$(22.9)$$

Because the RBE depends on the photon reference radiation, the reference has to be stated when reporting RBE values. Clinically, the RBE relative to 6 MV photons is of particular interest. However, cell experiments are often based on low-energy photons. For example, some laboratory experiments use 250 kVp X-rays as a reference, which depending on the endpoint, does show an RBE of ~1.15 relative to 6 MV photons [24].

For treatment planning considerations, not only the total dose but also fractionation needs to be taken into account. Fractionation effects [25,26] are beyond the scope of this chapter and are, by definition, not included in the RBE formalism. A biologically equivalent dose is used via a linear quadratic dose–response model when applying hyper- or hypofractionation [26,27].

22.3.2 The RBE at Low Doses

The definition of RBE as the ratio of two doses is valid for all dose levels but typically cannot be quantified accurately at very low doses. Equation 22.7 predicts an increase in RBE as dose decreases. The low-dose RBE refers to RBE_{max} (Equation 22.8) and is relevant particularly for radiation protection considerations where a conservative estimation defined by regulatory bodies rather than an exact RBE value is typically desired. Dose limits are based on defining either a radiation weighting factor or a radiation quality factor.

The "dose equivalent" has been defined for radiological protection purposes (Equation 22.10) [28]. Here, Q is the quality factor defined as a function of the unrestricted LET of charged particles in water [29]. Its maximum value is ~30 as defined by the International Commission for Radiological Protection (ICRP) [30]:

$$H = DQ, \tag{22.10}$$

$$Q(LET_\infty) = \begin{cases} 1 & LET_\infty < 10 \, \text{keV}/\mu\text{m} \\ 0.32 \, LET_\infty - 2.2 & LET_\infty \leq 100 \, \text{keV}/\mu\text{m} \\ 300/\sqrt{LET_\infty} & LET_\infty > 100 \, \text{keV}/\mu\text{m} \end{cases} \tag{22.11}$$

For a given volume, the quality factor is obtained by integrating over the dose-weighted contributions of charged particles:

$$Q = \frac{1}{D} \int_{LET_\infty = 0}^{\infty} Q(LET_\infty) \frac{dD}{dLET_\infty} \cdot dLET_\infty \tag{22.12}$$

A different multiplier to the average dose to an organ is the "radiation weighting factor," w_R, [31]. This leads to organ "equivalent dose" instead of "dose equivalent." From the available data, a set of continuous functions was suggested by the ICRP. For example, an energy-dependent bell-shaped curve was recommended for neutron radiation while the proton w_R is defined as 2 [32,33]. For the calculation of radiation weighting factors for neutrons, the latest recommendation shows a maximum of ~20 at energies ~1–2 MeV [34].

The use of the equivalent dose to assess the biological effectiveness at low doses in proton therapy patients is problematic because the definition is valid for an external radiation beam. For instance, the proton w_R of 2 is considering an external radiation source and not protons generated in the patient from proton or neutron nuclear interactions. Note that neutrons deposit their energy mainly primarily via secondary protons. This makes the scoring of equivalent doses in Monte Carlo simulations difficult, because the weighting factor would depend on the particle history. This problem can be overcome by using the quality factor (Equation 22.11) in the region of interest [35]. In contrast to neutrons generated in the patient, those neutrons that are generated in the treatment room (and thus act as external radiation source impinging on the patient) can be weighted with w_R values for conservative radiation protection purposes.

22.4 THE USE OF 1.1 AS A GENERIC RBE IN PROTON THERAPY

Since there is no simple relationship between dose and clinical endpoint, prescription doses and dose constraints are assigned empirically. The optimization of radiation therapy treatment plans is typically solely based on physical dose and relies on dose-based surrogates of biologic response, such as the underlying dose prescription and organ at risk (OAR) plan constraints, instead of directly optimizing clinical endpoints such as tumor control probability (TCP) and normal tissue complication probability (NTCP). Dose is a physical parameter so that differences in biological effect between modalities such as photons and protons need to be corrected for as they may lead to differences in radiation damage for the same absorbed dose.

Clinical experience is largely based on treating patients with photon beams. For consistency and to benefit from the pool of clinical results obtained with photon beams, prescription doses are defined as photon doses. Proton therapy prescriptions are based on physical dose times a generic constant RBE of 1.1. Thus, proton therapy patients receive a 10% lower prescribed dose than would be prescribed using photons. The currently used RBE was deduced in the early days of proton therapy as an average of measured values in vivo in the center of a spread-out Bragg peak (SOBP) (e.g., [36–38]). The average value of 1.1 relative to ^{60}Co was initially reported for the center of a target volume at 2 Gy(RBE) per fraction, and averaged over various endpoints such as skin reaction or lethal dose (LD_{50}). Based on an analysis of all published cell survival data in vitro, the estimated average RBE is about 1.15 in the center of a typical SOBP at 2 Gy(RBE) fraction size [39,40]. Because one aims at a conservative RBE definition for tumor control, this is in line with the clinical use of 1.1. The International Commission on Radiation Units and Measurements (ICRU) [22] as well as the International Atomic Energy Agency [41] judged that the continued use of a constant RBE of 1.1 is not unreasonable.

There are several practical advantages when using a constant RBE. For example, clinical dosimetry can be based on homogeneous dose distributions in the target. On the other hand, a generic value disregards the dependencies of the RBE on physical and biological properties.

22.5 VARIATIONS IN RBE: LABORATORY EXPERIMENTS

22.5.1 RBE Dependency on LET

For a given heavy charged particle, the RBE increases with increasing LET up to a maximum [10]. If the LET is further increased, far fewer tracks are required to deposit the same dose. This leads to saturation of the effect in small regions and eventually to a decrease of RBE with increasing LET. For protons, the maximum RBE occurs at extremely low proton energies where the contribution to the dose in a clinical scenario is negligible. Thus, one can safely assume that RBE increases with LET, with the slope depending on the biological endpoint. RBE values for cell survival for near-monoenergetic proton beams of <8.7 MeV are shown in Figure 22.3. Very low-energy protons contribute typically less than ≈1% to the total dose within an SOBP.

Due to protons slowing down, their LET increases, resulting in an increasing RBE with depth in a pristine Bragg curve or SOBP [40] are shown in Figure 22.4. The region of maximum LET is not occurring at the Bragg peak but downstream of the Bragg peak (Figure 22.5). The reason for the Bragg peak is a combination of LET and decreasing proton fluence.

The LET_d values in the entrance region of an SOBP up to the center of the SOBP are typically 0–3 keV/μm (relative to the photon LET_d), 3–6 keV/μm in the downstream half of an SOBP, 6–9 keV/μm in distal edge region, and 9–15 keV/μm in the dose falloff. The LET_d in the central region of an SOBP is typically between ~2.0 and ~3.0 keV/μm. Based on these typical LET_d values, the average RBE for cell survival in vitro at 2 Gy increases with depth from ~1.1 in the entrance region, to ~1.15 in the center, ~1.35 at the distal edge, and ~1.7 in the distal falloff [40]. This holds for typical clinical SOBP (range 10–25 cm; modulation width 5–20 cm). The rise in RBE results in an extension of the biologically effective range by 1–2 mm (see, e.g., [42–46]).

Phenomenological models describing cell survival in vitro and based on measured data typically predict a more or less linear relationship between RBE and LET_d as illustrated in Figure 22.6 [47]. Note that the RBE also depends on the tissue, e.g., the $(\alpha/\beta)_x$, and from model predictions one would expect a steeper slope as $(\alpha/\beta)_x$ decreases. There has been some indication that the relationship becomes nonlinear at high LET values, with an increasing slope in RBE [48,49].

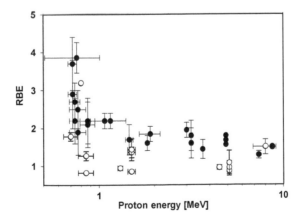

FIGURE 22.3 Experimental RBE values (relative to ^{60}Co; various dose levels) as a function of proton energy for cell inactivation measured *in vitro* for near monoenergetic protons. Open circles refer to human tumor cell lines at 2 Gy. (Adapted from Ref. [39].)

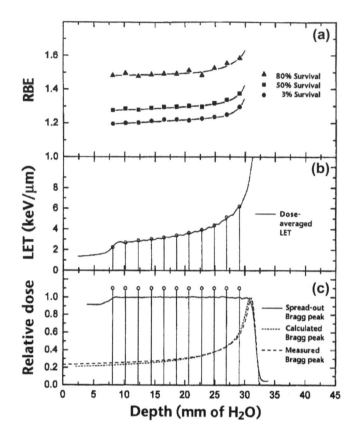

FIGURE 22.4 Upper: Proton RBE for different cell survival levels of V79 using a 70 MeV 2.5 cm SOBP beam. Middle: Dose-averaged LET as a function of depth in the SOBP. Lower: Depth-dose distribution at position of cell samples (open circles). (From Ref. [43], with permission.)

22.5.2 RBE Dependency on Dose

Due to the more pronounced shoulder in the X-ray survival curve compared to the proton survival curve (Figure 22.2), the RBE depends on dose. Specifically, there is experimental evidence that RBE increases with decreasing dose for cell survival [39,40]. The RBE is expected to increase more rapidly with decreasing dose for late-responding tissues (low α/β) compared to early responding tissues (high α/β) [39,40]. Figure 22.7 shows the RBE as a function of dose for clonogenic cell survival in vitro.

Most experimental RBE studies in vivo have employed large doses for which an RBE effect is expected to be minimal. In vitro, the dose dependency of the RBE is difficult to assess from experimental data in the clinically relevant region because of limitations of the available experimental cell survival assays at large doses (surviving fractions less than about 10^{-3} are very challenging to measure). Furthermore, most published studies only report cell survival data for one or two dose levels below 2 Gy, which increases the uncertainties in RBE at lower doses.

22.5.3 RBE Dependency on Endpoint

Proton RBE values, measured in vivo (various endpoints) and in vitro (using colony formation as the measure of cell survival), show significant variations [39] (Figure 22.8).

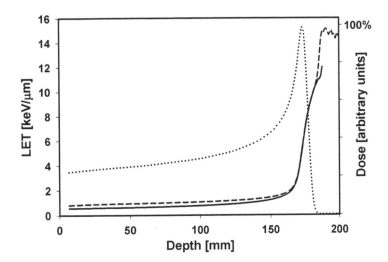

FIGURE 22.5 Dose (dotted line; right axis scale) and dose-averaged LET (left axis scale) as a function of depth in a water phantom for a 160 MeV beam. The dashed line shows the total dose-averaged LET (primary and secondary particles), while the solid line shows the dose-averaged LET for the primary protons only.

Differences between in vivo and in vitro data are most likely due to the different endpoints used in the experiments, i.e., a larger shoulder seen in survival curves for the endpoints used in vitro. In vitro studies have predominantly employed Chinese Hamster Ovary (CHO), and especially V79 cells, which exhibit large shoulders on their X-ray response curve and low α/β ratios. Most in vitro experiments use cell kill of one cell population (colony formation). On the other hand, most in vivo RBE studies have employed early reacting tissues having a high α/β ratio. The in vivo response reflects the more complex expression of radiation damage to several three-dimensional tissue systems and various biological processes (e.g., mutation). Nevertheless, most RBE values for endpoints other than clonogenic cell

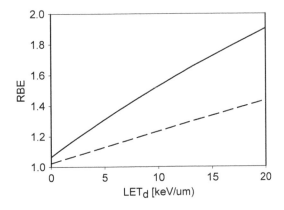

FIGURE 22.6 Proton RBE for clonogenic cell survival as a function of LET_d at 2 Gy photon dose for $(\alpha/\beta)_x$ of 2 Gy (solid) and 10 Gy (dashed) as predicted by an empirical model [47].

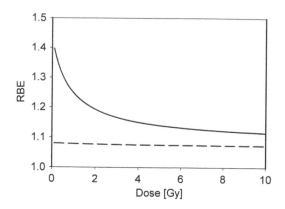

FIGURE 22.7   Proton RBE for clonogenic cell survival as a function of dose for an LET_d of 2.5 keV/µm for $(\alpha/\beta)_x$ of 2 Gy (solid) and 10 Gy (dashed) as predicted by an empirical model [47].

survival are not in disagreement with a clinical RBE of 1.1, but selected endpoints did show considerable deviations from this value [40].

Endpoint dependencies of RBE for clonogenic cell survival are often characterized by α/β. Based on the validity of the linear-quadratic model, one expects a higher RBE for low $(\alpha/\beta)_x$ [45]. Experimentally, the dependency of RBE on $(\alpha/\beta)_x$ shows only a weak trend [50]. Differences in interlaboratory experimental procedures, uncertainties associated with the reference and proton beam dosimetry, LET_d estimates, biological assays and nonlinear regression analysis of the measured data are likely to contribute to the lack of a definitive trend towards increasing RBE values with decreasing $(\alpha/\beta)_x$. Studies have indicated that the increase of RBE with decreasing $(\alpha/\beta)_x$ might be significant only at low $(\alpha/\beta)_x$ values (<~5 Gy) [21,50]. Figure 22.9 shows theoretical RBE values plotted as a function of the tissues' $(\alpha/\beta)_x$. Figure 22.3 shows the RBE measured in vitro for low-energy protons on human tumor cells and V79 Chinese hamster cells data. In these studies with low-energy beams, the data on human cells are significantly lower than the RBE values determined for hamster cells.

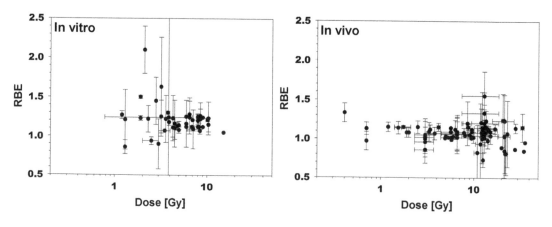

FIGURE 22.8   Experimental proton RBE values (relative to ^{60}Co) as a function of dose/fraction for cell inactivation measured in vitro (left) and in vivo (right). (From Ref. [39], with permission.)

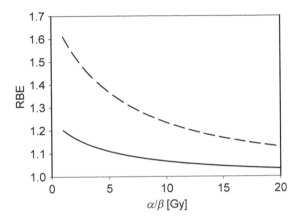

FIGURE 22.9   Proton RBE for clonogenic cell survival as a function of $(\alpha/\beta)_x$ for a photon dose of 2 Gy and LET_d values of 2 keV/μm (solid) and 10 keV/μm (dashed), as predicted by an empirical model [47].

Various other RBE values for endpoints other than cell survival have been measured, e.g., intestinal crypt regeneration assay in mice, induction of reactive oxygen species leading to oxidative stress and regulation of a variety of response pathways, DNA strand breaks in the form of single (SSB) or DSB, foci formation, repair proteins, gene expression, chromosome aberrations, mutations, micronuclei formation, apoptosis and cell cycle effects, and others. The magnitude of RBE variation with physical or biological parameters is usually small relative to our abilities to determine RBE values. The required number of animals to measure a 5% RBE difference (1.10 vs. 1.15) for one endpoint can be several hundred [39].

22.6 VARIATIONS IN RBE: PATIENTS

22.6.1 RBE Dependency on LET

According to in vitro experiments, if regions of high LET_d are within OARs with a low $(\alpha/\beta)_x$ we may severely underestimate the RBE. The increasing LET_d with increasing depth of a pristine Bragg curve or SOBP is of concern for critical structures immediately downstream of the target area. It could lead to an underestimation of the RBE-weighted dose and may additionally cause a slight shift in beam range. Furthermore, depending on the number of fields and intensity-modulated delivery, the LET distribution can be highly inhomogeneous [51,52]. One might expect higher LET values locally when using beam scanning, which emphasizes the importance to revisit the current clinical practice as the number of beam scanning facilities is on the rise.

LET distributions in a patient geometry can be calculated using Monte Carlo simulations [52]. Figures 22.4 and 22.5 show distributions of dose and LET_d in water while Figure 22.10 illustrates the distributions in a patient. LET_d values in patient geometries can be >10 keV/μm in the distal falloff, but typically only between 1.5 and 5 keV/μm in the target for typical beam arrangements [52]. Due to various sources of uncertainties and margins added in treatment planning, high LET regions may extend well into normal tissues, especially in or near low-density structures.

Despite the potential shortcomings associated with proton $RBE=1.1$, clinical evidence that this assumption results in significant patient under- or overdosage is weak, perhaps due to interpatient variability and other confounding factors in vivo. We have not (yet) clearly identified toxicities or recurrences that are a result of significant RBE effects [53,54]. On the other hand, it is difficult to assess RBE effects in critical structures from clinical data because the dose distributions differ between photon and proton irradiations. Nevertheless, treatment planners have to keep these effects in mind [55,56]. RBE variations can also be considered similar to physics-related range uncertainties [57].

Delivery uncertainties in proton therapy will be reduced in the future, which may lead to a reduction in margins, which in turn might expose RBE variations. Proton beam scanning and intensity-modulated proton therapy (IMPT) could lead to differences in RBE as a function of delivery and/or planning parameters [52,58,59].

22.6.2 RBE Dependency on Dose

Biophysical models suggest an increase in RBE as dose decreases, which becomes more pronounced as LET increases and $(\alpha/\beta)_x$ decreases [60]. Consequently, it is likely that hypofractionated regimens will result in lower RBE values. For large doses per fraction, RBE might decrease towards ~1.0 because of the disappearing shoulder in the dose–response curve. Several theoretical studies have addressed the issue of RBE spatial variations in patients [60–62] as well as the impact of RBE on fractionation in proton therapy [26,63,64].

22.6.3 RBE Dependency on Endpoint
22.6.3.1 Tumor Control Probability
While cell survival might be a valid surrogate for tumor control, there are various pathways leading to cell death, which is not caused by radiation-induced damage itself but a combination of damage and apoptosis or failure to complete mitosis. In terms of TCP, clinically more relevant than cell survival studies are measurements of TCD_{50}, i.e., the dose for 50% local control, using human tumor cells that have been implanted in immune-deficient animals. Very few studies in mice exist [39,40].

It has been speculated that medulloblastoma having a high $(\alpha/\beta)_x$ could be underdosed when using protons because of an RBE below 1.1 [65], but no evidence was found in clinical data [54]. This trend would also suggest an advantage of proton therapy when treating, for example, prostate carcinoma having low $(\alpha/\beta)_x$ [66]. The impact of $(\alpha/\beta)_x$ could affect the interpretation of clinical trials comparing photon and proton treatments.

There are differences in patient radiosensitivity due to, for instance, genomic factors. A subset of human cancers have defects in DNA repair pathways that may influence the RBE. For example, homologous recombination is required for the repair of DSBs in late S- and G2-phases of the cell cycle. It has been shown that in human cell lines and in CHO cells, defective homologous recombination increases the RBE [67–71].

When analyzing the impact of $(\alpha/\beta)_x$ on TCP, one needs to consider that not only the RBE but also TCP itself depends on $(\alpha/\beta)_x$ [72]. Assuming interpatient variability in linear-quadratic radiosensitivity parameters for a given tumor type, one might expect a lower TCP for patients with low $(\alpha/\beta)_x$. In proton therapy compared to photon therapy, the magnitude of such TCP variations in patients is potentially reduced due to an increase in RBE as $(\alpha/\beta)_x$ decreases.

22.6.3.2 Normal Tissue Complication Probability

Many endpoints have been studied in animal models, such as skin reactions or organ weight loss. The variety of endpoints does not allow a comprehensive analysis towards a clinical RBE for NTCP considerations. In patients, organ effects are dependent on the dose distribution, and the mean dose is not necessarily a valid approximation for both photon and proton radiation. Data analysis has been done on a voxel-by-voxel basis because proton dose distributions in critical structures are typically more heterogeneous compared to photon therapy. The difference in dose distribution between proton and photon therapy could have even more complex consequences, e.g., when interpreting side effects, we might have to consider the physiological interactions of different organs (e.g., lung and heart) [73]. There has been some progress in identifying fundamental differences between photon- and proton-induced radiation effects [74].

Considering typically lower $(\alpha/\beta)_x$ of healthy tissue for cell survival as well as lower doses than in the target, one might expect larger RBE values for normal tissue effects. This leads to a concern in tissues with a low $(\alpha/\beta)_x$ such as the spinal cord, which often has to be partially irradiated to achieve sufficient tumor coverage. The biggest variation in RBE can thus be expected in late-responding normal tissues as compared to early responding tissues. Considering NTCP models, patients with a lower $(\alpha/\beta)_x$ are predicted to have a lower complication probability, which is counteracted by an increase in RBE as $(\alpha/\beta)_x$ decreases. Consequently, toxicities in proton therapy would be more affected by variations in $(\alpha/\beta)_x$ compared to photon therapy [72].

Not all normal tissue complications can be parameterized with cell kill parameters such as $(\alpha/\beta)_x$. For instance, it has been shown that the proton RBE for asymptomatic radiographic changes in lung CT is significantly higher than 1.1 [75]. Thus, while for tumors the RBE for cell death might be relevant, other endpoints are presumably more relevant for most normal tissue complications. Organ-specific effects of interest are early effects such as erythema and late effects such as lung fibrosis, brain necrosis, or spinal cord injury. It is unclear to what extent the RBE for cell survival is a predictive surrogate for OAR RBE. Surviving cells with unrepaired or misrepaired damage can transmit changes to descendent cells, and malignant transformation can thus be initiated by gene mutation. For instance, no relation has been found between fibroblast radiosensitivity and the development of late normal tissue effects such as fibrosis [76,77]. The relationship between clinically observed late effects and DNA repair capacity has been discussed by Bentzen [78].

22.7 MODELING CELLULAR RADIATION EFFECTS

22.7.1 LET-Based Models

LET describes the energy deposition per path length and not the energy deposited in a cellular volume. Nevertheless, LET-based models are a valuable approximation because they can be based on few parameters. They follow the linear-quadratic dose–response curve and parameterize the change in α and β with dose-averaged LET_d relative to the reference

TABLE 22.2   Functional Form and Parameters of Some LET-Based RBE Models

ϕ	ψ	A_1	A_2	B_1	B_2	Reference
$\dfrac{LET_d}{(\alpha/\beta)_x}$	$\dfrac{LET_d}{(\alpha/\beta)_\gamma}$	0.843	0.413644 Gy/(keV/μm)	1.09	0.01612 Gy/(keV/ μm)	[45]
LET_d	1	0.1/α_x Gy^{-1}	0.02/α_x Gy^{-1} (keV/μm)$^{-1}$	1	0	[80]
$\dfrac{LET_d}{(\alpha/\beta)_x}$	1	1	0.434 Gy/(keV/μm)	1	0	[81]
$\dfrac{LET_d}{(\alpha/\beta)_x}$	$LET_d\sqrt{(\alpha/\beta)_\gamma}$	0.99064	0.35605 Gy/(keV/μm)	1.1012	−0.00387 Gy$^{-1/2}$/(keV/ μm)	[47]

radiation according to Equations 22.13 and 22.14, with parameters given in Table 22.2. Each of the model is fitted to a subset of the available in vitro experimental data from the literature. The largest data set was used by McNamara et al. [47]. By knowing α and β from photon data and as a function of LET, one can thus predict the cell survival curve and the RBE. For mixed radiation fields, e.g., considering proton energy distribution, dose-averaged means of α and $\sqrt{\beta}$ have to be applied.

A related phenomenological approach is to combine measured RBE values from various low-energy proton beams with the proton energy distribution as a function of location in the irradiated volume [79].

$$\alpha\left[(\alpha/\beta)_x, LET_d\right] = \alpha_x\left\{A_1 + A_2\phi\left[(\alpha/\beta)_x, LET_d\right]\right\} \qquad (22.13)$$

$$\beta\left[(\alpha/\beta)_x, LET_d\right] = \beta_x\left\{B_1 + B_2\psi\left[(\alpha/\beta)_x, LET_d\right]\right\} \qquad (22.14)$$

22.7.2 Microdosimetric Models

Microdosimetric quantities consider the stochastic nature and spatial distribution of particle interactions in micrometer-sized volumes [82]. A radiation field is characterized by energy deposition in a small subcellular volume, i.e., by the specific energy z (in Gy). Another important parameter is the frequency distribution of the energy deposited, the lineal energy y (in keV/μm). It is the energy deposited in a volume divided by the mean chord length and is related to the macroscopic quantity of LET. While the LET is defined for a track segment, the lineal energy considers energy deposition in a volume.

In microdosimetric models, the biological effect is described by experimentally determined response functions describing a cumulative probability that a subcellular target structure will respond to a specific target-averaged ionization density [83–87]. A microdosimetry spectrum, e.g., the measured dose distribution in lineal energy $d(y)$, can be convolved with a biological response function $r(y)$ to obtain biological effectiveness as a function of y [86–90]:

$$RBE(y) = \int r(y)d(y)dy \qquad (22.15)$$

The concept finds application in radiation protection [91]. The biological part (i.e., dose–response curves) and the physical part (i.e., microdosimetric energy loss spectra) can be measured directly [92,93]. Microdosimetric quantities (lineal energy spectra or event size distributions) can either be measured [94,95] with a tissue-equivalent proportional counter (TEPC) or determined from event-by-event Monte Carlo track structure simulations [96,97].

22.7.3 Track structure Models

Track structure models are based on the physical details of a particle track, including its secondary particles. Since the main goal is to calculate RBE values, biological phenomena are treated in a phenomenological way, as relative effects. Track structure theory ignores the mechanisms of damage initiation and repair mechanisms. Knowledge of both the response to photons and the spatial distribution of dose yields the spatial distribution of response around the path of a particle. Radiation is more effective per unit dose when the energy depositions are more concentrated in space leading to more complex damages. To predict relative damages, the cellular response to δ-electrons from protons is assumed to be identical to that due to δ-electrons from photons if the locally absorbed dose is the same.

The biological effect is calculated from the overlap of the track with the target, which is assumed to be an infinitely thin disc to ensure that there is no LET variation. A proton track is then characterized by its radial dose distribution originating from δ-electrons emitted in ionization events. The track's δ-electron halo, i.e., the dose as a function of distance to the track center, is approximated by a continuous radial distribution so that each particle track has an amorphous track with a certain diameter [98]. Generally, the dose deposited as a function of the distance, r, to the ion path can be written as a function of $1/r^2$ up to a limit determined by the maximal radial penetration of δ-rays. The track volume for protons is mostly empty with only a few δ-ray events on the scale of the cell nucleus size. The stochastic nature of particle-induced events is not taken into account when applying a continuous radial dose distribution [99].

The initial track structure model approach considers local energy depositions in sub-targets [100–102]. Each biological endpoint is characterized by a set of four parameters (n, D_0, σ_0, κ) [103]. The saturation value of the action cross section of the cell nucleus, σ_0, is not necessarily equal to the geometrical size of the target. A set of sensitive cellular subtargets with radius a_0 have to be inactivated to achieve a given effect. The radius of the subtargets is calculated using D_0 (radiosensitivity in the multitarget/single-hit formalism) and κ, both assumed to be independent fit parameters on experimental dose response. The parameters can be obtained by fitting a set of response curves for the considered endpoint measured in different radiation fields [101,104]. Photon dose–response data are needed to obtain the parameters n and D_0 (see Equation 22.5).

Two modes of interaction are considered, i.e., single-track action (called ion-kill) and multitrack action (called γ-kill). Ion-kill usually dominates at low fluence and high LET. The initial slope of the dose–response curve is attributed to the ion-kill mode, because at low doses, particles are so far apart that it is unlikely that their δ-rays overlap in the nucleus. If the cell is not inactivated in this mode, the remaining dose can still lead to an inactivation due to the overlapping of several particle tracks. There is typically a mixture of the ion-kill and γ-kill components so that the cell inactivation is a product of single-target/single-hit and multitarget/single-hit components.

Within this concept, one integrates the dose deposited in the subtarget over the radial dose distribution. Next, the effect produced by this average dose is calculated and the effects are integrated over impact parameters. To calculate the response to a mixed radiation field, a calculation for each energy and particle type must be done [101,105–107]. A dose–response relationship can be written as

$$\frac{N}{N_0} = \exp\left(-\sigma D \frac{dE}{dx}\right) \times \left(1 - \left[1 - \exp\left\{\frac{-D}{D_0}\left\langle 1 - \frac{\sigma}{\sigma_0}\right\rangle\right\}\right]^n\right) \qquad (22.16)$$

The track structure approach was implemented in a different fashion in the form of the local effect model (LEM) [108,109]. Here, the whole cell nucleus is assumed to be the critical target for which local effects are averaged, rather than averaging local doses in subtargets. Because the dose is not averaged over subvolumes, the $1/r^2$-dependent radial dose distribution shows infinite values for the local dose at $r=0$. Since in this model approach, local effects have to be determined, a cutoff for small radii must be introduced and a normalization constant adjusted, so that an integral over all track radii will yield an unrestricted LET. Consequently, a modified formulation of the linear-quadratic dose–response curve has to be used, because the standard linear-quadratic model (Equation 22.4) is a poor approximation for very high doses. An additional parameter, D_t, denotes the transition point from the shouldered low-dose region to an exponential tail for high doses, ensuring that there is an exponential asymptote at high doses.

The input parameters for the model are the geometrical cross section, i.e., the size of the cell nucleus (a thin disc covering an area A_{nucl}), and the photon dose–response curve with the parameters α, β, and D_t. The diameter of the cell nucleus can be determined microscopically. The parameter D_t is difficult to measure since values in the range of several hundreds or thousands of Gy can occur. Therefore, the transition point between the linear-quadratic and the exponential shape, D_t, must often be estimated from ion response curves.

For a proton radiation field, one can derive the mean number of lethal events by integrating the survival over the cell nucleus volume, assuming local photon dose response. Based on a photon survival parameterization, the mean number of lethal events produced per cell can be calculated and the effect be determined based on Poisson distribution of the mean.

Several tracks can contribute to the local dose, and the superposition of dose from different tracks leads to an enhanced efficiency for the production of lethal events. The average number of lethal events does not increase linearly with the particle fluence, and this leads to a shouldered response curve. The relative enhancement is highest when small doses are superimposed, so it is mostly the outer parts of the track contributing to the shoulder of the dose–response curve. Shouldered dose–response curves after charged particle irradiation can only be expected if the track diameters and particle fluences are sufficiently large that they obtain a significant fraction of energy deposition by superposition of local doses from several tracks. For cell inactivation, the total number of lethal events is determined, and the survival probability is calculated as

$$\frac{N}{N_0} = \exp\left[-Y\left(A_{\text{nucl}}, \sum_{j=0}^{\text{hits}}\left(D^{\text{local}}(r_j), \Phi(r_j, t_j, A_{\text{nucl}})\right)\right)\right] \tag{22.17}$$

The average number of lethal events, Y, is a function of the number of lethal particle hits, the nuclear area and the local dose caused by the radial dose distribution, D^{local} (Φ describes the overlap between particle track and cell nucleus).

The LEM has been modified several times to improve the agreement with experimental data [110,111]. Furthermore, the radial dose distribution can be modified to take into account the radical diffusion length in cells. In the latest versions [112–114], it is assumed that the final biological response of a cell to radiation is directly linked to the initial spatial distribution and density of DSB within subnuclear targets rather than the local dose distribution itself.

22.7.4 Lesion Induction and Repair Models

In lesion interaction models, e.g., the dual radiation action model (DRA) [115], nonlethal sublesions are produced proportional to the energy absorbed in a radiosensitive volume, z. This theory is linked to microdosimetry because the radiosensitive volume can be associated with a microdosimetric volume and the energy imparted in it, z_{1d}, i.e., the dose-mean specific energy for a single event. The model results in linear-quadratic dose–response relationship. The model was later revised from a site model to a distance model by introducing a sublesion interaction probability that depends on the distance between sublesions [116,117]. Related to the theory of DRA is the stochastic track structure-dependent approach [118,119]. It also contains a proximity concept where the probability of misrepair increases with decreasing distance of sublesions.

The lethal-potentially lethal (LPL) model [120] is based on lesions that are either repairable or nonrepairable. Nonrepairable lesions are responsible for the linear part of the dose–response curve, while repairable lesions include repair and binary misrepair. Lesions that cannot be repaired (lethal lesions) are formed proportional to dose and a parameter η_L. Repairable, or potentially lethal, lesions are formed proportional to dose and a parameter η_{PL}. These lesions are categorized into two groups: slow and fast repairing components. The slowly repairing lesions (repair rate ε_{PL}) can either turn lethal (fixed) or interact with another potentially lethal lesion (PLL) to form a lethal lesion (rate $2\varepsilon_{2PL}$; binary misrepair). The model converges to a linear-quadratic formulation at low doses. Note that, unlike in the DRA model, individual lesions can be lethal without the need of sublesion interaction by undergoing fixation. With t_r being the time after irradiation determining the repair rate, the dose–response curve for cell survival becomes

$$\frac{N}{N_0} = \exp\left(-[\eta_L + \eta_{PL}]D\right) \times \left(1 + \frac{\eta_{PL}D}{\varepsilon_{PL}/\varepsilon_{2PL}}\left[1 - \exp\{-\varepsilon_{PL}t_r\}\right]\right)^{\varepsilon_{PL}/\varepsilon_{2PL}} \tag{22.18}$$

Related to the LPL model is the repair–misrepair model, which assumes that there are linear repair processes and quadratic misrepair processes [121]. Further, the saturable repair model assumes that repair becomes saturated with increasing dose [122]. The saturable repair approach assumes that repair capability depends upon the damage complexity, the total number of lesions, the time for repair, and the availability of repair enzymes. This is different from other approaches, e.g., the LPL model, where repair capability solely depends on the damage concentration.

Another ansatz is the two-lesion kinetic model, which provides a direct link between the biochemical processing of DSBs and cell survival [123]. All DSBs are subdivided into simple and complex DSBs, each with a unique repair characteristic. It can be transformed into linear-quadratic formalism for therapeutic doses and dose rates [124].

The microdosimetric-kinetic model (MKM) [125–128] uses principles of the DRA, the LPL, and the repair–misrepair models. Further, it applies microdosimetric concepts via the dose mean lineal energy, y_D. It is assumed that the sensitivity of cells to low LET radiation is largely determined by their vulnerability to the formation of lethal unrepairable lesions from single PLLs. As in other models, differences in radiation action are due to differences in the energy deposition characteristics. The cell nucleus is divided into subvolumes, called domains, which can be considered spheres of unit density. The size of a domain is a unique property of a cell type, and the DNA content may vary from domain to domain. The MKM model assumes a linear-quadratic dose–response relationship in each domain with two types of lesions, lethal (proportional to z) and repairable. Typical domain diameters (d) are between 0.3 and 1 μm [126–128], with a few hundred to a few thousand domains per nucleus. The diameter of a domain is inversely related to the rate of repair of a PLL because it determines the maximum travel distance of a PLL. Domains are a mathematical concept to define the proximity of two lesions in order to have a chance of connecting. A PLL may be repaired or it may combine with another PLL in the same domain to form a lethal lesion. In the LET range typical for protons, the accumulation of lethal damage is well approximated by a Poisson distribution, and the MKM can be formulated similar to the expressions for the LET-based models such as shown in Equation 22.19 [126].

$$\alpha = \alpha_x\left(1 + \frac{0.229}{\rho d^2} \cdot \frac{LET_d}{[\alpha/\beta]_x}\right); \beta = \beta_x \tag{22.19}$$

The repair–misrepair fixation (RMF) model [129] assumes that it is the formation of lethal point mutations (e.g., misrepair or fixation of individual DSB) and chromosome damage (primarily exchanges) rather than the initial DSB that are the dominant mechanism of radiation effects from energy depositions. The RMF model assumes two repair terms and accounts for the track structure by including the particle- and energy-dependent number of DSBs per gray and cell, which can be estimated using Monte Carlo simulations. The radiosensitivity parameters are calculated from the yield in DSBs for protons and photons, the nuclear diameter (d), the radiosensitivity parameters of a reference radiation (α_x and β_x) as well as LET_d [130]. For doses comparable to and smaller than $(\alpha/\beta)_x$, the RMF is well approximated by the linear-quadratic model in which α and β are related to the RBE for DSB induction (RBE_{DSB}) in Equation 22.20. The term quadratic in RBE arises from the interactions of pairs of DSB formed by the same track, and the term proportional to RBE_{DSB} arises from the misrepair or fixation of individual DSB.

$$\alpha = \alpha_\gamma \cdot RBE_{DSB}\left(1 + \frac{2\bar{z}_F}{(\alpha/\beta)_\gamma} RBE_{DSB}\right)$$

(22.20)

$$\beta = \beta_\gamma \cdot RBE_{DSB} \cdot RBE_{DSB}$$

Here, the frequency-mean specific energy $\bar{z}_F \cong LET_d / \rho d^2$ [82], where d is the diameter of the cell nucleus and ρ is the mass density of the cell nucleus.

More recently, a general framework of cellular radiation response using cellular phenotypic characteristics was developed. The model is capable of describing a range of endpoints including DNA repair, genetic aberration, and cellular survival. To include lesion induction and repair characteristics, the model incorporates 11 parameters, nine characterizing DNA repair and two cell death rates. To predict RBE values, it incorporates the kinetics of different DNA repair processes, the spatial distribution of DSBs, and the resulting probability and severity of misrepair. The model describes a range of biological endpoints such as DNA repair kinetics, chromosome aberration, and mutation formation, as well as cell survival [131,132].

22.8 BIOLOGICAL OPTIMIZATION

Due to the uncertainties in RBE, treatment plan optimization based on RBE models is currently infeasible with clinically acceptable accuracy [60,133]. RBE variations are thus taken into account intuitively (for instance, by avoiding certain beam angles) but not quantitatively during the planning process. The use of robust planning techniques has been suggested to deal with RBE uncertainties [57,134].

Because for given dose and $(\alpha/\beta)_x$, the RBE increases steadily and more or less linearly with LET_d (see Figure 22.6), the changes of the latter can be used as a surrogate for RBE changes. The LET is based on physical properties and can be calculated using Monte Carlo simulations based on the treatment plan information [52,58] (Figure 22.10). IMPT allows the delivery of inhomogeneous dose distributions for each field using beam scanning [135]. Interestingly, LET distributions can be influenced in IMPT without significantly altering the dose constraints in treatment planning, i.e., dosimetrically equivalent plans can show differences in LET distributions (Figure 22.11) [52,57,136]. This can potentially be utilized to increase the efficacy of proton therapy by allowing biological dose optimization despite uncertainties in RBE values.

The LET-based planning concept was demonstrated in a multicriteria optimization framework [51]. Significant differences in LET_d distributions were observed in different base plans, in particular for OAR, while preserving target coverage. Subsequently, optimization

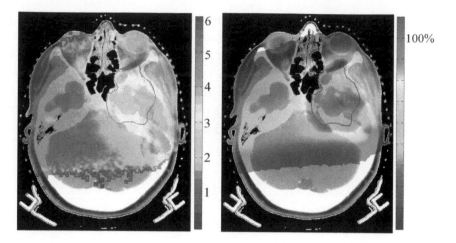

FIGURE 22.10 Dose distribution and distribution of dose-averaged LET (LET_d) for an IMPT treatment plan. The contour for the GTV is shown in blue. Right: dose in percent of prescribed dose. Left: LET_d distribution in keV/µm. The LET distribution is a potential measure of biological effectiveness. (See Ref. [52] for more details.)

using ($LET_d \times$ dose) was proposed [137]. This parameter is a well-defined physical quantity that (unlike RBE) does not depend on any model parameters. Monte Carlo dose calculation algorithms yield ($LET_d \times$ dose) at almost no additional computational cost, and also

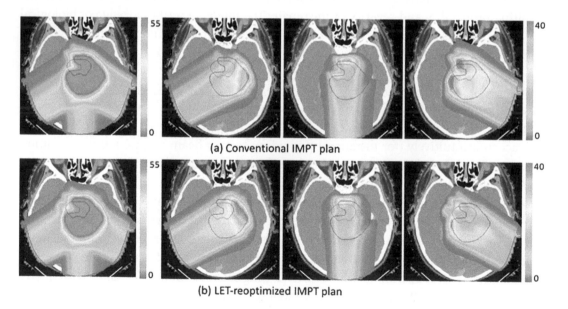

FIGURE 22.11 IMPT plans for an ependymoma patient in whom the target volume involves parts of the brain stem. The patient was treated with three posterior oblique beams. The left panel shows the total dose distribution while the other panels show the dose contributions of the three beams. (a) Conventional IMPT plan created based on a 2 mm CTV to PTV expansion. (b) LET reoptimized plan obtained after minimizing ($LET_d \times$ dose) in the brain stem while constraining the dose distribution to remain close to the conventional plan. (Adapted from Ref. [57].)

pencil beam algorithms can provide sufficient approximations. It can to first approximation be interpreted as a measure of the biological extra dose that comes with an elevated LET. From a mathematical perspective, ($LET_d \times$ dose) has the advantage that it is a linear function of the pencil beam fluence. Therefore, the same optimization algorithms that are well established for physical dose optimization can be applied. See Chapter 19 for more details on LET-based optimization.

22.9 OUT-OF-FIELD EFFECTS

The most common volume definitions in radiation therapy are illustrated in Figure 22.12. Target volumes might be defined as the gross tumor volume (GTV), the clinical target volume (CTV), and the planning target volume (PTV) [22]. For critical structures, OAR and the planning OAR volume (PRV) are defined [22]. When analyzing late effects from radiation, one can define three (overlapping) volumes in the patient:

- target (e.g., CTV), treated with the therapeutic dose.

- OAR in the tumor vicinity that are imaged (considered) in treatment planning, which may intersect with the beam path and are allowed to receive low-to-intermediate doses (in-field volume (IFV)).

- rest of the patient, which typically receives low doses (way below 1% of the target dose) are not considered, or even imaged, for treatment planning (out-of-field volume (OFV)).

The definitions of IFV and OFV are arbitrary. One might also distinguish between low and intermediate dose levels. However, there is a difference in dosimetry when analyzing biological effects of the IFV and the OFV. Organ doses in the IFV can typically be extracted

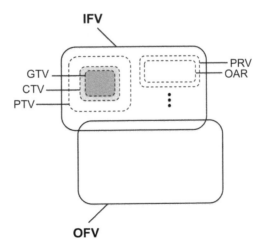

FIGURE 22.12 Volumes considered in treatment planning as defined by the ICRU [22] for the target (GTV; CTV; PTV,) and OAR; PRV). Also shown are the volume definitions introduced here for the analysis of scattered and secondary doses (IFV and OFV.)

from the treatment planning system. On the other hand, organ doses in the OFV can typically be calculated using Monte Carlo simulations, because doses within the human body are not directly measurable. Treatment planning systems cannot be applied for this purpose (even if a whole-body CT is available), because they are not commissioned for small doses and do not explicitly take into account the particle type and energy distribution of secondary radiation.

The term "integral dose" (denoting the total energy deposited) is often used to describe the total dose deposited in the patient including the tumor as well as all healthy tissues (target, IFV, and OFV). Proton therapy reduces the integral dose by a factor of 2–3 compared to photon therapy, which is expected to significantly reduce early and late side effects in the OFV when using protons in favor of photons [138,139]. Side effects do not simply scale with total energy deposited, as the distribution of dose also plays a role.

The dose in the high-to-medium dose region is predominantly caused by electromagnetic interactions of primary protons. However, protons undergo nuclear interactions in the treatment head and the patient itself, which can result in neutron radiation. Thus, far outside the main radiation field, the majority of the dose might be due to neutrons. Neutrons have a very long mean free path and deposit very little dose due to their low interaction probability. They deposit most of their dose via subsequent nuclear interactions resulting in protons.

The main concern with neutron radiation is the development of radiation-induced malignancies. The neutron RBE varies considerably as a function of dose, dose rate, and biological endpoint. Data derived from human data and exposure to high-energy neutrons are sparse, but the RBE at very low doses can be substantial (see Section 22.3.2). Because of uncertainties in RBE for carcinogenesis, one uses regulatory quantities. The majority of neutron RBE values are derived from experiments using neutrons in the 1–2 MeV regions. Although the majority of neutrons impinging on organs have energies of about 1–2 MeV, about two-third of the neutron dose in a typical proton therapy scenario is deposited via neutrons with energy above 100 MeV [140]. Neutron energy distributions in patients (or phantoms) can be calculated using Monte Carlo simulations (see, e.g., [141]). Figure 22.13 shows the simulated energy distribution of neutrons entering a water tank after being generated in a brass block by a 200 MeV proton beam.

Very few data are available on the biological effects of high-energy neutrons in the range from 20–250 MeV that would be most pertinent to proton therapy. A study in mice using a therapeutic proton beam showed an RBE for cancer induction, much lower than expected from the ICRP recommendations [142]. While a large integral dose should generally be avoided, most secondary cancers typically occur in the medium dose range close to the target volume [143].

Various dose–response relationships and low-dose risk models for carcinogenesis have been suggested by radiation protection bodies [144–147]. There is considerable uncertainty in the shape of the dose–response curve, and linearity may not hold for all cancers. Figure 22.14 shows the different regions of dose response. Risk modeling for second cancers depends on whether one considers IFV or OFV. In the OFV, linear dose–response models combined with regulatory quantities are applied. In the IFV, mutagenesis is in

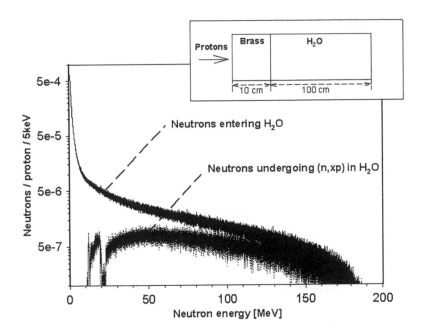

FIGURE 22.13 Results of a Monte Carlo simulation of a 200 MeV proton beam stopping in a Brass block and the generated neutrons entering a water tank. The simulated setup is shown in the upper right, with the protons indicated by an arrow and entering a 10 cm Brass block, which stops all primary particles. The figure shows the neutron energy distributions of those neutrons entering the water tank downstream of the Brass block as well as those neutrons causing a secondary proton in a nuclear interaction, the main mechanism of dose deposition. The unit is in neutrons per incident proton per 5 keV bin.

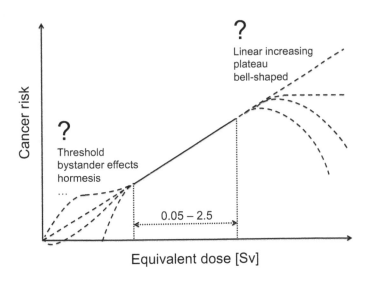

FIGURE 22.14 Dose–response relationships for carcinogenesis showing a linear dose response in the dose range of the atomic bomb exposures (~0.05–2.5 Sv [153]) and the uncertain relationships for higher and lower doses.

competition with cell survival causing the dose–response relationship to become non-linear, as sterilized cells will not mutate. In addition, tissues may respond to radiation by accelerated repopulation contributing to tissue sparing during fractionated radiation therapy. Other than in the OFV, average organ doses are not meaningful in the IFV, as dose-volume effects for inhomogeneous dose distributions might play a role. To consider such effects, one might use the concept of "organ equivalent dose" for cancer risk assessment [148–150].

22.10 SUMMARY AND CONCLUSION

Other than assuming a 10% difference in required prescription doses and dose constraints, the biological difference between proton and photon therapy is not considered quantitatively in treatment planning. Treatment planning decisions might be affected by RBE considerations in the context of beam angle selections. With the number of proton therapy patients increasing rapidly, one might be able to define prescription doses based on proton therapy alone, which, in theory, would make the RBE concept obsolete. On the other hand, due to the typically inhomogeneous dose distribution in OAR, the concept of RBE or assigning constraints based on proton mean dose might be flawed, and voxel-based RBE values have to be incorporated in TCP and NTCP models.

While the value of 1.1 is appropriate if a generic RBE is being applied as also recommended by the ICRU [22], the proton therapy community will for sure move toward variable RBE values in the future. Even with considerable uncertainties in RBE, we may be able to identify regions where adjustments to the value of 1.1 might benefit patient outcomes [151]. The key research areas will be on identifying biomarkers to identify patients with RBE values, either low or high compared to the overall patient population [70,71,152]. In the meantime, biological optimization based on LET can cause a decrease in patient-specific RBE values for OAR despite of patient-specific RBE uncertainties.

The issue of low-dose RBE and the risk for second malignances in proton therapy is not resolved, but with the move towards beam scanning will become less relevant.

REFERENCES

1. Paretzke HG. Radiation track structure theory. In *Kinetics of Nonhomogeneous Processes*, Freeman, GR Ed, John Wiley & Sons, 1987, 90–170.
2. Nikjoo H, Goodhead DT. Track structure analysis illustrating the prominent role of low-energy electrons in radiobiological effects of low-LET radiations. *Phys Med Biol.* 1991; 36: 229–38.
3. Prise KM, Ahnstroem G, Belli M, Carlsson J, Frankenberg D, Kiefer J, et al. A review of dsb induction data for varying quality radiations. *Int J Radiat Biol.* 1998; 74: 173–84.
4. Pastwa E, Neumann RD, Mezhevaya K, Winters TA. Repair of radiation-induced DNA double-strand breaks is dependent upon radiation quality and the structural complexity of double-strand breaks. *Radiat Res.* 2003; 159(2): 251–61.
5. Goodhead DT. Initial events in the cellular effects of ionizing radiations: Clustered damage in DNA. *Int J Radiat Biol.* 1994; 65: 7–17.
6. Belli M, Cera F, Cherubini R, Dalla Vecchia M, Haque AM, Ianzini F, et al. RBE-LET relationships for cell inactivation and mutation induced by low energy protons in V79 cells: further results at the LNL facility. *Int J Radiat Biol.* 1998; 74(4): 501–9.

7. Belli M, Goodhead DT, Ianzini F, Simone G, Tabocchini MA. Direct comparison of biological effectiveness of protons and alpha-particles of the same LET. II. Mutation induction at the HPRT locus in V79 cells. *Int J Radiat Biol.* 1992; 61: 625–9.

8. Jenner TJ, Belli M, Goodhead DT, Ianzini F, Simone G, Tabocchini MA. Direct comparison of biological effectiveness of protons and alpha-particles of the same LET. III. Initial yield of DNA double-strand breaks in V79 cells. *Int J Radiat Biol.* 1992; 61: 631–7.

9. Goodhead DT. Mechanisms for the biological effectiveness of high-LET radiations. *J Radiat Res.* 1999; 40 (Supplement): 1–13.

10. Goodhead DT. Radiation effects in living cells. *Canad J Phys.* 1990; 68: 872–86.

11. Nikjoo H, Uehara S, Wilson WE, Hoshi M, Goodhead DT. Track structure in radiation biology: theory and applications. *Int J Radiat Biol.* 1998;73:355–64.

12. Loebrich M, Cooper PK, Rydberg B. Non-random distribution of DNA double-strand breaks induced by particle irradiation. *Int J Radiat Biol.* 1996;70:493–503.

13. Fowler JF. Dose response curves for organ function or cell survival. *Br J Radiol.* 1983; 56: 497–500.

14. Fowler JF. The linear-quadratic formula and progress in fractionated radiotherapy. *Br J Radiol.* 1989; 62: 679–94.

15. Sachs RK, Hahnfeld P, Brenner DJ. The link between low-LET dose-response relations and the underlying kinetics of damage production/repair/misrepair. *Int J Radiat Biol.* 1997; 72: 351–74.

16. Joiner MC, Lambin P, Malaise EP, Robson T, Arrand JE, Skov KA, et al. Hypersensitivity to very-low single radiation doses: Its relationship to the adaptive response and induced radioresistance. *Mutat Res.* 1996; 358: 171–83.

17. Skarsgard LD, Wouters BG. Substructure in the cell survival response at low radiation dose: effect of different subpopulations. *Int J Radiat Biol.* 1997; 71: 737–49.

18. Andisheh B, Edgren M, Belkic D, Mavroidis P, Brahme A, Lind BK. A comparative analysis of radiobiological models for cell surviving fractions at high doses. *Technol Cancer Res Treat.* 2013; 12(2): 183–92.

19. Kirkpatrick JP, Brenner DJ, Orton CG. Point/Counterpoint. The linear-quadratic model is inappropriate to model high dose per fraction in radiosurgery. *Med Phys.* 2009; 36: 3381–4.

20. Gilbert CW, Hendry JH, Major D. The approximation in the formulation for survival $S = \exp\text{-}(aD+bD^2)$. *Int J Radiat Biol.* 1980; 37: 469–71.

21. Paganetti H, Gerweck LE, Goitein M. The general relation between tissue response to x-radiation (a/b-values) and the relative biological effectiveness (RBE) of protons: Prediction by the Katz track-structure model. *Int J Radiat Biol.* 2000;76:985–98.

22. ICRU. Prescribing, Recording, and Reporting Proton-Beam Therapy. International Commission on Radiation Units and Measurements, Bethesda, MD. 2007;Report No. 78.

23. Carabe-Fernandez A, Dale RG, Jones B. The incorporation of the concept of minimum RBE (RbEmin) into the linear-quadratic model and the potential for improved radiobiological analysis of high-LET treatments. *Int J Radiat Biol.* 2007; 83(1): 27–39.

24. Sinclair WK. The relative biological effectiveness of 22-MVp X-rays, cobalt-60 gamma-rays, and 200 kVp X-rays. VII. Summary of studies for five criteria of effect. *Radiat Res.* 1962; 16: 394–8.

25. Denekamp J, Waites T, Fowler JF. Predicting realistic RBE values for clinically relevant radiotherapy schedules. *Int J Radiat Biol.* 1997; 71: 681–94.

26. Carabe-Fernandez A, Dale RG, Hopewell JW, Jones B, Paganetti H. Fractionation effects in particle radiotherapy: implications for hypo-fractionation regimes. *Phys Med Biol.* 2010; 55: 5685–700.

27. Stuschke M, Pottgen C. Altered fractionation schemes in radiotherapy. *Front Radiat Ther Oncol.* 2010; 42: 150–6.

28. ICRU. Report of the RBE committee to the international commission on radiological protection and of radiological units and measurements. *Health Phys.* 1963; 9: 357–84.
29. ICRU. *Radiation Quantities and Units.* International Commision on Radiation Units and Measurements, Bethesda, MD, 1971, 19.
30. ICRP. Relative biological effectiveness (RBE), quality factor (Q), and radiation weighting factor (wR). *ICRP* (Pergamon Press). 2003;92.
31. ICRP. Recommendations of the International Commission on Radiological Protection. *ICRP Publication 60 Ann ICRP.* 1991; 21(1–3).
32. ICRP. *Relative Biological Effectiveness, Radiation Weighting and Quality Factor,* Contract No. 103. Oxford, UK: International Commission on Radiological Protection, 2003.
33. ICRP. The 2007 recommendations of the international commission on radiological protection. *ICRP Publication 103 Ann ICRP.* 2007; 37(2–4).
34. ICRP. Recommendations of the international commission on radiological protection. *ICRP Publication 103 Ann ICRP.* 2007; 37(2–4).
35. Xu XG, Paganetti H. Better radiation weighting factors for neutrons generated from proton treatment are needed. *Radiat Prot Dosim.* 2010; 138: 291–4.
36. Dalrymple GV, Lindsay IR, Hall JD, Mitchell JC, Ghidoni JJ, Kundel HL, et al. The relative biological effectiveness of 138-MeV protons as compared to cobalt-60 gamma radiation. *Radiat Res.* 1966; 28: 489–506.
37. Tepper J, Verhey L, Goitein M, Suit HD. In vivo determinations of RBE in a high energy modulated proton beam using normal tissue reactions and fractionated dose schedules. *Int J Radiat Oncol Biol Phys.* 1977; 2: 1115–22.
38. Urano M, Goitein M, Verhey L, Mendiondo O, Suit HD, Koehler A. Relative biological effectiveness of a high energy modulated proton beam using a spontaneous murine tumor in vivo. *Int J Radiat Oncol Biol Phys.* 1980; 6: 1187–93.
39. Paganetti H, Niemierko A, Ancukiewicz M, Gerweck LE, Loeffler JS, Goitein M, et al. Relative biological effectiveness (RBE) values for proton beam therapy. *Int J Radiat Oncol Biol Phys.* 2002; 53: 407–21.
40. Paganetti H. Relative biological effectiveness (RBE) values for proton beam therapy. Variations as a function of biological endpoint, dose, and linear energy transfer. *Phys Med Biol.* 2014; 59: R419–R72.
41. International Atomic Energy Agency I. *Relative Biological Effectiveness in Ion Beam Therapy.* IAEA Technical Report Series. 2008, 461.
42. Robertson JB, Williams JR, Schmidt RA, Little JB, Flynn DF, Suit HD. Radiobiological studies of a high-energy modulated proton beam utilizing cultured mammalian cells. *Cancer.* 1975; 35: 1664–77.
43. Wouters BG, Lam GKY, Oelfke U, Gardey K, Durand RE, Skarsgard LD. RBE measurement on the 70 MeV proton beam at TRIUMF using V79 cells and the high precision cell sorter assay. *Radiat Res.* 1996; 146: 159–70.
44. Paganetti H, Goitein M. Radiobiological significance of beam line dependent proton energy distributions in a spread-out Bragg peak. *Med Phys.* 2000;27:1119–26.
45. Carabe A, Moteabbed M, Depauw N, Schuemann J, Paganetti H. Range uncertainty in proton therapy due to variable biological effectiveness. *Phys Med Biol.* 2012; 57(5): 1159–72.
46. Gruen R, Friedrich T, Kraemer M, Zink K, Durante M, Engenhart-Cabilic R, et al. Physical and biological factors determining the effective proton range. *Med Phys.* 2013; 40: 111716.
47. McNamara AL, Schuemann J, Paganetti H. A phenomenological relative biological effectiveness (RBE) model for proton therapy based on all published in vitro cell survival data. *Phys Med Biol.* 2015; 60(21): 8399–416.
48. Abolfath R, Peeler CR, Newpower M, Bronk L, Grosshans D, Mohan R. A model for relative biological effectiveness of therapeutic proton beams based on a global fit of cell survival data. *Sci Rep.* 2017; 7(1): 8340.

49. Guan F, Bronk L, Titt U, Lin SH, Mirkovic D, Kerr MD, et al. Spatial mapping of the biologic effectiveness of scanned particle beams: Towards biologically optimized particle therapy. *Sci Rep.* 2015; 5: 9850.
50. Gerweck L, Kozin SV. Relative biological effectiveness of proton beams in clinical therapy. *Radiother Oncol.* 1999; 50: 135–42.
51. Giantsoudi D, Grassberger C, Craft D, Niemierko A, Trofimov A, Paganetti H. Linear energy transfer-guided optimization in intensity modulated proton therapy: Feasibility study and clinical potential. *Int J Radiat Oncol Biol Phys.* 2013; 87(1): 216–22.
52. Grassberger C, Trofimov A, Lomax A, Paganetti H. Variations in linear energy transfer within clinical proton therapy fields and the potential for biological treatment planning. *Int J Radiat Oncol Biol Phys.* 2011; 80(5): 1559–66.
53. Indelicato DJ, Flampouri S, Rotondo RL, Bradley JA, Morris CG, Aldana PR, et al. Incidence and dosimetric parameters of pediatric brainstem toxicity following proton therapy. *Acta Oncol.* 2014; 53(10): 1298–304.
54. Sethi RV, Giantsoudi D, Raiford M, Malhi I, Niemierko A, Rapalino O, et al. Patterns of failure after proton therapy in medulloblastoma; linear energy transfer distributions and relative biological effectiveness associations for relapses. *Int J Radiat Oncol Biol Phys.* 2014; 88(3): 655–63.
55. Giantsoudi D, Adams J, MacDonald SM, Paganetti H. Proton treatment techniques for posterior fossa tumors: Consequences for linear energy transfer and dose-volume parameters for the brainstem and organs at risk. *Int J Radiat Oncol Biol Phys.* 2017; 97(2): 401–10.
56. Fjaera LF, Li Z, Ytre-Hauge KS, Muren LP, Indelicato DJ, Lassen-Ramshad Y, et al. Linear energy transfer distributions in the brainstem depending on tumour location in intensity-modulated proton therapy of paediatric cancer. *Acta Oncol.* 2017; 56(6): 763–8.
57. Unkelbach J, Paganetti H. Robust proton treatment planning: Physical and biological optimization. *Semin Radiat Oncol.* 2018; 28(2): 88–96.
58. Grassberger C, Paganetti H. Elevated LET components in clinical proton beams. *Phys Med Biol.* 2011; 56: 6677–91.
59. Gridley DS, Pecaut MJ, Mao XW, Wroe AJ, Luo-Owen X. Biological effects of passive versus active scanning proton beams on human lung epithelial cells. *Technol Cancer Res Treat.* 2014.
60. Carabe A, Espana S, Grassberger C, Paganetti H. Clinical consequences of Relative Biological Effectiveness variations in proton radiotherapy of the prostate, brain and liver. *Phys Med Biol.* 2013; 58(7): 2103–17.
61. Tilly N, Johansson J, Isacsson U, Medin J, Blomquist E, Grusell E, et al. The influence of RBE variations in a clinical proton treatment plan for a hypopharynx cancer. *Phys Med Biol.* 2005; 50(12): 2765–77.
62. Frese MC, Wilkens JJ, Huber PE, Jensen AD, Oelfke U, Taheri-Kadkhoda Z. Application of constant vs. variable relative biological effectiveness in treatment planning of intensity-modulated proton therapy. *Int J Radiat Oncol Biol Phys.* 2011; 79(1): 80–8.
63. Dasu A, Toma-Dasu I. Impact of variable RBE on proton fractionation. *Med Phys.* 2013; 40: 011705.
64. Holloway RP, Dale RG. Theoretical implications of incorporating relative biological effectiveness into radiobiological equivalence relationships. *Br J Radiol.* 2013; 86.
65. Jones B, Wilson P, Nagano A, Fenwick J, McKenna G. Dilemmas concerning dose distribution and the influence of relative biological effect in proton beam therapy of medulloblastoma. *Br J Radiol.* 2012; 85: e912–e8.
66. Fowler J, Chappell R, Ritter M. Is a/b for prostate tumors really low? *Int J Radiat Oncol Biol Phys.* 2001; 50: 1021–31.
67. Grosse N, Fontana A, Hug EB, Lomax A, Coray A, Augsburger M, et al. Deficiency in homologous recombination renders mammalian cells more sensitive to proton versus photon irradiation. *Int J Radiat Oncol Biol Phys.* 2013; 88: 175–81.

68. Rostek C, Turner EL, Robbins M, Rightnar S, Xiao W, Obenaus A, et al. Involvement of homologous recombination repair after proton-induced DNA damage. *Mutagenesis.* 2008; 23(2): 119–29.

69. Fontana AO, Augsburger MA, Grosse N, Guckenberger M, Lomax AJ, Sartori AA, et al. Differential DNA repair pathway choice in cancer cells after proton- and photon-irradiation. *Radiother Oncol.* 2015; 116(3): 374–80.

70. Liu Q, Underwood TS, Kung J, Wang M, Lu HM, Paganetti H, et al. Disruption of SLX4-MUS81 Function increases the relative biological effectiveness of proton radiation. *Int J Radiat Oncol Biol Phys.* 2016; 95(1): 78–85.

71. Liu Q, Ghosh P, Magpayo N, Testa M, Tang S, Biggs P, et al. Lung cancer cell line screen links fanconi anemia pathway defects to increased relative biological effectiveness of proton radiation. *Int J Radiat Oncol Biol Phys.* 2015; 91(5): 1081–9.

72. Paganetti H. Relating the proton relative biological effectiveness to tumor control and normal tissue complication probabilities assuming interpatient variability in alpha/beta. *Acta Oncol.* 2017; 56(11): 1379–86.

73. Paganetti H, van Luijk P. Biological considerations when comparing proton therapy with photon therapy. *Semin Radiat Oncol.* 2013; 23(2): 77–87.

74. Girdhani S, Sachs R, Hlatky L. Biological effects of proton radiation: what we know and don't know. *Rad Res.* 2013; 179(3): 257–72.

75. Underwood TSA, Grassberger C, Bass R, Jimenez R, Meyersohn N, Yeap BY, et al. Asymptomatic Late-phase Radiographic Changes Among Chest-Wall Patients Are Associated With a Proton RBE Exceeding 1.1. *Int J Radiat Oncol Biolo Phys.* 2018; 101(4): 809–819.

76. Peacock J, Ashton A, Bliss J, Bush C, Eady J, Jackson C, et al. Cellular radiosensitivity and complication risk after curative radiotherapy. *Radiother Oncol.* 2000; 55(2): 173–8.

77. Russell NS, Grummels A, Hart AA, Smolders IJ, Borger J, Bartelink H, et al. Low predictive value of intrinsic fibroblast radiosensitivity for fibrosis development following radiotherapy for breast cancer. *Int J Radiat Biol.* 1998; 73(6): 661–70.

78. Bentzen SM. Preventing or reducing late side effects of radiation therapy: radiobiology meets molecular pathology. *Nat Rev Cancer.* 2006; 6(9): 702–13.

79. Belli M, Campa A, Ermolli I. A semi-empirical approach to the evaluation of the relative biological effectiveness of therapeutic proton bemas: The methodological framework. *Radiat Res.* 1997; 148: 592–8.

80. Wilkens JJ, Oelfke U. A phenomenological model for the relative biological effectiveness in therapeutic proton beams. *Phys Med Biol.* 2004; 49: 2811–25.

81. Wedenberg M, Lind BK, Hardemark B. A model for the relative biological effectiveness of protons: the tissue specific parameter alpha/beta of photons is a predictor for the sensitivity to LET changes. *Acta Oncol.* 2013; 52(3): 580–8.

82. ICRU. *Microdosimetry.* International Commission on Radiation Units and Measurements, Bethesda, MD, 1983, 36.

83. Hall EJ, Kellerer AM, Rossi HH, Yuk-Ming PL. The relative biological effectiveness of 160 MeV Protons. II. Biological data and their interpretation in terms of microdosimetry. *Int J Radiat Oncol Biol Phys.* 1978; 4: 1009–13.

84. Kliauga PJ, Colvett RD, Yuk-Ming PL, Rossi HH. The Relative Biological Effectiveness of 160 MeV Protons I. Microdosimetry. *Int J Radiat Oncol Biol Phys.* 1978; 4: 1001–8.

85. Menzel HG, Pihet P, Wambersie A. Microdosimetric specification of radiation quality in neutron radiation therapy. *Int J Radiat Biol.* 1990; 57: 865–83.

86. Morstin K, Bond VP, Baum JW. Probabilistic approach to obtain hit-size effectiveness functions which relate microdosimetry and radiobiology. *Radiat Res.* 1989; 120: 383–402.

87. Pihet P, Menzel HG, Schmidt R, Beauduin M, Wambersie A. Biological weighting function for RBE specification of neutron therapy beams. Intercomparison of 9 European centres. *Radiat Prot Dosim.* 1990; 31: 437–42.

88. Brenner DJ, Zaider M. Estimating RBEs at clinical doses from microdosimetric spectra. *Med Phys.* 1998; 25: 1055–7.
89. Loncol T, Cosgrove V, Denis JM, Gueulette J, Mazal A, Menzel HG, et al. Radiobiological effectiveness of radiation beams with broad LET spectra: microdosimetric analysis using biological weighting functions. *Radiat Prot Dosim.* 1994; 52: 347–52.
90. Paganetti H, Olko P, Kobus H, Becker R, Schmitz T, Waligorski MPR, et al. Calculation of RBE for proton beams using biological weighting functions. *Int J Radiat Oncol Biol Phys.* 1997; 37: 719–29.
91. Zaider M, Brenner DJ. On the microdosimetric definition of quality factors. *Radiat Res.* 1985; 103: 302–16.
92. Cosgrove VP, Delacroix S, Green S, Mazal A, Scott MC. Microdosimetric Studies on the ORSAY Proton Synchrocyclotron at 73 and 200 MeV. *Radiat Prot Dosim.* 1997; 70: 493–6.
93. Coutrakon G, Cortese J, Ghebremedhin A, Hubbard J, Johanning J, Koss P, et al. Microdosimetry spectra of the Loma Linda proton beam and relative biological effectiveness comparisons. *Med Phys.* 1997; 24: 1499–506.
94. Borak TB, Doke T, Fuse T, Guetersloh S, Heilbronn L, Hara K, et al. Comparisons of LET distributions for protons with energies between 50 and 200 MeV determined using a spherical tissue-equivalent proportional counter (TEPC) and a position-sensitive silicon spectrometer (RRMD-III). *Radiat Res.* 2004;162(6):687–92.
95. Kase Y, Yamashita W, Matsufuji N, Takada K, Sakae T, Furusawa Y, et al. Microdosimetric calculation of relative biological effectiveness for design of therapeutic proton beams. *J Radiat Res.* 2013; 54: 485–93.
96. Chen J. Microdosimetric characteristics of proton beams from 50 keV to 200 MeV. *Radiat Prot Dosim.* 2011; 143(2–4): 436–9.
97. Liamsuwan T, Uehara S, Nikjoo H. Microdosimetry of the full slowing down of protons using Monte Carlo track structure simulations. *Radiat Prot Dosim.* 2015; 166(1–4): 29–33.
98. Chen J, Kellerer AM. Calculation of radial dose distributions for heavy ions by a new analytical approach. *Radiat Prot Dosim.* 1997; 70: 55–8.
99. Schmollack JU, Klaumuenzer SL, Kiefer J. Stochastic radial dose distributions and track structure theory. *Radiat Res.* 2000; 153: 469–78.
100. Katz R, Ackerson B, Homayoonfar M, Sharma SC. Inactivation of Cells by Heavy Ion Bombardment. *Radiat Res.* 1971; 47: 402–25.
101. Katz R, Sharma SC. Response of cells to fast neutrons, stopped pions, and heavy ion beams. *Nucl Instrum Methods.* 1973; 111: 93–116.
102. Katz R, Sharma SC. Heavy particles in therapy: An application of track theory. *Phys Med Biol.* 1974; 19: 413–35.
103. Katz R, Sharma SC, Homayoonfar M. The structure of particle tracks. In: *Topics in Radiation Dosimetry*, Suppl. 1. Attix, F.H., Ed. Academic Press, 1972, 317–83.
104. Roth RA, Sharma SC, Katz R. Systematic evaluation of cellular radiosensitivity parameters. *Phys Med Biol.* 1976; 21: 491–503.
105. Curtis SB. The Katz cell-survival model and beams of heavy charged particles. *Nucl Tracks Radiat Meas (Int J Radiat Appl Instrum Part D).* 1989; 16: 97–103.
106. Katz R. Relative effectiveness of mixed radiation fields. *Radiat Res.* 1993; 133: 390.
107. Katz R, Fullerton BG, Roth RA, Sharma SC. Simplified RBE-dose calculations for mixed radiation fields. *Health Phys.* 1976; 30: 148–50.
108. Krämer M, Jäkel O, Haberer T, Kraft G, Schardt D, Weber U. Treatment planning for heavy ion radiotherapy: physical beam model and dose optimization. Physics in Medicine and Biology. 2000;45:3299–317.
109. Scholz M, Kraft G. Calculation of heavy ion inactivation probabilities based on track structure, X ray sensitivity and target size. *Radiat Protect Dosim.* 1994; 52: 29–33.
110. Elsasser T, Kramer M, Scholz M. Accuracy of the local effect model for the prediction of biologic effects of carbon ion beams in vitro and in vivo. *Int J Radiat Oncol Biol Phys.* 2008; 71(3): 866–72.

111. Elsasser T, Scholz M. Cluster effects within the local effect model. *Radiat Res.* 2007; 167(3): 319–29.

112. Elsasser T, Weyrather WK, Friedrich T, Durante M, Iancu G, Kramer M, et al. Quantification of the relative biological effectiveness for ion beam radiotherapy: direct experimental comparison of proton and carbon ion beams and a novel approach for treatment planning. *Int J Radiat Oncol Biol Phys.* 2010; 78(4): 1177–83.

113. Friedrich T, Scholz U, Elsasser T, Durante M, Scholz M. Calculation of the biological effects of ion beams based on the microscopic spatial damage distribution pattern. *Int J Radiat Biol.* 2012; 88(1–2): 103–7.

114. Tommasino F, Friedrich T, Scholz U, Taucher-Scholz G, Durante M, Scholz M. A DNA double-strand break kinetic rejoining model based on the local effect model. *Radiat Res.* 2013; 180(5): 524–38.

115. Kellerer AM, Rossi HH. The theory of dual radiation action. *Curr Top Radat Res.* 1974; Q8: S17–S67.

116. Kellerer AM, Rossi HH. A generalized formulation of dual radiation action. *Radiat Res.* 1978; 75: 471–88.

117. Zaider M, Rossi HH. On the application of microdosimetry to radiobiology. *Radiat Res.* 1988; 113(1): 15–24.

118. Brenner DJ. Track structure, lesion development, and cell survival. *Radiat Res.* 1990; 124: S29–S37.

119. Sachs RK, Chen AM, Brenner DJ. Review: Proximity effects in the production of chromosome aberrations by ionizing radiation. *Int J Radiat Biol.* 1997; 71(1): 1–19.

120. Curtis SB. Lethal and potentially lethal lesions induced by radiation -a unified repair model. *Radiat Res.* 1986; 106: 252–70.

121. Tobias CA, Blakely EA, Ngo FQH, Yang TCH. The repair-misrepair model of cell survival. In *Radiation Biology and Cancer Research*, Meyn, RE and Withers, HR, Eds. Raven, New York, 1980, 195–230.

122. Goodhead DT. Saturable Repair Models of Radiation Action in Mammalian Cells. *Radiat Res.* 1985; 104: S58–S67.

123. Stewart RD. Two-lesion kinetic model of double-strand break rejoining and cell killing. *Radiat Res.* 2001; 156: 365–78.

124. Guerrero M, Stewart RD, Wang JZ, Li XA. Equivalence of the linear-quadratic and two-lesion kinetic models. *Phys Med Biol.* 2002; 47: 3197–209.

125. Hawkins RB. A microdosimetric-kinetic theory of the dependence of the RBE for cell death on LET. *Med Phys.* 1998; 25: 1157–70.

126. Hawkins RB. A microdosimetric-kinetic model for the effect of non-Poisson distribution of lethal lesions on the variation of RBE with LET. *Radiat Res.* 2003; 160(1): 61–9.

127. Hawkins RB. The relationship between the sensitivity of cells to high-energy photons and the RBE of particle radiation used in radiotherapy. *Radiat Res.* 2009; 172(6): 761–76.

128. Hawkins RB. A microdosimetric-kinetic model of cell death from exposure to ionizing radiation of any LET, with experimental and clinical applications. *Int J Radiat Biol.* 1996; 69(6): 739–55.

129. Carlson DJ, Stewart RD, Semenenko VA, Sandison GA. Combined use of Monte Carlo DNA damage simulations and deterministic repair models to examine putative mechanisms of cell killing. *Radiat Res.* 2008; 169(4): 447–59.

130. Frese MC, Yu VK, Stewart RD, Carlson DJ. A mechanism-based approach to predict the relative biological effectiveness of protons and carbon ions in radiation therapy. *Int J Radiat Oncol Biol Phys.* 2012; 83(1): 442–50.

131. McMahon SJ, Schuemann J, Paganetti H, Prise KM. Mechanistic modelling of dna repair and cellular survival following radiation-induced dna damage. *Sci Rep.* 2016; 6: 33290.

132. McMahon SJ, McNamara AL, Schuemann J, Paganetti H, Prise KM. A general mechanistic model enables predictions of the biological effectiveness of different qualities of radiation. *Sci Rep*. 2017; 7: 10790.

133. Resch AF, Landry G, Kamp F, Cabal G, Belka C, Wilkens JJ, et al. Quantification of the uncertainties of a biological model and their impact on variable RBE proton treatment plan optimization. *Phys Med*. 2017; 36: 91–102.

134. Oden J, Eriksson K, Toma-Dasu I. Inclusion of a variable RBE into proton and photon plan comparison for various fractionation schedules in prostate radiation therapy. *Med Phys*. 2017; 44(3): 810–22.

135. Lomax A. Intensity modulation methods for proton radiotherapy. *Phys Med Biol*. 1999; 44: 185–205.

136. Fagerholm R, Schmidt MK, Khan S, Rafiq S, Tapper W, Aittomaki K, et al. The SNP rs6500843 in 16p13.3 is associated with survival specifically among chemotherapy-treated breast cancer patients. *Oncotarget*. 2015.

137. Unkelbach J, Botas P, Giantsoudi D, Gorissen BL, Paganetti H. reoptimization of intensity modulated proton therapy plans based on linear energy transfer. *Int J Radiat Oncol Biol Phys*. 2016; 96(5): 1097–106.

138. Merchant TE, Hua CH, Shukla H, Ying X, Nill S, Oelfke U. Proton versus photon radiotherapy for common pediatric brain tumors: comparison of models of dose characteristics and their relationship to cognitive function. *Pediatr Blood Cancer*. 2008; 51(1): 110–7.

139. Miralbell R, Lomax A, Cella L, Schneider U. Potential reduction of the incidence of radiation-induced second cancers by using proton beams in the treatment of pediatric tumors. *Int J Radiat Oncol Biol Phys*. 2002; 54: 824–9.

140. Zheng Y, Newhauser W, Fontenot J, Taddei P, Mohan R. Monte Carlo study of neutron dose equivalent during passive scattering proton therapy. *Phys Med Biol*. 2007; 52(15): 4481–96.

141. Jiang H, Wang B, Xu XG, Suit HD, Paganetti H. Simulation of organ-specific patient effective dose due to secondary neutrons in proton radiation treatment. *Phys Med Biol*. 2005; 50(18): 4337–53.

142. Gerweck LE, Huang P, Lu HM, Paganetti H, Zhou Y. Lifetime increased cancer risk in mice following exposure to clinical proton beam-generated neutrons. *Int J Radiat Oncol Biol Phys*. 2014; 89(1): 161–6.

143. Diallo I, Haddy N, Adjadj E, Samand A, Quiniou E, Chavaudra J, et al. Frequency distribution of second solid cancer locations in relation to the irradiated volume among 115 patients treated for childhood cancer. *Int J Radiat Oncol Biol Phys*. 2009; 74(3): 876–83.

144. BEIR. *Health Risks from Exposure to Low Levels of Ionizing Radiation*, BEIR VII, Phase 2. National Research Council, National Academy of Science, 2006.

145. EPA USEPA. Environmental Protection Agency—estimating radiogenic cancer risks. EPA. 1994, EPA 402-R-93-076 (Washington, DC).

146. NCRP. *Limitation of Exposure to Ionizing Radiation* (Supersedes NCRP Report No. 91). National Council on Radiation Protection and Measurements Report, 1993, 116.

147. UNSCEAR UNSCotEoAR. Sources and effects of ionizing radiation. United Nations Scientific Committee on the Effects of Atomic Radiation UNSCEAR; Report to the General Assembly, with scientific annexes. 2006;Vol. I: Report to the General Assembly, Scientific Annexes A and B.

148. Schneider U, Kaser-Hotz B. Radiation risk estimates after radiotherapy: application of the organ equivalent dose concept to plateau dose-response relationships. *Radiat Environ Biophys*. 2005; 44(3): 235–9.

149. Schneider U, Kaser-Hotz B. A simple dose-response relationship for modeling secondary cancer incidence after radiotherapy. *Z Med Phys*. 2005;15(1):31–7.

150. Schneider U, Lomax A, Besserer J, Pemler P, Lombriser N, Kaser-Hotz B. The impact of dose escalation on secondary cancer risk after radiotherapy of prostate cancer. *Int J Radiat Oncol Biol Phys*. 2007; 68(3): 892–7.

151. Paganetti H. Relating proton treatments to photon treatments via the relative biological effectiveness (RBE): Should we revise the current clinical practice? *Int J Radiat Oncol Biol Phys.* 2015; 91(5): 892–4.
152. Grosse N, Fontana AO, Hug EB, Lomax A, Coray A, Augsburger M, et al. Deficiency in homologous recombination renders Mammalian cells more sensitive to proton versus photon irradiation. *Int J Radiat Oncol Biol Phys.* 2014; 88(1): 175–81.
153. Hall EJ. Henry S. Kaplan Distinguished Scientist Award 2003: The crooked shall be made straight; dose response relationships for carcinogenesis. *Int J Radiat Biol.* 2004; 80: 327–37.

CHAPTER 23

Fully Exploiting the Benefits of Protons*

Using Risk Models for Normal Tissue Complications in Treatment Optimization

Peter van Luijk

Universtiy of Groningen

Marco Schippers

Paul Scherrer Institute

CONTENTS

---

* Colour figures available online at www.crcpress.com/9781138626508.

23.1 INTRODUCTION

For many cancers, radiotherapy is a common and effective treatment aiming at sterilizing tumor cells using radiation. At present, roughly 45%–50% of all patients receive radiotherapy at some stage of their treatment. For most cancers, however, normal tissues are inevitably coirradiated. This often leads to toxicity and a reduction of the quality of life (QoL) of the patient. For some tumors, such as lung cancer, the dose that can be administered safely to the tumor and consequently the efficacy of the treatment, are limited by the risk of severe toxicity.

Since the risk of toxicity depends on radiation dose and the amount of irradiated normal tissue, technological developments such as proton therapy are aimed at minimizing the dose and amount of normal tissue that is coirradiated. Protons indisputably have a superior physical dose distribution compared to classically used photons, since they lose most of their energy in a small region called the Bragg peak that can be positioned in the target volume (see previous chapters for details on the depth-dose distribution of protons.)

Optimal use of proton therapy, however, requires the (i) selection of patient populations for which use of protons may offer advantages, (ii) comparison with alternative treatment techniques, and (iii) individualized optimization of the selected technique (using e.g. optimization techniques described in Chapter 19). Each of these optimizations requires the ability to quantify the quality of a treatment (plan) by a figure of merit related to tumor control and normal tissue toxicity. Therefore, an important prerequisite of optimized use of proton therapy is, e.g., the availability of accurate risk models to estimate the risk of toxicity for given dose distributions.

In all current evaluation methods, the physical quantity "dose" plays a major role (see previous chapters for examples). This is partly due to the fact that this quantity can be calculated, adjusted, measured, and compared with great accuracy. However, it is the biological or clinical effect (tumor control or the occurrence of a complication) that determines the treatment outcome. "Dose" is only a surrogate for what is clinically important [1]. In this chapter, we will describe the attempts that are currently being made to determine the relationship(s) between "dose" and toxicity.

Since relatively little data is available on complications occurring after proton therapy, at present, this relation can only be estimated from data obtained after treatment with photons. However, the shapes of dose distributions that can be achieved using protons differ considerably from those obtained using photon-based techniques (see also Chapters 15 and 16). As such, extrapolating photon-based experience to proton therapy by using normal tissue complication probability (NTCP) models fitted to photon data is not trivial. In this chapter, an overview of various aspects of development and use of risk models for the prediction of responses to proton therapy as well as their limitations will be given.

23.2 RADIOTHERAPY: PATIENT CHARACTERIZATION FOR INDIVIDUALIZED TAILORING OF TREATMENT TECHNOLOGY

Different types and stages of tumor are treated using different treatment modalities such as fractionated photon therapy using a linear accelerator (linac), single fraction stereotactic radiotherapy, proton therapy, or brachytherapy using implanted radiation sources, all with the aim to maximize the probability of curing the patient while limiting toxicity to normal tissues. The choice of treatment modality is based on the clinical characteristics of the patient as well as the radiobiological characteristics of the tumor and surrounding normal tissues. As such, the choice for proton therapy is based on, e.g., differences in expected risks of normal tissue complications. Therefore, optimized selection of a treatment modality and specifically the choice for proton therapy is a very important field for the application of risk models for normal tissue damage.

Each treatment modality possesses many parameters that can be used to optimize its application. For example in external radiotherapy (linac-based photon therapy, proton therapy), the choice of number and shape of the beams and their angles with respect to the patient facilitate optimization of the dose to the tumor and normal tissues. Again based on the general location of the tumor and adjacent normal tissues, class solutions (treatment techniques) are developed for specific groups of patients. The development and optimization of these treatment techniques therefore represent a second level of treatment optimization where risk models for normal tissue damage are used.

Though these treatment techniques can serve as starting points for whole classes of patients, variability in patient anatomy necessitates individualized tailoring of this technique to each patient (treatment plan optimization). Factors determining the final treatment plan are, e.g., the size, shape, and position of the tumor and the exact anatomy of adjacent organs. In this final optimization step, imaging plays a pivotal role. The most used imaging modality in radiotherapy is X-ray computed tomography (CT) scanning. In addition, imaging modalities such as positron emission tomography (PET), single photon-emission CT (SPECT), or magnetic resonance imaging (MRI) are used to facilitate the identification of the most active parts of the tumor (e.g., ^{18}FDG PET), most functional parts of the lung (^{99m}Tc SPECT) or to improve the visibility of different brain regions (MRI). As an example, Figure 23.1A shows the CT scan of a patient with a lung tumor. To distinguish better between normal structures within the lung and extensions of the tumor, an ^{18}FDG PET scan was made and superimposed on the CT scan. The available images can be imported in a treatment planning system that contains a dose model of the treatment machine. Based on the CT and PET scan, the tumor and adjacent critical organs (e.g., the lung) are contoured. By varying treatment parameters such as beam angles, weights, and shapes, the treatment plan can be optimized (see also Chapters 15, 16, and 19). Figure 23.1B shows an example of the beam angles used in this patient. Based on the dose model, the treatment plan, and the patient geometry, the dose distribution in the patient can be calculated (Figure 23.1C).

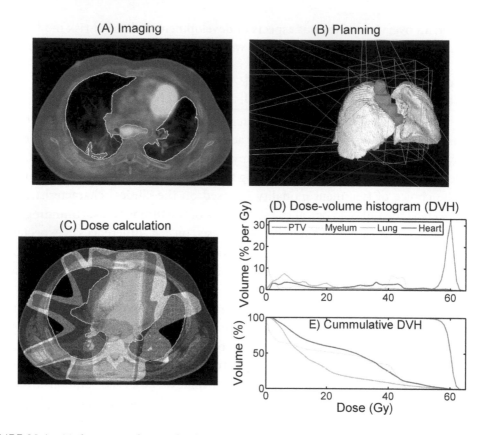

FIGURE 23.1   Tailoring modern radiotherapy technology to the patient requires characterization of the patient using imaging (panel A). The anatomy of the patient is characterized using a CT scan (gray scale). In addition to better distinguish between normal and tumor tissue, [18]FDG-PET scanning was used (color scale). The gross tumor volume is identified (green contour) and expanded with a margin to account for microscopic extensions, patient positioning, and tumor motion uncertainties to yield the planning target volume (red contour). Moreover, the critical organs are contoured (e.g., white contour for the lung). Subsequently in a treatment planning system, a treatment technique is applied to the patient (panel B) and the resulting dose distribution is calculated (color scale, panel C). Next, DVHs are constructed of dose deposited in the target volume and critical organs (panel D). In practice, these differential DVHs are converted in cumulative DVHs, giving the volume receiving more than a specific dose as a function of that dose (panel E). Based on these DVHs, the technique may be further adapted to the patient.

Direct comparison of 3D dose distributions of alternative treatment plans is very inconvenient. Therefore, present practice is to summarize the 3D dose distribution in a dose–volume histogram (DVH). For each (relative) dose level, this histogram gives the volume that receives that dose (Figure 23.1D). In practice, the cumulative DVH, giving the volume receiving more than a certain dose as a function of that dose, is reported by treatment planning systems (Figure 23.1E). Even though a DVH discards a lot of information as compared to the 3D dose distribution, at present, this DVH is the starting point for the derivation of figures of merit, based on which the treatment plan can be optimized. Classically, individual DVH points, the mean dose or generalized forms of the mean dose are used for this.

However, since the relation of these figures of merit to clinical outcome is not trivial, large efforts have been made to develop and use models explicitly describing the risk of normal tissue damage instead.

23.3 THE DEVELOPMENT OF RADIATION-INDUCED NORMAL TISSUE DAMAGE

To fully appreciate the complexity of events that risk models need to describe, a brief overview of the different events that lead from dose deposition to tissue damage and possibly clinical complications will be given.

In the presence of oxygen, photon irradiation leads to the formation of reactive oxygen species. These induce sparse DNA lesions such as single- and double-strand breaks dispersed through the cell nucleus. In comparison with photon irradiation, particle beams act more directly on the DNA and produce dense ionization tracks that cause clustered DNA damage, which are, especially in the Bragg peak region, more difficult to repair for a cell (see Chapter 22 for details.)

Cell death is primarily determined by nonrepairable double-strand breaks. Normal tissue damage, however, also depends strongly on other stress responses. Besides death, in normal tissue cells, DNA damage may also lead to, e.g., differentiation or proliferation of specific cells, cell-cycle arrest, initiating immune responses, changed metabolism, collagen and extracellular matrix deposition and other fibrotic reactions in response to activation of cytokines, and upregulation of profibrotic genes.

How these local effects are translated into clinical complications, however, strongly depends on organ or even endpoint-specific mechanisms, e.g., in the parotid gland, late loss of function is mostly due to a loss of repopulating primitive cells, whereas in the lung, late loss of function is due to excessive deposition of extracellular matrix material. Moreover, whether damage to tissue within the organ actually leads to a complication may depend on dose to other organs. Examples of these are dose to the heart, increasing the risk and severity of loss of pulmonary function after lung irradiation [2,3] and the risk of patient-rated xerostomia (dry mouth) depending on dose to the parotid gland as well as on dose to the submandibular gland [4]. Moreover, the relationship between DVH parameters and complication risk may also be influenced by other factors. For example, the relationship between dose and complication risk after irradiation for prostate cancer depends on whether or not abdominal surgery has been performed [5]. As such, each endpoint depends on a unique chain of events, leading from initial dose deposition and local damage to the final clinical endpoint. Therefore, NTCP models describing these endpoints need to be tailored to the organ and the endpoint studied.

23.4 METHODOLOGY OF MODEL DEVELOPMENT

For the application of new techniques, obtaining an NTCP estimate from an NTCP model is in fact extrapolating from current knowledge (the data). This extrapolation, however, heavily depends on the assumptions of the NTCP model. As such, it is critical to realize that the estimates obtained from the model are as good as the data and the assumptions it was built on. Therefore in this section, various stages of model development, such as the collection of data, describing the data with a model and testing the applicability of this model, are given.

23.4.1 Choice of Endpoint

Since the application of NTCP models is to optimize the treatment with respect to normal tissue toxicity, the exact type of toxicity (or "endpoint") to be described by the model has to be chosen carefully. Roughly, three classes of endpoints can be distinguished.

First, toxicity can be determined objectively in terms deterioration of organ function or changes in tissue, visualized by, e.g., imaging. This type of endpoint has the advantage that it depends on direct measurements of changes in organ function (e.g., lung capacity, saliva production), and therefore it allows an objective determination of radiation effects. Moreover, since this type of measurement can also be performed in, e.g., animal models, modeling this type of endpoints can benefit from mechanistic insight obtained from radiobiology studies. For most of these endpoints, however, it is unclear what the clinical impact is. Though, e.g., loss of 75% of parotid gland saliva production or density changes in the lung visible on CT clearly demonstrates damage to the parotid gland and lung, respectively, by itself, it does not establish a clinical impact that can be weighed against toxicity in other organs or tumor control.

Second, toxicity can be reported in terms of a clinical complication. As an example, the Common Terminology Criteria for Adverse Events (CTCAE v4.0) defines for each complication a grade between 1 and 5, where 1 indicates toxicity that remains asymptomatic or results in mild symptoms not requiring intervention, and 5 indicates death related to the complication. Though using this system complications can be classified according to clinical impact, this inherently also introduces some subjectivity, e.g., radiation pneumonitis is classified as grade 2 based on the need to administer steroids. Whether or not steroids are prescribed may, however, vary between physicians. Moreover, the observed clinical complication may originate from damage and/or functional problems in multiple organs [4]. As such, accurate prediction of clinical complications requires a good insight in the doses to which organs potentially contribute to the clinical symptoms. Consequently, modeling this type of endpoints requires the inclusion of a large number of dosimetric predictors and a large data set with sufficient number of events to support the model.

Finally, since the ultimate aim of (the optimization of) the radiotherapy treatment is to cure the patient while preserving or improving the QoL of the cancer patient, the impact of a treatment on QoL can also be used as an endpoint. QoL is generally measured using questionnaires (e.g., EORTC QLQ-C30 [6]).

The results of a number of studies clearly showed an inverse relationship between the presence and severity of radiation-induced side effects and the more general dimensions of QoL. Similar to scoring of clinical symptoms, changes in QoL may result from radiation effects in many organs as well as from the clinical status of the patient and/or comorbid diseases, e.g., in head and neck cancer, the clinical complication xerostomia is the most frequently reported late side effect, which is also considered by patients to affect their QoL. However, in a recent study, radiation-induced swallowing dysfunction appeared to be more important with regard to the effect on QoL as compared to xerostomia [7]. Thus, changes in QoL may result from multiple clinical complications, which in turn may result from dose to multiple organs. As such, including dosimetric information on all organs that may be involved in the reduction of QoL requires an enormous amount of data. Moreover, the QoL could ignore clinically relevant toxicity that has limited impact on the QoL. When basing treatment optimization of QoL, these toxicities would not be taken into account. In conclusion, choosing between these classes of endpoints is making a trade-off between specificity and relevance to the patient [8].

The optimization of a treatment will usually involve multiple endpoints. Normal tissue damage has to be weighed against the risk of not curing the patient. Moreover, for many disease sites, the patient is at risk for developing more than one type of toxicity. Simultaneous optimization of a treatment with respect to multiple endpoints requires that comparable figures of merit are used for each separately. A solution to this is to specify which amount of function loss, what grade of clinical complication, or what loss of health-related QoL is considered clinically relevant. Based on criteria such as loss of 75% of parotid saliva production, a grade 2 or higher radiation pneumonitis or loss of more than 10 points in QoL, these endpoints can be converted to a binary endpoint that specifies whether or not the complication occurred. Now the treatment can be optimized with respect to these binary endpoints, using tumor control probability and NTCP models. Though probabilities are suitable to compare nonequivalent types of endpoints, care should be taken to use binary endpoints that specify equally severe complications. Optimizing the occurrence of a grade 1 (asymptomatic) radiation pneumonitis against the probability of tumor control would compromise a lethal complication (not achieving tumor control) too much in favor of an asymptomatic complication.

23.4.2 Data Gathering

Since a model is in essence a summary of previously made observations, the quality of the data used for the development of the model is critical to its applicability. As such, when collecting data to facilitate the development of NTCP models to be used in treatment optimization with respect to some clinical endpoint, these data should at least contain the candidate predictors that can be expected to be related to this endpoint. It is important to recognize that, since many clinical endpoints (e.g., swallowing disorders, xerostomia, and radiation pneumonitis) are multiorgan endpoints, it is imperative that a data set used for their predictive modeling includes clinical and dosimetric parameters of all organs that are potentially involved. Moreover, since the risk of many complications depends on a combination of dosimetric and other clinical, treatment, or demographic factors, such as the addition of concurrent chemotherapy to radiation and age, it is imperative to collect these parameters and include them in the analysis as candidate predictors. Since models built on such multitude of predictive factors require an amount of data that is generally beyond the capacity of individual centers, data gathering requires extensive collaborations such as the QUANTEC collaboration [9].

Finally, since a model in principle summarizes the data used to construct it, the quality of a model critically depends on the quality of these data. As such, the quality of a model is influenced strongly by, e.g., the accuracy by which endpoints are scored, consistency in organ delineation, and the accuracy of the estimated dose distribution.

23.4.3 Fitting a Model to Binary Data

After the data has been collected, a model has to be selected to describe them. For models with a fixed set of predictors and parameters such as the Lyman–Kutcher–Burman (LKB) and critical-element and critical-volume models or logistic regression models with a known set of predictors (see next section for descriptions), parameter values can be determined by fitting the model to the data. The most appropriate fitting method for a model describing a process with binary outcome (the occurrence or absence of a complication) is the maximum-likelihood fit [10]. In this approach, the model parameters are selected for which the model prediction maximizes the probability that the experimental data would be observed. This in general corresponds to maximizing:

$$L(m)=\sum_i e_i \cdot \ln\left(NTCP_i(m)\right)+(1-e_i)\cdot \ln\left(1-NTCP_i(m)\right) \qquad (23.1)$$

L indicates the logarithm of the likelihood, e_i indicates the outcome (0 or 1) for patient i, and $NTCP_i(m)$ indicates the estimate of the $NTCP$ for the parameter set m. For testing the consistency between model and data (see later in this section), patients need to be grouped based on similarity of treatment with respect to the predictors in the model. The likelihood function to be optimized then becomes

$$L(m)=\sum_i r_i \cdot \ln\left(NTCP_i(m)\right)+(n_i-r_i)\cdot \ln\left(1-NTCP_i(m)\right) \qquad (23.2)$$

Here n_i and r_i indicate the number of patients and the number of complications in group i, respectively. As an example, Figure 23.2A shows the incidence of a reduction to 25% saliva production compared to the pretreatment level in patients treated in the head-and-neck region (data published in [11]). Using the maximum-likelihood method given by Equation 23.2 a logistic model (see Equations 23.6 and 23.7) was fitted to these data (curves).

23.4.4 Testing Model Validity

Though fitting a model to data will always produce a set of parameters that makes the model correspond to the data as good as possible, this does not imply that the model accurately describes the data. Different ways to assess the quality of the model put emphasis on different aspects such as correspondence between model and data or the ability to correctly distinguish between responding and nonresponding patients. As a first measure of how well the data is described by the model, the likelihood $L(m)$ (Equation 23.1 or 23.2) can be used. Although higher values of $L(m)$ indicate a better fit of the model to the data by itself, it does not provide a means to determine whether the model predictions are consistent to the data and can be used for a specific task. Derived from the likelihood, various information criteria have been developed.

These criteria also depend on the number of model parameters and provide a means to determine whether the addition of a new parameter in a model can be justified by the

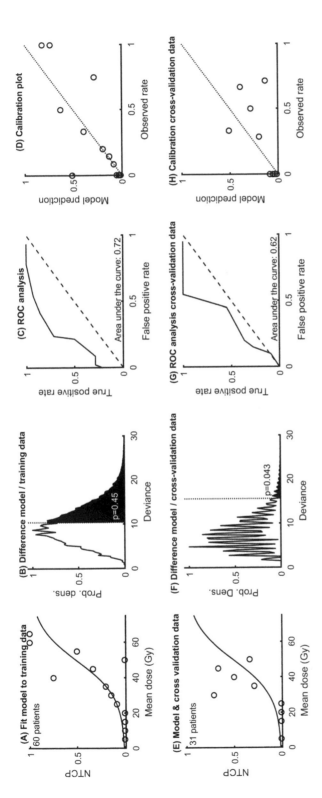

FIGURE 23.2 Overview of methods and their use in an example. Data on reduction >75% of parotid gland saliva production at 6 weeks and 1 year [11] are modeled using the mean parotid gland dose in a logistic model (Equation 23.2, panel A). The model is fitted to two-third of the data using the maximum likelihood method (Equation 23.2, panel A). Panel B shows the expected distribution of the deviance, determined by fitting to 10^5 alternative outcomes that were obtained using a Monte Carlo technique (Section 23.4.4). The dotted line indicates the deviance obtained in the real study. Based on these, the probability of finding larger deviances is 45%. As such, the differences between model and data are not significant ($p = 0.45$) and the model is not rejected by the data. To determine to what extent the model distinguishes between responders and nonresponders, an ROC analysis was performed (panel C). For internal cross-validation, the fitted model shown in panel A was compared to the unused one-third of the data (validation data set, panel D). New Monte Carlo data sets were generated to determine the deviance distribution for the validation data (panel E). The difference between the model fitted in panel A and the validation data, however, is significant. Finally, panel F shows that the performance of the model in the validation data is lower than in the training set.

improvement in the likelihood. Examples of these are Akaike's or the Bayesian information-tion criteria (AIC/BIC):

$$AIC = 2 \cdot k - 2 \cdot L \tag{23.3}$$

$$BIC = k \cdot \ln(n) - 2 \cdot L \tag{23.4}$$

Here k indicates the number of model parameters and L the natural logarithm of the likelihood obtained from the maximum likelihood fit. By comparing Equations 23.3 and 23.4, the difference between the BIC and the AIC can be seen to be in the penalization of the addition of a model parameter. In the BIC, this penalty depends on the logarithm of the number of samples, which leads to more conservative models, for data sets larger than six samples.

Many other statistical methods exist to quantify correspondence between model and data and to what extent the model separates patients with complications from patients without complications.

The most elementary test of the model is whether ranking of patients according to their predicted risk corresponds to ranking them according to the incidence of the complication. To this end, Spearman's rank correlation coefficient can be used [12]. Though a high correlation coefficient indicates that the model can be used to optimize the plan based on this model, it also does not test whether the probabilities are quantitatively correct. Treatment optimization based on risks of multiple endpoints (e.g., tumor control, xerostomia, and dysphagia), however, requires quantitatively correct NTCP values. Biases in the individual models would inadvertently influence plan selection [13]. Therefore, in addition to testing the ability of a model to rank plans with respect to a specific endpoint, checking the accuracy of the NTCP values obtained from the model is desired.

For large data sets with normal-distributed uncertainties, generally, the χ^2-test is used to test whether differences between model and data are larger than expected based on the statistical properties of the data. A similar method suitable for smaller data sets and binary outcome data, is based on the likelihood given by Equation 23.1 or 23.2 [10]. The data themselves are used as the best possible model, also called "the full model" [14]. Subsequently, it is tested whether the likelihood of the fitted model has significantly deteriorated as compared to the full model. The expected distribution of the difference between model and data is determined using a Monte Carlo method. The likelihood of obtaining the data if the model NTCP values are given by the data themselves can be obtained by substituting $NTCP_i = r_i/n_i$ in Equation 23.2 to yield:

$$L_0 = \sum_i r_i \cdot \ln\left(\frac{r_i}{n_i}\right) + (n_i - r_r)\ln\left(1 - \frac{r_i}{n_i}\right) \tag{23.5}$$

Now it can be tested whether the decrease of $L(m)$ with respect to L_0 significantly exceeds the amount expected based on the statistical noise in the data. In this analysis usually the deviance given by $-2 \cdot (L(m)-L_0)$ is used, since for large data sets, the distribution of the deviance resembles the χ^2 distribution [14]. However, the probability distribution of the deviance has to be known so that one can set a maximum decrease of the fitted models deviance with respect to the full model at which the model is considered to significantly deviate from the data. This distribution can be determined based on alternative study outcomes obtained using the fitted model and a Monte Carlo technique [10]. In this approach, for each data point, the NTCP is calculated using the fitted model. Subsequently, for each patient included in this data point, a random number between 0 and 1 is drawn and compared to the NTCP. If the random number is lower than the NTCP, for the simulated outcome, the patient is scored as showing the complication. Using this procedure, a large number of alternate data sets that could have occurred given the fitted NTCP model is obtained. Subsequently, L_0 can be recalculated and the NTCP model refitted for each of these data sets. This results in a large number of alternate values of the deviance and their distribution. Finally, from this distribution it can be determined which fraction of the fits to the Monte Carlo datasets resulted in a larger deviance than the fit to the experimental data. Since larger values indicate a larger difference between model and data, a small fraction (e.g., <5%) indicates that the difference between model and actual outcome is unlikely (e.g., probability <5%) to occur based on the statistical spread in the data, which is a good reason to reject the model. Figure 23.2B shows the deviance distribution for the model fit shown in Figure 23.2A. The difference between the model and data is not significant.

23.4.5 Model Performance

Besides testing whether the fitted model corresponds to the data, it is important to assess its performance. One such performance characteristic is the ability of a model to distinguish between responding and nonresponding patients. This aspect of performance, called discrimination, can be characterized by the receiver operating characteristic (ROC) curve [15]. The ROC plot is a plot of true-positive vs. false-positive rates for all possible NTCP threshold values (see, e.g., Figure 23.2C). For a random prediction (lowest performance), the area under this curve is 0.5. For better performing models, this area increases up to a maximum of 1.

A second performance characteristic is calibration, which specifies the quantitative accuracy of the risk estimates the model produces. An intuitive tool to assess calibration is the calibration plot in which the model estimates are plotted against the observed rate (Figure 23.2D.) In this example, it can be seen that the model plotted in Figure 23.2A performs rather well in patients observed to be at risks (<0.5), but seems to underestimate the risk in patients at high risk.

23.4.6 Cross-Validation

The fact that a model was developed based on a data set does not automatically imply that it is capable of predicting normal tissue toxicity in a new group of patients receiving the same type of treatment or can be extrapolated to other treatment modalities. Testing the ability of a model to predict the response in new patients can be done by cross-validation. Two stages of cross-validation can be distinguished. Internal cross-validation can be performed to test the predictive power of the model within the patient population and treatment type in which the model was developed. In internal cross-validation, the data set is split into a "training set" used to develop (i.e., determine predictive factors and/or fit model parameters) the model and a "validation set" is used to test the model performance in an independent set of data. In fact for the model fits shown in Figure 23.2A, only two-third of the available data was used as a training set. Every third patient was saved for a cross-validation. Figure 23.2E shows the original model curves copied from panel A, together with the cross-validation data. Here it can be seen that though the model follows the data, it appears to underestimate the risk at higher dose levels. Similar to Figure 23.2B, the significance of the difference between model and data can be tested. Since the deviance depends on, e.g., L_0 and thus on the dataset, the deviance distribution has to be redetermined by generating alternative outcomes and calculating the deviance using Equation 23.1 or 23.2, and the model fitted to the training set. The resulting distribution is shown in Figure 23.2F. The difference between the model and the cross-validation data sets is indeed significant. As expected, the performance of the model assessed by, e.g., the ROC curve, and the calibration plot is reduced (Figure 23.2G and H).

This approach to cross-validation may be sensitive to the specific selection of development and cross-validation data sets. This sensitivity can be reduced by using bootstrapping. In bootstrapping, multiple validation sets are sampled by randomly selecting patients from the original data set [12]. The ability of the model developed in the original data set to predict the outcomes in the bootstrapping data sets can now be assessed by the previously described methods, such as the ROC curve, calibration curve, or Spearman's correlation.

A successful internal cross-validation does not imply that it is accurate beyond the domain specified by the data set it was based on (e.g., other new treatment techniques or modalities). Therefore, any extrapolation to other patient populations or treatment techniques/modalities should be regarded as a hypothesis that needs to be tested prospectively in a new data set that was obtained in this new patient population or using the new treatment technique. To this end, again the previously described methods (see the "Testing Model Validity" section) can be applied to test consistency or performance of the model in new data sets. This procedure is called external cross-validation.

23.5 AVAILABLE MODELS FOR THE RISK OF NORMAL TISSUE DAMAGE

For a long time, efforts have been made to create predictive models for the benefit of treatment optimization. Only during the past two decades, evolution of treatment techniques and technology made the systematic collection of dose distribution and response data, and consequently, the development of models describing their relation possible. In the following subsections, a selection of these models will be described. In addition, in Table 23.1, a summary of these models, their properties, and typical applications are given.

23.5.1 The Sigmoid Curve: The Shape of the Population Distribution of the Tolerance

In 1924, it was recognized in the field of toxicology that the dose dependence of "poisoning can be described by an S-shaped curve, and that such a curve is properly conceived as an expression of the variation, either in sensitiveness or in resistance, of organisms, tissues,

TABLE 23.1 Overview of Models and Their Properties

Model	Parameters	Remarks
Logistic regression	Depending on model composition	Classical epidemiology approach
LKB	n volume effect (Figure 23.3) $D_{50}(1)$ tolerance to whole-organ irradiation m relative slope	Most generally used specialized NTCP model at present.
Critical element	$P(1,D)$ Dose response to whole-organ irradiation.	Designed for so-called "Serial organs."
Critical volume	Tolerance dose and slope of dose–response curve of the function subunit v, functional reserve σ, population spread of v_r	Designed for so-called "Parallel organs."
Relative seriality	$P(1,D)$ Dose–response curve to whole-organ irradiation. s relative seriality parameter	Designed for mixed behavior. Parameter s describes the extent to which the organ behaves serially or parallel.
Extended multivariate logistic regression modeling framework	Depending on model composition	Approach rather than a model. Most versatile allowing combining different types of predictive factors.

or cells toward a given poison" [16]. Though radiation can be regarded similar to poison, the normal tissues are generally not receiving a uniform dose. Therefore, one of the challenges in NTCP modeling is to recognize what predictive factor, derived from the dose distribution applied to the patient, best characterizes the variation of the resistance to the treatment (population tolerance distribution).

23.5.2 Logistic Regression Analysis: Identifying Predictive Factors

Classically, the description of the relation between (candidate) predictive factors and outcome are the domains of the epidemiologist. Though in this field, many possible modeling techniques are available, logistic regression analysis is still one of the most commonly used tools. The logistic regression model is given by

$$\text{NTCP} = \frac{1}{1+e^{-z}} \tag{23.6}$$

Here z is the predictive factor representing the stimulus of which the population tolerance distribution can be used to predict outcome. This predictive factor z can be a single variable (e.g., mean dose) or a function of multiple predictive variables. Though any type of function is possible, usually a linear function such as

$$z = \beta_0 + \beta_1 \cdot x_1 + \beta_2 \cdot x_2 + \cdots \tag{23.7}$$

is used. In Equation 23.7, the parameters β_i represent regression coefficients and x_i the corresponding predictive variables, such as points from the DVH, mean dose, or clinical parameters such as pretreatment pulmonary function. Though Equation 23.7 shows a linear function of independent predictive factors, each of these factors may depend on (combinations of) dose, volume, or clinical factors via a nonlinear function. The advantages of this type of model are the wide availability in statistics software packages of methods for building models and fitting them to data and the intrinsic capability of combining a heterogenic set of predictive factors, such as dose-related and patient factors.

23.5.3 Lyman-Kutcher-Burman Model

In the clinical physics, field efforts have been made to create mathematical formalisms that summarize 3D dose distributions in a single predictive factor. Since mechanisms underlying normal tissue damage vary by organ and even type of toxicity within the organ, the optimal strategy by which the 3D dose distribution is reduced to a single predictive factor is organ-dependent. One of the first and most often used models in this class is the LKB model [17]. In this model, the nonuniform dose distribution is reduced to a uniform dose distribution that is assumed to be equivalently effective as the original nonuniform one. To this end, the relative contribution of each subvolume of the organ to the uniformly irradiated volume is assumed to be proportional to a power function of its relative dose:

$$v_{eff} = \sum_i \left(\frac{D_i}{D_{max}} \right)^{\frac{1}{n}} \cdot v_i = \frac{1}{N} \sum_i \left(\frac{D_i}{D_{max}} \right)^{\frac{1}{n}} \tag{23.8}$$

Here D_i indicates the dose in relative subvolume of size v_i in the original, nonuniform dose distribution. If equally sized subvolumes (e.g., voxels in the 3D dose distribution) are used, the second formulation can be used. Here N indicates the number of subvolumes. Depending on the value of parameter n subvolumes contribute only to the effective volume if they contain a dose close to the maximum dose ($n \rightarrow 0$), proportionally to the relative dose ($n = 1$), or already starting at a low dose ($n \gg 1$). Irradiation of a volume v_{eff} to a dose D_{max} is assumed to result in the same clinical outcome as irradiation with the original dose distribution. Furthermore, the parameter D_{50}, also called the tolerance dose, is assumed to depend on the effective irradiated volume and the tolerance to irradiation of the whole organ $D_{50}(1)$ following a power law:

$$D_{50}(v) = \frac{D_{50}(1)}{v_{eff}^n} \tag{23.9}$$

Finally, by assuming that this $D_{50}(v)$ is normally distributed over the population with a relative uncertainty (i.e., standard deviation) m, the NTCP is given by

$$NTCP = \Phi \left(\frac{D_{max} - D_{50}(v)}{m \cdot D_{50}(v)} \right) \tag{23.10}$$

where Φ represents the cumulative normal distribution and is given by

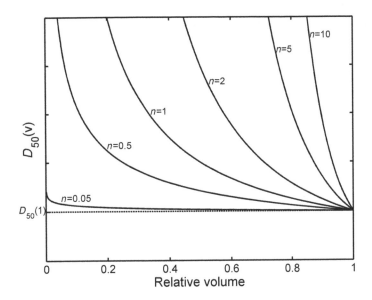

FIGURE 23.3 Dependence of the relation between the tolerance dose and irradiated volume on the value of parameter n of the Lyman model. For small values of n, the tolerance dose hardly depends on irradiated volume, in contrast to large values of n, leading to a strong volume dependency.

$$\Phi(x)=\frac{1}{\sqrt{2\pi}}\cdot\int\limits_{-\infty}^{x}e^{-y^2/2}\cdot dy \qquad (23.11)$$

Altogether, by specifying the parameter n, organ-specific behavior can be introduced (see Figure 23.3). For small values of n, the tolerance dose will hardly depend on irradiated volume, and the NTCP is determined mostly by the $D_{\max}$ of the dose distribution. For $n = 1$, however, the contribution of all subvolumes to the effective volume is proportional to dose and changes in irradiated volume will result in a proportional change in tolerance dose. At large values of n ($\gg 1$) any subvolume receiving even a low dose contributes fully to the effective volume. The tolerance dose, however, rises steeply if part of the organ is spared. Since this type of volume dependence might arise if subvolumes of the organ function independently and only a small functional subvolume is required for the organ to perform its function, this type of behavior is often characterized "parallel," as opposed "serial" behavior occurring for $n \ll 1$, where damaging even the smallest subvolume renders the organ dysfunctional.

At present, treatment optimization is still most often performed based on dose metrics such as DVH points and mean dose, rather than NTCP models. One of the figures of merit often used in such dose-based optimization is the generalized equivalent uniform dose (gEUD) [18]:

$$gEUD=\left(\frac{1}{N}\sum_{i}D_{i}^{a}\right)^{\frac{1}{a}} \qquad (23.12)$$

Here N represents the number of equivalently sized subvolumes in the organ at risk and a is a free parameter. The formulation of the EUD is very similar to the effective volume of the LKB model. In fact by substituting $a = 1/n$ in Equation 23.12, the LKB model can be reformulated in terms of gEUD:

$$\mathrm{NTCP} = \Phi\left(\frac{\mathrm{gEUD} - D_{50}(1)}{m \cdot D_{50}(1)}\right) \tag{23.13}$$

showing that the gEUD is in fact equivalent to the dosimetric predictor underlying the LKB model.

23.5.4 Functional Subunit-Based Models

Parallel and serial types of responses were modeled explicitly in a class of models, assuming that organs consist of independent substructures, called functional subunits (FSU), and that the organ response is determined by the organization and response of these functional subunits to radiation [19]. Initially, three of these models have been developed, assuming a serial (critical-element model [20]), parallel (critical-volume model [21]), or mixed (relative seriality [22]) organization. All three models start with the assumption that the risk of failure of this substructure (p) as a function of dose is given by a sigmoid curve. Many formulations (e.g., logistic, poison, cumulative normal distribution) for this type of curve are available. As such the logistic regression model of Equation 23.6 could be used with dose as a predictive factor. Alternatively it could be assumed that the tolerance dose of functional subunits is distributed according to the normal distribution. In this case, the dose–response curve is given by the cumulative normal distribution (see Equation 23.11). Finally, the Poisson distribution that describes the number of rare events (e.g., a lethal hit to a cell) occurring in a large sample size (e.g., all dose deposition events in the cell), could be used. Although the underlying assumptions differ, it is important to note that for all practical applications, these models perform similarly.

The critical-element model is based on the assumption that damage to any subvolume of the organ will result in failure of the entire organ [20]. Since the probability of preserving function in subvolume i is given by

$$1 - p(D_i) \tag{23.14}$$

and the probability of preserving function in all subvolumes is obtained by multiplication over all subvolumes, the risk of organ failure is given by

$$\mathrm{NTCP} = 1 - \prod_i (1 - p(D_i)) \tag{23.15}$$

Here D_i indicates the dose in FSU i. Similar to the Lyman model, the critical element model can be expressed in terms of the response to whole-organ irradiation:

$$\text{NTCP} = 1 - \prod_j \left(1 - P(1, D_j)\right)^{v_j} \qquad (23.16)$$

Here $P(1, D_j)$ equals the NTCP after irradiation of the whole organ to a dose D_j, and (D_j, v_j) represents the DVH.

In contrast, the critical-volume model is based on the assumption that the organ possesses some spare capacity. Organ failure is assumed to occur only if the damaged fraction of the organ exceeds this spare capacity. Therefore, to estimate the NTCP, first an estimate of the damaged volume has to be obtained:

$$v_d = \sum_{i=1}^{N} v_i \cdot p(D_i) \qquad (23.17)$$

Here v_i indicates the relative volume of subvolume i and N the total number of subvolumes. Assuming a normal distribution for the spare capacity, the risk of the damaged volume exceeding the spare capacity is then given by

$$\text{NTCP} = \Phi\left(\frac{v_d - v_r}{\sigma_r}\right) \qquad (23.18)$$

Here v_r and σ_r indicate the spare capacity and its population spread, respectively.

A model that combines both types of behaviors and allows the type of response (i.e., the degree of seriality) to be determined by a fit parameter is the relative seriality model:

$$\text{NTCP} = \left[1 - \prod_i \left(1 - p_{\text{FSU}}^s(D_i)\right)\right]^{1/s} \qquad (23.19)$$

Here the parameter s determines whether the volume dependency of the tolerance dose follows a parallel ($s \ll 1$) or serial ($s = 1$) behavior (Figure 23.4). Similar to the LKB and critical-element model, the relative seriality model can be expressed in the response after whole-organ irradiation:

$$\text{NTCP} = \left[1 - \prod_j \left(1 - P^s(1, D_j)\right)^{v_j}\right]^{1/s} \qquad (23.20)$$

During the last decade, the number of available models has increased enormously. Most of these recent models, however, share the most critical assumptions of the aforementioned models, and until now, their benefit over the previously described ones has not been demonstrated experimentally.

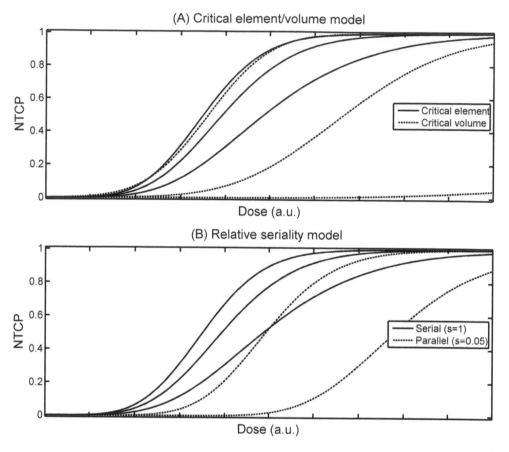

FIGURE 23.4 Behavior of FSU-based models illustrated by dose–response curves expected after irradiation of 100%, 67%, and 33% of the organ. The curves of the critical-element model and the relative seriality model in its serial limit demonstrate the characteristic small effect of irradiated volume on tolerance dose for low risk levels. In contrast, the critical volume model and the relative seriality model in its parallel regime show a strong dependence of tolerance dose on irradiated volume for all risk levels.

23.5.5 Including Clinical and Patient Characteristics: Logistic Regression Models

The LKB, critical element, critical volume, and relative seriality models are exclusively based on the dose distribution. Their formulation does not generally facilitate the inclusion of clinical parameters (e.g., addition of chemo or pre-existing morbidity), which would improve the accuracy of NTCP models. Alternatively, however, this can be achieved using a multivariate logistic regression model in which, e.g., the damaged volume for the critical-volume model and/or the EUD are used as predictors. Establishing the optimal combination of predictors resulting in the most accurate prediction of normal tissue toxicity is, however, not trivial. El Naqa et al. [12] gave an overview of methods that can be used to achieve this using a multivariable modeling approach combined with methods to test model robustness and prevent overfitting. The development of such a model involves (i) determining which parameter combinations result in a good description of the data, (ii) determining the number of parameters that can be supported by the data without overfitting, and (iii) determining the robustness of parameter selection.

First, determining combinations of candidate predictors that result in an accurate description of the data by testing all possible combinations is not feasible, especially if large numbers of candidates are considered. Therefore, two possible alternative approaches are sequentially adding the candidate that most improves the performance of the model or starting with all candidates and sequentially leaving out the one that leads to the least deterioration of model performance. Both approaches will result in a series of models with a number of predictors, ranging from a single parameter to all candidate predictors. Subsequently, the optimal model order needs to be determined. To this end various criteria, such as the AIC or BIC (Equations 23.3 and 23.4), can be used. Finally, the stability of the selection of predictive factors can be assessed by performing this procedure on alternative data sets, sampled by e.g., bootstrapping. For each bootstrap data set, the selection of predictive factors as described earlier, can be repeated, and for each factor, the number of times it is selected can be determined, giving an indication of which factors are generally selected, independent of statistical fluctuations in the data.

Since in many studies it was demonstrated that accurate prediction of normal tissue damage requires combining clinical and dosimetric factors, optimized multivariate modeling is a promising approach to NTCP modeling.

23.6 DEVELOPMENTS AND FUTURE DIRECTIONS

NTCP model development was initially inspired by improvements in radiotherapy technology, allowing sparing of normal tissues. Since both clinical data relating 3D-dose distributions to clinical outcome and knowledge on mechanisms underlying the irradiated-volume dependence of the risk of toxicity were sparse, the models described in the previous section were based on abstract assumptions. To test the validity of these assumptions to allow improvement of the accuracy of NTCP models, many radiobiological studies have been performed, leading to new insights in the development of normal tissue damage and providing directions for improving and developing predictive NTCP models.

All previously described models share a set of assumptions. First, the risk of damaging a subvolume (or functional subunit) is assumed to depend on dose to that subvolume alone. As such, damage in the tissue is assumed to be independent of events occurring elsewhere. Second, the radiation response of all subvolumes is assumed to be identical. Third, it is assumed that organ failure is the result of inactivation of a single type of target. To test these assumptions, during the past two decades, numerous studies have been performed on the lung, parotid gland, and spinal cord. Already in the 1990s, it was demonstrated in the mouse lung that the response of the lung to irradiation of the apex differed from that after irradiation of the base [23], indicating that either the functional consequence of dose may vary with the location it is deposited within the organ. Similarly, experiments in the rat spinal cord demonstrated that laterally located white matter is more radiosensitive than centrally located white matter [24], leading to large regional differences in tolerance dose for paralysis. Finally, in the rat parotid gland, the response to irradiation of the cranial parts of the gland differed from irradiation of the caudal parts of the glands [25], which was subsequently found to be due to a nonuniform distribution of stem cells within the gland [26]. Altogether, these results demonstrate that, in contrast to current model assumptions, the response of organs to irradiation is not uniform.

Moreover, in the rat spinal cord, the tolerance dose for irradiation of 8-mm cord length was found to be only 56% of the tolerance to a split-field dose distribution consisting of two segments of 4 mm [27], demonstrating irradiated volume is not generally the determinant of toxicity. In addition, the tolerance of the spinal cord to irradiation of a small subvolume (shower) was strongly reduced by a subtolerance dose (e.g., 20% of ED_{50}) administered to a larger, surrounding volume (bath) [27] (Figure 23.5A). Interestingly, this bath-and-shower effect was also observed in the rat parotid gland, showing that this is not an isolated finding, unique to the spinal cord [28] (Figure 23.5B). In addition, it has long been recognized that the response of the lung is not limited to its irradiated parts [29]. Taken together, these observations demonstrate that the occurrence of tissue damage does not depend only on the local dose. In fact, in the rat lung, it was demonstrated that radiation-induced loss of pulmonary strongly depended on the dose administered to the heart, showing that in fact the response of an organ may depend on the dose distribution in other organs [2].

Though the exact mechanisms underlying these nonlocal effects are not fully characterized; taken together, these observations all demonstrate that the functional response of an organ to irradiation is not trivially related to local dose. Moreover, even though this was demonstrated by looking at organ-specific effects, mostly in animal models, the fact that nonlocal effects were observed in virtually any organ that was studied, suggests that these nonlocal effects are likely to be the rule rather than an exception. More importantly, these effects have been shown to have a strong impact on tolerance doses and responses. This indicates that accounting for these effects in NTCP models will greatly enhance their accuracy.

Improvement of NTCP models by choosing predictors based on these nonlocal mechanisms requires an approach that is more organ-specific than has been done so far. Inclusion of full mathematical descriptions of biological mechanisms would likely lead to overly complex models with too many parameters. Therefore, an epidemiological approach (i.e., multivariable logistic regression analysis) that uses information on biological mechanisms to select candidate predictors may be preferable over mechanistic modeling approaches that were used classically. Moreover, to allow testing of candidate predictors in a clinical setting, the data gathering needs to be expanded from the widespread practice of collecting DVHs to actually storing 3D dose distributions to allow the extraction of these biology-based candidate predictors.

23.7 APPLICABILITY TO PROTON THERAPY

The development of NTCP models was inspired by the development of new techniques that allowed dose reductions in the normal tissues surrounding the target volume and the

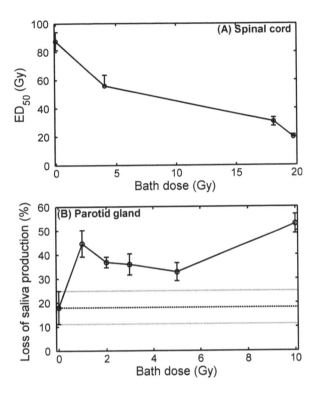

FIGURE 23.5 The effect of irradiation of a small subvolume can be strongly modulated by coirradiation of larger volumes to a low dose. Panel A shows the tolerance dose for irradiation of a 2 mm section of the cervical spinal cord of the rat in the center of a 20 mm section irradiated to a low dose (bath dose). Though the tolerance dose for irradiation of 20 mm of the spinal cord is 19.7 Gy, a dose of only 4 Gy reduces the tolerance dose in the 2 mm section from 87.5 to 56 Gy. Similarly, the reduction of the rat parotid saliva production after irradiation of the caudal 50% of the parotid gland to 50 Gy was increased by ~15%–30% when adding a bath dose to the cranial 50%. (Composed from Refs. [28,30,31].)

question of how to optimally put this new technology to use. As an example, the introduction of advanced techniques such as proton therapy provided the clinic with an unprecedented control over dose distribution. However, the most generally used dosimetric predictors, such as DVH points, mean lung doses were selected exclusively from available clinical data, which were mostly obtained in photon-based treatment. In such an analysis, only predictors showing both impact on outcome *and* strong variability in the data set are selected. Due to this variability criterion, however, this does not necessarily yield those parameters that best describe the biological processes underlying the toxicity with respect to which treatment is to be optimized. Consequently, optimization of novel treatment techniques using a selected predictor is not necessarily minimizing toxicity. Moreover, new treatment modalities, such as particle therapy, allow controlling more and other characteristics of the dose distribution than was possible with the modalities the models were based on. As such, when optimizing these new modalities based on a model solely based on data obtained in old treatment techniques, unique properties of the new treatment modality are left unused used to their full potential. Thus, when using NTCP models developed in photon-based data sets for the optimization of proton therapy treatment, it is important to realize that these models may not optimize unique features of proton therapy and NTCP estimates will be biased. As such, the NTCP model can only be used routinely for proton therapy after prospective testing against actual data on the effect of proton therapy.

These issues might be avoided by using independent mechanistic information obtained in e.g., preclinical studies, since this may offer the identification of predictive factors describing the underlying biological processes, independent of the extent to which their effect was the most prominent effect in existing photon-based data. As such, using these biology-based factors for treatment optimization is expected to better optimize the biological effect and use new treatment modalities such as proton therapy to their full potential.

23.8 ACKNOWLEDGMENTS

The authors wish to thank Dr. R.P. Coppes, Dr. C. Schilstra and Prof. Dr. J.A. Langendijk for critically reading the manuscript of this chapter. Moreover, they wish to thank numerous colleagues in the field for the fruitful discussions that contributed to the insights presented in this chapter.

REFERENCES

1. Goitein, M. 2007. *Radiation Oncology: A Physicist's-Eye View*. Springer, New York, NY.
2. Van Luijk, P., Novakova-Jiresova, A., Faber, H., Schippers, J.M., Kampinga, H.H., Meertens, H., and Coppes, R.P. 2005. Radiation damage to the heart enhances early radiation-induced lung function loss. *Cancer Res*, 65, 6509–6511.
3. van Luijk, P., Faber, H., Meertens, H., Schippers, J.M., Langendijk, J.A., Brandenburg, S., Kampinga, H.H. and Coppes, R.P. 2007. The impact of heart irradiation on dose-volume effects in the rat lung. *International Journal of Radiation Oncology Biology Physics*, 69, 552–559.
4. Jellema, A.P., Doornaert, P., Slotman, B.J., Leemans, C.R. and Langendijk, J.A. 2005. Does radiation dose to the salivary glands and oral cavity predict patient-rated xerostomia and sticky saliva in head and neck cancer patients treated with curative radiotherapy? *Radiotherapy and Oncology : Journal of the European Society for Therapeutic Radiology and Oncology*, 77, 164–71.
5. Peeters, S.T.H., Hoogeman, M.S., Heemsbergen, W.D., Hart, A.A.M., Koper, P.C.M. and Lebesque, J. V. 2006. Rectal bleeding, fecal incontinence, and high stool frequency after conformal radiotherapy for prostate cancer: Normal tissue complication probability modeling. *International Journal of Radiation Oncology Biology Physics*, 66, 11–19.
6. Aaronson, N.K., Ahmedzai, S., Bergman, B., Bullinger, M., Cull, A., Duez, N.J., Filiberti, A., Flechtner, H., Fleishman, S.B. and de Haes, J.C. 1993. The european organization for research and treatment of cancer QLQ-C30: A quality-of-life instrument for use in international clinical trials in oncology. *Journal of the National Cancer Institute*, 85, 365–76.
7. Langendijk, J.A., Doornaert, P., Verdonck-de Leeuw, I.M., Leemans, C.R., Aaronson, N.K. and Slotman, B.J. 2008. Impact of late treatment-related toxicity on quality of life among patients with head and neck cancer treated with radiotherapy. *Journal of Clinical Oncology*, 26, 3770–3776.
8. Bentzen, S., Dorr, W., Anscher, M., Denham, J., Hauerjensen, M., Marks, L. and Williams, J. 2003. Normal tissue effects: Reporting and analysis. *Seminars in Radiation Oncology*, 13, 189–202.

9. Marks, L.B., Ten Haken, R.K. and Martel, M.K. 2010. Guest editor's introduction to QUANTEC: A users guide. *International Journal of Radiation Oncology, Biology, Physics*, 76, S1–2.
10. Luijk, P. Van and Delvigne, T. 2003. Estimation of parameters of dose–volume models and their confidence limits. *Physics in Medicine*, 48, 1863–1884.
11. Burlage, F.R., Roesink, J.M., Kampinga, H.H., Coppes, R.P., Terhaard, C., Langendijk, J.A., van Luijk, P., Stokman, M.A. and Vissink, A. 2008. Protection of salivary function by concomitant pilocarpine during radiotherapy: A double-blind, randomized, placebo-controlled study. *International Journal of Radiation Oncology Biology Physics*, 70, 14–22.
12. El Naqa, I., Bradley, J., Blanco, A.I., Lindsay, P.E., Vicic, M., Hope, A. and Deasy, J.O. 2006. Multivariable modeling of radiotherapy outcomes, including dose–volume and clinical factors. *International Journal of Radiation Oncology, Biology, Physics*, 64, 1275–1286.
13. Langer, M., Morrill, S.S. and Lane, R. 1998. A test of the claim that plan rankings are determined by relative complication and tumor-control probabilities. *International Journal of Radiation Oncology, Biology, Physics*, 41, 451–457.
14. Collett, D. 2003. *Modelling Binary Data*. Chapman & Hall/CRC.
15. Lind, P.A., Marks, L.B., Hollis, D., Fan, M., Zhou, S.-M., Munley, M.T., Shafman, T.D., Jaszczak, R.J. and Coleman, R.E. 2002. Receiver operating characteristic curves to assess predictors of radiation-induced symptomatic lung injury. *International Journal of Radiation Oncology, Biology, Physics*, 54, 340–7.
16. Shackell, L.F., Williamson, W., Deitchman, M.M., Katzman, G.M. and Kleinman, B.S. 1924. The relation of dosage to effect. *Journal of Pharmacology and Experimental Therapeutics*, 24.
17. Lyman, J.T. 1985. Complication probability as assessed from dose-volume histograms. *Radiation research. Supplement*, 8, S13–9.
18. Qiuwen Wu, Mohan, R. and Niemierko, A. IMRT optimization based on the generalized equivalent uniform dose (EUD). In Proceedings of the 22nd Annual International Conference of the IEEE Engineering in Medicine and Biology Society (Cat. No.00CH37143), Volume 1. IEEE, 710–713.
19. Rodney Withers, H., Taylor, J.M.G. and Maciejewski, B. 1988. Treatment volume and tissue tolerance. *International Journal of Radiation Oncology, Biology, Physics*, 14, 751–759.
20. Niemierko, A. and Goitein, M. 1991. Calculation of normal tissue complication probability and dose-volume histogram reduction schemes for tissues with a critical element architecture. *Radiotherapy and Oncology : Journal of the European Society for Therapeutic Radiology and Oncology*, 20, 166–76.
21. Niemierko, A. and Goitein, M. 1993. Modeling of normal tissue response to radiation: the critical volume model. *International Journal of Radiation Oncology, Biology, Physics*, 25, 135–45.
22. Källman, P., Agren, A. and Brahme, A. 1992. Tumour and normal tissue responses to fractionated non-uniform dose delivery. *International Journal of Radiation Biology*, 62, 249–62.
23. Travis, E.L., Liao, Z.X. and Tucker, S.L. (1997) Spatial heterogeneity of the volume effect for radiation pneumonitis in mouse lung. *International Journal of Radiation Oncology, Biology, Physics*, 38, 1045–1054.
24. Bijl, H.P., Van Luijk, P., Coppes, R.P., Schippers, J.M., Konings, A.W.T. and Van Der Kogel, A.J. (2005) Regional differences in radiosensitivity across the rat cervical spinal cord. *International Journal of Radiation Oncology, Biology, Physics*, 61, 543–551.
25. Konings, A.W.T., Cotteleer, F., Faber, H., Van Luijk, P., Meertens, H. and Coppes, R.P. (2005) Volume effects and region-dependent radiosensitivity of the parotid gland. *International Journal of Radiation Oncology, Biology, Physics*, 62, 1090–1095.
26. van Luijk, P., Pringle, S., Deasy, J.O., Moiseenko, V.V., Faber, H., Hovan, A., Baanstra, M., van der Laan, H.P., Kierkels, R.G.J., van der Schaaf, A., Witjes, M.J., Schippers, J.M., Brandenburg, S., Langendijk, J.A., Wu, J. and Coppes, R.P. 2015. Sparing the region of the salivary gland containing stem cells preserves saliva production after radiotherapy for head and neck cancer. *Science Translational Medicine*, 7, 305ra147.

27. Bijl, H.P., Van Luijk, P., Coppes, R.P., Schippers, J.M., Konings, A.W.T. and Van Der Kogel, A.J. 2003. Unexpected changes of rat cervical spinal cord tolerance caused by inhomogeneous dose distributions. *International Journal of Radiation Oncology, Biology, Physics*, 57, 274–281.

28. van Luijk, P., Faber, H., Schippers, J.M., Brandenburg, S., Langendijk, J.A., Meertens, H. and Coppes, R.P. 2009. Bath and shower effects in the rat parotid gland explain increased relative risk of parotid gland dysfunction after intensity-modulated radiotherapy. *International Journal of Radiation Oncology, Biology, Physics*, 74, 1002–1005.

29. Morgan, G.W., Pharm, B. and Breit, S.N. 1995. Radiation and the lung: A reevaluation of the mechanisms mediating pulmonary injury. *International Journal of Radiation Oncology, Biology, Physics*, 31, 361–369.

30. Bijl, H.P., Van Luijk, P., Coppes, R.P., Schippers, J.M., Konings, A.W.T. and Van Der Kogel, A.J. 2006. Influence of adjacent low-dose fields on tolerance to high doses of protons in rat cervical spinal cord. *International Journal of Radiation Oncology, Biology, Physics*, 64, 1204–1210.

31. Van Luijk P., Bijl H.P., Konings, A.W.T., Van der Kogel, A.J. and Schippers, J.M. (2005) Data on dose-volume effects in the rat spinal cord do not support existing NTCP models. *International Journal of Radiation Oncology, Biology, Physics*, 61, 892–900.

Index

G

Gadolinium orthosilicate (GSO) 649
Gamma (Anger) camera 660, 663
GAMOS Monte Carlo framework 250
Gantry 99–102, 132–133, 196–197, 466
    achromaticity 100
    bending magnets 101
    corkscrew gantry 99
    gantry angle 132–133, 352–353, 355–357, 388–389, 392, 618
    gantry dipole 197
    isocentricity 350–352, 383, 388–389, 392
    superconducting 102
GATE Monte Carlo framework 250
Gating See Respiratory gating
Geant4 Monte Carlo code 224, 250, 252, 266, 333, 402
Gel dosimeter 281–283, 294
Generalized equivalent uniform dose (gEUD) 585, 729–730
Gesellschaft für Schwerionenforschung, GSI (Germany) 171, 183, 571, 631, 655
gEUD See Generalized equivalent uniform dose
gPMC (GPU based Monte Carlo code) 448
GPU See Graphics processing unit 402, 436–437, 447–448
Gray(RBE) 686
Gross tumor volume (GTV) 457, 602, 703
GSI See Gesellschaft für Schwerionenforschung
GTV See Gross tumor volume
Gustav Werner Institute, Uppsala (Sweden) 10, 70, 81

H

Half life 209, 647, 653
Hard scatter See Nuclear interactions
Harmonic number 78
Harvard Cyclotron Laboratory, Cambridge (USA) 11, 17, 70, 81, 145, 526, 616, 635
Head and Neck cancer 496, 526–527, 620, 695
Heavy Ion Medical Accelerator Chiba (HIMAC), Japan 655
Heidelberg Ion-Beam Therapy Center (HIT) (Germany) 619, 657
High-gradient insulator 104
High-voltage switching circuit 104
Highland approximation 38, 41–42, 409, 423
HIMAC See Heavy Ion Medical Accelerator
History of proton therapy 9–26
HIT See Heidelberg Ion-Beam Therapy Center
Hitachi America Ltd. 92–93, 96, 98, 157–159, 189, 197, 410

Hodoscope 662, 664
Hokkaido University, Japan 620–621
Hounsfield Unit conversion 361–363, 384, 438–439, 524–526, 625
    density override 459
    stoichiometric method 362, 439
Hypofractionation 124, 685

I

IBA See Ion Beam Application S.A.
IFV See In-field volume
IGRT See Image-guidance
iGTV See Internal gross tumor volume
Image guidance 123, 538, 615–636, 644
    AlignRT 631
    c-arm imager 619, 624–625
    cone-beam CT (CBCT) 538, 616, 642–627, 631
    CT on rails 627–630
    digitally reconstructed radiograph (DRR) 616, 619, 620, 634
    DRR See digitally reconstructed radiograph
    dual KV imaging 627
    electronic portal imaging (EPID) 615, 644
    EPID See electronic portal imaging
    fiducial marker 382–383, 473, 535, 538–539, 558, 620–622
    flat panel imager 616–617, 620–621, 625
    fluoroscopic image guidance 620–621
    four-dimensional cone beam CT (4D CBCT) 627
    marker tracking 630–631
    optical tracking 630–633
    planar radiography 617, 620, 630–631
    radiographic localization 617–621, 644
    robotic arm 619, 624–625, 634
    RPM system 631
    surface imaging 631–633
    volumetric localization 622–630
Image guided radiation therapy (IGRT) See Image guidance
Image reconstruction 649
Image registration 512, 620
Imaging and Radiation Oncology Core (IROC), Houston (USA) 336, 372, 382
Immobilization See Patient immobilization
IMPT See Intensity-modulated proton therapy
IMRT See Intensity-modulated photon therapy
Incoherent nuclear interation 46
Inductance 189
Inelastic cross section 211–212
Inelastic nuclear interactions 46, 209, 214
In-field volume 703
Injector 93